T0388668

Advances in Experimental Medicine and Biology

Volume 1131

Editorial Board
IRUN R. COHEN, *The Weizmann Institute of Science, Rehovot, Israel*
ABEL LAJTHA, *N.S. Kline Institute for Psychiatric Research, Orangeburg, NY, USA*
JOHN D. LAMBRIS, *University of Pennsylvania, Philadelphia, PA, USA*
RODOLFO PAOLETTI, *University of Milan, Milan, Italy*
NIMA REZAEI, *Children's Medical Center Hospital, Tehran University of Medical Sciences, Tehran, Iran*

More information about this series at http://www.springer.com/series/5584

Md. Shahidul Islam
Editor

Calcium Signaling

Second Edition

Volume 2

Editor
Md. Shahidul Islam
Department of Clinical Science
and Education
Södersjukhuset, Karolinska Institutet
Stockholm, Sweden

Department of Emergency Care
and Internal Medicine
Uppsala University Hospital
Uppsala, Sweden

Calcium signaling contains a unique selection of chapters that cover a wide range of contemporary topics in this ubiquitous and diverse system of cell signaling.

ISSN 0065-2598 ISSN 2214-8019 (electronic)
Advances in Experimental Medicine and Biology
ISBN 978-3-030-12456-4 ISBN 978-3-030-12457-1 (eBook)
https://doi.org/10.1007/978-3-030-12457-1

1st edition: © Springer Science+Business Media Dordrecht 2012
© Springer Nature Switzerland AG 2020
This work is subject to copyright. All rights are reserved by the Publisher, whether the whole or part of the material is concerned, specifically the rights of translation, reprinting, reuse of illustrations, recitation, broadcasting, reproduction on microfilms or in any other physical way, and transmission or information storage and retrieval, electronic adaptation, computer software, or by similar or dissimilar methodology now known or hereafter developed.
The use of general descriptive names, registered names, trademarks, service marks, etc. in this publication does not imply, even in the absence of a specific statement, that such names are exempt from the relevant protective laws and regulations and therefore free for general use.
The publisher, the authors, and the editors are safe to assume that the advice and information in this book are believed to be true and accurate at the date of publication. Neither the publisher nor the authors or the editors give a warranty, express or implied, with respect to the material contained herein or for any errors or omissions that may have been made. The publisher remains neutral with regard to jurisdictional claims in published maps and institutional affiliations.

This Springer imprint is published by the registered company Springer Nature Switzerland AG.
The registered company address is: Gewerbestrasse 11, 6330 Cham, Switzerland

Contents of Volume 2

Contents of Volume 1

Chapter 23
Review: Structure and Activation Mechanisms of CRAC Channels

Carmen Butorac, Adéla Krizova, and Isabella Derler

Abstract Ca^{2+} release activated Ca^{2+} (CRAC) channels represent a primary pathway for Ca^{2+} to enter non-excitable cells. The two key players in this process are the stromal interaction molecule (STIM), a Ca^{2+} sensor embedded in the membrane of the endoplasmic reticulum, and Orai, a highly Ca^{2+} selective ion channel located in the plasma membrane. Upon depletion of the internal Ca^{2+} stores, STIM is activated, oligomerizes, couples to and activates Orai. This review provides an overview of novel findings about the CRAC channel activation mechanisms, structure and gating. In addition, it highlights, among diverse STIM and Orai mutants, also the disease-related mutants and their implications.

Keywords Calcium · CRAC channel · STIM1 · Orai1 · STIM-Orai interaction · Orai gating · Gain-of-function mutants · Electrophysiology · FRET · Structural resolution

Abbreviations

2-APB	2-aminoethoxydiphenyl borate
aa	amino acid
ADP	adenosine diphosphate
ATP	adenosine triphosphate
ANSGA	4-point mutation in hinge region aa position 261-265
ATPase	adenosine triphosphatase
BK	large conductance, Ca^{2+}-activated potassium channels
cCAD	*C. elegans* CAD
cOrai	*C. elegans* Orai
cSTIM	*C. elegans* STIM

C. Butorac · A. Krizova · I. Derler (✉)
Institute of Biophysics, Johannes Kepler University of Linz, Linz, Austria
e-mail: isabella.derler@jku.at

© Springer Nature Switzerland AG 2020

M. S. Islam (ed.), *Calcium Signaling*, Advances in Experimental Medicine and Biology 1131, https://doi.org/10.1007/978-3-030-12457-1_23

Ca^{2+}	calcium ion
CAD	Ca^{2+} Release-Activated Ca^{2+} activating domain
CaM	calmodulin
CAR	Ca^{2+} accumulating region
CC	coiled-coil
Ccb9	coiled-coil domain containing region b9
cEF	canonical EF hand
CFP	cyan fluorescent protein
CRAC	Ca^{2+} release-activated Ca^{2+}
CRACR2A	calcium release activated channel regulator 2A
Cs^{+}	cesium ion
Δ	represents deletion mutants
dOrai	*Drosophila melanogaster* Orai
DVF	divalent-free
EGTA	ethylene glycol tetraacetic acid
ER	endoplasmic reticulum
ERK1/2	extracellular-signal-regulated kinases 1 and 2
ETON	extended transmembrane Orai1 N-terminal
FCDI	fast calcium dependent inactivation
FIRE	FRET-derived interaction in a restricted environment
FRAP	fluorescence recovery after photobleaching
FRET	fluorescence resonance energy transfer
GoF	gain of function
HEK	Human Embryonic Kidney
I/V	current voltage relationship
I_{Ca2+}	CRAC current
I_{Na+}	sodium current in sodium divalent free solution
ID	inhibitory domain
IH	inhibitory helix
IP3	inositol-triphosphate
K^{+}	potassium ion
Kir	inward-rectifier potassium channels
L1-L3	Loop 1-3 (of Orai channels)
L-type	long-lasting Calcium channel
LRET	luminescence resonance energy transfer
MD simulations	molecular dynamics simulations
Na^{+}-DVF	sodium divalent free
nEF	non-canonical EF hand
nAChR	nicotinic acetylcholine receptors
NMR	nuclear magnetic resonance
OASF	Orai – activating small fragment
Orai 1-3	Orai proteins (also as O1-3)
P/S	proline/serine
PIP_2	phosphatidylinositol 4,5-bisphosphate

PM	plasma membrane
RASSF4	Ras association domain family member 4
RNAi	RNA interference
SAM	sterile α-motif
S	signal peptide
SCDI	slow calcium dependent inactivation
SCID	Severe Combined Immune Deficiency
SOAP	STIM – Orai association pocket
SOAR	STIM – Orai activating region
SOC	store operated channel
SOCE	store-operated calcium entry
SPCA2	Secretory Pathway Ca^{2+}-ATPase
STIM	stromal interaction molecule
TM	transmembrane helices
TRP	transient receptor potential ion channel (C-canonical, M-melastatin, V-vallinoid)
WT	wild-type

23.1 Introduction

Ca^{2+} is a very important second messenger in eukaryotic cells. Sustaining Ca^{2+}-homeostasis within the cell is indispensable for immune cell function and activity of neurons. Perturbations in Ca^{2+} levels can lead to severe diseases such as cancer or immune deficiencies [1, 2]. In many cell types, one well-known Ca^{2+} entry pathway is the Ca^{2+} release activated Ca^{2+} (CRAC) channel [3]. This pathway is composed of two molecular key players STIM1, which belongs to the stromal interaction molecule (STIM) family (STIM1 and STIM2) and Orai1, which belongs to the Orai family (Orai1, Orai2, Orai3). Both of these have been identified via a systematic RNA interference (RNAi) screen [4]. Intact communication of STIM and Orai proteins maintains proper cell function, especially of immune cells and neurons, while abnormal up- or downregulation of these proteins can lead to defects in signaling pathways [5]. Additionally, several mutations in STIM1 and Orai1 are currently known to lead to either gain- [6] or loss-of-function [7] and have been associated with diseases like Severe Combined Immune Deficiency (SCID), Stormorken Syndrome and tubular aggregate myopathy, highlighting their clinical relevance [2, 8]. A list of most currently known STIM and Orai mutants and their functional effects are summarized in Tables 23.1 and 23.2, which also states whether they are disease-related.

Upon binding of IP_3 to receptors in the membrane of the endoplasmic reticulum (ER), Ca^{2+} is released from the ER, which initiates the activation of STIM1 [3]. STIM1 is a single transmembrane spanning protein within the ER membrane that senses the ER-lumenal Ca^{2+} concentration via its N-terminal EF-hand motif. It is uniformly distributed within the ER membrane in the resting state [9]. Once

Table 23.1 Mutations in STIM1

Mutation	Domain	Orai coupling	Current	Related disease	References
H72Q	EF-hand	Yes	CONSTITUTIVE	Tubular aggregate myopathy	[218]
D76A	EF-hand	Yes	Constitutive		[11]
N80T	EF-hand	Yes	Constitutive	Tubular aggregate myopathy	[219]
G81D	EF-hand	Yes	Constitutive	Tubular aggregate myopathy	[220]
D84G	EF-hand	Yes	Constitutive	Tubular aggregate myopathy	[218]
E87A	EF-hand	Yes	Constitutive		[23]
L96V	EF-hand	Yes	Constitutive	Tubular aggregate myopathy	[219]
F108I/L	EF-hand	Yes	Constitutive	Tubular aggregate myopathy	[219]
H109R/N	EF-hand	Yes	Constitutive	Tubular aggregate myopathy	[218]
I115F	EF-hand	Yes	Constitutive	Tubular aggregate myopathy; York platelet syndrome	[221, 222]
E136X	SAM	No	Inactive	Combined immune deficiency	[223]
P165Q	SAM	Yes	Store-operated	Late-onset immunodeficiency	[224]
I220W	TM	Yes	Constitutive		[40]
C227W	TM	Yes	Constitutive		[40]
L248S	CC1	Yes	Constitutive		[86]
L251S	CC1	Yes	Constitutive		[86]
R304W	CC1	Yes	Constitutive	Stormorken Syndrome	[8]
Y316A	CC1	Yes	Constitutive		[87]
E318/319/320/322A	CC1	Yes	Constitutive		[225]
I364A	CC2	Yes	Store-operated (enhanced)		[114]
A369K	CC2	Yes	Constitutive		[74]
A376K	CC2	Yes	Constitutive		[74]
A380R	CC2	Yes	Constitutive		[16]
K382/384/385/386E	CC2	No	Inactive		[16]
F394H	CC2	Reduced	Inactive		[52]
R426L	CC3	No	Inactive		[86]
R429C	CC3	No	Inactive	Combined immune deficiency	[97]

STIM1 has been activated, it homomerizes and oligomerizes into ER-PM-junctions, where it can bind to and activate Orai1 via the cytosolic C-terminus [2, 10–12]. Orai1 represents a highly Ca^{2+} selective pore in the plasma membrane [12, 13]. A milestone was reached in 2012, when Hou et al. [14] managed to crystallize the Orai channel of *Drosophila melanogaster* (dOrai), revealing the hexameric stoichiometry of this channel complex. Recently, the structure of a dOrai mutant representing a potential open Orai state was also resolved [15]. Furthermore, structural resolutions

Table 23.2 Mutations in Orai1

Mutation	Domain	Pm-localization	STIM1 coupling	Current	Related disease	References
L74I	N-term.	Yes	Yes	Store-operated (enhanced)		[113]
Y80S	N-term.	Yes	Yes	Store-operated (enhanced)		[113]
L74/W76E/ R/S	N-term.	Yes	Yes	Store-operated (reduced)		[64]
K85E	N-term.	Yes	Reduced	Inactive		[109]
R91W	TM1	Yes		Inactive	SCID	[2]
S97C	TM1	Yes		Constitutive	Tubular aggregate myopathy	[226]
G98C/D/P	TM1	Yes	Yes	Constitutive		[69, 156]
G98R	TM1	No	No	Inactive	CID; autoimmunity; EDA	[227]
G98S	TM1	Yes		Constitutive	Tubular aggregate myopathy	[71]
F99C/G/M/ S/T/Y/W	TM1	Yes	Yes	Constitutive		[69]
V102A/C/ G/S/T	TM1	Yes	Yes	Constitutive		[67]
V102I/L/M/V	TM1	Yes	Yes	Store-operated		[67]
A103E	TM1	No	No	Inactive	Immunodeficiency	[228]
E106Q	TM1	Yes	Yes	Inactive		[154]
V107M	TM1	Yes		Constitutive	Tubular aggregate myopathy	[71]
H134S/A/C/ T/V/Q/E/M	TM2	Yes	Yes	Constitutive		[68, 73]
H134K/W	TM2	Yes	Yes	Inactive		[73]
A137V	TM2	Yes		Constitutive	Colorectal tumor	[68, 229]
L138F	TM2	Yes		Constitutive	Tubular aggregate myopathy	[72]
M139V	TM2	Yes		Constitutive	Stomach carcinoma	[68, 229]
S141C	TM2	Yes		Constitutive		[73]
S159L	Loop2	Yes		Constitutive	Uterine carcinoma	[68, 229]
L174D	TM3	Yes	Reduced	Inactive		[65]

(continued)

Table 23.2 (continued)

Mutation	Domain	Pm-localization	STIM1 coupling	Current	Related disease	References
W176C	TM3	Yes	Yes	Constitutive		[162]
A177D	TM3	Yes		Constitutive		[68, 229]
V181SfsX8	TM3	No	No	Inactive	CID; autoimmunity; EDA	[227]
G183A	TM3	Yes		Inactive		[162]
G183D	TM3	No	No	Inactive	Glioblastoma	[68, 229]
T184M	TM3	Yes	Yes	Store-operated	Tubular aggregate myopathy	[71]
F187C	TM3	Yes		Constitutive		[73]
E190C	TM3	Yes	Yes	Constitutive		[73]
E190Q	TM3	Yes		Reduced		[13]
L194P	TM3	No	No	Inactive	CID; autoimmunity; EDA; immunodeficiency	[227, 228]
A235C	TM4	Yes	Yes	Constitutive		[73]
S239C	TM4	Yes		Constitutive		[73]
G247S	TM4	Yes		Constitutive	Neck carcinoma	[68, 229]
F250C	TM4	Yes		Constitutive		[73]
P245L	TM4	Yes	Yes	Constitutive	Stormorken-like syndrome	[8]
L273S	C-term.	Yes	No	Inactive		[89]
L273D	C-term.	Yes	No	Inactive		[127]
L276D	C-term.	Yes	No	Inactive		[230]

of STIM1 N- and C-terminal fragments as well as an interacting complex formed by C-terminal fragments of STIM1 and Orai1 are currently available [16, 17]. This structural data together with functional and simulation studies improve our understanding of the STIM/Orai coupling and activation mechanism.

23.2 Composition and Structure of the Molecular Key Players of CRAC Channels

23.2.1 Structure of STIM

The STIM protein family, includes STIM1 and STIM2, two isoforms which share ~61% sequence identity [18, 19]. They are expressed in the endoplasmic reticulum [9, 11, 20, 21] and, at lower levels, in the plasma membrane [19, 20, 22]. STIM1 located in the plasma membrane has been suggested to control the extent of store-operated Ca^{2+} entry [23]. The activity of arachidonic acid regulated channels (ARC) has been proposed to solely depend on plasma membrane resident STIM1 [24–26]. According to recent reports, STIM proteins are also expressed in the acidic stores of lysosome-related organelles and the dense granules of the human platelets [27–29]. However, their roles there have so far remained elusive. In this review, we focus particularly on the role, function and structure of STIM proteins expressed in the ER, which are essential for the activation of CRAC channels [30].

The family of STIM proteins is further enriched in splice variants of STIM1 (STIM1L) and STIM2 (STIM 2.1 or STIM2β, STIM 2.2 or STIM2α, STIM2.3) [31–33] (see sequence alignment – Fig. 23.1).

Briefly, key domains in the STIM proteins (Fig. 23.2a) are the N-terminus embedded in the ER lumen, containing the Ca^{2+} – sensing region, a single TM spanning region of ~20-amino acids and the long cytosolic C-terminus, which binds to Orai channels in the plasma membrane [34]. Whereas the N-terminus of STIM is well conserved, the C- terminus is relatively varied among diverse species [35]. So far, only one luminal and two different C-terminal portions of STIM1 have been crystallized [16, 17, 36, 37], while a structural resolution of full-length STIM1 is still lacking.

23.2.2 STIM1

The Ca^{2+}-sensor protein STIM1 is composed of 685 amino acids (Fig. 23.2a) including the ER signal peptide (aa 1-22), a luminal EF hand (a canonical aa 63-96 and a non-canonical aa 97-128 EF hand), a sterile α-motif (SAM aa 132-200) followed by an α-helical TM domain (aa 212-234) and the C-terminus (aa 238-685). The first third of the cytosolic segment includes the three highly conserved

```
                        Isoform of STIM2 with Long Peptide Insertion
STIM1             ------------------------------------------------------------ 0
STIM1 isoform     ------------------------------------------------------------ 0
STIM2             ------------------------------------------------------------ 0
STIM2 isoform 1   MNAAGIRAPEAAGADGTRLAPGGSPCLRRRGRPEESPAAVVAPRGAGELQAAGAPLRFYP 60
STIM2 isoform 2   ------------------------------------------------------------ 0

                                      Signal Peptide
STIM1             ---------------MDVC---VRLALWLLWGLLLHQG---QSLS--HSHSEKATGTSS- 36
STIM1 isoform     ---------------MDVC---VRLALWLLWGLLLHQG---QSLS--HSHSEKATGTSS- 36
STIM2             ---------------------------MLVLGLLVAGAADGCELVPRHLRGRRATGSAAT 33
STIM2 isoform 1   ASPRRLHRASTPGPAWGWLLRRRRWAALLVLGLLVAGAADGCELVPRHLRGRRATGSAAT 120
STIM2 isoform 2   ---------------------------MLVLGLLVAGAADGCELVPRHLRGRRATGSAAT 33

                                                        Canonical EF Hand
STIM1             -------GANSEESTAAEFCRIDKPLCHSEDEKLSFEAVRNIHKLMDDDANGDVDVEESD 89
STIM1 isoform     -------GANSEESTAAEFCRIDKPLCHSEDEKLSFEAVRNIHKLMDDDANGDVDVEESD 89
STIM2             AASSPAAAAGDSPALMTDPCMSLSPPCFTEEDRFSLEALQTIHKQMDDDKDGGIEVEESD 93
STIM2 isoform 1   AASSPAAAAGDSPALMTDPCMSLSPPCFTEEDRFSLEALQTIHKQMDDDKDGGIEVEESD 180
STIM2 isoform 2   AASSPAAAAGDSPALMTDPCMSLSPPCFTEEDRFSLEALQTIHKQMDDDKDGGIEVEESD 93

                              Non-canonical EF Hand
STIM1             EFLREDLNYHDPTVKHSTFHGEDKLISVEDLWKAWKSSEVYNWTVDEVVQWLITYVELPQ 149
STIM1 isoform     EFLREDLNYHDPTVKHSTFHGEDKLISVEDLWKAWKSSEVYNWTVDEVVQWLITYVELPQ 149
STIM2             EFIREDMKYKDATNKHSHLHREDKHITIEDLWKRWKTSEVHNWTLEDTLQWLIEFVELPQ 153
STIM2 isoform 1   EFIREDMKYKDATNKHSHLHREDKHITIEDLWKRWKTSEVHNWTLEDTLQWLIEFVELPQ 240
STIM2 isoform 2   EFIREDMKYKDATNKHSHLHREDKHITIEDLWKRWKTSEVHNWTLEDTLQWLIEFVELPQ 153

                       SAM – Sterile α Motif
STIM1             YEETFRKLQLSGHAMPRLAVTNTTMTGTVLKMTDRSHRQKLQLKALDTVLFGPPLLTRHN 209
STIM1 isoform     YEETFRKLQLSGHAMPRLAVTNTTMTGTVLKMTDRSHRQKLQLKALDTVLFGPPLLTRHN 209
STIM2             YEKNFRDNNVKGTTLPRIAVHEPSFMISQLKISDRSHRQKLQLKALDVVLFGPLTRPPHN 213
STIM2 isoform 1   YEKNFRDNNVKGTTLPRIAVHEPSFMISQLKISDRSHRQKLQLKALDVVLFGPLTRPPHN 300
STIM2 isoform 2   YEKNFRDNNVKGTTLPRIAVHEPSFMISQLKISDRSHRQKLQLKALDVVLFGPLTRPPHN 213

                           TM Domain
STIM1             HLKDFMLVVSIVIGVGGCWFAYIQNRYSKEHMKKMMKDLEGLHRAEQSLHDLQERLHKAQ 269
STIM1 isoform     HLKDFMLVVSIVIGVGGCWFAYIQNRYSKEHMKKMMKDLEGLHRAEQSLHDLQERLHKAQ 269
STIM2             WMKDFILTVSIVIGVGGCWFAYTQNKTSKEHVAKMMKDLESLQTAEQSLMDLQERLEKAQ 273
STIM2 isoform 1   WMKDFILTVSIVIGVGGCWFAYTQNKTSKEHVAKMMKDLESLQTAEQSLMDLQERLEKAQ 360
STIM2 isoform 2   WMKDFILTVSIVIGVGGCWFAYTQNKTSKEHVAKMMKDLESLQTAEQSLMDLQERLEKAQ 273

                                       Coiled-coil 1
STIM1             EEHRTVEVEKVHLEKKLRDEINLAKQEAQRLKELREGTENERSRQKYAEEELEQVREALR 329
STIM1 isoform     EEHRTVEVEKVHLEKKLRDEINLAKQEAQRLKELREGTENERSRQKYAEEELEQVREALR 329
STIM2             EENRNVAVEKQNLERKMMDEINYAKEEACRLRELREGAECELSRRQYAEQELEQVRMALK 333
STIM2 isoform 1   EENRNVAVEKQNLERKMMDEINYAKEEACRLRELREGAECELSRRQYAEQELEQVRMALK 420
STIM2 isoform 2   EENRNVAVEKQNLERKMMDEINYAKEEACRLRELREGAECELSRRQYAEQELEQVRMALK 333

                                         Coiled-coil 2    Exon 9 VASSYLIQ insertion
STIM1             KAEKELESHSSWYAPEALQKWLQLTHEVEVQYYNIKKQNAEKQLLVAKEG--------AE 381
STIM1 isoform     KAEKELESHSSWYAPEALQKWLQLTHEVEVQYYNIKKQNAEKQLLVAKEG--------AE 381
STIM2             KAEKEFELRSSWSVPDALQKWLQLTHEVEVQYYNIKRQNAEMQLAIAKDE--------AE 385
STIM2 isoform 1   KAEKEFELRSSWSVPDALQKWLQLTHEVEVQYYNIKRQNAEMQLAIAKDEVAASYLIQAE 480
STIM2 isoform 2   KAEKEFELRSSWSVPDALQKWLQLTHEVEVQYYNIKRQNAEMQLAIAKDE--------AE 385

                                         Coiled-coil 3
STIM1             KIKKKRNTLFGTFHVAHSSSLDDVDHKILTAKQALSEVTAALRERLHRWQQIEILCGFQI 441
STIM1 isoform     KIKKKRNTLFGTFHVAHSSSLDDVDHKILTAKQALSEVTAALRERLHRWQQIEILCGFQI 441
STIM2             KIKKKRSTVFGTLHVAHSSSLDEVDHKILEAKKALSELTTCLRERLFRWQQIEKICGFQI 445
STIM2 isoform 1   KIKKKRSTVFGTLHVAHSSSLDEVDHKILEAKKALSELTTCLRERLFRWQQIEKICGFQI 540
STIM2 isoform 2   KIKKKRSTVFGTLHVAHSSSLDEVDHKILEAKKALSELTTCLRERLFRWQQIEKICGFQI 445

                                                   Inhibitory Domain
STIM1             VNNPGIHSLVAALNIDPSWMGSTRPNPAHFIMTDDVDDMDEEIVSPLSMQSPSLQSSVRQ 501
STIM1 isoform     VNNPGIHSLVAALNIDPSWMGSTRPNPAHFIMTDDVDDMDEEIVSPLSMQSPSLQSSVRQ 501
STIM2             AHNSGLPSLTSSLYSDHSWVVMPRVSIPPYPIAGGVDDLDEDTPPIVS-QFP-------G 497
STIM2 isoform 1   AHNSGLPSLTSSLYSDHSWVVMPRVSIPPYPIAGGVDDLDEDTPPIVS-QFP-------G 592
STIM2 isoform 2   AHNSGLPSLTSSLYSDHSWVVMPRVSIPPYPIAGGVDDLDEDTPPIVS-QFP-------G 497

                               STIM1 splice variant     P/S Rich Region STIM2
STIM1             RLTEPQHGLGSQRDLTHSDSESSLHMSDRQRVAPKPPQMSRAADEALNAMTSNGSHRLIE 561
STIM1 isoform     RLTEPQHGLGSQRGSSLKANRLSSKGFDPFRFGVLPPHE--------------------- 540
STIM2             TMAKPPGSLARSSSLCR----------SRRSIVPSSPQPQRAQLAPHAPHPSHPRHPHHP 547
STIM2 isoform 1   TMAKPPGSLARSSSLCR----------SRRSIVPSSPQPQRAQLAPHAPHPSHPRHPHHP 642
```

Fig. 23.1 Sequence alignment of STIM isoforms

```
STIM2 isoform 2 TMAKPPGSLARSSSLCR----------SRRSIVPSSPQPQRAQLAPHAPHPSHPRHPHHP 547
                                   STIM2 missing alternate in-frame exon
STIM1           GVHPGSLVEKLP------DSPALAKKAL----------------------LALNHGLDKA 593
STIM1 isoform   ------------------------------------------------------------ 540
STIM2           QHTPHSLPSPDPDILSVSSCPALYRNEEEEEAIYFSAEKQWEVPDTASECDSLNSSIGRK 607
STIM2 isoform 1 QHTPHSLPSPDPDILSVSSCPALYRNEEEEEAIYFSAEKQWEVPDTASECDSLNSSIGRK 702
STIM2 isoform 2 QHTPHSLPSPDPDILSVSSCPALYRNEEEEEAIYFSAEKQCIHLGL-GACKSE------- 599
                                 P/S Rich Region STIM1
STIM1           H-------SLMELSPSAPPGGS-PHLDSSRSHSPSSPDPDT--PSPVGD----------- 632
STIM1 isoform   ------------------------------------------------------------ 540
STIM2           QSPPLSLEIYQTLSPRKISRDEVSLEDSSRGDSPVTVDVSWGSPDCVGLTETKSMIFSPA 667
STIM2 isoform 1 QSPPLSLEIYQTLSPRKISRDEVSLEDSSRGDSPVTVDVSWGSPDCVGLTETKSMIFSPA 762
STIM2 isoform 2 ------------------------------------------------------------ 599

STIM1           -------------------------------------SRALQ----ASRNTRIPHLAGKK 651
STIM1isoform    ------------------------------------------------------------ 540
STIM2           SKVYNGILEKSCSMNQLSSGIPVPKPRHTSCSSAGNDSKPVQEAPSVARISSIPHDLC-- 725
STIM2isoform1   SKVYNGILEKSCSMNQLSSGIPVPKPRHTSCSSAGNDSKPVQEAPSVARISSIPHDLC-- 820
STIM2isoform2   ------------------------------------------------------------ 599
                               K Rich Region
STIM1           AVAEEDNGSIGEETDSSPGRKKFPLKIFKKPLKK     685
STIM1isoform    ----------------------------------     540
STIM2           --------HNGEKS-KKPSKI---KSLFKKKSK-     746
STIM2isoform1   --------HNGEKS-KKPSKI---KSLFKKKSK-     841
STIM2isoform2   ----------------------------------     599
```

Fig. 23.1 (continued)

coiled-coil (CC1-CC3) helices (aa 238-437), CC1 aa: 238-343, CC2 aa: 345-391, CC3 aa: 393-437. Afterwards, the inhibitory domain (aa 470-491), containing the CRAC modulatory domain (aa 475-483), a proline/serine-rich (aa 600-629) and a positively charged lysine-rich region (aa 672-685) follow (Fig. 23.2a).

The structure of the N-terminal Ca^{2+} sensor, the STIM1 EF-SAM complex has been resolved via NMR in the Ca^{2+}-bound state [37]. The EF hand domain consists of two EF hand motifs, a canonical and a non-canonical one. They both possess typical helix-loop-helix structures, α1-loop1-α2 for the canonical and α3-loop2-α4 for the non-canonical (Fig. 23.2b, left) motif, respectively. While the canonical EF hand motif coordinates a single Ca^{2+} ion, the non-canonical one lacks this ability and contributes to structural stabilization of the canonical EF hand [26, 37]. The canonical EF hand includes several negatively charged residues that have been proposed to contribute to Ca^{2+} binding [36] (Fig. 23.2b, left).

The EF hand domain is linked to the SAM domain via a short helical structure. The SAM region forms a five α-helix bundle structure (Fig. 23.2b, left) and represents another essential region of the STIM1 protein, which is vital for the regulation of store-operated Ca^{2+} entry [16, 26]. Deletion of the SAM domain has been shown to alter inducible punctae formation [38].

The EF-hand and SAM domains form a complex via hydrophobic interactions, under resting conditions. Here, two hydrophobic amino acids L195 and L199, positioned within the last α-helical segment of the SAM domain, couple to the hydrophobic cleft of the EF-hand domain [26, 37].

The TM domain of STIM1 links the EF-SAM domain to STIM1 C-terminus, in order to enable signal transmission from the N- to the C-terminus [39]. This helical portion includes three glycines (G223, G225, G226) which provide it with

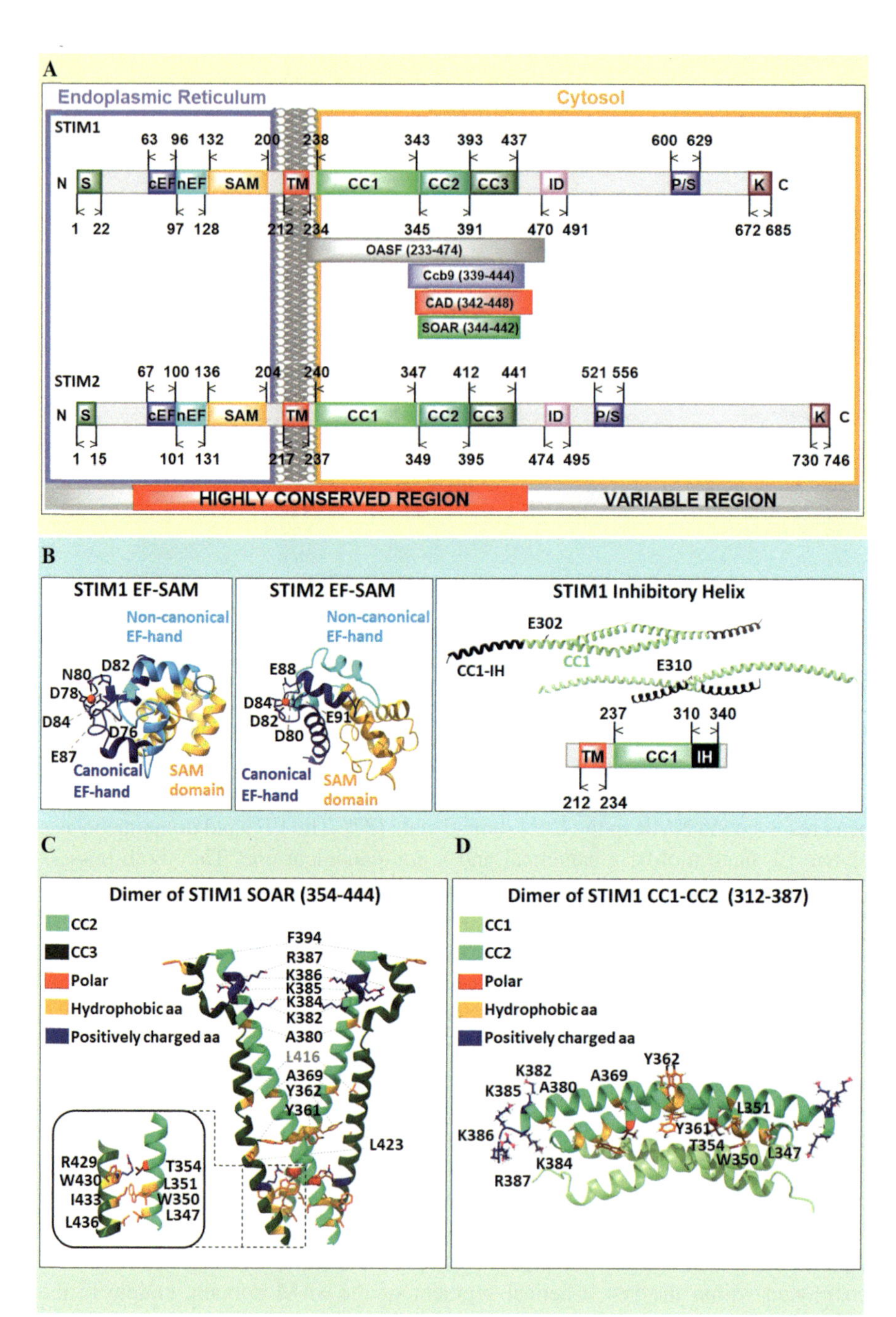

Fig. 23.2 Stromal interaction molecule 1 and 2 (STIM1, STIM2) (**a**) Scheme showing a comparison of a full-length human STIM1 (top) and STIM2 (bottom) with respect to regions critical for the regulation of the STIM1/Orai1 signaling cascade (upper part). Important fragments, such as

the flexibility required for conformational changes upon STIM1 activation [40]. Structural alterations will be addressed in more detail in the following sections.

STIM1 C-terminus has been shown to include minimal fragments such as the Orai1 activating small fragment: OASF (aa 233-474) [41], Ccb9 (aa 339-444) [42], the CRAC-activating domain: CAD (aa 342-448) [43] and the STIM Orai activating region: SOAR (aa 344-442) [44], which is sufficient to activate Orai channels. Structural predictions have revealed that the long putative CC1 domain contains three α-helical regions $CC1_{\alpha 1}$ (aa 238-271), $CC1_{\alpha 2}$ (aa 278-304), $CC1_{\alpha 3}$ (aa 308-337), respectively [45]. Currently structural resolutions of two distinct portions of STIM1 C-terminus are available: a STIM1-C-terminal, SOAR-like fragment aa 354-444 [17] (Fig. 23.2c) and a SOAR overlapping fragment aa 312-387 (Fig. 23.2d) [16]. Both fragments incorporate parts of STIM1 CC2 (aa: 345-391) and CC3 (aa: 393-437) that are supposed to be critical for oligomerization of STIM1 proteins, coupling to and activation of CRAC channels [41]. It is noteworthy that both structures of STIM1 C-terminal fragments reveal a dimeric assembly, while the two monomers assemble in an antiparallel manner with a cross point at residue Y361 (Fig. 23.2c, d). Within the crystal structure of the SOAR-like fragments arranged as dimers, the two monomers together exhibit an overall V-shaped conformation (Fig. 23.2c). The structure of each monomer resembles the capital letter "R", that is constituted by four α-helices $S_{\alpha 1}$-$S_{\alpha 4}$ ($S_{\alpha 1}$ (aa 345-391), $S_{\alpha 2}$ (aa 393-398), $S_{\alpha 3}$ (aa 400-403) and $S_{\alpha 4}$ (aa 408-437)). Within the NMR structure of the SOAR dimer, each monomer forms a bend between the two helical portions $CC1_{\alpha 3}$ and CC2 [16, 17] (Fig. 23.2d).

The inhibitory (aa 470-491) [46] or CRAC modulatory domain (aa 474-485) [47] contains seven negatively charged residues which are critical for the maintenance of typical biophysical characteristics of CRAC channels, as explained later in this review.

Downstream of the C-terminal coiled-coil regions STIM1 possesses a Ser/Pro-rich region (aa 600-629) which is important for proper targeting of STIM1 into clusters close to the cell membrane upon Ca^{2+} store depletion [48].

Fig. 23.2 (continued) OASF, Ccb9, CAD and SOAR are further represented as insets. (**b**) The left and middle panels show high resolution EF-SAM domains structures of human STIM1 and STIM2, each loaded with a Ca^{2+} ion (red spheres), respectively. The residues with proposed Ca^{2+} binding ability are highlighted. The right panel displays the portion of the STIM1 CC1- inhibitory helix with the critical residues highlighted. (**c**) The crystallographic structure of the STIM1 SOAR dimer, forming a V-shape, consists of CC2 and CC3 domains. Each monomer resembles the capital letter ‚R". Residues that represent potential interaction sites within the dimer and those mediating coupling to Orai1 are highlighted. Left inset: Magnified view of amino acids involved in dimer interactions between the N-terminal portion of the first SOAR monomer and the C-terminal portion of the second SOAR monomer. (**d**) The NMR structure consists of a dimer of STIM1 $CC1_{\alpha 3}$-CC2 monomers that couple in an antiparallel manner. Each monomer bends with a sharp kink between the two coiled-coil domains

Furthermore, a lysine-rich region at the end of STIM1 C-terminus [34, 49] also participates in correct STIM1 targeting. However, it is dispensable for CRAC channel activation [50]. It resembles a phosphatidyl inositol-4,5-bisphosphat (PIP_2) binding domain [50, 51]. The role of PIP_2 in STIM1 regulation will be explained in the section entitled "Lipid mediated regulation of the STIM1/Orai1 complex.".

23.2.3 STIM2

The STIM2 protein contains 746 amino acids (Fig. 23.2a) and critical domains are analogous to those for STIM1 as described above. Briefly, this protein includes the ER signal peptide (aa 1-15), a luminal EF hand (a canonical aa 67-100 and a non-canonical aa 101-131 EF hand) (Fig. 23.2b, middle), a sterile α-motif (SAM aa 136-204) followed by an α helical TM domain (aa 217-237) and the large cytosolic part, which includes the three coiled-coil (CC1-CC3) helices (aa 240-441; CC1 aa: 240-347, CC2 aa: 349-395, CC3 aa: 412-441), a proline/serine-rich (aa 521-556) and a positively charged lysine-rich region (aa 730-746).

Despite the many similarities and $\sim$61% sequence identity of STIM1 and STIM2 isoforms, STIM2 possesses several notable differences compared to STIM1, providing evidence for distinctions in its structure and function.

STIM1 and STIM2 EF-SAM domains possess significant differences, despite having a sequence similarity of $\sim$85%. Specifically, the SAM domain of STIM2 (Fig. 23.2b middle), in contrast to that of STIM1, consists of an additional third non-polar residue V201 that packs into its hydrophobic cleft. Thus, the hydrophobic core of STIM2 – unlike STIM1 – shows not only an enhancement in size but also in stability in the presence of bound Ca^{2+}. Additional support for the stability of STIM2 is provided by the existence of a possible ionic bond between D200 in the SAM domain and K103 in the EF hand domain. Thus, due to the presence of more stable hydrophobic and electrostatic interactions, the STIM2 SAM domain oligomerizes more efficiently than that in STIM1 [26, 37].

A comparison of STIM1 and STIM2 C-termini exhibits considerable differences within the last third, termed as variable region. Homology modelling of the STIM1 C-terminal SOAR region predicts an almost identical structure. Nevertheless, one main difference represents the non-conserved residue F394 in SOAR of STIM1 [52], corresponding to L398 in STIM2. Chimera and single point mutation studies by Wang et al. [52] revealed clear functional differences in the dependence of the introduced residue at this position, which will be outlined in detail in the section entitled "Activation mechanisms of the STIM1/Orai signaling machinery".

The lysine-rich region at the very end of STIM2 is notably larger than that of STIM1 and possesses an enhanced affinity for PIP_2, as explained in more detail in the section entitled "Lipid-mediated regulation of the STIM1/Orai1 complex" [51].

23.2.4 Orai Proteins

The Orai protein family includes three highly conserved isoforms, Orai1-3, expressed in the plasma membrane. Orai, as the pore-forming subunit of CRAC channels, is unique among the huge diversity of Ca^{2+} ion channels due to its high degree of Ca^{2+} selectivity as well as its hexameric structure. [2, 12, 14, 53] In native cells, Orai proteins are assumed to form homo- as well as heteromeric assemblies [54, 55]. Each Orai subunit is composed of four transmembrane (TM) domains that are connected via one intracellular and two extracellular loops and flanked by a cytosolic N- and C-terminal strand (Fig. 23.3a, b) [56–58]. The TM1 segment (aa 92-106) is fully conserved among the protein family, while TM2 (aa 118-140), TM3 (aa 174-197) and TM4 (aa 236-258) share ~81-87% sequence identity. Flanking strands and connecting loops possess major isoform specific structural differences (Fig. 23.2) [59]. Specifically, the cytosolic N- (aa 1-90) and C-terminal (aa 265-301) strands show only 34% and 46% sequence identity, respectively [60]. Furthermore, especially the cytosolic extension of TM2 has been suggested as longer in Orai3 than in Orai1. Thus, Orai3 in contrast to Orai1, contains a shorter flexible loop2 portion connecting TM2 and TM3 [61].

In 2012, Hou et al. [14] published the crystal structure of *Drosophila melanogaster* dOrai, that exhibits a hexameric stoichiometry (Fig. 23.3c). In accord with the dOrai crystal structure, the results of concatemeric studies confirmed that the functional state of human Orai1 represents a hexameric assembly [62, 63]. The Orai pore is formed by the six TM1 domains arranged as a ring in the center of the channel complex. Its detailed composition (Fig. 23.3d) is explained in the section entitled "Ca^{2+} ion conduction pathway". The other TM regions, TM2 – TM4 are arranged around the pore in two rings, whereas TM2 and TM3 form the second and TM4 the third ring. The TM1 domains extend further into the cytosol [14] culminating into a conserved, helical region, that has been termed the so-called ETON (Extended TM Orai1 NH_2-terminal; aa 73-90 in hOrai1) region [64] (Fig. 23.3d, e) for hOrai homologues. It is approximately 20 Å long and includes the last 20 amino acids of the Orai N-terminus. Both, TM2 and TM3 have been shown to expand into the cytosol by a couple of helical turns [14]. TM4 is divided by a kink at P245 in hOrai (corresponding to P288 in dOrai) into two regions. Helical extensions of TM4 exposed to the cytosol, represent the Orai C-termini and are connected to TM4 via a highly conserved hinge region (aa 261-265). Within a hexamer, three dimers of Orai subunits exhibit antiparallel oriented C-termini, which respectively form an angle of 152° [14, 65]. Thereby, they capture a belt-like assembly surrounding the intracellular side of the channel [15]. The structural properties of the Orai C-termini provide in addition to the sixfold, a threefold symmetry to the Orai channel. (Fig. 23.3c). Unfortunately, structural resolution of the more flexible Orai segments including the intracellular, both extracellular loops and approximately the first ~70 residues of the N- terminal strands are still missing. Nevertheless, flexible loop portions have been integrated into a hOrai1 model that is based on the dOrai X-ray structure [66].

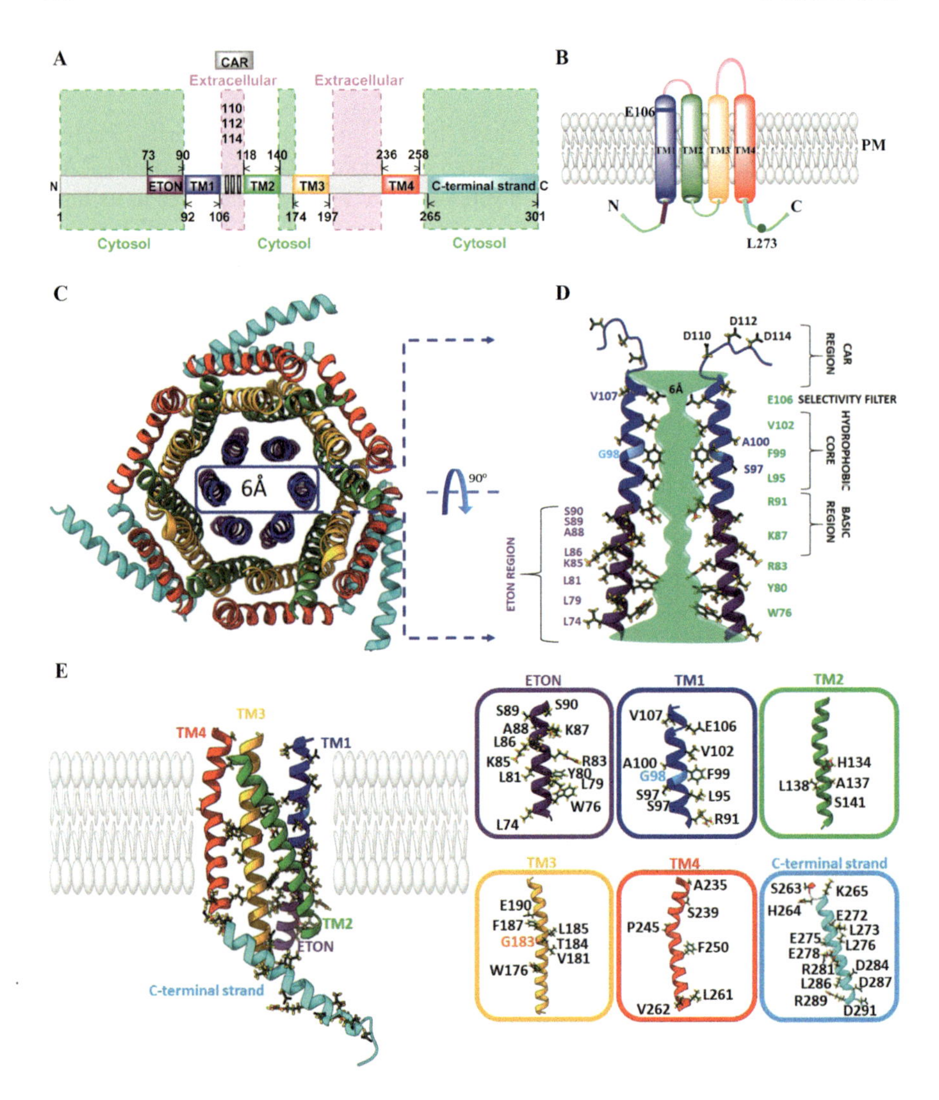

Fig. 23.3 Orai1 channel (**a**) Scheme depicting the full-length human Orai1 channel with highlighted regions and residues that are essential for Orai1 function. (**b**) Scheme of overall full-length human Orai1 subunit structure representing distinct extracellular, cytosolic and transmembrane-spanning regions. The selectivity filter of the channel, E106, is depicted as a blue line; the residue L273 that causes impaired STIM1 binding upon single point mutation is shown as a green circle. (**c**) The cartoon shows the Orai1 channel possessing a hexameric assembly based on the *Drosophila* dOrai X-ray structure. While the inner ring surrounding the pore is formed by TM1, other TM domains within the six subunits are constituted into concentric rings around the pore. (**d**) The cartoon displays two opposite TM1 domains with the cytosolic and extracellular helical extension and highlights the essential residues lining the pore of the channel depicted in teal. Residues of the N-terminal region, ETON, are depicted in purple. On the C-terminal side is the CAR region with its residues depicted in black. The selectivity filter, hydrophobic core and basic region of the pore are highlighted. (**e**) The cartoon of one Orai1 subunit with four TM segments along with N- and C- terminal helices depicted in distinct colors (same as applied within **a**–**d**) displays the residues in more detail that are known to manifest proper Orai1 channel function and maintain the closed state of the Orai1 channel

```
Orai1    MHPEPAPPPSRSSPELPPSGGSTTSGSRRSRRRSGDGEPPGAPPPPPSAVTYPDWIGQSY 60
Orai2    --------------------------MSAELNVPIDPSAPACPEPGHKGMDYRDWVRRSY 34
Orai3    -----------------MKGGEGDAGEQAPLNP--------EGESPAGSATYREFVHRGY 35
                    ETON region              TM1            CAR
Orai1    SEVMSLNEHSMQALSWRKLYLSRAKLKASSRTSALLSGFAMVAMVEVQLDADHDYPPGLL 120
Orai2    LELVTSNHHSVQALSWRKLYLSRAKLKASSRTSALLSGFAMVAMVEVQLETQYQYPRPLL 94
Orai3    LDLMGASQHSLRALSWRRLYLSRAKLKASSRTSALLSGFAMVAMVEVQLESDHEYPPGLL 95
             TM2
Orai1    IAFSACTTVLVAVHLFALMISTCILPNIEAVSNVHNLNSVKESPHERMHRHIELAWAFST 180
Orai2    IAFSACTTVLVAVHLFALLISTCILPNVEAVSNIHNLNSISESPHERMHPYIELAWGFST 154
Orai3    VAFSACTTVLVAVHLFALMVSTCLLPHIEAVSNIHNLNSVHQSPHQRLHRYVELAWGFST 155
             TM3
Orai1    VIGTLLFLAEVVLLCWVKFLPLKKQPGQPRPT---SKPPASGA----------------- 220
Orai2    VLGILLFLAEVVLLCWIKFLPVDARRQPG-P------PPG-------------------- 187
Orai3    ALGTFLFLAEVVLVGWVKFVPIGAPLDTPTPMVPTSRVPGTLAPVATSLSPASNLPRSSA 215
                                              TM4
Orai1    --------------AANVSTSGITPGQAAAIASTTIMVPFGLIFIVFAVHFYRSLVSHKT 266
Orai2    --------------------PGSHTGWQAALVSTIIMVPVGLIFVVFTIHFYRSLVRHKT 227
Orai3    SAAPSQAEPACPPRQACGGGGAHGPGWQAAMASTAIMVPVGLVFVAFALHFYRSLVAHKT 275
             C-terminal strand
Orai1    DRQFQELNELAEFARLQDQLDHRGDHPLTPGSHYA           301
Orai2    ERHNREIEELHKLKVQLDGHER-SLQVL-------           254
Orai3    DRYKQELEELNRLQGELQAV---------------           295
```

Fig. 23.4 Sequence alignment of Orai isoforms

It is noteworthy that several residues (Fig. 23.4e) within Orai TM domains have recently been reported as important for maintaining the entire channel complex in the closed state, since their single or multiple point mutations have resulted in constitutively active channels. These are known as gain-of-function mutations [65, 67–73]. One recently described gain-of-function point mutant Orai1 H134A (dOrai H206A) [68] has been utilized for further crystallographic studies representing one open conformation of the channel [15], in contrast to the potential quiescent state of the dOrai structure published in 2012 [14].

In accordance with the dOrai crystal structure, the novel dOrai H206A (corresponds to H134A in hOrai1) mutant structure exhibits also a hexameric assembly [15]. Main conformational alterations have been observed in the basic region of the pore as well as the TM4 regions. Unlike to the dOrai closed pore, the open state of the dOrai mutant displays an extension of approximately 10Å at dOrai K159 (corresponding to K87 in hOrai1) within the basic pore region. While in the quiescent state TM4 is separated into two regions by a kink at P288 (corresponds to P245) and connected to the C-terminus via another kink, the so-called nexus (in hOrai1 aa 261-265) [65], the open state of the Orai mutant exhibits a straight helix formed by the two TM4 segments and the C-terminus thus, expanding approximately 45Å into the cytosol [15].

It is important to note that the Orai crystal structures available so far display the channel's potential open and closed conformations only in the absence of STIM1. However, physiologically, Orai channels are activated via their coupling to STIM1, which potentially leads to a distinct conformation that differ from the structure of specific gain-of-function point mutants. Thus, the structural resolution of a STIM1-Orai1 complex would represent the ultimate goal to provide insights in Orai conformational changes from the closed to the open state at a more physiological level.

23.3 Activation Mechanisms of the STIM1/Orai Signaling Machinery

23.3.1 STIM1 in the Resting State

The STIM1 conformation in the resting state is controlled by STIM1 N-terminal, transmembrane and C-terminal domains, which is explained in detail in the following. Inactive STIM1 proteins exist as dimers [41, 74] that are uniformly distributed throughout the ER membrane. They are stably connected to microtubules [75, 76] and their average diffusion velocity is 0.22 ±0.07 μm/s [75].

In the quiescent state, the EF-hand domain at the luminal N-terminus of STIM1 binds Ca^{2+} via negatively charged residues (aspartates and glutamates) [77]. Mutations of some of these residues (D76A, D76/78N, E87Q) lead to the loss of Ca^{2+} binding ability and, hence, to constitutive STIM1 activity [9, 11, 78]. The resting conformation of the EF-SAM domain represents a compact structure consisting of mainly α-helices (Fig. 23.2b). They are stabilized by hydrophobic interactions between the EF-hand and the SAM domain (see previous section) [36]. Several studies have revealed that the EF-SAM domain exists as a stable monomer in the quiescent state, while it forms dimers upon dissociation of Ca^{2+}, findings that are in line with STIM1 oligomerization upon store-depletion [36, 37]. In contrast, Huang et al. [77] have shown that the EF-hand domains isolated from STIM1 tend to form dimers already in the presence of Ca^{2+}, in accord with the STIM1 dimeric state under resting conditions.

STIM2 already leaves the quiescent state upon smaller changes in the luminal Ca^{2+} concentration than STIM1, although Orai channel activation by STIM2 is slower compared to STIM1 [37]. This underlies differences in sequence and structure of the EF-SAM domain of STIM1 and STIM2 that lead to distinct affinities for Ca^{2+} with a K_d of 200 μM and 500 μM for STIM1 and STIM2, respectively [36, 37, 79]. While the canonical EF-hand of STIM2 possesses a lower Ca^{2+} affinity than that of STIM1, hydrophobic and electrostatic interactions between the EF-hands and the SAM domains are more stable in STIM2 than in STIM1. Thus, the STIM1 EF-SAM domain undergoes faster unfolding and aggregation upon dissociation of Ca^{2+} than that of STIM2 [26, 37]. Furthermore Subedi et al. [80] recently revealed that STIM2 is already located in ER-PM junctions in the resting state and assists STIM1 in unfolding and coupling to Orai1 upon minimal store-depletion. In addition, a splice variant STIM1L has been shown to be located in the ER-PM junctions in the resting state, which has been found to underlie a 106 aa long extension in its C-terminus. The latter enables the anchoring of STIM1L to the cytoskeleton [81].

In addition to the EF-SAM domain, the conformation of the TM domains also controls the resting state of STIM proteins. Chemical crosslinking and computational modeling have revealed that the TM domains of two STIM1 proteins mainly interact at their C-terminal portions (aa 221-232) whereas their N-terminal parts (aa 212-220) are kept apart via a crossing angle of approximately 45° in the resting state [40] (Fig. 23.5a). A tryptophan-scanning approach was used by Zhou et al. [40]

to identify two residues (I220, C227) within the transmembrane domain that keep STIM1 in the closed state, as their mutation (I220W, C227W) leads to constitutive STIM1 activity.

The dimeric state of STIM1 C-termini under resting conditions [38, 41, 82–85] is mediated by CAD/SOAR and further supported by CC1 [74]. Based on the crystal structure of human SOAR, STIM1 dimer formation is maintained via hydrophobic, hydrogen bond and stacking interactions between C-terminal residues of one SOAR (R429, W430, I433, L436 in $\alpha4$) and N-terminal residues of another one (L347, W350, L351, T354, Y361 in $\alpha1$). Their importance is evident from the results of mutagenesis studies, as interference with the dimer formation at those positions impairs the SOAR binding to Orai. [17] Moreover, the NMR structure of the overlapping SOAR domain resolves the interactions of $CC1_{\alpha3}$ (aa 320-331) and CC2 (aa 355-369) within the monomers, respectively [16] (Figs. 23.2c, d and 23.5a). These NMR data provide valuable information on the SOAR structure. Nevertheless, one still has to keep in mind that the STIM1 fragment (aa 312-387), which has been structurally resolved, is unable to bind to and activate Orai1 [41].

The cytosolic part of STIM1 is also involved in controlling the inactive state via an intramolecular clamp between CC1 and SOAR (Fig. 23.5a), particularly between $CC1_{\alpha1}$ and CC3 [45, 86]. Single point mutations within $CC1_{\alpha1}$ (L248S, L251S, L258S) and CC3 (L416S, L423S) engineered in a double labeled OASF fragment resulted in decreased FRET values, indicating that these residues are involved in maintaining the tight conformation of STIM1 [83, 86]. Additionally, Y316 in CC3 has been shown to contribute to the inactive state of STIM1, as a mutation to an alanine (Y316A) leads to STIM1 cluster formation without store-depletion [87]. In contrast, mutating the arginine at position 426 in CC3 to a leucine (R426L) strengthened the CC1-CC3 clamp, thereby stabilizing the quiescent state of STIM1 [86]. Moreover, four other residues, L258 and L261 in $CC1_{\alpha1}$ and V419 and L416 in CC3 have been proposed to support this intramolecular clamp, keeping STIM1 in the closed state [40].

The gain-of-function point mutant STIM1 R304W, which has been associated with Stormorken Syndrome, suggests that residue R304 in CC1 supports the maintenance of STIM1 in the closed state [8]. A recent study on this mutant revealed that this specific point mutation induces an extension of the $CC1_{\alpha2}$ helix and lead to increased CC1 homomerization [88], which is in line with the observed destabilizing effect of $CC1_{\alpha2}$ leading to an open state as mentioned in the next section.

In contrast to STIM1, the CC1-CC3 interaction within STIM2 has been found to be much weaker, thus providing one potential explanation for the observed clustering of STIM2 within the ER-PM junction in unstimulated cells [80].

In summary, several parts of STIM1 are involved in keeping the protein in the resting state, including the EF-hand domain as long as Ca^{2+} is bound, specific residues within the crossing TM domains of a STIM1 dimer as well as the C-terminus via the intramolecular clamp between CC1 and SOAR.

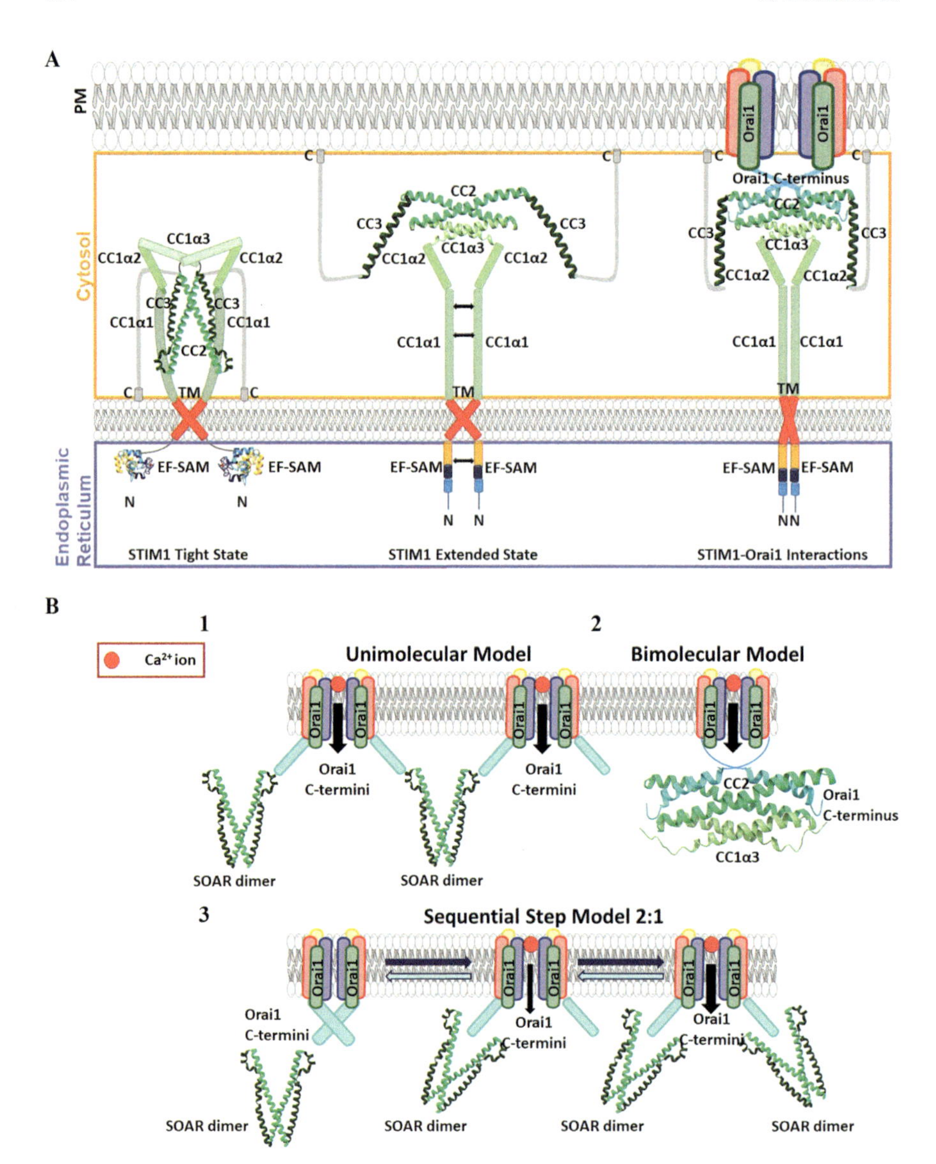

Fig. 23.5 Hypothetical model of the binding of STIM1 to Orai1 (**a**) As long as Ca^{2+} ions are bound to the EF-hand of STIM1, the latter stays in an inactive, tight state (left part). However, upon store depletion, when the EF-hand domain of STIM1 loses the Ca^{2+} ion, conformational changes are induced, and STIM1 proteins reassemble into the active, extended state (middle part). Subsequently, the crossing angle of the STIM1 TM helices is altered and the inhibitory, intramolecular clamp between CC1α1 and CC3 is released, which finally results in a STIM1 interaction with the Orai channel in the plasma membrane (right). (**b**) Three different stoichiometries of the functional STIM1-Orai1 complex have been proposed (according to

23.3.2 STIM1 Oligomerization and Conformational Changes Upon Activation

STIM1 activation is initiated upon store-depletion and induces conformational changes in STIM1 that propagate from the N-terminus via the TM domain to the C-terminus. This leads to STIM1 oligomerization, punctae formation and finally coupling to and activation of Orai. Initial studies employing FRET microscopy have revealed an increase in the STIM1 homomeric interaction following store-depletion which decreases upon store-refilling [50, 89]. It is worth mentioning, that STIM1 oligomerization occurs prior to STIM1-Orai1 coupling [50].

The first step in the STIM1 activation cascade is the dissociation of the EF-hand bound Ca^{2+} ion upon store-depletion, which leads to the structural destabilization of the EF-SAM motif [79, 90]. Hydrophobic regions become exposed, leading to the dimerization of neighboring STIM1 N-termini. In accordance with this observation, STIM1 mutants lacking the entire C-terminus have been reported to form multimers upon the loss of Ca^{2+} binding [74]. Oligomerization upon Ca^{2+} store-depletion is in line with the constitutive punctae formation seen in the STIM1 EF-hand mutants (D76A, E87A), that are defect in Ca^{2+} binding [9, 23]. In addition, oxidative stress has been hypothesized to reduce the Ca^{2+}-binding affinity of STIM1, as native CRAC currents have been shown to activate in the presence of micromolar concentrations of H_2O_2 [91–94].

Structural rearrangements within the STIM1 N-terminus lead to conformational changes in the TM domains, allowing the transmission of the activation signal to the C-terminus. STIM1-TM domains have been found to possess distinct conformations in the closed versus the open state. NMR and LRET experiments have revealed that store-depletion decreases the crossing angle of interacting TM regions thereby bringing the N-terminal parts (aa 214-220) closer together [40]. In line with this observation, cysteine crosslinking experiments have shown that the TM helices

◄

Fig. 23.5 (continued) [16, 217]). (1) In the unimolecular SOAR-Orai coupling model (STIM1-Orai1 1:1 ratio) one of the SOAR monomers of a SOAR dimer interacts with a single Orai1 channel subunit and induces gating (upper part left). The second monomer of the dimer binds to another Orai channel, thus enabling clustering of Orai channels. (2) Bimolecular coupling model shows (STIM1-Orai1 1:1 ratio), based on the NMR [16] structure using isolated partial SOAR fragments and SOAR-coupling C-terminal strand of Orai, a dimer of STIM1 which binds to the crossing C-termini of two adjacent Orai1 subunits in the hexameric Orai1 channel, thus inducing gating (left). (3) The sequential step model shows, first, the resting state of the Orai channel which possesses the assembly with an antiparallel configuration of neighboring Orai1 C-terminal strands providing the binding site for a SOAR dimer (lower part, left). In the next step, Orai1 C-terminal strands extend towards the cytosol leading to only a partial pore opening (middle). Finally, the dissociated Orai1 C terminal strands of the partially active channel represent a new binding site for the SOAR dimer with a final 2:1 stoichiometry (right)

of two STIM1 proteins move closer together after store-depletion and undergo a change in the orientation towards each other, as the crosslinking of residues L216, S219, G223, G226, A230 and Q233, when mutated to a cysteine, has only been visible in the active state [40, 95]. Due to their high flexibility, three glycines (G223, G225, G226) within the transmembrane domain likely facilitate the conformational changes necessary to transmit the activation signal from the luminal side to STIM1 C-terminus [40, 96]. It is worth mentioning that STIM1 TM rearrangements are not only triggered by luminal Ca^{2+} but can also be governed by alterations in the C-terminus, as has been shown via the L251S mutation. The latter induces constitutive activity of STIM1 and already allows in the presence of Ca^{2+}a TM cysteine crosslinking pattern [95], as observed under wild-type conditions in the absence of Ca^{2+}. Thus, while the activation signal within STIM1 is transmitted from the N- to the C-terminus, conformational changes within STIM1 have been assumed to occur in a bidirectional manner.

The final step in the activation cascade of STIM1 represents the signal transmission from the TM domain to the C-terminus inducing conformational changes together with oligomerization. Here, one main step represents the release of the intramolecular clamp between CC1 and CC3, thereby exposing SOAR [40, 45, 83]. By labeling STIM1 fragments on both sides a reduction in intramolecular FRET/LRET has been observed upon the release of the intramolecular clamp either upon coupling to Orai1, single point mutations such as L248S, L251S, L258S or the deletion of CC1 domain [83, 86]. Along the same lines, ER-targeted CC1 domains are unable to couple to SOAR/CAD fragments when mutated (e.g. at L251S) [40, 45]. Moreover, while CC1 occurs as a monomer in solution and binds to CAD, artificially crosslinked CC1 fragments form dimers that exhibit impaired interaction with cytosolic CAD/SOAR [83] that is compatible with a di-/homomerization of the CC1 region and contributes to the switch into an extended conformation [83]. Additionally, FIRE (FRET-derived Interaction in a Restricted Environment) experiments have shown that $CC1_{\alpha2}$ plays a destabilizing role, as the deletion of this region significantly delays the activation of Orai1 by STIM1 $\Delta CC1_{\alpha2}$. Breakup of the CC1-CC3 clamp leads to enhanced CC1 homomerization and SOAR exposure, while CC3 is involved in formation of higher order STIM1 oligomers [16, 40, 41, 45, 83, 86]. Mutation of residue R429 in CC3 to a cysteine (R429C), associated with combined immune deficiency [97], does not affect the STIM1 transition to the active state, but reduces STIM1 oligomerization and its coupling to Orai1 [98]. A C-terminal region overlapping with CC3 (aa 421-450) forms the STIM1 homomerization domain (SHD), the deletion of which leads to drastically reduced STIM1 oligomerization and loss of Orai1 activation [41].

In contrast to STIM1, clustering of STIM2 is more pronounced, probably due to the more relaxed and open conformation of OASF in STIM2 than in STIM1. Herewith, STIM2 helps STIM1 to transit into the active state, thus enabling STIM1/Orai1 activation [80]. Furthermore, the accessory protein STIMATE, a

STIM-activating enhancer located in the ER membrane, has been found to interact with STIM1 CC1 in order to support the release of SOAR upon store-depletion [99].

A STIM2 splice variant, STIM2.1 has been found to couple to, but not gate Orai1, and it functions as a negative regulator for STIM1 and STIM2 wild-type. This inhibitory function arises due to an insertion of 8 aa in the CAD region of STIM2, which potentially interferes with stable CAD formation. [33]

In addition to Ca^{2+} dependent regulation of STIM1 function, STIM1 glycosylation sites within the SAM domain have been found to control its function depending on the properties of the inserted amino acids. Specifically, a STIM1 N131D N171Q mutation has been found to create gain-of-function, thus enabling not constitutive but faster Orai channel activation, suggesting that these mutations facilitate STIM1's transition from the closed to the open state. Intriguingly, this STIM1 DQ mutant reduced the amount of Orai protein in the cell. This suggests that an alternative mechanism leads to enhanced Orai activation, likely due to enhanced oligomerization rates or potentially increased STIM1 dimer fusion. [100]

Further phosphorylation sites in Orai1 possess regulatory roles. Specifically, the N-terminus includes two serines, S27 and S30, which, when mutated, enhance CRAC channel activation upon mutation. This suggests that protein kinase C (PKC) inhibits CRAC channel activation upon phosphorylation of these sites [101]. The residue Y361 in STIM1 CC2 has been found to be phosphorylated by the proline-rich kinase 2 upon store-depletion. Knocking out this phosphorylation site (Y361F) still leads to STIM1 punctae formation, but coupling to Orai is completely abolished [102]. Additional phosphorylation sites within the STIM1 C-terminus (S468, S668) regulate store-operated Ca^{2+} currents during meiosis and mitosis [20, 103, 104]. Furthermore, STIM1 has been reported to be regulated by the extracellular-signal-regulated kinases 1 and 2 (ERK1/2) [105], involving the phosphorylation of specific sites within the STIM1 C-terminal serine/proline-rich region.

In summary, upon store-depletion and dissociation of the Ca^{2+} ion from the STIM1 EF-hand domain, the activation signal is transmitted via the TM domain to the C-terminus. Subsequently, the intramolecular clamp between CC1 and CC3 in the C-terminus is released, thereby exposing SOAR, the binding domain for Orai. In addition, STIMATE, glycosylation sites and several phosphorylation sites are involved in the regulation of the STIM's active state.

23.4 STIM/Orai Coupling

Upon store-triggered conformational changes of STIM1, this Ca^{2+} sensor protein interacts with the pore forming CRAC channel component, Orai. In the following section, the main coupling domains within STIM1 and Orai that establish their assembly and activation will be explained in detail.

23.4.1 Main Coupling Domains Within STIM1

The STIM1 C-terminus located in the cytoplasm contains sites that are essential for direct coupling to Orai1. In the STIM1 C-terminus, OASF, SOAR, CAD or Ccb9 (Fig. 23.2a) are sufficient for Orai activation [41–44]. All of these fragments comprise the CC2 (aa 345-391) as well as parts of the extended CC3 (aa 393-450) domain, the latter including the STIM1 homomerization domain (SHD, aa 421-450) [41]. Park et al. [43] showed a direct interaction between CAD and Orai1. Additional studies with STIM1 C-terminal and Orai1 cytosolic fragments have revealed a strong interaction between CAD and Orai1 C-terminus and, a weak one with the N-terminus, and no interaction with the loop2 [43, 61, 64]. By employing a slightly longer Orai1 loop2 fragment, our recent study revealed interactions with a STIM1-C-terminal fragment [61].

The NMR structure of the so-called SOAP (STIM-Orai association pocket) clearly shows that two STIM1 C-terminal fragments (aa312-387) bind to two Orai C-termini in an antiparallel manner [16]. The residues involved in this association are L347, L351 of one CC2, Y362, L373 and A376 of the second CC2, but also a positively charged cluster (K382, K384, K385 and K386) [16]. This findings are in line with those reported in previous publications which have shown that mutations within this region (L347R, L351R, L373S, A376K) disrupt STIM1 binding to Orai1 [16, 106].

A more recent discovery shows that the α2 domain of SOAR (aa 393-398), located between CC2 and CC3, plays a role in the coupling as well as activation process of SOCE, as the non-conserved residue F394 therein seems to represent an important interaction site within STIM1 [52]. It is worth mentioning that residue F394 is the only non-conserved residue between STIM1 and STIM2 (L398) within α2 of SOAR, which may also contribute to the distinct behavior of the two homologues. F394 in STIM1 has been hypothesized to interact either with the Orai1 N-terminus or the Orai1 hinge plate [65], a region between TM3 and TM4 that forms hydrophobic interactions (L174, L261) at the more cytosolic side of Orai1 [52, 107].

In summary, SOAR, located within the STIM1 C-terminus, represents an essential binding site for Orai channels. While the binding of STIM1 C-terminal residues to the Orai1 C-terminus has been clearly defined, residues mediating potential interactions with other Orai cytosolic sites are currently elusive.

23.4.2 Main Coupling Domains Within Orai1

STIM1 coupling to and activation of Orai1 involves cytosolic domains of the Orai protein for either direct or indirect interaction. The main STIM1 binding site within Orai1 represents the C-terminus. The Orai1 C-terminus contains two hydrophobic residues (L273, L276) which have been identified as important for STIM1 coupling, as their mutation to more hydrophilic amino acids (S/D) completely abolished

STIM1 binding to Orai1 [48, 89]. Interestingly, only double, but not single point mutations in the C-termini of Orai2 and Orai3 are sufficient to abolish STIM1 binding [106]. Besides L273 and L276, the resolution of the SOAP structure uncovered additional residues within Orai1 C-terminus that are involved in STIM1 binding: R281, L286 and R289 [16]. The crystal structure of dOrai reveals that the C-termini of two adjacent subunits are crossing each other in an antiparallel manner at a crossing angle of 152° [14]. Crosslinking of cysteines at positions L273 and L276 impaired STIM1 mediated Orai1 channel activation, thus locking it into the inactive state [108]. The Orai1 C-terminus is connected to TM4 by a bent region [14], the so-called nexus (aa261-265). It is assumed to alter its configuration upon STIM1 coupling and initiates signal transmission to the pore [15, 65]. Indeed, point mutations within the nexus (L261A V262N H264G K265A; Orai1 ANSGA) can lead to a constitutively active channel that is accompanied by a reduction in STIM1 binding. This suggests that an altered Orai1 C-terminus conformation affects both STIM1 coupling and the channel's active state [65]. Thus, there is evidence that the orientation of the C-termini for proper STIM1 binding is the key to preserving store-operated Orai1 channel activation.

The Orai1 C-terminus representing the main and indispensable binding site for STIM1 [43, 48, 89], but there is also evidence for the involvement of the N-terminus in coupling, although to a weaker extent [43, 64]. These findings are based on in vitro interaction studies with N-terminal fragments. Overexpression studies of Orai1 N- and C-terminal truncation or single point mutants have allowed researchers to clearly distinguish between STIM1 binding either to the N- or C-terminus, as a deletion of or single point mutation within the C-terminus has led to total loss of STIM1 binding. This suggests that if there is STIM1 binding to other cytosolic domains of Orai1, STIM1 coupling to the Orai C-terminus is an inalienable prerequisite for this. Nevertheless, both N-terminal truncation ($\Delta N_{1-76/78}$) and point mutations (L74/W76EE/RR/SS, R83/K87A and K85E) within the extended transmembrane Orai1 N-terminal (ETON) region (aa 73-90), lead to reduced STIM1 binding, pointing to a potential direct or indirect involvement of the Orai N-terminus in STIM1 coupling. [64, 109]. It is worth mentioning that although the entire ETON region is conserved among the Orai homologues, Orai1 activation is lost upon N-terminal truncation up to residue 76 (Orai1 ΔN_{1-76}), whereas Orai3 activation is only lost upon deletion of the first 57 residues (Orai3 ΔN_{1-57}), corresponding to residue 82 in Orai1 [110]. These observations suggest an inhibitory role for Orai1 loop2 [61].

Thus, the loop2 of Orai1 has also been assumed to play a role in the gating process. Exchanging the loop2 in Orai1 with that of Orai3 allows the recovery of activation for some inactive Orai1 N-terminal mutants (e.g. Orai1 ΔN_{1-78}, Orai1 L74E W76E) [61] (Fig. 23.6a), indicating that loop2 has distinct properties in the two homologues. The observed effects are accompanied by enhanced STIM1 coupling to the N-truncated Orai1-Orai3-loop2 chimera. Indeed, experiments employing the FIRE system have revealed an interaction between a STIM1 C-terminal and the Orai1 loop2 fragment; this interaction, however, has the same extent as with the analogous Orai3-loop2 segment [61]. Further investigation of the role of loop2 revealed that it already impacts CRAC channels at the level of Orai independent of STIM1. Specifically, diverse gain-of-function Orai mutants, that lose their function upon N-terminal truncation, recover their activity upon swapping Orai3-loop2, even without co-expression of STIM1, except for Orai1 ANSGA (Fig. 23.6m). MD simulations revealed that the helical extension of TM2 into the cytosol is shorter in Orai1 than in Orai3, resulting in a longer and more flexible loop2 segment in Orai1 as compared to that in Orai3 [61]. This higher flexibility may facilitate interactions between loop2 and the truncated N-terminus in Orai1, causing this inhibitory effect within Orai1 $\Delta N_{1\text{-}78}$, but not within Orai1 $\Delta_{1\text{-}78}$ Orai3-L2. This altered Orai conformation at the cytosolic side is thought to affect STIM1 binding either in a direct or indirect manner. It is probable that the loop2 and the N-terminus function as individual binding sites or form a new binding site, but this requires further proof. Cysteine crosslinking experiments have shown that STIM1 induced Orai1 currents are drastically diminished upon the crosslinking of K78 in the N-terminus and E166 in loop2 [61]. Thus, in wild-type Orai1 a permissive communication between the N-terminus and loop2 is required to maintain STIM1 mediated Orai activation. In contrast, a non-functional Orai1 ANSGA N-truncation mutant does not regain function (Fig. 23.6m) upon swapping Orai3-loop2. This suggests that this mutant employs a distinct signal propagation pathway compared to wild-type Orai1 and other constitutive Orai1 mutants, which probably bypasses the loop2. A recent study on STIM and Orai in C. *elegans* revealed a distinct gating mechanism compared to that in mammals, where the binding of cSTIM or cCAD to the loop2 of cOrai is sufficient to gate the channel [111], indicating that the gating mechanism of SOCE has evolved over time.

In addition, although STIM1 and Orai1 are sufficient for CRAC channel activation [84], accessory proteins further adjust their interplay. Certain regulatory molecules have been reported to modulate CRAC channel function via coupling to Orai N-terminus, and especially to the ETON region [43, 112–115], as it has been shown for calmodulin (CaM). It adjusts CRAC channel inactivation, as outlined in detail in "Ca^{2+} dependent inactivation of Orai channels" [116]. CRACR2A has been reported to regulate the clustering of STIM1 and Orai1 upon store-depletion via positively charged residues in the N-terminus [115]. Furthermore, SPCA2, a Golgi secretory pathway Ca^{2+}-ATPase, has been found to bind directly to Orai1, whereas the region of interaction has been narrowed down to Orai1 N-terminus (aa 48-91) [117]. It mediates Ca^{2+} entry across Orai1 channels independent of STIM1 or ER-Ca^{2+}-store-depletion. Moreover, the lipid cholesterol has been shown to regulate CRAC channel function [113]. Detailed mechanisms for all these regulatory proteins are summarized in the following studies and reviews [59, 118–120], while more detailed insights on lipid mediated CRAC channel regulation are provided at the end of this review.

In summary, Orai1 C-terminus represents the main and best-characterized binding site for STIM1. Other Orai cytosolic fragments are essential for CRAC channel activation and maintenance of authentic CRAC channel hallmarks (as outlined in "Maintenance of CRAC channel hallmarks") but it remains unclear whether they directly interact with the STIM1 C-terminus.

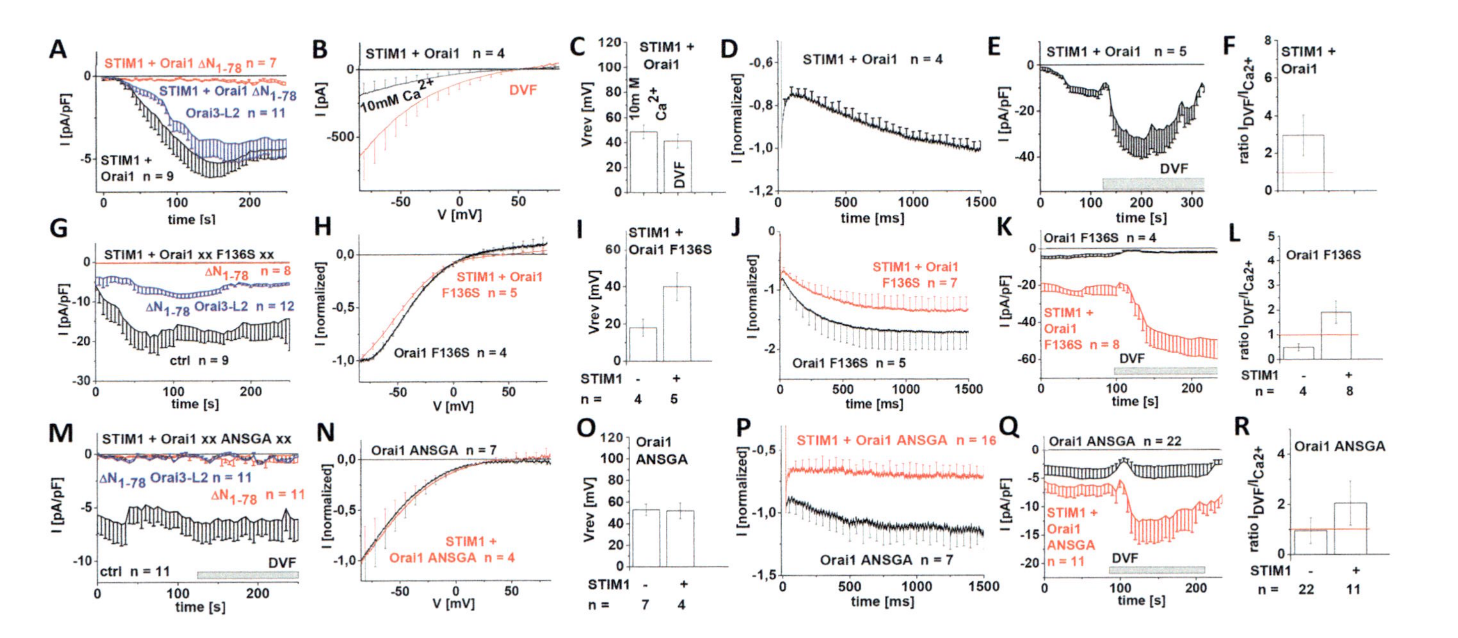

Fig. 23.6 Alterations of CRAC channel hallmarks of wild-type Orai1 versus two constitutively active Orai mutants (**a**) (**g**) (**m**) Respective time courses of whole-cell inward currents at -74 mV of N-truncated mutants and N-truncated chimeras containing Orai3-loop2 compared to the corresponding wild-type Orai1 (**a**), Orai1 F136S (**g**) and Orai1 ANSGA (**m**) in the presence of STIM1. (**b**) I/V relationship of Orai1 in the presence of STIM1 and (**h**) (**n**) normalized I/V relationships of constitutively active mutants Orai1 F136S (**h**) and Orai1 ANSGA (**n**) in the absence as compared to the presence of STIM1. (**c**) (**i**) (**o**) Block diagram showing the reversal potentials of the respective mutants in (**b**), (**h**), (**n**). (**d**) (**j**) (**p**) Inactivation characteristics of currents shown in (**b**), (**h**), (**n**). (**e**) (**k**) (**q**) Respective time courses of whole-cell inward currents at -74 mV of Orai1 in the presence of STIM1 (**e**) and constitutively active mutants Orai1 F136S (**k**) and Orai1 ANSGA (**q**) in the absence as compared to the presence of STIM1. At t = 0 s inward currents in 10 mM extracellular Ca^{2+} solution are activated upon passive store-depletion via 20mM EGTA, as shown after reaching a steady state level. After approximately 100s, the Na^{+}-DVF solution was perfused. (**f**) (**l**) (**r**) Block diagram exhibiting the ratio of I_{DVF} versus I_{Ca2+} of tested mutants with and without STIM1 (**e**), (**k**), (**q**)

23.5 Potential CRAC Channel Activation Mechanism via STIM1-Orai1 Coupling

The Orai channel activation cascade is initiated via the coupling of STIM1-SOAR to Orai1 C-terminus, which has been assumed to induce a conformational change in the C-terminus and the connected TM4 as outlined in the previous chapter. In line, comparison of the crystallographic resolution of a closed and an open conformation of dOrai suggests a straightening of TM4-C-terminus together with a widening of the pore in the basic region upon Orai pore opening [15]. However, a straightened TM4-C-terminus has also been observed for an Orai1 R91S mutant, which remains inactive in the absence of STIM1, in line with a lack of pore widening in the basic region. Thus, a straightening of TM4-C-terminus is not sufficient for pore opening. Additional conformational changes within the Orai channel complex are required to enable pore widening of the basic region. [15] Further investigations are required to determine how the signal of STIM1 binding to Orai1 C-terminus is transmitted to the pore helices located toward the N-terminal side of each subunit.

On the one hand, it is probable that Orai1 channel activation requires, both STIM1 coupling to Orai1 C-terminus and coupling to the cytosolic loop2 and/or the N-terminus, as are required for STIM1 mediated activation [61, 64, 110, 121]. Palty et al. [122] clearly demonstrated via local enrichment of STIM1 SOAR, that both the Orai1 N- and C-terminus contribute synergistically to STIM1 mediated Orai1 activation. Specifically, they have shown that Orai1 tethered to CAD remains active upon partial deletion (Δ1-76 or Δ276-301) of both the N- and the C-terminal strand. In contrast, while single point mutations (K85E or L273S) in either the N- or the C-terminus leave the fusion construct active, mutations within both cytosolic strands lead to total loss of function [122]. Apparently, the local attachment of CAD/SOAR is able to circumvent the need for two intact cytosolic strands as is usually seen with full-length STIM1 or cytosolic STIM1 fragments. Thus, Palty et al. [122] even hypothesized, as an alternative to a potential stepwise STIM1 coupling, which initially starts with STIM1 binding to the Orai1 C-terminus and continues to the N-terminus, that the N- and C-terminal binding sites assemble into a distinct binding pocket for STIM1 that controls the gating and selectivity of Orai1 channels. It is still unclear whether and how STIM1 bridges the potential signal transmission from the Orai1 C-terminus via the loop2 to the N-terminus.

On the other hand, it is possible that STIM1 coupling to Orai1 C-terminus solely initiates signal transmission emanating from the nexus via all the Orai1 TM domains, thus enabling Orai1 pore opening. Indeed, mutations within the nexus can lead to a channel that is constitutively active, but has reduced STIM1 affinity [65]. The nexus comprises the flexible hinge region (SHK, aa 263-265) and the hinge plate (LV, aa 261-262). Here, two residues L261 (TM4) and L174 (TM3) form potential hydrophobic interactions to mediated close proximity of TM3 and TM4 [65]. They may contribute to the transmission of the gating signal from the C-terminus towards the pore, as mutation of one of these residues to an aspartate (L261D, L174D) drastically decreased or completely abolished channel activation, respectively [65]. Additionally, cysteine crosslinking experiments with L174C and L261C revealed enhanced activation, suggesting that these two residues change their orientation towards each other upon Orai1 activation [65]. Whether the hinge plate represents another target for STIM1 coupling as has been suggested by Zhou et al. [107] still requires further investigations. Conformational changes within the C-terminus and the nexus region might be potentially transferred to loop2 and the N-terminus [61], in line with the observations that their permissive communication is essential for Orai channel activation.

In summary, STIM1 SOAR binding to Orai1 C-terminus represents the trigger for pore opening. However, how the signal is further propagated to the pore and whether STIM1 requires further cytosolic coupling sites still needs to be clarified. Potential rearrangements within the Orai1 TM domains upon pore opening will be highlighted in the following sections, based on several gain-of-function point mutants.

23.6 Structure and Stoichiometry of the STIM/Orai Complex

Upon Ca^{2+} store-depletion, the direct interaction between STIM1 and Orai1 is initiated and results in the formation of an oligomeric, heteromeric complex (Fig. 23.5a). However, the exact stoichiometry of these coupling subunits is still unclear. According to the published X-ray dOrai structures and Orai concatemeric studies [62, 63], Orai is known to form a hexamer. Therefore, one might assume that, within a STIM1-Orai complex, six STIM1 molecules bind to the hexameric complex. Along those lines, NMR studies by Stathopulos et al. [16] indicate that the antiparallel oriented Orai1 C-terminal strands are attached to a dimer of STIM1 C-terminal fragments which would fits with the 1:1 STIM1:Orai1 stoichiometry, termed as the bimolecular binding model (Fig. 23.5b2). In contrast to this bimolecular model, a recent study by Zhou et al. [107] revealed a completely distinct STIM1-Orai1 interaction model. Through super-resolution microscopy and FRAP studies, the authors have discovered that SOAR dimers induced substantial clustering of Orai channels. A single point mutant, STIM1 F394H [52], impaired coupling to and activation of Orai1. Introduction of this single point mutation within one subunit of a SOAR dimer still enables the full activation of Orai1 channels, however, but lacked the ability to crosslink Orai1 channels [123]. This outcome suggests, in contrast to the bimolecular model, a unimolecular interaction, with an overall 1:1 stoichiometry (Fig. 23.5b1). However, in the latter model two monomers of one STIM1 dimer each interact with one Orai subunit of two distinct Orai channels, enabling local clustering of Orai channels.

Alternatively, several studies have demonstrated that maximal Orai1 activation requires a ratio of 2:1 of STIM1:Orai1 subunits [124–127]. Here, twelve STIM1 proteins would be required to activate one Orai hexamer. A recent study by Palty et al. [128] which employ a constitutively active Orai1 mutant (Orai1 P245L) proposes a sequential activation mechanism of Orai1 via STIM1, whereby the authors assumed a final 2:1 STIM1:Orai1 stoichiometry (Fig. 23.5b3). They supposed that, in a first step, one C-terminal strand of a STIM1 dimer is coupled to an Orai1 C-terminus, thereby inducing partial activation and leading to conformational rearrangements within the Orai channel. The latter enables binding of the second C-terminus of the STIM1 dimer to the same Orai subunit to trigger full Orai1 activation [128]. At this point, however, it remains unclear whether the second STIM1 monomer of the dimer also binds to the C-terminus or some other cytosolic site in the Orai protein.

The amount of bound STIM1 has been reported to alter the Orai channels' biophysical characteristics. Scrimgeour et al. [125, 126] reported that a lower number of STIM1 proteins bound to the Orai1 channel corresponds to a reduced extent of typical fast Ca^{2+} dependent inactivation of CRAC channels. The decreased STIM1:Orai1 ratio has been shown to lead to a slower increment in divalent cation selectivity together with reduced 2-aminoethoxydiphenyl borate (2-APB) current potentiation. [125, 126] The amount of bound STIM1 also determines Orai1 channel Ca^{2+} selectivity, as it has been shown for Orai1 and Orai1 V102A proteins tethered to a single CAD/SOAR domain, as opposed to a tandem one [67].

Overall, the stoichiometry of an active STIM1-Orai1 channel complex is currently still controversial. Thus far, three distinct scenarios have been proposed including a uni- and a bimolecular model as well as a sequential activation model (Fig. 23.5b). Functional studies have suggested that Orai1 channel activation can occur at distinct STIM1:Orai1 ratios, while the biophysical characteristics might be distinct depending on the respective ratios. Full clarification of the stoichiometry of a functional CRAC channel would require additional functional studies together with the structural resolution of the STIM1-Orai1 channel complex (Figs. 23.5a and 23.7).

23.7 Orai Channel Activation

23.7.1 CRAC/Orai Channel Permeability

CRAC channels are characterized by two main properties: high Ca^{2+} selectivity and low single channel conductance [13, 129, 130]. Among a myriad of Ca^{2+} ion channels, this type of store-operated Ca^{2+} channel is one of the most selective ones. Its high Ca^{2+} selectivity is characterized by a 1000 times higher permeation for Ca^{2+} than for the more prevalent Na^{+} ion [131]. This appears in the strong inward rectification of the CRAC channel's whole cell currents, which exhibit a reversal potential higher than +50 mV [129, 130]. At the structural level, high Ca^{2+} selectivity is determined by a ring of glutamates (E106) at the external pore mouth, forming the selectivity filter together with three negatively charged residues (D110/112/114) in the first extracellular loop, the so-called CAR region [66]. Despite the fact that CRAC channels exhibit a 20-fold lower binding affinity for Ca^{2+} than for, for example, Ca_V channels [132], their high level of Ca^{2+} selectivity is justified by a slow rate of ion flux [133]. Additionally, the Ca^{2+} selectivity of

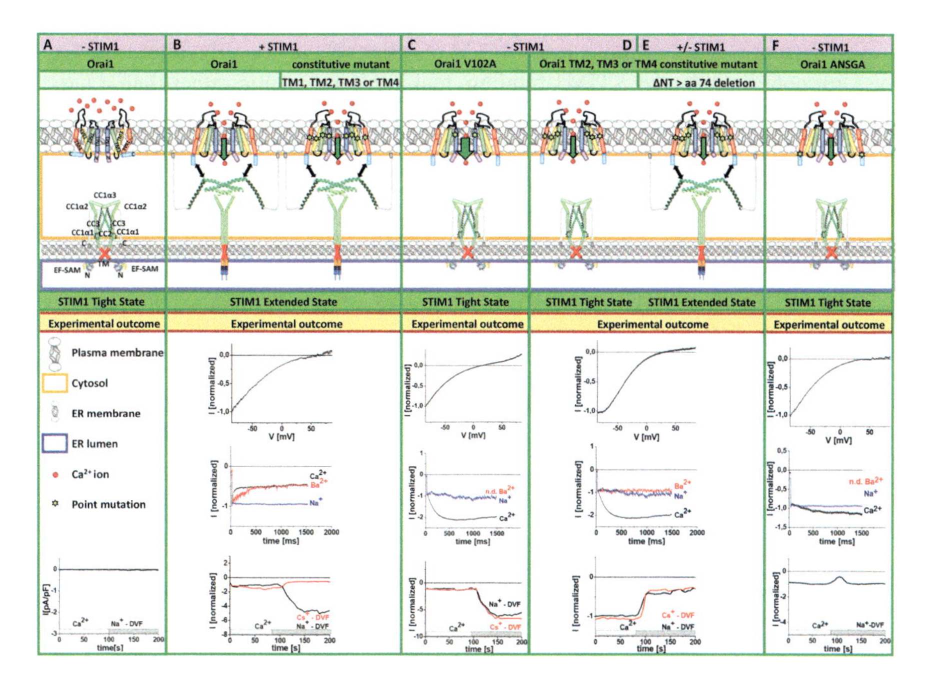

Fig. 23.7 Hypothetical models for gating of Orai1 channel The cartoon depicts how Orai channel activation via STIM1 or point mutation inducing constitutive activity or both is related to CRAC channel hallmarks. (**a**) Orai1 wild-type in the absence of STIM1 shows no activation. (**b**) STIM1 mediated activation of Orai1 wild-type or most constitutive Orai1 mutants leads to maintenance of CRAC channel hallmarks, with V_{rev} > 50mV, strong inactivation in a Ca^{2+} containing solution and a ratio of $I_{DVF}/I_{Ca2+} > 1$. Inactivation is preserved in a Ba^{2+} containing solution, while it is lost in a Na^{+} containing divalent free solution. Furthermore, the STIM1 activated Orai proteins are impermeable to Cs^{+}. (**c–f**) In the absence of STIM1 or the deletion of at least the first 74 N-terminal amino acids of Orai1, constitutively active mutants loose the typical CRAC channel characteristics shown in (**b**). The mutants show differences and can be categorized in three groups (**c**, **d**, **f**). Orai1 V102A/C (**c**) show a V_{rev} of +20 mV, reactivation, but maintain a ratio of $I_{DVF}/I_{Ca2+} > 1$. Orai1 TM2/TM3/TM4 point mutants (**d**) show constitutive activity, a V_{rev} of 30–40mV, reactivation and a ratio of $I_{DVF}/I_{Ca2+} < 1$. (**e**) Deletion of the N-terminus of the constitutively active mutants in (**d**) leads to mutants with comparable properties as described in (**d**). Orai1 ANSGA (**f**) exhibits a V_{rev} of +50mV and loss of inactivation, but no reactivations and a ratio of $I_{DVF}/I_{Ca2+} = 1$

Orai is maintained via the conserved glutamate E190 in TM3. Its substitution to an alanine or glutamine (E190A/Q) results in increased permeation of Cs^{+}, an increased pore diameter to 7Å and decreased activation via the compound 2-APB [134]. It has been suggested that E190 allosterically affects the pore, probably through alterations in the intramolecular TM interactions. Novel MD simulations of dOrai revealed that the mutation E262Q (corresponding to E190Q in hOrai1) severely impacts the hydration pattern of the pore and the dynamics of a positively charged residue K270 (corresponds to K198 in hOrai1) in TM3. It has been suggested that high Ca^{2+} selectivity relies on a counterbalancing effect of E262 and K270 which is lost for the E262Q mutant [135].

Furthermore, the Ca^{2+} selectivity of CRAC channels is not only determined by the Orai pore, but also by STIM1 binding, making it a dynamic rather than a fixed property of the channel [64, 67].

The low unitary conductance represents the second, more poorly understood property of CRAC channels and thus far precludes these Ca^{2+} channels from direct single channel current recordings [136]. Only non-stationary noise analysis has enabled researchers to determine CRAC channel open probabilities and unitary conductance [137]. To date, promising studies at the single molecule level have only been obtained via optical recordings on Orai proteins fused to a genetically encoded Ca^{2+} indicator, which allowed to monitor single channel opening events [138]. Upon co-expression of CAD/SOAR, the Orai activity was resolved as periodic fluctuations with multiple conductance states [138].

Besides Ca^{2+}, CRAC channels also conduct small monovalent ions such as Na^{+}, Li^{+} or K^{+}, however, only as long as divalent ions are lacking in the monovalent solution. One typical CRAC channel hallmark represents the 5–6 fold enhancement of currents in a divalent free Na^{+}- versus a Ca^{2+}-containing solution. The anomalous mole fraction behaviour for Ca^{2+} versus Na^{+} currents has revealed that Na^{+} currents are blocked by the addition of Ca^{2+} in the μM range. [46, 131, 137, 139–145] Unlike other Ca^{2+} ion channels such as L-type and TRPV6, Cs^{+} ions are impermeable across CRAC channels. This is thought to be due to the very narrow pore diameters of CRAC channels, which are in the range of 3.8–3.9 Å and are significantly narrower than for Ca_V channels. [134, 137, 146] Intriguingly, the dOrai crystal structure displays a diameter of 6 Å at the narrowest part of the pore [14] (Fig. 23.3c, d), which seems to vary little compared to the novel structure of the gain-of-function mutant dOrai H206A [15]. The variance in pore diameter between structural and experimental [134] data can probably be explained by distinct pore diameters in the absence or the presence of STIM1. In order to clarify this, the structural resolution of an Orai complex bound to STIM1 proteins is required.

STIM1 possesses the ability to narrow the pore diameter of Orai channels, as it has been shown for a constitutive Orai1 mutant, Orai1 V102A/C. This mutant exhibits non-selective currents together with an enhanced pore diameter of around 6.9 Å in the absence of STIM1, while Orai1 V102A/C currents become selective and exhibit a pore diameter of 4.9 Å in the presence of STIM1, which is comparable to that observed for the STIM1-bound Orai1 pore. [67]

23.7.2 *Ca^{2+} Ion Conduction Pathway*

The Ca^{2+} permeation pathway of Orai channels is formed by TM1 together with the N-terminal helical extension, the so-called ETON region (aa 73-90) (see Fig. 23.3c, d). An initial idea of the pore conformation has been obtained by conclusive electrophysiological studies, cysteine scanning mutagenesis experiments and the characterization of the inhibition of Orai currents by oxidative crosslinking of either cysteines or block via cadmium [134, 147, 148]. The crystal structure of dOrai, elucidated in 2012, finally confirmed the concept of the Orai pore [14] and showed that the conduction pathway is formed by an association of six TM1 helices in the center of the Orai channel. The Orai pore (see Fig. 23.3d) is composed of a Ca^{2+} accumulating region (CAR) at the external side [66], the selectivity filter, followed by a hydrophobic cavity and finally a basic region at the cytosolic side [14, 149, 150]. Their detailed roles will be explained in the following section.

The external CAR region (see Fig. 23.3d) is constituted by the loops connecting TM1 and TM2 (TM1-TM2 loops). Substituted cysteine accessibility method (SCAM), disulphide crosslinking and Cd^{+} block experiments have shown that the 1st loops are highly flexible portions within the Orai channel complex [148, 151]. Each first loop segment contains three negatively charged amino acids (D110, D112, D114) that attract Ca^{2+} ions [66] thereby reducing the energetic barrier for Ca^{2+} ions to enter the pore. While the mutation of all three residues to alanines changes the ion selectivity and widens the pore diameter [134], single cysteine or alanine point mutations do not affect Ca^{2+} selectivity [66, 148]. Molecular dynamics simulations have revealed that in particular, D110 and D112 act as transient Ca^{2+} binding sites before the Ca^{2+} ions bind to E106, the selectivity filter, which is 1.2 nm away from the CAR region. Indeed, Ca^{2+} permeation is reduced upon substitution of D110 to an alanine (D110A), especially at lower extracellular Ca^{2+} concentrations. The reason for this might be that Ca^{2+} binding shifts to D112 and D114, leading to an increased energy barrier between CAR and the selectivity filter. Ca^{2+} accumulation at the pore is fine-tuned still further by electrostatic interactions between TM1-TM2 loop1 and the TM3-TM4 connecting loop3 [66].

In contrast to the three glutamines found in the first loop of Orai1, Orai2 and Orai3 contain a mixture of glutamates, glutamines and aspartates. This difference leaves homomeric Orai assemblies unaffected, but heteromeric Orai1/Orai3 channels exhibit reduced Ca^{2+} selectivity and higher Cs^{+} permeation [152]. These findings suggest that the acidic Ca^{2+} coordination site in the first loop regulates not only the Ca^{2+} permeation, but also the Ca^{2+} selectivity of Orai channels. Interestingly, co-expression of Orai1 and Orai2 maintained the Ca^{2+} selectivity of store-operated currents at a level comparable to that seen in wild-type conditions [153].

Originating from the CAR region, Ca^{2+} ions are guided toward to the selectivity filter, which is exclusively formed by E106 at the more extracellular side of TM1 [13, 14, 154, 155]. While single point mutants Orai1 E106A/Q lose function, Orai1 E106D exhibits reduced Ca^{2+} selectivity and a widened pore diameter of about 5.4 Å [134]. The coordination of the Ca^{2+} ions is established by the close proximity of the residues E106 and a high rigidity of the residues aa 99-104 in the TM1 helix [147]. MD simulations showed that a single Ca^{2+} ion is bound to E106 under equilibrium conditions [66]. The novel crystal structure of the constitutively open dOrai H206A (equal to Orai1 H134A) mutant, representing an open channel state, revealed that either one or two ions directly coordinate the side chains of the glutamate ring (E178 in dOrai, corresponds to E106 in hOrai1) [15]. This ring of E106 residues forming the selectivity filter establishes a negative electrostatic potential and has been suggested to act as a potential gate for Ca^{2+} ions to enter the channel [67].

After passing the selectivity filter, Ca^{2+} ions enter a wide, hydrophobic cavity, that is composed of V102, F99 and L95 [14] (see Fig. 23.3d). Furthermore, this area includes a glycine G98 that has been suggested to function as a flexible gating hinge [156], as is known for many other ion channels [157–159]. Cysteine crosslinking studies suggest that V102 and L95 point directly into the ion conduction pathway [147]. In support of this observation, Cd^{+} blockage experiments showed the significant inhibition of V102C, G98C, L95C mutant currents [151]. Single point mutations of Orai1 V102 (V102C/A/G) and Orai1 F99 (F99Y/M/S/T/W/C/G) lead to constitutively active function, thus somehow locking Orai1 in an open state [69]. In line with these findings, molecular dynamics simulations for the dOrai channel showed that a F171Y and V174A mutation (corresponding to F99Y and V102A in human Orai1, respectively) provided a more favourable permeation pathway than the WT channel, based on potentials of mean force for the translocation of a single Na^{+} ion [160] and enhanced hydration of the hydrophobic zone within the pore [67, 69].

The flexible glycine G98 within this hydrophobic cavity [156] is assumed to provide flexibility for the upstream pore-lining region, enabling Ca^{2+} ions to easily pass through the pore after passing the selectivity filter. Single point mutations of G98 (G98 S/T/Q/H/N/K/E/D) induce constitutive activity [69]. Recent studies suggest that, while F99 points into the pore in the closed state, it moves out and G98 moves into the pore upon STIM1 coupling, thus decreasing the hydrophobic barrier for Ca^{2+} permeation [69]. Diverse constitutive Orai1 mutants that contain substitutions at the positions G98, F99 or V102 exhibited non-selective currents in the absence of STIM1, but became selective in the presence of STIM1. Thus, the potential rotation within the hydrophobic portion in TM1 is assumed to govern the Ca^{2+} selectivity of the Orai channel. [69]

The attachment of Tb^{3+} to the pore close to the selectivity filter has revealed alterations in the luminescence upon STIM1 coupling [67, 161]. In line with these findings, a comparison of the dOrai crystal structure in a closed versus an open conformation showed an altered positioning of Ba^{2+} around the selectivity filter, indicating that the latter undergoes subtle changes upon Orai pore opening [15]. Altogether, these studies suggest that the hydrophobic cavity together with E106 contribute as one gate of the CRAC channel and that ion selectivity and gating of Orai channels are closely coupled [67, 161].

The hydrophobic cavity is followed by a basic region that is already located in the elongated Orai1 helix in the cytosol connecting TM1 to the Orai1 N-terminus, the so-called ETON region [64]. The ETON region potentially forms an electrostatic barrier via the three positively charged amino acids R91, K87 and R83 (see Fig. 23.3d). The crystal structure suggests that all three residues point into the pore, in line with the results of crosslinking experiments with Orai1 R91C [147, 151]. These residues are suggested to inhibit Ca^{2+} entry in the closed channel state probably either simply via electrostatic repulsion in this region or due to bound anions [14]. MD simulations on wild-type Orai1 versus a constitutively open Orai1 mutant, where a Ca^{2+} ion has been pulled through the pore, suggested that the side chains of R91 are moved more out of the pore in the Ca^{2+} permeable state [68]. Severe Combined Immune Deficiency (SCID) is associated with a substitution of R91 by a hydrophobic amino acid which assumedly generates a robust hydrophobic barrier [14], hindering Ca^{2+} flow through the elongated pore. Currently, it is being debated whether the ETON region, with the positively charged residues pointing into the pore, functions as a potential second gate or merely a barrier. Furthermore, it still remains to be determined whether it contributes to STIM1 coupling [64].

In summary, Ca^{2+} permeation through the Orai1 pore starts upon the attraction of Ca^{2+} ions by the external CAR region. Ca^{2+} ions are then moved toward the selectivity filter that is formed by the single glutamate E106. This functions together with the hydrophobic region as one gate of the Orai channel. Subsequently, Ca^{2+} ions pass through an electrostatic barrier and finally, they are guided into the cytosol potentially via the repulsion of positively charged residues at the end of the pore [136].

23.7.3 Gating of Orai Channels

Gating of Orai channels implicitly requires the coupling of STIM1 to the Orai C-terminus. However, how this activation is finally transmitted to the pore helix at the N-terminal side of each Orai subunit is currently unclear. As described above it has been hypothesized that other cytosolic Orai domains are also involved in channel activation by STIM1 either via direct binding or indirectly. Recent studies additionally suggest that Orai TM domains are involved in signal propagation to the pore. Within this section, we aim to provide a picture of how the activation signal of STIM1 binding propagates across Orai TM helices to the pore.

Site-directed mutagenesis studies have revealed that the closed state of the Orai channel is not only maintained by the pore-lining helix TM1, but also by the other three TM helices and the hinge region that connects TM4 and the C-terminus of the Orai subunit, even if these do not line the pore [39, 70, 73]. A series of point mutations at positions throughout all TM domains, identified individually [68–71] as well as by a systematic screen [73], have been shown to lead to constitutive Orai activation. Some of those gain-of-function mutants have also been associated with diseases such as cancer, Stormorken syndrome, tubular aggregate myopathy (TAM) and hypocalcemia [68, 72] (see Table 23.2). Additionally, mutagenesis studies have revealed that a cluster of hydrophobic contacts between TM1-TM3 contribute to proper pore opening [73]. Below, we provide an overview of the diversity of mutants that are currently known.

Along the pore-lining helix TM1, S97, G98, F99, A100, V102 and V107 (Fig. 23.3d) have been discovered to contribute to the establishment of the closed state [67, 73, 156] as their mutation to small and/or polar residues leads to constitutive activity. Furthermore, the V107M mutation, which is associated with tubular aggregate myopathy (TAM) or Stormorken syndrome, results in constitutively active Orai1 mutant channels [71]. As mentioned in the previous section, studies on diverse other constitutive Orai1 TM1 mutants (e.g. Orai1 V102A/C) together with Cd^+ block experiments [67, 69] have suggested that Orai1 pore opening is accompanied by a rotation of the pore helix TM1 [69, 70, 122, 136] at least at position G98 and F99.

In TM2, H134, F136 (Fig. 23.6g) A137 [68], L138 [72], M139 [68] and S141 [73] are suggested to keep the Orai1 channel in the closed state, as single point mutation of these residues (H134A/C, A137V, L138F, S141C) lead to constitutively active currents [68, 73]. In addition, Orai1 M139V enhances Ca^{2+} levels independent of STIM1 [68]. MD simulations together with Cd^+ block experiments provide a potential explanation for facilitated Ca^{2+} transport across the constitutive Orai1 mutants, containing substitutions at H134. In two independent studies enhanced water permeation was observed through the pore of constitutive Orai1 H134A/S mutants in MD simulations [68, 73]. Furthermore, pulling a Ca^{2+} ion through the pore of hOrai1 H134A revealed an unaffected movement of the R91 side chain, in contrast to Orai1 wild-type [68]. In addition, Cd^+ crosslinking experiments showed that Orai1 H134S exhibits no rotation of TM1 around G98/F99, as G98 points already into the pore region in the absence of STIM1 [68, 73]. The crystal structure of the dOrai H206A analogue of the Orai1 H134A mutant displays a strongly dilated pore together with a fully extended TM4-C-terminus region [15].

In TM3, W176, V181, G183, T184, L185, F187, E190 have been demonstrated to keep the Orai1 channel in the closed state, as their substitution to a cysteine or several other small, polar residues can lead to constitutive activity and/or altered selectivity [61, 73, 110, 162]. A mutation of W176 to a cysteine (W176C) induces constitutive, but fewer Ca^{2+} selective currents, in the absence or presence of STIM1 [162]. While a single substitution of V181 or L185 to an alanine results in small, constitutively active Orai currents [61], a double point mutation L185A F250A of two opposing residues in TM3 and TM4 lead to strongly enhanced constitutive activity that was not further enhanced upon STIM1 coupling [110]. Interestingly, Orai3 F160A, unlike the analogue Orai1 L185A, has revealed much higher constitutive currents, indicating isoform specific signal propagation [61, 110]. These critical residues of the two Orai isoforms in TM3 are located opposite to TM4. At this point, further examination is required to determine whether mutations in TM3 that lead to constitutive activity, induce conformational changes such as the movement of TM3 closer to or farther apart from the adjacent TM4. Moreover, the single point mutant T184M, associated with TAM and Stormorken syndrome, exhibits constitutive activity.

A235, S239, P245 and F250 in TM4 have also been discovered to contribute to the maintenance of the closed state [70, 73]. P245 is located at the kink of TM4 (Fig. 23.3e). A GoF point mutation P245L, associated with the Stormorken-like syndrome [8], and the analogue Orai3 P254L mutant [110] have been shown to cause constitutive activation. Interestingly, substitution of this proline by any other amino acids leads to constitutive activation suggesting that only the proline keeps Orai1 in the closed state [122]. Since this proline is located at and responsible for the kink in TM4, it might be assumed that this kink contributes to keeping Orai1 in the closed state.

The nexus region connecting TM4 and the C-terminus also contributes to the maintenance of the closed state of the Orai channel. A double (L261A V262N) and a fourfold point mutation (L261A V262N H264G K265A corresponds to Orai1 ANSGA) in this kink region, which is 80% conserved among all Orai homologues, resulted in constitutively active Orai1 channels with high selectivity and a V_{rev} of +60 mV (Fig. 23.6n, o, 5) [65].

The gain-of-function point mutants mentioned above exhibit a partially open state, that probably represents one step in the structural change that occurs upon STIM1 binding [122]. At this point, it is worth noting that these diverse GoF mutants exhibit distinct biophysical characteristics in the absence of STIM1, potentially in line with slightly distinct partially active states, as outlined in the section entitled "Authentic CRAC channel hallmarks are tuned by STIM1 and Orai1 N-terminus".

In addition, atomic packing analysis revealed a close density of residues around TM1-TM2/3 situated within the same membrane plane as E106 and V102. Mutational analysis of hydrophobic residues in this area, such as L96, M101, M104, F187, yielded a finding that they contribute to proper communication with the pore helix, because their substitution by an alanine impairs Orai activation [73].

In summary, several residues throughout all Orai TM helices contribute to the maintenance of the closed state of Orai channels and are likely displaced upon STIM1 binding. Thus, Orai pore opening is governed by a concerted, overall conformational change in the Orai channel complex rather than the sole rearrangement of TM1 helices [73]. Furthermore, the establishment of the open state requires an intact communication of the residues in a TM1-TM2/3 hydrophobic clamp [73]. It is noticeable that several residues, not lining the pore, are all located within the same membrane plane of the Orai1 channel complex and affect gating and/or permeation of Orai channels indirectly.

A long the same line, a comparison of the crystal structures of dOrai and a mutant (H206) form, exhibiting either a potentially closed or open form, indicated that conformational rearrangements take place throughout all TM domains. Hence, STIM1 coupling probably initially releases the belt formed by the crossing Orai C-termini, which likely provides to all TM helices more freedom to move. Subsequently, a signal is propagated to the pore, sequentially altering the conformations of TM4, TM3 and TM2, which finally leads to a pore dilation in the basic region to around 10 Å. However, as these assumptions are only based on the comparison of two dOrai states, more structural resolutions – especially of a human Orai channel complexed together with STIM1 – are needed to draw firm conclusions. This would provide information on how STIM1 binds to the channel and rearranges the orientation of the TM helices to finally stabilize the open state [65, 136].

23.7.4 Regulatory Residues in Orai TM Domains

In addition to the huge variety of sites within the Orai TM domains that are known to control Orai channel gating, a few other positions have been found to possess distinct functional roles in Orai1.

Unlike the constitutive Orai1 G183C mutant, the G183A mutation leads to fully abolished store-operated activation and an altered sensitivity to 2-APB [162]. Furthermore, this glycine in TM3 ensures proper plasma membrane insertion, as several amino acid substitutions at this position have resulted in loss of plasma membrane expression [68]. The analogue Orai3 G158C exhibits altered kinetics in response to 2-APB and an impaired closing of the channel upon 2-APB washout [163]. These effects of G158C in Orai3 have been attributed to a potential TM2-TM3 interaction with an endogenous cysteine C101, which probably controls the activation state of Orai3 channels [163].

Oxidative stress has also been reported to control Orai channel function via TM3. While H_2O_2 blocks the activation of Orai1 channels, it leaves Orai3 channels unaffected. This isoform specific sensitivity to H_2O_2 underlies a cysteine C195, which is located at the extracellular side of TM3 in Orai1 and is lacking in Orai3. [164]. This inhibitory mechanism underlies a reduction in subunit interactions within the oxidized Orai1 channel. Specifically, the oxidized C195 has been proposed to form inhibitory interactions with S239 in TM4, which locks Orai1 into the closed state [165].

23.7.5 Ca^{2+} Dependent Inactivation of Orai Channels

Ca^{2+} dependent inactivation (CDI) governs the inhibition of Orai/CRAC channel activity. It represents an important feedback mechanism to regulate cytosolic Ca^{2+} levels and can be separated into fast (FCDI) and slow Ca^{2+}-dependent inactivation (SCDI) [166]. FCDI can typically be monitored as a decrease in CRAC currents during a hyperpolarizing voltage step that takes place over tens of milliseconds [3, 131, 167, 168]. In contrast, SCDI requires several minutes for full completion [3, 168].

The CDI of CRAC channels is regulated by several components: STIM1 and Orai as well as the accessory proteins, CaM and SARAF [118]. Here, the cytosolic regions of STIM1 and Orai1 especially contribute to the inactivation of Orai channels.

The extent of inactivation has been suggested to depend on the STIM1-Orai ratio [125, 126]. Moreover, the CRAC Modulatory Domain (CMD), an acidic cluster (aa 475-483) in the STIM1 C-terminus is required for the FCDI of Orai/CRAC channels [47, 116, 169]. Mutation of these negatively charged amino acids in the CMD to alanines has led to a reduction in inactivation of all Orai isoforms as well as native CRAC currents in RBL-2H3 cells [47, 169]. Moreover, studies of several gain-of-function Orai mutants have shown that FCDI, which is typical for STIM1 mediated Orai currents, can only be established in the presence of STIM1 and an intact N-terminus, which includes the whole ETON region [73, 110].

Fast and slow CDI of STIM1-Orai mediated currents have also been shown to be regulated by SARAF, another protein localized in the ER [166, 170, 171]. So far, only its role in SCDI is understood, while the mechanisms underlying the regulation of FCDI via SARAF are still elusive [166, 170, 172]. SARAF only binds to the CAD/SOAR domain of STIM1 in the presence of Orai1 [170]. STIM1-mediated regulation by SARAF occurs via a domain downstream from SOAR, the so-called C-terminal inhibitory domain (CTID). Here, SARAF interacts with different portions of CTID to accomplish SCDI of Orai channels [170]. Briefly, in the presence of SARAF, STIM1-induced Orai1 currents exhibit SCDI that is reduced when PIP_2 is depleted, the polybasic cluster is lacking, or STIM1 proteins are targeted to PIP_2-poor regions [172]. The interplay between STIM1 and SARAF is further supported by septin 4 and ESyt1, which helps to keep STIM1/Orai1 complexes in PIP_2-rich regions [166, 172].

All three isoforms (Orai1, Orai2 and Orai3) display fast CDI that occurs within the first 100 ms of a voltage step. However, it is three times weaker for Orai1 and Orai2 compared to Orai3 [129, 152, 169]. Orai1 currents exhibit a late reactivation phase after FCDI, while Orai2 and Orai3 channels subsequently show a slow inactivation phase over a 2s voltage step [129, 152]. Unlike overexpressed Orai subunits, the native CRAC currents in RBL-2H3 mast cells [47, 167] display FCDI without the subsequent, typical reactivation phase known for Orai1 currents. This suggests that other cellular components may contribute to the inactivation of native CRAC channels.

Several cytosolic domains in the Orai channel have been reported to maintain its FCDI.

Within the Orai1 N-terminus, a proline/arginine-rich region aa 1 – 47 was found to induce reactivation [173]. Furthermore, CaM has been suggested to regulate inactivation of Orai channels via its binding to the N-terminus [116, 173, 174]. However, recent studies have found no alteration in Orai channel's inactivation via overexpression of CaM or CaM_{MUT} [175]. Specific residues (W76, Y80, R83) that line the pore have been proposed to adjust inactivation via the side chain properties of the respective amino acids [145].

Chimeric and mutagenesis studies have shown that the intracellular loop2 of Orai1 connecting TM2 and TM3 is also involved in the modulation of fast and slow inactivation [173, 176]. Specifically, the mutation of the amino acid stretch 151-VSNV-154 to a sequence of four alanines compromises fast inactivation. A short peptide (aa 153-157) including this region has been shown to reduce CRAC currents, suggesting that loop2 probably functions either as a blocking particle or affects inactivation in an allosteric manner.

The Orai C-terminus also modulates Orai channel's fast inactivation. Orai2 and Orai3 chimeras, in which the C-terminus was exchanged with that of Orai1, exhibited diminished fast inactivation [169], potentially due to three glutamates that are only present in the C-termini of Orai2 and Orai3 [169].

Thus, Orai channels utilize diverse regions to regulate CDI. The cytosolic N- and C-termini and loop2 seem to regulate Orai inactivation/gating in a cooperative manner [173]. Moreover, the FCDI is adjusted by negatively charged residues within the outer pore vestibule of Orai channels [134]. The interpretation of potential inactivation sites may require a strict control of STIM/Orai stoichiometry [124, 126].

In summary, the FCDI of Orai channels is controlled by negatively charged residues in STIM1 as well as all three cytosolic portions of Orai channels, the two cytosolic strands as well as the loop2 region. Orai channel current's SCDI underlies the regulation via STIM1 and SARAF. However, whether Orai domains are also involved in SCDI is still unclear.

23.7.6 Authentic CRAC Channel Hallmarks are Tuned by STIM1 and Orai1 N-terminus

STIM1 plays several roles in CRAC channel activation. It triggers not only CRAC channel gating but also adjusts authentic CRAC channel hallmarks [110]. The latter include high Ca^{2+} selectivity ($V_{rev} \sim +50mV$) (Fig. 23.6b, c), fast Ca^{2+} dependent inactivation (FCDI) (Fig. 23.6d) and enhancement of currents upon the switch from a Ca^{2+} containing to a Na^{+} containing divalent free solution (Fig. 23.6e, f).

The high Ca^{2+} selectivity of CRAC channels is triggered upon functional coupling of STIM1 to the Orai channel. This finding has been proven by studies with a series of constitutively active mutants throughout all TM helices [67, 70, 73, 110] (see section "Gating of Orai channels"). In the absence of STIM1, these mutants exhibit constitutive, non-selective currents, whereas constitutive Orai1 TM mutants (e.g., Orai1 V102A/C) show a lower V_{rev} and are less Ca^{2+} selective than most other mutants that contain substitutions in the outer TM regions TM2/TM3/TM4 [110] (Fig. 23.7d). Contrary to the absence of STIM1, in its presence the Ca^{2+} selectivity of all constitutive mutants is enhanced to levels comparable to wild-type STIM1-Orai-mediated currents [73, 110] (Fig. 23.7b). As a representative for most constitutive TM2/TM3/TM4 mutants, the typical CRAC channel hallmarks of the constitutively active Orai1 TM2 mutants (Orai1 F136S) are shown in Fig. 23.6g–l. The ability of STIM1 to fine-tune Ca^{2+} selectivity is not unique to constitutive mutant channels. Wild-type Orai1 currents have also been shown to display increased ion selectivity with a progressive amount of bound STIM1 proteins [67]. Exceptions to this are some other non-selective constitutive TM1 mutants (e.g. S97C/M/V/L/I) that cannot regain Ca^{2+} selectivity in the presence of STIM1, potentially due to constrained motions of TM helices that trap the channel in a non-selective state [73, 156]. In addition, constitutive Orai1 H134A/S/C (mutations in TM2) [68, 73] and the TM4-C-terminus hinge mutant (Orai1 ANSGA) [65] (Fig. 23.6n, o) have revealed comparable V_{rev} in both the absence and presence of STIM1, suggesting that these mutants possess similar conformational alterations and properties as those observed for activated endogenous CRAC and STIM1-induced Orai1 currents.

Both the Orai channels' Ca^{2+} selectivity and their fast Ca^{2+}-dependent inactivation (FCDI) is maintained only in the presence of STIM1 (Fig. 23.6d, 23.7b), which has been visualized with constitutively active mutants. Constitutive Orai1 TM mutants display strong reactivation in the absence of STIM1 upon the application of a hyperpolarizing voltage step [67, 70, 73, 110] as shown for Orai1 F136S exemplarily (Fig. 23.6j). It is worth mentioning, that the extent of reactivation in the absence of STIM1 is less pronounced for Orai1 ANSGA (Fig. 23.6p) compared to all other Orai1 TM mutants [73, 110]. The FCDI of all constitutive Orai mutant currents is maintained only in the presence of STIM1, but, it lacks the typical reactivation phase after FCDI within the first 100ms [110] (Fig. 23.6d, j, p), potentially due to altered STIM1 coupling (see "Main coupling domains within Orai1").

Another common CRAC channel characteristic represents the increase in I_{DVF} compared to I_{Ca2+} (Fig. 23.6e, f). For CRAC channels, it has been hypothesized that this prominent enhancement in current densities correlates with the extent of inactivation [137]. In a Ca^{2+}-containing solution, CRAC channels display FCDI, but in a divalent free Na^{+}-containing solution, CRAC channel currents exhibit no inactivation, similar to constitutively active mutants in the presence of STIM1 [110]. This possibly explains the increased current levels observed in a solution containing Na^{+}-compared to Ca^{2+}- as a charge carrier [13]. This correlation has been suggested to potentially underlie an altered open probability in a Ca^{2+}- versus Na^{+}- containing solution [137]. Indeed, diverse studies [46, 137, 144, 167] have suggested that Ca^{2+} dependent inactivation is likely to reduce the open probability when Ca^{2+} rather than Na^{+} is the charge carrier. In accordance with these findings, the extent of the FCDI of Orai1 and Orai3, which is much more pronounced for Orai3, also correlates with larger current enhancements for Orai3 in a Na^{+} versus Ca^{2+} containing solution compared to Orai1 [129].

Interestingly, in the absence of STIM1 the currents of several constitutively active mutants, especially those in the outer TM regions, TM2-TM4, exhibit decreased I_{DVF} versus I_{Ca2+} in a reversible manner, a feature that was reported for the first time by Derler et al. [110] and is here exemplarily shown for Orai1 F136S (Fig. 23.6k, l; 5D). A potential explanation might be a distinct inactivation behaviour found in Ca^{2+}- versus Na^{+}-containing solutions. While hyperpolarizing voltage steps have shown loss of inactivation, but strong reactivation in the presence of Ca^{2+}, in a Na^{+}-containing divalent-free solution, these mutants yielded only loss of inactivation, and lack reactivation. Thus, this difference in the "inactivation behavior" in a Ca^{2+}-versus Na^{+}-containing solution might account for reduced I_{DVF} versus I_{Ca2+}. One exception represents Orai1 H134A, as it shows enhancements in I_{DVF} versus I_{Ca2+} [68], thus, probably matches the STIM1 bound Orai1 open state even more than other Orai1 TM2-TM4 constitutively active mutants.

Intriguingly, in contrast to the TM2/3/4 mutants, the poorly Ca^{2+}-selective Orai1 V102A/C mutants displayed strongly increased Na^{+} divalent-free versus Ca^{2+} currents, both, in the absence and presence of STIM1 [67]. Other investigations of various constitutively active TM mutants have revealed differences in Cs^{+} permeation. As mentioned in the section entitled “CRAC/Orai channel permeability”, Cs^+ is known to be impermeable for CRAC or STIM1/Orai1 channels [67]. The non-selective Orai1 V102A/C allow permeation of Cs^+ [110], due to an enhanced pore diameter. In the presence of STIM1, Cs^+ permeation across Orai1 V102A/C is abolished, as STIM1 coupling narrows the pore, thus restoring the pore dimensions of the wild-type Orai1 channel. In contrast to Orai1 V102A/C, other constitutively active TM3/TM4 mutants are impermeable to Cs^+ in the absence of STIM1, suggesting that their pore diameters are comparable in the absence or presence of STIM1 [110]. Thus, differences in current size in a Ca^{2+} containing versus Na^+ containing divalent free solution of different constitutively active mutants are probably determined by the differences in pore diameter, which still requires further clarification.

However, at this point, it is unclear why Orai1 V102A/C and other Orai TM mutants exhibit an identical inactivation behaviour, despite so far discussed differences in V_{rev} and I_{DVF}/I_{Ca2+}, [110, 148] (Fig. 23.7c, d). All constitutive Orai mutants exhibit reactivation in the absence of STIM1 and inactivation in the presence of STIM1. It is probable that an enhanced pore diameter, as is the case for Orai1 V102A, does not additionally enhance the extent of reactivation, but this supposition requires further evaluation and direct comparisons.

The Orai1 ANSGA mutant represents another exception, with respect to the enhancements in currents in a Na^+ divalent free solution, in the absence of STIM1. The current ratio I_{DVF}/I_{Ca2+} amounts to 1 (Fig. 23.6q, r; 5F), showing that currents in a Ca^{2+}-containing solution exhibit levels comparable to those in divalent free Na^+-containing solution only for this mutant.

Maintenance of authentic CRAC channel hallmarks not only requires STIM1, but also an intact Orai N-terminus. Using Orai1 V102A it has been impressively shown that an N-terminal deletion up to aa 78 still preserves constitutive activity, while the Ca^{2+} selectivity in the presence of STIM1 is lost. Only a deletion up to aa 74 fully maintains Ca^{2+} selectivity in line with the preserved store-operated activation of wild-type Orai1 upon deletion up to aa 74/75 in the presence of STIM1 [64, 67, 110, 121]. Other constitutive Orai1 mutants lead to loss of function upon deletion up to aa 78, while the function can be restored via the swap of Orai3 loop2 (Fig. 23.6g) as in the wild-type Orai1 (Fig. 23.6a) and in line with preserved function of analogue Orai3 N-terminal deletion mutants [61, 110]. N-truncation Orai1-Orai3-loop2 chimeras and analogue Orai3 N-truncation mutants have revealed that the reversal potential is not altered in the presence of STIM1 to more positive values, underlining the fact that residues upstream of aa 78, including the entire ETON region, are required to maintain the Ca^{2+} selectivity of Orai channels [110].

Furthermore, N-truncated constitutive mutants in the outer TM regions, TM2-TM4, have been employed to clarify the role of the N-terminus in maintaining the other CRAC channel hallmarks (FCDI and the ratio of I_{DVF}/I_{Ca2+}). We have discovered that a deletion of the first 74 residues already leads to abolished FCDI and enhancement of I_{DVF} versus I_{Ca2+} also in the presence of STIM1, suggesting that residues upstream of position 74 are required for full maintenance of authentic CRAC channel characteristics [110]. Moreover, this holds not only for the constitutively active mutants, but also for wild-type Orai proteins. Thus, in contrast to store-operated activation which requires only the last half of the ETON region (aa 80-90), for maintenance of CRAC channel hallmarks the whole ETON region is indispensable. Furthermore, we have shown that the interplay of the N-terminus with loop2 governs the maintenance of authentic CRAC channel characteristics, as a swap of loop2 between Orai1 and Orai3 N-truncation mutants alters CRAC channel properties in various ways. It is noteworthy that the Orai1 ANSGA mutant loses function upon N-terminal deletion up to aa 78. However, the function cannot be recovered via the switch of the Orai3 loop2 for this mutant function (Fig. 23.6m). This suggests that conformational alterations within the hinge region disrupt the communication with the loop2 region pointing to distinct conformational alterations via ANSGA compared to the diversity of other constitutively active mutants [61].

In summary, authentic CRAC channel hallmarks are only maintained in the presence of STIM1 and an intact Orai N-terminus. Otherwise, typical CRAC channel characteristics are fully abolished, as shown with diverse constitutively active Orai mutants. The latter, however, exhibit differences in their biophysical properties and thus can be categorized into three groups: Orai1 V102A, most TM2/TM3/TM4 constitutively active mutants and the Orai1 ANSGA mutant (Fig. 23.7). Obviously, these different constitutively active mutants adjust distinct partially open states. At this point, however, it remains unclear if STIM1 mediated Orai activation induces comparable structural alterations at some point as observed for the constitutively active mutants before reaching the final open state. Moreover, it is elusive, whether STIM1 binding and the Orai N-terminus establish these authentic CRAC channel properties via direct or indirect communication which should be ultimately addressed at a structural level.

23.8 Lipid Mediated Regulation of the STIM1/Orai Channel Complex

STIM1 and Orai1 proteins can fully reconstitute CRAC channel currents [84]. However, several components [115, 117, 171, 177–190] in the cell have been found to modulate the interplay, interaction and activation of the STIM1-Orai channel complex. Some regulatory proteins, such as SARAF, STIMATE or CRACR2A have already been briefly mentioned in this review, while a detailed presentation of these can be found in the following reviews [59, 118–120]. Here, we provide an overview of lipid mediated regulation of the STIM1/Orai channel complex. Among the lipids, PIP_2 and cholesterol have been found to predominantly regulate CRAC channels.

PIP_2 represents a minor phospholipid that is found in the inner leaflet of the plasma membrane and plays an essential role in the modulation of several ion channels [191–193]. Furthermore, it reportedly regulates STIM proteins. Both, STIM1 and STIM2 contain a lysine-rich domain at the very end of their C-termini, which resembles a PIP_2 binding site [50]. This domain is larger in STIM2 [51].

The STIM1-Orai1 clustering, STIM1 targeting to ER-PM junctions and stable interaction between STIM1 and the plasma membrane have been shown to be affected upon reduction of PIP_2 or PIP_3 levels [50, 183–185]. Nevertheless, store-operated currents are maintained upon PIP_2 depletion [183, 185]. Enhanced PIP_2 levels in liposomes enabled preferential binding of STIM2 [51], suggesting that PIP_2 contributes to STIM2 recruitment to the plasma membrane. In contrast, the reduction in PIP_2 levels affects STIM1 translocation only mildly, although CRAC currents are strongly impaired [183]. Recently, a novel regulatory protein RASSF4 was found to control SOCE and ER-PM junctions by affecting steady state PIP_2 levels in the plasma membrane. Knock-down of RASSF4 leads to reduced PIP_2 levels and in consequence to decreased STIM1-Orai1 mediated currents [194, 195].

Cholesterol represents a class that is distinct from other membrane lipids, as it cannot form membranes on its own and it penetrates less deeply into the hydrophobic layer of membrane leaflets [196]. Moreover, cholesterol is a highly mobile lipid, and drug-induced modulation of the cholesterol levels in the membrane can have a large impact on virus replication and trafficking pathways [196]. Membrane rafts have been suggested to contain high concentrations of cholesterol as well as sphingolipids [196].

Among the diversity of ion channels (nAChR, Kir, BK, TRPV [197]), the CRAC channel components, STIM1 and Orai1 have also been recently found to be regulated by cholesterol [112–114]. Enzymatically or chemically induced cholesterol depletion enhances SOCE or Ca^{2+} currents from STIM1/Orai1 expressing HEK cells and increases endogenous CRAC currents in mast cells [113]. The latter also leads to enhanced mast cell degranulation as pathophysiologically observed in hypocholesterolemia patients associated with the Smith-Lemli-Opitz syndrome [198]. In support of these observations, both, STIM1 and Orai1 have been shown to bind cholesterol which affects STIM1 mediated Orai1 currents in an inhibitory manner [113, 114].

Cholesterol can interact with membrane proteins via specific cholesterol binding motifs, such as the cholesterol recognition amino acid consensus motif (CRAC) [197, 199, 200]. The CRAC motif, with the consensus sequence -L/V-$(X)_{(1\text{-}5)}$-Y-$(X)_{(1\text{-}5)}$-R/K-, is located at the protein-lipid interface of a transmembrane domain [200]. Cholesterol as part of the membrane is located either in the outer or inner leaflet of the membrane, and its head groups (-OH group in cholesterol) are aligned in a row with the head groups of the other lipids located in the upper or lower leaflet. Due to considerable variability of the consensus sequence for cholesterol binding motifs, other criteria such as its close proximity to a TM region, formation of an α-helical structure and/or its incorporation in a protein located within lipid rafts is required [196, 201, 202]. The STIM1 C-terminus and Orai1 N-terminus have been shown to contain a putative cholesterol binding motif. The cholesterol binding motif within Orai1 N-terminus is located in the conserved ETON region that extends from aa 74-83 [113]. In STIM1 C-terminus it is located within the SOAR domain, specifically at aa 357 – 366 in CC2 [114]. Mutations of those cholesterol binding motifs (L74I, Y80S in Orai1, I364A in STIM1), that lead to abolished cholesterol binding, have been shown to strongly enhance Orai1 currents mediated by STIM1 or STIM1-C-terminal fragments, and these findings are in line with the enhancements observed upon chemically induced cholesterol depletion. A lack of cholesterol binding does not affect the selectivity of CRAC currents and retains the association of STIM1 to Orai1. Overall, these findings clearly indicate that the CRAC channel function can be modulated by lipids such as PIP_2 and cholesterol.

23.9 Perspectives

Over the past decade, considerable progress has been made in understanding the key molecular components of CRAC channels. Currently, two 3D atomic structures of the Orai protein are available; one showing the closed and one the open state. Nevertheless, several questions remain to be answered in the CRAC channel field.

With regard to STIM1, structural resolutions are currently only available for cytosolic C-terminal fragments. Furthermore, the two different, available structures of STIM1 C-terminal portions have revealed significant differences. Hence, further structural studies are needed, particularly on longer STIM1 C-terminal fragments in order to clarify the intra- and intermolecular interactions that govern the activation of STIM1. Mutations within STIM1 proteins locking them in the active, extended conformation e.g. L251S probably helps to obtain structures that mimic the active state of cytosolic STIM1.

Unlike STIM1, the dOrai structure has already been resolved in two conformations, which are assumed to represent the closed and the open state, respectively. Nevertheless, the structure of the gain-of-function mutant represents an open state in the absence of STIM1. Thus, it is unclear whether this structure mimics the STIM1-bound Orai open conformation. The novel structure indicates a huge structural rearrangement within TM4-C-terminus, but a great deal of energy would be required to stabilize this conformation. Thus, the provision of additional structures for other gain-of-function mutants (e.g. Orai1 P245X) or the STIM1-Orai complex as well as a structure for a human Orai would further clarify the pore opening mechanisms of CRAC channels.

The detailed molecular mechanism of how STIM1 gates Orai channels into the open state is still only partially understood. Thus far, only the NMR structure of a complex of C-terminal fragments of STIM1 and Orai1 is available and this has been a key to characterizing the main STIM1-Orai1 interaction sites. There is clear evidence that other cytosolic strands within Orai are involved in the STIM1 mediated gating process. However, whether this occurs via direct or indirect interaction still remains to be resolved. In addition, structural resolutions of a STIM1-Orai1 complex might also provide further insight into essential binding sites of STIM1 within Orai. These studies may reveal the structural alterations that take place upon activation of the Orai channel including re-arrangements of the cytosolic strands that are linked with those in the TM regions. Here, single particle cryo-EM [203, 204] represents another valuable approach, that can be used, as this has been successfully used with ion channels such as TRPV1 [205], TRPML3 [206] and the IP_3-receptor [207] to obtain resolutions that are similar to those obtained using crystallography.

As a complement to structural resolutions, the precise and dynamic resolution of the CRAC channel function was obtained using light [40, 208–210]. This sophisticated optogenetics approach can be capitalized upon to control and understand the STIM1-Orai communication more effectively and elucidate the structure-function relationship in more detail.

The current achievements hat have been made in the understanding of the communication between STIM and Orai proteins have allowed to enhance our knowledge of physiological down-stream signaling for the CRAC channels [211, 212]. The native CRAC channel system becomes even more complex when considering the accessory cellular components that regulate the STIM/Orai communication, such as SARAF [171], STIMATE [213] and cholesterol [112–114] and potentially enables fine-tuned interference with drugs. Even though several CRAC channel blockers are currently available [214–216], their precise sites of action have remained elusive. In addition to the classical pore blockers, the complexity of the CRAC channel machinery allows a variety of interventions to be made, which can interfere with the coupling of CRAC channel components or their conformational changes from the closed to the open state in order. Hence, novel CRAC channel modulators with significant therapeutic potential in immune deficiency, autoimmune or allergic disorders can potentially be uncovered.

Acknowledgements This work was supported by the Austrian Science Fund (FWF projects P27641 and P30567 to I.D.).

References

1. Misceo D, Holmgren A, Louch WE, Holme PA, Mizobuchi M, Morales RJ, De Paula AM, Stray-Pedersen A, Lyle R, Dalhus B, Christensen G, Stormorken H, Tjonnfjord GE, Frengen E (2014) A dominant STIM1 mutation causes Stormorken syndrome. Hum Mutat 35(5):556–564. https://doi.org/10.1002/humu.22544
2. Feske S, Gwack Y, Prakriya M, Srikanth S, Puppel SH, Tanasa B, Hogan PG, Lewis RS, Daly M, Rao A (2006) A mutation in Orai1 causes immune deficiency by abrogating CRAC channel function. Nature 441(7090):179–185
3. Parekh AB, Putney JW Jr (2005) Store-operated calcium channels. Physiol Rev 85(2):757–810. https://doi.org/10.1152/physrev.00057.2003
4. Zhang SL, Yeromin AV, Zhang XH, Yu Y, Safrina O, Penna A, Roos J, Stauderman KA, Cahalan MD (2006) Genome-wide RNAi screen of Ca^{2+} influx identifies genes that regulate Ca^{2+} release-activated Ca^{2+} channel activity. Proc Natl Acad Sci U S A 103(24):9357–9362. https://doi.org/10.1073/pnas.0603161103
5. Berna-Erro A, Woodard GE, Rosado JA (2012) Orais and STIMs: physiological mechanisms and disease. J Cell Mol Med 16(3):407–424. https://doi.org/10.1111/j.1582-4934.2011.01395.x
6. Morin G, Bruechle NO, Singh AR, Knopp C, Jedraszak G, Elbracht M, Bremond-Gignac D, Hartmann K, Sevestre H, Deutz P, Herent D, Nurnberg P, Romeo B, Konrad K, Mathieu-Dramard M, Oldenburg J, Bourges-Petit E, Shen Y, Zerres K, Ouadid-Ahidouch H, Rochette J (2014) Gain-of-function mutation in STIM1 (P.R304W) is associated with stormorken syndrome. Hum Mutat 35(10):1221–1232. https://doi.org/10.1002/humu.22621

7. Thompson JL, Mignen O, Shuttleworth TJ (2009) The Orai1 severe combined immune deficiency mutation and calcium release-activated Ca^{2+} channel function in the heterozygous condition. J Biol Chem 284(11):6620–6626. https://doi.org/10.1074/jbc.M808346200
8. Nesin V, Wiley G, Kousi M, Ong EC, Lehmann T, Nicholl DJ, Suri M, Shahrizaila N, Katsanis N, Gaffney PM, Wierenga KJ, Tsiokas L (2014) Activating mutations in STIM1 and ORAI1 cause overlapping syndromes of tubular myopathy and congenital miosis. Proc Natl Acad Sci U S A 111(11):4197–4202. https://doi.org/10.1073/pnas.1312520111 1312520111 [pii]
9. Zhang SL, Yu Y, Roos J, Kozak JA, Deerinck TJ, Ellisman MH, Stauderman KA, Cahalan MD (2005) STIM1 is a Ca^{2+} sensor that activates CRAC channels and migrates from the Ca^{2+} store to the plasma membrane. Nature 437(7060):902–905. doi:nature04147 [pii] https://doi.org/10.1038/nature04147
10. Roos J, DiGregorio PJ, Yeromin AV, Ohlsen K, Lioudyno M, Zhang S, Safrina O, Kozak JA, Wagner SL, Cahalan MD, Velicelebi G, Stauderman KA (2005) STIM1, an essential and conserved component of store-operated Ca^{2+} channel function. J Cell Biol 169(3):435–445. https://doi.org/10.1083/jcb.200502019
11. Liou J, Kim ML, Heo WD, Jones JT, Myers JW, Ferrell JE Jr, Meyer T (2005) STIM is a Ca^{2+} sensor essential for Ca^{2+}-store-depletion-triggered Ca^{2+} influx. Curr Biol 15(13):1235–1241
12. Vig M, Peinelt C, Beck A, Koomoa DL, Rabah D, Koblan-Huberson M, Kraft S, Turner H, Fleig A, Penner R, Kinet JP (2006) CRACM1 is a plasma membrane protein essential for store-operated Ca^{2+} entry. Science 312(5777):1220–1223
13. Prakriya M, Feske S, Gwack Y, Srikanth S, Rao A, Hogan PG (2006) Orai1 is an essential pore subunit of the CRAC channel. Nature 443(7108):230–233. https://doi.org/10.1038/nature05122
14. Hou X, Pedi L, Diver MM, Long SB (2012) Crystal structure of the calcium release-activated calcium channel Orai. Science 338(6112):1308–1313. https://doi.org/10.1126/science.1228757
15. Hou X, Burstein SR, Long S (2018) Structures reveal opening of the store-operated calcium channel Orai. bioRxiv. https://doi.org/10.1101/284034
16. Stathopulos PB, Schindl R, Fahrner M, Zheng L, Gasmi-Seabrook GM, Muik M, Romanin C, Ikura M (2013) STIM1/Orai1 coiled-coil interplay in the regulation of store-operated calcium entry. Nat Commun 4:2963. https://doi.org/10.1038/ncomms3963
17. Yang X, Jin H, Cai X, Li S, Shen Y (2012) Structural and mechanistic insights into the activation of Stromal interaction molecule 1 (STIM1). Proc Natl Acad Sci U S A 109(15):5657–5662. https://doi.org/10.1073/pnas.1118947109
18. Soboloff J, Rothberg BS, Madesh M, Gill DL (2012) STIM proteins: dynamic calcium signal transducers. Nat Rev Mol Cell Biol 13(9):549–565. https://doi.org/10.1038/nrm3414
19. Cahalan MD (2009) STIMulating store-operated Ca^{2+} entry. Nat Cell Biol 11(6):669–677. https://doi.org/10.1038/ncb0609-669
20. Manji SS, Parker NJ, Williams RT, van Stekelenburg L, Pearson RB, Dziadek M, Smith PJ (2000) STIM1: a novel phosphoprotein located at the cell surface. Biochim Biophys Acta 1481(1):147–155
21. Soboloff J, Spassova MA, Tang XD, Hewavitharana T, Xu W, Gill DL (2006) Orai1 and STIM reconstitute store-operated calcium channel function. J Biol Chem 281(30):20661–20665. https://doi.org/10.1074/jbc.C600126200
22. Ambily A, Kaiser WJ, Pierro C, Chamberlain EV, Li Z, Jones CI, Kassouf N, Gibbins JM, Authi KS (2014) The role of plasma membrane STIM1 and Ca^{2+}entry in platelet aggregation. STIM1 binds to novel proteins in human platelets. Cell Signal 26(3):502–511. https://doi.org/10.1016/j.cellsig.2013.11.025
23. Spassova MA, Soboloff J, He LP, Xu W, Dziadek MA, Gill DL (2006) STIM1 has a plasma membrane role in the activation of store-operated Ca^{2+} channels. Proc Natl Acad Sci U S A 103(11):4040–4045. https://doi.org/10.1073/pnas.0510050103
24. Shuttleworth TJ (2009) Arachidonic acid, ARC channels, and Orai proteins. Cell Calcium 45(6):602–610. https://doi.org/10.1016/j.ceca.2009.02.001

25. Thompson JL, Shuttleworth TJ (2012) A plasma membrane-targeted cytosolic domain of STIM1 selectively activates ARC channels, an arachidonate-regulated store-independent Orai channel. Channels (Austin) 6(5):370–378. https://doi.org/10.4161/chan.21947
26. Novello MJ, Zhu J, Feng Q, Ikura M, Stathopulos PB (2018) Structural elements of stromal interaction molecule function. Cell Calcium 73:88–94. https://doi.org/10.1016/j.ceca.2018.04.006
27. Varga-Szabo D, Braun A, Nieswandt B (2011) STIM and Orai in platelet function. Cell Calcium 50(3):270–278. https://doi.org/10.1016/j.ceca.2011.04.002
28. Braun A, Vogtle T, Varga-Szabo D, Nieswandt B (2011) STIM and Orai in hemostasis and thrombosis. Front Biosci (Landmark Ed) 16:2144–2160
29. Zbidi H, Jardin I, Woodard GE, Lopez JJ, Berna-Erro A, Salido GM, Rosado JA (2011) STIM1 and STIM2 are located in the acidic Ca^{2+} stores and associates with Orai1 upon depletion of the acidic stores in human platelets. J Biol Chem 286(14):12257–12270. https://doi.org/10.1074/jbc.M110.190694
30. Soboloff J, Spassova MA, Dziadek MA, Gill DL (2006) Calcium signals mediated by STIM and Orai proteins – a new paradigm in inter-organelle communication. Biochim Biophys Acta 1763(11):1161–1168. https://doi.org/10.1016/j.bbamcr.2006.09.023
31. Rosado JA, Diez R, Smani T, Jardin I (2015) STIM and Orai1 variants in store-operated calcium entry. Front Pharmacol 6:325. https://doi.org/10.3389/fphar.2015.00325
32. Rana A, Yen M, Sadaghiani AM, Malmersjo S, Park CY, Dolmetsch RE, Lewis RS (2015) Alternative splicing converts STIM2 from an activator to an inhibitor of store-operated calcium channels. J Cell Biol 209(5):653–669. https://doi.org/10.1083/jcb.201412060
33. Miederer AM, Alansary D, Schwar G, Lee PH, Jung M, Helms V, Niemeyer BA (2015) A STIM2 splice variant negatively regulates store-operated calcium entry. Nat Commun 6:6899. https://doi.org/10.1038/ncomms7899
34. Hewavitharana T, Deng X, Soboloff J, Gill DL (2007) Role of STIM and Orai proteins in the store-operated calcium signaling pathway. Cell Calcium 42(2):173–182. https://doi.org/10.1016/j.ceca.2007.03.009
35. Cai X (2007) Molecular evolution and functional divergence of the Ca^{2+} sensor protein in store-operated Ca^{2+} entry: stromal interaction molecule. PLoS One 2(7):e609. https://doi.org/10.1371/journal.pone.0000609
36. Stathopulos PB, Zheng L, Li GY, Plevin MJ, Ikura M (2008) Structural and mechanistic insights into STIM1-mediated initiation of store-operated calcium entry. Cell 135(1):110–122. https://doi.org/10.1016/j.cell.2008.08.006
37. Zheng L, Stathopulos PB, Schindl R, Li GY, Romanin C, Ikura M (2011) Auto-inhibitory role of the EF-SAM domain of STIM proteins in store-operated calcium entry. Proc Natl Acad Sci U S A 108(4):1337–1342. https://doi.org/10.1073/pnas.1015125108
38. Baba Y, Hayashi K, Fujii Y, Mizushima A, Watarai H, Wakamori M, Numaga T, Mori Y, Iino M, Hikida M, Kurosaki T (2006) Coupling of STIM1 to store-operated Ca^{2+} entry through its constitutive and inducible movement in the endoplasmic reticulum. Proc Natl Acad Sci U S A 103(45):16704–16709. https://doi.org/10.1073/pnas.0608358103
39. Derler I, Jardin I, Romanin C (2016) Molecular mechanisms of STIM/Orai communication. Am J Physiol Cell Physiol 310(8):C643–C662. https://doi.org/10.1152/ajpcell.00007.2016
40. Ma G, Wei M, He L, Liu C, Wu B, Zhang SL, Jing J, Liang X, Senes A, Tan P, Li S, Sun A, Bi Y, Zhong L, Si H, Shen Y, Li M, Lee MS, Zhou W, Wang J, Wang Y, Zhou Y (2015) Inside-out Ca^{2+} signalling prompted by STIM1 conformational switch. Nat Commun 6:7826. https://doi.org/10.1038/ncomms8826
41. Muik M, Fahrner M, Derler I, Schindl R, Bergsmann J, Frischauf I, Groschner K, Romanin C (2009) A cytosolic homomerization and a modulatory domain within STIM1 C terminus determine coupling to ORAI1 channels. J Biol Chem 284(13):8421–8426. https://doi.org/10.1074/jbc.C800229200
42. Kawasaki T, Lange I, Feske S (2009) A minimal regulatory domain in the C terminus of STIM1 binds to and activates ORAI1 CRAC channels. Biochem Biophys Res Commun 385(1):49–54. https://doi.org/10.1016/j.bbrc.2009.05.020

43. Park CY, Hoover PJ, Mullins FM, Bachhawat P, Covington ED, Raunser S, Walz T, Garcia KC, Dolmetsch RE, Lewis RS (2009) STIM1 clusters and activates CRAC channels via direct binding of a cytosolic domain to Orai1. Cell 136(5):876–890. https://doi.org/10.1016/j.cell.2009.02.014
44. Yuan JP, Zeng W, Dorwart MR, Choi YJ, Worley PF, Muallem S (2009) SOAR and the polybasic STIM1 domains gate and regulate Orai channels. Nat Cell Biol 11(3):337–343. https://doi.org/10.1038/ncb1842
45. Fahrner M, Muik M, Schindl R, Butorac C, Stathopulos P, Zheng L, Jardin I, Ikura M, Romanin C (2014) A coiled-coil clamp controls both conformation and clustering of stromal interaction molecule 1 (STIM1). J Biol Chem 289(48):33231–33244. https://doi.org/10.1074/jbc.M114.610022
46. Mullins FM, Lewis RS (2016) The inactivation domain of STIM1 is functionally coupled with the Orai1 pore to enable Ca^{2+}-dependent inactivation. J Gen Physiol 147(2):153–164. https://doi.org/10.1085/jgp.201511438
47. Derler I, Fahrner M, Muik M, Lackner B, Schindl R, Groschner K, Romanin C (2009) A Ca2(+)release-activated Ca2(+) (CRAC) modulatory domain (CMD) within STIM1 mediates fast Ca2(+)-dependent inactivation of ORAI1 channels. J Biol Chem 284(37):24933–24938. doi:C109.024083 [pii] https://doi.org/10.1074/jbc.C109.024083
48. Li Z, Lu J, Xu P, Xie X, Chen L, Xu T (2007) Mapping the interacting domains of STIM1 and Orai1 in Ca^{2+} release-activated Ca^{2+} channel activation. J Biol Chem 282(40):29448–29456. https://doi.org/10.1074/jbc.M703573200
49. Hogan PG, Lewis RS, Rao A (2010) Molecular basis of calcium signaling in lymphocytes: STIM and ORAI. Annu Rev Immunol 28:491–533. https://doi.org/10.1146/annurev.immunol.021908.132550
50. Liou J, Fivaz M, Inoue T, Meyer T (2007) Live-cell imaging reveals sequential oligomerization and local plasma membrane targeting of stromal interaction molecule 1 after Ca^{2+} store depletion. Proc Natl Acad Sci U S A 104(22):9301–9306. https://doi.org/10.1073/pnas.0702866104
51. Ercan E, Momburg F, Engel U, Temmerman K, Nickel W, Seedorf M (2009) A conserved, lipid-mediated sorting mechanism of yeast Ist2 and mammalian STIM proteins to the peripheral ER. Traffic 10(12):1802–1818. doi:TRA995 [pii] https://doi.org/10.1111/j.1600-0854.2009.00995.x
52. Wang X, Wang Y, Zhou Y, Hendron E, Mancarella S, Andrake MD, Rothberg BS, Soboloff J, Gill DL (2014) Distinct Orai-coupling domains in STIM1 and STIM2 define the Orai-activating site. Nat Commun 5:3183. https://doi.org/10.1038/ncomms4183
53. Bogeski I, Al-Ansary D, Qu B, Niemeyer BA, Hoth M, Peinelt C (2010) Pharmacology of ORAI channels as a tool to understand their physiological functions. Expert Rev Clin Pharmacol 3(3):291–303. https://doi.org/10.1586/ecp.10.23
54. Vanden Abeele F, Dubois C, Shuba Y, Prevarskaya N (2015) Disrupting the dynamic equilibrium of ORAI channels determines the phenotype of malignant cells. Mol Cell Oncol 2(2):e975631. https://doi.org/10.4161/23723556.2014.975631
55. Dubois C, Vanden Abeele F, Lehen'kyi V, Gkika D, Guarmit B, Lepage G, Slomianny C, Borowiec AS, Bidaux G, Benahmed M, Shuba Y, Prevarskaya N (2014) Remodeling of channel-forming ORAI proteins determines an oncogenic switch in prostate cancer. Cancer Cell 26(1):19–32. https://doi.org/10.1016/j.ccr.2014.04.025
56. Ji W, Xu P, Li Z, Lu J, Liu L, Zhan Y, Chen Y, Hille B, Xu T, Chen L (2008) Functional stoichiometry of the unitary calcium-release-activated calcium channel. Proc Natl Acad Sci U S A 105(36):13668–13673. doi:0806499105 [pii] https://doi.org/10.1073/pnas.0806499105
57. Mignen O, Thompson JL, Shuttleworth TJ (2008) Orai1 subunit stoichiometry of the mammalian CRAC channel pore. J Physiol 586(2):419–425. doi:jphysiol.2007.147249 [pii] https://doi.org/10.1113/jphysiol.2007.147249
58. Penna A, Demuro A, Yeromin AV, Zhang SL, Safrina O, Parker I, Cahalan MD (2008) The CRAC channel consists of a tetramer formed by Stim-induced dimerization of Orai dimers. Nature 456(7218):116–120. https://doi.org/10.1038/nature07338

59. Hogan PG, Rao A (2015) Store-operated calcium entry: mechanisms and modulation. Biochem Biophys Res Commun 460(1):40–49. https://doi.org/10.1016/j.bbrc.2015.02.110
60. Shuttleworth TJ (2012) Orai3 – the 'exceptional' Orai? J Physiol 590(Pt 2):241–257. https://doi.org/10.1113/jphysiol.2011.220574
61. Fahrner M, Pandey SK, Muik M, Traxler L, Butorac C, Stadlbauer M, Zayats V, Krizova A, Plenk P, Frischauf I, Schindl R, Gruber HJ, Hinterdorfer P, Ettrich R, Romanin C, Derler I (2018) Communication between N terminus and loop2 tunes Orai activation. J Biol Chem 293(4):1271–1285. https://doi.org/10.1074/jbc.M117.812693
62. Yen M, Lokteva LA, Lewis RS (2016) Functional analysis of Orai1 concatemers supports a hexameric stoichiometry for the CRAC channel. Biophys J 111(9):1897–1907. https://doi.org/10.1016/j.bpj.2016.09.020
63. Cai X, Zhou Y, Nwokonko RM, Loktionova NA, Wang X, Xin P, Trebak M, Wang Y, Gill DL (2016) The Orai1 store-operated calcium channel functions as a hexamer. J Biol Chem 291(50):25764–25775. https://doi.org/10.1074/jbc.M116.758813
64. Derler I, Plenk P, Fahrner M, Muik M, Jardin I, Schindl R, Gruber HJ, Groschner K, Romanin C (2013) The extended transmembrane Orai1 N-terminal (ETON) region combines binding interface and gate for Orai1 activation by STIM1. J Biol Chem 288(40):29025–29034. https://doi.org/10.1074/jbc.M113.501510
65. Zhou Y, Cai X, Loktionova NA, Wang X, Nwokonko RM, Wang X, Wang Y, Rothberg BS, Trebak M, Gill DL (2016) The STIM1-binding site nexus remotely controls Orai1 channel gating. Nat Commun 7:13725. https://doi.org/10.1038/ncomms13725
66. Frischauf I, Zayats V, Deix M, Hochreiter A, Jardin I, Muik M, Lackner B, Svobodova B, Pammer T, Litvinukova M, Sridhar AA, Derler I, Bogeski I, Romanin C, Ettrich RH, Schindl R (2015) A calcium-accumulating region, CAR, in the channel Orai1 enhances Ca^{2+} permeation and SOCE-induced gene transcription. Sci Signal 8(408):ra131. https://doi.org/10.1126/scisignal.aab1901
67. McNally BA, Somasundaram A, Yamashita M, Prakriya M (2012) Gated regulation of CRAC channel ion selectivity by STIM1. Nature 482(7384):241–245. https://doi.org/10.1038/nature10752
68. Frischauf I, Litvinukova M, Schober R, Zayats V, Svobodova B, Bonhenry D, Lunz V, Cappello S, Tociu L, Reha D, Stallinger A, Hochreiter A, Pammer T, Butorac C, Muik M, Groschner K, Bogeski I, Ettrich RH, Romanin C, Schindl R (2017) Transmembrane helix connectivity in Orai1 controls two gates for calcium-dependent transcription. Sci Signal 10(507). https://doi.org/10.1126/scisignal.aao0358
69. Yamashita M, Yeung PS, Ing CE, McNally BA, Pomes R, Prakriya M (2017) STIM1 activates CRAC channels through rotation of the pore helix to open a hydrophobic gate. Nat Commun 8:14512. https://doi.org/10.1038/ncomms14512
70. Palty R, Stanley C, Isacoff EY (2015) Critical role for Orai1 C-terminal domain and TM4 in CRAC channel gating. Cell Res 25(8):963–980. https://doi.org/10.1038/cr.2015.80
71. Bohm J, Bulla M, Urquhart JE, Malfatti E, Williams SG, O'Sullivan J, Szlauer A, Koch C, Baranello G, Mora M, Ripolone M, Violano R, Moggio M, Kingston H, Dawson T, DeGoede CG, Nixon J, Boland A, Deleuze JF, Romero N, Newman WG, Demaurex N, Laporte J (2017) ORAI1 mutations with distinct channel gating defects in tubular aggregate myopathy. Hum Mutat 38(4):426–438. https://doi.org/10.1002/humu.23172
72. Endo Y, Noguchi S, Hara Y, Hayashi YK, Motomura K, Miyatake S, Murakami N, Tanaka S, Yamashita S, Kizu R, Bamba M, Goto Y, Matsumoto N, Nonaka I, Nishino I (2015) Dominant mutations in ORAI1 cause tubular aggregate myopathy with hypocalcemia via constitutive activation of store-operated Ca(2)(+) channels. Hum Mol Genet 24(3):637–648. https://doi.org/10.1093/hmg/ddu477
73. Yeung PS, Yamashita M, Ing CE, Pomes R, Freymann DM, Prakriya M (2018) Mapping the functional anatomy of Orai1 transmembrane domains for CRAC channel gating. Proc Natl Acad Sci U S A 115(22):E5193–E5202. https://doi.org/10.1073/pnas.1718373115
74. Covington ED, Wu MM, Lewis RS (2010) Essential role for the CRAC activation domain in store-dependent oligomerization of STIM1. Mol Biol Cell 21(11):1897–1907. https://doi.org/10.1091/mbc.E10-02-0145

75. Grigoriev I, Gouveia SM, van der Vaart B, Demmers J, Smyth JT, Honnappa S, Splinter D, Steinmetz MO, Putney JW Jr, Hoogenraad CC, Akhmanova A (2008) STIM1 is a MT-plus-end-tracking protein involved in remodeling of the ER. Curr Biol 18(3):177–182. https://doi.org/10.1016/j.cub.2007.12.050
76. Honnappa S, Gouveia SM, Weisbrich A, Damberger FF, Bhavesh NS, Jawhari H, Grigoriev I, van Rijssel FJ, Buey RM, Lawera A, Jelesarov I, Winkler FK, Wuthrich K, Akhmanova A, Steinmetz MO (2009) An EB1-binding motif acts as a microtubule tip localization signal. Cell 138(2):366–376. https://doi.org/10.1016/j.cell.2009.04.065
77. Huang Y, Zhou Y, Wong HC, Chen Y, Wang S, Castiblanco A, Liu A, Yang JJ (2009) A single EF-hand isolated from STIM1 forms dimer in the absence and presence of Ca^{2+}. FEBS J 276(19):5589–5597. https://doi.org/10.1111/j.1742-4658.2009.07240.x
78. Mercer JC, Dehaven WI, Smyth JT, Wedel B, Boyles RR, Bird GS, Putney JW Jr (2006) Large store-operated calcium selective currents due to co-expression of Orai1 or Orai2 with the intracellular calcium sensor, Stim1. J Biol Chem 281(34):24979–24990. https://doi.org/10.1074/jbc.M604589200
79. Brandman O, Liou J, Park WS, Meyer T (2007) STIM2 is a feedback regulator that stabilizes basal cytosolic and endoplasmic reticulum Ca^{2+} levels. Cell 131(7):1327–1339. https://doi.org/10.1016/j.cell.2007.11.039
80. Subedi KP, Ong HL, Son GY, Liu X, Ambudkar IS (2018) STIM2 induces activated conformation of STIM1 to control Orai1 function in ER-PM junctions. Cell Rep 23(2):522–534. https://doi.org/10.1016/j.celrep.2018.03.065
81. Darbellay B, Arnaudeau S, Bader CR, Konig S, Bernheim L (2011) STIM1L is a new actin-binding splice variant involved in fast repetitive Ca^{2+} release. J Cell Biol 194(2):335–346. https://doi.org/10.1083/jcb.201012157
82. Huang GN, Zeng W, Kim JY, Yuan JP, Han L, Muallem S, Worley PF (2006) STIM1 carboxyl-terminus activates native SOC, I(crac) and TRPC1 channels. Nat Cell Biol 8(9):1003–1010. https://doi.org/10.1038/ncb1454
83. Zhou Y, Srinivasan P, Razavi S, Seymour S, Meraner P, Gudlur A, Stathopulos PB, Ikura M, Rao A, Hogan PG (2013) Initial activation of STIM1, the regulator of store-operated calcium entry. Nat Struct Mol Biol 20(8):973–981. https://doi.org/10.1038/nsmb.2625
84. Zhou Y, Meraner P, Kwon HT, Machnes D, Oh-hora M, Zimmer J, Huang Y, Stura A, Rao A, Hogan PG (2010) STIM1 gates the store-operated calcium channel ORAI1 in vitro. Nat Struct Mol Biol 17(1):112–116. https://doi.org/10.1038/nsmb.1724
85. Williams RT, Senior PV, Van Stekelenburg L, Layton JE, Smith PJ, Dziadek MA (2002) Stromal interaction molecule 1 (STIM1), a transmembrane protein with growth suppressor activity, contains an extracellular SAM domain modified by N-linked glycosylation. Biochim Biophys Acta 1596(1):131–137
86. Muik M, Fahrner M, Schindl R, Stathopulos P, Frischauf I, Derler I, Plenk P, Lackner B, Groschner K, Ikura M, Romanin C (2011) STIM1 couples to ORAI1 via an intramolecular transition into an extended conformation. EMBO J 30(9):1678–1689. https://doi.org/10.1038/emboj.2011.79
87. Yu J, Zhang H, Zhang M, Deng Y, Wang H, Lu J, Xu T, Xu P (2013) An aromatic amino acid in the coiled-coil 1 domain plays a crucial role in the auto-inhibitory mechanism of STIM1. Biochem J 454(3):401–409. https://doi.org/10.1042/BJ20130292
88. Fahrner M, Stadlbauer M, Muik M, Rathner P, Stathopulos P, Ikura M, Muller N, Romanin C (2018) A dual mechanism promotes switching of the Stormorken STIM1 R304W mutant into the activated state. Nat Commun 9(1):825. https://doi.org/10.1038/s41467-018-03062-w
89. Muik M, Frischauf I, Derler I, Fahrner M, Bergsmann J, Eder P, Schindl R, Hesch C, Polzinger B, Fritsch R, Kahr H, Madl J, Gruber H, Groschner K, Romanin C (2008) Dynamic coupling of the putative coiled-coil domain of ORAI1 with STIM1 mediates ORAI1 channel activation. J Biol Chem 283(12):8014–8022. doi:M708898200 [pii] https://doi.org/10.1074/jbc.M708898200

90. Stathopulos PB, Li GY, Plevin MJ, Ames JB, Ikura M (2006) Stored Ca^{2+} depletion-induced oligomerization of stromal interaction molecule 1 (STIM1) via the EF-SAM region: an initiation mechanism for capacitive Ca^{2+} entry. J Biol Chem 281(47):35855–35862. https://doi.org/10.1074/jbc.M608247200
91. Grupe M, Myers G, Penner R, Fleig A (2010) Activation of store-operated I(CRAC) by hydrogen peroxide. Cell Calcium 48(1):1–9. doi: S0143-4160(10)00085-0 [pii] https://doi.org/10.1016/j.ceca.2010.05.005
92. Hawkins BJ, Irrinki KM, Mallilankaraman K, Lien YC, Wang Y, Bhanumathy CD, Subbiah R, Ritchie MF, Soboloff J, Baba Y, Kurosaki T, Joseph SK, Gill DL, Madesh M (2010) S-glutathionylation activates STIM1 and alters mitochondrial homeostasis. J Cell Biol 190(3):391–405. doi:jcb.201004152 [pii] https://doi.org/10.1083/jcb.201004152
93. Droge W (2002) Free radicals in the physiological control of cell function. Physiol Rev 82(1):47–95. https://doi.org/10.1152/physrev.00018.2001
94. Rhee SG (2006) Cell signaling. H2O2, a necessary evil for cell signaling. Science 312(5782):1882–1883. doi: 312/5782/1882 [pii] https://doi.org/10.1126/science.1130481
95. Hirve N, Rajanikanth V, Hogan PG, Gudlur A (2018) Coiled-coil formation conveys a STIM1 signal from ER lumen to cytoplasm. Cell Rep 22(1):72–83. https://doi.org/10.1016/j.celrep.2017.12.030
96. Dong H, Sharma M, Zhou HX, Cross TA (2012) Glycines: role in alpha-helical membrane protein structures and a potential indicator of native conformation. Biochemistry 51(24):4779–4789. https://doi.org/10.1021/bi300090x
97. Fuchs S, Rensing-Ehl A, Speckmann C, Bengsch B, Schmitt-Graeff A, Bondzio I, Maul-Pavicic A, Bass T, Vraetz T, Strahm B, Ankermann T, Benson M, Caliebe A, Folster-Holst R, Kaiser P, Thimme R, Schamel WW, Schwarz K, Feske S, Ehl S (2012) Antiviral and regulatory T cell immunity in a patient with stromal interaction molecule 1 deficiency. J Immunol 188(3):1523–1533. https://doi.org/10.4049/jimmunol.1102507
98. Maus M, Jairaman A, Stathopulos PB, Muik M, Fahrner M, Weidinger C, Benson M, Fuchs S, Ehl S, Romanin C, Ikura M, Prakriya M, Feske S (2015) Missense mutation in immunodeficient patients shows the multifunctional roles of coiled-coil domain 3 (CC3) in STIM1 activation. Proc Natl Acad Sci U S A 112(19):6206–6211. https://doi.org/10.1073/pnas.1418852112
99. Hooper R, Soboloff J (2015) STIMATE reveals a STIM1 Transitional State. Nat Cell Biol 17(10):1232–1234. https://doi.org/10.1038/ncb3245
100. Kilch T, Alansary D, Peglow M, Dorr K, Rychkov G, Rieger H, Peinelt C, Niemeyer BA (2013) Mutations of the Ca^{2+}-sensing stromal interaction molecule STIM1 regulate Ca^{2+} influx by altered oligomerization of STIM1 and by destabilization of the Ca^{2+} channel Orai1. J Biol Chem 288(3):1653–1664. https://doi.org/10.1074/jbc.M112.417246
101. Kawasaki T, Ueyama T, Lange I, Feske S, Saito N (2010) Protein kinase C-induced phosphorylation of Orai1 regulates the intracellular Ca^{2+} level via the store-operated Ca^{2+} channel. J Biol Chem 285(33):25720–25730. doi:M109.022996 [pii] https://doi.org/10.1074/jbc.M109.022996
102. Yazbeck P, Tauseef M, Kruse K, Amin MR, Sheikh R, Feske S, Komarova Y, Mehta D (2017) STIM1 phosphorylation at Y361 recruits Orai1 to STIM1 puncta and induces Ca^{2+} entry. Sci Rep 7:42758. https://doi.org/10.1038/srep42758
103. Yu F, Sun L, Machaca K (2009) Orai1 internalization and STIM1 clustering inhibition modulate SOCE inactivation during meiosis. Proc Natl Acad Sci U S A 106(41):17401–17406. doi:0904651106 [pii] https://doi.org/10.1073/pnas.0904651106
104. Smyth JT, Petranka JG, Boyles RR, DeHaven WI, Fukushima M, Johnson KL, Williams JG, Putney JW Jr (2009) Phosphorylation of STIM1 underlies suppression of store-operated calcium entry during mitosis. Nat Cell Biol 11(12):1465–1472. doi:ncb1995 [pii] https://doi.org/10.1038/ncb1995
105. Pozo-Guisado E, Campbell DG, Deak M, Alvarez-Barrientos A, Morrice NA, Alvarez IS, Alessi DR, Martin-Romero FJ (2010) Phosphorylation of STIM1 at ERK1/2 target sites modulates store-operated calcium entry. J Cell Sci 123(Pt 18):3084–3093. doi:jcs.067215 [pii] https://doi.org/10.1242/jcs.067215

106. Frischauf I, Muik M, Derler I, Bergsmann J, Fahrner M, Schindl R, Groschner K, Romanin C (2009) Molecular determinants of the coupling between STIM1 and Orai channels: differential activation of Orai1-3 channels by a STIM1 coiled-coil mutant. J Biol Chem 284(32):21696–21706. https://doi.org/10.1074/jbc.M109.018408
107. Zhou Y, Cai X, Nwokonko RM, Loktionova NA, Wang Y, Gill DL (2017) The STIM-Orai coupling interface and gating of the Orai1 channel. Cell Calcium 63:8–13. https://doi.org/10.1016/j.ceca.2017.01.001
108. Tirado-Lee L, Yamashita M, Prakriya M (2015) Conformational changes in the Orai1 C-terminus evoked by STIM1 binding. PLoS One 10(6):e0128622. https://doi.org/10.1371/journal.pone.0128622
109. Lis A, Zierler S, Peinelt C, Fleig A, Penner R (2010) A single lysine in the N-terminal region of store-operated channels is critical for STIM1-mediated gating. J Gen Physiol 136(6):673–686. https://doi.org/10.1085/jgp.201010484
110. Derler I, Butorac C, Krizova A, Stadlbauer M, Muik M, Fahrner M, Frischauf I, Romanin C (2018) Authentic CRAC channel activity requires STIM1 and the conserved portion of the Orai N terminus. J Biol Chem 293(4):1259–1270. https://doi.org/10.1074/jbc.M117.812206
111. Kim KM, Wijerathne T, Hur JH, Kang UJ, Kim IH, Kweon YC, Lee AR, Jeong SJ, Lee SK, Lee YY, Sim BW, Lee JH, Baig C, Kim SU, Chang KT, Lee KP, Park CY (2018) Distinct gating mechanism of SOC channel involving STIM-Orai coupling and an intramolecular interaction of Orai in Caenorhabditis elegans. Proc Natl Acad Sci U S A 115(20):E4623–e4632. https://doi.org/10.1073/pnas.1714986115
112. Bohorquez-Hernandez A, Gratton E, Pacheco J, Asanov A, Vaca L (2017) Cholesterol modulates the cellular localization of Orai1 channels and its disposition among membrane domains. Biochim Biophys Acta 1862(12):1481–1490. https://doi.org/10.1016/j.bbalip.2017.09.005
113. Derler I, Jardin I, Stathopulos PB, Muik M, Fahrner M, Zayats V, Pandey SK, Poteser M, Lackner B, Absolonova M, Schindl R, Groschner K, Ettrich R, Ikura M, Romanin C (2016) Cholesterol modulates Orai1 channel function. Sci Signal 9(412):ra10. https://doi.org/10.1126/scisignal.aad7808
114. Pacheco J, Dominguez L, Bohorquez-Hernandez A, Asanov A, Vaca L (2016) A cholesterol-binding domain in STIM1 modulates STIM1-Orai1 physical and functional interactions. Sci Rep 6:29634. https://doi.org/10.1038/srep29634
115. Srikanth S, Jung HJ, Kim KD, Souda P, Whitelegge J, Gwack Y (2010) A novel EF-hand protein, CRACR2A, is a cytosolic Ca^{2+} sensor that stabilizes CRAC channels in T cells. Nat Cell Biol 12(5):436–446. https://doi.org/10.1038/ncb2045
116. Mullins FM, Park CY, Dolmetsch RE, Lewis RS (2009) STIM1 and calmodulin interact with Orai1 to induce Ca^{2+}-dependent inactivation of CRAC channels. Proc Natl Acad Sci U S A 106(36):15495–15500. https://doi.org/10.1073/pnas.0906781106
117. Feng M, Grice DM, Faddy HM, Nguyen N, Leitch S, Wang Y, Muend S, Kenny PA, Sukumar S, Roberts-Thomson SJ, Monteith GR, Rao R (2010) Store-independent activation of Orai1 by SPCA2 in mammary tumors. Cell 143(1):84–98. https://doi.org/10.1016/j.cell.2010.08.040
118. Lopez JJ, Albarran L, Gomez LJ, Smani T, Salido GM, Rosado JA (2016) Molecular modulators of store-operated calcium entry. Biochim Biophys Acta 1863(8):2037–2043. https://doi.org/10.1016/j.bbamcr.2016.04.024
119. Shaw PJ, Qu B, Hoth M, Feske S (2013) Molecular regulation of CRAC channels and their role in lymphocyte function. Cell Mol Life Sci 70(15):2637–2656. https://doi.org/10.1007/s00018-012-1175-2
120. Srikanth S, Gwack Y (2013) Molecular regulation of the pore component of CRAC channels, Orai1. Curr Top Membr 71:181–207. https://doi.org/10.1016/B978-0-12-407870-3.00008-1
121. McNally BA, Somasundaram A, Jairaman A, Yamashita M, Prakriya M (2013) The C- and N-terminal STIM1 binding sites on Orai1 are required for both trapping and gating CRAC channels. J Physiol 591(Pt 11):2833–2850. https://doi.org/10.1113/jphysiol.2012.250456
122. Palty R, Isacoff EY (2016) Cooperative binding of stromal interaction molecule 1 (STIM1) to the N and C termini of calcium release-activated calcium modulator 1 (Orai1). J Biol Chem 291(1):334–341. https://doi.org/10.1074/jbc.M115.685289

123. Zhou Y, Nwokonko RM, Cai X, Loktionova NA, Abdulqadir R, Xin P, Niemeyer BA, Wang Y, Trebak M, Gill DL (2018) Cross-linking of Orai1 channels by STIM proteins. Proc Natl Acad Sci U S A 115(15):E3398–E3407. https://doi.org/10.1073/pnas.1720810115
124. Hoover PJ, Lewis RS (2011) Stoichiometric requirements for trapping and gating of Ca^{2+} release-activated Ca^{2+} (CRAC) channels by stromal interaction molecule 1 (STIM1). Proc Natl Acad Sci U S A 108(32):13299–13304. https://doi.org/10.1073/pnas.1101664108
125. Scrimgeour NR, Wilson DP, Barritt GJ, Rychkov GY (2014) Structural and stoichiometric determinants of Ca^{2+} release-activated Ca^{2+} (CRAC) channel Ca^{2+}-dependent inactivation. Biochim Biophys Acta 1838(5):1281–1287. https://doi.org/10.1016/j.bbamem.2014.01.019
126. Scrimgeour N, Litjens T, Ma L, Barritt GJ, Rychkov GY (2009) Properties of Orai1 mediated store-operated current depend on the expression levels of STIM1 and Orai1 proteins. J Physiol 587(Pt 12):2903–2918. https://doi.org/10.1113/jphysiol.2009.170662
127. Li Z, Liu L, Deng Y, Ji W, Du W, Xu P, Chen L, Xu T (2011) Graded activation of CRAC channel by binding of different numbers of STIM1 to Orai1 subunits. Cell Res 21(2):305–315. https://doi.org/10.1038/cr.2010.131
128. Palty R, Fu Z, Isacoff EY (2017) Sequential steps of CRAC channel activation. Cell Rep 19(9):1929 1939. https://doi.org/10.1016/j.celrep.2017.05.025
129. Lis A, Peinelt C, Beck A, Parvez S, Monteilh-Zoller M, Fleig A, Penner R (2007) CRACM1, CRACM2, and CRACM3 are store-operated Ca^{2+} channels with distinct functional properties. Curr Biol 17(9):794–800. https://doi.org/10.1016/j.cub.2007.03.065
130. Hoth M, Penner R (1992) Depletion of intracellular calcium stores activates a calcium current in mast cells. Nature 355(6358):353–356
131. Hoth M, Penner R (1993) Calcium release-activated calcium current in rat mast cells. J Physiol 465:359–386
132. Sather WA, McCleskey EW (2003) Permeation and selectivity in calcium channels. Annu Rev Physiol 65:133–159. https://doi.org/10.1146/annurev.physiol.65.092101.142345
133. Yamashita M, Prakriya M (2014) Divergence of Ca^{2+} selectivity and equilibrium Ca^{2+} blockade in a Ca^{2+} release-activated Ca^{2+} channel. J Gen Physiol 143(3):325–343. https://doi.org/10.1085/jgp.201311108
134. Yamashita M, Navarro-Borelly L, McNally BA, Prakriya M (2007) Orai1 mutations alter ion permeation and Ca^{2+}-dependent fast inactivation of CRAC channels: evidence for coupling of permeation and gating. J Gen Physiol 130(5):525–540. https://doi.org/10.1085/jgp.200709872
135. Alavizargar A, Berti C, Ejtehadi MR, Furini S (2018) Molecular dynamics simulations of Orai reveal how the third transmembrane segment contributes to hydration and Ca^{2+} selectivity in calcium release-activated calcium channels. J Phys Chem B 122(16):4407–4417. https://doi.org/10.1021/acs.jpcb.7b12453
136. Yeung PS, Yamashita M, Prakriya M (2016) Pore opening mechanism of CRAC channels. Cell Calcium. https://doi.org/10.1016/j.ceca.2016.12.006
137. Prakriya M, Lewis RS (2006) Regulation of CRAC channel activity by recruitment of silent channels to a high open-probability gating mode. J Gen Physiol 128(3):373–386. https://doi.org/10.1085/jgp.200609588
138. Dynes JL, Amcheslavsky A, Cahalan MD (2015) Genetically targeted single-channel optical recording reveals multiple Orai1 gating states and oscillations in calcium influx. Proc Natl Acad Sci U S A. https://doi.org/10.1073/pnas.1523410113
139. Lepple-Wienhues A, Cahalan MD (1996) Conductance and permeation of monovalent cations through depletion-activated Ca^{2+} channels (ICRAC) in Jurkat T cells. Biophys J 71(2):787–794. https://doi.org/10.1016/S0006-3495(96)79278-0
140. Bakowski D, Parekh AB (2002) Permeation through store-operated CRAC channels in divalent-free solution: potential problems and implications for putative CRAC channel genes. Cell Calcium 32(5-6):379–391
141. Prakriya M, Lewis RS (2002) Separation and characterization of currents through store-operated CRAC channels and Mg2+-inhibited cation (MIC) channels. J Gen Physiol 119(5):487–507

142. Su Z, Shoemaker RL, Marchase RB, Blalock JE (2004) Ca^{2+} modulation of Ca^{2+} release-activated Ca^{2+} channels is responsible for the inactivation of its monovalent cation current. Biophys J 86(2):805–814. https://doi.org/10.1016/S0006-3495(04)74156-9
143. Mullins FM, Yen M, Lewis RS (2016) Correction: Orai1 pore residues control CRAC channel inactivation independently of calmodulin. J Gen Physiol 147(3):289. https://doi.org/10.1085/jgp.20151143701262016c
144. Mullins FM, Yen M, Lewis RS (2016) Orai1 pore residues control CRAC channel inactivation independently of calmodulin. J Gen Physiol 147(2):137–152. https://doi.org/10.1085/jgp.201511437
145. Prakriya M, Lewis RS (2015) Store-operated calcium channels. Physiol Rev 95(4):1383–1436. https://doi.org/10.1152/physrev.00020.2014
146. McCleskey EW, Almers W (1985) The Ca channel in skeletal muscle is a large pore. Proc Natl Acad Sci U S A 82(20):7149–7153
147. Zhou Y, Ramachandran S, Oh-Hora M, Rao A, Hogan PG (2010) Pore architecture of the ORAI1 store-operated calcium channel. Proc Natl Acad Sci U S A 107(11):4896–4901. https://doi.org/10.1073/pnas.1001169107
148. McNally BA, Prakriya M (2012) Permeation, selectivity and gating in store-operated CRAC channels. J Physiol 590(Pt 17):4179–4191. https://doi.org/10.1113/jphysiol.2012.233098
149. Rothberg BS, Wang Y, Gill DL (2013) Orai channel pore properties and gating by STIM: implications from the Orai crystal structure. Sci Signal 6(267):pe9. https://doi.org/10.1126/scisignal.2003971
150. Derler I, Fahrner M, Carugo O, Muik M, Bergsmann J, Schindl R, Frischauf I, Eshaghi S, Romanin C (2009) Increased hydrophobicity at the N terminus/membrane interface impairs gating of the severe combined immunodeficiency-related ORAI1 mutant. J Biol Chem 284(23):15903–15915. https://doi.org/10.1074/jbc.M808312200
151. McNally BA, Yamashita M, Engh A, Prakriya M (2009) Structural determinants of ion permeation in CRAC channels. Proc Natl Acad Sci U S A 106(52):22516–22521. https://doi.org/10.1073/pnas.0909574106
152. Schindl R, Frischauf I, Bergsmann J, Muik M, Derler I, Lackner B, Groschner K, Romanin C (2009) Plasticity in Ca^{2+} selectivity of Orai1/Orai3 heteromeric channel. Proc Natl Acad Sci U S A 106(46):19623–19628. doi:0907714106 [pii] https://doi.org/10.1073/pnas.0907714106
153. Vaeth M, Yang J, Yamashita M, Zee I, Eckstein M, Knosp C, Kaufmann U, Karoly Jani P, Lacruz RS, Flockerzi V, Kacskovics I, Prakriya M, Feske S (2017) ORAI2 modulates store-operated calcium entry and T cell-mediated immunity. Nat Commun 8:14714. https://doi.org/10.1038/ncomms14714
154. Vig M, Beck A, Billingsley JM, Lis A, Parvez S, Peinelt C, Koomoa DL, Soboloff J, Gill DL, Fleig A, Kinet JP, Penner R (2006) CRACM1 multimers form the ion-selective pore of the CRAC channel. Curr Biol 16(20):2073–2079. https://doi.org/10.1016/j.cub.2006.08.085
155. Yeromin AV, Zhang SL, Jiang W, Yu Y, Safrina O, Cahalan MD (2006) Molecular identification of the CRAC channel by altered ion selectivity in a mutant of Orai. Nature 443(7108):226–229. https://doi.org/10.1038/nature05108
156. Zhang SL, Yeromin AV, Hu J, Amcheslavsky A, Zheng H, Cahalan MD (2011) Mutations in Orai1 transmembrane segment 1 cause STIM1-independent activation of Orai1 channels at glycine 98 and channel closure at arginine 91. Proc Natl Acad Sci U S A 108(43):17838–17843. https://doi.org/10.1073/pnas.1114821108
157. Webster SM, Del Camino D, Dekker JP, Yellen G (2004) Intracellular gate opening in Shaker K^+ channels defined by high-affinity metal bridges. Nature 428(6985):864–868. https://doi.org/10.1038/nature02468
158. Payandeh J, Scheuer T, Zheng N, Catterall WA (2011) The crystal structure of a voltage-gated sodium channel. Nature 475(7356):353–358. https://doi.org/10.1038/nature10238
159. Zhao Y, Yarov-Yarovoy V, Scheuer T, Catterall WA (2004) A gating hinge in Na^+ channels; a molecular switch for electrical signaling. Neuron 41(6):859–865

160. Dong H, Fiorin G, Carnevale V, Treptow W, Klein ML (2013) Pore waters regulate ion permeation in a calcium release-activated calcium channel. Proc Natl Acad Sci U S A 110(43):17332–17337. https://doi.org/10.1073/pnas.1316969110
161. Gudlur A, Quintana A, Zhou Y, Hirve N, Mahapatra S, Hogan PG (2014) STIM1 triggers a gating rearrangement at the extracellular mouth of the ORAI1 channel. Nat Commun 5:5164. https://doi.org/10.1038/ncomms6164
162. Srikanth S, Yee MK, Gwack Y, Ribalet B (2011) The third transmembrane segment of orai1 protein modulates Ca^{2+} release-activated Ca^{2+} (CRAC) channel gating and permeation properties. J Biol Chem 286(40):35318–35328. https://doi.org/10.1074/jbc.M111.265884
163. Amcheslavsky A, Safrina O, Cahalan MD (2013) Orai3 TM3 point mutation G158C alters kinetics of 2-APB-induced gating by disulfide bridge formation with TM2 C101. J Gen Physiol 142(4):405–412. https://doi.org/10.1085/jgp.201311030
164. Bogeski I, Kummerow C, Al-Ansary D, Schwarz EC, Koehler R, Kozai D, Takahashi N, Peinelt C, Griesemer D, Bozem M, Mori Y, Hoth M, Niemeyer BA (2010) Differential redox regulation of ORAI ion channels: a mechanism to tune cellular calcium signaling. Sci Signal 3(115):ra24. doi:3/115/ra24 [pii] https://doi.org/10.1126/scisignal.2000672
165. Alansary D, Schmidt B, Dorr K, Bogeski I, Rieger H, Kless A, Niemeyer BA (2016) Thiol dependent intramolecular locking of Orai1 channels. Sci Rep 6:33347. https://doi.org/10.1038/srep33347
166. Cao X, Choi S, Maleth JJ, Park S, Ahuja M, Muallem S (2015) The ER/PM microdomain, PI(4,5)P(2) and the regulation of STIM1-Orai1 channel function. Cell Calcium 58(4):342–348. https://doi.org/10.1016/j.ceca.2015.03.003
167. Zweifach A, Lewis RS (1995) Rapid inactivation of depletion-activated calcium current (ICRAC) due to local calcium feedback. J Gen Physiol 105(2):209–226
168. Prakriya M, Lewis RS (2003) CRAC channels: activation, permeation, and the search for a molecular identity. Cell Calcium 33(5-6):311–321
169. Lee KP, Yuan JP, Zeng W, So I, Worley PF, Muallem S (2009) Molecular determinants of fast Ca^{2+}-dependent inactivation and gating of the Orai channels. Proc Natl Acad Sci U S A 106(34):14687–14692. doi:0904664106 [pii] https://doi.org/10.1073/pnas.0904664106
170. Jha A, Ahuja M, Maleth J, Moreno CM, Yuan JP, Kim MS, Muallem S (2013) The STIM1 CTID domain determines access of SARAF to SOAR to regulate Orai1 channel function. J Cell Biol 202(1):71–79. https://doi.org/10.1083/jcb.201301148
171. Palty R, Raveh A, Kaminsky I, Meller R, Reuveny E (2012) SARAF inactivates the store operated calcium entry machinery to prevent excess calcium refilling. Cell 149(2):425–438. https://doi.org/10.1016/j.cell.2012.01.055
172. Maleth J, Choi S, Muallem S, Ahuja M (2014) Translocation between PI(4,5)P2-poor and PI(4,5)P2-rich microdomains during store depletion determines STIM1 conformation and Orai1 gating. Nat Commun 5:5843. https://doi.org/10.1038/ncomms6843
173. Frischauf I, Schindl R, Bergsmann J, Derler I, Fahrner M, Muik M, Fritsch R, Lackner B, Groschner K, Romanin C (2011) Cooperativeness of Orai cytosolic domains tunes subtype-specific gating. J Biol Chem 286(10):8577–8584. doi:M110.187179 [pii] https://doi.org/10.1074/jbc.M110.187179
174. Bergsmann J, Derler I, Muik M, Frischauf I, Fahrner M, Pollheimer P, Schwarzinger C, Gruber HJ, Groschner K, Romanin C (2011) Molecular determinants within N terminus of Orai3 protein that control channel activation and gating. J Biol Chem 286(36):31565–31575. https://doi.org/10.1074/jbc.M111.227546
175. Litjens T, Harland ML, Roberts ML, Barritt GJ, Rychkov GY (2004) Fast Ca^{2+}-dependent inactivation of the store-operated Ca^{2+} current (ISOC) in liver cells: a role for calmodulin. J Physiol 558(Pt 1):85–97. https://doi.org/10.1113/jphysiol.2004.065870
176. Srikanth S, Jung HJ, Ribalet B, Gwack Y (2010) The intracellular loop of Orai1 plays a central role in fast inactivation of Ca^{2+} release-activated Ca^{2+} channels. J Biol Chem 285(7):5066–5075. doi:M109.072736 [pii] https://doi.org/10.1074/jbc.M109.072736
177. Howie D, Nolan KF, Daley S, Butterfield E, Adams E, Garcia-Rueda H, Thompson C, Saunders NJ, Cobbold SP, Tone Y, Tone M, Waldmann H (2009) MS4A4B

is a GITR-associated membrane adapter, expressed by regulatory T cells, which modulates T cell activation. J Immunol 183(7):4197–4204. doi:jimmunol.0901070 [pii] https://doi.org/10.4049/jimmunol.0901070

178. Walsh CM, Doherty MK, Tepikin AV, Burgoyne RD (2010) Evidence for an interaction between Golli and STIM1 in store-operated calcium entry. Biochem J 430(3):453–460. doi:BJ20100650 [pii] https://doi.org/10.1042/BJ20100650
179. Krapivinsky G, Krapivinsky L, Stotz SC, Manasian Y, Clapham DE 2011 POST, partner of stromal interaction molecule 1 (STIM1), targets STIM1 to multiple transporters. Proc Natl Acad Sci U S A 108 (48):19234-19239. doi:1117231108 [pii] https://doi.org/10.1073/pnas.1117231108
180. Deb BK, Pathak T, Hasan G (2016) Store-independent modulation of Ca^{2+} entry through Orai by Septin 7. Nat Commun 7. https://doi.org/10.1038/ncomms11751
181. Sharma S, Quintana A, Findlay GM, Mettlen M, Baust B, Jain M, Nilsson R, Rao A, Hogan PG (2013) An siRNA screen for NFAT activation identifies septins as coordinators of store-operated Ca^{2+} entry. Nature 499(7457):238–242. https://doi.org/10.1038/nature12229
182. Srivats S, Balasuriya D, Pasche M, Vistal G, Edwardson JM, Taylor CW, Murrell-Lagnado RD (2016) Sigma1 receptors inhibit store-operated Ca^{2+} entry by attenuating coupling of STIM1 to Orai1. J Cell Biol 213(1):65–79. https://doi.org/10.1083/jcb.201506022
183. Korzeniowski MK, Popovic MA, Szentpetery Z, Varnai P, Stojilkovic SS, Balla T (2009) Dependence of STIM1/Orai1-mediated calcium entry on plasma membrane phosphoinositides. J Biol Chem 284(31):21027–21035. doi:M109.012252 [pii] https://doi.org/10.1074/jbc.M109.012252
184. Chvanov M, Walsh CM, Haynes LP, Voronina SG, Lur G, Gerasimenko OV, Barraclough R, Rudland PS, Petersen OH, Burgoyne RD, Tepikin AV (2008) ATP depletion induces translocation of STIM1 to puncta and formation of STIM1-ORAI1 clusters: translocation and re-translocation of STIM1 does not require ATP. Pflugers Arch 457(2):505–517. https://doi.org/10.1007/s00424-008-0529-y
185. Walsh CM, Chvanov M, Haynes LP, Petersen OH, Tepikin AV, Burgoyne RD (2010) Role of phosphoinositides in STIM1 dynamics and store-operated calcium entry. Biochem J 425(1):159–168. doi:BJ20090884 [pii] https://doi.org/10.1042/BJ20090884
186. Galan C, Woodard GE, Dionisio N, Salido GM, Rosado JA Lipid rafts modulate the activation but not the maintenance of store-operated Ca^{2+} entry. Biochim Biophys Acta 1803(9):1083–1093. doi:S0167-4889(10)00176-X [pii] https://doi.org/10.1016/j.bbamcr.2010.06.006
187. Jardin I, Salido GM, Rosado JA (2008) Role of lipid rafts in the interaction between hTRPC1, Orai1 and STIM1. Channels (Austin) 2(6). doi:7055 [pii]
188. Liao Y, Plummer NW, George MD, Abramowitz J, Zhu MX, Birnbaumer L (2009) A role for Orai in TRPC-mediated Ca^{2+} entry suggests that a TRPC:Orai complex may mediate store and receptor operated Ca^{2+} entry. Proc Natl Acad Sci U S A 106(9):3202–3206. doi:0813346106 [pii] https://doi.org/10.1073/pnas.0813346106
189. Martin AC, Willoughby D, Ciruela A, Ayling LJ, Pagano M, Wachten S, Tengholm A, Cooper DM (2009) Capacitative Ca^{2+} entry via Orai1 and stromal interacting molecule 1 (STIM1) regulates adenylyl cyclase type 8. Mol Pharmacol 75(4):830–842. doi:mol.108.051748 [pii] https://doi.org/10.1124/mol.108.051748
190. Korade Z, Kenworthy AK (2008) Lipid rafts, cholesterol, and the brain. Neuropharmacology 55(8):1265–1273. doi:S0028-3908(08)00064-6 [pii] https://doi.org/10.1016/j.neuropharm.2008.02.019
191. Toth BI, Oberwinkler J, Voets T (2016) Phosphoinositide regulation of TRPM channels – TRPM3 joins the club! Channels (Austin) 10(2):83–85. https://doi.org/10.1080/19336950.2015.1113719
192. Hille B, Dickson EJ, Kruse M, Vivas O, Suh BC (2015) Phosphoinositides regulate ion channels. Biochim Biophys Acta 1851(6):844–856. https://doi.org/10.1016/j.bbalip.2014.09.010
193. Zaydman MA, Cui J (2014) PIP2 regulation of KCNQ channels: biophysical and molecular mechanisms for lipid modulation of voltage-dependent gating. Front Physiol 5:195. https://doi.org/10.3389/fphys.2014.00195

194. Dickson EJ (2017) RASSF4: Regulator of plasma membrane PI(4,5)P2. J Cell Biol 216(7):1879–1881. https://doi.org/10.1083/jcb.201706042
195. Chen YJ, Chang CL, Lee WR, Liou J (2017) RASSF4 controls SOCE and ER-PM junctions through regulation of PI(4,5)P2. J Cell Biol 216(7):2011–2025. https://doi.org/10.1083/jcb.201606047
196. Schroeder C (2010) Cholesterol-binding viral proteins in virus entry and morphogenesis. Subcell Biochem 51:77–108. https://doi.org/10.1007/978-90-481-8622-8_3
197. Levitan I, Singh DK, Rosenhouse-Dantsker A (2014) Cholesterol binding to ion channels. Front Physiol 5:65. https://doi.org/10.3389/fphys.2014.00065
198. Kovarova M, Wassif CA, Odom S, Liao K, Porter FD, Rivera J (2006) Cholesterol deficiency in a mouse model of Smith-Lemli-Opitz syndrome reveals increased mast cell responsiveness. J Exp Med 203(5):1161–1171. https://doi.org/10.1084/jem.20051701
199. Fantini J, Barrantes FJ (2013) How cholesterol interacts with membrane proteins: an exploration of cholesterol-binding sites including CRAC, CARC, and tilted domains. Front Physiol 4:31. https://doi.org/10.3389/fphys.2013.00031
200. Baier CJ, Fantini J, Barrantes FJ (2011) Disclosure of cholesterol recognition motifs in transmembrane domains of the human nicotinic acetylcholine receptor. Sci Rep 1:69. https://doi.org/10.1038/srep00069
201. Epand RM (2006) Cholesterol and the interaction of proteins with membrane domains. Prog Lipid Res 45(4):279–294. doi:S0163-7827(06)00015-4 [pii] https://doi.org/10.1016/j.plipres.2006.02.001
202. Epand RM (2008) Proteins and cholesterol-rich domains. Biochim Biophys Acta 1778(7-8):1576–1582
203. Cheng Y (2015) Single-particle Cryo-EM at crystallographic resolution. Cell 161(3):450–457. https://doi.org/10.1016/j.cell.2015.03.049
204. Bai XC, McMullan G, Scheres SH (2015) How cryo-EM is revolutionizing structural biology. Trends Biochem Sci 40(1):49–57. https://doi.org/10.1016/j.tibs.2014.10.005
205. Liao M, Cao E, Julius D, Cheng Y (2013) Structure of the TRPV1 ion channel determined by electron cryo-microscopy. Nature 504(7478):107–112. https://doi.org/10.1038/nature12822
206. Zhou X, Li M, Su D, Jia Q, Li H, Li X, Yang J (2017) Cryo-EM structures of the human endolysosomal TRPML3 channel in three distinct states. Nat Struct Mol Biol 24(12):1146–1154. https://doi.org/10.1038/nsmb.3502
207. Fan G, Baker ML, Wang Z, Baker MR, Sinyagovskiy PA, Chiu W, Ludtke SJ, Serysheva II (2015) Gating machinery of InsP3R channels revealed by electron cryomicroscopy. Nature 527(7578):336–341. https://doi.org/10.1038/nature15249
208. Ishii T, Sato K, Kakumoto T, Miura S, Touhara K, Takeuchi S, Nakata T (2015) Light generation of intracellular Ca^{2+} signals by a genetically encoded protein BACCS. Nat Commun 6:8021. https://doi.org/10.1038/ncomms9021
209. He L, Zhang Y, Ma G, Tan P, Li Z, Zang S, Wu X, Jing J, Fang S, Zhou L, Wang Y, Huang Y, Hogan PG, Han G, Zhou Y (2015) Near-infrared photoactivatable control of Ca^{2+} signaling and optogenetic immunomodulation. elife 4. https://doi.org/10.7554/eLife.10024
210. Ma G, Wen S, Huang Y, Zhou Y (2017) The STIM-Orai pathway: light-operated Ca^{2+} entry through engineered CRAC channels. Adv Exp Med Biol 993:117–138. https://doi.org/10.1007/978-3-319-57732-6_7
211. Parekh AB (2008) Store-operated channels: mechanisms and function. J Physiol 586(13):3033. https://doi.org/10.1113/jphysiol.2008.156885
212. Kar P, Parekh AB (2015) Distinct spatial Ca^{2+} signatures selectively activate different NFAT transcription factor isoforms. Mol Cell 58(2):232–243. https://doi.org/10.1016/j.molcel.2015.02.027
213. Jing J, He L, Sun A, Quintana A, Ding Y, Ma G, Tan P, Liang X, Zheng X, Chen L, Shi X, Zhang SL, Zhong L, Huang Y, Dong MQ, Walker CL, Hogan PG, Wang Y, Zhou Y (2015) Proteomic mapping of ER-PM junctions identifies STIMATE as a regulator of Ca influx. Nat Cell Biol. https://doi.org/10.1038/ncb3234

214. Parekh AB (2010) Store-operated CRAC channels: function in health and disease. Nat Rev Drug Discov 9(5):399–410. https://doi.org/10.1038/nrd3136
215. Putney JW (2010) Pharmacology of store-operated calcium channels. Mol Interv 10(4):209–218. https://doi.org/10.1124/mi.10.4.4
216. Jairaman A, Prakriya M (2013) Molecular pharmacology of store-operated CRAC channels. Channels (Austin) 7(5):402–414. https://doi.org/10.4161/chan.25292
217. Zhou Y, Wang X, Wang X, Loktionova NA, Cai X, Nwokonko RM, Vrana E, Wang Y, Rothberg BS, Gill DL (2015) STIM1 dimers undergo unimolecular coupling to activate Orai1 channels. Nat Commun 6:8395. https://doi.org/10.1038/ncomms9395
218. Bohm J, Chevessier F, Maues De Paula A, Koch C, Attarian S, Feger C, Hantai D, Laforet P, Ghorab K, Vallat JM, Fardeau M, Figarella-Branger D, Pouget J, Romero NB, Koch M, Ebel C, Levy N, Krahn M, Eymard B, Bartoli M, Laporte J (2013) Constitutive activation of the calcium sensor STIM1 causes tubular-aggregate myopathy. Am J Hum Genet 92(2):271–278. https://doi.org/10.1016/j.ajhg.2012.12.007
219. Bohm J, Chevessier F, Koch C, Peche GA, Mora M, Morandi L, Pasanisi B, Moroni I, Tasca G, Fattori F, Ricci E, Penisson-Besnier I, Nadaj-Pakleza A, Fardeau M, Joshi PR, Deschauer M, Romero NB, Eymard B, Laporte J (2014) Clinical, histological and genetic characterisation of patients with tubular aggregate myopathy caused by mutations in STIM1. J Med Genet 51(12):824–833. https://doi.org/10.1136/jmedgenet-2014-102623
220. Walter MC, Rossius M, Zitzelsberger M, Vorgerd M, Muller-Felber W, Ertl-Wagner B, Zhang Y, Brinkmeier H, Senderek J, Schoser B (2015) 50 years to diagnosis: autosomal dominant tubular aggregate myopathy caused by a novel STIM1 mutation. Neuromuscul Disord 25(7):577–584. https://doi.org/10.1016/j.nmd.2015.04.005
221. Hedberg C, Niceta M, Fattori F, Lindvall B, Ciolfi A, D'Amico A, Tasca G, Petrini S, Tulinius M, Tartaglia M, Oldfors A, Bertini E (2014) Childhood onset tubular aggregate myopathy associated with de novo STIM1 mutations. J Neurol 261(5):870–876. https://doi.org/10.1007/s00415-014-7287-x
222. Markello T, Chen D, Kwan JY, Horkayne-Szakaly I, Morrison A, Simakova O, Maric I, Lozier J, Cullinane AR, Kilo T, Meister L, Pakzad K, Bone W, Chainani S, Lee E, Links A, Boerkoel C, Fischer R, Toro C, White JG, Gahl WA, Gunay-Aygun M (2015) York platelet syndrome is a CRAC channelopathy due to gain-of-function mutations in STIM1. Mol Genet Metab 114(3):474–482. https://doi.org/10.1016/j.ymgme.2014.12.307
223. Picard C, McCarl CA, Papolos A, Khalil S, Luthy K, Hivroz C, LeDeist F, Rieux-Laucat F, Rechavi G, Rao A, Fischer A, Feske S (2009) STIM1 mutation associated with a syndrome of immunodeficiency and autoimmunity. N Engl J Med 360(19):1971–1980. https://doi.org/10.1056/NEJMoa0900082
224. Schaballie H, Rodriguez R, Martin E, Moens L, Frans G, Lenoir C, Dutre J, Canioni D, Bossuyt X, Fischer A, Latour S, Meyts I, Picard C (2015) A novel hypomorphic mutation in STIM1 results in a late-onset immunodeficiency. J Allergy Clin Immunol 136(3):816–819.e814. https://doi.org/10.1016/j.jaci.2015.03.009
225. Korzeniowski MK, Manjarres IM, Varnai P, Balla T (2010) Activation of STIM1-Orai1 involves an intramolecular switching mechanism. Sci Signal 3(148):ra82. https://doi.org/10.1126/scisignal.2001122
226. Garibaldi M, Fattori F, Riva B, Labasse C, Brochier G, Ottaviani P, Sacconi S, Vizzaccaro E, Laschena F, Romero NB, Genazzani A, Bertini E, Antonini G (2017) A novel gain-of-function mutation in ORAI1 causes late-onset tubular aggregate myopathy and congenital miosis. Clin Genet 91(5):780–786. https://doi.org/10.1111/cge.12888
227. Lian J, Cuk M, Kahlfuss S, Kozhaya L, Vaeth M, Rieux-Laucat F, Picard C, Benson MJ, Jakovcevic A, Bilic K, Martinac I, Stathopulos P, Kacskovics I, Vraetz T, Speckmann C, Ehl S, Issekutz T, Unutmaz D, Feske S (2017) ORAI1 mutations abolishing store-operated Ca^{2+} entry cause anhidrotic ectodermal dysplasia with immunodeficiency. J Allergy Clin Immunol. https://doi.org/10.1016/j.jaci.2017.10.031

228. McCarl CA, Picard C, Khalil S, Kawasaki T, Rother J, Papolos A, Kutok J, Hivroz C, Ledeist F, Plogmann K, Ehl S, Notheis G, Albert MH, Belohradsky BH, Kirschner J, Rao A, Fischer A, Feske S (2009) ORAI1 deficiency and lack of store-operated Ca^{2+} entry cause immunodeficiency, myopathy, and ectodermal dysplasia. J Allergy Clin Immunol 124(6):1311–1318.e1317. https://doi.org/10.1016/j.jaci.2009.10.007
229. Cancer Genome Atlas Network. https://cancergenome.nih.gov/
230. Navarro-Borelly L, Somasundaram A, Yamashita M, Ren D, Miller RJ, Prakriya M (2008) STIM1-Orai1 interactions and Orai1 conformational changes revealed by live-cell FRET microscopy. J Physiol 586(Pt 22):5383–5401. https://doi.org/10.1113/jphysiol.2008.162503

Chapter 24
Targeting Transient Receptor Potential Channels by MicroRNAs Drives Tumor Development and Progression

Giorgio Santoni, Maria Beatrice Morelli, Matteo Santoni, Massimo Nabissi, Oliviero Marinelli, and Consuelo Amantini

Abstract Transient receptor potential (TRP) cation channel superfamily plays important roles in a variety of cellular processes such polymodal cellular sensing, adhesion, polarity, proliferation, differentiation and apoptosis. The expression of TRP channels is strictly regulated and their de-regulation can stimulate cancer development and progression.

In human cancers, specific miRNAs are expressed in different tissues, and changes in the regulation of gene expression mediated by specific miRNAs have been associated with carcinogenesis. Several miRNAs/TRP channel pairs have been reported to play an important role in tumor biology. Thus, the TRPM1 gene regulates melanocyte/melanoma behaviour via TRPM1 and microRNA-211 transcripts. Both miR-211 and TRPM1 proteins are regulated through microphthalmia-associated transcription factor (MIFT) and the expression of miR-211 is decreased during melanoma progression. Melanocyte phenotype and melanoma behaviour strictly depend on dual TRPM1 activity, with loss of TRPM1 protein promoting melanoma aggressiveness and miR-211 expression supporting tumour suppressor. TRPM3 plays a major role in the development and progression of human clear cell renal cell carcinoma (ccRCC) with von Hippel-Lindau (VHL) loss. TRPM3, a direct target of miR-204, is enhanced in ccRCC with inactivated or deleted VHL. Loss of VHL inhibits miR-204 expression that lead to increased oncogenic autophagy. Therefore,

G. Santoni (✉) · M. B. Morelli · M. Nabissi
School of Pharmacy, Experimental Medicine Section, University of Camerino, Camerino, Italy
e-mail: giorgio.santoni@unicam.it

M. Santoni
Clinic and Oncology Unit, Macerata Hospital, Macerata, Italy

O. Marinelli
School of Pharmacy, Experimental Medicine Section, University of Camerino, Camerino, Italy

School of Biosciences and Veterinary Medicine, University of Camerino, Camerino, Italy

C. Amantini
School of Biosciences and Veterinary Medicine, University of Camerino, Camerino, Italy

© Springer Nature Switzerland AG 2020

M. S. Islam (ed.), *Calcium Signaling*, Advances in Experimental Medicine and Biology 1131, https://doi.org/10.1007/978-3-030-12457-1_24

the understanding of specific TRP channels/miRNAs molecular pathways in distinct tumors could provide a clinical rationale for target therapy in cancer.

Keywords TRP channels · miRNAs · Channelopathies · Tumor progression · Target therapy · Calcium/calcineurin signaling · TRPV · TRPA1 · TRPP · TRPM

Abbreviations

BRAF	proto-oncogene protein B-raf
$BRAF^{V600}$	BRAF harbouring somatic missense mutations at the amino acid residue V600
BRN2	POU-domain transcription factor (POU3F2)
ccRCC	human clear cell renal cell carcinoma
CRC	colorectal cancer
EC	endometrial cancer
EOC	epithelial ovarian cancer
ETS-1	erythroblastosis virus E26 oncogene homolog 1
FGR2	fibroblast growth factor receptor type 2
HCC	hepatocellular carcinoma cells
LUAD	lung adenocarcinoma
MIFT	microphthalmia-associated transcription factor
miR	MicroRNAs
mRNA	messenger RNA
MTSS1	metastasis suppressor gene 1
NCX1	Na^+/Ca^{2+} exchanger-1
NFAT5	nuclear factor of activated T-cells 5
NFATC3	nuclear factor of activated T-cells isoform c3
NSCLC	non-small cell lung carcinoma
OC	ovarian cancer
PCa	prostate cancer
PKD	Polycystic kidney disease
pri-miRs	primary miRNAs
TrkB	Tropomyosin receptor kinase B
TRPA	Transient receptor potential ankyrin
TRPC	Transient receptor potential canonical
TRPM	Transient receptor potential melastatin
TRPP	Transient receptor potential polycystic
TRPV	Transient receptor potential vanilloid
UTR	untranslated region
VHL	von Hippel-Lindau

24.1 Introduction

Ion channels belonging to the Transient Receptor Potential (TRP) family are expressed in every living cell, where they participate in controlling a lot of biological processes and physiological functions, such as cell excitation, electrical activity, cellular osmolarity, as well as growth and death. They show common features in the structure such us the presence of six transmembrane segments with intracellular N- and C-termini and varying degrees of sequence homology. They are grouped into seven subfamilies: TRPC ("C" for canonical), TRPV ("V" for vanilloid), TRPM ("M" for melastatin), TRPN ("N" for no mechanoreceptor potential C), TRPA ("A" for ankyrin), TRPP ("P" for polycystic) and TRPML ("ML" for mucolipin). The majority of TRPs is permeable to Ca^{2+} and these channels are considered as multiple signal integrators. In fact they play critical roles in chemosensation, mechanosensation, thermosensation and nociception sensing stimuli from both external and local environments [1]. Expression of TRP channels is tightly regulated and their expression deregulation can trigger abnormal processes, leading to pathologies, called channelopathies. Several transcription factors play a critical role in controlling the transcriptome of TRP channels by acting on the $5'$-flanking gene region. Microribonucleic acids (miRNAs), a small non-coding ribonucleic acids (RNAs) of approximately 22 bp, induce RNA interference by base-pairing with the $3'$ untranslated region (UTR) of mRNA, which triggers either mRNA translational repression or RNA degradation [2, 3]. In this manner, miRNAs function as sequence-specific inhibitors of gene expression. miRNAs are initially transcribed as precursor transcripts called primary miRNAs (pri-miRNAs). pri-miRNAs are at first processed in the nucleus to precursor miRNAs (pre-miRNAs) by the class 2 RNase III enzyme Drosha, then, after the transport into the cytoplasm, they become mature miRNAs by the action of Dicer, an RNase III type protein. Finally they are integrated into the Argonaute protein to produce the effector RNA-induced silencing complex (RISC). RISCs target mRNAs recognized through partial sequence complementarity promoting either translational repression or mRNA degradation [4]. Over 1000 different miRNAs are encoded by the human genome; approximately 20–30% of all genes are targeted by miRNAs, and a single miRNA may target up to 200 genes [5]. In human cancers, specific miRNAs are expressed in different tissues, and changes in the control of gene expression have been associated with carcinogenesis [6], including in endometrial, colorectal, prostate cancers and melanomas [7–11]. Furthermore, miRNAs cooperatively exert their function with certain transcription factors in the regulation of mutual sets of target genes, allowing coordinated modulation of gene expression both transcriptionally and post-transcriptionally [12]. In addition, these small noncoding RNAs regulate the expression of different genes involved also in cardiac excitability, pain, brain edema etc. [3]. Future studies might decode other miRNAs/TRP deregulation in human diseases (Table 24.1).

Table 24.1 Expression of miRs and TRP channels in cancers

Tumors	TRP gene	miR	Refs.
Breast cancer	TRPC5	miR-320a (−)	[28]
Renal cell cancer	TRPM3	miR-204 (−)	[54, 56]
Melanoma	TRPM1	miR-211 (−)	[7, 8, 17–20]
Endometrial cancer	TRPM3	miR-204 (−)	[9]
Prostate cancer	TRPM8	miR-26a (+)	[11]
Colon-rectal cancer	TRPV6	miR-122 (+)	[30]
	TRPC1 (?)	miR-135a (+)	[73]
Epithelial ovarian cancer	TRPM1	miR-211 (−)	[26]
	TRPM3	mir-204 (−)	[62]
	TRPC1	miR-135b (+)	[76]
Hepatocarcinoma	TRPC6 (?)	miR-30 (?)	[64]
Lung adenocarcinoma	TRPA1	miR-142 (?)	[31]
Non small cell lung carcinoma	TRPP2	miR-106b (−)[a]	[29]

TRP: Transient receptor potential channels; (?): correlation has been suggested; (−): down-regulated; (+): up-regulated
[a]Cisplatin cell resistant vs sensitive

24.1.1 *miR-211 and Its Target Genes in Melanoma Progression*

Malignant melanoma has increased the frequency of its occurrence in the last years [13]. Surgical removal of superficial primary tumors is satisfactory in term of survival. However, metastatic melanoma shows a poor survival rate [14]. Transient receptor potential melastatin channel 1 (TRPM1) transcripts are over-expressed in benign nevi, dysplastic nevi and melanomas *in situ*; it is variably expressed in invasive melanomas and is absent in most melanoma metastasis [15]. TRPM1 is regulated by a microphthalmia-associated transcription factor (MITF) [16]. The expression of miR-211, which decreases during melanoma progression [7], is driven by the TRPM1 promoter sequences in a MITF-dependent manner [17]. The gene encoding miR-211 is located within the sixth intron of the TRPM1 gene, and both miR-211 and TRPM1 channel protein are regulated by MITF. The miR-211 directly targets potassium calcium-activated channel subfamily M alpha 1 (KCNMA1) which is often associated with both cell proliferation and migration/invasion in various cancers [7]. Moreover, it has been demonstrated that miR-211 expression is greatly decreased in melanoma cells compared to normal melanocytes [18]. Levy and co-workers have demonstrated by over-expression and knockdown of either TRPM1 channel protein or miR-211, respectively, that miR-211, rather than TRPM1 channel protein, modified the melanoma invasiveness [17]. They also identified that miR-211 regulates insulin-like growth factor 2 (IGF2R), transforming growth factor beta receptor II (TGFBR2), and nuclear factor of activated T-cells 5 (NFAT5) signal transduction pathways to have a suppressive effect on a tumour. These data support the hypothesis that TRPM1 gene regulates melanocyte/melanoma behaviour via generation of two transcripts including TRPM1 protein, and miR-211: the loss of TRPM1 protein is an excellent marker of melanoma aggressiveness while the miR-211 expression is linked to the tumour suppressor functions. Therefore, different RNA transcription program (i.e. TRPM1 mRNA and/or miR-211) decides on the melanocyte phenotype and melanoma behaviour. Clarifying this phenomenon requires future in vitro studies with targeted modulation of TRPM1 expression and clinic-pathologic correlation using large clinical cohorts of melanoma patients [8]. Furthermore, another miR-211 target, BRN2, also known as POU-domain transcription factor (POU3F2) has been identified. In melanocyte, miR-211 modulates BRN2 expression by repressing its translation [18]. In melanoma miR-211 is expressed at low levels and is related to an over-expression of BRN2, that mediates the de-differentiated and invasive phenotype [18]. The identification of miR-211 and its TRPM1 and BRN2 target genes, could suggest a new therapeutic strategy for the treatment of metastatic melanomas.

24.1.2 miR-211/TRPM1 in BRAF^{V600} Malignant Melanoma

About 50% of malignant melanomas showed proto-oncogene protein B-raf (BRAF) somatic missense mutations at the amino acid residue V600 (BRAF^{V600}) [19]. Inhibition of BRAF^{V600} with vemurafenib induces a rapid regression of metastatic BRAF^{V600} melanomas. Recently, by studying the secretome of melanoma-derived extracellular CD81^{+} and TSG-101^{+} vesicles in vemurafenib-treated cells, an increased expression of several miRNAs including miR-211-5p was evidenced (Table 24.2). In melanomas harboring Braf^{V600} mutation, the expression of miR-211-5p because of BRAF inhibition was induced by increased MIFT expression. The later transcriptional factor that up-regulates the TRPM1 gene expression induces miR-211-5p expression, resulting in activation of survival pathway through the Bcl-2 and Melan-A anti-apoptotic molecules [20]. Bcl-2 is a direct target of

Table 24.2 Vemurofenib treatment increases the microRNA levels in exosomes-derived and BRAF^{V600} melanoma cells, compared to not treated cells

	Melanoma cells		Exosomes	
Type	Fold changes	p-value	Fold changes	p-value
microRNA-211-5p	4.07	0.03		
microRNA-34a-5p	1.91	0.04		
microRNA-15b-5p	−2.11	0.01		
microRNA-1307-3p	−1.72	0.01		
microRNA-1301-3p	−2.22	0.02		
microRNA-1307-5p	−2,25	0.03		
microRNA-339-5p	−2.23	0.01		
microRNA-574-3p			−1.40	0.00
microRNA-9-5p			−1.40	0.04
microRNA-7-5p			−1.50	0.04

Table shows the microRNA changes (fold expression) from sequencing analysis in exosome-secreted from and melanoma cells. Fold change is utilized as microRNA up- or down-regulation. Data are presented as the ±SEM. $p < 0.05$

MIFT and modulation of Bcl-2 regulates Melan-A [21]. Inhibition of $BRAF^{V600}$ leads to down-regulation of pERK1/2 that increases MIFT expression [22].

MIFT transcriptionally activates TRPM1 and simultaneously up-regulates the intronic miR-211-5p. Recently has been reported that MIFT induces miR-211 target genes such as AP1S2, SOX11, IGFBP5 and SERINC3 that increase melanoma cell invasion. In addition, also a role for miR-211 as metabolic regulator in melanoma cells, by targeting the hypoxia inducible factor 1a (HIF-1α) has been reported [23].

24.1.3 miR-211 in Ovarian Cancer

The miR-211 is located on intron 6 of the TRPM1 gene at 15q13-q21, a locus frequently lost in neoplasms [24, 25]. It has been demonstrated that miR-211 expression is significantly down-regulated in ovarian cancer. The miR-211 negatively regulates the activity of CDK6 and Cyclin D1 by directly binding to 3'UTR sequences of the related mRNAs repressing their translation into proteins [26]. The cyclin D controls the CDK6 activity and has been reported to regulate angiogenesis, growth factor-stimulated proliferation and the promotion of G1 phase progression. Moreover, miR-211 suppresses the expression of PHF19, promoting apoptosis and inhibiting cell migration [27]. Overall, the cyclin D1/CDK6 and PHF19 are key players in epithelial ovarian cancer (EOC) tumorigenesis and TRPM1/miR-211 might provides new data in the diagnosis, prognosis and therapy for EOC [26, 27].

24.2 TRPC5/miR-320a and TRPP2/miR-106-5p in Breast and Lung Cancer Drug Sensitivity

Over-expression of the transient receptor potential canonical 5 (TRPC5) channel and the nuclear factor of activated T-cells isoform c3 (NFATC3) are essential for breast cancer chemoresistance. However, the mechanism by which TRPC5 and NFATC3 are regulated are unknown. The miR-320a was found to be downregulated in chemoresistant breast cancer cells. It directly targeted TRPC5 and NFATC3 and downregulation of miR-320a triggered TRPC5 and NFATC3 over-expression. In chemoresistant breast cancer cells, downregulation of miR-320a was associated with promoter methylation of the miR-320a coding sequence [28]. Furthermore, the transcription factor v-ets erythroblastosis virus E26 oncogene homolog 1 (ETS-1), which inhibits the miR-320a expression, was found to be activated in chemoresistant breast cancer cells and such activation was associated with hypomethylation of the ETS-1 promoter [28]. Finally, downregulation of miR-320a and enhanced expression of TRPC5, NFATC3, and ETS-1 were verified in clinically chemoresistant breast cancer samples. Low expression of miR-320a was also found to be a significant unfavourable predictor for clinic outcome.

In addition, a role for the miR-106b-5p and the transient receptor potential polycystic channel 2 (TRPP2) channel in the sensitivity of non-small cell lung carcinoma (NSCLC) to cisplatin treatment has been reported [29]. Treatment of NSCL patients with cisplatin is hindered by cisplatin resistance. Yu e co-workers, have demonstrated in human lung adenocarcinoma MDRA549/cisplatin (A549/DDP) and its progenitor A549 cell line, that miR-106b-5p was decreased in A549/DDP cells. The miR-106b-5p affected the tolerance of cancer lung cells to cisplatin treatment, by negatively regulating the TRPP2 channel [29]. Up-regulation of the miR-106b-5p or down-regulation of the TRPP2 channel expression increased the sensitivity of A549/DDP cells to cisplatin treatment, suggesting that mR-106b-5p may represent a clinical strategy in the treatment of NSCLC [29].

24.3 TRPV6/miR-122 in Colorectal Liver Metastasis

Control of liver metastasis is an important goal in the treatment of colorectal cancer (CRC). In liver metastasis of primary CRCs, the most abundant miRNA, compared with primary tumors, is miR-122 [30].

The expression levels of transient receptor potential vanilloid channel 6 (TRPV6) channels, the cationic amino acid transporter 1 (CAT1), a negative target gene of miR-122, were found to be lower in liver metastases than in primary tumors. The expression levels of TRPV6 evaluated in 132 formalin-fixed paraffin-embedded primary tumors and their corresponding metastatic liver tumors isolated by using laser capture microdissection, were negatively correlated with synchronous liver metastasis and tumor stage. Results from the analysis on 121 CRC patients without synchronous liver metastasis, demonstrated that patients with low TRPV6 expression showed significantly shorter liver metastasis-free survival, but not disease-free survival. Over-expression of miR-122 and concomitant suppression of TRPV6 in the primary CRC appears to play important roles in the development of colorectal liver metastasis.

Thus, expression of TRPV6 in the primary CRC represents a novel biomarker to predict the risk of postoperative liver metastasis of CRC patients [30].

24.4 TRPA1 Channel-Targeting Exosomal miR-142-3p

The transient receptor potential ankyrin channel A1 (TRPA1) channel has been suggested to play an important role in lung cancers [31, 32]. Recent studies have demonstrated the capability of the TRPA1 to form a complex with the fibroblast growth factor receptor type 2 (FGR2) in lung adenocarcinoma (LUAD), a diffuse lung cancer that metastasizes in different organs and brain [33]. As in other lung tumors, in LUAD, FGFR2 is a major factor responsible of tumor progression [34, 35]. In this regard, the TRPA1 channel through the ankyrin repeats, has been demonstrated to bind the terminal prolin-rich region of FGR2. This binding that it is induced, independently by external stimulation, inhibits TRPA1 channel activity, resulting in FGFR2 signaling activation that leads to increased cell proliferation and metastatic spread invasion [33, 34]. In addition, Berrout and coworkers also demonstrated that the dormant state of LUAD cells observed in the brain upon astrocytes encounter, may be related to a crosstalk between cancer cells and astrocytes. Previously, has been reported that the miR-142-3p targeting the TRPA1 channel can suppress NSC lung cancer progression [35, 36]. In regard to LUAD, astrocytes have been found to be able to transfer micro-vesicles called exosomes containing miRNA (e.g. miR-142-3p) specifically targeting the TRP channel. The binding of miRNA-142-3p to the 3'-UTR of TRPA1 triggers the depletion of TRPA1 expression in metastatic LUAD cells and subsequently abrogation of the FGFR2-driven cell proliferation and invasion of lung cancer cells in the brain.

24.5 TRPM8 and miRNAs in Prostate Cancer

Prostate cancer (PCa) is the second most frequent tumor and the 60 leading cause of cancer related death among males worldwide [37]. TRPM8 is an androgen-responsive gene and essential for the survival of PCa cells [38]. It is involved in the regulation of the intracellular Ca^{2+} concentration and exhibited an elevated expression in PCa cells. TRPM8 shows a significant association with age, serum prostate specific antigen concentration, tumor state, Gleason score or metastasis at prostatectomy. The analysis of a possible correlation between the expression of selected miRNAs and the TRPM8 gene, have evidenced a moderate inverse correlation between high TRPM8 expression and low miR-26a expression. It was found that miR-26a expression was decreased in PCa tissues and cell lines, with androgen-independent prostate cancer showing lower miR-26a expression compared to androgen-dependent prostate cancer [39]. Over-expression of miR-26a enhances apoptosis, and this upregulation is triggered by cytochrome *c* oxidase subunit II inhibition. In addition, a low miR-26a density resulted in an evidently poor prognosis. Further research is warranted to confirm a direct regulatory effect of miRNA on their potential target genes and to the development of miRNA-based therapy [11].

24.6 miR-17/TRPP Channels in Polycystic Kidney Disease and Cancers

The miR-17 and related miRNAs are derived from three miRNA clusters: miR-17 $\sim$ 92, miR-106a-363 and miR-106b $\sim$ 25 clusters. The genomic organization and coding sequences of these miRNA clusters are evolutionarily conserved in vertebrates. Based on their seed sequence, miRNAs derived from these three clusters can be classified into four families: the miR-17, miR-18, miR-19 and miR-25 families. Since members of each family have an identical seed sequence, they are predicted to target the same mRNAs. Interestingly, miR-17 $\sim$ 92 and related clusters are enriched in developing tissues and are essential for heart and lung development [40]. The miR-17 and related miRNAs are also implicated in the pathogenesis of Polycystic kidney disease (PKD) [41], a most common genetic cause of chronic kidney failure characterized by the presence of numerous, progressively enlarging fluid-filled cysts in the renal parenchyma [41]. By bioinformatics analysis has been reported that miR-17 directly targets the 3′UTR of TRPP2 (PKD2) and post-transcriptionally represses its expression [42]. Dysregulated miRNA expression is observed in PKD, with miR-17 $\sim$ 92, that is upregulated in a mouse model of PKD. Kidney-specific transgenic over-expression of miR-17 $\sim$ 92 produces kidney cysts in mice. Conversely, kidney-specific inactivation of miR-17 $\sim$ 92 in a mouse model of PKD retards kidney cyst growth, improves renal function, and prolongs survival. miR-17 $\sim$ 92 may mediate these effects by promoting proliferation and through post-transcriptional repression of the TRPP1 and TRPP2 genes (Pkd1 and Pkd2, respectively) and of the hepatocyte nuclear factor-1β [41]. The cysts arise from renal tubules and are lined by abnormally functioning and hyperproliferative epithelial cells.

In addition, two major lines of evidence also implicate miR-17 and related miRs in the pathogenesis of various cancers [43, 44]. First, these miRs are amplified in numerous human cancers, promote proliferation [45] and cause tumor growth in vivo [44]. Second, the oncogenic transcription factor c-Myc has been demonstrated to bind to the miR-17 $\sim$ 92 promoter and to induce its transcription [46]. Further studies should be required to completely address the oncogenic role of miR-17 and TRPP channels.

24.6.1 miR-204/TRPM3 and Cancer Survival and Apoptosis

At present, several data on the role of miR-204 in cancers have been provided. Roldo and coworkers showed the upregulation of miR-204 in insulinoma [47]. Similar results were also reported by Zanette in the acute lymphoblastic leukemia [48]. By contrast, decreased miR-204 expression was reported in glioblastoma [49], gastric, bladder and lung cancers, suggesting that miR-204 may also be a tumor suppressor gene. In hepatocarcinoma, miR-204 has been found to inhibit the expression of long non-coding RNA (IncRNA) for homeobox A distal transcript antisense RNA (HOTTIP), through interference with the argonaute-2 pathway [50]. BCL-2 represents a target for miR-204, and apoptosis represents the suppressive mechanism regulated by miR-204 by binding to the 3′-UTR of BCL-2 [51]. The miR-204 suppression has been reported to inhibit the transition from epithelial to mesenchymal, IL-11, SOX4 and SIX1 target gene expression [52] and bone metastasis in breast cancer cells by reduction of the 68-kDa Src-associated protein in mitosis (SAM68) activity [53].

The miR-204-encoding gene is located in the sixth intron of TRPM3, and expression of mature miR-204 and pri-miR-204 strictly correlates in vitro and in vivo with that of TRPM3 gene [54].

24.6.2 Loss of miR-204 Triggers TRPM3-Mediated Oncogenic Autophagy in Clear Cell Renal Cell Carcinoma

Autophagy is an important homeostatic process for lysosome degradation of damaged organelles and proteins. Interestingly, alterations of the crosstalk between miRs and ion channels belonging to the TRP family, alter the homeostatic control and trigger oncogenic autophagy to survive to stressful stimuli [55]. Among TRP ion channels, TRPM3 plays a major role in the development and progression of clear cell renal cell carcinoma (ccRCC) with von Hippel-Lindau (VHL) loss mutation. TRPM3 expression is enhanced in human ccRCC with inactivated or deleted VHL. Loss of VHL inhibits the expression of miR-204 that in turn leads to increased oncogenic autophagy in ccRCC, resulting in an augmented expression of TRPM3, a direct target of miR-204 [54, 56, 57]. Binding of miR-204 to the 3′UTR of TRPM3 inhibits the TRPM3 translation. Similarly, binding of miR-204 to the 3′-UTR of caveolin 1 (CAV1) inhibits the CAV1 expression required for TRPM3 expression (Fig. 24.1). TRPM3 activation by stimulating of Ca^{2+} influx rise, triggers oncogenic autophagy through increased autophagosomes and CAMKK2

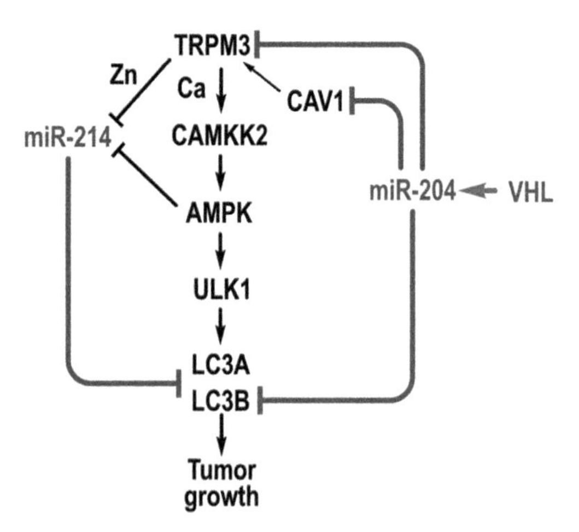

Fig. 24.1 Robust control of the autophagic network by microRNAs and calcium- and zinc-activated pathways. Calcium and zinc entering the cell through the TRPM3 channel stimulate oncogenic autophagy mediated by LC3A and LC3B through a dual mechanism. Calcium stimulates phagophore initiation through Ca^{2+}-dependent activation of CAMKK2 and AMPK, and the resulting phosphorylation of ULK1. Calcium and zinc also inhibit miR-214, which directly targets LC3A and LC3B. The VHL tumor suppressor inhibits expression of TRPM3 directly and indirectly through the effect of miR-204 on CAV1. In addition, miR-204 directly targets LC3B. AMPK, AMP-activated protein kinase; CAMKK2, calcium/calmodulin-dependent protein kinase kinase 2, β; CAV1, caveolin 1; LC3A, microtubule-associated protein 1 light chain 3 α; LC3B, microtubule-associated protein 1 light chain 3 β; TRPM3, transient receptor potential melastatin 3; ULK1, unc-51 like autophagy activating kinase 1; VHL, Von Hippel-Lindau [52]. (Courtesy Hall et al., Cancer Cell. 2014; 26(5): 738–753)

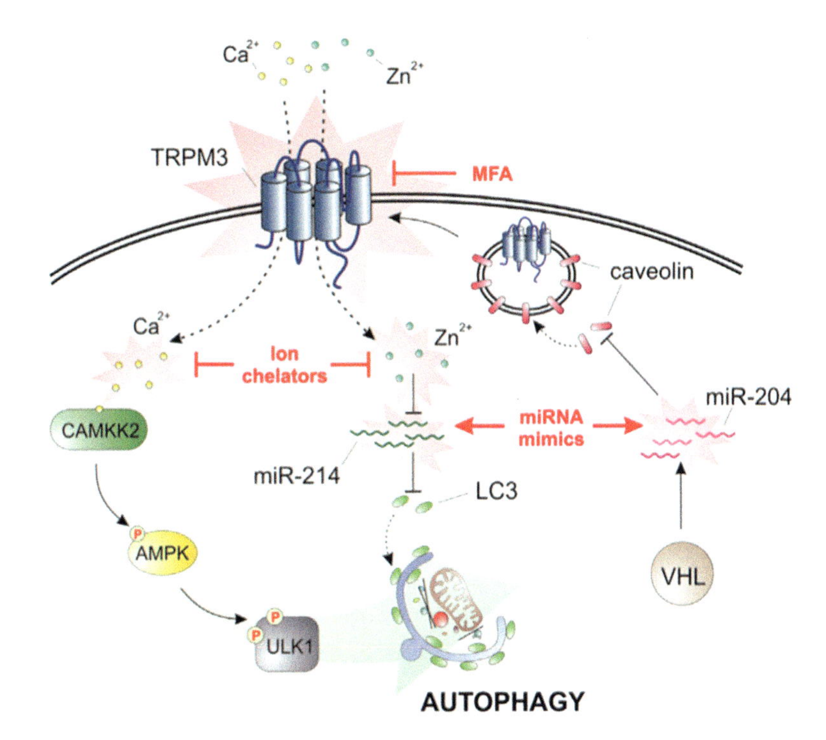

Fig. 24.2 Novel autophagy pathway important in ccRCC, centered on the ion channel TRPM3. The possible strategies to manipulate this pathway at several steps are indicated in red [53]. (Courtesy of Cecconi and Jaattela, Cancer Cell. 2014;26(5):599–600)

and AMPK activation, resulting in ULK1 phosphorylation; by contrast miR-204 inhibits oncogenic autophagy [56]. In addition, TRPM3-induced Ca^{2+} and Zn^{2+} influx and CAMKK2-AMPK pathway activation, inhibit the expression of miR-214 directly targeting the LC3A and LC3B proteins (Fig. 24.2) [58, 59]. Overall, the inhibition of TRPM3-induced oncogenic autophagy in VHL-mutated ccRCC may provide a clinical rationale therapy leading to ccRCC regression. On the other hand, the signal transducer and activator of transcription (STAT)-3 has been demonstrated to down-regulate the miR-204 expression in nasopharyngeal carcinoma [60] and endometrial carcinoma [9].

24.6.3 TRPM3/TrkB/miR-204 Interplay in the Endometrial Cancer

Endometrial carcinoma (EC) is the most common gynecological malignancy worldwide [61]. A recent study identifies a novel TrkB–STAT3–miR-204-5p signaling axis playing an important role in EC growth through the accumulation of the key

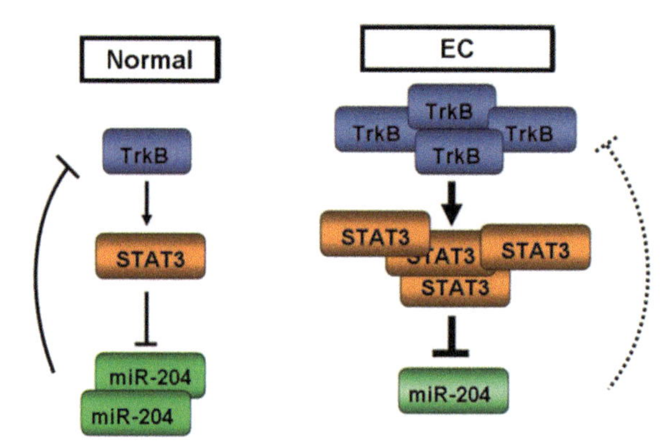

Fig. 24.3 Left: In normal cells, a recurrent auto-regulatory circuit involving the expression of TrkB induces phosphorylation of STAT3 to negatively regulate the expression of miR-204-5p. MiR-204-5p, in turn, represses TrkB expression. The expression of miR-204 within this circuit maintains endometrial cells in a normal differentiated state. Right: In endometrial cancer cells, this circuit becomes dysregulated due to increased activity of the TrkB–STAT3 component of the circuit, which constitutively represses miR-204-5p. In the absence of sufficient miR-204 tumor suppressor activity, TrkB is left uncontrolled, thereby leading to carcinogenesis [7]. (Courtesy of Bao et al. Mol Cancer. 2013; 12: 155)

tumor oncogene, TrkB [9]. The TrkB oncogene is a novel target of miR-204-5p. In normal cells, a recurrent auto-regulatory circuit involving the expression of TrkB induces phosphorylation of STAT3 that negatively regulates the miR-204-5p expression. The miR-204-5p in turn, represses the TrkB expression. The expression of miR-204 in this circuit maintains the endometrial cells in a normal differentiated state. On the other hand, in EC cells, this circuit becomes dysregulated due to increased activity of the TrkB–STAT3 component, which constitutively represses the miR-204-5p. In the absence of sufficient miR-204 tumor suppressor activity, the TrkB oncogene is left uncontrolled, thereby leading to carcinogenesis (Fig. 24.3).

Ectopic over-expression or knockdown of TrkB expression caused changes in miR expression in EC cells. qRT-PCR showed that elevated TrkB repressed miR-204-5p expression in EC cells. Furthermore, TrkB over-expression in Ishikawa^{TrkB} cells increased JAK2 and STAT3 phosphorylation, which was aborted by TrkB knockdown in HEC-1B^{shTrkB} cells. Moreover, by ChIP assays, phospho-STAT3 direct binding to STAT3-binding sites near the TRPM3 promoter region, upstream of miR-204-5p, has been reported [9]. The miR-204-5p suppresses the clonogenic growth, migration and invasion of EC cells and also inhibits the growth of tumor xenografts bearing human EC cells. Interestingly, lower miR-204-5p expression was associated with lymph node metastasis and lowered the survival in EC patients. Finally, it has been also recently reported in EOC [62] that IL-6 treated EOC via IL-6R, triggers STAT3 activation that in turn represses miR-204 near to the TRPM3 promoter. This effect is required for IL-6-induced cisplatin resistance [62]. Collectively, the reestablishment of miR-204-5p expression could be explored as a potential new therapeutic target for this disease.

24.6.4 TRPC6 and miR-30 in Hepatocellular Carcinoma Invasion

TRPC6 channel is a critical component of calcium/calcineurin signaling together with protein phosphatase 3 family member 3CA/B, R1 (PPP3CA/B, PPP3R1) and NFATC3. This channel is highly expressed in several types of cancer [63]. Recently, a role for TRPC6 in driven TGFβ-mediated migration and invasion of human hepatocellular carcinoma cells (HCC) by forming a complex with the Na^{+}/Ca^{2+} exchanger-1 (NCX1) has been reported [64]. This complex-mediating Ca^{2+} signaling regulates the effects of TGFβ on the migration, invasion in HepG2 and Huh7 cells, and intrahepatic metastasis of human HCC cells in nude mice. TGFβ upregulates TRPC6 and NCX1 expression and induces the formation and activation of the TRPC6/NCX1 molecular complex, generating a positive feedback between TRPC6/NCX1 and Smad signaling. The expression of both TRPC6 and NCX1 was markedly increased in native human HCC tissues, and their expression levels positively correlated with advancement of HCC in patients. These data reveal the relevance of TRPC6/NCX1 molecular complex in HCC suggesting it as potential targets for therapy [64]. TRPC6 is a target gene of miR-30 [65] found to inhibit cell proliferation and invasion in different tumors [66]. However, at present the mechanisms regulating the miR-30/TRPC6 molecular interaction in HCC are still unknown. In parallel to TRPC6 upregulation that increases cell proliferation, TRPC1 silencing suppressed proliferation. Thus, it may be suggested that miR-30 by targeting TRPC1 and TRPC6 channels may take a part in the mechanism regulating HCC cell proliferation [67].

24.6.5 MiR-135 by Targeting TRPC1 Promotes Cancer Invasion and Chemotherapy Resistance

The miR-135 family comprises two members, miR-135a and miR-135b. miR-135a functions as a tumor suppressor gene in gastric [68], prostate [69] and renal cancers [70], malignant glioma [71] and colon cancer [72]. An in vitro study has demonstrated that in SW480 and SW620 CRC cell lines miR-135a promotes mobility and invasion via the metastasis suppressor gene 1 (MTSS1) [73]. In contrast, inhibition of miR-135a reduced their invasive capability. The miR-135a-mediated cell mobility and invasion were reduced after MTSS1 knocked-down by small interfering RNA, indicating that miR-135a promotes the invasion of CRC cells, partially through targeting MTSS1 [73].

By luciferase reporter assay, TRPC1 was identified as other target gene of miR-135a [63]. In cultured podocytes, TGFβ stimulation and adriamycin treatment promote miR-135a expression and TRPC1 down-regulation. Ectopic expression of miR-135a led to severe podocyte injury and disarray of podocyte cytoskeleton, which was reversed by TRPC1 [63]. Thus, in the view of the important role of miR-

135a in cytoskeleton stability, a contribute of this miRNA in cancer migration and metastatic invasion could be suggested. A role of TRPC1 in cancer development and progression, such as a role in apoptosis of hepatocellular carcinoma [67], metastasis of nasopharyngeal carcinoma [74], proliferation of NSCL carcinoma [75] as well as proliferation and tumorigenesis in ovarian cancer (OC) has been reported. A marked decrease in TRPC1 mRNA levels in human OC and cisplatin-resistant OC cells was observed [76]. TRPC1 directly interacts with several proteins/genes (MORC4, EGFR, STAT3, PDCD4, MET, OGDHL, BCL2, PTEN, SPARCL1, PIK3C3) implicated in drug resistance of OC. In the same way 5 miRs (miR-135b, miR-186, miR-26a, miR-497 and miR-548b-3p) targeting TRPC1, controlling drug resistance in OC [76] have been identified, with increased miR-135b and miR-186 expression that significantly correlates with the reduction of TRPC1 expression. The miR-135a regulates HOXA10 expression in epithelial OC, which correlates with platinum resistance [77]; moreover, a role for the TRPC1/SPARCL1 in the regulation of the autophagy and drug resistance in OC has been suggested.

24.7 Conclusion

MicroRNA are single stranded 19–25 nucleotides short RNAs that modulate gene expression at posttranslational level by targeting mRNAs and through binding of the 3′-untranslated region (UTR) of mRNAs. In the last years, miRNAs have attracted great interest from the oncologists for their versatility to regulate every phase of the carcinogenesis processes. A growing body of evidence suggests that miRNAs are aberrantly expressed in many human cancers (Table 24.1). Some high expressed miRNAs may function as oncogenes by repressing tumor suppressors, whereas other miRNAs are down-regulated and negatively regulate oncogenes, thus functioning as tumor suppressor. miRNAs have a pivotal role in tumorigenesis and the understanding of their functions may help to provide new cancer therapies.

In the last years it was demonstrated that several members of TRP family are target of miRNAs. Considering that these not selective cation channels fulfill several roles in cell physiology and in pathology such as regulating tumorigenesis and tumor progression, they became promising therapeutic targets in cancer treatment. Indeed, the interruption of one or more of the above described signaling network could be more effective than a single target gene to overcome cancer.

Interestingly, RNA molecules are not only retained in the cytoplasm of the cells, but they can also be released into the extracellular milieu, often in extracellular vesicles. These extracellular vesicles can transfer functional RNA between cells. In addition, different types of vesicles such as apoptotic bodies, microvesicles and exosomes contain distinct RNA molecules, especially miRNAs.

MiRNAs dysregulation was also involved in cancer chemoresistance, by regulating specific TRP-mediated pathway developed as consequence of high selection pressure in response to a disadvantageous microenvironment.

Future studies should be required to identify the TRP channels and miRNAs expression and their role played in distinct phase of tumor development and progression. Much work still remains to do; we are only at the beginning to develop strategies to treat cancer manipulating the TRP/miRNA interactive network.

References

1. Zheng J (2013) Molecular mechanism of TRP channels. Compr Physiol 3:221–242
2. Bartel DP (2004) MicroRNAs: genomics, biogenesis, mechanism, and function. Cell 12:281–297
3. Wang Z (2013) miRNA in the regulation of ion channel/transporter expression. Compr Physiol 3:599–653
4. Hyun YJ, Changchun X (2015) MicroRNA mechanisms of action: what have we learned from mice? Front Genet 6:328
5. Krek A, Grun D, Poy MN, Wolf R, Rosenberg L, Epstein EJ, MacMenamin P, da Piedade I, Gunsalus KC, Stoffel M, Rajewsky N (2005) Combinatorial microRNA target predictions. Nat Genet 12:495–500
6. Lu J, Getz G, Miska EA, Alvarez-Saavedra E, Lamb J, Peck D, Sweet-Cordero A, Ebert BL, Mak RH, Ferrando AA (2005) MicroRNA expression profiles classify human cancers. Nature 12:834–838
7. Mazar J, DeYoung K, Khaitan D, Meister E, Almodovar A, Goydos J, Ray A, Perera RJ (2010) The regulation of miRNA-211 expression and its role in melanoma cell invasiveness. PLoS One 5:e13779
8. Guo H, Carlson JA, Slominski A (2012) Role of TRPM in melanocytes and melanoma. Exp Dermatol 21(9):650–654
9. Bao W, Wang HH, Tian FJ, He XY, Qiu MT, Wang JY (2013) A TrkB-STAT3-miR-204-5p regulatory circuitry controls proliferation and invasion of endometrial carcinoma cells. Mol Cancer 12:155
10. Banno K, Yanokura M, Kisu I, Yamagami W, Susumu N, Aoki D (2013) MicroRNAs in endometrial cancer. Int J Clin Oncol 12:186–192
11. Erdmann K, Kaulke K, Thomae C, Huebner D, Sergon M, Froehner M, Wirth MP, Fuessel S (2014) Elevated expression of prostate cancer-associated genes is linked to down-regulation of microRNAs. BMC Cancer 14:82
12. Shalgi R, Lieber D, Oren M, Pilpel Y (2007) Global and local architecture of the mammalian microRNA-transcription factor regulatory network. PLoS Comput Biol 3(7):e131
13. Australian Institute of Health and Welfare (AIHW) (2008) Australia Cancer Incidence and Mortality (ACIM) books: incidence numbers and rates from 1982 to 2005, and mortality numbers and rates from 1968 to 2006
14. Thompson JF, Scolyer RA, Kefford RF (2005) Cutaneous melanoma. Lancet 365(9460):687–701
15. Deeds J, Cronin F, Duncan LM (2000) Patterns of melastatin mRNA expression in melanocytic tumors. Hum Pathol 31(11):1346–1356
16. Miller AJ, Du J, Rowan S, Hershey CL, Widlund HR, Fisher DE (2004) Transcriptional regulation of the melanoma prognostic marker melastatin (TRPM1) by MITF in melanocytes and melanoma. Cancer Res 64(2):509–516
17. Levy C, Khaled M, Iliopoulos D, Janas MM, Schubert S, Pinner S, Chen PH, Li S, Fletcher AL, Yokoyama S, Scott KL, Garraway LA, Song JS, Granter SR, Turley SJ, Fisher DE, Novina CD (2010) Intronic miR-211 assumes the tumor suppressive function of its host gene in melanoma. Mol Cell 40:841–849

18. Boyle GM, Woods SL, Bonazzi VF, Stark MS, Hacker E, Aoude LG, Dutton-Regester K, Cook AL, Sturm RA, Hayward NK (2011) Melanoma cell invasiveness is regulated by miR-211 suppression of the BRN2 transcription factor. Pigment Cell Melanoma Res 24:525–537
19. Davies H, Bignell GR, Cox C, Stephens P, Edkins S, Clegg S, Teague J, Woffendin H, Garnett MJ, Bottomley W, Davis N, Dicks E, Ewing R, Floyd Y, Gray K, Hall S, Hawes R, Hughes J, Kosmidou V, Menzies A, Mould C, Parker A, Stevens C, Watt S, Hooper S, Wilson R, Jayatilake H, Gusterson BA, Cooper C, Shipley J, Hargrave D, Pritchard-Jones K, Maitland N, Chenevix-Trench G, Riggins GJ, Bigner DD, Palmieri G, Cossu A, Flanagan A, Nicholson A, Ho JW, Leung SY, Yuen ST, Weber BL, Seigler HF, Darrow TL, Paterson H, Marais R, Marshall CJ, Wooster R, Stratton MR, Futreal PA (2002) Mutation of the BRAF gene in human cancer. Nature 417:949–954
20. Lunavat TR, Cheng L, Einarsdottir BO, Olofsson Bagge R, Veppil Muralidharan S, Sharples RA, Lässer C, Gho YS, Hill AF, Nilsson JA, Lötvall J (2017) BRAF(V600) inhibition alters the microRNA cargo in the vesicular secretome of malignant melanoma cells. Proc Natl Acad Sci U S A 114(29):E5930–E5939
21. De Luca T, Pelosi A, Trisciuoglio D, D'Aguanno S, Desideri M, Farini V, Di Martile M, Bellei B, Tupone MG, Condiloro A, Rcgazzo G, Rizzo MG, Del Bufalo D (2016) miR-211 and MIFT modulation by Bcl-2 protein in melanoma cells. Mol Carcinog 55:2304–2312
22. Wellbrock C, Arozarena I (2015) Microphthalmia-associated transcription factor in melanoma development and MAP-kinase pathway targed therapy. Pigment Cell Melanoma Res 28:390–406
23. Margue C, Philippidou D, Reinsbach SE, Schmitt M, Behrmann I, Kreis S (2013) New target genes of MITF-induced microRNA-211 contribute to melanoma cell invasion. PLoS One 8(9):e73473
24. Natrajan R, Louhelainen J, Williams S, Laye J, Knowles MA (2003) High-resolution deletion mapping of 15q13.2-q21.1 in transitional cell carcinoma of the bladder. Cancer Res 63:7657–7662
25. Poetsch M, Kleist B (2006) Loss of heterozygosity at 15q21.3 correlates with occurrence of metastases in head and neck cancer. Mod Pathol 19:1462–1469
26. Xia B, Yang S, Liu T, Lou G (2015) miR-211 suppresses epithelial ovarian cancer proliferation and cell-cycle progression by targeting Cyclin D1 and CDK6. Mol Cancer 14:57
27. Tao F, Tian X, Ruan S, Shen M, Zhang Z (2018) miR-211 sponges lncRNA MALAT1 to suppress tumor growth and progression through inhibiting PHF19 in ovarian carcinoma. FASEB J 6:fj201800495RR
28. He DX, Gu XT, Jiang L, Jin J, Ma X (2014) A methylation-based regulatory network for microRNA 320a in chemoresistant breast cancer. Mol Pharmacol 86(5):536–547
29. Yu S, Qin X, Chen T, Zhou L, Xu X, Feng J (2017) MicroRNA-106b-5p regulates cisplatin chemosensitivity by targeting polycystic kidney disease-2 in non-small-cell lung cancer. Anti-Cancer Drugs 28(8):852–860
30. Iino I, Kikuchi H, Myyazaki S, Hiramatsu Y, Ohta M, Kamiya K, Kusama Y, Baba S, Setou M, Konno H (2013) Effect of mir-122 and its target gene cationic aminoacid transporter 1 on colorectal liver metastasis. Cancer Sci 104:624–630
31. Shapiro D, Deering-Rice CE, Romero EG, Hughen RW, Light AR, Veranth JM, Reilly CA (2013) Activation of transient receptor potential ankyrin-1 (TRPA1) in lung cells by wood smoke particulate material. Chem Res Toxicol 26(5):750–758
32. Zygmunt PM, Hogestatt ED (2014) Trpa1. Handb Exp Pharmacol 222:583–630
33. Berrout J, Kyriakopoulou E, Moparthi L, Hogea AS, Berrout L, Ivan C, Lorger M, Boyle J, Peers C, Muench S, Gomez JE, Hu X, Hurst C, Hall T, Umamaheswaran S, Wesley L, Gagea M, Shires M, Manfield I, Knowles MA, Davies S, Suhling K, Gonzalez YT, Carragher N, Macleod K, Abbott NJ, Calin GA, Gamper N, Zygmunt PM, Timsah Z (2017) TRPA1-FGFR2 binding event is a regulatory oncogenic driver modulated by miRNA-142-3p. Nat Commun 8(1):947
34. Turner N, Grose R (2010) Fibroblast growth factor signalling: from development to cancer. Nat Rev Cancer 10:116–129

35. Timsah Z, Berrout J, Suraokar M, Behrens C, Song J, Lee JJ, Ivan C, Gagea M, Shires M, Hu X, Vallien C, Kingsley CV, Wistuba I, Ladbury JE (2015) Expression pattern of FGFR2, Grb2 and Plcγ1 acts as a novel prognostic marker of recurrence recurrence-free survival in lung adenocarcinoma. Am J Cancer Res 5(10):3135–3148
36. Peng X, Liu WL (2015) MiR-142-3p functions as a potential tumour suppressor directly targeting HMGB1 in non-small-cell lung carcinoma. Int J Clin Exp Pathol 8(9):10800
37. Jemal A, Bray F, Center MM, Ferlay J, Ward E, Forman D (2011) Global cancer statistics. CA Cancer J Clin 61(2):69–90
38. Zhang L, Barritt GJ (2004) Evidence that TRPM8 is an androgen-dependent Ca^{2+} channel required for the survival of prostate cancer cells. Cancer Res 64(22):8365–8373
39. Zhang J, Liang J, Huang J (2016) Downregulated microRNA-26a modulates prostate cancer cell proliferation and apoptosis by targeting COX-2. Oncol Lett 12(5):3397–3402
40. Tong MH, Mitchell DA, McGowan SD, Evanoff R, Griswold MD (2012) Two miRNA clusters, Mir-17-92 (Mirc1) and Mir-106b-25 (Mirc3), are involved in the regulation of spermatogonial differentiation in mice. Biol Reprod 86(3):72
41. Patel V, Williams D, Hajarnis S, Hunter R, Pontoglio M, Somlo S, Igarashi P (2013) miR-17~92 miRNA cluster promotes kidney cyst growth in polycystic kidney disease. Proc Natl Acad Sci U S A 110(26):10765–10770
42. Sun H, Li QW, Lv XY, Ai JZ, Yang QT, Duan JJ, Bian GH, Xiao Y, Wang YD, Zhang Z, Liu YH, Tan RZ, Yang Y, Wei YQ, Zhou Q (2010) MicroRNA-17 post-transcriptionally regulates polycystic kidney disease-2 gene and promotes cell proliferation. Mol Biol Rep 37:2951–2958
43. Mendell JT (2008) miRiad roles for the miR-17-92 cluster in development and disease. Cell 133(2):217–222
44. Conkrite K, Sundby M, Mukai S, Thomson JM, Mu D, Hammond SM, MacPherson D (2011) miR-17~92 cooperates with RB pathway mutations to promote retinoblastoma. Genes Dev 25(16):1734–1745
45. Cloonan N, Brown MK, Steptoe AL, Wani S, Chan WL, Forrest AR, Kolle G, Gabrielli B, Grimmond SM (2008) The miR-17-5p microRNA is a key regulator of the G1/S phase cell cycle transition. Genome Biol 9(8):R127
46. O'Donnell KA, Wentzel EA, Zeller KI, Dang CV, Mendell JT (2005) c-Myc-regulated microRNAs modulate E2F1 expression. Nature 435(7043):839–843
47. Roldo C, Missiaglia E, Hagan JP, Falconi M, Capelli P, Bersani S, Calin GA, Volinia S, Liu CG, Scarpa A, Croce CM (2006) MicroRNA expression abnormalities in pancreatic endocrine and acinar tumors are associated with distinctive pathologic features and clinical behavior. J Clin Oncol 24(29):4677–4684
48. Zanette DL, Rivadavia F, Molfetta GA, Barbuzano FG, Proto-Siqueira R, Silva WA Jr, Falcão RP, Zago MA (2007) miRNA expression profiles in chronic lymphocytic and acute lymphocytic leukemia. Braz J Med Biol Res 40(11):1435–1440
49. Xin J, Zheng L-M, Sun D-K, Li X-F, Xu P, Tian L-Q (2018) miR-204 functions as a tumor suppressor gene, at least partly by suppressing CYP27A1 in glioblastoma. Oncol Lett 16:1439–1448
50. Ge Y, Yan X, Jin Y, Yang X, Yu X, Zhou L, Han S, Yuan Q, Yang M (2015) fMiRNA-192 and miRNA-204 directly suppress lncRNA HOTTIP and interrupt GLS1-mediated glutaminolysis in hepatocellular carcinoma. Terracciano L ed. PLoS Genet 11(12):e1005726
51. Kuwano Y, Nishida K, Kajita K, Satake Y, Akaike Y, Fujita K, Kano S, Masuda K, Rokutan K (2015) Transformer 2beta and miR-204 regulate apoptosis through competitive binding to 3′ UTR of BCL2 mRNA. Cell Death Differ 22(5):815–825
52. Imam JS, Plyler JR, Bansal H, Prajapati S, Bansal S, Rebeles J, Chen HI, Chang YF, Panneerdoss S, Zoghi B, Buddavarapu KC, Broaddus R, Hornsby P, Tomlinson G, Dome J, Vadlamudi RK, Pertsemlidis A, Chen Y, Rao MK (2012) Genomic loss of tumor suppressor miRNA-204 promotes cancer cell migration and invasion by activating AKT/mTOR/Rac1 signaling and actin reorganization. PLoS One 7(12):e52397
53. Wang L, Tian H, Yuan J, Wu H, Wu J, Zhu X (2015) CONSORT. Sam68 is directly regulated by MiR-204 and promotes the self-renewal potential of breast cancer cells by activating the Wnt/Beta-catenin signaling pathway. Medicine 94(49):e2228

54. Hall DP, Cost NG, Hegde S, Kellner E, Mikhaylova O, Stratton Y, Ehmer B, Abplanalp WA, Pandey R, Biesiada J, Harteneck C, Plas DR, Meller J, Czyzyk-Krzeska MF (2014) TRPM3 and miR-204 establish a regulatory circuit that controls oncogenic autophagy in clear cell renal cell carcinoma. Cancer Cell 26(5):738–753
55. Cecconi F, Jäättelä M (2014) Targeting ions-induced autophagy in cancer. Cancer Cell 26(5):599–600
56. Cost NG, Czyzyk-Krzeska MF (2015) Regulation of autophagy by two products of one gene: TRPM3 and miR-204. Mol Cell Oncol 2(4):e1002712
57. Harteneck C (2005) Function and pharmacology of TRPM cation channels. Naunyn Schmiedeberg's Arch Pharmacol 371:307–314
58. Chow TF, Youssef YM, Lianidou E, Romaschin AD, Honey RJ, Stewart R, Pace KT, Youssef GM (2010) Differential expression profiling of microRNAs and their potential involvement in renal cell carcinoma pathogenesis. Clin Biochem 43:150–158
59. Osanto S, Qin Y, Buermans HP, Berkers J, Lerut E, Goeman JJ, van Poppel H (2012) Genome-wide microRNA expression analysis of clear cell renal cell carcinoma by next generation deep sequencing. PLoS One 7:e38298
60. Ma L, Deng X, Wu M, Zhang G, Huang J (2014) Down-regulation of miRNA-204 by LMP-1 enhances CDC42 activity and facilitates invasion of EBV-associated nasopharyngeal carcinoma cells. FEBS Lett 588(9):1562–1570
61. Siegel R, Naishadham D, Jemal A (2013) Cancer statistics, 2013. CA Cancer J Clin 12:11–30
62. Zhu X, Shen H, Yin X, Long L, Chen X, Feng F, Liu Y, Zhao P, Xu Y, Li M, Xu W, Li Y (2017) IL-6R/STAT3/miR-204 feedback loop contributes to cisplatin resistance of epithelial ovarian cancer cells. Oncotarget 8(24):39154–39166
63. Shapovalov G, Ritaine A, Skryma R, Prevarskaya N (2016) Role of TRP ion channels in cancer and tumorigenesis. Semin Immunopathol 38(3):357–369
64. Xu J, Yang Y, Xie R, Liu J, Nie X, An J, Wen G, Liu X, Jin H, Tuo B (2018) The NCX1/TRPC6 complex mediates TGFβ-driven migration and invasion of human hepatocellular carcinoma cells. Cancer Res 78(10):2564–2576
65. Wu J, Zheng C, Wang X, Yun S, Zhao Y, Liu L, Lu Y, Ye Y, Zhu X, Zhang C, Shi S, Liu Z (2015) MicroRNA-30 family members regulate calcium/calcineurin signaling in podocytes. J Clin Invest 125(11):4091–4106
66. Liu Y, Zhou Y, Gong X, Zhang C (2017) MicroRNA-30a-5p inhibits the proliferation and invasion of gastric cells by targeting insulin-like growth factor 1 receptor. Exp Ther Med 14:173–180
67. Selli C, Erac Y, Tosun M (2015) Simultaneous measurement of cytosolic and mitochondrial calcium levels: observations in TRPC1-silenced hepatocellular carcinoma cells. J Pharmacol Toxicol Methods 72:29–34
68. Zhang C, Chen X, Chen X, Wang X, Ji A, Jiang L, Sang F, Li F (2016) miR-135a acts as a tumor suppressor in gastric cancer in part by targeting KIFC1. Onco Targets Ther 9:3555–3563
69. Wan X, Pu H, Huang W, Yang S, Zhang Y, Kong Z, Yang Z, Zhao P, Li A, Li T, Li Y (2016) Androgen-induced miR-135a acts as a tumor suppressor through downregulating RBAK and MMP11, and mediates resistance to androgen deprivation therapy. Oncotarget 7:51284–51300
70. Yamada Y, Hidaka H, Seki N, Yoshino H, Yamasaki T, Itesako T, Nakagawa M, Enokida H (2013) Tumor-suppressive microRNA-135a inhibits cancer cell proliferation by targeting the c-MYC oncogene in renal cell carcinoma. Cancer Sci 104:304–312
71. Wu S, Lin Y, Xu D, Chen J, Shu M, Zhou Y, Zhu W, Su X, Zhou Y, Qiu P, Yan G (2012) MiR-135a functions as a selective killer of malignant glioma. Oncogene 31:3866–3874
72. Nagel R, le Sage C, Diosdad B, van der Waal M, Oude Vrielink JA, Bolijn A, Meijer GA, Agami R (2008) Regulation of the adenomatous polyposis coli gene by the miR-135 family in colorectal cancer. Cancer Res 68:5795–5802
73. Zhou W, Li X, Liu F, Xiao Z, He M, Shen S, Liu S (2012) MiR-135a promotes growth and invasion of colorectal cancer via metastasis suppressor 1 in vitro. Acta Biochim Biophys Sin Shanghai 44:838–846

74. He B, Liu F, Ruan J, Li A, Chen J, Li R, Shen J, Zheng D, Luo R (2012) Silencing TRPC1 expression inhibits invasion of CNE2 nasopharyngeal tumor cells. Oncol Rep 27:1548–1554
75. Tajeddine N, Gailly P (2012) TRPC1 protein channel is major regulator of epidermal growth factor receptor signaling. J Biol Chem 287:16146–16157
76. Liu X, Zou J, Su J, Lu Y, Zhang J, Li L, Yin F (2016) Downregulation of transient receptor potential cation channel, subfamily C, member 1 contributes to drug resistance and high histological grade in ovarian cancer. Int J Oncol 48(1):243–252
77. Tang W, Jiang Y, Mu X, Xu L, Cheng W, Wang X (2014) MiR-135a functions as a tumor suppressor in epithelial ovarian cancer and regulates HOXA10 expression. Cell Signal 26:1420–1426

Chapter 25
Calcium Channels and Calcium-Regulated Channels in Human Red Blood Cells

Lars Kaestner, Anna Bogdanova, and Stephane Egee

Abstract Free Calcium (Ca^{2+}) is an important and universal signalling entity in all cells, red blood cells included. Although mature mammalian red blood cells are believed to not contain organelles as Ca^{2+} stores such as the endoplasmic reticulum or mitochondria, a 20,000-fold gradient based on a intracellular Ca^{2+} concentration of approximately 60 nM vs. an extracellular concentration of 1.2 mM makes Ca^{2+}-permeable channels a major signalling tool of red blood cells. However, the internal Ca^{2+} concentration is tightly controlled, regulated and maintained primarily by the Ca^{2+} pumps PMCA1 and PMCA4. Within the last two decades it became evident that an increased intracellular Ca^{2+} is associated with red blood cell clearance in the spleen and promotes red blood cell aggregability and clot formation. In contrast to this rather uncontrolled deadly Ca^{2+} signals only recently it became evident, that a temporal increase in intracellular Ca^{2+} can also have positive effects such as the modulation of the red blood cells O_2 binding properties or even be vital for brief transient cellular volume adaptation when passing constrictions like small capillaries or slits in the spleen. Here we give an overview of Ca^{2+} channels and Ca^{2+}-regulated channels in red blood cells, namely the Gárdos channel, the non-selective voltage dependent cation channel, Piezo1, the NMDA receptor, VDAC, TRPC channels, $Ca_V2.1$, a Ca^{2+}-inhibited channel novel to red blood cells and i.a. relate these channels to the molecular unknown sickle cell disease conductance P_{sickle}. Particular attention is given to correlation of functional measurements with

L. Kaestner (✉)
Theoretical Medicine and Biosciences, Saarland University, Homburg, Germany

Experimental Physics, Saarland University, Saarbrücken, Germany
e-mail: lars_kaestner@me.com

A. Bogdanova
Red Blood Cell Research Group, Institute of Veterinary Physiology, Vetsuisse Faculty and the Zürich Center for Integrative Human Physiology (ZIHP), University of Zürich, Zürich, Switzerland

S. Egee
CNRS, UMR8227 LBI2M, Sorbonne Université, Roscoff, France

Laboratoire d'Excellence GR-Ex, Paris, France

© Springer Nature Switzerland AG 2020

M. S. Islam (ed.), *Calcium Signaling*, Advances in Experimental Medicine and Biology 1131, https://doi.org/10.1007/978-3-030-12457-1_25

molecular entities as well as the physiological and pathophysiological function of these channels. This view is in constant progress and in particular the understanding of the interaction of several ion channels in a physiological context just started. This includes on the one hand channelopathies, where a mutation of the ion channel is the direct cause of the disease, like Hereditary Xerocytosis and the Gárdos Channelopathy. On the other hand it applies to red blood cell related diseases where an altered channel activity is a secondary effect like in sickle cell disease or thalassemia. Also these secondary effects should receive medical and pharmacologic attention because they can be crucial when it comes to the life-threatening symptoms of the disease.

Keywords Gárdos channel · Non-selective voltage dependent cation channel · Piezo1 · NMDA receptor · VDAC · TRPC channel · $Ca_V2.1$ · Calcium-inhibited channel · P_{sickle} · Anaemia

25.1 Introduction to Calcium in Red Blood Cells

Free Calcium (Ca^{2+}) is an important and universal second messenger in all cells [1, 2], red blood cells (RBCs) included [3–5]. This results in the abundance of Ca^{2+}-binding proteins in RBCs with differing Ca^{2+} sensitivities as outlined in Fig. 25.1. Mature mammalian RBCs are believed to not contain organelles as Ca^{2+} stores such as the endoplasmic reticulum or mitochondria [6]. Compared to other cell types, where the Ca^{2+} liberated from stores within intracellular organelles can be used in the regulation of free cytosolic Ca^{2+} concentration and thereby Ca^{2+} signalling, in mammalian erythrocytes the control of free intracellular Ca^{2+} concentration must be done by regulation of membrane transport. A 20,000-fold gradient based on an intracellular Ca^{2+} concentration of approximately 60 nM vs. an extracellular concentration of 1.2 mM makes Ca^{2+}-permeable channels a major signalling tool of RBCs. As historically RBCs served as the model cell to investigate membrane transport, it is well known that the internal Ca^{2+} concentration is tightly controlled, regulated and maintained primarily by the Ca^{2+} pumps PMCA1 and PMCA4 [4, 7]. The Ca^{2+} pumping in turn is regulated by multiple factors, such as the Ca^{2+} concentration itself [8], calmodulin [9], calpain [10], phospholipids and various kinases [11] or even self-association [12]. Within the last two decades it became evident that an increased intracellular Ca^{2+} is associated with RBC clearance in the spleen and promotes RBCs aggregability and clot formation [3, 13–16]. There was a long debate within the community whether this process should be called eryptosis [17], which is no longer recommended [18]. In contrast to this rather uncontrolled deadly Ca^{2+} signals (resulting in Ca^{2+} overload), within the recent years it became evident that a temporal increase in intracellular Ca^{2+} can also have positive effects such as the modulation of the RBCs O_2 binding properties [19] or even be vital for brief transient cellular volume adaptation when passing constrictions like small capillaries or slits in the spleen [5, 20, 21] as depicted in Fig. 25.2. The perilous balance of Ca^{2+} in RBCs was recently reviewed [3].

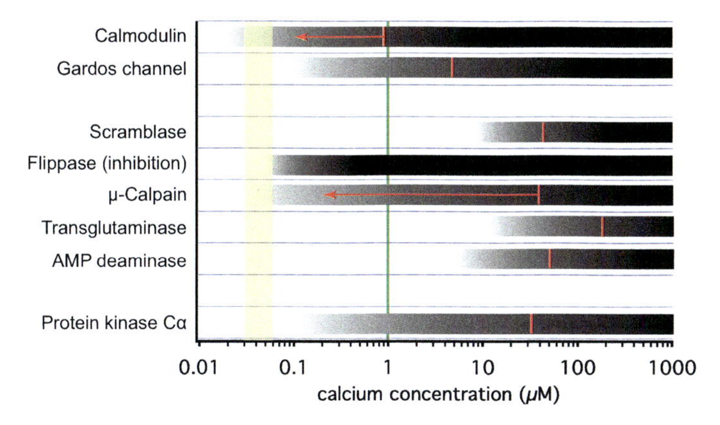

Fig. 25.1 Overview of concentration dependence of Ca^{2+}activated proteins in RBCs. The yellow column indicates the estimated range of RBCs'resting free Ca^{2+} [96]. The gray/black bars indicate the activation of the proteins with the intensity of darkness related to the activation level (details see below). The red lines depict the half activation concentration. For orientation the green line provides the in vivo k_D for Fluo-4 [101], probably the most appropriate Ca^{2+} fluorophore to be used in RBCs [102]. The universal intermediate messenger calmodulin has a dissociation constant for Ca^{2+} of 920 nM [103], which can be shifted down to 100 nM, indicated by the red arrow. The Gárdos channel has an open probability of EC_{50} of 4.7 μM with a Hill slope factor of approximately 1 [104]. Values were measured in excised patches at a membrane potential of 0 mV. The curve of the opening frequency is almost superimposable (EC_{50} of 4.3 μM) [104] keeping the values given in the figure valid also for whole cell and hence population based investigations. The values for half maximal activation of the scramblase was determined by different studies with varying methodologies and a slightly different result. Values varied between approximately 30 μM determined in liposomes [105] and 70 μM measured in RBC ghosts [106]. The flippase displays almost full inhibition already at a Ca^{2+} concentration of 400 nM [107]. μ-Calpain, a protein that cleaves cytoskeleton and membrane proteins depicts an half activation at 40 μM Ca^{2+} [108] but can be activated and then shifting half-maximal activation down to 200 nM [109]. Transglutaminase mediating polymerisation of RBC membrane proteins in its native form has a dissociation constant for Ca^{2+} of 190 μM [110]. Adenosine monophosphate (AMP) deaminase is an enzyme that converts AMP into inosin monophosphate and is directly stimulated by Ca^{2+} at a half maximal concentration of 50 μM free Ca^{2+} [111]. The binding of Ca^{2+} to the C2-domain of PKCα was determined in vitro to be 35 μM with a Hill coefficient of 0.9 [112]. Although the Ca^{2+} dependence of the membrane binding was measured to be one order of magnitude lower [112], the initial Ca^{2+} binding is the crucial step for PKCα activation and therefore the relevant number in this compilation. (This figure is reproduced from Bogdanova et al. 2013 [3])

25.2 The Gárdos Channel – A Calcium-Activated Potassium Channel

The Gárdos channel is one of the Ca^{2+} sensors in RBCs transferring Ca^{2+} uptake into K^+ and water loss and mediating thereby Ca^{2+}-dependent volume regulation and changes in RBC rheology. It is also annotated as KCNN4, $K_{Ca}3.1$, IK1 or SK4. It is the first channel we describe in this chapter because it was the first channel found in RBCs [22, 23], i.e. utilising the patch-clamp technique as a direct read-out of channel activity. However, its name goes back to the effect of Ca^{2+} dependent

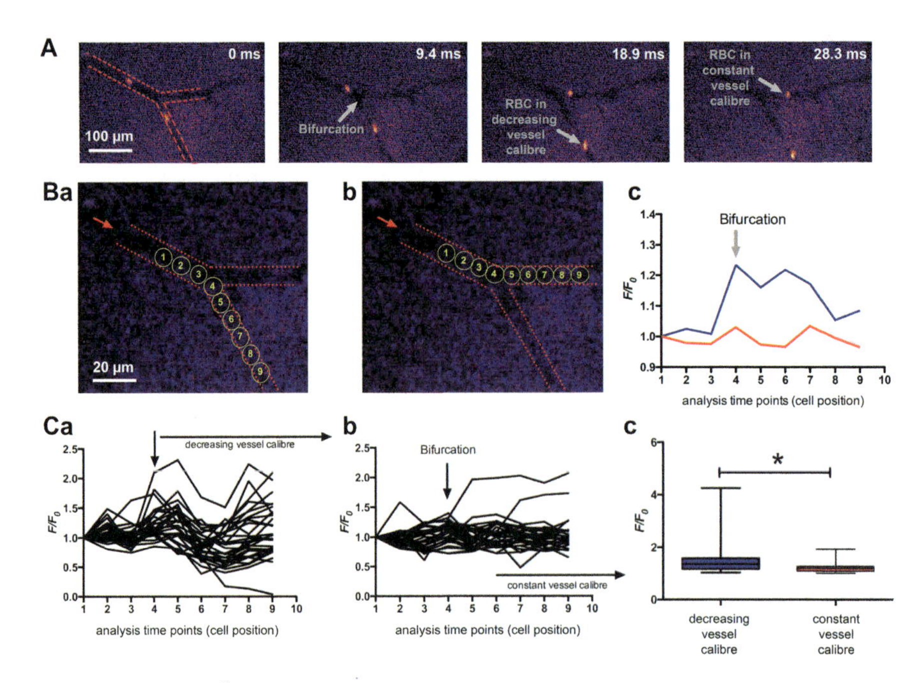

Fig. 25.2 In vivo Ca^{2+}-signalling of mouse RBCs when passing through capillaries. Mouse RBCs were ex vivo stained with Fluo-4 and then re-injected into the mouse circulation. Fluorescence imaging of capillaries was performed in the dorsal skinfold chamber. (A) shows representative snapshots of RBCs passing a bifurcation. For the 30 ms sequence only every second recorded image is presented. For a better orientation the vessel walls are indicated by red dashed lines in the leftmost image and further annotations (grey) are added in the other images. (B) depicts the positions where fluorescence intensity (F/F_0) was analysed for a decreasing vessel calibre (Ba) and for a constant vessel calibre (Bb) of the same example section as in (A). The dashed red lines mark the vessel walls, the red arrow indicates the blood flow direction and the yellow circles depict the analysis positions which are plotted in the following diagrams. Example fluorescence traces of the two cells analysed as pointed out in (Ba) and (Bb) are shown in (Bc). (C) depicts the analysis of 30 cells passing through a capillary with decreasing vessel calibre and 28 cells passing through a capillary with constant vessel calibre. Analysis was performed at 3 vessel-bifurcations in 2 mice. The fluorescence intensity (F/F_0) traces of all measured RBCs passing through a capillary with decreasing vessel calibre is plotted in (Ca), while the traces of all measured RBCs passing a capillary with constant vessel calibre is plotted in (Cb). The statistical analysis of the maximal fluorescence intensity (F/F_0) of RBCs from both groups is depicted in (Cc). The increase in Ca^{2+}, while passing through a vessel with decreasing calibre is significant ($p = 0.014$; *). (This figure is reproduced from Danielczok et al. 2017 [5])

K^+ efflux found in RBCs by G. Gárdos once metabolic pathways are poisoned [24, 25]. As the molecular identity of this transport in RBCs was not known, it was referred to as Gárdos effect and later, when it turned out to be ion channel mediated, the involved transport protein was called Gárdos channel. Even after its molecular identification in 2003 [26] in the RBC field it remained to be referred to as Gárdos channel. Figure 25.3 provides the milestones in the Gárdos channel research. A comprehensive review of the Gárdos channel structure and function was published

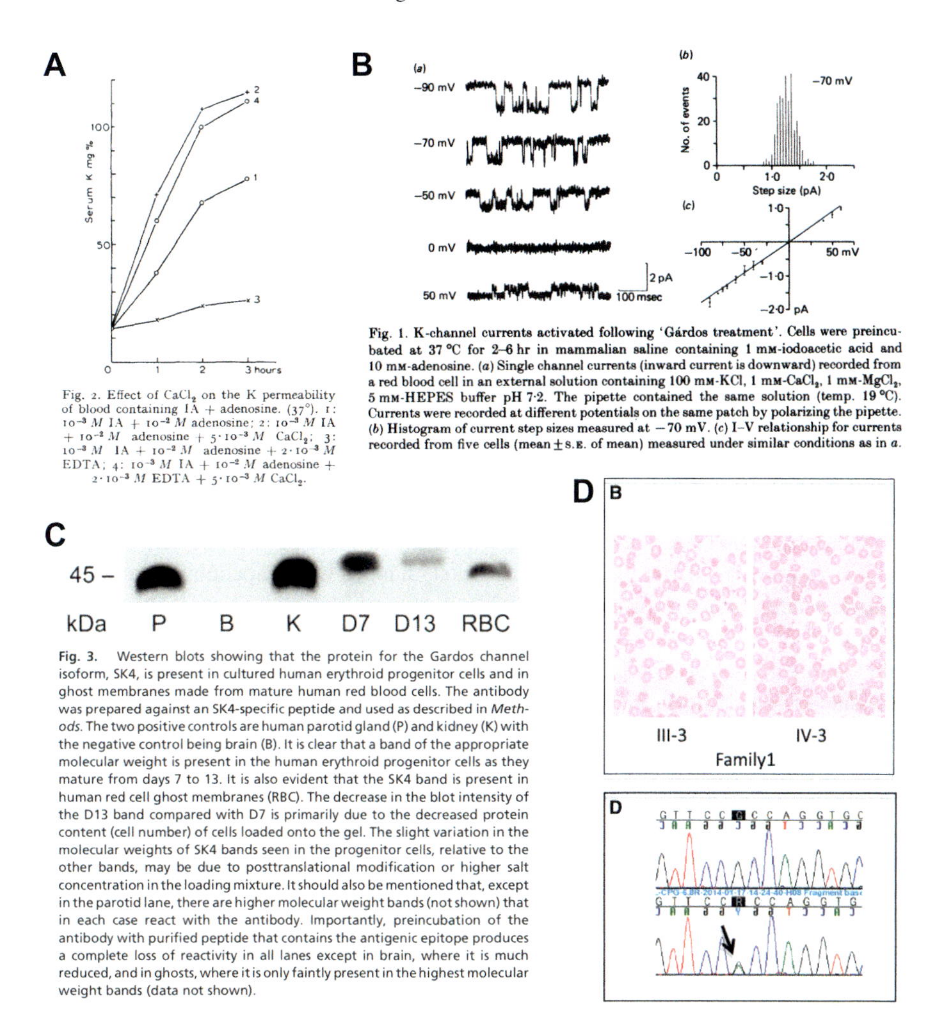

Fig. 25.3 Milestones in Gárdos channel research. (**a**) Reprint from Gárdos 1958 [25]; a one page article mentioning a dependence of K^+-efflux from Ca^{2+} in the category "preliminary notes" was retrospectively sufficient for the naming of an ion channel. IA stands for iodoacedic acid. (**b**) Reprint from Hamill 1981 [23]; the first published single channel current traces from the Gárdos channel appeared in a poster Abstract for the Physiological Society (London) meeting. Numerous patch-clamp based characterisations followed this initial recording [22, 28, 104, 103–107]. (**c**) Reprint from Hoffman et al. 2003 [26] providing the molecular identification of the Gárdos channel to be KCNN4 ($K_{Ca}3.1$, IK1, SK4). (**d**) Reprint from Rapetti-Maus et al. 2015 [29] providing the first report of a Gárdos channel mutation and its link to a pathophysiological setting. Upper panel: Blood film smears for two patients with the Gárdos channel mutation p.Arg352His (mother and son). Lower panel KCNN4 transcript sequencing; upper line presents the wild type sequence and the lower line the transcript with mutation c.1055G.A (p.Arg352His)

some 15 years ago [27]. An interesting property of the channel is its temperature dependence. With decreasing temperature, a continuous decrease of Gárdos channel conductance is observed. The Arrhenius plot of the unitary channel conductance between 0 °C and 47 °C is strictly linear and has a slope which corresponds to an activation energy of 29.6 ± 0.4 kJ/mol. Nevertheless, simultaneously, altered gating kinetics results in an increase of channel opening probability at reduced temperatures. At saturating concentration of intracellular Ca^{2+} (10 μM), reducing the temperature from 35 to 30 °C results in a change of the opening and closing kinetic of the Gárdos channel. Brief channel openings and closing are progressively replaced by longer openings and shorter closing states [28]. More importantly, reducing the Ca^{2+} concentration at the intracellular face of the channel at half the EC_{50} for Ca^{2+} and temperature close to 0 °C drastically increases open probability indicating that even at very low Ca^{2+} concentrations the Gárdos channel may be activated. Knowing that blood samples for analyses and RBC concentrates for blood transfusion are kept refrigerated, one has to keep in mind this peculiar property of the Gárdos channel that may have deleterious effects respective to the cell volume.

The physiological function of the Gárdos channel was a speculative topic for many decades although a link to RBC cell volume threatening was clear, since activation of the Gárdos channel results in cellular K^+ loss associated with Cl^- and osmotically obliged water loss may lead to rapid cell shrinkage. In this context the channel was believed to be a 'suicide mechanism' triggered by the intracellular increase in Ca^{2+}. This was proposed to happen in the process of clot formation, in thrombotic events as well as during RBC clearance. With the finding of the first mutations in the Gárdos channel [29, 30] (see also Fig. 25.3D) and the associated pathophysiology, a physiological function of the Gárdos channel was doubtless proven. Although the initial reports link the mutation of the Gárdos channel to Hereditary Xerocytosis [29, 30], further studies revealed that 'Gárdos channelopathy' is its own disease or at least an own variant of Hereditary Stomatocytosis [31]. Any prolonged Gárdos channel activation lead to changes in cell volume which eventually affect rheological, stiffness and rigidity properties that compromise their survival within the circulation especially during their passage within the slits of the spleen. The mutations reported so far resulted in a gain of function that could be treated with a Gárdos channel inhibitor. There are numerous Gárdos channel inhibitors available some of them already clinically tested for other diseases, like clotrimazole for topical applications or senicapoc [32]. The Gárdos channel also shows an increased activity in other haemolytic diseases, such as sickle cell disease. Therefore one has tried to use the Gárdos channel as a pharmacological target to treat sickle cell disease [33] overlooking the fact that upstream of the signalling cascade is an increase in intracellular Ca^{2+} through a pathway named P_{sickle} for which we are still seeking molecular identity (see below) and triggers numerous other pathophysiological processes in the RBCs (compare Fig. 25.1) leaving the Gárdos channel only a minor portion to account for the cellular symptoms of sickle cell disease [34]. However, what failed in sickle cell disease may still work out well in Gárdos Channelopathy [31, 35].

25.3 Non-selective Cation Channels Permeable for Calcium

25.3.1 Non-selective Voltage Dependent Cation Channel and Piezo1

RBCs contain a variety of non-selective ion channels that are permeable to Ca^{2+}. First there is a non-selective voltage dependent cation channel initially described by Christophersen and Bennekou [36, 37] and later to be reported to be Ca^{2+} permeable [38]. However, the molecular identity of this channel is still not quite clear [39, 40]. Although it was proposed to be a conductive state of the voltage-dependent anion channel (VDAC) [41], recent reports rather make a link to the Piezo1 [42, 43]. Figure 25.4 provides a comparison of the non-selective voltage dependent cation channel and Piezo1. The unique hysteresis-like open probability was also modelled successfully [44].

PIEZO1 and in particular mutations of the channel have been associated with the RBC-related disease Hereditary Xerocytosis [45, 46]. Therefore it seems obvious that this channel, originally described as a mechanosensitive channel, is present in the RBC membrane. Furthermore, knock-out approaches in zebrafish [21] and mice [20] gave further evidence for the conserved abundance of Piezo1 in RBCs as well as for its function (see below). Piezo1 and its mutations were mainly characterised in heterologous overexpressing cell lines and initial measurements of Piezo1 in RBC have been rather episodic [47]. However, the discovery of the pharmacological activation of Piezo1 by Yoda1 [48] lead to the development of high throughput patch-clamp assays as potential diagnostic tools that were recently implemented [49].

The interplay of Piezo1 (in particular its property to mediate Ca^{2+} entry) with the above mentioned Gardos channel [5, 20, 21] provides additional evidence for a functional Piezo1 in RBCs. Furthermore, the interplay between the channels was proposed to be vital for the RBCs to maintain their ion homeostasis [50]. In pathophysiology, Piezo1 seems to play a mayor role in an increased RBC Ca^{2+}-homeostasis. The reported mutations of Piezo1 [45, 46, 51–54] are mostly gain of function mutations suggesting an easier and more pronounced Ca^{2+} entry. The consequent increase in intracellular Ca^{2+} is most likely the trigger for the early removal of the RBC from the circulation and hence the reason for the anaemic symptoms [55]. One of the clinical treatments to handle severe anaemias is splenectomy. Interestingly splenectomy introduces comorbidity, namely thrombotic events, in a subpopulation of Hereditary Xerocytosis patients. This could be explained by an active participation of RBCs in the thrombus formation due to increased intracellular Ca^{2+} [13, 14, 56–58]. Surprisingly such thrombotic events were not reported for splenectomised Gárdos Channelopathy patients [31].

It is worthwhile to mention that in sickle cells an increased conductance carrying also Ca^{2+} was reported and named P_{sickle} [59, 60]. It is likely that P_{sickle} resembles the superposition of several ion channel entities. The sensitivity of P_{sickle} to GsMTx-4 [61], a toxin that inhibits Piezo1 [62] points to this mechanosensitive channel, while the increased abundance of *N*-methyl-D-aspartate (NMDA) receptors in sickle cells [63] is a very strong indicator for these ionotrope glutamate receptors (see below).

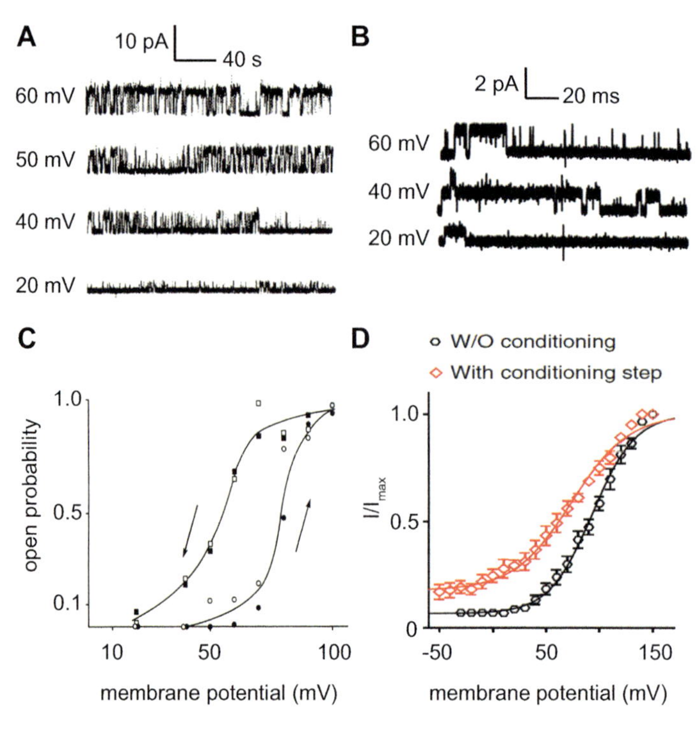

Fig. 25.4 Comparison of the non-selective voltage activated cation channel recorded in RBCs and Piezo1 recorded in overexpressing Neuro2A cells. (**a**) Current traces of the non-selective voltage activated cation channel in inside-out patches of RBCs in symmetrical KCl-solution in mM (500 KCl, 5 MOPS, 4 NMDG, 1 EGTA, 0.02 Ca^{2+}, pH = 7.4). (**b**) Current traces of Piezo1 in outside-out patches of overexpressing Neuro2A cells in symmetrical NaCl-solution in mM (140 NaCl, 10 HEPES, 5 EGTA, pH = 7.4). The difference in ion strength between panels **a** and **b** could explain the different current amplitudes shown in the single channel openings. (**c**) The open state probability as function of the membrane potential. Filled symbols represent data from the experiment shown panel **a**. Open symbols represent data from an identical experiment performed on the same patch after 15 min. In both series, the open probability was calculated from 3 min of continuous recording at each potential. The curves were drawn by eye. (**d**) Tail currents from individual cells were normalized to their maximum and fitted to a Boltzmann relationship. Pooled data are shown as mean ± SEM. (Panels **a** and **c** are reproduced from Kaestner et al. 2000 [38] and panels **b** and **d** from Moroni et al. 2018 [42])

25.3.2 N-Methyl D-Aspartate Receptor

There is clear evidence for the abundance of erythroid N-methyl D-aspartate (NMDA)-receptors in RBCs in particular in the young cell population [19, 64, 65] and its abnormally high prevalence and activity in sickle cell patients [34, 63]. Inhibition of these receptors by oral administration of memantine, the pore-targeting antagonist of NMDARs, results in a decrease of the intracellular Ca^{2+} (Fig. 25.5). In general, cells stemming from myeloid lineage are expressing

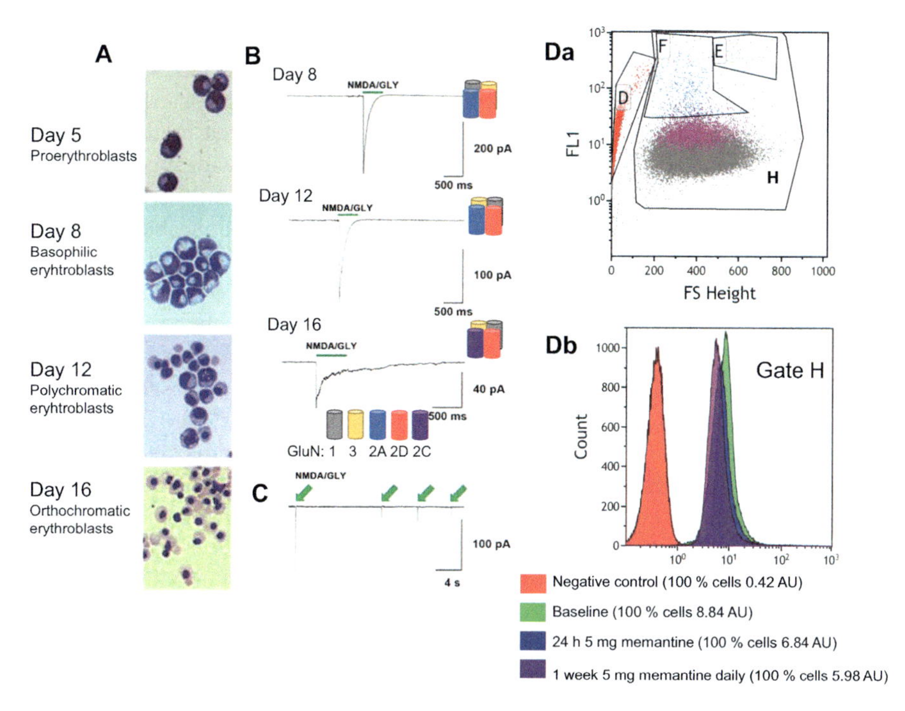

Fig. 25.5 NMDA receptors structure and function. (A) Erythroid NMDA receptors are expressed early on in differentiating erythroid progenitor cells starting from proerythroblasts (CD34 + -derived cells on day 5 in culture) to orthochromatic erythroblasts (day 16 in culture). (B) Electrophysiological recordings of currents mediated by treatment of these cells with agonists NMDA and glycine reveal the change in subunit composition of the receptor during differentiation. Whereas proerythroblasts and polychromatic erythroblasts are equipped with receptors built by glycine-binding GluN1 and 3A/B and glutamate/NMDA-binding Glun2A and 2D (see cartoons), late orthochromatic erythroblasts and reticulocytes contain receptors in which GluN2A is replaced by GluN2C. As a result channels with high current amplitudes and short times to inactivation in early progenitors turn into slowly inactivating channels with lower amplitude in late progenitors and circulating RBCs. For details see [67]. (C) Repeated activation of eNMDARs triggers inactivation of the channels. (D) Changes in the intracellular Ca^{2+} concentration in RBCs upon systemic administration of memantine in patients with sickle cell disease. (Da) Dot plot showing the heterogeneity of free Ca^{2+} levels in RBCs (gates F, E, and H) and RBC-derived vesicles (gate D). (Db) shows a histogram of Ca^{2+}-dependent fluorescence (all cells in population, gate H) in unstained cells (negative control, red), at baseline before the onset of treatment with 5 mg memantine a day (green) after 24 h (blue) and 1 week (pink) of therapy with 5 mg Memantin Mepha daily. The treatment was performed within the MemSID clinical trial (ClinicalTrials.gov Identifier: NCT02615847 approved by SWISSMEDIC (# 2015DR2096) and Cantonal ethic committee of canton Zurich (#2015–0297)). (Panels A–C are reproduced from Hänggi et al. 2015 [67] and panel D from Makhro et al. 2017 [66])

particular type of ionotropic glutamate receptors making immune responses, clotting and RBC function sensitive to the changes in ambient glutamate levels [66]. Subunit composition of erythroid NMDA receptors (eNMDARs) as well as the number of receptor copies changes in the course of differentiation (Fig. 25.5A–C).

Receptor abundance declines from thousand copies in proerythroblasts to about 30 in reticulocytes and about 5 (on average) in mature RBCs [19, 67]. At the same time high amplitude currents with short inactivation time carried by the GluN2A/2D-containing receptors in proerythroblasts are replaced by the currents mediated by the receptors built by the GluN2C/2D subunits with much smaller amplitude and prolonged inactivation time in ortho/polychromatic erythroblasts and reticulocytes [67], (Fig. 25.5A–C). eNMDARs are highly permeable for Ca^{2+} [19, 64] and are actively involved in Ca^{2+}-driven signaling during differentiation and maintenance of intracellular Ca^{2+} in mature RBCs [19, 67]. Clearance of eNMDARs in reticulocytes released into the circulation most likely occurs by way of 'shedding' when receptors are released together with other membrane proteins from the membrane in the form of vesicles. Whereas no direct measurements of eNMDARs in vesicles were performed so far, inability to clear eNMDARs from membranes of RBCs of patients with sickle cell disease is an indirect evidence for this hypothesis. In erythroid precursor cells obtained by differentiation of peripheral CD34+ cells of sickle cell disease patients the number of eNMDAR copies was like that in cells of healthy patients [67]. However, circulating RBCs of patients were presented with abnormally high abundance and activity levels of eNMDARs. As a result, basal Ca^{2+} levels in RBCs of sickle cell disease patients were exceeding that in cells of healthy subjects. Pharmacological inhibition of the receptors decreased Ca^{2+} levels and resulted in rehydration and reduction in oxidative load [67]. First pilot clinical trial MemSID in which patients with sickle cell disease were treated with the antagonist of NMDA receptors, memantine, revealed that these receptors may be an attractive pharmacological target for this group of patients [68, 69] (Fig. 25.5D). Among physiological factors that may control eNMDARs are endurance exercises that are associated with glutamate release into the circulation [70].

25.3.3 Voltage-Dependent Anion Channel

Another multifunctional channel with a clear molecular identity in RBCs that also conducts Ca^{2+} is the Voltage-Dependent Anion Channel (VDAC) [41, 71]. VDACs have originally been characterized as mitochondrial porins [72]. Three different isoforms of VDAC have been identified so far: VDAC1, VDAC2 and VDAC3. Showing the expression of 'Porin 31HL' in the plasmalemma of human B lymphocytes, gave first evidence on the multitopological localisation of VDAC [73]. The existence in the membrane of RBCs of a 32 kDa associated voltage dependent anion channel (VDAC) in a peripheral benzodiazepine receptor-like PBR protein complex to 18 kDa protein TSPO 'translocator proteins' and 30 kDa ANT 'adenine nucleotide transporter' proteins has been demonstrated [41, 74]. It has a nanomolar affinity for PK11195, Ro5–4864 and Diazepam ligands [75,

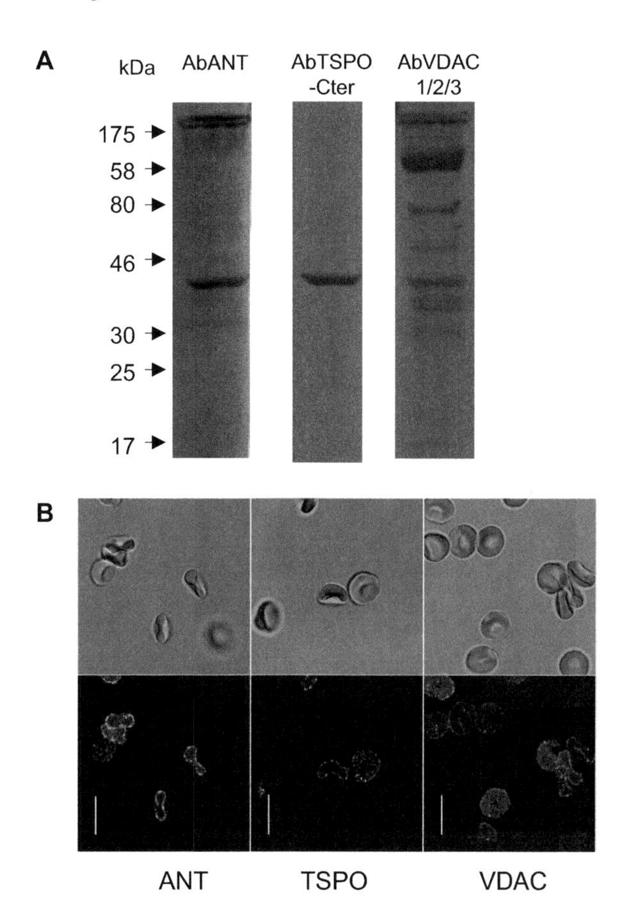

Fig. 25.6 VDAC, ANT and TSPO detection in human red blood cell ghosts. (**a**) Samples (15 μg of protein) of whole lysates were subjected to SDS-PAGE (10% acrylamide) and analysed by Western blotting using polyclonal anti-ANT (1:1000 dilution), polyclonal goat anti-TSPO raised against the C terminus of human TSPO (1:1000) or rabbit polyclonal anti VDAC 1–2-3 (1:100 dilution). Multiple bands at different molecular weights is consistent with the oligomerisation of VDAC proteins. ANT and TSPO proteins are also clearly visible. (**b**) Immunofluorescence experiments were performed on smears Dilution were 1/5 for primary and 1/20 for secondary antibodies. Scale bars represent 10 μm. (This figure is reproduced from Bouyer et al. 2011 [41])

76]. All blood cells have a population of receptors with micromolar affinity for PK11195 ranging from approximately 750,000 sites for lymphocytes to over one hundred sites for RBCs [77, 78]. These indications are corroborated by analysis of messenger RNA expression data provided by GeneAtlas U133A where 3 isoforms of VDAC, 2 isoforms of ANT and 2 isoforms of TSPO were found in erythroid progenitors from CD34+ to CD71+ (Fig. 25.6). VDAC is a protein that has remarkably well-preserved structural and functional characteristics, despite major variations in the sequence [79]. Although it is also present in the plasma membrane, most of the information we have on its structure function comes from studies on mitochondrial proteins [80, 81]. The maximum conductances reach 4–5 nS in the presence of 1 M NaCl or KCl, 350–450 pS for more physiological concentrations (NaCl or 150 mM KCl). Conductance and selectivity are voltage dependent; at low voltages, close to -10 mV, the channel is stable and remains open, whereas at positive or negative potentials higher than 40 mV, VDAC has multiple sub states of different permeabilities and selectivities, as well as closing episodes of which frequency increases with voltage [82, 83]. The highest levels are permeable to small ions (Na^+, K^+, Cl^-, etc.) but also to large anions (glutamate, ATP) and large cations (acetylcholine, dopamine, Tris, etc.). They have a preference for anions (2:1) when saline solutions are composed of ions of equal mobility such as NaCl or KCl. More importantly, at low conductances, VDAC is more permeable to small ions with, apparently, a marked preference for cations and higher permeability to Ca^{2+} ions than in large conductances [82–84]. VDAC may have different oligomerization states: mono-, di-, tri, tetra-, hexamers or even more. Indeed, atomic force microscopy revealed the presence of VDAC1 monomers as well as dimers and larger oligomers showcasing the interaction of the pore with itself, however, dimers are more frequent. Very little is known about the activation and regulation mechanisms of the channel. Nevertheless, when the pores dimerize, the selectivity for Ca^{2+} increases. Various studies support the function of VDAC (more precisely VDAC1 the most studied yet) in the transport of Ca^{2+} and in cellular Ca^{2+} homeostasis. Lipids-reconstituted bilayer incorporating VDAC1 in the presence of different $CaCl_2$ concentration gradients showed well-defined voltage-dependent channel conductance, as observed with either NaCl or KCl solution, with higher permeability to Ca^{2+} once VDAC is in the low conductance state. It is obvious that the permeability ratios of VDAC1 for Ca^{2+} is very low compared to Cl^- (P_{Ca2+}/P_{Cl-} is 0.02–0.38) [83] but considering the tremendous electrochemical gradient for Ca^{2+} between intra- and extracellular face of RBCs (see Introduction) a short activation may represent a significant input of Ca^{2+} into the cell.

25.3.4 Transient Receptor Potential Channels of Canonical Type

Yet another type of non-selective cation channels that are believed to be abundant in RBCs are Transient Receptor Potential channels of Canonical type (TRPC channels). Indications point to a different expression pattern of isoforms in precursor cells compared to mature RBCs and also differences between mammalian species seem likely [85–88]. In humans it is believed that TRPC6 is abundant in RBCs [39, 89, 90]. So far a dedicated physiological function of TRPC6 in mature RBCs remains elusive.

25.4 Voltage-Activated Calcium Channels and Their Regulation

Evidence for the existence of a number of voltage-activated Ca^{2+} channels in RBCs has been reported [91, 92], and the most convincing evidence is for $Ca_V2.1$, based on molecular biology data (Western blot) [93] and, presumably, $Ca_V2.1$-specific pharmacological interactions (ω-agatoxinTK) [93, 94] – both are shown in Fig. 25.7A, B. Nevertheless, so far, we and others have failed to obtain direct functional evidence for the existence of $Ca_V2.1$ or other voltage-activated Ca^{2+} channels in RBCs by patch-clamp techniques [95]. However, also RBCs although non-excitable cells meet the condition of voltage jumps necessary to activate voltage-activated channels such as $Ca_V2.1$ [95]. In particular when the Gárdos channel (see Sect. 25.2) is activated, the resting membrane potential changes from approximately -10 mV to approximately -70 mV [96]. Not hyperpolarisation but depolarisation is required to activate $Ca_V2.1$ [97]. Nevertheless, hyperpolarisation is a requirement to switch $Ca_V2.1$ channels from the inactivated state to the closed state, which is a prerequisite to subsequently transition to the open state [98] (Fig. 25.7C). Closing of the Gárdos channels after their initial activation could well provide the necessary conditions for subsequent depolarisation to activate $Ca_V2.1$ [95]. Such a proposed mechanism is sensible also in the context of other voltage-activated channels in the RBC membrane (compare Sect. 25.3).

25.5 Evidence for a Calcium-Inhibited Channel

There is also evidence for a non-selective cation channel in RBCs that is activated when extracellular Ca^{2+} is removed [99]. Original recordings and an I–V curve are shown in Fig. 25.8A, B. There are two conceptual question related to this recent report (a) if the channel is abundant in almost all RBCs why it was not reported before (in four decades of patch-clamping RBCs) and (b) since divalent cations in general and Ca^{2+} in particular support seal formation, a removal of Ca^{2+} could impair the seal quality/tightness. Under these circumstances it is almost impossible to discriminate a leak in the seal from an ion channel. However, here are also two arguments in favour of the existence of this channel: (A) The suspicion of the phenomenon described in (b) could have prevented scientists to report about the channel (a). The non-ohmic behaviour of the I-V curve (Fig. 25.8B) is in favour of a channel rather than a leak. (B) A channel activated by the removal of Ca^{2+} is an ideal explanation of the dissipation of the monovalent cation gradients when

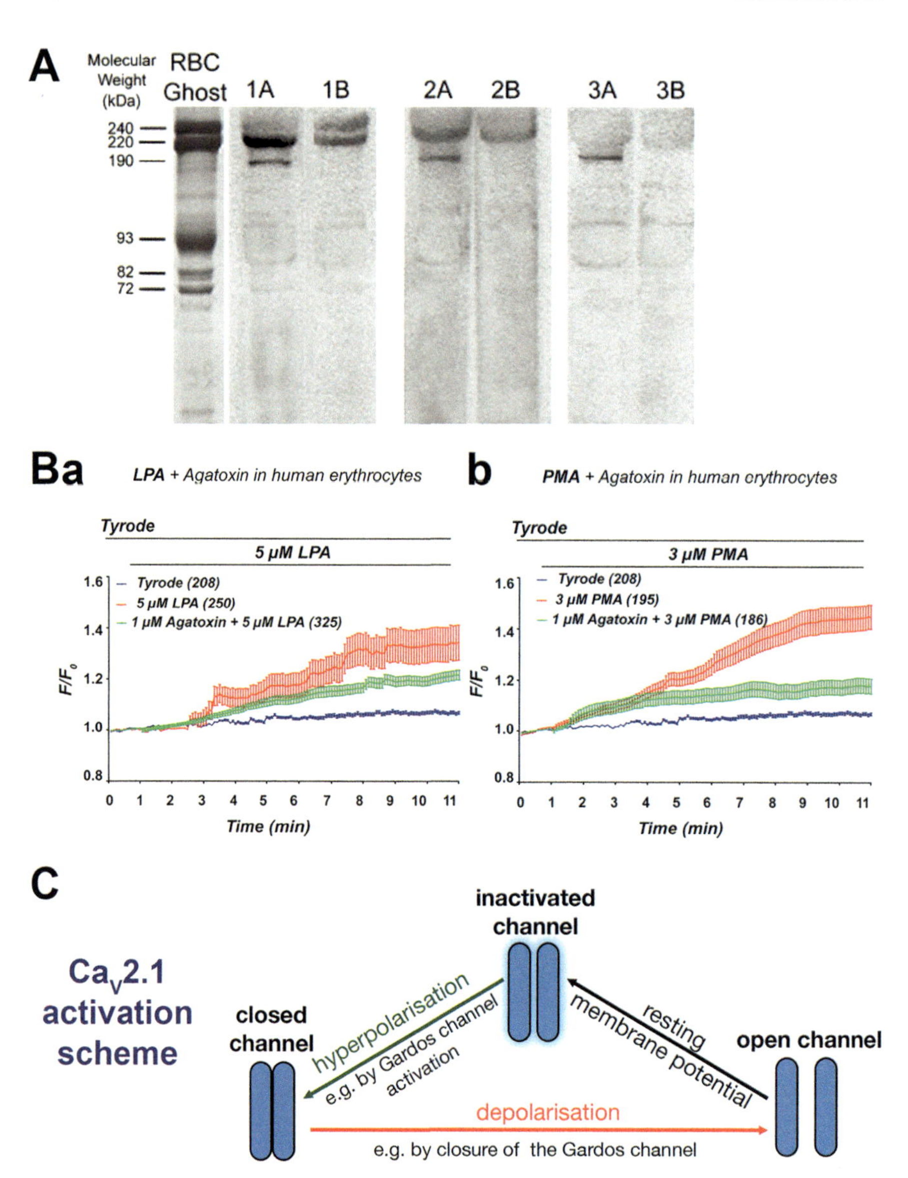

Fig. 25.7 **Ca_V2.1 in human RBCs.** (A) Western blot analysis of the 1A subunit of voltage-gated calcium channels in human erythrocyte ghosts from 3 different donors. RBC membrane proteins were separated electrophoretically at high protein loads (80 μg). After electrophoretic transfer to nitrocellulose paper, blots were stained with antibodies directed against residues 865 to 881 of the α_{1A} subunit of the rat brain voltage-gated calcium channel (lanes 1A, 2A, 3A). Because several nonspecific bands were also visualized, competition of Ca_V2.1 antibody staining using its specific peptide was also performed. For this purpose, 10 μg of the Ca_V2.1 antibody was preincubated with 10 μg of antibody-specific peptide for 1 h at 22 °C and then further incubated with the blot for 2 h at 22 °C (lanes 1B, 2B, 3B). Polypeptides with M_r of approximately 190,000 and 220,000 are characteristic of the major splicing variants of the α_{1A} subunit of Ca_V2.1 in the brain. An SDS-PAGE Coomassie blue–stained gel of RBC membranes (left lane; “RBC Ghost”) serves as an

cells are placed in tubes containing Ca^{2+}-chelating anticoagulants as exemplified in Fig. 25.8C. This experimental result is a showcase of the cation gradient dissipation associated with RBC storage lesions [100].

Fig. 25.7 (continued) approximate molecular weight marker. (B) Kinetics of lysophosphatidic acid (LPA; phospholipid released by activated platelets)-induced (Ba) and Phorbol 12-myristate 13-acetate (PMA; protein kinase C activator)-induced (Bb) Ca^{2+} entry in the presence and absence of ω-agatoxin TK. Average traces of single cells derived from live cell imaging experiments are presented as self-ratio values. Labelled lines above the traces indicate the stimulation regime. The traces are the mean values of 3 independent experiments, and the numbers in brackets at the end of the colour legend refer to the number of cells measured. (C) Activation scheme for the $Ca_V2.1$ channel modulated by underlying Gardos channel activity. Closing of the Gardos channels after their initial activation could provide the necessary conditions for subsequent depolarisation to activate $Ca_V2.1$. Since the hypothetical switching behaviour of the Gardos channel would be crucial for the activation of $Ca_V2.1$, we propose three principle modes by which this switching could occur: (i) Because channel activity is a stochastic event and because the number of Gardos channels per RBC is rather low (in the single digit numbers [114, 118]), depolarisation could be the result of stochastic Gardos channel closures. This hypothesis is supported by the rather sparse whole cell patch-clamp recordings of Gardos channel activity in human RBCs [31, 119–121]. Whole cell current traces do not show a smooth appearance but rather a flickering pattern similar to that observed with single channel recordings, especially at higher (positive and negative) membrane potentials. (ii) When looking at Gardos channel-induced changes in the membrane potential of cell populations, a gradual Ca^{2+} concentration-dependent effect can be seen [122], i.e., the hyperpolarisation observed in RBC suspensions is a gradual Ca^{2+} concentration-dependent effect. However, the abovementioned study [122] as well as another report [123] showed that the activation of the Gardos channel at the cellular level is an all-or-none response. This means that the gradual change in membrane potential would be the result of the summation of cells with open or closed Gardos channels. Taking into consideration that the Ca^{2+} pump [124] continuously operates in response to any increase in intracellular Ca^{2+} levels, one would imagine that the state of the Gardos channels is exclusively modulated by variations in intracellular Ca^{2+} concentrations. Hence, the switching behaviour of the Gardos channel would be the direct consequence of continuous variations in RBC intracellular Ca^{2+} concentrations. (iii) Localized interactions between the Gárdos channel and $Ca_V2.1$ in RBCs could occur in lipid rafts or nanodomains, as is the case with closely related ion transporters in other cell types, for example, within the fuzzy space or dyadic cleft in myocytes [125]. Although RBCs do not possess membrane-constricted subspaces, there are indications for functional compartments in the immediate vicinity of the plasma membrane [126]. Colocalization of ion channels is common in excitable cells [127, 128]. For RBCs, it is still unknown if the different ion channels colocalize or cluster to allow their interaction in nanodomains. However, in support of this idea is the observation that local activation of mechanosensitive channels (most likely Piezo 1) by patch-clamp micropipettes resulted in local activation (single-channel recordings) of the Gardos channel [124]. (Panels A, B and C are reproduced from Andrews et al. 2000 [93], Wagner-Britz et al. 2013 [94] and Kaestner et al. 2018 [95], respectively)

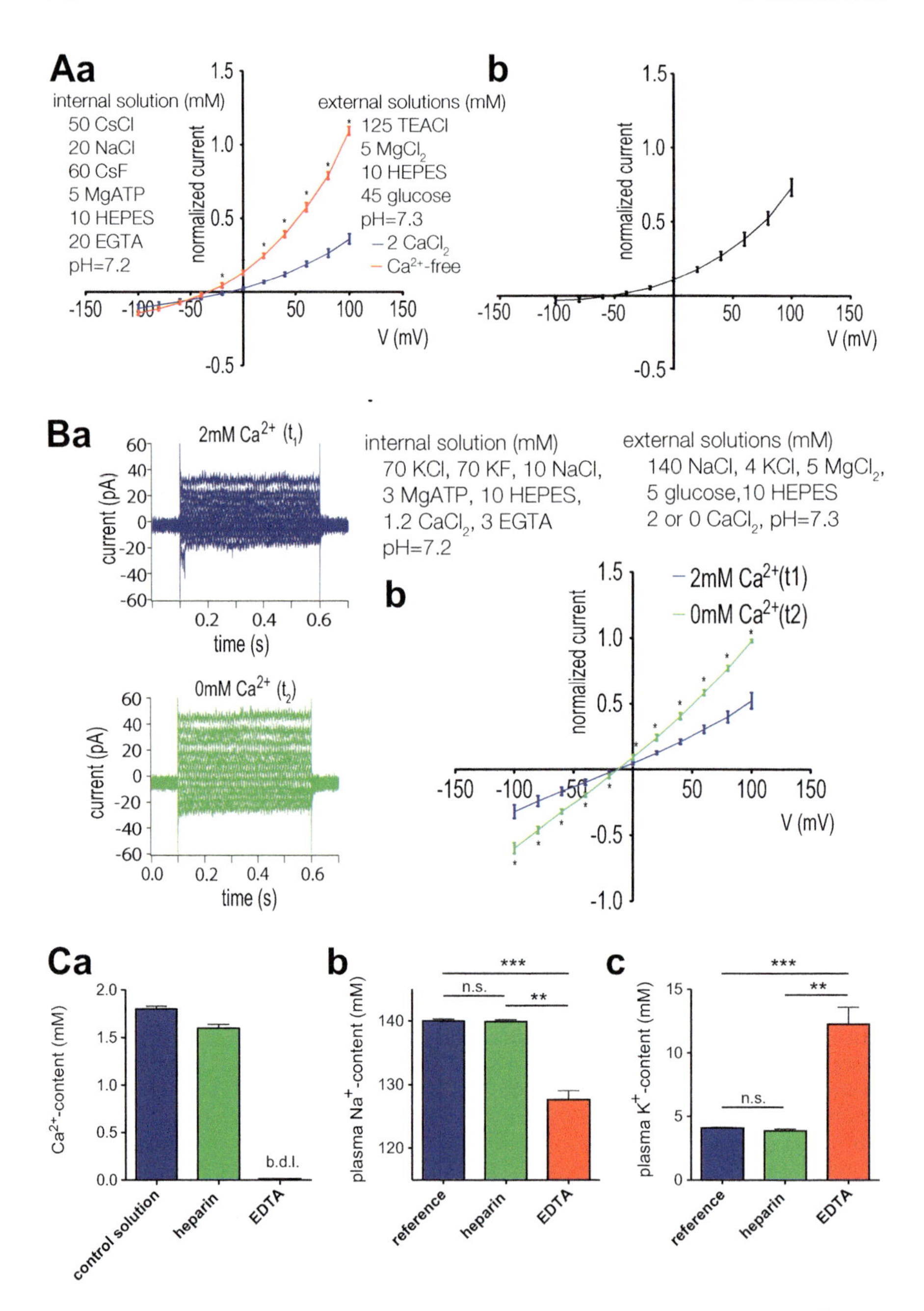

Fig. 25.8 Evidence for non-selective cation channel activated by the removal of Ca^{2+}. (A) Whole-cell patch clamp recordings in a Cs^{+}-based internal and a TEACl-based external solutions (Aa) I/V curves with 2 mM $CaCl_2$ (blue) and 0 mM $CaCl_2$ (red) in the external solution ($n = 5$). (Ab) I/V curve of the Ca^{2+} blocked current – the current recorded in 2 mM $CaCl_2$-external solution was subtracted from the current recorded in 0 mM $CaCl_2$ -external solution. Currents were elicited

25.6 Summary

The importance of Ca^{2+} in the membrane transport regulation and mediation of RBCs was early recognised. However, only in the recent years it became evident how this ion transport is related to ion channels and a correlation to molecular entities could be performed. This process is everything but finished and in particular the understanding of the interaction of several ion channels in a physiological context just started. This includes also pathophysiological conditions, on the one hand channelopathies, where a mutation of the ion channel is the direct cause of the disease, like the above described Hereditary Xerocytosis [45] and the Gárdos Channelopathy [31]. On the other hand it applies to RBC related diseases where an altered channel activity is a secondary effect like in sickle cell disease [34, 63] or thalassemia. Also these secondary effects should receive medical and pharmacologic attention because they can be crucial when it comes to the life-threatening symptoms of the disease [55]. An overview of the involvement of Ca^{2+} and Ca^{2+}-conducting channels as general components in anaemias is summarised in Fig. 25.9. However, this scheme can only be regarded as a current snapshot of our knowledge about Ca^{2+} and Ca^{2+}-conducting channels in RBCs. Further investigations on a better match between functional and molecular knowledge will arise as well as a better understanding of the activity of Ca^{2+} and Ca^{2+}-conducting channels within the signalling networks in RBCs.

Fig. 25.8 (continued) by voltage steps from −100 mV to 100 mV for 500 ms in 20 mV increments at $V_h = -30$ mV. Detailed solutions composition is given next to the graphs. Data are presented as mean ± SEM, with n being the number of cells. Significance is assessed with a paired Student's t test and set at $p < 0.05$. For better visualization, a significance anywhere below $p < 0.05$ is denoted with one star. (B) Whole-cell patch clamp recordings in physiological (a K^+-based internal and a Na^+-based external) solutions (Ba) Raw current traces from a representative RBC in an external solution containing 2 mM $CaCl_2$ at t_1 (dark blue) and 0 mM $CaCl_2$ at t_2 (green). Detailed solutions composition is given next to the current traces. (Bb) I/V curves in 2 mM $CaCl_2$ (t_1) (dark blue) and 0 mM $CaCl_2$ (t_2) (green) external solutions ($n = 7$). (C) Blood plasma ion content of healthy adults in heparin and EDTA. (Ca) Ca^{2+} content of a control aqueous non-buffered 1.8 mM $CaCl_2$ solution filled in heparin and EDTA vacutainers. The abbreviation b.d.l. denotes 'below detection limit'. (Cb) Na^+ plasma content, (Cc) K^+ plasma content of blood anticoagulated with heparin and EDTA. Measurements in heparin and EDTA were performed on blood of healthy adults ($n = 3$) collected in heparin and EDTA vacutainers, respectively and reference values were taken from Liappis, 1972 [130]. Error bars represent SEM and stars denote significances as follows: n.s for not significant, ** for $p < 0.01$ and *** for $p < 0.001$. (This figure is reproduced from Petkova-Kirowa et al. 2018 [99])

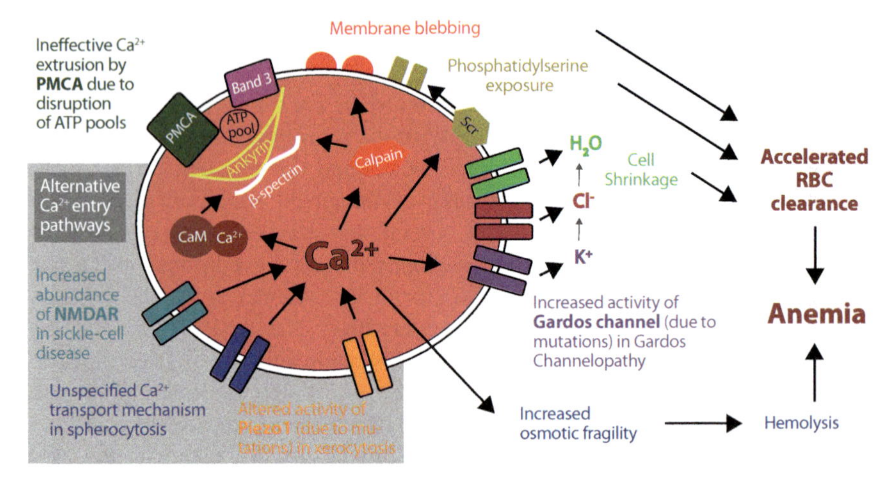

Fig. 25.9 Proposed mechanisms leading to increased intracellular Ca^{2+}levels in diseased RBCs and accordingly to accelerated clearance of cells from the blood stream. Alternative or cumulating Ca^{2+} entry pathways are highlighted with grey background: increased abundance of NMDA-receptors (NMDAR), e.g., in sickle cell disease, altered activity of Piezo1, e.g. in Hereditary Xerocytosis, increased activity of Gárdos Channel, e.g. in Gárdos Channelopathy, or unspecified Ca^{2+} transport mechanisms. Additionally, ineffective extrusion of Ca^{2+} due to disruption of ATP pools fueling the plasma membrane Ca^{2+} ATPase (PMCA) can contribute. Several downstream processes follow Ca^{2+} overload in RBCs, e.g.: activation of calmodulin by formation of the Ca^{2+}-calmodulin complex (Ca-CaM) and activation of calpain, thereby loosening the cytoskeletal structure; activation of the scramblase (Scr) leading to exposure of phosphatidylserine on the outer leaflet of the membrane; activation of the Gárdos channel followed by the efflux of K^+, Cl^- and H_2O and consecutive cell shrinkage. (This figure is reproduced from Hertz et al. 2017 [55])

References

1. Berridge MJ (1994) The biology and medicine of calcium signalling. Mol Cell Endocrinol 98:119–124
2. Berridge MJ (2006) Calcium microdomains: organization and function. Cell Calcium 40:405–412
3. Bogdanova A, Makhro A, Wang J et al (2013) Calcium in red blood cells-a perilous balance. Int J Mol Sci 14:9848–9872. https://doi.org/10.3390/ijms14059848
4. Lew VL, Tsien RY, Miner C, Bookchin RM (1982) Physiological [Ca^{2+}]i level and pump-leak turnover in intact red cells measured using an incorporated Ca chelator. Nature 298:478–481
5. Danielczok JG, Terriac E, Hertz L et al (2017) Red blood cell passage of small capillaries is associated with transient Ca^{2+}-mediated adaptations. Front Physiol 8:979. https://doi.org/10.3389/fphys.2017.00979
6. Hammer K, Ruppenthal S, Viero C et al (2010) Remodelling of Ca^{2+} handling organelles in adult rat ventricular myocytes during long term culture. J Mol Cell Cardiol 49:427–437. https://doi.org/10.1016/j.yjmcc.2010.05.010
7. Schatzmann HJ (1983) The red cell calcium pump. Annu Rev Physiol 45:303–312. https://doi.org/10.1146/annurev.ph.45.030183.001511
8. Scharff O, Foder B (1977) Low Ca^{2+} concentrations controlling two kinetic states of Ca^{2+}-ATPase from human erythrocytes. Biochim Biophys Acta 483:416–424

9. Kosk-Kosicka D, Bzdega T (1990) Effects of calmodulin on erythrocyte Ca2(+)-ATPase activation and oligomerization. Biochemistry 29:3772–3777
10. Wang KK, Roufogalis BD, Villalobo A (1990) Calpain I activates Ca^{2+} transport by the human erythrocyte plasma membrane calcium pump. Adv Exp Med Biol 269:175–180
11. Wang KK, Villalobo A, Roufogalis BD (1992) The plasma membrane calcium pump: a multiregulated transporter. Trends Cell Biol 2:46–52
12. Kosk-Kosicka D, Bzdega T (1988) Activation of the erythrocyte Ca^{2+}-ATPase by either self-association or interaction with calmodulin. J Biol Chem 263:18184–18189
13. Andrews DA, Low PS (1999) Role of red blood cells in thrombosis. Curr Opin Hematol 6:76–82
14. Kaestner L, Tabellion W, Lipp P, Bernhardt I (2004) Prostaglandin E2 activates channel-mediated calcium entry in human erythrocytes: an indication for a blood clot formation supporting process. Thromb Haemost 92:1269–1272. https://doi.org/10.1267/THRO04061269
15. Lang KS, Duranton C, Poehlmann H et al (2003) Cation channels trigger apoptotic death of erythrocytes. Cell Death Differ 10:249–256. https://doi.org/10.1038/sj.cdd.4401144
16. Kaestner L, Minetti G (2017) The potential of erythrocytes as cellular aging models. Cell Death Differ 24:1475–1477. https://doi.org/10.1038/cdd.2017.100
17. Lang KS, Lang PA, Bauer C et al (2005) Mechanisms of suicidal erythrocyte death. Cell Physiol Biochem 15:195–202
18. Galluzzi L, Vitale I, Aaronson SA et al (2018) Molecular mechanisms of cell death: recommendations of the Nomenclature Committee on Cell Death 2018. Cell Death Differ:1–56. https://doi.org/10.1038/s41418-017-0012-4
19. Makhro A, Hanggi P, Goede JS et al (2013) N-methyl D-aspartate (NMDA) receptors in human erythroid precursor cells and in circulating red blood cells contribute to the intracellular calcium regulation. Am J Physiol Cell Physiol. https://doi.org/10.1152/ajpcell.00031.2013
20. Cahalan SM, Lukacs V, Ranade SS et al (2015) Piezo1 links mechanical forces to red blood cell volume. Elife 4:e07370. https://doi.org/10.7554/eLife.07370
21. Faucherre A, Kissa K, Nargeot J et al (2014) Piezo1 plays a role in erythrocyte volume homeostasis. Haematologica 99:70–75. https://doi.org/10.3324/haematol.2013.086090
22. Hamill OP (1983) Potassium and chloride channels in red blood cells. In: Sakmann B, Neher E (eds) Single channel recording. Plenum Press, New York/London, pp 451–471
23. Hamill OP (1981) Potassium channel currents in human red blood cells. J Physiol Lond 319:97P–98P
24. Gardos G (1956) The permeability of human erythrocytes to potassium. Acta Physiol Hung 10:185–189
25. Gardos G (1958) The function of calcium in the potassium permeability of human erythrocytes. Biochim Biophys Acta 30:653–654
26. Hoffman JF, Joiner W, Nehrke K et al (2003) The hSK4 (KCNN4) isoform is the Ca^{2+}-activated K^{+} channel (Gardos channel) in human red blood cells. Proc Natl Acad Sci U S A 100:7366–7371. https://doi.org/10.1073/pnas.1232342100
27. Maher AD, Kuchel PW (2003) The Gárdos channel: a review of the Ca^{2+}-activated K^{+} channel in human erythrocytes. Int J Biochem Cell Biol 35:1182–1197
28. Grygorczyk R (1987) Temperature dependence of Ca^{2+}-activated K^{+} currents in the membrane of human erythrocytes. Biochim Biophys Acta 902:159–168
29. Rapetti-Mauss R, Lacoste C, Picard V et al (2015) A mutation in the Gardos channel is associated with hereditary xerocytosis. Blood 126:1273–1280. https://doi.org/10.1182/blood-2015-04-642496
30. Glogowska E, Lezon-Geyda K, Maksimova Y et al (2015) Mutations in the Gardos channel (KCNN4) are associated with hereditary xerocytosis. Blood 126:1281–1284. https://doi.org/10.1182/blood-2015-07-657957
31. Fermo E, Bogdanova A, Petkova-Kirova P et al (2017) "Gardos Channelopathy": a variant of hereditary stomatocytosis with complex molecular regulation. Sci Rep 7:1744. https://doi.org/10.1038/s41598-017-01591-w
32. Mankad VN (2001) Exciting new treatment approaches for pathyphysiologic mechanisms of sickle cell disease. Pediatr Pathol Mol Med 20:1–13

33. Ataga KI, Reid M, Ballas SK et al (2011) Improvements in haemolysis and indicators of erythrocyte survival do not correlate with acute vaso-occlusive crises in patients with sickle cell disease: a phase III randomized, placebo-controlled, double-blind study of the Gardos channel blocker senicapoc (ICA-17043). Br J Haematol 153:92–104. https://doi.org/10.1111/j.1365-2141.2010.08520.x
34. Bogdanova A, Makhro A, Kaestner L (2015) Calcium handling in red blood cells of sicke cell disease patients. In: Lewis ME (ed) Sickle cell disease: *Genetics, Management and Prognosis*, Nova Publishing, pp 29–59
35. Rapetti-Mauss R, Soriani O, Vinti H et al (2016) Senicapoc: a potent candidate for the treatment of a subset of Hereditary Xerocytosis caused by mutations in the Gardos channel. Haematologica 2016:149104. https://doi.org/10.3324/haematol.2016.149104
36. Christophersen P, Bennekou P (1991) Evidence for a voltage-gated, non-selective cation channel in the human red cell membrane. Biochim Biophys Acta 1065:103–106
37. Bennekou P (1993) The voltage-gated non-selective cation channel from human red cells is sensitive to acetylcholine. Biochim Biophys Acta 1147:165–167
38. Kaestner L, Christophersen P, Bernhardt I, Bennekou P (2000) The non-selective voltage-activated cation channel in the human red blood cell membrane: reconciliation between two conflicting reports and further characterisation. Bioelectrochemistry 52:117–125
39. Kaestner L (2011) Cation channels in erythrocytes – historical and future perspective. Open Biol J 4:27–34
40. Bouyer G, Thomas S, Egée S (2012) Patch-clamp analysis of membrane transport in erythrocytes. In: Patch clamp technique. InTech, Rijeka, pp 171–202
41. Bouyer G, Cueff A, Egée S et al (2011) Erythrocyte peripheral type benzodiazepine receptor/voltage-dependent anion channels are upregulated by Plasmodium falciparum. Blood 118:2305–2312. https://doi.org/10.1182/blood-2011-01-329300
42. Moroni M, Servin-Vences MR, Fleischer R et al (2018) Voltage gating of mechanosensitive PIEZO channels. Nat Commun 9:1096. https://doi.org/10.1038/s41467-018-03502-7
43. Kaestner L, Egee S (2018) Commentary: voltage gating of mechanosensitive PIEZO channels. Front Physiol 9:1565
44. Andersson T (2010) Exploring voltage-dependent ion channels in silico by hysteretic conductance. Math Biosci 226:16–27. https://doi.org/10.1016/j.mbs.2010.03.004
45. Zarychanski R, Schulz VP, Houston BL et al (2012) Mutations in the mechanotransduction protein PIEZO1 are associated with hereditary xerocytosis. Blood 120:1908–1915. https://doi.org/10.1182/blood-2012-04-422253
46. Bae C, Gnanasambandam R, Nicolai C et al (2013) Xerocytosis is caused by mutations that alter the kinetics of the mechanosensitive channel PIEZO1. Proc Natl Acad Sci U S A 110:E1162–E1168. https://doi.org/10.1073/pnas.1219777110
47. Kaestner L (2015) Channelizing the red blood cell: molecular biology competes with patch-clamp. Front Mol Biosci 2:46. https://doi.org/10.3389/fmolb.2015.00046
48. Syeda R, Xu J, Dubin AE et al (2015) Chemical activation of the mechanotransduction channel Piezo1. Elife. https://doi.org/10.7554/eLife.07369
49. Rotordam GM, Fermo E, Becker N et al (2019) A novel gain-of-function mutation of Piezo1 is functionally affirmed in red blood cells by high-throughput patch clamp. Haematologica 104(5). https://doi.org/10.3324/haematol.2018.201160
50. Lew VL, Tiffert T (2017) On the mechanism of human red blood cell longevity: roles of calcium, the sodium pump, PIEZO1, and Gardos channels. Front Physiol 8:977. https://doi.org/10.3389/fphys.2017.00977
51. Albuisson J, Murthy SE, Bandell M et al (2013) Dehydrated hereditary stomatocytosis linked to gain-of-function mutations in mechanically activated PIEZO1 ion channels. Nat Commun 4:1884. https://doi.org/10.1038/ncomms2899
52. Andolfo I, Alper SL, De Franceschi L et al (2013) Multiple clinical forms of dehydrated hereditary stomatocytosis arise from mutations in PIEZO1. Blood 121:3925–3935. https://doi.org/10.1182/blood-2013-02-482489
53. Archer NM, Shmukler BE, Andolfo I et al (2014) Hereditary xerocytosis revisited. Am J Hematol 89:1142–1146. https://doi.org/10.1002/ajh.23799

54. Glogowska E, Schneider ER, Maksimova Y et al (2017) Novel mechanisms of PIEZO1 dysfunction in hereditary xerocytosis. Blood 130:1845–1856. https://doi.org/10.1182/blood-2017-05-786004
55. Hertz L, Huisjes R, Llaudet-Planas E et al (2017) Is increased intracellular calcium in red blood cells a common component in the molecular mechanism causing anemia? Front Physiol 8:673. https://doi.org/10.3389/fphys.2017.00673
56. Steffen P, Jung A, Nguyen DB et al (2011) Stimulation of human red blood cells leads to Ca^{2+}-mediated intercellular adhesion. Cell Calcium 50:54–61. https://doi.org/10.1016/j.ceca.2011.05.002
57. Kaestner L, Steffen P, Nguyen DB et al (2012) Lysophosphatidic acid induced red blood cell aggregation in vitro. Bioelectrochemistry 87:89–95. https://doi.org/10.1016/j.bioelechem.2011.08.004
58. Chung SM, Bae ON, Lim KM et al (2007) Lysophosphatidic acid induces thrombogenic activity through phosphatidylserine exposure and procoagulant microvesicle generation in human erythrocytes. Arterioscler Thromb Vasc Biol 27:414–421
59. Lew VL, Ortiz OE, Bookchin RM (1997) Stochastic nature and red cell population distribution of the sickling-induced Ca^{2+} permeability. J Clin Invest 99:2727–2735. https://doi.org/10.1172/JCI119462
60. Browning JA, Robinson HC, Ellory JC, Gibson JS (2007) Deoxygenation-induced non-electrolyte pathway in red cells from sickle cell patients. Cell Physiol Biochem 19:165–174. https://doi.org/10.1159/000099204
61. Ma Y-L, Rees DC, Gibson JS, Ellory JC (2012) The conductance of red blood cells from sickle cell patients: ion selectivity and inhibitors. J Physiol Lond 590:2095–2105. https://doi.org/10.1113/jphysiol.2012.229609
62. Bae C, Sachs F, Gottlieb PA (2011) The mechanosensitive ion channel Piezo1 is inhibited by the peptide GsMTx4. Biochemistry 50:6295–6300. https://doi.org/10.1021/bi200770q
63. Hanggi P, Makhro A, Gassmann M et al (2014) Red blood cells of sickle cell disease patients exhibit abnormally high abundance of N-methyl D-aspartate receptors mediating excessive calcium uptake. Br J Haematol 167:252–264. https://doi.org/10.1111/bjh.13028
64. Makhro A, Wang J, Vogel J et al (2010) Functional NMDA receptors in rat erythrocytes. Am J Physiol Cell Physiol 298:C1315–C1325. https://doi.org/10.1152/ajpcell.00407.2009
65. Bogdanova A, Makhro A, Goede J et al (2009) NMDA receptors in mammalien erythrocytes. Clin Biochem 42:1858–1859
66. Makhro A, Kaestner L, Bogdanova A (2017) NMDA receptor activity in circulating red blood cells: methods of detection. Methods Mol Biol 1677:265–282, 484. https://doi.org/10.1007/978-1-4939-7321-7_15
67. Hanggi P, Telezhkin V, Kemp PJ et al (2015) Functional plasticity of the N-methyl-d-aspartate receptor in differentiating human erythroid precursor cells. Am J Physiol Cell Physiol 308:C993–C1007. https://doi.org/10.1152/ajpcell.00395.2014
68. Hegemann I, Sasselli C, Valeri F, Makhro A, Müller R, Bogdanova A, Manz MG, Gassmann M, Goede JS. Memantine treatment is well tolerated by sickle cell patients and improves erythrocyte stability: phase II study MemSID (submitted)
69. Bogdanova A, Makhro A, Hegemann I, Seiler E, Bogdanov N, Simionato G, Kaestner L, Claveria V, Saselli C, Torgeson P, Manz M, Goede J, Gassmann M. Improved maturation and increased stability of red blood cells of sickle cell patients on memantine treatment (submitted)
70. Makhro A, Haider T, Wang J et al (2016) Comparing the impact of an acute exercise bout on plasma amino acid composition, intraerythrocytic Ca^{2+} handling, and red cell function in athletes and untrained subjects. Cell Calcium 60:235–244. https://doi.org/10.1016/j.ceca.2016.05.005
71. Thomas SLY, Bouyer G, Cueff A et al (2011) Ion channels in human red blood cell membrane: actors or relics? Blood Cells Mol Dis 46:261–265. https://doi.org/10.1016/j.bcmd.2011.02.007
72. Schein SJ, Colombini M, Finkelstein A (1976) Reconstitution in planar lipid bilayers of a voltage-dependent anion-selective channel obtained from paramecium mitochondria. J Membr Biol 30:99–120

73. Thinnes FP, Flörke H, Winkelbach H et al (1994) Channel active mammalian porin, purified from crude membrane fractions of human B lymphocytes or bovine skeletal muscle, reversibly binds the stilbene-disulfonate group of the chloride channel blocker DIDS. Biol Chem Hoppe Seyler 375:315–322
74. Marginedas-Freixa I, Hattab C, Bouyer G et al (2016) TSPO ligands stimulate ZnPPIX transport and ROS accumulation leading to the inhibition of P. falciparum growth in human blood. Sci Rep 6:33516. https://doi.org/10.1038/srep33516
75. McEnery MW, Snowman AM, Trifiletti RR, Snyder SH (1992) Isolation of the mitochondrial benzodiazepine receptor: association with the voltage-dependent anion channel and the adenine nucleotide carrier. Proc Natl Acad Sci U S A 89:3170–3174
76. Le Fur G, Vaucher N, Perrier ML et al (1983) Differentiation between two ligands for peripheral benzodiazepine binding sites, [3H]RO5-4864 and [3H]PK 11195, by thermodynamic studies. Life Sci 33:449–457
77. Olson JM, Ciliax BJ, Mancini WR, Young AB (1988) Presence of peripheral-type benzodiazepine binding sites on human erythrocyte membranes. Eur J Pharmacol 152:47–53
78. Canat X, Carayon P, Bouaboula M et al (1993) Distribution profile and properties of peripheral-type benzodiazepine receptors on human hemopoietic cells. Life Sci 52:107–118
79. Shoshan-Barmatz V, De Pinto V, Zweckstetter M et al (2010) VDAC, a multi-functional mitochondrial protein regulating cell life and death. Mol Asp Med 31:227–285. https://doi.org/10.1016/j.mam.2010.03.002
80. Moran O, Sorgato MC (1992) High-conductance pathways in mitochondrial membranes. J Bioenerg Biomembr 24:91–98
81. Benz R (1994) Permeation of hydrophilic solutes through mitochondrial outer membranes: review on mitochondrial porins. Biochim Biophys Acta 1197:167–196
82. Hodge T, Colombini M (1997) Regulation of metabolite flux through voltage-gating of VDAC channels. J Membr Biol 157:271–279
83. Gincel D, Silberberg SD, Shoshan-Barmatz V (2000) Modulation of the voltage-dependent anion channel (VDAC) by glutamate. J Bioenerg Biomembr 32:571–583
84. Báthori G, Csordás G, Garcia-Perez C et al (2006) Ca^{2+}-dependent control of the permeability properties of the mitochondrial outer membrane and voltage-dependent anion-selective channel (VDAC). J Biol Chem 281:17347–17358. https://doi.org/10.1074/jbc.M600906200
85. Tong Q, Hirschler-Laszkiewicz I, Zhang W et al (2008) TRPC3 is the erythropoietin-regulated calcium channel in human erythroid cells. J Biol Chem 283:10385–10395. https://doi.org/10.1074/jbc.M710231200
86. Hirschler-Laszkiewicz I, Tong Q, Conrad K et al (2009) TRPC3 activation by erythropoietin is modulated by TRPC6. J Biol Chem 284:4567–4581. https://doi.org/10.1074/jbc.M804734200
87. Kucherenko YV, Bhavsar SK, Grischenko VI et al (2010) Increased cation conductance in human erythrocytes artificially aged by glycation. J Membr Biol 235:177–189. https://doi.org/10.1007/s00232-010-9265-2
88. Danielczok J, Hertz L, Ruppenthal S et al (2017) Does erythropoietin regulate TRPC channels in red blood cells? Cell Physiol Biochem 41:1219–1228. https://doi.org/10.1159/000464384
89. Foller M, Kasinathan RS, Koka S et al (2008) TRPC6 contributes to the Ca^{2+} leak of human erythrocytes. Cell Physiol Biochem 21:183–192
90. Dietrich A, Gudermann T (2014) TRPC6: physiological function and pathophysiological relevance. Handb Exp Pharmacol 222:157–188. https://doi.org/10.1007/978-3-642-54215-2_7
91. Pinet C, Antoine S, Filoteo AG et al (2002) Reincorporated plasma membrane Ca^{2+}-ATPase can mediate B-type Ca^{2+} channels observed in native membrane of human red blood cells. J Membr Biol 187:185–201. https://doi.org/10.1007/s00232-001-0163-5
92. Romero PJ, Romero EA, Mateu D et al (2006) Voltage-dependent calcium channels in young and old human red cells. Cell Biochem Biophys 46:265–276. https://doi.org/10.1385/CBB:46:3:265
93. Andrews DA, Yang L, Low PS (2002) Phorbol ester stimulates a protein kinase C-mediated agatoxin-TK-sensitive calcium permeability pathway in human red blood cells. Blood 100:3392–3399

94. Wagner-Britz L, Wang J, Kaestner L, Bernhardt I (2013) Protein kinase Cα and P-type Ca channel CaV2.1 in red blood cell calcium signalling. Cell Physiol Biochem 31:883–891. https://doi.org/10.1159/000350106
95. Kaestner L, Wang X, Hertz L, Bernhardt I (2018) Voltage-activated ion channels in non-excitable cells – a viewpoint regarding their physiological justification. Front Physiol 9:450
96. Tiffert T, Bookchin RM, Lew VL (2003) Calcium homeostasis in normal and abnormal human red cells. In: Bernhardt I, Ellory C (eds) Red cell membrane transport in health and disease. Springer, Berlin, pp 373–405
97. Catterall WA (2011) Voltage-gated calcium channels. Cold Spring Harb Perspect Biol 3:a003947. https://doi.org/10.1101/cshperspect.a003947
98. Catterall WA (2000) Structure and regulation of voltage-gated Ca^{2+} channels. Annu Rev Cell Dev Biol 16:521–555. https://doi.org/10.1146/annurev.cellbio.16.1.521
99. Petkova-Kirova P, Hertz L, Makhro A et al (2018) A previously unrecognized Ca^{2+}-inhibited non-selective cation channel in red blood cells. HemaSphere, 2:e146
100. Flatt JF, Bawazir WM, Bruce LJ (2014) The involvement of cation leaks in the storage lesion of red blood cells. Front Physiol 5:214. https://doi.org/10.3389/fphys.2014.00214
101. Lipp P, Kaestner L (2014) Detecting calcium in cardiac muscle: fluorescence to dye for. Am J Physiol Heart Circ Physiol 307:H1687–H1690. https://doi.org/10.1152/ajpheart.00468.2014
102. Kaestner L, Tabellion W, Weiss E et al (2006) Calcium imaging of individual erythrocytes: problems and approaches. Cell Calcium 39:13–19. https://doi.org/10.1016/j.ceca.2005.09.004
103. Jarrett HW, Kyte J (1979) Human erythrocyte calmodulin. Further chemical characterization and the site of its interaction with the membrane. J Biol Chem 254:8237–8244
104. Leinders T, van Kleef RG, Vijverberg HP (1992) Single Ca^{2+}-activated K^+ channels in human erythrocytes: Ca^{2+} dependence of opening frequency but not of open lifetimes. Biochim Biophys Acta 1112:67–74
105. Stout JG, Zhou Q, Wiedmer T, Sims PJ (1998) Change in conformation of plasma membrane phospholipid scramblase induced by occupancy of its Ca^{2+} binding site. Biochemistry 37:14860–14866. https://doi.org/10.1021/bi9812930
106. Woon LA, Holland JW, Kable EP, Roufogalis BD (1999) Ca^2+ sensitivity of phospholipid scrambling in human red cell ghosts. Cell Calcium 25:313–320
107. Bitbol M, Fellmann P, Zachowski A, Devaux PF (1987) Ion regulation of phosphatidylserine and phosphatidylethanolamine outside-inside translocation in human erythrocytes. Biochim Biophys Acta 904:268–282
108. Murakami T, Hatanaka M, Murachi T (1981) The cytosol of human erythrocytes contains a highly Ca^{2+}-sensitive thiol protease (calpain I) and its specific inhibitor protein (calpastatin). J Biochem 90:1809–1816
109. Salamino F, De Tullio R, Mengotti P et al (1993) Site-directed activation of calpain is promoted by a membrane-associated natural activator protein. Biochem J 290(Pt 1):191–197
110. Bergamini CM, Signorini M (1993) Studies on tissue transglutaminases: interaction of erythrocyte type-2 transglutaminase with GTP. Biochem J 291(Pt 1):37–39
111. Almaraz L, García-Sancho J, Lew VL (1988) Calcium-induced conversion of adenine nucleotides to inosine monophosphate in human red cells. J Physiol Lond 407:557–567
112. Kohout SC, Corbalán-García S, Torrecillas A et al (2002) C2 domains of protein kinase C isoforms alpha, beta, and gamma: activation parameters and calcium stoichiometries of the membrane-bound state. Biochemistry 41:11411–11424. https://doi.org/10.1021/bi026401k
113. Schwarz W, Grygorczyk R, Hof D (1989) Recording single-channel currents from human red cells. Methods Enzymol 173:112–121
114. Grygorczyk R, Schwarz W, Passow H (1984) Ca^{2+}-activated K^+ channels in human red cells. Comparison of single-channel currents with ion fluxes. Biophys J 45:693–698. https://doi.org/10.1016/S0006-3495(84)84211-3
115. Grygorczyk R, Schwarz W (1985) Ca^{2+}-activated K^+ permeability in human erythrocytes: modulation of single-channel events. Eur Biophys J 12:57–65

116. Leinders T, van Kleef RG, Vijverberg HP (1992) Distinct metal ion binding sites on Ca^{2+}-activated K^+ channels in inside-out patches of human erythrocytes. Biochim Biophys Acta 1112:75–82
117. Dunn PM (1998) The action of blocking agents applied to the inner face of Ca^{2+}-activated K^+ channels from human erythrocytes. J Membr Biol 165:133–143
118. Wolff D, Cecchi X, Spalvins A, Canessa M (1988) Charybdotoxin blocks with high affinity the Ca-activated K^+ channel of Hb A and Hb S red cells: individual differences in the number of channels. J Membr Biol 106:243–252
119. Qadri SM, Kucherenko Y, Lang F (2011) Beauvericin induced erythrocyte cell membrane scrambling. Toxicology 283:24–31. https://doi.org/10.1016/j.tox.2011.01.023
120. Kucherenko Y, Zelenak C, Eberhard M et al (2012) Effect of casein kinase 1α activator pyrvinium pamoate on erythrocyte ion channels. Cell Physiol Biochem 30:407–417. https://doi.org/10.1159/000339034
121. Kucherenko YV, Wagner-Britz L, Bernhardt I, Lang F (2013) Effect of chloride channel inhibitors on cytosolic Ca^{2+} levels and Ca^{2+}-activated K^+ (Gardos) channel activity in human red blood cells. J Membr Biol 246:315–326. https://doi.org/10.1007/s00232-013-9532-0
122. Baunbaek M, Bennekou P (2008) Evidence for a random entry of Ca^{2+} into human red cells. Bioelectrochemistry 73:145–150. https://doi.org/10.1016/j.bioelechem.2008.04.006
123. Seear RV, Lew VL (2011) IKCa agonist (NS309)-elicited all-or-none dehydration response of human red blood cells is cell-age dependent. Cell Calcium 50:444–448. https://doi.org/10.1016/j.ceca.2011.07.005
124. Schatzmann HJ (1973) Dependence on calcium concentration and stoichiometry of the calcium pump in human red cells. J Physiol Lond 235:551–569
125. Lines GT, Sande JB, Louch WE et al (2006) Contribution of the Na^+/Ca^{2+} exchanger to rapid Ca^{2+} release in cardiomyocytes. Biophys J 91:779–792. https://doi.org/10.1529/biophysj.105.072447
126. Chu H, Puchulu-Campanella E, Galan JA et al (2012) Identification of cytoskeletal elements enclosing the ATP pools that fuel human red blood cell membrane cation pumps. Proc Natl Acad Sci U S A 109:12794–12799. https://doi.org/10.1073/pnas.1209014109
127. Bers DM (2002) Cardiac excitation-contraction coupling. Nature 415:198–205. https://doi.org/10.1038/415198a
128. Rasband MN, Shrager P (2000) Ion channel sequestration in central nervous system axons. J Physiol Lond 525(Pt 1):63–73. https://doi.org/10.1111/j.1469-7793.2000.00063.x
129. Dyrda A, Cytlak U, Ciuraszkiewicz A et al (2010) Local membrane deformations activate Ca^{2+}-dependent K^+ and anionic currents in intact human red blood cells. PLoS One 5:e9447. https://doi.org/10.1371/journal.pone.0009447
130. Liappis N (1972) Sodium-, potassium- and chloride-concentrations in the serum of infants, children and adults. Monatsschr Kinderheilkd 120:138–142

Chapter 26
Regulation of Multifunctional Calcium/Calmodulin Stimulated Protein Kinases by Molecular Targeting

Kathryn Anne Skelding and John A. P. Rostas

Abstract Multifunctional calcium/calmodulin-stimulated protein kinases control a broad range of cellular functions in a multitude of cell types. This family of kinases contain several structural similarities and all are regulated by phosphorylation, which either activates, inhibits or modulates their kinase activity. As these protein kinases are widely or ubiquitously expressed, and yet regulate a broad range of different cellular functions, additional levels of regulation exist that control these cell-specific functions. Of particular importance for this specificity of function for multifunctional kinases is the expression of specific binding proteins that mediate molecular targeting. These molecular targeting mechanisms allow pools of kinase in different cells, or parts of a cell, to respond differently to activation and produce different functional outcomes.

Keywords CaMKK · CaMKI · CaMKII · CaMKIV · Casein kinase I · Targeting

Abbreviations

α-KAP	αCaMKII anchoring protein
AMPK	AMP-activated protein kinase

K. A. Skelding
Priority Research Centre for Cancer Research, Innovation and Translation, School of Biomedical Sciences and Pharmacy, Faculty of Health and Medicine, The University of Newcastle, Callaghan, NSW, Australia

Hunter Medical Research Institute, New Lambton Heights, NSW, Australia

J. A. P. Rostas
Hunter Medical Research Institute, New Lambton Heights, NSW, Australia

Priority Research Centre for Brain and Mental Health, and Priority Research Centre for Stroke and Brain Injury, School of Biomedical Sciences and Pharmacy, Faculty of Health and Medicine, The University of Newcastle, Callaghan, NSW, Australia
e-mail: John.Rostas@newcastle.edu.au

© Springer Nature Switzerland AG 2020

M. S. Islam (ed.), *Calcium Signaling*, Advances in Experimental Medicine and Biology 1131, https://doi.org/10.1007/978-3-030-12457-1_26

ATP	Adenosine triphosphate
Ca^{2+}	Calcium ions
CaM	Calmodulin
CaMK	Calcium/calmodulin stimulated protein kinases
CaMKI	Calcium/calmodulin stimulated protein kinase I
CaMKII	Calcium/calmodulin stimulated protein kinase II
CaMKIV	Calcium/calmodulin stimulated protein kinase IV
CaMKK	Calcium/calmodulin stimulated protein kinase kinase
CK1	Casein kinase 1
CLK2	CDC-like kinase 2
GSK-3	Glycogen synthase kinase 3
PKA	cAMP-dependent protein kinase
PKB	Protein kinase B; Akt
PSD	Post-synaptic density
S	Serine
T	Threonine

26.1 Introduction

Calcium is a major second messenger in all cells, and is integral in many important signalling pathways. Changes in intracellular Ca^{2+} regulate many biological processes, including neurotransmitter release, gene expression, and the cell cycle. Though free Ca^{2+} can activate a number of proteins directly (for example myosin, phospholipase A_2, and protein kinase C), it regulates the activity of many enzymes indirectly via a number of low molecular weight Ca^{2+} binding proteins, the most abundant of which is calmodulin (CaM). CaM consists of two globular lobes, each of which contain two Ca^{2+}-binding sites. Binding of Ca^{2+} dramatically changes the conformation of CaM, allowing Ca^{2+}/CaM to interact with a variety of other proteins, including several classes of protein kinases.

Many proteins that bind Ca^{2+}/CaM do so via an α-helical region consisting of approximately 20 amino acids, which contain positively charged amino acids among hydrophobic residues. There are two classes of CaM binding motif [1]. The ***IQ motif*** (IQXXXRGXXXR) is present in proteins that bind CaM in the absence of Ca^{2+}. The majority, if not all, of the proteins that contain this motif are not enzymes, and appear to limit the concentration of diffusible CaM during periods of low intracellular Ca^{2+}. The second class of motifs are related to each other, and indicate CaM binding in the presence of Ca^{2+}. These motifs include ***1-12***, ***1-14***, ***1-5-10***, and ***1-8-14*** (named based on the conserved hydrophobic residues within these motifs). However, several identified/putative CaM binding sites have sequence motifs that are called ***unclassified*** because they do not conform to either of the preceding sequence motifs.

Ca^{2+}/CaM stimulated protein kinases are classified based on their substrate specificity, and there are two main types: ***restricted*** kinases, which only phosphorylate one, or a small number, of substrates, and ***multifunctional*** kinases, which have broad substrate specificity and regulate multiple functions in the same and different cell types.

There are three main families of restricted Ca^{2+}/CaM stimulated protein kinases: phosphorylase kinase (PhK), elongation factor 2 kinase (eEF2K) and myosin light chain kinase (MLCK). These kinase families do not share a common protein domain structure and, though they are all regulated by multiple mechanisms, only MLCK appears to be controlled by molecular targeting. For a description of the molecular structures and control mechanisms of the restricted kinases, the reader is referred to our previous review [2]. This review will focus on the multifunctional kinases and the molecular regulatory mechanisms that allow these widely expressed kinases to produce functional responses that are both stimulus- and cell type-specific.

Multifunctional kinases control a plethora of functions within multiple cell types and, consequently, there are multiple levels of control that regulate the functions of these protein kinases. The most basic method of controlling kinase function is via the regulation of Ca^{2+} dynamics, specifically the frequency, amplitude and duration of oscillations in the intracellular concentration of Ca^{2+}. This is most commonly controlled by ion channels and many kinases can be directly regulated by intracellular Ca^{2+} fluxes. For example, the multifunctional Ca^{2+}/CaM-stimulated protein kinase II (CaMKII) can translate the frequency of Ca^{2+} spikes into corresponding amounts of kinase activity [3]. However, several additional mechanisms exist that produce extra forms of control of kinase activity. Modulation of the response to changes in Ca^{2+} can be controlled by expression of different splice forms of the kinase which are differentially sensitive to the various control mechanisms, phosphorylation, or by the kinase becoming autonomously active (i.e. no longer require Ca^{2+}/CaM for activity) [4–6]. Another level of control has also been identified that provides both temporal, tissue-specific and cellular site-specific control of kinase function. This mechanism is termed 'molecular targeting' and involves the interactions between kinases and specific binding proteins. This chapter will examine the role of molecular regulation and targeting in controlling the function of multifunctional Ca^{2+}/CaM stimulated protein kinases (Table 26.1). We will focus particularly on CaMKII because it has been most extensively studied with respect to molecular targeting.

26.2 Multifunctional Calcium/Calmodulin Stimulated Protein Kinases

26.2.1 Two Families of Kinases with Homologous Domain Structures

The multifunctional calcium/calmodulin stimulated protein kinases can be classified into two broad families based on the homology of their domain structures

Table 26.1 Structural and functional properties of multifunctional Ca^{2+}/CaM-stimulated kinases

	Multifunctional Ca^{2+}/CaM-stimulated kinases				
	CaMKK	CaMKI	CaMKIV	CaMKII	CK1
Genes	2 (*CAMKK1*, *CAMKK2*)	4 (*CAMK1A*, *CAMK1B*, *CAMK1G*, *CAMK1D*)	1 (*CAMK4*)	4 (*CAMK2A*, *CAMK2B*, *CAMK2G*, *CAMK2D*)	7 (*CSNK1A1*, *CSNK1B*, *CSNK1G1*, *CSNK1G2*, *CSNK1G3*, *CSNK1D*, *CSNK1E*)
Splice Variants	Multiple	Multiple	At least 1	Multiple	Multiple
Subunit (kDa)	54–68	38–53	65–67	50–60	37–51
Structure	Monomer	Monomer	Monomer	Multimer	Monomer
CaM Binding Motif	Unclassified	1-14 motif	1-8-14 motif	1-5-10 motif	Unclassified
Contains an Autoinhibitory Region?	Yes	Yes	Yes	Yes	Yes (proposed)
Expression	Neuronal, immune, testis	Ubiquitous	Neuronal, immune, testis	Ubiquitous	Ubiquitous
Regulated by Targeting?	Yes	Yes	Yes	Yes	Yes
Requirement for Activation	Ca^{2+}/CaM	Ca^{2+}/CaM	Ca^{2+}/CaM,and phosphorylation	Ca^{2+}/CaM	Constitutively active
Capable of Autonomous Activity?	Yes	No	Yes	Yes	Yes
Number of Phosphorylation Sites	Multiple	T174 to T180 (depending on isoform)	T196 and S332 (CaMKIVα)	Multiple	Multiple
Regulation by Phosphorylation	Phosphorylation by PKA (α) or CDK, GSK (β) inhibitors kinase activity. Autophosphorylation at S74 may regulate targeting	Phosphorylation by CaMKK activates kinase activity	Phosphorylation by CaMKK activates kinase activity	Autophosphorylation regulates both kinase activity (inhibitory and activating) and targeting	Autophosphorylation inhibits kinase activity

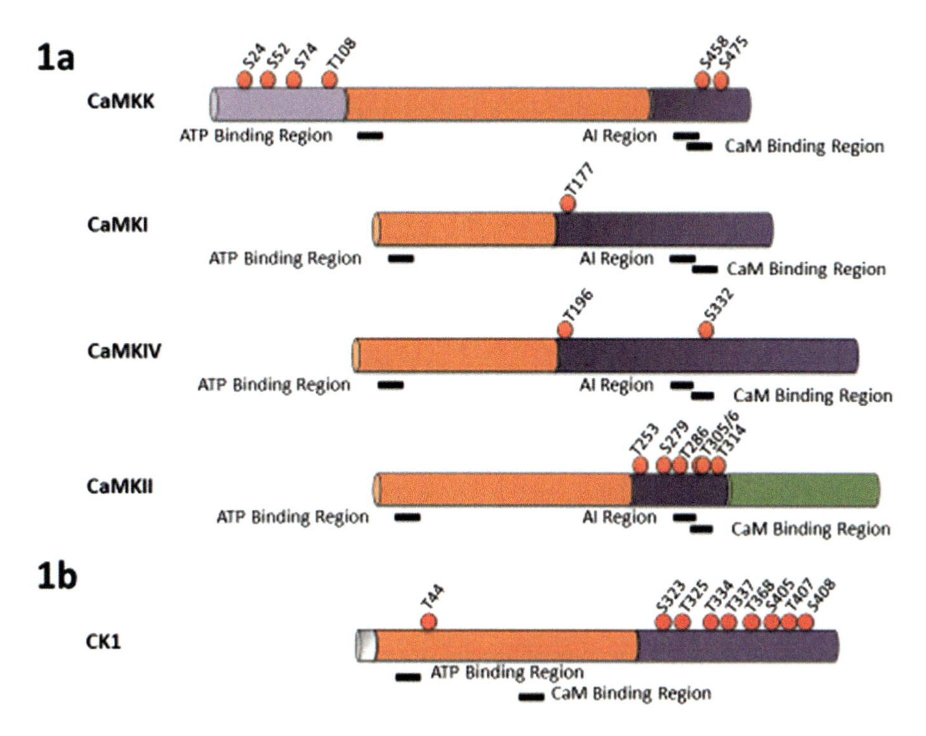

Fig. 26.1 Schematic representation, drawn to relative scale, of the domain structures of the multifunctional Ca^{2+}/CaM stimulated protein kinases. Domain structures of the α isoforms of the four homologous kinases: CaMKK, CaMKI, CaMKIV and CaMKII (**a**) and the ε isoform of casein kinase I (CK1) (**b**). Each kinase has a similar domain structure illustrated as catalytic (orange), regulatory (purple), and putative regulatory (lilac) domains. CaMKII also has an association domain (green). The major characterised phosphorylation sites (red balls) are also indicated. Black bars indicate the locations of ATP binding regions, autoinhibitory (AI) pseudosubstrate regions and calcium/calmodulin (CaM) binding regions

(Fig. 24.1a, b). Each of these kinases in both families has a regulatory domain (purple) and a catalytic domain (orange) linked to its N-terminus. Calcium/calmodulin stimulated protein kinase II (CaMKII) is the only member of these families that is multimeric (Table 26.1) and its multimeric structure is held together by interactions between the association domains (green) of the subunits which are attached to the C-terminus of their regulatory domains (Fig. 26.1).

The four kinases which form one family (Fig. 24.1a) – calcium/calmodulin stimulated protein kinase kinase (CaMKK) and calcium/calmodulin stimulated protein kinase I, IV and II (CaMKI, CaMKIV and CaMKII) – all have the following common features in their regulatory domains: (i) an autoinhibitory (AI) pseudosubstrate region which acts like a substrate and binds to the active site of the kinase, thereby inhibiting it until the binding of calcium/calmodulin causes the displacement of the autoinhibitory pseudosubstrate region and activates the kinase; (ii) a calcium/calmodulin binding region that partially overlaps with the autoinhibitory region; and (iii) one or more phosphorylation sites (red balls) that modify the activity of the kinase but are not necessary for kinase activity. The autoinhibitory and calcium/calmodulin binding regions provide an obvious area of homology and point for alignment for the kinases in Fig. 24.1a. There is no consensus in the literature on the precise extents of the regulatory domains. We have chosen to define the extent of the regulatory domains in CaMKI and CaMKIV based on CaMKII whose regulatory mechanisms have been characterised most extensively. For the purposes of Fig. 24.1a, we have placed the N-terminal boundary of the regulatory domain of CaMKII to include T253, whose phosphorylation has been shown to regulate molecular targeting in vivo, and the C-terminal boundary to include the first of the four splice sites in the C terminal part of CaMKII because the functional differences between the alternative spice variants in this region are all related to regulatory properties. The N termini of the regulatory domains of CaMKI and CaMKIV have been chosen to include T196 in CaMKIV and T177 in CaMKI because phosphorylation of these sites has been shown to regulate kinase activity. As CaMKK, CaMKI and CaMKIV are monomeric we have assumed that the regulatory domains of these kinases extend to their C termini. As the N terminal domain of CaMKK contains three phosphorylation sites that regulate CaMKK activity, we have designated this domain as a putative second regulatory domain and coloured it lilac to differentiate it from the well-established C-terminal regulatory domain.

The multiple variants of casein kinase I form another family of homologous kinases with a variable C-terminal regulatory domain (purple) and a conserved catalytic domain (orange) attached to its N-terminus (Fig. 24.1b). There is also a short N-terminal extension (white) to the catalytic domain whose function is not known.

26.2.2 Calcium/Calmodulin Stimulated Protein Kinase Kinase (CaMKK)

Calcium/calmodulin stimulated protein kinase kinase (CaMKK) is a multifunctional protein kinase encoded by two genes (*CAMKK1* and *CAMKK2*) that produce CaMKKα and CaMKKβ, respectively. Several splice isoforms of the monomeric enzyme are generated [7], which are between 54 and 68 kDa in size. CaMKK is primarily expressed in the brain, but is also present in the thymus, spleen, and testis [5, 8]. The expression pattern of CaMKKβ appears to parallel that of CaMKIV [9]. CaMKK phosphorylates CaMKI and CaMKIV, but can also phosphorylate other proteins in their activation loop, such as AMP activated protein kinase (AMPK) [10] and protein kinase B (PKB/Akt) [11].

CaMKK, CaMKI and CaMKIV have been shown to form a signalling pathway termed the Ca^{2+}/CaM-dependent kinase cascade, which has been implicated in several cellular processes, including axonal and dendritic outgrowth and elongation, adiposity regulation, glucose homeostasis, hematopoietic stem cell maintenance, cell proliferation, apoptosis, and normal immune cell function [5, 12–17]. An unusual aspect of this cascade is that binding of Ca^{2+}/CaM to both CaMKK and its substrates (CaMKI and CaMKIV) is required for phosphorylation of their activation loops [18]. Whilst unusual, this mechanism has been noted in other signalling pathways, including the AMP-kinase cascade [19].

CaMKK requires Ca^{2+}/CaM for maximal activity [18], however, CaMKKβ exhibits partially autonomous activity in the absence of Ca^{2+}/CaM, whereas CaMKKα is completely dependent on the binding of Ca^{2+}/CaM for activity [9, 20]. The major site of autosphorylation of CaMKKα is S24 [21], but CaMKK can also autophosphorylate at S74 in the presence of Ca^{2+}/CaM, however, it is very slow, substochiometric, and does not appear to affect catalytic activity [9, 22]. A possible role for this phosphorylation in regulating molecular targeting has not been investigated.

The activity of CaMKKα is regulated by phosphorylation at multiple sites. cAMP-dependent protein kinase (PKA) can phosphorylate S52, S74, T108, S458, and S475 on CaMKKα [21], and S100, S495 and S511 on CaMKKβ [5]. Binding of Ca^{2+}/CaM blocks phosphorylation at S52, S74, T108, and S458, but enhances phosphorylation at S475 [21]. CaMKKα activity is negatively regulated by phosphorylation on S74, T108 and S458 [23–25] and phosphorylation of S458 blocks Ca^{2+}/CaM binding. Cyclin dependent kinase 5 (CDK5) phosphorylates CaMKKβ at S137, thereby priming CaMKKβ for phosphorylation by glycogen synthase kinase 3β (GSK-3β) at S129 and S133 [26]. Phosphorylation at these sites decreases the autonomous activity and the half-life of CaMKKβ. In addition, phosphorylation of CaMKKβ by GSK-3β and CDK5 is critical for its role in neurite development [26]. Furthermore, CaMKK binding partners can alter CaMKK kinetics. Whilst 14-3-3 protein binding to CaMKKβ does not inhibit the catalytic activity of phosphorylated CaMKK, it slows down its dephosphorylation by affecting the structure of several regions of CaMKKβ outside of the 14-3-3- binding motifs [27].

Evidence indicates that molecular targeting plays a role in regulating CaMKK function, as CaMKKα is known to translocate to the nucleus [28]. Furthermore, inhibition of this translocation prevents type-II monocytic cells from being activated [28]. However, whether this targeting is regulated by phosphorylation has not been identified.

26.2.3 Calcium/Calmodulin Stimulated Protein Kinase I (CaMKI)

The CaMKI family is composed of four members, which are encoded by four different genes: CAMK1 (CaMKIα), PNCK (CaMKIβ/Pnck), CAMK1G (CaMKIγ/CLICK3) and CAMK1D (CaMKIδ/CKLiK). Each gene produces at least one splice variant and all members of this family are monomeric with sizes between 38 and 53 kDa. The various isoforms of CaMKI are widely expressed in rat brain and other tissues, with CaMKIα being found in most mammalian cells [29]. CaMKI has been implicated in a variety of cellular functions, including the control of synapsin in nerve terminals, growth cone motility and axon outgrowth, aldosterone synthase expression, the visual signalling process, and the cell cycle [30–35].

In addition to activation by binding of Ca^{2+}/CaM, CaMKI is activated through phosphorylation by CaMKK, and phosphorylation of the conserved T (T174 to T180, depending on the isoform) in the activation loop by CaMKK is required for maximal CaMKI activity [36] in a substrate dependent manner [37], suggesting that targeting may also be involved in the regulation of CaMKI. Additionally, once CaMKIδ is phosphorylated by CaMKK, it becomes resistant to protein phosphatases, which induces a 'primed' state, where it can more readily be activated in response to Ca^{2+} signals than other CaMKI enzymes [38].

Subcellular localisation regulates CaMKI function, and several CaMKI isoforms can translocate to the nucleus. For example, the translocation of CaMKIα to the nucleus (mediated by interacting with a CRM1 complex) is enhanced by Ca^{2+}/CaM, suggesting that nuclear export may be enhanced by activation of the kinase [39]. Furthermore, nuclear translocation of CaMKIδ is triggered by stimuli that produce an influx of intracellular Ca^{2+} (potassium depolarisation or glutamate stimulation) [40]. The mechanisms and functions involved, however, remain to be determined.

26.2.4 Calcium/Calmodulin Stimulated Protein Kinase IV (CaMKIV)

CaMKIV is encoded by one gene (CAMK4) which encodes the α isoform and produces at least one splice variant (β); both are monomeric and are 65 and 67 kDa, respectively in size. The two isoforms, CaMKIVα and CaMKIVβ, are identical, except CaMKIVβ contains a 28 aa N-terminal extension, of unknown function. As mentioned previously, the CaMKIV expression pattern is similar to that of CaMKKβ, with CaMKIV primarily expressed in the brain, but also present in immune cells and in the testis and ovary [8, 14, 41, 42]. The *CAMK4* gene also encodes calspermin, a Ca^{2+}/CaM binding protein of unknown function that is expressed exclusively in spermatids in the testes [42]. CaMKIV has been implicated in the regulation of cyclic AMP element binding protein (CREB), homeostatic plasticity, neurite outgrowth, fear memory, immune and inflammatory responses, tau accumulation, neuropathic pain, and cell cycle control [31, 43–51].

CaMKIV requires Ca^{2+}/CaM to become active, as well as phosphorylation of the conserved T in the activation loop by CaMKK [6], which generates an autonomously active kinase. Phosphorylation of S332 within the CaM binding region prevents CaM binding [52]. Phosphorylation at T196 enhances glucokinase promoter activity [53], suggesting that CaMKIV phosphorylation can regulate insulin secretion and glucose homeostasis. Additionally, a subfraction of hyper-phosphorylated CaMKIV (identified by its shift in electrophoretic mobility) has been identified specifically localised to the nuclear matrix in spermatids [42], however the precise molecular function regulated by this is yet to be elucidated. The subcellular distribution of CaMKIV is dynamic as CaMKIV can translocate between the cytoplasm and nucleus, however, catalytic activity is required for this translocation, as catalytically inactive CaMKIV remains in the cytoplasm [54]. CaMKIV translocates to the nucleus in neocortical neurons after disruption of sensory input [55], whereas CaMKIV translocates to the cytoplasm in luteinised ovarian granulosa cells [41]. This indicates that targeting plays a role in regulating CaMKIV function, but the mechanisms involved require further investigation.

26.2.5 Calcium/Calmodulin Stimulated Protein Kinase II (CaMKII)

Ca^{2+}/CaM stimulated protein kinase II (CaMKII) is encoded by four genes (α, β, γ, and δ) [56], which produce over 30 isoforms ranging in size from 50 to 60 kDa. There are four main variable regions (V1-4) in the regulatory and association domains at which the alternative splicing occurs [57]. The V1 region, which we have included at the C-terminal of the regulatory domain, is the primary site for divergence among the four CaMKII genes. The functional differences between the splice variants are all regulatory e.g. altered sensitivity to intracellular Ca^{2+} and altered molecular targeting by interaction with different CaMKII binding proteins (Table 26.2). One or more members of this family are found in virtually every tissue, and mediate diverse physiological functions. CaMKIIα is expressed most abundantly in neurons (where it can account for up to 1% of total protein [58]), and is involved in regulating many aspects of normal neuronal function, including neurotransmitter synthesis and release, cellular morphology and neurite extension, cortical neuron migration, long-term plasticity, learning, memory consolidation, and memory erasure following retrieval [59–68]. Non-neuronal CaMKII has been implicated in the regulation of other biological processes, such as fertilisation, cell proliferation, osteogenic differentiation, the maintenance of vascular tone, normal cardiac function and heart failure, and cancer cell invasion and migration [69–79].

The three-dimensional structure of CaMKII is in fact highly unusual [80], and in contrast to the other CaMKI family members, CaMKII associates into a multimeric form through interactions between the association domains of its subunits (Fig. 24.1a). The CaMKIIα crystal is an asymmetric unit that consists of two autoinhibited catalytic domains in a symmetric dimer held together by interactions between anti-parallel coiled-coil structures formed by the regulatory domains. The regulatory domains are joined by a hinge to the C-terminus of the catalytic domain [80]. The regulatory domain functions like a gate (with T286 as its hinge), so that it is positioned to block the protein substrate and adenosine triphosphate (ATP) binding sites when CaMKII is autoinhibited and is 'open' when Ca^{2+}/CaM is bound or CaMKII is autophosphorylated at T286. Therefore, CaMKII

Table 26.2 Identified CaMKII targeting proteins

Binding protein		CaMKII used in study		Refs.
Protein	Function	Phosphorylation state	Isoform	
α-actinin-1	Microfilament protein	Non-phosphorylated	α (β, γ, δ nt)	[161]
α-actinin-2	Microfilament protein	Autophosphorylated (radiolabelled CaMKII)[a]	α (β, γ, δ nt)	[162]
F-actin	Actin cytoskeleton	Asp286-CaMKII does not bind.	α, β (γ, δ nt)	[163]
BAALC 1-6-8	Marker of human hematopoietic progenitor cells; proposed lipid raft targeting protein in neurons	Non-phosphorylated and Asp286-CaMKII both bind, Asp253-CaMKII binds more strongly through a different binding site	α binds, β does not (γ, δ nt)	[70, 164]
Calcium channel α-subunit isoforms (L-type)	Calcium influx	Non-phosphorylated CaMKII binds α1, α2a, α3, and α4, while pThr286 only binds α1 and α2a	δ2 (α, β, γ nt)	[165]
Calcium channel, N-type	Pre-synaptic calcium influx	Non-phosphorylated	Not specified	[166]
Camguk/CASK	Synaptic protein targeting and synaptic plasticity	Non-phosphorylated CaMKII binds Camguk, pThr305/306 decreases binding	Not specified	[167]
CARMA1	Regulator of NFκB activation of lymphocytes	Autophosphorylated (radiolabelled CaMKII)[a]	γ (α, β, δ nt)	[168]
Cdk5 activators p35 and p39	Proline-directed serine/threonine kinase	Non-phosphorylated	α (β, γ, δ nt)	[161]
Cytoplasmic polyadenylation element binding protein (CPEB)	Regulates protein synthesis and initiates mRNA polyadenylation and translation	pThr286	Not specified	[169]

Densin-180	Dendritic scaffolding protein	pThr286 enhances binding compared to non-phosphorylated CaMKII	α (β, γ, δ nt)	[170, 171]
Desmin	Muscle intermediate filament	Non-phosphorylated CaMKII binds desmin, Asp286 and Asp253 increases binding	Purified from rat brain	[70, 172, 173]
MAP-2	Microtubule assembly	Non-phosphorylated binds MAP-2, Asp286 increases binding	Purified from rat brain	[70, 172, 173]
Myelin Basic Protein	Major myelin sheath protein	Autophosphorylated (radiolabelled CaMKII)[a]	γ (α, β, δ nt)	[174]
		Non-phosphorylated binds, Asp286 slightly decreases binding, Asp253 completely blocks binding	α (β, γ, δ nt)	[70]
NR2A/B	Voltage sensitive ionotropic glutamate receptor involved in synaptic plasticity	pThr286 enhances binding compared to non-phosphorylated CaMKII	α (β, γ, δ nt)	[84, 175–177]
PP2A	Serine/threonine protein phosphatase	Non-phosphorylated	α (β, γ, δ nt)	[178]
Projectin	Integral protein of insect flight muscle	pThr286 decreases binding compared to non-phosphorylated CaMKII	Purified from rat brain	[179]
Rad	GTP binding protein	Autophosphorylated (radiolabelled CaMKII)[a]	Not specified	[180]
SCP3	PP2C-type protein phosphatase	Non-phosphorylated	γ (G-2 variant) (α, β, δ nt)	[181]

(continued)

Table 26.2 (continued)

Binding protein		CaMKII used in study		Refs.
Protein	Function	Phosphorylation state	Isoform	
STOP	Microtubule associated protein	pThr286	Purified from rat brain	[182]
Synapsin 1	Synaptic vesicle binding protein	Autophosphorylated (radiolabelled CaMKII)[a]	α (β, γ, δ nt)	[183]
Syntaxin 1A	Component of exocytotic molecular machinery	Only pThr286 bound syntaxin 1A (non-phosphorylated CaMKII did not interact)	α (β, γ, δ nt)	[59]
Tau	Microtubule assembly	Non-phosphorylated and Asp253 CaMKII bind tau, Asp286 increases binding	Purified from rat brain	[70, 173, 184]
Tyrosine hydroxylase (rat)	Catecholamine biosynthesis	Non-phosphorylated	α (β, γ, δ nt)	[185]
Tyrosine hydroxylase isoform 2 (human)	Catecholamine biosynthesis	Non-phosphorylated and Asp286 CaMKII bind, Asp253 increases binding	α (β, γ, δ nt)	[70]
Tyrosine hydroxylase isoform 2 (human), phosphorylated at Ser19 and Ser40	Catecholamine biosynthesis	Binding is enhanced for non-phosphorylated, Asp286, and Asp253 CaMKII when compared to non-phosphorylated tyrosine hydroxylase	α (β, γ, δ nt)	[70]

nt not tested

[a]When autophosphorylated, CaMKII will be predominantly (though not always exclusively) phosphorylated at Thr286 Asp286-CaMKII and Asp253-CaMKII are phosphomimic mutants of CaMKII in which an aspartatic acid has been substituted for threonine at positions 286 and 253, respectively

is comprised of six mutually inhibited dimers. Homomers of α, β, γ, and δ all exhibit the same basic structure. Whilst heteromultimers are known to exist [81], their structures are unknown.

The biological properties of CaMKII are regulated by a variety of post-translational modifications (including phosphorylation and oxidation by reactive oxygen species in pathological conditions) and targeting to specific subcellular locations through interactions with other proteins. These two control mechanisms can also influence one another, as the interaction between CaMKII and some binding partners can be modified by the phosphorylation state of the kinase, as well as by phosphorylation of the binding partner [4, 70].

Purified CaMKII requires the presence of Ca^{2+}/CaM for enzyme activity. Binding of two CaM molecules to two adjacent subunits within a holoenzyme allows autophosphorylation of one or both of these subunits to occur at T286 [82]. Autophosphorylation of T286 in CaMKIIα (T287 in CaMKIIβ, γ, and δ) occurs quickly, greatly enhances the affinity of CaMKII for Ca^{2+}/CaM by more than 100 fold, changes enzyme activity, and alters targeting to specific subcellular sites. CaMKII phosphorylation at T286 allows the enzyme to remain active even after CaM has dissociated from it (autonomous activity), and can also regulate the function of the enzyme by increasing the binding of CaMKII to specific subcellular sites [83–85]. However, phosphorylation of T286 is not required for kinase activity.

Once the kinase activity of CaMKII is Ca^{2+}-independent (autonomous), and Ca^{2+}/CaM is no longer bound to the kinase, secondary sites that are within the CaM-binding site can be phosphorylated (T305/306 in CaMKIIα, and T306/307 in CaMKIIβ, γ, and δ) but these phosphorylation changes occur slowly because the rate of dissociation of Ca^{2+}/CaM is greatly decreased by phosphorylation at T286 [86, 87]. Once these sites are phosphorylated, CaM can no longer bind so CaMKII becomes insensitive to changes in Ca^{2+}/CaM but CaMKII remains active until T286/7 is dephosphorylated [88].

CaMKII can also be directly activated by Zn^{2+} independent of Ca^{2+}/CaM resulting in autophosphorylation of T306, T286 and S279. The level of phosphorylation of S279 in vitro increases with increasing Zn^{2+} concentration and inhibits CaMKII activity [89]. This may be particularly relevant in certain pathological conditions that involve elevated intracellular Zn^{2+} concentrations [90].

However, not all CaMKII phosphorylation sites modulate Ca^{2+}/CaM binding and kinase activity. A phosphorylation site on CaMKII at T253 was identified in vivo [91]. T253 has previously been overlooked as a phosphorylation site of interest because: (i) T253 phosphorylation has no direct effect on the kinase activity or Ca^{2+}/CaM binding of purified CaMKII in vitro [91]; (ii) T253 autophosphorylation of purified CaMKII in vitro occurs relatively slowly (at least 10 times slower than the phosphorylation of T286) under standard assay conditions [91]; and (iii) the overall stoichiometry of T253 phosphorylation in tissues and cells in vivo is relatively low [91] . However, subsequent studies have shown that the rate and extent of T253 phosphorylation is very sensitive to the molecular environment so that: (i) the rate of T253 phosphorylation in vivo can be very fast depending on the cell type and stimulus [70, 91–93]; (ii) by inducing interaction with specific binding proteins, T253 phosphorylation can lead to allosteric activation of CaMKII [94]; and (iii) T253 phosphorylation only occurs in particular subpopulations of CaMKII at specific cellular locations and the stoichiometry of T253 phosphorylation at these locations can be high [91]. Therefore, T253 phosphorylation exerts its regulatory effects on CaMKII through modifying molecular targeting.

Other sites, such as S279 and S314, are phosphorylated both in vitro [89, 95, 96] and in vivo [97–99], but the stoichiometry of phosphorylation is relatively low and phosphorylation of S314 does not affect CaMKII activity in vitro. Although these sites have not been investigated for their effects on targeting, it is possible that, as with T253, their major functional role is to regulate molecular targeting rather than to directly modify enzyme activity.

Other forms of post-translational modification have also been demonstrated to alter CaMKII kinase activity. Specifically, a pair of methionine residues in CaMKII (M281/282), present in β, γ, and δ, but not α, isoforms, can become oxidised during periods of elevated reactive oxygen species and produce a conformational change in CaMKII similar to that produced following T286 phosphorylation leading to an autonomous activation of CaMKII [100]. Consequently, one might expect that M281/282 oxidation would also modify molecular targeting in a way that is similar to T286 phosphorylation, but this has not yet been investigated.

A detailed discussion of the mechanisms involved in molecular targeting is given in Section 2 with a specific focus on CaMKII.

26.2.6 Casein Kinase 1 (CK1)

The casein kinase 1 (CK1) family of multifunctional serine/threonine protein kinases are abundantly expressed in all eukaryotic organisms [101], and in a variety of tissues [102]. All organisms contain several isoforms [103], and 7 isoforms have been identified in vertebrates (α1, β, γ1, γ2, γ3, δ, ε), [104–106]. In addition, several splice variants exhibiting different biochemical and cellular properties also exist [107]. CK1 phosphorylates a variety of proteins that are involved in many cellular processes, including cell division, neurite outgrowth, differentiation, antiviral responses, circadian rhythms and metabolism [108–111].

CK1 phosphorylates a wide range of substrates [112–117] and shows a strong preference for 'primed' pre-phosphorylated substrates at N-3 (e.g. pS/T-X-X-S/T). However, CK1 can also phosphorylate unprimed substrates that contain a cluster of acidic amino acids in the N-3 position. Furthermore, CK1 purified from erythrocytes and Xenopus oocytes is able to phosphorylate tyrosine residues in vitro [118, 119], though it has not been determined whether this activity occurs in vivo.

The different isoforms/variants of CK1 are highly conserved within their catalytic domains but vary significantly in the length and structure of their regulatory domains, which contain multiple phosphorylation sites [120–122]. The catalytic domain of CK1 is similar to other kinases, with a smaller N-terminal lobe, a large C-terminal lobe, and an intermediate catalytic cleft where ATP and substrates bind.

CK1δ, ε, and γ3 have large C-terminal domains, which have been suggested to function as pseudosubstrates which, when phosphorylated, inhibit kinase activity [120, 123, 124]. The CK1 family of kinases have been described as monomeric, constitutively active enzymes. However, CK1δ has been suggested to form dimers [124], with this dimerisation potentially inhibiting its activity. This hypothesis has yet to be proven.

The precise mechanisms involved in regulating CK1 function have not been fully elucidated. CK1 was initially identified as being stimulated by Ca^{2+}/CaM [125, 126]. However, because CK1 is constitutively active, mechanisms controlled by second messengers can modify its activity but are not required for its initial activation, unlike many other kinases. The function of CK1 is regulated via a combination of phosphorylation and targeting to specific subcellular locations and via interactions with specific binding proteins. In addition, CK1ε can undergo limited proteolysis, which produces a protease-resistant core kinase with increased activity [121].

CK1 does not require phosphorylation on its activation loop for activity. CK1 can be autophosphorylated, or phosphorylated by PKA, PKB/Akt, CLK2 (CDC-like kinase 2) and PKC. Phosphorylation of CK1δ by these kinases at S370 decreases substrate phosphorylation efficiency [127]. Additionally, CK1ε can autophosphorylate at T44 in vivo, and phosphomimic mutation of this site activates TCF/LEF-driven transcription in breast cancer cell lines [128]. Several inhibitory autophosphorylation sites have been identified for CK1δ and ε (CK1δ at S331, S370, S382, D383, S384, S411 and CK1ε at S323, T325, T324, T337, S368, S405, T407, S408) [122, 123, 129–131]. Although CK1 can autophosphorylate in vivo, it is actively maintained in its dephosphorylated, active state by protein phosphatases [132, 133].

Similarly to CaMKII, there is a large body of evidence demonstrating that targeting is an important regulatory mechanism for controlling CK1 function in vivo. Studies in yeast [134–136], and more recently in higher organisms [137, 138], have demonstrated that the function of constitutively active CK1 is regulated by its subcellular localisation. For example, CK1α exhibits a cell cycle dependent subcellular distribution, interacting with cytosolic vesicles and nuclear structures during interphase, and the mitotic spindle during mitosis [139–142], with this localisation being controlled by the activity of CK1 [137]. In addition, domain swapping experiments performed with yeast CK1 demonstrate the interaction between CK1 and its substrates is controlled by subcellular distribution. Yeast encode four homologues of CK1, 3 of which localise to the plasma membrane, with the fourth being located at the nucleus [135, 143]. However, these homologues functionally complement each other when the localisation signals are switched [135], strongly supporting a role for subcellular targeting in regulating the function of CK1. Additionally, phosphorylation of CK1δ at T347 regulates CK1δ activity towards the PER protein, which regulates the circadian clock [144]. Furthermore, alternative splicing of isoforms influences substrate binding and subcellular location [107, 145–147] highlighting another layer of complexity in the regulation of CK1.

In addition, the function of CK1 varies depending on the proteins with which it is associated. For example, CK1ε only phosphorylates CRY when both proteins are bound to PGR [148]. The phosphorylation state of CK1 can also affect its ability to interact with proteins. For example, the interaction between CK1α and 14-3-3 is dependent upon phosphorylation of CK1α [149]. By contrast, dephosphorylation of CK1ε increases activity towards the SV40 large T antigen, IκB and Ets-1 [121]. These examples highlight the importance of molecular targeting in mediating cell and tissue specific CK1 function in vivo.

26.3 Mechanisms of Molecular Targeting

Molecular targeting is an important regulatory mechanism for all the multifunctional calcium/calmodulin stimulated protein kinases but it has been most extensively studied in CaMKII. This is because the high level of expression of CaMKII in neurons, and the morphological complexity of neurons, made it easier to identify the heterogeneity of the responses of pools of CaMKII molecules in different cellular locations. Therefore, in this section, we will discuss the mechanisms involved in molecular targeting using CaMKII as the specific example, but the general principles outlined here apply to all the multifunctional kinases.

A wide range of proteins with varied functions have been shown to bind CaMKII (Table 26.2). The expression patterns of CaMKII binding proteins vary with cell type and also subcellular localisation. Additionally, phosphorylation of either the binding partner or CaMKII can alter their ability to bind CaMKII [4, 70]. As cells contain multiple pools of CaMKII (with different post-translational modifications and hence affinity for binding partners) these pools of CaMKII can respond differently to cellular stimulation, depending on the binding partner with which they are associated.

Fig. 26.2 shows a schematic diagram depicting how molecular targeting can generate different functional responses to cellular stimulation by multiple pools of CaMKII. Non-phosphorylated CaMKII, which has a protein binding site represented by a square on its surface, interacts with three different proteins in different cellular locations (A). When cellular stimulation produces a rise in intracellular Ca^{2+} and activation of CaMKII by Ca^{2+}/CaM, the different molecular environments provided by the three binding proteins cause the three pools of CaMKII to respond with a different pattern of autophosphorylation (A). This difference may be due to proximity to different ion channels or conformational changes induced in CaMKII by the interaction with the binding protein. The autophosphorylation induces a conformational change in the CaMKII that hides the binding site through which non-phosphorylated CaMKII interacted with its binding partners causing the phosphorylated CaMKII to dissociate from those binding proteins. Each pattern of autophosphorylation exposes a different binding site on the surface of CaMKII (pT286 – trapezium; pT253 – triangle; pT286+pT253 – circle) which allows that pool of CaMKII to interact with a different binding protein (purple) that has a matching complementary binding site (B and C). These events may involve translocation of CaMKII between protein complexes in the same region of the cell

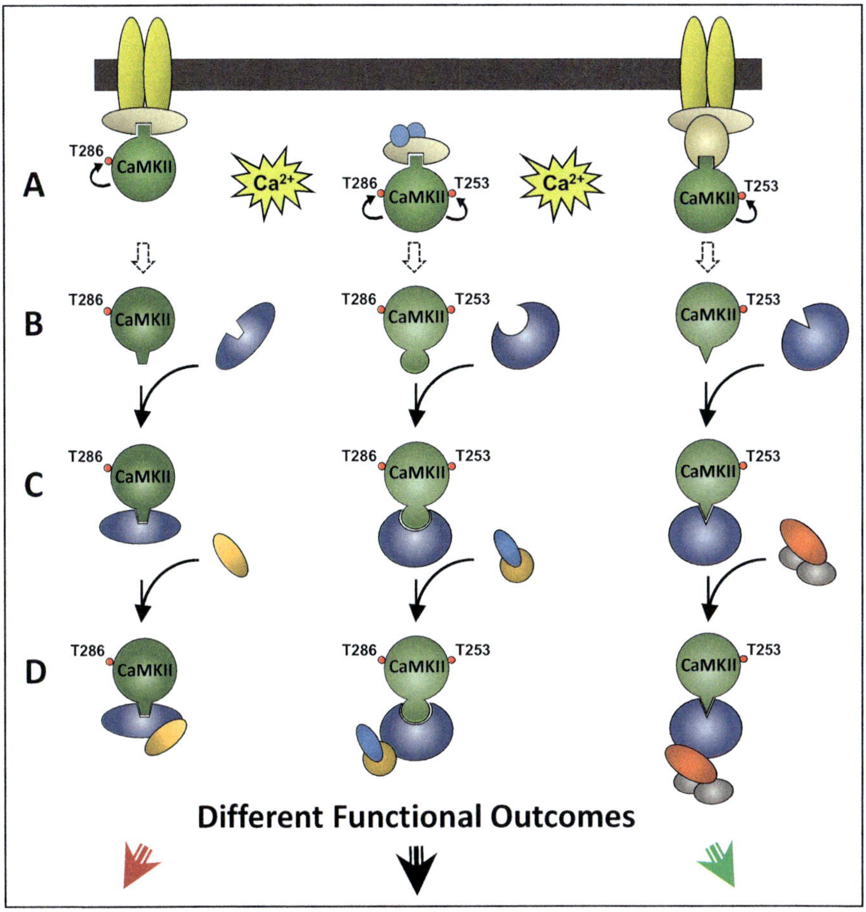

Fig. 26.2 Schematic diagram depicting how molecular targeting can create multiple pools of CaMKII that can generate different functional responses to cellular stimulation. (**a**) Non-phosphorylated CaMKII, which has a protein binding site represented by a square on its surface, interacts with three different proteins in different cellular locations. When cellular stimulation produces a rise in intracellular Ca^{2+} and activation of CaMKII by Ca^{2+}/CaM, the different molecular environments provided by the three binding proteins cause the three pools of CaMKII to respond with different patterns of autophosphorylation (curved arrows and red balls on CaMKII), namely phosphorylation at T286 (left) alone, T253 (right) alone or both T286 and T253 (middle). (**b**) This autophosphorylation induces a conformational change in CaMKII that hides the binding site through which non-phosphorylated CaMKII interacted with its binding partners causing the phosphorylated CaMKII to dissociate from those binding proteins. Each pattern of autophosphorylation exposes a different binding site on the surface of CaMKII (pT286 – trapezium; pT253 – triangle; pT286+pT253 – circle). (**c**) This allows that pool of CaMKII to interact with a different binding protein (dark blue) that has a matching complementary binding site. (**d**) Each CaMKII-binding protein complex then interacts with a different group of additional proteins targeting that pool of CaMKII to different substrates, possibly at different cellular locations, resulting in the activation of different downstream molecular events producing different functional outcomes

or in different subcellular compartments. Each CaMKII-binding protein complex then interacts with a different group of additional proteins targeting that pool of CaMKII to different substrates, possibly at different cellular locations (D), resulting in the activation of different downstream molecular events producing different functional outcomes. These targeted pools of CaMKII may be activated in their new location by Ca^{2+}/CaM following a subsequent rise in intracellular Ca^{2+}, or may be autonomously active due to phosphorylation at T286 or allosteric activation of pT253-CaMKII [150].

We have shown that phosphorylation of CaMKII at T253 is essential for CaMKII-mediated ischaemia-induced cell death [94]. Brain regions with enhanced sensitivity to ischaemic damage show enhanced ischaemia/excitotoxicty-induced phosphorylation of CaMKII at T253, but not T286 or T305/306, and this difference in response is intrinsic to the tissue and independent of blood perfusion, since excitotoxic stimulation of brain slices from different regions in vitro faithfully mimics the responses to cerebral ischaemia in vivo. [93]. Brain regions with enhanced sensitivity to ischaemia also express different patterns of CaMKII binding proteins compared to regions that are more resistant [70]. The interaction of pT253-CaMKII with one or more specific binding proteins activates the pT253-CaMKII by an induced conformational change that is independent of Ca^{2+}-calmodulin [150]. The binding protein targets the pT253-CaMKII to proteins associated with cell death pathway(s) allowing the sustained CaMKII activity to activate cell death responses. The specificity of these targeting mechanisms provides an opportunity for therapeutic intervention in conditions such as stroke through selectively disrupting the interaction between pT253-CaMKII and its specific binding protein. Using a peptide or other small molecule that mimics the pT253-induced binding site on CaMKII for the specific binding protein, the pT253-CaMKII could be displaced, thereby preventing the sustained CaMKII activity and its targeting to cell death pathways and consequently reducing the amount of cell death [150]. Importantly, such a small molecule inhibitor would specifically act on the key pool of pT253-CaMKII molecules without inhibiting the CaMKII molecules in other pools within the same cell or in other cells. Such an approach offers the potential of neuroprotective therapy with enhanced tissue specificity and reduced side effects. The fact that each CaMKII-mediated functional outcome depends on the interaction of a particular pool of CaMKII with a specific binding protein that targets it to molecular pathway involved, means that developing specific inhibitors of the interaction between CaMKII and particular targeting proteins has the potential of producing highly selective therapeutic agents in many clinical conditions.

In addition to differences in phosphorylation state, interaction with specific binding proteins can also be regulated by other factors. Several binding proteins interact with only particular isoforms or splice variants of CaMKII (Table 26.2). All four CaMKII genes undergo alternative splicing in their variable regions [151], which produces some variability in the kinase properties in vitro. However, the number of splice variants is much greater than the differences observed in enzyme activity and the splicing occurs in parts of the molecule well away from the catalytic domain and the autoinhibitory and Ca^{2+}/CaM binding regions suggesting that the primary function of many of the splicing isoforms is not to alter enzyme activity. The αCaMKII-anchoring protein (αKAP), provides an unusual example of targeting. RNA splicing of CaMKIIα produces αKAP as a truncated, enzymatically inactive protein, which is mostly comprised of the association domain of CaMKIIα, and an N-terminal hydrophobic anchor sequence. αKAP is found in skeletal muscle and the heart, and is expressed at low levels in the lung, kidney, and testis [152]. αKAP can form heteromultimers with full length CaMKII, thereby targeting the active kinase subunits to the sarcoplasmic reticulum membrane in rat skeletal muscle [153].

A small number of splice variants of CaMKII contain a consensus nuclear localisation sequence and others contain specific binding sites for individual proteins (for example, the binding sequence for actin is specific to the CaMKIIβ isoform [154]). The fact that the association domain and the C-terminal part of the regulatory domain contain all the main sequence variations between isoforms of CaMKII suggests that these regions contains part or all of the binding sites for other proteins.

The change in interaction between CaMKII and its specific binding protein partners as a consequence of stimulus induced phosphorylation highlights the mobility of CaMKII involved in targeting. This was first identified in the brain where it was recognised that CaMKII is highly concentrated in certain cellular locations, such as the post synaptic density (PSD) and the cytoskeleton, and that the concentration at such sites can change by translocation from the cytoplasm. Translocation to the PSD can occur slowly during the normal maturation phase of brain development by a process that is sensitive to thyroid hormone [155]. Translocation can also occur rapidly to the cytoskeleton [156] and also to the PSD in response to excitotoxic stimulation, hypoxia or post-mortem delay [85, 157, 158]. Phosphorylation of CaMKII at T305/306 decreases the amount of CaMKII bound to the PSD and stimulates the translocation from the PSD to the cytosol [159]. By contrast, phosphorylation at either T286 or T253 stimulates translocation from the cytosol to the PSD and enhances binding to the PSD through different binding proteins [91]. Once located at the PSD CaMKII can bind to, and move between, a number of proteins and phosphorylate a variety of substrates [160].

Figure 26.3 shows a schematic representation of how stimulus-induced short distance translocation of CaMKII between different binding proteins in protein complexes (such as found in the PSD or the cytoskeleton) can modify CaMKII-mediated functional responses to stimulation. Panel A shows the effects of two sequential stimulus-induced rises in intracellular Ca^{2+} on CaMKII bound to a protein as part of a protein complex. Non-phosphorylated CaMKII is bound to one protein in the protein complex. The initial stimulus-induced rise in intracellular Ca^{2+} allows the non-phosphorylated CaMKII to be activated by Ca^{2+}/CaM and to phosphorylate itself and a nearby protein substrate in the protein complex (this functional response is denoted as response ①). The phospho-CaMKII dissociates from the protein to which the non-phosphorylated CaMKII was bound and binds to a neighbouring protein. When a subsequent stimulus-induced rise in intracellular Ca^{2+} activates the translocated phospho-CaMKII, it is able to phosphorylate two different sites on neighbouring proteins (this different functional response is denoted as response ②).

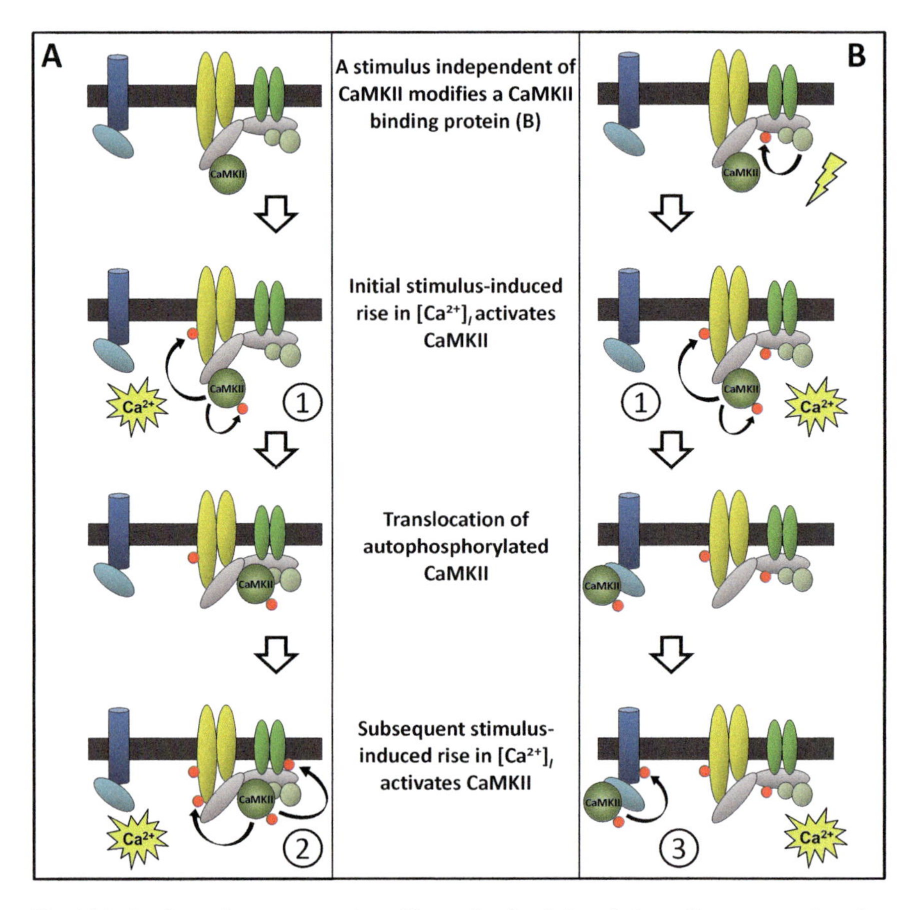

Fig. 26.3 A schematic representation of how stimulus-induced short distance translocation of CaMKII and cross talk between different signalling pathways can alter CaMKII-mediated responses. (**a**) The effects of two sequential stimulus-induced rises in intracellular Ca^{2+} on CaMKII bound to a protein as part of a protein complex. Non-phosphorylated CaMKII is bound to one protein in the protein complex. The initial stimulus-induced rise in intracellular Ca^{2+} allows the non-phosphorylated CaMKII to be activated by Ca^{2+}/CaM and to phosphorylate itself and a nearby protein substrate in the protein complex (curved arrows and red balls). This is functional response ①. The pCaMKII dissociates from the protein to which the non-phosphorylated CaMKII was bound and binds to a neighbouring protein. When a subsequent stimulus-induced rise in intracellular Ca^{2+} activates the translocated pCaMKII, it is able to phosphorylate two different sites on neighbouring proteins (functional response ②) (**b**) How cross talk between different signalling pathways can alter CaMKII-mediated responses. Non-phosphorylated CaMKII is bound to the same protein as in A when a different stimulus (yellow lightning bolt) activates a CaMKII-independent signalling pathway that phosphorylates the protein that binds pCaMKII following functional response ① in panel A. The initial stimulus-induced rise in intracellular Ca^{2+} produces the same functional response ① as in A. However, the pCaMKII now translocates to another pCaMKII binding protein in a nearby complex because the prior CaMKII-independent phosphorylation has prevented pCaMKII binding to the same protein as in A. When a subsequent stimulus-induced rise in intracellular Ca^{2+} activates the translocated pCaMKII, it is able to phosphorylate a different site on a neighbouring protein in the complex (functional response ③)

Therefore, the CaMKII-mediated functional outcome of the second stimulus can be altered by a previous stimulus of the same type. Panel B shows how cross talk between different signalling pathways can alter CaMKII-mediated responses. Non-phosphorylated CaMKII is bound to the same protein as in Panel A when a different stimulus (yellow lightning bolt) activates a CaMKII-independent signalling pathway that phosphorylates the protein which binds phospho-CaMKII following functional response ① in panel A′, inhibiting the ability of that protein to bind phospho-CaMKII. Following this change, the initial stimulus-induced rise in intracellular Ca^{2+} allows the non-phosphorylated CaMKII to be activated by Ca^{2+}/CaM and to produce the same functional response ① as in Panel A. However, when the phospho-CaMKII dissociates from the protein to which the non-phosphorylated CaMKII was bound, it can no longer bind to the same phospho-CaMKII binding protein as in Panel A because it has been phosphorylated in response to the earlier stimulus, so it binds to another phospho-CaMKII binding protein in a nearby complex. When a subsequent stimulus-induced rise in intracellular Ca^{2+} activates the translocated phospho-CaMKII, it is able to phosphorylate a different site on a neighbouring protein in the complex (this different functional response is denoted as response ③).

26.4 Conclusions

Multifunctional Ca^{2+}/CaM stimulated protein kinases are abundant, expressed in most tissues and are responsible for regulating a broad range of physiological functions. We have examined the multiple molecular control mechanisms that regulate the functional responses of these multifunctional kinases. We have focussed particularly on the mechanism of molecular targeting whereby the interaction of these kinases with specific binding proteins enables these enzymes to produce tissue-specific and stimulus-specific responses despite having broad substrate specificities.

As different cells express different complements of binding proteins, the microenvironment in which the kinase is located affects its response to different stimuli and the functional outcome from the stimulus. Understanding molecular targeting will allow us to have a better understanding of molecular mechanisms underlying normal and pathological cellular events. In addition, if cell specific mechanisms controlling kinase function can be identified, then drugs can be designed that will selectively disrupt interactions between widely expressed kinases and their substrates only in specific tissues or in response to specific stimuli. In the past, such widely expressed, multifunctional kinases have not been regarded as suitable drug targets for fear that lack of tissue- or stimulus-specificity would produce unwanted side-effects. But, with our developing understanding of how targeting can produce cell- and stimulus-specific responses in widely expressed kinases, the potential has arisen to achieve specific therapeutic outcomes for a variety of conditions and pathologies by developing drugs that disrupt specific molecular targeting interactions, as proposed above in the case of pT253-CaMKII and ischaemia.

References

1. Rhoads AR, Friedberg F (1997) Sequence motifs for calmodulin recognition. FASEB J 11(5):331–340
2. Skelding KA, Rostas JA (2012) The role of molecular regulation and targeting in regulating calcium/calmodulin stimulated protein kinases. Adv Exp Med Biol 740:703–730
3. De Koninck P, Schulman H (1998) Sensitivity of CaM kinase II to the frequency of Ca^{2+} oscillations. Science 279(5348):227–230
4. Skelding KA, Rostas JA (2009) Regulation of CaMKII in vivo: the importance of targeting and the intracellular microenvironment. Neurochem Res 34(10):1792–1804
5. Racioppi L, Means AR (2012) Calcium/calmodulin-dependent protein kinase kinase 2: roles in signaling and pathophysiology. J Biol Chem 287(38):31658–31665
6. Selbert MA, Anderson KA, Huang QH, Goldstein EG, Means AR, Edelman AM (1995) Phosphorylation and activation of Ca^{2+}-calmodulin-dependent protein kinase IV by Ca^{2+}-calmodulin-dependent protein kinase Ia kinase. Phosphorylation of threonine 196 is essential for activation. J Biol Chem 270(29):17616–17621
7. Hsu LS, Chen GD, Lee LS, Chi CW, Cheng JF, Chen JY (2001) Human Ca^{2+}/calmodulin-dependent protein kinase kinase beta gene encodes multiple isoforms that display distinct kinase activity. J Biol Chem 276(33):31113–31123
8. Ohmstede CA, Jensen KF, Sahyoun NE (1989) Ca^{2+}/calmodulin-dependent protein kinase enriched in cerebellar granule cells. Identification of a novel neuronal calmodulin-dependent protein kinase. J Biol Chem 264(10):5866–5875
9. Anderson KA, Means RL, Huang QH, Kemp BE, Goldstein EG, Selbert MA et al (1998) Components of a calmodulin-dependent protein kinase cascade. Molecular cloning, functional characterization and cellular localization of Ca^{2+}/calmodulin-dependent protein kinase kinase beta. J Biol Chem 273(48):31880–31889
10. Hawley SA, Selbert MA, Goldstein EG, Edelman AM, Carling D, Hardie DG (1995) 5′-AMP activates the AMP-activated protein kinase cascade, and Ca^{2+}/calmodulin activates the calmodulin-dependent protein kinase I cascade, via three independent mechanisms. J Biol Chem 270(45):27186–27191
11. Yano S, Tokumitsu H, Soderling TR (1998) Calcium promotes cell survival through CaM-K kinase activation of the protein-kinase-B pathway. Nature 396(6711):584–587
12. Anderson KA, Means AR (2002) Defective signaling in a subpopulation of CD4(+) T cells in the absence of Ca^{2+}/calmodulin-dependent protein kinase IV. Mol Cell Biol 22(1):23–29
13. Neal AP, Molina-Campos E, Marrero-Rosado B, Bradford AB, Fox SM, Kovalova N et al (2010) CaMKK-CaMKI signaling pathways differentially control axon and dendrite elongation in cortical neurons. J Neurosci 30(8):2807–2809
14. Kitsos CM, Sankar U, Illario M, Colomer-Font JM, Duncan AW, Ribar TJ et al (2005) Calmodulin-dependent protein kinase IV regulates hematopoietic stem cell maintenance. J Biol Chem 280(39):33101–33108
15. Kahl CR, Means AR (2004) Regulation of cyclin D1/Cdk4 complexes by calcium/calmodulin-dependent protein kinase I. J Biol Chem 279(15):15411–15419
16. Anderson KA, Ribar TJ, Illario M, Means AR (1997) Defective survival and activation of thymocytes in transgenic mice expressing a catalytically inactive form of Ca^{2+}/calmodulin-dependent protein kinase IV. Mol Endocrinol 11(6):725–737
17. Wayman GA, Lee YS, Tokumitsu H, Silva AJ, Soderling TR (2008) Calmodulin-kinases: modulators of neuronal development and plasticity. Neuron 59(6):914–931

18. Tokumitsu H, Soderling TR (1996) Requirements for calcium and calmodulin in the calmodulin kinase activation cascade. J Biol Chem 271(10):5617–5622
19. Hawley SA, Davison M, Woods A, Davies SP, Beri RK, Carling D et al (1996) Characterization of the AMP-activated protein kinase kinase from rat liver and identification of threonine 172 as the major site at which it phosphorylates AMP-activated protein kinase. J Biol Chem 271(44):27879–27887
20. Edelman AM, Mitchelhill KI, Selbert MA, Anderson KA, Hook SS, Stapleton D et al (1996) Multiple Ca^{2+}-calmodulin-dependent protein kinase kinases from rat brain. Purification, regulation by Ca^{2+}-calmodulin, and partial amino acid sequence. J Biol Chem 271(18):10806–10810
21. Okuno S, Kitani T, Fujisawa H (2001) Regulation of Ca^{2+}/calmodulin-dependent protein kinase kinase alpha by cAMP-dependent protein kinase: I. Biochemical analysis. J Biochem 130(4):503–513
22. Tokumitsu H, Takahashi N, Eto K, Yano S, Soderling TR, Muramatsu M (1999) Substrate recognition by Ca^{2+}/Calmodulin-dependent protein kinase kinase. Role of the arg-pro-rich insert domain. J Biol Chem 274(22):15803–15810
23. Davare MA, Saneyoshi T, Guire ES, Nygaard SC, Soderling TR (2004) Inhibition of calcium/calmodulin-dependent protein kinase kinase by protein 14-3-3. J Biol Chem 279(50):52191–52199
24. Wayman GA, Tokumitsu H, Soderling TR (1997) Inhibitory cross-talk by cAMP kinase on the calmodulin-dependent protein kinase cascade. J Biol Chem 272(26):16073–16076
25. Matsushita M, Nairn AC (1999) Inhibition of the Ca^{2+}/calmodulin-dependent protein kinase I cascade by cAMP-dependent protein kinase. J Biol Chem 274(15):10086–10093
26. Green MF, Scott JW, Steel R, Oakhill JS, Kemp BE, Means AR (2011) Ca^{2+}/calmodulin-dependent protein kinase kinase beta is regulated by multisite phosphorylation. J Biol Chem 286(32):28066–28079
27. Psenakova K, Petrvalska O, Kylarova S, Lentini Santo D, Kalabova D, Herman P et al (2018) 14-3-3 protein directly interacts with the kinase domain of calcium/calmodulin-dependent protein kinase kinase (CaMKK2). Biochim Biophys Acta. ePub ahead of print
28. Guest CB, Deszo EL, Hartman ME, York JM, Kelley KW, Freund GG (2008) Ca^{2+}/calmodulin-dependent kinase kinase alpha is expressed by monocytic cells and regulates the activation profile. Plos One 3(2):e1606
29. Picciotto MR, Zoli M, Bertuzzi G, Nairn AC (1995) Immunochemical localization of calcium/calmodulin-dependent protein kinase I. Synapse 20(1):75–84
30. Rasmussen CD (2000) Cloning of a calmodulin kinase I homologue from Schizosaccharomyces pombe. J Biol Chem 275(1):685–690
31. Skelding KA, Rostas JA, Verrills NM (2011) Controlling the cell cycle: the role of calcium/calmodulin-stimulated protein kinases I and II. Cell Cycle 10(4):631–639
32. Nairn AC, Greengard P (1987) Purification and characterization of Ca^{2+}/calmodulin-dependent protein kinase I from bovine brain. J Biol Chem 262(15):7273–7281
33. Wayman GA, Kaech S, Grant WF, Davare M, Impey S, Tokumitsu H et al (2004) Regulation of axonal extension and growth cone motility by calmodulin-dependent protein kinase I. J Neurosci 24(15):3786–3794
34. Condon JC, Pezzi V, Drummond BM, Yin S, Rainey WE (2002) Calmodulin-dependent kinase I regulates adrenal cell expression of aldosterone synthase. Endocrinology 143(9):3651–3657
35. Jusuf AA, Sakagami H, Kikkawa S, Terashima T (2016) Expression of beta subunit 2 of Ca(2)+/calmodulin-dependent protein kinase I in the developing rat retina. Kobe J Med Sci 61(4):E115–E123
36. Haribabu B, Hook SS, Selbert MA, Goldstein EG, Tomhave ED, Edelman AM et al (1995) Human calcium-calmodulin dependent protein-kinase-I – CDNA cloning, domain-structure and activation by phosphorylation at threonine-177 by calcium-calmodulin dependent protein-kinase-I kinase. EMBO J 14(15):3679–3686

37. Hook SS, Kemp BE, Means AR (1999) Peptide specificity determinants at P-7 and P-6 enhance the catalytic efficiency of Ca^{2+}/calmodulin-dependent protein kinase I in the absence of activation loop phosphorylation. J Biol Chem 274(29):20215–20222
38. Senga Y, Ishida A, Shigeri Y, Kameshita I, Sueyoshi N (2015) The phosphatase-resistant isoform of CaMKI, Ca(2)(+)/calmodulin-dependent protein kinase Idelta (CaMKIdelta), remains in its "Primed" form without Ca(2)(+) stimulation. Biochemistry 54(23):3617–3630
39. Stedman DR, Uboha NV, Stedman TT, Nairn AC, Picciotto MR (2004) Cytoplasmic localization of calcium/calmodulin-dependent protein kinase I-alpha depends on a nuclear export signal in its regulatory domain. FEBS Lett 566(1-3):275–280
40. Sakagami H, Kamata A, Nishimura H, Kasahara J, Owada Y, Takeuchi Y et al (2005) Prominent expression and activity-dependent nuclear translocation of Ca^{2+}/calmodulin-dependent protein kinase I delta in hippocampal neurons. Eur J Neurosci 22(11):2697–2707
41. Wu JY, Gonzalez-Robayana IJ, Richards JS, Means AR (2000) Female fertility is reduced in mice lacking Ca^{2+}/calmodulin-dependent protein kinase IV. Endocrinology 141:4777–4783
42. Wu JY, Means AR (2000) Ca^{2+}/calmodulin-dependent protein kinase IV is expressed in spermatids and targeted to chromatin and the nuclear matrix. J Biol Chem 275(11):7994–7999
43. Kimura Y, Corcoran EE, Eto K, Gengyo-Ando K, Muramatsu MA, Kobayashi R et al (2002) A CaMK cascade activates CRE-mediated transcription in neurons of Caenorhabditis elegans. EMBO Rep 3(10):962–966
44. Bleier J, Toliver A (2017) Exploring the role of CaMKIV in homeostatic plasticity. J Neurosci 37(48):11520–11522
45. Takemura M, Mishima T, Wang Y, Kasahara J, Fukunaga K, Ohashi K et al (2009) Ca^{2+}/calmodulin-dependent protein kinase IV-mediated LIM kinase activation is critical for calcium signal-induced neurite outgrowth. J Biol Chem 284(42):28554–28562
46. Wei F, Qiu CS, Liauw J, Robinson DA, Ho N, Chatila T et al (2002) Calcium calmodulin-dependent protein kinase IV is required for fear memory. Nat Neurosci 5(6):573–579
47. Racioppi L, Means AR (2008) Calcium/calmodulin-dependent kinase IV in immune and inflammatory responses: novel routes for an ancient traveller. Trends Immunol 29(12):600–607
48. Gu R, Ding M, Shi D, Huang T, Guo M, Yu L et al (2018) Calcium/calmodulin-dependent protein kinase IV mediates IFN-gamma-induced immune behaviors in skeletal muscle cells. Cell Physiol Biochem 46(1):351–364
49. Shi D, Gu R, Song Y, Ding M, Huang T, Guo M et al (2018) Calcium/calmodulin-dependent protein kinase IV (CaMKIV) mediates acute skeletal muscle inflammatory response. Inflammation 41(1):199–212
50. Wei YP, Ye JW, Wang X, Zhu LP, Hu QH, Wang Q et al (2018) Tau-induced Ca^{2+}/calmodulin-dependent protein kinase-IV activation aggravates nuclear Tau hyperphosphorylation. Neurosci Bull 34(2):261–269
51. Zhao X, Shen L, Xu L, Wang Z, Ma C, Huang Y (2016) Inhibition of CaMKIV relieves streptozotocin-induced diabetic neuropathic pain through regulation of HMGB1. BMC Anesthesiol 16(1):27
52. Swulius MT, Waxham MN (2008) Ca^{2+}/calmodulin-dependent protein kinases. Cell Mol Life Sci 65(17):2637–2657
53. Murao K, Li J, Imachi H, Muraoka T, Masugata H, Zhang GX et al (2009) Exendin-4 regulates glucokinase expression by CaMKK/CaMKIV pathway in pancreatic beta-cell line. Diabetes Obes Metab 11(10):939–946
54. Lemrow SM, Anderson KA, Joseph JD, Ribar TJ, Noeldner PK, Means AR (2004) Catalytic activity is required for calcium/calmodulin-dependent protein kinase IV to enter the nucleus. J Biol Chem 279(12):11664–11671
55. Lalonde J, Lachance PE, Chaudhuri A (2004) Monocular enucleation induces nuclear localization of calcium/calmodulin-dependent protein kinase IV in cortical interneurons of adult monkey area V1. J Neurosci 24(2):554–564

56. Miller SG, Kennedy MB (1985) Distinct forebrain and cerebellar isozymes of type II Ca^{2+}/calmodulin-dependent protein kinase associate differently with the postsynaptic density fraction. J Biol Chem 260(15):9039–9046
57. Hudmon A, Schulman H (2002) Neuronal CA^{2+}/calmodulin-dependent protein kinase II: the role of structure and autoregulation in cellular function. Annu Rev Biochem 71:473–510
58. Lisman J, Schulman H, Cline H (2002) The molecular basis of CaMKII function in synaptic and behavioural memory. Nat Rev Neurosci 3(3):175–190
59. Ohyama A, Hosaka K, Komiya Y, Akagawa K, Yamauchi E, Taniguchi H et al (2002) Regulation of exocytosis through Ca^{2+}/ATP-dependent binding of autophosphorylated Ca^{2+}/calmodulin-activated protein kinase II to syntaxin 1A. J Neurosci 22(9):3342–3351
60. Giese KP, Fedorov NB, Filipkowski RK, Silva AJ (1998) Autophosphorylation at Thr286 of the alpha calcium-calmodulin kinase II in LTP and learning. Science 279(5352):870–873
61. Miller S, Yasuda M, Coats JK, Jones Y, Martone ME, Mayford M (2002) Disruption of dendritic translation of CaMKIIalpha impairs stabilization of synaptic plasticity and memory consolidation. Neuron 36(3):507–519
62. Soderling TR, Derkach VA (2000) Postsynaptic protein phosphorylation and LTP. Trends Neurosci 23(2):75–80
63. Taha S, Hanover JL, Silva AJ, Stryker MP (2002) Autophosphorylation of alphaCaMKII is required for ocular dominance plasticity. Neuron 36(3):483–491
64. Cao X, Wang H, Mei B, An S, Yin L, Wang LP et al (2008) Inducible and selective erasure of memories in the mouse brain via chemical-genetic manipulation. Neuron 60(2):353–366
65. von Hertzen LS, Giese KP (2005) Alpha-isoform of Ca^{2+}/calmodulin-dependent kinase II autophosphorylation is required for memory consolidation-specific transcription. Neuroreport 16(12):1411–1414
66. Vigil FA, Giese KP (2018) Calcium/calmodulin-dependent kinase II and memory destabilization: a new role in memorymaintenance. J Neurochem.. ePub ahead of print
67. Li X, Goel P, Wondolowski J, Paluch J, Dickman D (2018) A glutamate homeostat controls the presynaptic inhibition of neurotransmitter release. Cell Rep 23(6):1716–1727
68. Nicole O, Bell DM, Leste-Lasserre T, Doat H, Guillemot F, Pacary E (2018) A novel role for CAMKIIbeta in the regulation of cortical neuron migration: implications for neurodevelopmental disorders. Mol Psychiatry. ePub ahead of print.
69. Jones KT (2007) Intracellular calcium in the fertilization and development of mammalian eggs. Clin Exp Pharmacol Physiol 34(10):1084–1089
70. Skelding KA, Suzuki T, Gordon S, Xue J, Verrills NM, Dickson PW et al (2010) Regulation of CaMKII by phospho-Thr253 or phospho-Thr286 sensitive targeting alters cellular function. Cell Signal 22(5):759–769
71. Hoffman A, Carpenter H, Kahl R, Watt LF, Dickson PW, Rostas JAP et al (2014) Dephosphorylation of CaMKII at T253 controls the metaphase-anaphase transition. Cellular Signal 26(4):748–756
72. Shin MK, Kim MK, Bae YS, Jo I, Lee SJ, Chung CP et al (2008) A novel collagen-binding peptide promotes osteogenic differentiation via Ca^{2+}/calmodulin-dependent protein kinase II/ERK/AP-1 signaling pathway in human bone marrow-derived mesenchymal stem cells. Cell Signal 20(4):613–624
73. Munevar S, Gangopadhyay SS, Gallant C, Colombo B, Sellke FW, Morgan KG (2008) CaMKIIT287 and T305 regulate history-dependent increases in alpha agonist-induced vascular tone. J Cell Mol Med 12(1):219–226
74. Maier LS, Bers DM (2007) Role of Ca^{2+}/calmodulin-dependent protein kinase (CaMK) in excitation-contraction coupling in the heart. Cardiovasc Res 73(4):631–640
75. Chi M, Evans H, Gilchrist J, Mayhew J, Hoffman A, Pearsall EA et al (2016) Phosphorylation of calcium/calmodulin-stimulated protein kinase II at T286 enhances invasion and migration of human breast cancer cells. Sci Rep 6:33132
76. Liu Z, Han G, Cao Y, Wang Y, Gong H (2014) Calcium/calmodulindependent protein kinase II enhances metastasis of human gastric cancer by upregulating nuclear factorkappaB and Aktmediated matrix metalloproteinase9 production. Mol Med Rep 10(5):2459–2464

77. Abdul Majeed ABB, Pearsall E, Carpenter H, Brzozowski J, Dickson PW, Rostas JAP et al (2014) CaMKII kinase activity, targeting and control of cellular functions: effect of single and double phosphorylation of CaMKIIalpha. Calcium Signal 1:36–51
78. Sun X, Cao H, Zhan L, Yin C, Wang G, Liang P et al (2018) Mitochondrial fission promotes cell migration by Ca^{2+} /CaMKII/ERK/FAK pathway in hepatocellular carcinoma. Liver Int 38(7):1263–1272
79. Yu G, Cheng CJ, Lin SC, Lee YC, Frigo DE, Yu-Lee LY et al (2018) Organelle-derived acetyl-CoA promotes prostate cancer cell survival, migration, and metastasis via activation of calmodulin kinase II. Cancer Res 78(10):2490–2502
80. Rosenberg OS, Deindl S, Sung RJ, Nairn AC, Kuriyan J (2005) Structure of the autoinhibited kinase domain of CaMKII and SAXS analysis of the holoenzyme. Cell 123(5):849–860
81. Kolb SJ, Hudmon A, Ginsberg TR, Waxham MN (1998) Identification of domains essential for the assembly of calcium/calmodulin-dependent protein kinase II holoenzymes. J Biol Chem 273(47):31555–31564
82. Hanson PI, Meyer T, Stryer L, Schulman H (1994) Dual role of calmodulin in autophosphorylation of multifunctional cam kinase may underlie decoding of calcium signals. Neuron 12(5):943–956
83. Meyer T, Hanson PI, Stryer L, Schulman H (1992) Calmodulin trapping by calcium-calmodulin-dependent protein kinase. Science 256(5060):1199–1202
84. Strack S, Colbran RJ (1998) Autophosphorylation-dependent targeting of calcium/ calmodulin-dependent protein kinase II by the NR2B subunit of the N-methyl- D-aspartate receptor. J Biol Chem 273(33):20689–20692
85. Strack S, Choi S, Lovinger DM, Colbran RJ (1997) Translocation of autophosphorylated calcium/calmodulin-dependent protein kinase II to the postsynaptic density. J Biol Chem 272(21):13467–13470
86. Hanson PI, Schulman H (1992) Inhibitory autophosphorylation of multifunctional Ca^{2+}/calmodulin-dependent protein kinase analyzed by site-directed mutagenesis. J Biol Chem 267(24):17216–17224
87. Patton BL, Miller SG, Kennedy MB (1990) Activation of type II calcium/calmodulin-dependent protein kinase by Ca^{2+}/calmodulin is inhibited by autophosphorylation of threonine within the calmodulin-binding domain. J Biol Chem 265:11204–11212
88. Kato K, Iwamoto T, Kida S (2013) Interactions between alphaCaMKII and calmodulin in living cells: conformational changes arising from CaM-dependent and -independent relationships. Mol Brain 6:37
89. Lengyel I, Fieuw-Makaroff S, Hall AL, Sim AT, Rostas JA, Dunkley PR (2000) Modulation of the phosphorylation and activity of calcium/calmodulin-dependent protein kinase II by zinc. J Neurochem 75(2):594–605
90. Mizuno D, Kawahara M (2013) The molecular mechanisms of zinc neurotoxicity and the pathogenesis of vascular type senile dementia. Int J Mol Sci 14(11):22067–22081
91. Migues PV, Lehmann IT, Fluechter L, Cammarota M, Gurd JW, Sim ATR et al (2006) Phosphorylation of CaMKII at Thr253 occurs in vivo and enhances binding to isolated postsynaptic densities. J Neurochem 98(1):289–299
92. Gurd JW, Rawof S, Zhen Huo J, Dykstra C, Bissoon N, Teves L et al (2008) Ischemia and status epilepitcus result in enhanced phosphorylation of calcium and calmodulin-stimulated protein kinase II on threonine 253. Brain Res 1218:158–165
93. Skelding KA, Spratt NJ, Fluechter L, Dickson PW, Rostas JAP (2012) alpha CaMKII is differentially regulated in brain regions that exhibit differing sensitivities to ischemia and excitotoxicity. J Cerebr Blood F Met 32(12):2181–2192
94. Rostas JA, Hoffman A, Murtha LA, Pepperall D, McLeod DD, Dickson PW et al (2017) Ischaemia- and excitotoxicity-induced CaMKII-Mediated neuronal cell death: the relative roles of CaMKII autophosphorylation at T286 and T253. Neurochem Int. 104:6–10
95. Hanson PI, Kapiloff MS, Lou LL, Rosenfeld MG, Schulman H (1989) Expression of a multifunctional Ca^{2+}/calmodulin-dependent protein kinase and mutational analysis of its autoregulation. Neuron. 3(1):59–70

96. Colbran RJ, Soderling TR (1990) Calcium calmodulin-independent autophosphorylation sites of calcium calmodulin-dependent protein kinase-II – studies on the effect of phosphorylation of threonine-305/306 and serine-314 on calmodulin binding using synthetic peptides. J Biol Chem 265(19):11213–11219
97. Jaffe H, Vinade L, Dosemeci A (2004) Identification of novel phosphorylation sites on postsynaptic density proteins. Biochem Biophys Res Commun 321(1):210–218
98. Molloy SS, Kennedy MB (1991) Autophosphorylation of type II Ca^{2+}/calmodulin-dependent protein kinase in cultures of postnatal rat hippocampal slices. Proc Natl Acad Sci U S A 88(11):4756–4760
99. Collins MO, Yu L, Coba MP, Husi H, Campuzano I, Blackstock WP et al (2005) Proteomic analysis of in vivo phosphorylated synaptic proteins. J Biol Chem 280(7):5972–5982
100. Erickson JR, Joiner ML, Guan X, Kutschke W, Yang J, Oddis CV et al (2008) A dynamic pathway for calcium-independent activation of CaMKII by methionine oxidation. Cell 133(3):462–474
101. Tuazon PT, Traugh JA (1991) Casein kinase I and II–multipotential serine protein kinases: structure, function, and regulation. Adv Second Messenger Phosphoprotein Res 23:123–164
102. Nakajo S, Hagiwara T, Nakaya K, Nakamura Y (1987) Tissue distribution of casein kinases. Biochem Int 14(1):701–707
103. Manning G, Whyte DB, Martinez R, Hunter T, Sudarsanam S (2002) The protein kinase complement of the human genome. Science 298(5600):1912–1934
104. Rowles J, Slaughter C, Moomaw C, Hsu J, Cobb MH (1991) Purification of casein kinase I and isolation of cDNAs encoding multiple casein kinase I-like enzymes. Proc Natl Acad Sci U S A 88(21):9548–9552
105. Tapia C, Featherstone T, Gomez C, Taillon-Miller P, Allende CC, Allende JE (1994) Cloning and chromosomal localization of the gene coding for human protein kinase CK1. FEBS Lett 349(2):307–312
106. Fish KJ, Cegielska A, Getman ME, Landes GM, Virshup DM (1995) Isolation and characterization of human casein kinase I epsilon (CKI), a novel member of the CKI gene family. J Biol Chem 270(25):14875–14883
107. Burzio V, Antonelli M, Allende CC, Allende JE (2002) Biochemical and cellular characteristics of the four splice variants of protein kinase CK1alpha from zebrafish (Danio rerio). J Cell Biochem 86(4):805–814
108. Vielhaber E, Virshup DM (2001) Casein kinase I: from obscurity to center stage. IUBMB Life 51(2):73–78
109. Zhang L, Li H, Chen Y, Gao X, Lu Z, Gao L et al (2017) The down-regulation of casein kinase 1 alpha as a host defense response against infectious bursal disease virus infection. Virology 512:211–221
110. Bischof J, Muller A, Fander M, Knippschild U, Fischer D (2011) Neurite outgrowth of mature retinal ganglion cells and PC12 cells requires activity of CK1delta and CK1epsilon. PLoS One. 6(6):e20857
111. Zhang B, Butler AM, Shi Q, Xing S, Herman PK (2018) P-body localization of the Hrr25/CK1 protein kinase is required for the completion of meiosis. Mol Cell Biol. ePub ahead of print.
112. Pulgar V, Marin O, Meggio F, Allende CC, Allende JE, Pinna LA (1999) Optimal sequences for non-phosphate-directed phosphorylation by protein kinase CK1 (casein kinase-1) – a re-evaluation. Eur J Biochem 260(2):520–526
113. Marin O, Bustos VH, Cesaro L, Meggio F, Pagano MA, Antonelli M et al (2003) A noncanonical sequence phosphorylated by casein kinase 1 in beta-catenin may play a role in casein kinase 1 targeting of important signaling proteins. P Natl Acad Sci USA 100(18):10193–10200
114. Flotow H, Graves PR, Wang AQ, Fiol CJ, Roeske RW, Roach PJ (1990) Phosphate groups as substrate determinants for casein kinase I action. J Biol Chem 265(24):14264–14269
115. Flotow H, Roach PJ (1991) Role of acidic residues as substrate determinants for casein kinase-I. J Biol Chem 266(6):3724–3727

116. Meggio F, Perich JW, Reynolds EC, Pinna LA (1991) A synthetic beta-casein phosphopeptide and analogs as model substrates for casein kinase-1, a ubiquitous, phosphate directed protein-kinase. Febs Letters 283(2):303–306
117. Bustos VH, Marin O, Meggio F, Cesaro L, Allende CC, Allende JE et al (2005) Generation of protein kinase Ck1alpha mutants which discriminate between canonical and non-canonical substrates. Biochem J 391(Pt 2):417–424
118. Pulgar V, Tapia C, Vignolo P, Santos J, Sunkel CE, Allende CC et al (1996) The recombinant alpha isoform of protein kinase CK1 from Xenopus laevis can phosphorylate tyrosine in synthetic substrates. Eur J Biochem 242(3):519–528
119. Braun S, Raymond WE, Racker E (1984) Synthetic tyrosine polymers as substrates and inhibitors of tyrosine-specific protein-kinases. J Biol Chem 259(4):2051–2054
120. Graves PR, Roach PJ (1995) Role of COOH-terminal phosphorylation in the regulation of casein kinase I delta. J Biol Chem 270(37):21689–21694
121. Cegielska A, Gietzen KF, Rivers A, Virshup DM (1998) Autoinhibition of casein kinase I epsilon (CKI epsilon) is relieved by protein phosphatases and limited proteolysis. J Biol Chem 273(3):1357–1364
122. Gietzen KF, Virshup DM (1999) Identification of inhibitory autophosphorylation sites in casein kinase I epsilon. J Biol Chem 274(45):32063–32070
123. Zhai L, Graves PR, Robinson LC, Italiano M, Culbertson MR, Rowles J et al (1995) Casein kinase I gamma subfamily. Molecular cloning, expression, and characterization of three mammalian isoforms and complementation of defects in the Saccharomyces cerevisiae YCK genes. J Biol Chem 270(21):12717–12724
124. Longenecker KL, Roach PJ, Hurley TD (1998) Crystallographic studies of casein kinase I delta toward a structural understanding of auto-inhibition. Acta Crystallogr D Biol Crystallogr 54(Pt 3):473–475
125. Kuret J, Schulman H (1984) Purification and characterization of a Ca^{2+}/calmodulin-dependent protein kinase from rat brain. Biochemistry 23(23):5495–5504
126. Brooks CL, Landt M (1984) Calcium-ion and calmodulin-dependent kappa-casein kinase in rat mammary acini. Biochem J 224(1):195–200
127. Giamas G, Hirner H, Shoshiashvili L, Grothey A, Gessert S, Kuhl M et al (2007) Phosphorylation of CK1delta: identification of Ser370 as the major phosphorylation site targeted by PKA in vitro and in vivo. Biochem J 406(3):389–398
128. Foldynova-Trantirkova S, Sekyrova P, Tmejova K, Brumovska E, Bernatik O, Blankenfeldt W et al (2010) Breast cancer-specific mutations in CK1epsilon inhibit Wnt/beta-catenin and activate the Wnt/Rac1/JNK and NFAT pathways to decrease cell adhesion and promote cell migration. Breast Cancer Res 12(3):R30
129. Graves PR, Haas DW, Hagedorn CH, DePaoli-Roach AA, Roach PJ (1993) Molecular cloning, expression, and characterization of a 49-kilodalton casein kinase I isoform from rat testis. J Biol Chem 268(9):6394–6401
130. Carmel G, Leichus B, Cheng X, Patterson SD, Mirza U, Chait BT et al (1994) Expression, purification, crystallization, and preliminary x-ray analysis of casein kinase-1 from Schizosaccharomyces pombe. J Biol Chem 269(10):7304–7309
131. Schittek B, Sinnberg T (2014) Biological functions of casein kinase 1 isoforms and putative roles in tumorigenesis. Mol Cancer 13:231
132. Rivers A, Gietzen KF, Vielhaber E, Virshup DM (1998) Regulation of casein kinase I epsilon and casein kinase I delta by an in vivo futile phosphorylation cycle. J Biol Chem 273(26):15980–15984
133. Swiatek W, Tsai IC, Klimowski L, Pepler A, Barnette J, Yost HJ et al (2004) Regulation of casein kinase I epsilon activity by Wnt signaling. J Biol Chem 279(13):13011–13017
134. Wang PC, Vancura A, Mitcheson TG, Kuret J (1992) Two genes in Saccharomyces cerevisiae encode a membrane-bound form of casein kinase-1. Mol Biol Cell 3(3):275–286
135. Vancura A, Sessler A, Leichus B, Kuret J (1994) A prenylation motif is required for plasma membrane localization and biochemical function of casein kinase I in budding yeast. J Biol Chem 269(30):19271–19278

136. Ho Y, Mason S, Kobayashi R, Hoekstra M, Andrews B (1997) Role of the casein kinase I isoform, Hrr25, and the cell cycle-regulatory transcription factor, SBF, in the transcriptional response to DNA damage in Saccharomyces cerevisiae. Proc Natl Acad Sci U S A 94(2):581–586
137. Milne DM, Looby P, Meek DW (2001) Catalytic activity of protein kinase CK1 delta (casein kinase 1delta) is essential for its normal subcellular localization. Exp Cell Res 263(1):43–54
138. Yin H, Laguna KA, Li G, Kuret J (2006) Dysbindin structural homologue CK1BP is an isoform-selective binding partner of human casein kinase-1. Biochemistry 45(16):5297–5308
139. Gross SD, Hoffman DP, Fisette PL, Baas P, Anderson RA (1995) A phosphatidylinositol 4,5-bisphosphate-sensitive casein kinase I alpha associates with synaptic vesicles and phosphorylates a subset of vesicle proteins. J Cell Biol 130(3):711–724
140. Gross SD, Simerly C, Schatten G, Anderson RA (1997) A casein kinase I isoform is required for proper cell cycle progression in the fertilized mouse oocyte. J Cell Sci 110:3083–3090
141. Brockman JL, Gross SD, Sussman MR, Anderson RA (1992) Cell cycle-dependent localization of casein kinase I to mitotic spindles. Proc Natl Acad Sci U S A 89(20):9454–9458
142. Elmore ZC, Guillen RX, Gould KL (2018) The kinase domain of CK1 enzymes contains the localization cue essential for compartmentalized signaling at the spindle pole. Mol Biol Cell. ePub ahead of print. mbcE18020129
143. Wang X, Hoekstra MF, DeMaggio AJ, Dhillon N, Vancura A, Kuret J et al (1996) Prenylated isoforms of yeast casein kinase I, including the novel Yck3p, suppress the gcs1 blockage of cell proliferation from stationary phase. Mol Cell Biol 16(10):5375–5385
144. GWL E, Edison, Virshup DM (2017) Site-specific phosphorylation of casein kinase 1 delta (CK1delta) regulates its activity towards the circadian regulator PER2. PLoS One 12(5):e0177834
145. Zhang J, Gross SD, Schroeder MD, Anderson RA (1996) Casein kinase I alpha and alpha L: alternative splicing-generated kinases exhibit different catalytic properties. Biochemistry 35(50):16319–16327
146. Takano A, Hoe HS, Isojima Y, Nagai K (2004) Analysis of the expression, localization and activity of rat casein kinase 1epsilon-3. Neuroreport 15(9):1461–1464
147. Kannanayakal TJ, Tao H, Vandre DD, Kuret J (2006) Casein kinase-1 isoforms differentially associate with neurofibrillary and granulovacuolar degeneration lesions. Acta Neuropathol 111(5):413–421
148. Eide EJ, Vielhaber EL, Hinz WA, Virshup DM (2002) The circadian regulatory proteins BMAL1 and cryptochromes are substrates of casein kinase Iepsilon. J Biol Chem 277(19):17248–17254
149. Clokie S, Falconer H, Mackie S, Dubois T, Aitken A (2009) The interaction between casein kinase Ialpha and 14-3-3 is phosphorylation dependent. FEBS J. 276(23):6971–6984
150. Rostas JAP, Spratt NJ, Dickson PW, Skelding KA (2017) The role of Ca^{2+}-calmodulin stimulated protein kinase II in ischaemic stroke – A potential target for neuroprotective therapies. Neurochem Int 107:33–42
151. Hudmon A, Schulman H (2002) Structure-function of the multifunctional Ca^{2+}/calmodulin-dependent protein kinase II. Biochem J 364(Pt 3):593–611
152. Sugai R, Takeuchi M, Okuno S, Fujisawa H (1996) Molecular cloning of a novel protein containing the association domain of calmodulin-dependent protein kinase II. J Biochem 120(4):773–779
153. Bayer KU, Harbers K, Schulman H (1998) alphaKAP is an anchoring protein for a novel CaM kinase II isoform in skeletal muscle. EMBO J 17(19):5598–5605
154. O'Leary H, Lasda E, Bayer KU (2006) CaMKIIbeta association with the actin cytoskeleton is regulated by alternative splicing. Mol Biol Cell 17(11):4656–4665
155. Wang X, Rostas JA (1996) Effect of hypothyroidism on the subcellular distribution of Ca^{2+}/calmodulin-stimulated protein kinase II in chicken brain during posthatch development. J Neurochem 66(4):1625–1632
156. Lin YC, Redmond L (2008) CaMKIIbeta binding to stable F-actin in vivo regulates F-actin filament stability. Proc Natl Acad Sci U S A 105(41):15791–15796

157. Kolb SJ, Hudmon A, Waxham MN (1995) Ca^{2+}/calmodulin kinase II translocates in a hippocampal slice model of ischemia. J Neurochem 64(5):2147–2156
158. Suzuki T, Okumuranoji K, Tanaka R, Tada T (1994) Rapid translocation of cytosolic Ca^{2+}/calmodulin-dependent protein-kinase-II into postsynaptic density after decapitation. J Neurochem 63(4):1529–1537
159. Elgersma Y, Fedorov NB, Ikonen S, Choi ES, Elgersma M, Carvalho OM et al (2002) Inhibitory autophosphorylation of CaMKII controls PSD association, plasticity, and learning. Neuron 36(3):493–505
160. Bayer KU, Schulman H (2001) Regulation of signal transduction by protein targeting: the case for CaMKII. Biochem Biophys Res Commun 289(5):917–923
161. Dhavan R, Greer PL, Morabito MA, Orlando LR, Tsai LH (2002) The cyclin-dependent kinase 5 activators p35 and p39 interact with the alpha-subunit of Ca^{2+}/calmodulin-dependent protein kinase II and alpha-actinin-1 in a calcium-dependent manner. J Neurosci 22(18):7879–7891
162. Robison AJ, Bartlett RK, Bass MA, Colbran RJ (2005) Differential modulation of Ca^{2+}/calmodulin-dependent protein kinase II activity by regulated interactions with N-methyl-D-aspartate receptor NR2B subunits and alpha actinin. J Biol Chem 280(47):39316–39323
163. Khan S, Conte I, Carter T, Bayer KU, Molloy JE (2016) Multiple CaMKII binding modes to the actin cytoskeleton revealed by single-molecule imaging. Biophys J 111(2):395–408
164. Wang X, Tian QB, Okano A, Sakagami H, Moon IS, Kondon H et al (2005) BAALC 1-6-8 protein is targeted to postsynaptic lipid rafts by its N-terminal myristoylation and palmitoylation, and interacts with alpha, but not beta, subunit of Ca^{2+}/calmodulin-dependent protein kinase II. J Neurochem 92:647–659
165. Grueter CE, Abiria SA, Wu Y, Anderson ME, Colbran RJ (2008) Differential regulated interactions of calcium/calmodulin-dependent protein kinase II with isoforms of voltage-gated calcium channel beta subunits. Biochemistry 47(6):1760–1767
166. Hell JW, Appleyard SM, Yokoyama CT, Warner C, Catterall WA (1994) Differential phosphorylation of two size forms of the N-type calcium channel alpha 1 subunit which have different COOH termini. J Biol Chem 269(10):7390–7396
167. Lu CS, Hodge JJ, Mehren J, Sun XX, Griffith LC (2003) Regulation of the Ca^{2+}/CaM-responsive pool of CaMKII by scaffold-dependent autophosphorylation. Neuron 40(6):1185–1197
168. Ishiguro K, Green T, Rapley J, Wachtel H, Giallourakis C, Landry A et al (2006) Ca^{2+}/calmodulin-dependent protein kinase II is a modulator of CARMA1-mediated NF-kappaB activation. Mol Cell Biol 26(14):5497–5508
169. Atkins CM, Nozaki N, Shigeri Y, Soderling TR (2004) Cytoplasmic polyadenylation element binding protein-dependent protein synthesis is regulated by calcium/calmodulin-dependent protein kinase II. J Neurosci 24(22):5193–5201
170. McNeill RB, Colbran RJ (1995) Interaction of autophosphorylated Ca^{2+}/calmodulin-dependent protein kinase II with neuronal cytoskeletal proteins. Characterization of binding to a 190-kDa postsynaptic density protein. J Biol Chem 270(17):10043–10049
171. Robison AJ, Bass MA, Jiao Y, MacMillan LB, Carmody LC, Bartlett RK et al (2005) Multivalent interactions of calcium/calmodulin-dependent protein kinase II with the postsynaptic density proteins NR2B, densin-180, and alpha-actinin-2. J Biol Chem 280(42):35329–35336
172. Jefferson AB, Schulman H (1991) Phosphorylation of microtubule-associated protein-2 in GH3 cells. Regulation by cAMP and by calcium. J Biol Chem 266(1):346–354
173. Yamamoto H, Fukunaga K, Goto S, Tanaka E, Miyamoto E (1985) Ca^{2+}, calmodulin-dependent regulation of microtubule formation via phosphorylation of microtubule-associated protein 2, tau factor, and tubulin, and comparison with the cyclic AMP-dependent phosphorylation. J Neurochem 44(3):759–768
174. Shoju H, Sueyoshi N, Ishida A, Kameshita I (2005) High level expression and preparation of autonomous Ca^{2+}/calmodulin-dependent protein kinase II in Escherichia coli. J Biochem 138(5):605–611

175. Bayer KU, De Koninck P, Leonard AS, Hell JW, Schulman H (2001) Interaction with the NMDA receptor locks CaMKII in an active conformation. Nature 411(6839):801–805
176. Gardoni F, Caputi A, Cimino M, Pastorino L, Cattabeni F, Di Luca M (1998) Calcium/calmodulin-dependent protein kinase II is associated with NR2A/B subunits of NMDA receptor in postsynaptic densities. J Neurochem 71(4):1733–1741
177. Gardoni F, Schrama LH, van Dalen JJ, Gispen WH, Cattabeni F, Di Luca M (1999) AlphaCaMKII binding to the C-terminal tail of NMDA receptor subunit NR2A and its modulation by autophosphorylation. FEBS Lett 456(3):394–398
178. Yamashita T, Inui S, Maeda K, Hua DR, Takagi K, Fukunaga K et al (2006) Regulation of CaMKII by alpha4/PP2Ac contributes to learning and memory. Brain Res 1082(1):1–10
179. Fahrmann M, Erfmann M, Beinbrech G (2002) Binding of CaMKII to the giant muscle protein projectin: stimulation of CaMKII activity by projectin. Biochim Biophys Acta 1569(1–3):127–134
180. Moyers JS, Bilan PJ, Zhu J, Kahn CR (1997) Rad and Rad-related GTPases interact with calmodulin and calmodulin-dependent protein kinase II. J Biol Chem 272(18):11832–11839
181. Gangopadhyay SS, Gallant C, Sundberg EJ, Lane WS, Morgan KG (2008) Regulation of Ca^{2+}/calmodulin kinase II by a small C-terminal domain phosphatase. Biochem J 412(3):507–516
182. Baratier J, Peris L, Brocard J, Gory-Faure S, Dufour F, Bosc C et al (2006) Phosphorylation of microtubule-associated protein STOP by calmodulin kinase II. J Biol Chem 281(28):19561–19569
183. Benfenati F, Valtorta F, Rubenstein JL, Gorelick FS, Greengard P, Czernik AJ (1992) Synaptic vesicle-associated Ca^{2+}/calmodulin-dependent protein kinase II is a binding protein for synapsin I. Nature. 359(6394):417–420
184. Bennecib M, Gong CX, Grundke-Iqbal I, Iqbal K (2001) Inhibition of PP-2A upregulates CaMKII in rat forebrain and induces hyperphosphorylation of tau at Ser 262/356. Febs Letters 490(1–2):15–22
185. Lehmann IT, Bobrovskaya L, Gordon SL, Dunkley PR, Dickson PW (2006) Differential regulation of the human tyrosine hydroxylase isoforms via hierarchical phosphorylation. J Biol Chem 281(26):17644–17651

Chapter 27
Readily Releasable Stores of Calcium in Neuronal Endolysosomes: Physiological and Pathophysiological Relevance

Koffi L. Lakpa, Peter W. Halcrow, Xuesong Chen, and Jonathan D. Geiger

Abstract Neurons are long-lived post-mitotic cells that possess an elaborate system of endosomes and lysosomes (endolysosomes) for protein quality control. Relatively recently, endolysosomes were recognized to contain high concentrations (400–600 μM) of readily releasable calcium. The release of calcium from this acidic organelle store contributes to calcium-dependent processes of fundamental physiological importance to neurons including neurotransmitter release, membrane excitability, neurite outgrowth, synaptic remodeling, and cell viability. Pathologically, disturbances of endolysosome structure and/or function have been noted in a variety of neurodegenerative disorders including Alzheimer's disease (AD) and HIV-1 associated neurocognitive disorder (HAND). And, dysregulation of intracellular calcium has been implicated in the neuropathogenesis of these same neurological disorders. Thus, it is important to better understand mechanisms by which calcium is released from endolysosomes as well as the consequences of such release to inter-organellar signaling, physiological functions of neurons, and possible pathological consequences. In doing so, a path forward towards new therapeutic modalities might be facilitated.

Keywords Endosomes · Lysosomes · Endolysosomes · Calcium · Store-operated calcium entry · N-type calcium channels · Neurodegenerative diseases · HIV-1 associated neurocognitive disorder · Alzheimer's disease · Neurons

K. L. Lakpa · P. W. Halcrow · X. Chen · J. D. Geiger (✉)
Department of Biomedical Sciences, University of North Dakota School of Medicine and Health Sciences, Grand Forks, ND, USA
e-mail: Jonathan.geiger@und.edu

© Springer Nature Switzerland AG 2020

M. S. Islam (ed.), *Calcium Signaling*, Advances in Experimental Medicine and Biology 1131, https://doi.org/10.1007/978-3-030-12457-1_27

27.1 An Evolutionary Perspective on Calcium, Intracellular Organelles and Endolysosomes

Intracellular calcium regulates many essential functions of neurons including neurotransmitter release, excitability, synaptic plasticity, and cell viability [1]. Levels of intraneuronal calcium are very tightly regulated both temporally and spatially by various mechanisms including calcium release from intracellular stores, calcium influx across plasma membranes, and its association with a whole host of calcium binding proteins. Because of its importance both physiologically and pathologically, we start our story about the presence and functional significance of readily releasable stores of calcium in neuronal endolysosomes with a brief evolutionary perspective about calcium and intracellular organelles.

Calcium is well-known to be important for signal transduction in most cells including neurons. Indeed, calcium has been referred to as a universal second messenger in eukaryotic cells. The approximate 10,000-fold gradient of extracellular to intracellular calcium originated evolutionarily because of the gradual rise in calcium levels from about 100 nM during the period when the basic building blocks of life developed in thermal ducts under the ocean floor to about 1 mM during the Pre-Cambrian period when multicellular life evolved [2, 3]. Due to the toxic nature of millimolar levels of calcium, evolutionary pressure was applied such that cellular survival dictated that semipermeable membranes appeared and a variety of mechanisms were formed to maintain appropriate calcium gradients across plasma membranes [3]. Simultaneously, embedded in the plasma membranes were newly developed calcium pumps and calcium binding proteins which helped with calcium homeostasis [3]. Together, in neurons, these evolutionary changes provide unique and complex spatial and temporal handling of calcium that is essential for not only proper cellular signaling but also neuronal cell life and death.

It was also during this billion-year evolutionary period that intracellular organelles began appearing including mitochondria resulting from symbiotic relationships with bacteria and the development of functional endocytic machinery [4]. Mitochondria are integral to the maintenance of cellular energetics and they are important ‘sinks’ for intracellular calcium [5]. However, when too much calcium is up-taken into mitochondria cellular energetics are compromised and the resulting calcium overload can lead to a cascade of events including increased oxidative stress and cell death. It has also become increasingly appreciated that organelles including endoplasmic reticulum, endosomes and lysosomes (hereafter referred to as endolysosomes) have readily releasable and functionally important pools of intracellular calcium. Although less well known, the approximate 500 μM levels of calcium in endolysosomes are similar to the calcium concentrations present in endoplasmic reticulum [6]. This is a very important concept because endoplasmic reticulum is commonly referred to as the principal intracellular store of readily releasable calcium. Furthermore, as the field of inter-organellar signaling as well as physical and chemical crosstalk between organelles has grown over the past decade it is prudent of us to now posit that this relatively new and highly complicated area of modern cell biology is key to our understanding of the regulation and dysregulation of calcium [7].

With this as a very quick trip across 1 billion years of evolutionary biology, here we embark on a brief but focused summary of findings that neuronal endolysosomes contain readily releasable stores of calcium and once released this calcium can lead to calcium influx into cells, calcium release from other organelles, and calcium dysregulation-induced neurotoxicity. The relevance of such an important upstream store of calcium to the regulation of physiological functions and pathophysiological events is obvious and will be discussed with particular relevance to the pathogenesis of two neurodegenerative disorders; Alzheimer's disease (AD) and HIV-1 associated neurocognitive disorder (HAND).

27.2 Endolysosomes Contain Readily Releasable Pools of Calcium

Neurons are long-lived post-mitotic cells that possess an elaborate endolysosome system for quality control especially for proteins. Endolysosomes are well known to be acidic organelles that contain high levels of cations including calcium, iron, zinc and copper. However, for the cation calcium it was not until fairly recently that these organelles were described as being 'acidic calcium stores' because the luminal pH of endolysosomes is acidic and endolysosomes contain high (400–600 μM) levels of readily releasable calcium [8, 9].

Neuronal calcium signals display spectacular spatiotemporal complexity and understanding how calcium signals are generated spatially and temporally is necessary to understand calcium-dependent cellular processes. Endolysosome calcium levels are maintained by a variety of uptake and efflux mechanisms. Essential for uptake of calcium into endolysosomes, proton gradients are established mainly by vacuolar H^+-ATPase (v-ATPase) that pumps H^+ into the lumen and this helps regulate Ca^{2+} levels [9–11]. Four main mechanisms for calcium release from endolysosomes have been described including: (1) Calcium release through two-pore channels (TPCs) triggered by nicotinic acid adenine dinucleotide phosphate (NAADP) [12–17]; (2) Elevation of endolysosome pH with, for example, the selective v-ATPase inhibitor bafilomycin (BAF) or the alkaline lysosomotropic agents NH_4Cl and chloroquine [8, 18, 19]; (3) Involvement of TRPML1 mucolipin-type channels and P2X4 receptors [20–22]; and (4) Selective disruption of endolysosome membranes with Gly-Phe-β-naphtylamide (GPN) [23, 24]. Of physiological significance, calcium released from endolysosomes has been shown to contribute to a variety of calcium-dependent neuronal processes including neurotransmitter release, neuronal excitability, neurite outgrowth, synaptic remodeling, and cell viability [25–27].

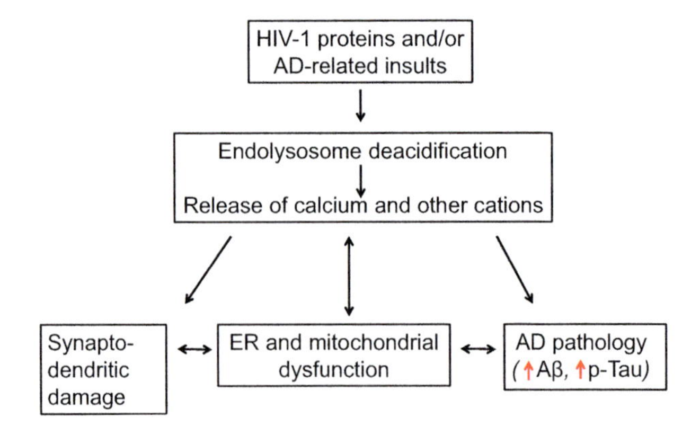

Fig. 27.1 HIV-1 proteins and other neurotoxic insults can cause deacidification of endolysosomes. Increasing endolysosome pH can release calcium and other cations from endolysosomes. Calcium released from readily releasable stores in endolysosomes can increase the release of calcium from other intracellular stores and can increase the influx of extracellular calcium. Such increases in pH and calcium levels can cause endoplasmic reticulum (ER) and mitochondrial dysfunction, Alzheimer's disease (AD)-like pathology, and synaptodendritic damage

Endolysosomes can release calcium transiently and in a highly localized and distinct fashion [17, 28, 29]. Endolysosome calcium can affect the release of calcium from organellar stores as well as through plasma membrane-based calcium influx mechanisms. The inter-organellar signaling and signaling with the plasma membrane is explained at least in part by findings that endolysosomes are highly mobile in cells, are highly dynamic metabolically, have high rates of biogenesis, and can interact physically and functionally with other intracellular organelles (Fig. 27.1).

At least three models of acidic store-induced calcium signaling mechanisms have been described [9]. (1) Acidic stores of calcium might communicate with endoplasmic reticulum calcium stores such that calcium released from endolysosomes can enhance endoplasmic reticulum calcium loading [30] and calcium-induced calcium release [13, 15]. (2) Changes in endolysosome pH may release calcium from a subgroup of acidic calcium stores and the released calcium may affect other subgroups of acidic stores through mechanisms such as vesicular fusion of late endosomes and lysosomes [9, 15, 31]. (3) Calcium released from acidic calcium stores might depolarize plasma membranes, evoke calcium-dependent currents, and stimulate calcium influx across plasma membranes [12].

27.3 Acidic Store-Operated Calcium Entry in Neurons

Acidic store-operated calcium entry (aSOCE) is a unique mechanism that links readily releasable calcium in endolysosomes with influx of extracellular calcium into neurons. This is a novel means by which intraneuronal stores of calcium can contribute to spatial and temporal integration of calcium signaling. In support of this novel mechanism, we found that calcium could be released from endolysosomes following stimulation of a number of different mechanisms, that the calcium release could be independent of other organellar stores of calcium, that release of calcium from endolysosomes triggered calcium influx, and that the calcium influx was regulated by N-type calcium channels and lysosome exocytosis (Fig. 27.2).

Capacitative influx of calcium into cells was described over 30 years ago [32]. Such calcium influx mechanisms, that are now commonly referred to as store-operated calcium entry (SOCE), are principally initiated by a reduction in

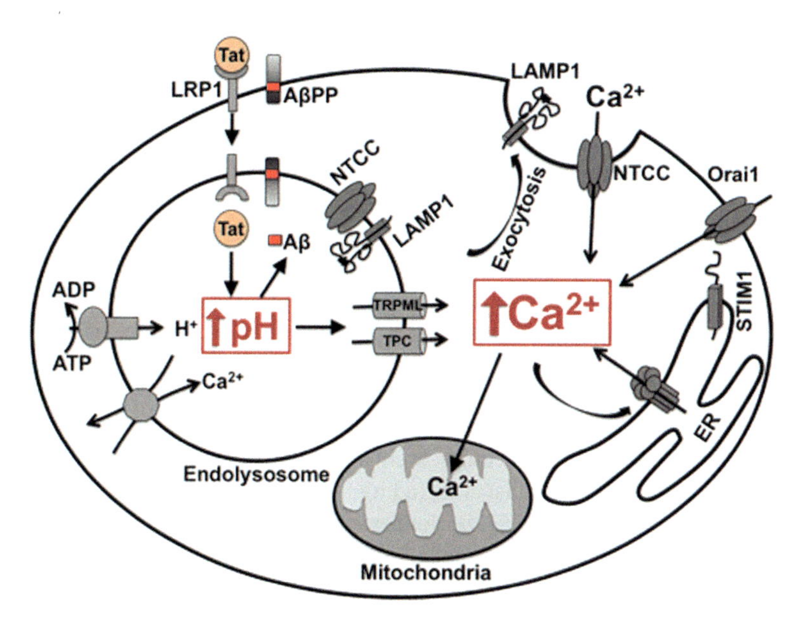

Fig. 27.2 HIV-1 Tat de-acidifies endolysosomes, increases amyloidogenesis, and releases calcium from readily releasable stores in endolysosomes. Calcium released from endolysosomes can affect mitochondrial and endoplasmic reticulum (ER) calcium stores, and increase store operated calcium (SOCE) mechanisms. Mechanistically, following de-acidification endolysosome calcium is released through TRPML1 and two pore channels (TPCs). The calcium signals can be amplified by releasing calcium from other organelles including mitochondria and ER, and by activating ER-based SOCE involving STIM1 and Orai channels as well as acidic store operated calcium entry involving N-type calcium channels (NTCCs)

endoplasmic reticulum calcium stores followed by influx of extracellular calcium in a variety of cells including neurons in order to refill the depleted stores of calcium. Mechanistically, depleting endoplasmic reticulum calcium stores drives the oligomerization and translocation of stromal interaction molecule 1 (STIM1) proteins to endoplasmic reticulum junctions close to the plasma membrane. Such STIM1 translocation induces the clustering of calcium release-activated calcium modulator 1 (Orai1) channels and/or transient receptor potential (TRP) cation channels into plasma membranes thereby enabling extracellular calcium entry [33].

Conceptually, but not mechanistically, we observed similar store-operated calcium entry involving endolysosomes in neurons. Using primary cultures of rat cortical neurons, we found that calcium was released from endolysosomes following treatment with the two-pore channel agonist NAADP-AM, the v-ATPase inhibitor BAF, and the lysosomotropic agent GPN; all of which de-acidify endolysosomes [34]. However, when these experiments were conducted in the absence of extracellular calcium, de-acidification of endolysosomes with NAADP-AM, BAF and GPN increased only slightly levels of free cytosolic calcium. When these same experiments were conducted in the presence of extracellular calcium, NAADP-AM, BAF and GPN all increased significantly the levels of free cytosolic calcium. Although it is not well understood currently, the relatively small release of calcium from endolysosomes causes a much larger influx of extracellular calcium and this might be due to plasma membrane depolarization as is accompanied by NAADP-induced endolysosome calcium release [34, 35]. Besides neurons, phenomena similar to aSOCE have been described in other cell types where NAADP has been found to induce endolysosome calcium release and large influxes of calcium across plasma membranes [12, 36–39]. These observations suggested to us that endolysosome de-acidification by three completely different mechanisms led directly or indirectly to an enhanced influx of calcium into neurons. Accordingly, we next tested more specifically the extent to which a store-operated mechanism might control the observed calcium influx across the plasma membrane.

Using approaches similar to those used by others and us, we began studying store-operated calcium mechanisms including the classical endoplasmic reticulum-based capacitative SOCE. Indeed, we confirmed that in the absence of extracellular calcium and following depletion of endoplasmic reticulum calcium with the SERCA pump inhibitor thapsigargin (TG) there was a significant increase in levels of free intracellular calcium only when calcium was re-introduced to the extracellular medium. With this positive control for the functional presence of endoplasmic reticulum-based SOCE in our cultured neurons, we conducted similar experiments with agents that de-acidify endolysosomes and release calcium from endolysosome stores. Even after depleting ER pools of calcium with TG, application of NAADP-AM, BAF and GPN still caused increased influx of extracellular calcium and still induced increased levels of intracellular calcium. Thus, in these neurons there appeared to be at least two separate and functional store-operated calcium mechanisms; one governed by endoplasmic reticulum and the other by endolysosomes.

In testing the distinctive nature of the two store-operated calcium mechanisms governed by endoplasmic reticulum or endolysosomes, we used pharmacological and molecular/genetic strategies. Using siRNA to knock-down protein expression levels of STIM1, a protein that is central to SOCE, and the SOCE blockers SKF-96365 and 2-APB we were able to block significantly TG-induced release of calcium from endoplasmic reticulum, but we were unable to block significantly NAADP-AM-, BAF- and GPN-induced calcium influx. However, we were able to block significantly NAADP-AM-, BAF- and GPN-induced calcium influx with the selective N-type calcium channel (NTCC) blocker (ω-conotoxin). The selective and specific nature of this inhibition by ω-conotoxin was confirmed further by showing that NAADP-AM-, BAF- and GPN-induced calcium influx was not blocked by inhibitors of L-type (nimodipine, verapamil) and P/Q-type (ω-agatoxin) calcium channels. Moreover, we confirmed these pharmacological findings by showing that siRNA knockdown of NTCCs attenuated significantly NAADP-AM-, BAF- and GPN-induced calcium influx, but did not affect TG-induced SOCE. Together, the above results demonstrated that calcium released from endolysosomes can be distinct from calcium released from endoplasmic reticulum through SOCE mechanisms and that the calcium released from endolysosomes is capable of activating cell surface calcium channels to stimulate calcium influx. These findings support and extend earlier findings that calcium released from endolysosomes did not stimulate endoplasmic reticulum-dependent SOCE in MDCK epithelial cells [23]. Accordingly, this new mechanism was termed by us as "acidic store-operated calcium entry" (aSOCE) [34].

27.4 Role of Lysosome Exocytosis in Acidic Store-Operated Calcium Entry (aSOCE)

Multiple mechanisms might control aSOCE involving NTCCs. One such mechanism might involve lysosome exocytosis because we have shown using a quantitative biotinylation of surface proteins assay that NAADP-AM, BAF and GPN all increased cell surface protein expression levels of NTCCs and lysosome-associated glycoprotein 1 (LAMP1). Next, we addressed the possibility that lysosome exocytosis and NTCCs were linked directly by conducting co-immunoprecipitation studies and found a physical interaction between NTCCs and LAMP1. Because LAMP1 is critical for lysosome exocytosis [40], those observations suggested to us that lysosome exocytosis might be a functional partner in aSOCE especially because aSOCE was inhibited following siRNA knockdown of protein expression levels of LAMP1. Thus, de-acidification of endolysosomes might be of central importance because NAADP-AM, BAF and GPN through very different initial mechanisms all appeared to enhance lysosome exocytosis and the recycling of NTCCs to the plasma membrane where they participated in calcium influx generally and aSOCE more specifically. Physically, this makes sense as well because of findings that de-acidification of endolysosomes changes cellular distribution patterns of these organelles from a mostly peri-nuclear pattern to one where the endolysosomes migrate close to the plasma membrane [41]. Thus, functionally and physically there is evidence favoring endolysosomes and endolysosome exocytosis in calcium entry.

27.5 Physical Interactions and Functional Relevance of Inter-organellar Signaling

In addition to physical interactions between endolysosomes and plasma membranes, it is becoming increasingly clear that endolysosomes physically and functionally interact as well with other intracellular organelles including mitochondria and endoplasmic reticulum. Such recognition has led to an appreciation for dynamic physical and chemical communications between intracellular organelles including those regulated by pH and calcium.

Physical interactions between mitochondria and endoplasmic reticulum were first described about 60 years ago and the functional significance of mitochondria-associated membranes was first characterized about 30 years ago [42]. Even today, there continues to be work focused on the physical and functional interactions between organelles [43] as well as the role that organellar interactions plays in the pathogenesis of neurodegenerative diseases [44, 45]. As it relates to endolysosomes, it is now known that there are extensive physical interactions between endolysosomes and mitochondria and that these inter-organellar communications participate in lipid and metabolite exchange as well as mitochondrial quality control [46]. Conversely, mitochondrial dysfunction has been found to negatively affect lysosome structure and function through reactive oxygen species-dependent mechanisms [47]. Extensive membrane contact sites have been described between lysosomes and endoplasmic reticulum, that these contact sites were evolutionarily conserved, and that calcium released from lysosomes was sufficient to stimulate endoplasmic reticulum-dependent calcium-induced calcium release [48, 49, 50]. However, only recently was it shown that endolysosomes maintain their 1000-fold calcium concentration gradient in cells in part by refilling endolysosome stores of calcium from IP_3-regulated stores of calcium in endoplasmic reticulum [51]. Some of the differences in findings as to calcium movements between organelles might be because of cell-specific mechanisms. In addition, the difficult nature of understanding inter-organellar calcium dynamics is highlighted by work showing that STIM1 and STIM2 are expressed in endolysosomes, at least in platelets, and that depletion of acidic organellar stores of calcium can increase protein-protein interactions between STIM proteins with Orai1 and TRPC channels to induce SOCE [52]. It is further complicated by findings that calcium released through endolysosome-resident TRML1 channels can cause calcium release from endoplasmic reticulum and calcium influx [53] and that NAADP has been implicated in this "cross-talk" [54].

27.6 Possible Role of Endolysosomes and aSOCE in Pathogenesis of Alzheimer's Disease and HIV-1 Associated Neurocognitive Disorder (HAND)

Disturbances in endolysosome structure and/or function have been noted in a variety of neurodegenerative disorders including Alzheimer's disease (AD) and HIV-1 neurocognitive disorder (HAND) [55–59]. AD is a devastating age-related neurodegenerative disease that is the commonest cause of dementia in people over the age of 65. People with HAND, on the other hand, exhibit neurological complications ranging from mild (mild cognitive impairment) to severe (dementia). In the current era of anti-retroviral therapeutics HIV-1 infected individuals are living almost full life-spans, but are now experiencing a prevalence rate of over 50% for HAND [60, 61]. Clinically and pathologically people living with neuroHIV-1 are exhibiting AD-like symptoms including learning and memory deficits as well as increased amyloidogenesis. Although the pathogenesis of HAND is not fully understood, HIV-1 proteins including the HIV-1 transactivator of transcription protein Tat have been implicated by others and us to be causative virotoxins in HAND [62–69]. Among the HIV-1 viral proteins, HIV-1 Tat is present in brains of HIV-1 infected individuals and its levels stay elevated in CSF even when HIV-1 viral levels are immeasurable [70]. Others and we have shown that HIV-1 Tat directly excites neurons [65, 71, 72], disturbs neuronal calcium homeostasis [64, 73], disrupts synaptic integrity [74, 75], and induces neurotoxicity [68, 76].

Endolysosome dysfunction has been implicated in the development of at least two pathological hallmarks of AD and HAND; Aβ accumulation and neurofibrillary tangle formation. Endolysosomes are very important for amyloidogenic processing of AβPP to Aβ because amyloid β precursor protein (AβPP) is first endocytosed, the amyloidogenic enzymes BACE-1 and γ-secretase are almost exclusively located in endosomes and lysosomes, the acidic environment of endolysosomes is favorable for amyloidogenic metabolism of AβPP, and Aβ can be either accumulated in or released by exocytosis from endolysosomes [77–83]. Tau is a microtubule-associated protein, and when hyperphosphorylated it aggregates and contributes to the formation of neurofibrillary tangles. Tau aggregates can be degraded by cathepsin D in autophagosomes-lysosomes [84, 85], and endolysosome dysfunction contributes to tau aggregation and neurofibrillary tangle formation [86, 87]. On the other hand, transcriptional activation of lysosome biogenesis can clear aggregated tau [88]. Thus, endolysosomes are important sites for development of these neurological disorders.

Dysregulation of intracellular calcium has also been implicated in the neuropathogenesis of these same neurological disorders. And it is clear (see above) that de-acidification of endolysosomes releases calcium from these acidic stores [28, 89, 90]. We found that HIV-1 Tat protein elevated endolysosome pH and disturbed the structure and function of endolysosomes [74], a prominent and early pathological feature of HAND [57, 58]. Clearly, endolysosome calcium stores contribute to neuronal calcium signaling and function [91–93] and calcium release from endolysosomes triggers calcium release from endoplasmic reticulum [11, 17] and through plasma membranes via aSOCE (see above).

HIV-1 proteins including HIV-1 Tat, and anti-retroviral therapeutic drugs contribute to the development of AD-like pathology including increases in Aβ levels [94–99]. HIV-1 Tat enters neurons via receptor-mediated endocytosis [100–102]. The Tat-induced de-acidification of endolysosomes and resulting effects on calcium dyshomeostasis likely results from the ability of HIV-1 Tat to decrease the levels and activity of vacuolar-ATPase as well as compensatory increases in cathepsin D and LAMP-1 [103]. The consequences of such alterations in calcium dynamics and homeostasis are synaptic disruption and neurotoxicity [104–106].

Endolysosomes contain physiologically important levels of calcium that is readily releasable by a number of stimuli and insults. The calcium can exit through a number of channels including TRPML and two pore channels. Once released the calcium can signal other organelles to release calcium and for greater influx of calcium through plasma membrane-resident calcium channels especially N-type calcium channels. These effects on endolysosome structure and function have clear implications to the pathogenesis of AD and HAND; neurological disorders that show overlap in terms of clinical and pathological features. We are excited to be part of this emerging area of cell biology focused on inter-organellar signaling and look forward to further studies elucidating physiological and pathological consequences of calcium release from endolysosome stores.

Acknowledgements The authors gratefully acknowledge the funding provided by the NIH for our work; P30GM103329, R01MH100972, R01MH105329, R01MH119000, R01NS065957, and R01DA032444.

References

1. Berridge MJ, Bootman MD, Roderick HL (2003) Calcium signalling: dynamics, homeostasis and remodelling. Nat Rev Mol Cell Biol 4:517–529
2. Jaiswal JK (2001) Calcium – how and why? J Biosci 26(3):357–363
3. Case RM, Eisner D, Gurney A, Jones O, Muallem S, Verkhratsky A (2007) Evolution of calcium homeostasis: from birth of the first cell to an omnipresent signalling system. Cell Calcium 42(4–5):345–350
4. Wideman JG, Leung KF, Field MC, Dacks JB (2014) The cell biology of the Endocytic system from an evolutionary perspective. Cold Spring Harb Perspect Biol [Internet] 6(4):a016998. Apr 1 [cited 2018 July 16]. Available from: http://www.ncbi.nlm.nih.gov/pubmed/24478384
5. Carafoli E (2010) The fateful encounter of mitochondria with calcium: how did it happen? Biochim Biophys Acta Bioenerg [Internet] 1797(6–7):595–606. June 1 [cited 2018 July 16]. Available from: https://www.sciencedirect.com/science/article/pii/S0005272810001301#fig1
6. Patel S, Cai X (2015) Evolution of acidic Ca^{2+}stores and their resident Ca^{2+}–permeable channels [internet]. Cell Calcium 57:222–230. [cited 2018 July 14]. Available from: https://www.sciencedirect.com/science/article/pii/S0143416014002012

7. Raffaello A, Mammucari C, Gherardi G, Rizzuto R (2016) Calcium at the Center of Cell Signaling: interplay between endoplasmic reticulum, mitochondria, and lysosomes. Trends Biochem Sci 41:1035–1049. [cited 2017 May 17]. Available from: http://www.cell.com/trends/biochemical-sciences/pdf/S0968-0004(16)30147-5.pdf
8. Christensen KA, Myers JT, JA S (2002) pH-dependent regulation of lysosomal calcium in macrophages. J Cell Sci 115(Pt 3):599–607
9. Morgan AJ, Platt FM, Lloyd-Evans E, Galione A (2011) Molecular mechanisms of endolysosomal Ca^{2+} signalling in health and disease. Biochem J [Internet] 439(3):349–374. Available from: http://www.biochemj.org/content/439/3/349.abstract
10. Moreno SNJ, Docampo R (2009) The role of acidocalcisomes in parasitic protists. J Eukaryot Microbiol 56:208–213
11. Patel S, Docampo R (2010) Acidic calcium stores open for business: expanding the potential for intracellular Ca^{2+}signaling. Trends Cell Biol 20:277–286
12. Brailoiu E, Churamani D, Cai X, Schrlau MG, Brailoiu GC, Gao X et al (2009) Essential requirement for two-pore channel 1 in NAADP-mediated calcium signaling (a). J Cell Biol 186(2):201–209
13. Calcraft PJ, Ruas M, Pan Z, Cheng X, Arredouani A, Hao X et al (2009) NAADP mobilizes calcium from acidic organelles through two-pore channels. Nature 459(7246):596–600
14. Zong X, Schieder M, Cuny H, Fenske S, Gruner C, Rötzer K et al (2009) The two-pore channel TPCN2 mediates NAADP-dependent Ca^{2+} –release from lysosomal stores. Pflugers Arch Eur J Physiol [Internet] 458(5):891–899. Sept 26 [cited 2018 July 14]. Available from: http://www.ncbi.nlm.nih.gov/pubmed/19557428
15. Ruas M, Rietdorf K, Arredouani A, Davis LC, Lloyd-Evans E, Koegel H et al (2010) Purified TPC isoforms form NAADP receptors with distinct roles for Ca^{2+}signaling and Endolysosomal trafficking. Curr Biol [Internet] 8(20):703–709. Apr 27 [cited 2018 July 27]. Available from: http://www.ncbi.nlm.nih.gov/pubmed/20346675
16. Schieder M, Rötzer K, Brüggemann A, Biel M, Wahl-Schott CA (2010) Characterization of two-pore channel 2 (TPCN2)-mediated Ca^{2+} currents in isolated lysosomes. J Biol Chem [Internet] 285(28):21219–21222. July 9 [cited 2018 July 14]. Available from: http://www.ncbi.nlm.nih.gov/pubmed/20495006
17. Zhu MX, Ma J, Parrington J, Calcraft PJ, Galione A, Evans AM (2010) Calcium signaling via two-pore channels: local or global, that is the question. AJP Cell Physiol [Internet] 298(3):C430–C441. Available from: http://ajpcell.physiology.org/cgi/doi/10.1152/ajpcell.00475.2009
18. Camacho M, Machado JD, Alvarez J, Borges R (2008) Intravesicular calcium release mediates the motion and exocytosis of secretory organelles: a study with adrenal chromaffin cells. J Biol Chem 283(33):22383–22389
19. Machado JD, Camacho M, Alvarez J, Borges R (2009) On the role of intravesicular calcium in the motion and exocytosis of secretory organelles. Commun Integr Biol 2(2):71–73
20. Starkus JG, Fleig A, Penner R (2010) The calcium-permeable non-selective cation channel TRPM2 is modulated by cellular acidification. J Physiol 588(8):1227–1240
21. Kiselyov K, Colletti GA, Terwilliger A, Ketchum K, CWP L, Quinn J et al (2011) TRPML: transporters of metals in lysosomes essential for cell survival? Cell Calcium 50:288–294
22. Cao Q, Zhong XZ, Zou Y, Murrell-Lagnado R, Zhu MX, Dong XP (2015) Calcium release through P2X4 activates calmodulin to promote endolysosomal membrane fusion. J Cell Biol 209(6):879–894
23. Haller T, Dietl P, Deetjen P, Völkl H (1996) The lysosomal compartment as intracellular calcium store in MDCK cells: a possible involvement in InsP3-mediated Ca^{2+}release. Cell Calcium 19(2):157–165
24. McGuinness L, Bardo SJ, Emptage NJ (2007) The lysosome or lysosome-related organelle may serve as a Ca^{2+}store in the boutons of hippocampal pyramidal cells. Neuropharmacology 52(1):126–135
25. Repnik U, Česen MH, Turk B (2013) The endolysosomal system in cell death and survival. Cold Spring Harb Perspect Biol [Internet] 5(1):a008755. Jan 1 [cited 2017 Oct 16]. Available from: http://www.ncbi.nlm.nih.gov/pubmed/23284043

26. Ferguson SM (2018) Neuronal lysosomes. Neurosci Lett [Internet]. Available from: http://linkinghub.elsevier.com/retrieve/pii/S030439401830260X
27. Goo MS, Sancho L, Slepak N, Boassa D, Deerinck TJ, Ellisman MH et al (2017) Activity-dependent trafficking of lysosomes in dendrites and dendritic spines. J Cell Biol 216(8):2499–2513
28. Galione A, Morgan AJ, Arredouani A, Davis LC, Rietdorf K, Ruas M et al (2010) NAADP as an intracellular messenger regulating lysosomal calcium-release channels. Biochem Soc Trans [Internet] 38(6):1424–1431. Available from: http://biochemsoctrans.org/lookup/doi/10.1042/BST0381424
29. Shen D, Wang X, Li X, Zhang X, Yao Z, Dibble S et al (2012) Lipid storage disorders block lysosomal trafficking by inhibiting a TRP channel and lysosomal calcium release. Nat Commun 3:731
30. Macgregor A, Yamasaki M, Rakovic S, Sanders L, Parkesh R, Churchill GC et al (2007) NAADP controls cross-talk between distinct Ca^{2+} Stores in the Heart. J Biol Chem [Internet] 282(20):15302–15311. May 18 [cited 2018 July 16]. Available from: http://www.ncbi.nlm.nih.gov/pubmed/17387177
31. Galionc A, Parrington J, Funnell T (2011) Physiological roles of NAADP-mediated Ca^{2+} signaling. Sci China Life Sci [Internet] 54(8):725–732. Available from: http://link.springer.com/10.1007/s11427-011-4207-5
32. Putney JW (1986) A model for receptor-regulated calcium entry. Cell Calcium 7(1):1–12
33. Putney JW (2009) Capacitative calcium entry: from concept to molecules. Immunol Rev 231:10–22
34. Hui L, Geiger NH, Bloor-Young D, Churchill GC, Geiger JD, Chen X (2015) Release of calcium from endolysosomes increases calcium influx through N-type calcium channels: evidence for acidic store-operated calcium entry in neurons. Cell Calcium [Internet] 58:617–627. [cited 2017 May 31]. Available from: http://ac.els-cdn.com/S0143416015001529/1-s2.0-S0143416015001529-main.pdf?_tid=d155fae0-460a-11e7-a743-00000aab0f02&acdnat=1496239973_5d092efe55666657b3cd6c5dc4881cf0
35. Arredouani A, Ruas M, Collins SC, Parkesh R, Clough F, Pillinger T et al (2015) Nicotinic acid adenine dinucleotide phosphate (NAADP) and endolysosomal two-pore channels modulate membrane excitability and stimulus-secretion coupling in mouse pancreatic β cells. J Biol Chem [Internet] 290(35):21376–21392. Aug 28 [cited 2018 July 14]. Available from: http://www.ncbi.nlm.nih.gov/pubmed/26152717
36. Moccia F, Lim D, Nusco GA, Ercolano E, Santella L (2003) NAADP activates a Ca^{2+} current that is dependent on F-actin cytoskeleton. FASEB J 17(13):1907–1909
37. Moccia F, Billington RA, Santella L (2006) Pharmacological characterization of NAADP-induced Ca^{2+} signals in starfish oocytes. Biochem Biophys Res Commun 348(2):329–336
38. Naylor E, Arredouani A, Vasudevan SR, Lewis AM, Parkesh R, Mizote A et al (2009) Identification of a chemical probe for NAADP by virtual screening. Nat Chem Biol 5(4):220–226
39. Churchill GC, O'Neill JS, Masgrau R, Patel S, Thomas JM, Genazzani AA et al (2003) Sperm deliver a new second messenger: NAADP. Curr Biol 13(2):125–128
40. Yogalingam G, Bonten EJ, van de Vlekkert D, Hu H, Moshiach S, Connell SA et al (2008) Neuraminidase 1 is a negative regulator of Lysosomal exocytosis. Dev Cell 15(1):74–86
41. Li X, Rydzewski N, Hider A, Zhang X, Yang J, Wang W et al (2016) A molecular mechanism to regulate lysosome motility for lysosome positioning and tubulation. Nat Cell Biol 18(4):404–417
42. Herrera-Cruz MS, Simmen T (2017) Over six decades of discovery and characterization of the architecture at mitochondria-associated membranes (MAMs). Adv Exp Med Biol 997:13–31
43. Wu Y, Whiteus C, Xu CS, Hayworth KJ, Weinberg RJ, Hess HF et al (2017) Contacts between the endoplasmic reticulum and other membranes in neurons. Proc Natl Acad Sci [Internet] 114(24):E4859–E4867. Available from: http://www.pnas.org/lookup/doi/10.1073/pnas.1701078114

44. Schon EA, Area-Gomez E (2013) Mitochondria-associated ER membranes in Alzheimer disease. Mol Cell Neurosci [Internet] 55:26–36. July [cited 2018 July 27]. Available from: http://www.ncbi.nlm.nih.gov/pubmed/22922446
45. Joshi AU, Kornfeld OS, Mochly-Rosen D (2016) The entangled ER-mitochondrial axis as a potential therapeutic strategy in neurodegeneration: a tangled duo unchained. Cell Calcium 60:218–234
46. Soto-Heredero G, Baixauli F, Mittelbrunn M (2017) Interorganelle communication between mitochondria and the Endolysosomal system. Front Cell Dev Biol [Internet] 5:95. Available from: http://journal.frontiersin.org/article/10.3389/fcell.2017.00095/full
47. Demers-Lamarche J, Guillebaud G, Tlili M, Todkar K, Bélanger N, Grondin M et al (2016) Loss of mitochondrial function impairs lysosomes$_{*}$. J Biol Chem [Internet] 291(19):10263–10276. May 6 [cited 2017 Dec 8]. Available from: http://www.ncbi.nlm.nih.gov/pubmed/26987902
48. Kilpatrick BS, Eden ER, Schapira AH, Futter CE, Patel S (2013) Direct mobilisation of lysosomal Ca^{2+} triggers complex Ca^{2+} signals. J Cell Sci [Internet] 126(Pt 1):60–66. Available from: http://www.ncbi.nlm.nih.gov/pubmed/23108667%5Cnhttp://jcs.biologists.org/content/joces/126/1/60.full.pdf
49. Penny CJ, Kilpatrick BS, Eden ER, Patel S (2015) Coupling acidic organelles with the ER through Ca^{2+} microdomains at membrane contact sites. Cell Calcium 58:387–396
50. Hariri H, Ugrankar R, Liu Y, Henne WM (2016) Inter-organelle ER-endolysosomal contact sites in metabolism and disease across evolution. Communicative and Integrative Biology 9(3):e1156278
51. Garrity AG, Wang W, Collier CM, Levey SA, Gao Q, Xu H (2016) The endoplasmic reticulum, not the pH gradient, drives calcium refilling of lysosomes. Elife [Internet] 5:e15887. May 23 [cited 2017 May 11]. Available from: http://www.ncbi.nlm.nih.gov/pubmed/27213518
52. Zbidi H, Jardin I, Woodard GE, Lopez JJ, Berna-Erro A, Salido GM et al (2011) STIM1 and STIM2 are located in the acidic Ca^{2+} stores and associates with Orai1 upon depletion of the acidic stores in human platelets. J Biol Chem 286(14):12257–12270
53. Kilpatrick BS, Yates E, Grimm C, Schapira AH, Patel S (2016) Endo-lysosomal TRP mucolipin-1 channels trigger global ER Ca^{2+} release and Ca^{2+} influx. J Cell Sci [Internet]. 129(20):3859–3867. [cited 2018 July 27]. Available from: http://www.ncbi.nlm.nih.gov/pubmed/27577094
54. Ronco V, Potenza DM, Denti F, Vullo S, Gagliano G, Tognolina M et al (2015) A novel Ca^{2+}–mediated cross-talk between endoplasmic reticulum and acidic organelles: implications for NAADP-dependent Ca^{2+}signalling. Cell Calcium [Internet] 57(2):89–100. Feb 1 [cited 2018 July 14]. Available from: https://www.sciencedirect.com/science/article/pii/S0143416015000020
55. Tate BA, Mathews PM (2006) Targeting the role of the endosome in the pathophysiology of Alzheimer's disease: a strategy for treatment. Sci Aging Knowl Environ [Internet] 2006(10):re2–re2. June 28 [cited 2018 July 27]. Available from: http://www.ncbi.nlm.nih.gov/pubmed/16807486
56. Boland B, Kumar A, Lee S, Platt FM, Wegiel J, Yu WH et al (2008) Autophagy induction and Autophagosome clearance in neurons: relationship to Autophagic pathology in Alzheimer's disease. J Neurosci [Internet] 28(27):6926–6937. July 2 [cited 2018 July 27]. Available from: http://www.ncbi.nlm.nih.gov/pubmed/18596167
57. Gelman BB, Soukup VM, Holzer CE 3rd, Fabian RH, Schuenke KW, Keherly MJ et al (2005) Potential role for white matter lysosome expansion in HIV-associated dementia. J Acquir Immune Defic Syndr 39(4):422–425
58. Spector SA, Zhou D (2008) Autophagy: an overlooked mechanism of HIV-1 pathogenesis and neuroAIDS? Autophagy [Internet] 4(5):704–706. July [cited 2018 July 27]. Available from: http://www.ncbi.nlm.nih.gov/pubmed/18424919

59. Cysique LA, Hewitt T, Croitoru-Lamoury J, Taddei K, Martins RN, Chew CS et al (2015) APOE ε4 moderates abnormal CSF-abeta-42 levels, while neurocognitive impairment is associated with abnormal CSF tau levels in HIV+ individuals – a cross-sectional observational study. BMC Neurol [Internet] 15(1):51. Dec 1 [cited 2018 July 27]. Available from: http://www.ncbi.nlm.nih.gov/pubmed/25880550
60. Ellis RJ, Rosario D, Clifford DB, McArthur JC, Simpson D, Alexander T et al (2010) Continued high prevalence and adverse clinical impact of human immunodeficiency virus–associated sensory neuropathy in the era of combination antiretroviral therapy. Arch Neurol [Internet] 67(5):552. May 1 [cited 2018 July 27]. Available from: http://www.ncbi.nlm.nih.gov/pubmed/20457954
61. Heaton RK, Clifford DB, Franklin DR, Woods SP, Ake C, Vaida F et al (2010) HIV-associated neurocognitive disorders persist in the era of potent antiretroviral therapy: charter study. Neurol Int 75(23):2087–2096. Dec 7 [cited 2017 Sept 13]. Available from: http://www.ncbi.nlm.nih.gov/pubmed/21135382
62. Sabatier JM, Vives E, Mabrouk K, Benjouad A, Rochat H, Duval A et al (1991) Evidence for neurotoxic activity of tat from human immunodeficiency virus type 1. J Virol [Internet] 65(2):961 967. Feb [cited 2018 July 27]. Available from: http://www.ncbi.nlm.nih.gov/pubmed/1898974
63. Weeks BS, Lieberman DM, Johnson B, Roque E, Green M, Loewenstein P et al (1995) Neurotoxicity of the human immunodeficiency virus type 1 tat transactivator to PC12 cells requires the tat amino acid 49-58 basic domain. J Neurosci Res [Internet] 42(1):34–40. Sept 1 [cited 2018 July 27]. Available from: http://doi.wiley.com/10.1002/jnr.490420105
64. Haughey NJ, Holden CP, Nath A, Geiger JD (1999) Involvement of inositol 1,4,5-trisphosphate-regulated stores of intracellular calcium in calcium dysregulation and neuron cell death caused by HIV-1 protein tat. J Neurochem [Internet] 73(4):1363–1374. Oct [cited 2018 July 27]. Available from: http://www.ncbi.nlm.nih.gov/pubmed/10501179
65. Nath A, Haughey NJ, Jones M, Anderson C, Bell JE, Geiger JD (2000) Synergistic neurotoxicity by human immunodeficiency virus proteins tat and gp120: protection by memantine. Ann Neurol [Internet] 47(2):186–194. Feb [cited 2018 July 27]. Available from: http://www.ncbi.nlm.nih.gov/pubmed/10665489
66. Pérez A, Probert AW, Wang KK, Sharmeen L (2001) Evaluation of HIV-1 tat induced neurotoxicity in rat cortical cell culture. J Neurovirol [Internet] 7(1):1–10. Feb 1 [cited 2018 July 27]. Available from: http://www.ncbi.nlm.nih.gov/pubmed/11519477
67. King JE, Eugenin EA, Buckner CM, Berman JW (2006) HIV tat and neurotoxicity. Microbes Infect [Internet] 8(5):1347–1357. Apr [cited 2018 July 27]. Available from: http://www.ncbi.nlm.nih.gov/pubmed/16697675
68. Buscemi L, Ramonet D, Geiger JD (2007) Human immunodeficiency virus type-1 protein tat induces tumor necrosis factor-alpha-mediated neurotoxicity. Neurobiol Dis [Internet] 26(3):661–670. June [cited 2018 June 1]. Available from: http://www.ncbi.nlm.nih.gov/pubmed/17451964
69. Agrawal L, Louboutin J-P, Reyes BAS, Van Bockstaele EJ, Strayer DS (2012) HIV-1 tat neurotoxicity: a model of acute and chronic exposure, and neuroprotection by gene delivery of antioxidant enzymes. Neurobiol Dis [Internet] 45(2):657–670. Feb [cited 2018 July 27]. Available from: http://www.ncbi.nlm.nih.gov/pubmed/22036626
70. Johnson TP, Patel K, Johnson KR, Maric D, Calabresi PA, Hasbun R et al (2013) Induction of IL-17 and nonclassical T-cell activation by HIV-tat protein. Proc Natl Acad Sci U S A [Internet] 110(33):13588–13593. Aug 13 [cited 2017 July 20]. Available from: http://www.ncbi.nlm.nih.gov/pubmed/23898208
71. DSK M, Jsnudsen BE, Geiger JD, Brownstone RM, Nath A (1995) Human lmmunodehclency V m s lype 1 tat activates non-N-Methyla-aspartate excitatory ammo acid receptors and causes neurotoxicity. Ann Neurol [Internet] 37(3):373–380. [cited 2017 Nov 20]. Available from: https://med-und.illiad.oclc.org/illiad/illiad.dll?Action=10&Form=75&Value=116096

72. Nath A, Psooy K, Martin C, Knudsen B, Magnuson DS, Haughey N et al (1996) Identification of a human immunodeficiency virus type 1 tat epitope that is neuroexcitatory and neurotoxic. J Virol [Internet] 3(70):1475–1480. [cited 2018 June 13]. Available from: https://www.ncbi.nlm.nih.gov/pmc/articles/PMC189968/pdf/701475.pdf
73. Haughey NJ, Mattson MP (2002) Calcium dysregulation and neuronal apoptosis by the HIV-1 proteins tat and gp120. J Acquir Immune Defic Syndr [Internet] 31(Suppl 2):S55–S61. Oct 1 [cited 2018 July 27]. Available from: http://www.ncbi.nlm.nih.gov/pubmed/12394783
74. Kim HJ, Martemyanov KA, Thayer SA (2008) Human immunodeficiency virus protein tat induces synapse loss via a reversible process that is distinct from cell death. J Neurosci [Internet] 28(48):12604–12613. Nov 26 [cited 2018 July 27]. Available from: http://www.ncbi.nlm.nih.gov/pubmed/19036954
75. Fitting S, Xu R, Bull C, Buch SK, El-Hage N, Nath A et al (2010) Interactive comorbidity between opioid drug abuse and HIV-1 tat: chronic exposure augments spine loss and sublethal dendritic pathology in striatal neurons. Am J Pathol [Internet] 177(3):1397–1410. Sept [cited 2018 July 27]. Available from: http://www.ncbi.nlm.nih.gov/pubmed/20651230
76. Hui L, Chen X, Haughey NJ, Geiger JD (2012) Role of Endolysosomes in HIV-1 tat-induced neurotoxicity. ASN Neuro [Internet] 4(4):AN20120017. Apr 3 [cited 2018 July 27]. Available from: http://www.ncbi.nlm.nih.gov/pubmed/22591512
77. Rajendran L, Annaert W (2012) Membrane trafficking pathways in Alzheimer's disease. Traffic [Internet] 13(6):759–770. June [cited 2018 July 27]. Available from: http://www.ncbi.nlm.nih.gov/pubmed/22269004
78. Morel E, Chamoun Z, Lasiecka ZM, Chan RB, Williamson RL, Vetanovetz C et al (2013) Phosphatidylinositol-3-phosphate regulates sorting and processing of amyloid precursor protein through the endosomal system. Nat Commun [Internet] 4(1):2250. Dec 2 [cited 2018 July 27]. Available from: http://www.ncbi.nlm.nih.gov/pubmed/23907271
79. Jiang S, Li Y, Zhang X, Bu G, Xu H, Zhang Y (2014) Trafficking regulation of proteins in Alzheimer's disease. Mol Neurodegener [Internet] 9:6. Jan 11 [cited 2018 July 27]. Available from: http://www.ncbi.nlm.nih.gov/pubmed/24410826
80. Nixon RA (2005) Endosome function and dysfunction in Alzheimer's disease and other neurodegenerative diseases. Neurobiol Aging [Internet] 26(3):373–382. Mar [cited 2018 July 27]. Available from: http://www.ncbi.nlm.nih.gov/pubmed/15639316
81. Rajendran L, Schneider A, Schlechtingen G, Weidlich S, Ries J, Braxmeier T et al (2008) Efficient inhibition of the Alzheimer's disease -Secretase by membrane targeting. Science (80-) [Internet] 320(5875):520–523. Apr 25 [cited 2018 July 27]. Available from: http://www.ncbi.nlm.nih.gov/pubmed/18436784
82. Shimizu H, Tosaki A, Kaneko K, Hisano T, Sakurai T, Nukina N (2008) Crystal structure of an active form of BACE1, an enzyme responsible for amyloid protein production. Mol Cell Biol [Internet] 28(11):3663–3671. June 1 [cited 2018 July 27]. Available from: http://www.ncbi.nlm.nih.gov/pubmed/18378702
83. Sannerud R, Declerck I, Peric A, Raemaekers T, Menendez G, Zhou L et al (2011) ADP ribosylation factor 6 (ARF6) controls amyloid precursor protein (APP) processing by mediating the endosomal sorting of BACE1. Proc Natl Acad Sci [Internet] 108(34):E559–E568. Aug 23 [cited 2018 July 27]. Available from: http://www.ncbi.nlm.nih.gov/pubmed/21825135
84. Hamano T, Gendron TF, Causevic E, Yen S-H, Lin W-L, Isidoro C et al (2008) Autophagic-lysosomal perturbation enhances tau aggregation in transfectants with induced wild-type tau expression. Eur J Neurosci [Internet] 27(5):1119–1130. Mar [cited 2018 July 27]. Available from: http://www.ncbi.nlm.nih.gov/pubmed/18294209
85. Chesser AS, Pritchard SM, GVW J (2013) Tau clearance mechanisms and their possible role in the pathogenesis of Alzheimer disease. Front Neurol [Internet] 4:122. Sept 3 [cited 2018 July 27]. Available from: http://www.ncbi.nlm.nih.gov/pubmed/24027553

86. Jo C, Gundemir S, Pritchard S, Jin YN, Rahman I, GVW J (2014) Nrf2 reduces levels of phosphorylated tau protein by inducing autophagy adaptor protein NDP52. Nat Commun [Internet] 5(1):3496. Dec 25 [cited 2018 July 27]. Available from: http://www.nature.com/articles/ncomms4496
87. Bi X, Liao G (2007) Erratum: Autophagic-lysosomal dysfunction and neurodegeneration in Niemann-pick type C mice: lipid starvation or indigestion? (autophagy) [internet]. Autophagy 3:646–648. [cited 2018 July 27]. Available from: http://www.ncbi.nlm.nih.gov/pubmed/17921694
88. Polito VA, Li H, Martini-Stoica H, Wang B, Yang L, Xu Y et al (2014) Selective clearance of aberrant tau proteins and rescue of neurotoxicity by transcription factor EB. EMBO Mol Med [Internet] 6(9):1142–1160. Sept 1 [cited 2018 July 27]. Available from: http://www.ncbi.nlm.nih.gov/pubmed/25069841
89. Masgrau R, Churchill GC, Morgan AJ, Ashcroft SJH (2003) Galione a. NAADP: a new second messenger for glucose-induced Ca^{2+} responses in clonal pancreatic beta cells. Curr Biol [Internet] 13(3):247–251. Feb 4 [cited 2018 July 27]. Available from: http://www.ncbi.nlm.nih.gov/pubmed/12573222
90. Mitchell KJ, Lai FA, Rutter GA (2003) Ryanodine receptor type I and nicotinic acid adenine dinucleotide phosphate receptors mediate Ca^{2+} release from insulin-containing vesicles in living pancreatic β-cells (MIN6). J Biol Chem [Internet] 278(13):11057–11064. Mar 28 [cited 2018 July 27]. Available from: http://www.ncbi.nlm.nih.gov/pubmed/12538591
91. Haas E, Bhattacharya I, Brailoiu E, Damjanovic M, Brailoiu GC, Gao X et al (2009) Regulatory role of G protein-coupled estrogen receptor for vascular function and obesity. Circ Res [Internet] 104(3):288–291. Feb 13 [cited 2018 July 27]. Available from: http://www.ncbi.nlm.nih.gov/pubmed/19179659
92. Pandey V, Chuang C-C, Lewis AM, Aley PK, Brailoiu E, Dun NJ et al (2009) Recruitment of NAADP-sensitive acidic Ca^{2+} stores by glutamate. Biochem J [Internet] 422(3):503–512. Sept 15 [cited 2018 July 27]. Available from: http://www.ncbi.nlm.nih.gov/pubmed/19548879
93. Dickinson GD, Churchill GC, Brailoiu E, Patel S (2010) Deviant nicotinic acid adenine dinucleotide phosphate (NAADP)-mediated Ca^{2+} signaling upon lysosome proliferation. J Biol Chem [Internet] 285(18):13321–13325. Apr 30 [cited 2018 July 27]. Available from: http://www.ncbi.nlm.nih.gov/pubmed/20231291
94. Rempel HC, Pulliam L (2005) HIV-1 tat inhibits neprilysin and elevates amyloid β. AIDS [Internet] 19(2):127–135. Jan 28 [cited 2018 July 27]. Available from: http://www.ncbi.nlm.nih.gov/pubmed/15668537
95. Giunta B, Hou H, Zhu Y, Rrapo E, Tian J, Takashi M et al (2009) HIV-1 tat contributes to Alzheimer's disease-like pathology in PSAPP mice. Int J Clin Exp Pathol [Internet] 5(2):433–443. [cited 2018 July 27]. Available from: http://www.ncbi.nlm.nih.gov/pubmed/19294002
96. Aksenov MY, Aksenova MV, Mactutus CF, Booze RM (2010) HIV-1 protein-mediated amyloidogenesis in rat hippocampal cell cultures. Neurosci Lett [Internet] 475(3):174–178. May 21 [cited 2018 July 27]. Available from: http://www.ncbi.nlm.nih.gov/pubmed/20363291
97. Chen X, Hui L, Geiger NH, Haughey NJ, Geiger JD (2013) Endolysosome involvement in HIV-1 transactivator protein-induced neuronal amyloid beta production. Neurobiol Aging [Internet] 34(10):2370–2378. Oct [cited 2017 Aug 8]. Available from: http://www.ncbi.nlm.nih.gov/pubmed/23673310
98. Kim J, Yoon JH, Kim YS (2013) HIV-1 tat interacts with and regulates the localization and processing of amyloid precursor protein. Chauhan A, editor. PLoS One [Internet] 8(11):e77972. Nov 29 [cited 2018 July 27]. Available from: http://www.ncbi.nlm.nih.gov/pubmed/24312169
99. Fields JA, Dumaop W, Crews L, Adame A, Spencer B, Metcalf J et al (2015) Mechanisms of HIV-1 tat neurotoxicity via CDK5 translocation and hyper-activation: role in HIV-associated neurocognitive disorders. Curr HIV Res [Internet] 13(1):43–54. [cited 2018 July 27]. Available from: http://www.ncbi.nlm.nih.gov/pubmed/25760044

100. Liu Y, Jones M, Hingtgen CM, Bu G, Laribee N, Tanzi RE et al (2000) Uptake of HIV-1 tat protein mediated by low-density lipoprotein receptor-related protein disrupts the neuronal metabolic balance of the receptor ligands. Nat Med [Internet] 6(12):1380–1387. Dec 1 [cited 2018 July 27]. Available from: http://www.ncbi.nlm.nih.gov/pubmed/11100124
101. Deshmane SL, Mukerjee R, Fan S, Sawaya BE (2011) High-performance capillary electrophoresis for determining HIV-1 tat protein in neurons. Kashanchi F, editor. PLoS One [Internet] 6(1):e16148. Jan 7 [cited 2018 July 27]. Available from: http://dx.plos.org/10.1371/journal.pone.0016148
102. Vendeville A, Rayne F, Bonhoure A, Bettache N, Montcourrier P, Beaumelle B (2004) HIV-1 tat enters T cells using coated pits before translocating from acidified endosomes and eliciting biological responses. Mol Biol Cell [Internet] 15(5):2347–2360. May [cited 2018 July 27]. Available from: http://www.ncbi.nlm.nih.gov/pubmed/15020715
103. Mangieri LR, Mader BJ, Thomas CE, Taylor CA, Luker AM, Tse TE et al (2014) ATP6V0C knockdown in Neuroblastoma cells alters autophagy-lysosome pathway function and metabolism of proteins that accumulate in neurodegenerative disease. Srinivasula SM, editor. PLoS One [Internet] 9(4):e93257. Apr 2 [cited 2018 July 27]. Available from: http://dx.plos.org/10.1371/journal.pone.0093257
104. Bendiske J, Caba E, Brown QB, Bahr BA (2002) Intracellular deposition, microtubule destabilization, and transport failure: an “early” pathogenic Cascade leading to synaptic decline. J Neuropathol Exp Neurol [Internet] 61(7):640–650. July 1 [cited 2018 July 27]. Available from: https://academic.oup.com/jnen/article-lookup/doi/10.1093/jnen/61.7.640
105. Bendiske J, Bahr BA (2003) Lysosomal activation is a compensatory response against protein accumulation and associated synaptopathogenesis–an approach for slowing Alzheimer disease? J Neuropathol Exp Neurol [Internet] 62(5):451–463. May [cited 2018 July 27]. Available from: http://www.ncbi.nlm.nih.gov/pubmed/12769185
106. Kanju PM, Parameshwaran K, Vaithianathan T, Sims CM, Huggins K, Bendiske J et al (2007) Lysosomal dysfunction produces distinct alterations in synaptic alpha-amino-3-hydroxy-5-methylisoxazolepropionic acid and N-methyl-D-aspartate receptor currents in hippocampus. J Neuropathol Exp Neurol [Internet] 66(9):779–788. Sept [cited 2018 July 27]. Available from: http://www.ncbi.nlm.nih.gov/pubmed/17805008

Chapter 28
At the Crossing of ER Stress and MAMs: A Key Role of Sigma-1 Receptor?

Benjamin Delprat, Lucie Crouzier, Tsung-Ping Su, and Tangui Maurice

Abstract Calcium exchanges and homeostasis are finely regulated between cellular organelles and in response to physiological signals. Besides ionophores, including voltage-gated Ca^{2+} channels, ionotropic neurotransmitter receptors, or Store-operated Ca^{2+} entry, activity of regulatory intracellular proteins finely tune Calcium homeostasis. One of the most intriguing, by its unique nature but also most promising by the therapeutic opportunities it bears, is the sigma-1 receptor (Sig-1R). The Sig-1R is a chaperone protein residing at mitochondria-associated endoplasmic reticulum (ER) membranes (MAMs), where it interacts with several partners involved in ER stress response, or in Ca^{2+} exchange between the ER and mitochondria. Small molecules have been identified that specifically and selectively activate Sig-1R (Sig-1R agonists or positive modulators) at the cellular level and that also allow effective pharmacological actions in several pre-clinical models of pathologies. The present review will summarize the recent data on the mechanism of action of Sig-1R in regulating Ca^{2+} exchanges and protein interactions at MAMs and the ER. As MAMs alterations and ER stress now appear as a common track in most neurodegenerative diseases, the intracellular action of Sig-1R will be discussed in the context of the recently reported efficacy of Sig-1R drugs in pathologies like Alzheimer's disease, Parkinson's disease, Huntington's disease, or amyotrophic lateral sclerosis.

Keywords Sigma-1 receptor · Calcium · Mitochondria · ER stress · UPR · MAMs · Neurodegenerative disease · Alzheimer's disease · Amyotrophic lateral sclerosis · Addiction · Pain

---

B. Delprat (✉) · L. Crouzier · T. Maurice
MMDN, University of Montpellier, EPHE, INSERM, U1198, Montpellier, France
e-mail: benjamin.delprat@inserm.fr

T.-P. Su
Cellular Pathobiology Section, Integrative Neuroscience Branch, Intramural Research Program, National Institute on Drug Abuse, NIH, DHHS, IRP, NIDA/NIH, Baltimore, MD, USA

© Springer Nature Switzerland AG 2020

M. S. Islam (ed.), *Calcium Signaling*, Advances in Experimental Medicine and Biology 1131, https://doi.org/10.1007/978-3-030-12457-1_28

28.1 Introduction: Physiopathology of Sigma-1 Receptor (Sig-1R)

The Sigma-1 receptor (Sig-1R), discovered in the mid 1970s [1] was identified as a 223-amino acid protein only in the mid-1990s [2, 3]. Although its involvement in physiopathology started to be documented earlier, its cellular role was precised only 10 years ago [4] and cellular biology studies continue to precise its intracellular partners and functions. It shares no homology with any other known protein, except some steroid related/emopamyl-binding enzymes [2, 5, 6]. The protein was initially viewed as a receptor since, very early, specific and selective small molecules have been identified binding to Sig-1Rs and triggering (for so-called agonists) or preventing (for so-called antagonists) biological responses. However, its activation by physiological triggers, including ER stress or oxidative stress [4, 7, 8], and its mode of action, relying on modifications of protein-protein interactions rather than coupling to second messenger systems, suggested more a chaperone-like identity than a classical receptor nature [4]. Indeed, the present review will detail the effects of Sig-1R at intracellular organelles and show that it offers a unique opportunity to finely tune its activity, thereby impacting numerous physiopathological pathways, through a very classical pharmacological approach involving agonists, positive modulators or antagonists.

The Sig-1Rs are expressed in numerous organs, including liver, heart, lung, gonads and the nervous system, and numerous cell types, including, in the latter, neurons and glial cells (astrocytes, microglia, oligodendrocytes, Schwann cells) and vascular cells [9–11]. The particular density of Sig-1R in the nervous system is coherent with its importance in numerous psychiatric and neurological conditions. Sig1-Rs have indeed been involved in epilepsy [12, 13], stroke [14–16], drug abuse [17–19], pain [20, 21], and neurodegenerative pathologies. Interestingly, the last field of research is currently very active and recent evidence show both that Sig-1Rs play a role in the physiopathology of several neurodegenerative disease and that Sig-1R agonists have effective neuroprotective effects in preclinical models that deserve translation in clinical trials and better understanding of the mechanism of action of Sig-1R drugs against neurodegeneration. In parallel to the accumulating evidences that small molecules acting as Sig-1R agonist have pharmacological action in preclinical models of neurodegenerative diseases, and thus therapeutic potential, arguments are also brought confirming that Sig-1R exerts its cell homeostatic and cytoprotective activities mainly by directly targeting ER/mitochondria communication. We will here detail these arguments.

28.2 Sig-1R at the MAM

The ER of a cell spread almost all over a cell either in close proximity or in direct contacts with other subcellular components including the Golgi, mitochondria, nucleus, and plasma membrane. Through those close encounters, the ER plays many critical functions in the cell. One such important contact site for the ER is the mitochondria-associated ER membrane, termed the MAM [22], which harbors not only the lipid exchange [22], mitochondrial DNA exchange, Ca^{2+} signaling between the ER and mitochondria [4, 23–27] but also plays a role in the ER-nucleus signaling for cellular survival [28]. Recent evidences also indicate that the MAM is the origin of the isolated membrane for autophagy [29]. In addition, the MAM is critical in the formation of inflammasome [30, 31].

The MAM contains a plethora of functional proteins [24, 32]. Among those is the Sig-1R which is an ER molecular chaperone with two transmembrane regions from cellular biology studies [4, 33] but only one from the X-ray crystallographic study [34]. At the MAM, the Sig-1R chaperones the inositol-1,4,5 trisphosphate receptor type 3 (IP_3R3) which would otherwise degrade after the stimulation of IP_3, ensuring thus proper Ca^{2+} signaling from the ER into mitochondria [4, 35]. At the MAM, the Sig-1R also chaperones inositol-requiring enzyme 1 (IRE-1), one of the ER stress sensors, to facilitate the signaling of the unfolded protein response from the ER into nucleus to call for the transcriptional activation of antioxidant proteins and chaperones [28]. The Sig-1R was also found to attenuate free radical formation around the MAM area to reduce the activation of caspase that would have degraded the guanine nucleotide exchange factor to inactivate Rac GTPase that is essential for dendritic spine formation [36]. The Sig-1R also plays a role, likely at the MAM, in binding and transferring myristic acid to p35 to facilitate the p35 degradation by proteasome at the plasma membrane, thereby diverting p35 from forming p25 that would otherwise stun the axon elongation [37]. However, it remains to be totally clarified how those molecular actions of the Sig-1R at the MAM may contribute to the overall cellular and physiological functions of the MAM in general in a cell.

Upon the stimulation of Sig-1R agonists, Sig-1Rs dissociate from innate co-chaperone binding immunoglobulin protein (BiP) and translocate to other parts of cell to interact with and regulate the function of receptors, ion channels, and other functional proteins at the plasma membrane, mitochondria, ER reticular network, and nucleus [38, 39]. Thus, due to the nature and dynamics of Sig-1Rs, the receptor plays multiple physiological roles in living systems.

One of the important physiological roles of the Sig-1R is to regulate Ca^{2+} signaling not only at the MAM but also at the ER reticular network and plasma membrane. Sig-1Rs at the MAM facilitate Ca^{2+} influx from the ER into mitochondria by chaperoning IP_3R3 at the MAM [4]. At the ER reticular network, the supranormal release of Ca^{2+} from the ER in medium spiny neurons of the YAC128 transgenic Huntington's disease mice was attenuated by a Sig-1R agonist [40]. The release of Ca^{2+} from the ER reticular network is mainly controlled by the IP_3R type 1 and the ryanodine receptor. Thus, this report suggests an inhibitory effect of Sig-1Rs on the IP_3R1 or the ryanodine receptor, which is in contrast to the facilitative effect of Sig-1Rs on IP_3R3. More experiments are needed as such.

The following three studies showed inconsistent effects of Sig-1Rs on the $[Ca^{2+}]_i$ in nevertheless different systems. Also, the site of action of Sig-1Rs were not identified. By using cultured cortical neurons, Sig-1R agonists were found to attenuate the ischemia-induced increase of $[Ca^{2+}]_i$ [41]. A recent study showed an increased $[Ca^{2+}]_i$ when Sig-1Rs were activated by agonists in cultured embryonic mouse spinal neurons from ALS-causing mutants [42]. As well, methamphetamine-induced increase of $[Ca^{2+}]_i$ was shown to be attenuated by Sig-1R agonists in dopaminergic neurons [43]. Again, sites of action of Sig-1Rs in those three studies were not identified. It remains to be seen if those actions of Sig-1Rs were at the MAM, the ER reticular network, or the plasma membrane. Possibility exists that results were manifestation of concerted actions at all of those sites.

At the plasma membrane, Sig-1Rs showed a presynaptic action in inhibiting N-type Ca^{2+} channels in cholinergic interneurons in rat striatum, resulting in a decrease in presynaptic $[Ca^{2+}]_i$ [44]. The Sig-1R co-immunoprecipitated and co-localized with the N-type Ca^{2+} channel [24]. In rat brain microvascular endothelial cells, the store-operated calcium entry (SOCE) was attenuated by a Sig-1R agonist cocaine [45]. The mechanism of this interesting action of Sig-1R was reported in an elegant study in the same year. The Sig-1R was shown to bind stromal interaction molecule 1 (STIM1) at the ER when extracellular Ca^{2+} was depleted, and, as a result slowed down the recruitment of STIM1 to the ER-plasma membrane junction where STIM1 binds Orai1 [46]. The resultant inhibition of SOCE was seen when Sig-1Rs were overexpressed or when cells were treated with Sig-1R agonists. The Sig-1R antagonists or shSig-1R treatment enhanced the SOCE [46].

The calcium channels on the plasma membrane were examined in autonomous neurons taken from neonatal rat intracardiac and superior cervical ganglia (SCG). It was found that Sig-1Rs depressed high-voltage activated calcium currents from all calcium channel subtypes found on the cell body of these neurons, which includes N-, L-, P/Q-, and R-type calcium channels [47]. This study suggests that the activation of sigma receptors on sympathetic and parasympathetic neurons may modulate cell-to-cell signaling in autonomic ganglia and thus the regulation of cardiac function by the peripheral nervous system [47]. No direct interaction between Sig-1Rs and those calcium channels were demonstrated in this study however. The effect of Sig-1Rs on the potassium chloride-induced Ca^{2+} influx was examined by using retinal ganglion cell line (RGC)-5 and rat primary RGCs by the whole-cell patch clamp technique [48]. Sig-1R agonists inhibited the calcium influx and the Sig-1R antagonist reversed the inhibitory effect of Sig-1R agonist [48]. The Sig-1R was found to co-immunoprecipitate with the L-type calcium channels in this study.

Calcium homeostasis is critical to cellular physiology and plays an important role in many central nervous system (CNS) diseases, in particular the neurodegenerative diseases [49–54]. Since Sig-1Rs play critical role in calcium signaling at several loci of a cell, Sig-1Rs may be related to neurodegenerative diseases which show dysfunctional calcium homeostasis. However, a direct link between Sig-1R-regulated calcium signaling and a neurodegenerative disease has only been recently demonstrated in Huntington's disease as mentioned above [40].

Nevertheless, because Sig-1Rs reside mainly at the MAM which is increasingly recognized as an important loci related to many neurodegenerative diseases

[25, 49, 52, 53, 55, 56], it is possible that the Sig-1R at the MAM may participate in those diseases in a manner either directly related to calcium signaling or *via* other yet-to-be-revealed mechanisms at the MAM.

A study has specifically examined the role of Sig-1Rs at the MAM on ALS. Using primary motor neuron cultures, the study found that the pharmacological or genetic inactivation of Sig-1Rs led to motor neuron axonal degeneration [57]. They also found that the disruption of Sig-1R function in motor neurons disturbed ER-mitochondria contacts and affected intracellular calcium signaling, and was accompanied by activation of ER stress and defects in mitochondrial dynamics and transport (direct quotes) [57]. It is interesting to note that several other studies have implicated Sig-1Rs in ALS although they did not directly examine if the action of Sig-1Rs was at the MAM [58–61].

Sig-1Rs have been related to Alzheimer's disease (AD). Several Sig-1R agonists, including PRE-084, MR-22, afobasole, ANAVEX1-41, ANAVEX2-73 or dehydroepiandrosterone, prevented amyloid-$\beta_{25\text{-}35}$ ($A\beta_{25\text{-}35}$)-induced toxicity in rat neuronal cultures [62, 63] and/or $A\beta_{25\text{-}35}$-induced toxicity and learning impairments in mice in vivo [64–69]. Among the biochemical markers of toxicity in both in vitro and in vivo models, the Sig-1R drugs appear particularly effective in alleviating oxidative stress. Similar data were obtained in transgenic animal models of AD by Fisher et al. [70] who reported that AF710B, a mixed M1 mAChR/Sig-1R agonist, administered for 2 months in female 3xTg-AD mice, attenuated memory impairments and neurotoxicity. The drug also diminished soluble and insoluble Aβ species accumulation, the number of plaques and Tau hyperphosphorylation [70], thus confirming the neuroprotection and potentially disease-modifying effects of the drug. Moreover, invalidation of the Sig-1R expression, using Sig-1R knockout mice or a repeated treatment with the Sig-1R antagonist NE-100, increased learning deficits and neurotoxicity in $A\beta_{25\text{–}35}$-injected mice or after cross-breeding with $APP_{Swe,Ldn}$ mice [71]. Therefore, it appeared that the absence of Sig-1R could worsen Aβ toxicity and behavioral deficits while it activation by therapeutic drugs showed neuroprotection.

The role of the MAM in the pathogenesis of AD remains an important area of research. Interestingly, a study has related the toxicity of Aβ to the increase of $[Ca^{2+}]_i$ and the overload of mitochondrial calcium [72]. Nanomolar concentrations of Aβ was shown to increase MAM-associated proteins and caused an increase of the MAM [55]. Importantly, knockdown of Sig-1Rs resulted in neurodegeneration [55, 71]. Moreover, a direct examination of the effects of Sig-1R drugs in isolated mitochondria exposed to β-amyloid peptide showed that agonists decreased $A\beta_{1\text{–}42}$-induced increase in reactive oxygen species (ROS) and attenuated $A\beta_{1\text{–}42}$-induced alterations in mitochondrial respiration related to decreases in complex I and IV activity [73]. The Sig-1R agonists increased complex I activity, in a Ca^{2+}-dependent and Sig-1R antagonist-sensitive manner in physiological conditions. These observations identified direct consequences on mitochondria of Sig-1R activity. However, further research on the involvement of the MAM in Alzheimer's disease is certainly warranted.

Moreover, Sig-1R agonists, and particularly PRE-084, have been shown to be neuroprotective in mouse models of Parkinson's disease [74], Huntington's disease [75], amyotrophic lateral sclerosis (ALS) [76, 77], multiple sclerosis [78] or retinal neurodegeneration [79, 80], notably. Several recent reviews addressed the different progresses made so far [17, 51, 81–86]. For instance, a study examined the role of Sig-1Rs in Parkinsonism in a mouse model of intrastriatal lesion by 6-hydroxydopamine and found that the Sig-1R agonist significantly improved the fore-limb use [74]. At the molecular level, the study found that the agonist increased the density of dopaminergic fibers at the most denervated striatal regions and also caused an increase of neurotrophic factor brain-derived neurotrophic factor (BDNF) [74]. Interestingly, the agonist treatment induced a wider intracellular distribution of Sig-1Rs [74]. Because Sig-1Rs can translocate to other parts of neuron upon the stimulation by an agonist, it is tempting to speculate that the action of Sig-1Rs in the improvement of Parkinsonism may occur at the MAM as well other parts of neuron.

The Sig-1R has been shown to relate to Huntington's disease. Most of the evidence cam from studies using a drug called pridopidine which was effective against Huntington's disease in preclinical models and in phase two clinical trial at the secondary end-point level [87]. Originally thought to be a dopamine D2 ligand, pridopidine was nevertheless found to have a 100-fold higher affinity at the Sig-1R than at the dopamine D2 receptor [88]. In an in vivo radioligand binding assay, behaviorally relevant doses of pridopidine blocked about 57–85% of radiotracer binding to Sig-1Rs while blocked only negligible fraction of D2 receptor [89]. Recently, pridopidine was found to attenuate the phencyclidine-induced memory impairment through the Sig-1R-mediated mechanism as the effect of pridopidine was blocked by a Sig-1R antagonist NE-100 [90].

The exact mechanism and therefore the cellular site of action of Sig-1Rs underlying the action of pridopidine against Huntington's disease are however not fully clarified. However, couple of studies provide some interesting results. In Q175 knock-in (Q175 KI) *vs* Q25 WT mouse models, the effect of pridopidine versus sham treatment on genome-wide expression profiling in the rat striatum was analyzed and compared to the pathological expression profile. Then a broad, unbiased pathway analysis was conducted, followed by testing the enrichment of relevant pathways [87]. Results showed that pridopidine upregulated the BDNF pathway ($P = 1.73E\text{-}10$), and its effect on BDNF secretion was Sig-1R-dependent [87]. It remains to be investigated how Sig-1Rs may upregulate BDNF at the molecular level. As mentioned before, the action of pridopidine was examined in a mouse model of Huntington's disease with a specific focus on intracellular calcium signaling [40]. Results showed that pridopidine attenuates spine loss of medium spiny neurons and the effect was absent with the neuronal deletion of Sig-1Rs [40]. Pridopidine suppressed supranormal ER Ca^{2+} release, restored ER calcium levels and reduced excessive SOCE entry into spines. Interestingly, normalization of ER Ca^{2+} levels by pridopidine was prevented by Sig-1R deletion [40]. Whether those effects of Sig-1Rs originate at the MAM or beyond are not clear at present.

28.3 Sig-1R and ER Stress

The ER is an essential organelle of the cell that plays important role in protein folding and quality control [91, 92], lipid synthesis [93] and Ca^{2+} homeostasis [94]. During the life of the cell, different factors may perturb these functions, leading to a cellular state referred to as 'ER stress'. These stressors may be intrinsic, *i.e.*, cancer [95–98], neurodegenerative disease [99, 100], or diabetes [101, 102], or extrinsic, *i.e.*, micro-environmental stress [103], exposure to ER stressors [104], temperature [105] or reactive oxygen species production [106, 107]. Nevertheless, every time the ER is stressed, it triggers an adaptive response. This adaptive response is called the unfolded protein response (UPR). This UPR will help the cells to counter the stress by attenuating protein synthesis, clearing the unfolded proteins and enhancing the ability of the ER to fold proteins.

The UPR is an intracellular signal transduction mechanism that protects cells from ER stress. Three ER-resident transmembrane proteins function as stress sensors: RNA-activated protein kinase (PKR)-like endoplasmic reticular kinase (PERK); activating transcription factor 6 (ATF6); and IRE1. In basal state, these three transmembrane proteins are bound to BiP, an ER resident chaperone and are inactive [108, 109]. Upon a stress, the folding capacity of the ER is surpassed, leading to the dissociation of BiP from PERK, ATF6 and IRE1. This dissociation allows the activation of the three sensors [110]. Their activations transduce the unfolded protein stress signal across ER membrane and lead to UPR activation [111]. PERK is transmembrane ER resident protein of 1116 amino acids with two functional domains, a luminal and a cytosolic Ser/Thr kinase domain [112]. The dissociation of BiP from the luminal domain leads to oligomerization [108] and trans-autophosphorylation [113]. Activation of the PERK pathway leads to attenuation of general protein translation by phosphorylation of the α subunit of eukaryotic translation initiation factor 2 (eIF2α) [114]. Phosphorylated eIF2α inhibits eukaryotic translation initiation 2B activity, thus leading to a decrease of protein synthesis [115]. The blockage of the translation during ER stress diminishes the protein load on the ER folding machinery and is a prerequisite to a reestablishment of the ER homeostasis. In contrast to its attenuation of translation, eIF2α phosphorylation can selectively enhance the translation of mRNAs containing inhibitory upstream open reading frames in their $5'$ untranslated region, such as activating transcription factor 4 (ATF4) [116]. The production of ATF4 induces the expression of a plethora of adaptive genes involved in amino acid transport, metabolism, protection from oxidative stress, protein homeostasis and autophagy [117]. Finally, ATF4 favors the expression of CAAT/enhancer-binding protein (C/EBP) homologous protein (CHOP), which will result in the expression of genes that are involved in protein synthesis and the UPR. If the expression of CHOP is sustained, the increased protein synthesis will lead to oxidative stress and cell death [118].

The second ER stress sensor is ATF6. ATF6 is also an ER resident transmembrane protein of 670 amino acids and two functional domains, an N-terminal cytosolic containing basic leucine zipper and a C-terminal luminal domain. When BiP dissociates from ATF6, this one is exported to the Golgi where it will be cleaved by site-1 and site-2 proteases [119]. This cleavage releases a fragment of 400 amino acids corresponding to ATF6 cytosolic N-terminal domain. This released fragment of ATF6 will then translocate to the nucleus in order to act as a transcription factor. ATF6 will bind to the promoter of UPR-inducible genes, resulting in an upregulations of proteins, which role is to adjust ER protein folding, including ER chaperones and X-box-binding protein-1 (XBP-1) [120, 121].

The last ER stress sensor is IRE1. IRE1 is, like the two other sensors, an ER resident transmembrane proteins of 977 amino acids with three functional domains, an N-terminal luminal domain, a C-terminal Ser/Thr kinase domain and a C-terminal RNase L domain. The dissociation with BiP triggers oligomerization and activation of its cytosolic kinase domain. This activation facilitates the unconventional splicing of XBP-1 mRNA and subsequent translation of an active transcription factor, XBP1s [111, 121]. XBP1s is a basic leucine zipper transcription factor [122]. XBP1s controls the expression of several targets including chaperones, foldases and components of the ER-associated degradation (ERAD) pathway, in order to stop the ER stress and restore homeostasis [123]. The ERAD system destroys unfolded proteins through degradation in the cytosol [124]. Indeed, unfolded ER proteins are retro-translocated across the ER membrane into the cytosol in order to be degraded by the proteasome, following ubiquitination by ubiquitin-conjugating enzymes [125, 126]. Finally, the RNase activity of IRE1 may also target other genes *via* a mechanism named regulated IRE1-dependent decay (RIDD) [127]. RIDD is a conserved mechanism in eukaryotes by which IRE1 cleaves its target substrates [128]. The cleaved transcripts are degraded by exoribonucleases [129]. Therefore, RIDD seems to be required for the maintenance of ER homeostasis by diminishing ER protein load *via* mRNA degradation. Notably, it has been recently suggested that the physiological activity of RIDD may increase with the severity of the ER stress [130].

Interestingly, a substantial number of proteins involved in UPR are localized in MAMs [23]. Indeed, two of the three major proteins involved in UPR, PERK [131] and IRE1 [28], are enriched in MAMs. Intriguingly, some ER chaperones involved in UPR are also enriched in MAMs. For example, Calnexin, a type I integral membrane protein which helps in folding newly synthetize proteins which is essential in mitigating ER stress, is expressed in MAMs [132]. Another chaperone expressed in MAMs is the Sig-1R. The first evidence of the role of Sig-1R in ER stress came from the observation that under Ca^{2+} depletion or when stimulated by its ligand, Sig-1R dissociates from BiP, thus allowing a sustained Ca^{2+} efflux from the ER via IP_3R [4]. In addition, under ER stress following treatment with tunicamycin or thapsigargin, Sig-1R is upregulated, suggesting that it is protective against ER stress. Interestingly, overexpression on Sig-1R suppressed ER stress-induced activation of PERK and ATF6. IRE1 is expressed in MAMs and Sig-1R regulates the stability of IRE1 [28]. This enhanced stability favors the phosphorylation level of IRE1 under ER stress. Notably, the Sig-1R knock down potentiates the apoptosis of cells under ER stress. They showed that increased apoptosis was due to a diminution of the Xbp1 splicing [28]. In addition, activation of Sig-1R increases Bcl-2 expression, allowing Bcl-2/IP3R interaction, leading to increased mitochondrial Ca^{2+} uptake and ATP production [133] (see Penke et al. for review [134]).

It is well known that Sig-1R plays an important protective role in retinal disease [85]. Using *in situ* hybridization, Ola et al. [135] detected the Sig-1R mRNA in retinal ganglion cells, cells of the inner nuclear layer, photoreceptor and retinal pigment epithelium. The mRNA expression was confirmed by immunohistochemistry. Ha et al. [136] described that in Müller cells of the retina from Sig-1R KO mice, the expression of PERK, IRE1 and ATF4 was decreased, whereas the expression of BiP, CHOP and ATF6 was increased. Intriguingly, no difference was detected in whole brain or whole retina. Similar to what was described by Yang et al. [133], Ha et al. saw a decrease expression of Bcl-2 associated with decrease in NFkB and pERK1/2 [136]. Moreover, Wang et al. [137] demonstrated that loss of Sig-1R in a model of retinitis pigmentosa (rd10), aggravates the degeneration of the photoreceptors. They revealed that at P28, the expression level of Xbp1 and CHOP is increased in the rd10 mice without Sig-1R expression.

Since Sig-1R is a receptor that can be activated or inhibited, different groups determined the effect of its activation or inhibition in following ER stress. Ha et al. [138] treated RGC-5 cells with (+)-pentazocine, a potent Sig-1R agonist. RGC-5 cells are a rat retinal ganglion cell line [139]. They showed that whereas the protein level of PERK, ATF4, ATF6 IRE1 and CHOP was upregulated during oxidative stress, in the presence of (+)-pentazocine, their expression level decreased, suggesting that Sig-1R plays a pivotal role in the UPR response. These results confirmed the initial observation of Wang et al. [140] that stimulation of Sig-1R protects against oxidative stress. Indeed, in human cell line FHL124, H_2O_2 treatment induces apoptosis, associated to an increase level of BiP, ATF6 and p-eIF2α. Application of (+)-pentazocine suppressed the induction of BiP and p-eIF2α. Another agonist, fluvoxamine, alleviates induction of CHOP, cleaved caspase 3 and 4 in cancer neuronal cell SK-N-SH [141]. In another experiment, Omi et al. [142] showed that treatment of neuronal cell line Neuro2a induces overexpression of Sig-1R. This expression is mediated by ATF4, a downstream element of PERK activation. Interestingly, this overexpression is achieved without activating UPR. Intriguingly, the increased translation of ATF4 is dependent of the presence/function of Sig-1R since if the concomitant treatment of Neuro2a cells with Fluvoxamine and NE-100, a Sig-1R antagonist, abolished the ATF4 translation. This result was confirmed by the use of mouse embryonic fibroblasts (MEF) from Sig-1R KO mice. Indeed, fluvoxamine treatment of these MEF did not increase ATF4 expression. Morihara et al. [143] treated mice with Sig-1R agonist aniline derivative compound (Comp-AD) following ischemic stroke, since it is well known that Sig-1R protects against ischemic stroke but the role of ER stress was unknown. So, treatment of mice after 90 min of transient middle cerebral artery, diminished the expression level of p-PERK and p-IRE1, suggesting that activation of Sig-1R protects against ischemic stroke *via* the attenuation of ER stress.

If Sig-1R activation suppresses effectively ER stress, it should be expected that inhibition of Sig-1R should do the contrary. This was demonstrated by Ono et al. [144] using Sig-1R antagonist, by Hong et al. [145] using Sig-1R KO, and Alam et al. [146] using siRNA to knock down Sig-1R. Ono et al. [144] showed that NE-100 protects ER stress induced cell death in hippocampal HT22 cells after tunicamycin treatment. Indeed, NE-100 application attenuated the upregulation of CHOP. Interestingly, NE-100 treatment alone was capable of upregulate the expression of both ATF6 and BiP. Total ablation of Sig-1R in dopaminergic neurons of substantia nigra in mice led to an elevation of the expression level of p-eIF2α and CHOP [145]. In cardiomyocytes treated with tunicamycin, the downregulation of Sig-1R by siRNA led to an increase of CHOP expression. They also showed that Sig-1R downregulation diminished IRE1 phosphorylation and Xbp1 splicing [146].

Mutations of Sig-1R in human may lead to Juvenile [58] and classic ALS [147] or distal hereditary neuropathy (dHMN) [148–151]. Interestingly, E102Q mutation, which induces juvenile ALS, leads to ER stress [59, 152]. Indeed, over-expression of Sig-1R mutant in MCF7 cells induced an aggregation of the mutant protein into the ER in contrast to the overexpression of the wild-type Sig-1R, which is localized in the ER, the nuclear envelope end ER-Golgi intermediate compartment. Using ER stress response element (ERSE) reporter assay, they detected an increase in ER stress in MCF7 cells. There was an increase expression level of p-eIF2α, BiP, HSP70, GADD. Moreover, they showed a co-localization of ubiquitin-positive Sig-1R mutant aggregates with 20S proteasome subunit, suggesting possible interference with the ubiquitin proteasome system machinery [59]. In parallel, they demonstrated that the proteasome activity was greatly reduced. In order to confirm the results observed in transfected cells, they generated immortalized primary lymphoblastoid cells (PLCs) from blood samples of ALS patients. In PLCs, they also showed an aggregation of mutant Sig-1R in the ER associated with an increase level of BiP and p-eIF2α together with an increase of HSP70, GADD and ubiquitin conjugates [59].

28.4 Sig-1R in the Nucleus

Although Sig-1R is known to be particularly enriched in MAMs, observation of Sig-1Rs at the nuclear envelope (NE) and within nucleoplasms have also been reported. First, after stimulation by agonists such as cocaine, Sig-1Rs were found to translocate from ER to the NE, where they bind NE protein emerin and recruit chromatin-remodeling molecules [37]. These partners include lamin A/C, barrier-to-autointegration factor, and histone deacetylase (HDAC), to form a complex with the gene repressor specific protein 3 (Sp3). The dynamics of the interaction was confirmed when knockdown of Sig-1Rs attenuated the complex formation [37]. These observations were confirmed and developed by Mavlyutov et al. [153] who expressed APEX2 peroxidase fused to Sig1R-GFP in a Sig1R-null NSC34 neuronal cell line generated with CRISPR-Cas9. They observed that Sig1R actually resides in the nucleoplasmic reticulum, a specialized nuclear compartment formed via NE invagination into the nucleoplasm. A major consequence for this localization appears to be related to neurodegenerative pathologies since accumulation of Sig-1R may be common to neuronal nuclear inclusions in various proteinopathies [154]. Sig-1R immunoreactivity was shown to be co-localized with neuronal nuclear inclusions in TDP-43 proteinopathy, five polyglutamine diseases and intranuclear inclusion body disease, as well as in intranuclear Marinesco bodies in aged normal controls [154]. These authors interestingly proposed that Sig-1Rs might shuttle between the nucleus and the cytoplasm and likely play an important role in neurodegenerative diseases, characterized by neuronal nuclear inclusions, and known to particularly rely on ER-related degradation machinery as a common pathway for the degradation of aberrant proteins [154].

28.5 Conclusion

The Sig-1R protein is not specifically a MAM or ER protein, and direct interactions have been described at or close to the plasma membrane with potassium or sodium ion channels, ether-a-gogo-related gene (ERG) ionophores and metabotropic neurotransmitter receptors [39]. One of the complexity seen with Sig-1R is the multiplicity of its intracellular partners and consequent target pathways affected by its activation, as summarized in Fig. 28.1. This multiplicity of actions within different types of cells, in the brain as well as in other tissues, explain its involvement in numerous physiopathological processes and its value as a potential therapeutic target. Moreover, the well-known observation that bearing a relatively simple pharmacophore, Sig-1R binds small molecules, and even steroids or peptides, of diverse nature with high affinity, contributed to poorly considered it as a pertinent pharmacological target for therapeutic intervention. The data we discussed in this review allow to realize that we are now accumulating evidence on the mechanisms of action of Sig-1R, on its major role at MAMs and the ER, on its efficacy to maintain, and putatively restore cellular integrity and Ca^{2+} homeostasis (Fig. 28.1). The recent progression of several molecules in clinical phases in Alzheimer's disease or Huntington's disease strengthened the validity of Sig-1R as a pharmacological target. Moreover, their efficacies in preclinical models of different pathologies outlined the importance of MAM and ER alterations in neurodegenerative processes.

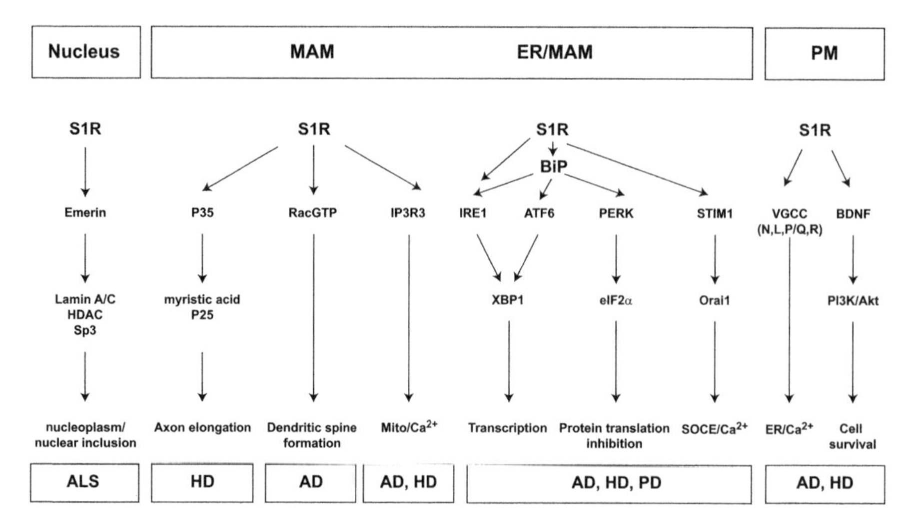

Fig. 28.1 Schematic diagram of the role of 1R in different compartment of the cell and putative major impacts in neurodegenerative diseases (AD, PD, HD, ALS)
In the nucleus, S1R interacts with Emerin [37]. This interaction leads to the recruitment of chromatin-remodeling molecules such as Lamin A/C and HDAC. This interaction is necessary for the creation of a supercomplex with Sp3. In the MAM, S1R interacts with P35 [37]. This will leads to the myristoylation of P25 and axon elongation. S1R interacts with RacGTP in order to foster dendritic spine formation [155]. S1R interacts with IP_3R3 in order to allow the proper Ca^{2+} efflux from the ER to the mitochondria [4]. In the ER, S1R interacts with BiP [4]. Upon stimulation, the dissociation of S1R with BiP induces activation of IRE1 [28] and ATF6 [136, 138, 144]. This will induce the splicing of XBP1 and the transcription of chaperones. The activation of PERK [144, 156] will induce the phosphorylation of eIF2α, to stop protein translation. Finally, S1R interacts with STIM1 and this interaction will regulate Ca^{2+} fluxes into the ER thought the STIM1/Orai1 axe [46]. In the plasma membrane, S1R interacts with voltage-gated calcium channels (for review, [157]) in order to modulate Ca^{2+} homeostasis. S1R favors BDNF secretion [74, 87]. BDNF will activate the PI3K/Akt pathway in order to improve cell survival

References

1. Martin WR, Eades CG, Thompson JA, Huppler RE, Gilbert PE (1976) The effects of morphine- and nalorphine- like drugs in the nondependent and morphine-dependent chronic spinal dog. J Pharmacol Exp Ther 197(3):517–532
2. Hanner M, Moebius FF, Flandorfer A, Knaus HG, Striessnig J, Kempner E et al (1996) Purification, molecular cloning, and expression of the mammalian $sigma_1$-binding site. Proc Natl Acad Sci U S A 93(15):8072–8077
3. Kekuda R, Prasad PD, Fei YJ, Leibach FH, Ganapathy V (1996) Cloning and functional expression of the human type 1 sigma receptor (hSigmaR1). Biochem Biophys Res Commun 229(2):553–558
4. Hayashi T, Su TP (2007) Sigma-1 receptor chaperones at the ER-mitochondrion interface regulate Ca^{2+} signaling and cell survival. Cell 131(3):596–610
5. Maurice T, Gregoire C, Espallergues J (2006) Neuro(active)steroids actions at the neuromodulatory $sigma_1$ (σ_1) receptor: biochemical and physiological evidences, consequences in neuroprotection. Pharmacol Biochem Behav 84(4):581–597

6. Moebius FF, Reiter RJ, Hanner M, Glossmann H (1997) High affinity of sigma 1-binding sites for sterol isomerization inhibitors: evidence for a pharmacological relationship with the yeast sterol C8-C7 isomerase. Br J Pharmacol 121(1):1–6
7. Meunier J, Hayashi T (2010) Sigma-1 receptors regulate Bcl-2 expression by reactive oxygen species-dependent transcriptional regulation of nuclear factor kappaB. J Pharmacol Exp Ther 332(2):388–397
8. Pal A, Fontanilla D, Gopalakrishnan A, Chae YK, Markley JL, Ruoho AE (2012) The sigma-1 receptor protects against cellular oxidative stress and activates antioxidant response elements. Eur J Pharmacol 682(1-3):12–20
9. Alonso G, Phan V, Guillemain I, Saunier M, Legrand A, Anoal M et al (2000) Immunocytochemical localization of the sigma_1 receptor in the adult rat central nervous system. Neuroscience 97(1):155–170
10. Palacios G, Muro A, Vela JM, Molina-Holgado E, Guitart X, Ovalle S et al (2003) Immunohistochemical localization of the sigma_1-receptor in oligodendrocytes in the rat central nervous system. Brain Res 961(1):92–99
11. Tagashira H, Bhuiyan S, Shioda N, Hasegawa H, Kanai H, Fukunaga K (2010) Sigma1-receptor stimulation with fluvoxamine ameliorates transverse aortic constriction-induced myocardial hypertrophy and dysfunction in mice. Am J Physiol Heart Circ Physiol 299(5):H1535–H1545
12. Meurs A, Clinckers R, Ebinger G, Michotte Y, Smolders I (2007) Sigma 1 receptor-mediated increase in hippocampal extracellular dopamine contributes to the mechanism of the anticonvulsant action of neuropeptide Y. Eur J Neurosci 26(11):3079–3092
13. Vavers E, Svalbe B, Lauberte L, Stonans I, Misane I, Dambrova M et al (2017) The activity of selective sigma-1 receptor ligands in seizure models in vivo. Behav Brain Res 328:13–18
14. Harukuni I, Bhardwaj A, Shaivitz AB, DeVries AC, London ED, Hurn PD et al (2000) sigma_1-receptor ligand 4-phenyl-1-(4-phenylbutyl)-piperidine affords neuroprotection from focal ischemia with prolonged reperfusion. Stroke 31(4):976–982
15. Lesage AS, De Loore KL, Peeters L, Leysen JE (1995) Neuroprotective sigma ligands interfere with the glutamate-activated NOS pathway in hippocampal cell culture. Synapse 20(2):156–164
16. Shen YC, Wang YH, Chou YC, Liou KT, Yen JC, Wang WY et al (2008) Dimemorfan protects rats against ischemic stroke through activation of sigma-1 receptor-mediated mechanisms by decreasing glutamate accumulation. J Neurochem 104(2):558–572
17. Cai Y, Yang L, Niu F, Liao K, Buch S (2017) Role of sigma-1 receptor in cocaine abuse and neurodegenerative disease. Adv Exp Med Biol 964:163–175
18. Maurice T, Martin-Fardon R, Romieu P, Matsumoto RR (2002) Sigma_1 (σ_1) receptor antagonists represent a new strategy against cocaine addiction and toxicity. Neurosci Biobehav Rev 26(4):499–527
19. Su TP, Hayashi T (2001) Cocaine affects the dynamics of cytoskeletal proteins via sigma_1 receptors. Trends Pharmacol Sci 22(9):456–458
20. Diaz JL, Zamanillo D, Corbera J, Baeyens JM, Maldonado R, Pericas MA et al (2009) Selective sigma-1 (σ_1) receptor antagonists: emerging target for the treatment of neuropathic pain. Cent Nerv Syst Agents Med Chem 9(3):172–183
21. Zamanillo D, Romero L, Merlos M, Vela JM (2013) Sigma 1 receptor: a new therapeutic target for pain. Eur J Pharmacol 716(1-3):78–93
22. Vance JE (1990) Phospholipid synthesis in a membrane fraction associated with mitochondria. J Biol Chem 265(13):7248–7256
23. Carreras-Sureda A, Pihan P, Hetz C (2017) The unfolded protein response: at the intersection between endoplasmic reticulum function and mitochondrial bioenergetics. Front Oncol 7:55
24. Hayashi T, Rizzuto R, Hajnoczky G, Su TP (2009) MAM: more than just a housekeeper. Trends Cell Biol 19(2):81–88
25. Raturi A, Simmen T (2013) Where the endoplasmic reticulum and the mitochondrion tie the knot: the mitochondria-associated membrane (MAM). Biochim Biophys Acta 1833(1): 213–224

26. Rizzuto R, Duchen MR, Pozzan T (2004) Flirting in little space: the ER/mitochondria Ca^{2+} liaison. Sci STKE 2004(215):re1
27. Walter L, Hajnoczky G (2005) Mitochondria and endoplasmic reticulum: the lethal interorganelle cross-talk. J Bioenerg Biomembr 37(3):191–206
28. Mori T, Hayashi T, Hayashi E, Su TP (2013) Sigma-1 receptor chaperone at the ER-mitochondrion interface mediates the mitochondrion-ER-nucleus signaling for cellular survival. PLoS One 8(10):e76941
29. Hamasaki M, Furuta N, Matsuda A, Nezu A, Yamamoto A, Fujita N et al (2013) Autophagosomes form at ER-mitochondria contact sites. Nature 495(7441):389–393
30. Sutterwala FS, Haasken S, Cassel SL (2014) Mechanism of NLRP3 inflammasome activation. Ann N Y Acad Sci 1319:82–95
31. Zhou R, Yazdi AS, Menu P, Tschopp J (2011) A role for mitochondria in NLRP3 inflammasome activation. Nature 469(7329):221–225
32. Poston CN, Krishnan SC, Bazemore-Walker CR (2013) In-depth proteomic analysis of mammalian mitochondria-associated membranes (MAM). J Proteome 79:219–230
33. Fontanilla D, Hajipour AR, Pal A, Chu UB, Arbabian M, Ruoho AE (2008) Probing the steroid binding domain-like I (SBDLI) of the sigma-1 receptor binding site using N-substituted photoaffinity labels. Biochemistry 47(27):7205–7217
34. Schmidt HR, Zheng S, Gurpinar E, Koehl A, Manglik A, Kruse AC (2016) Crystal structure of the human $sigma_1$ receptor. Nature 532(7600):527–530
35. Wu Z, Bowen WD (2008) Role of sigma-1 receptor C-terminal segment in inositol 1,4,5-trisphosphate receptor activation: constitutive enhancement of calcium signaling in MCF-7 tumor cells. J Biol Chem 283(42):28198–28215
36. Tsai SY, Hayashi T, Harvey BK, Wang Y, Wu WW, Shen RF et al (2009) Sigma-1 receptors regulate hippocampal dendritic spine formation via a free radical-sensitive mechanism involving Rac1xGTP pathway. Proc Natl Acad Sci U S A 106(52):22468–22473
37. Tsai SY, Pokrass MJ, Klauer NR, Nohara H, Su TP (2015) Sigma-1 receptor regulates Tau phosphorylation and axon extension by shaping p35 turnover via myristic acid. Proc Natl Acad Sci U S A 112(21):6742–6747
38. Maurice T, Su TP (2009) The pharmacology of sigma-1 receptors. Pharmacol Ther 124(2):195–206
39. Su TP, Su TC, Nakamura Y, Tsai SY (2016) The sigma-1 receptor as a puripotent modulator in living systems. Trends Pharmacol Sci 37(4):262–278
40. Ryskamp D, Wu J, Geva M, Kusko R, Grossman I, Hayden M et al (2017) The sigma-1 receptor mediates the beneficial effects of pridopidine in a mouse model of Huntington disease. Neurobiol Dis 97(Pt A):46–59
41. Katnik C, Guerrero WR, Pennypacker KR, Herrera Y, Cuevas J (2006) Sigma-1 receptor activation prevents intracellular calcium dysregulation in cortical neurons during in vitro ischemia. J Pharmacol Exp Ther 319(3):1355–1365
42. Tadic V, Malci A, Goldhammer N, Stubendorff B, Sengupta S, Prell T et al (2017) Sigma 1 receptor activation modifies intracellular calcium exchange in the G93A(hSOD1) ALS model. Neuroscience 359:105–118
43. Sambo DO, Lin M, Owens A, Lebowitz JJ, Richardson B, Jagnarine DA et al (2017) The sigma-1 receptor modulates methamphetamine dysregulation of dopamine neurotransmission. Nat Commun 8(1):2228
44. Zhang K, Zhao Z, Lan L, Wei X, Wang L, Liu X et al (2017) Sigma-1 receptor plays a negative modulation on N-type calcium channel. Front Pharmacol 8:302
45. Brailoiu GC, Deliu E, Console-Bram LM, Soboloff J, Abood ME, Unterwald EM et al (2016) Cocaine inhibits store-operated Ca^{2+} entry in brain microvascular endothelial cells: critical role for sigma-1 receptors. Biochem J 473(1):1–5
46. Srivats S, Balasuriya D, Pasche M, Vistal G, Edwardson JM, Taylor CW et al (2016) Sigma1 receptors inhibit store-operated Ca^{2+} entry by attenuating coupling of STIM1 to Orai1. J Cell Biol 213(1):65–79

47. Zhang H, Cuevas J (2002) Sigma receptors inhibit high-voltage-activated calcium channels in rat sympathetic and parasympathetic neurons. J Neurophysiol 87(6):2867–2879
48. Tchedre KT, Huang RQ, Dibas A, Krishnamoorthy RR, Dillon GH, Yorio T (2008) Sigma-1 receptor regulation of voltage-gated calcium channels involves a direct interaction. Invest Ophthalmol Vis Sci 49(11):4993–5002
49. Erpapazoglou Z, Mouton-Liger F, Corti O (2017) From dysfunctional endoplasmic reticulum-mitochondria coupling to neurodegeneration. Neurochem Int 109:171–183
50. Joshi AU, Kornfeld OS, Mochly-Rosen D (2016) The entangled ER-mitochondrial axis as a potential therapeutic strategy in neurodegeneration: a tangled duo unchained. Cell Calcium 60(3):218–234
51. Nguyen L, Lucke-Wold BP, Mookerjee S, Kaushal N, Matsumoto RR (2017) Sigma-1 receptors and neurodegenerative diseases: towards a hypothesis of sigma-1 receptors as amplifiers of neurodegeneration and neuroprotection. Adv Exp Med Biol 964:133–152
52. Ottolini D, Cali T, Negro A, Brini M (2013) The Parkinson disease-related protein DJ-1 counteracts mitochondrial impairment induced by the tumour suppressor protein p53 by enhancing endoplasmic reticulum-mitochondria tethering. Hum Mol Genet 22(11):2152–2168
53. Ouyang YB, Giffard RG (2012) ER-mitochondria crosstalk during cerebral ischemia: molecular chaperones and ER-mitochondrial calcium transfer. Int J Cell Biol 2012:493934
54. Zundorf G, Reiser G (2011) Calcium dysregulation and homeostasis of neural calcium in the molecular mechanisms of neurodegenerative diseases provide multiple targets for neuroprotection. Antioxid Redox Signal 14(7):1275–1288
55. Hedskog L, Pinho CM, Filadi R, Ronnback A, Hertwig L, Wiehager B et al (2013) Modulation of the endoplasmic reticulum-mitochondria interface in Alzheimer's disease and related models. Proc Natl Acad Sci U S A 110(19):7916–7921
56. Prudent J, McBride HM (2017) The mitochondria-endoplasmic reticulum contact sites: a signalling platform for cell death. Curr Opin Cell Biol 47:52–63
57. Bernard-Marissal N, Medard JJ, Azzedine H, Chrast R (2015) Dysfunction in endoplasmic reticulum-mitochondria crosstalk underlies SIGMAR1 loss of function mediated motor neuron degeneration. Brain 138(Pt 4):875–890
58. Al-Saif A, Al-Mohanna F, Bohlega S (2011) A mutation in sigma-1 receptor causes juvenile amyotrophic lateral sclerosis. Ann Neurol 70(6):913–919
59. Dreser A, Vollrath JT, Sechi A, Johann S, Roos A, Yamoah A et al (2017) The ALS-linked E102Q mutation in sigma receptor-1 leads to ER stress-mediated defects in protein homeostasis and dysregulation of RNA-binding proteins. Cell Death Differ 24(10):1655–1671
60. Luty AA, Kwok JB, Dobson-Stone C, Loy CT, Coupland KG, Karlstrom H et al (2010) Sigma nonopioid intracellular receptor 1 mutations cause frontotemporal lobar degeneration-motor neuron disease. Ann Neurol 68(5):639–649
61. Prause J, Goswami A, Katona I, Roos A, Schnizler M, Bushuven E et al (2013) Altered localization, abnormal modification and loss of function of Sigma receptor-1 in amyotrophic lateral sclerosis. Hum Mol Genet 22(8):1581–1600
62. Behensky AA, Yasny IE, Shuster AM, Seredenin SB, Petrov AV, Cuevas J (2013) Afobazole activation of sigma-1 receptors modulates neuronal responses to amyloid-β_{25-35}. J Pharmacol Exp Ther 347(2):468–477
63. Marrazzo A, Caraci F, Salinaro ET, Su TP, Copani A, Ronsisvalle G (2005) Neuroprotective effects of sigma-1 receptor agonists against beta-amyloid-induced toxicity. Neuroreport 16(11):1223–1226
64. Antonini V, Marrazzo A, Kleiner G, Coradazzi M, Ronsisvalle S, Prezzavento O et al (2011) Anti-amnesic and neuroprotective actions of the sigma-1 receptor agonist (−)-MR22 in rats with selective cholinergic lesion and amyloid infusion. J Alzheimers Dis 24(3):569–586
65. Lahmy V, Meunier J, Malmstrom S, Naert G, Givalois L, Kim SH et al (2013) Blockade of Tau hyperphosphorylation and $A\beta_{1-42}$ generation by the aminotetrahydrofuran derivative ANAVEX2-73, a mixed muscarinic and $sigma_1$ receptor agonist, in a nontransgenic mouse model of Alzheimer's disease. Neuropsychopharmacology 38(9):1706–1723

66. Meunier J, Ieni J, Maurice T (2006) The anti-amnesic and neuroprotective effects of donepezil against amyloid β_{25-35} peptide-induced toxicity in mice involve an interaction with the σ_1 receptor. Br J Pharmacol 149(8):998–1012
67. Villard V, Espallergues J, Keller E, Alkam T, Nitta A, Yamada K et al (2009) Antiamnesic and neuroprotective effects of the aminotetrahydrofuran derivative ANAVEX1-41 against amyloid β_{25-35}-induced toxicity in mice. Neuropsychopharmacology 34(6):1552–1566
68. Villard V, Espallergues J, Keller E, Vamvakides A, Maurice T (2011) Anti-amnesic and neuroprotective potentials of the mixed muscarinic receptor/sigma$_1$ (σ_1) ligand ANAVEX2-73, a novel aminotetrahydrofuran derivative. J Psychopharmacol 25(8):1101–1117
69. Yang R, Chen L, Wang H, Xu B, Tomimoto H, Chen L (2012) Anti-amnesic effect of neurosteroid PREGS in Aβ_{25-35}-injected mice through σ_1 receptor- and α_7 nAChR-mediated neuroprotection. Neuropharmacology 63(6):1042–1050
70. Fisher A, Bezprozvanny I, Wu L, Ryskamp DA, Bar-Ner N, Natan N et al (2016) AF710B, a Novel M1/σ_1 agonist with therapeutic efficacy in animal models of Alzheimer's disease. Neurodegener Dis 16(1-2):95–110
71. Maurice T, Strehaiano M, Duhr F, Chevallier N (2018) Amyloid toxicity is enhanced after pharmacological or genetic invalidation of the sigma1 receptor. Behav Brain Res 339:1–10
72. Sanz-Blasco S, Valero RA, Rodriguez-Crespo I, Villalobos C, Nunez L (2008) Mitochondrial Ca^{2+} overload underlies Aβ oligomers neurotoxicity providing an unexpected mechanism of neuroprotection by NSAIDs. PLoS One 3(7):e2718
73. Goguadze N, Zhuravliova E, Morin D, Mikeladze D, Maurice T (2019) Sigma-1 receptor agonists induce oxidative stress in mitochondria and enhance complex I activity in physiological condition but protect against pathological oxidative stress. Neurotox Res 35(1):1–18
74. Francardo V, Bez F, Wieloch T, Nissbrandt H, Ruscher K, Cenci MA (2014) Pharmacological stimulation of sigma-1 receptors has neurorestorative effects in experimental parkinsonism. Brain 137(Pt 7):1998–2014
75. Hyrskyluoto A, Pulli I, Tornqvist K, Ho TH, Korhonen L, Lindholm D (2013) Sigma-1 receptor agonist PRE084 is protective against mutant huntingtin-induced cell degeneration: involvement of calpastatin and the NF-κB pathway. Cell Death Dis 4:e646
76. Mancuso R, Olivan S, Rando A, Casas C, Osta R, Navarro X (2012) Sigma-1R agonist improves motor function and motoneuron survival in ALS mice. Neurotherapeutics 9(4): 814–826
77. Peviani M, Salvaneschi E, Bontempi L, Petese A, Manzo A, Rossi D et al (2014) Neuroprotective effects of the Sigma-1 receptor (S1R) agonist PRE-084, in a mouse model of motor neuron disease not linked to SOD1 mutation. Neurobiol Dis 62:218–232
78. Oxombre B, Lee-Chang C, Duhamel A, Toussaint M, Giroux M, Donnier-Marechal M et al (2015) High-affinity sigma1 protein agonist reduces clinical and pathological signs of experimental autoimmune encephalomyelitis. Br J Pharmacol 172(7):1769–1782
79. Smith SB, Duplantier J, Dun Y, Mysona B, Roon P, Martin PM et al (2008) In vivo protection against retinal neurodegeneration by sigma receptor 1 ligand (+)-pentazocine. Invest Ophthalmol Vis Sci 49(9):4154–4161
80. Zhao L, Chen G, Li J, Fu Y, Mavlyutov TA, Yao A et al (2017) An intraocular drug delivery system using targeted nanocarriers attenuates retinal ganglion cell degeneration. J Control Release 247:153–166
81. Francardo V, Schmitz Y, Sulzer D, Cenci MA (2017) Neuroprotection and neurorestoration as experimental therapeutics for Parkinson's disease. Exp Neurol 298(Pt B):137–147
82. Mancuso R, Navarro X (2017) Sigma-1 receptor in motoneuron disease. Adv Exp Med Biol 964:235–254
83. Maurice T, Goguadze N (2017) Role of σ_1 receptors in learning and memory and Alzheimer's disease-type dementia. Adv Exp Med Biol 964:213–233
84. Maurice T, Goguadze N (2017) Sigma-1 (σ_1) receptor in memory and neurodegenerative diseases. Handb Exp Pharmacol 244:81–108
85. Smith SB, Wang J, Cui X, Mysona BA, Zhao J, Bollinger KE (2018) Sigma 1 receptor: a novel therapeutic target in retinal disease. Prog Retin Eye Res 67:130–149

86. Weng TY, Tsai SA, Su TP (2017) Roles of sigma-1 receptors on mitochondrial functions relevant to neurodegenerative diseases. J Biomed Sci 24(1):74
87. Geva M, Kusko R, Soares H, Fowler KD, Birnberg T, Barash S et al (2016) Pridopidine activates neuroprotective pathways impaired in Huntington disease. Hum Mol Genet 25(18):3975–3987
88. Sahlholm K, Arhem P, Fuxe K, Marcellino D (2013) The dopamine stabilizers ACR16 and (−)-OSU6162 display nanomolar affinities at the sigma-1 receptor. Mol Psychiatry 18(1): 12–14
89. Sahlholm K, Sijbesma JW, Maas B, Kwizera C, Marcellino D, Ramakrishnan NK et al (2015) Pridopidine selectively occupies sigma-1 rather than dopamine D2 receptors at behaviorally active doses. Psychopharmacology 232(18):3443–3453
90. Sahlholm K, Valle-Leon M, Fernandez-Duenas V, Ciruela F (2018) Pridopidine reverses phencyclidine-induced memory impairment. Front Pharmacol 9:338
91. Braakman I, Bulleid NJ (2011) Protein folding and modification in the mammalian endoplasmic reticulum. Annu Rev Biochem 80:71–99
92. Hebert DN, Molinari M (2007) In and out of the ER: protein folding, quality control, degradation, and related human diseases. Physiol Rev 87(4):1377–1408
93. Fagone P, Jackowski S (2009) Membrane phospholipid synthesis and endoplasmic reticulum function. J Lipid Res 50(Suppl):S311–S316
94. Meldolesi J, Pozzan T (1998) The endoplasmic reticulum Ca^{2+} store: a view from the lumen. Trends Biochem Sci 23(1):10–14
95. Corazzari M, Gagliardi M, Fimia GM, Piacentini M (2017) Endoplasmic reticulum stress, unfolded protein response, and cancer cell fate. Front Oncol 7:78
96. Hanahan D, Weinberg RA (2011) Hallmarks of cancer: the next generation. Cell 144(5): 646–674
97. Hetz C, Papa FR (2018) The unfolded protein response and cell fate control. Mol Cell 69(2):169–181
98. Jain BP (2017) An overview of unfolded protein response signaling and its role in cancer. Cancer Biother Radiopharm 32(8):275–281
99. Hetz C, Saxena S (2017) ER stress and the unfolded protein response in neurodegeneration. Nat Rev Neurol 13(8):477–491
100. Xiang C, Wang Y, Zhang H, Han F (2017) The role of endoplasmic reticulum stress in neurodegenerative disease. Apoptosis 22(1):1–26
101. Ariyasu D, Yoshida H, Hasegawa Y (2017) Endoplasmic reticulum (ER) stress and endocrine disorders. Int J Mol Sci 18(2):382
102. Harding HP, Ron D (2002) Endoplasmic reticulum stress and the development of diabetes: a review. Diabetes 51(Suppl 3):S455–S461
103. Giampietri C, Petrungaro S, Conti S, Facchiano A, Filippini A, Ziparo E (2015) Cancer microenvironment and endoplasmic reticulum stress response. Mediat Inflamm 2015:417281
104. Foufelle F, Fromenty B (2016) Role of endoplasmic reticulum stress in drug-induced toxicity. Pharmacol Res Perspect 4(1):e00211
105. Liu Y, Sakamoto H, Adachi M, Zhao S, Ukai W, Hashimoto E et al (2012) Heat stress activates ER stress signals which suppress the heat shock response, an effect occurring preferentially in the cortex in rats. Mol Biol Rep 39(4):3987–3993
106. Bhandary B, Marahatta A, Kim HR, Chae HJ (2012) An involvement of oxidative stress in endoplasmic reticulum stress and its associated diseases. Int J Mol Sci 14(1):434–456
107. Cao SS, Kaufman RJ (2014) Endoplasmic reticulum stress and oxidative stress in cell fate decision and human disease. Antioxid Redox Signal 21(3):396–413
108. Bertolotti A, Zhang Y, Hendershot LM, Harding HP, Ron D (2000) Dynamic interaction of BiP and ER stress transducers in the unfolded-protein response. Nat Cell Biol 2(6):326–332
109. Shen J, Chen X, Hendershot L, Prywes R (2002) ER stress regulation of ATF6 localization by dissociation of BiP/GRP78 binding and unmasking of Golgi localization signals. Dev Cell 3(1):99–111

110. Almanza A, Carlesso A, Chintha C, Creedican S, Doultsinos D, Leuzzi B et al (2018) Endoplasmic reticulum stress signalling – from basic mechanisms to clinical applications. FEBS J 286:241–278
111. Schroder M, Kaufman RJ (2005) The mammalian unfolded protein response. Annu Rev Biochem 74:739–789
112. Harding HP, Zhang Y, Ron D (1999) Protein translation and folding are coupled by an endoplasmic-reticulum-resident kinase. Nature 397(6716):271–274
113. McQuiston A, Diehl JA (2017) Recent insights into PERK-dependent signaling from the stressed endoplasmic reticulum. F1000Res 6:1897
114. Schroder M (2006) The unfolded protein response. Mol Biotechnol 34(2):279–290
115. Rowlands AG, Panniers R, Henshaw EC (1988) The catalytic mechanism of guanine nucleotide exchange factor action and competitive inhibition by phosphorylated eukaryotic initiation factor 2. J Biol Chem 263(12):5526–5533
116. Lu PD, Harding HP, Ron D (2004) Translation reinitiation at alternative open reading frames regulates gene expression in an integrated stress response. J Cell Biol 167(1):27–33
117. Quiros PM, Prado MA, Zamboni N, D'Amico D, Williams RW, Finley D et al (2017) Multi-omics analysis identifies ATF4 as a key regulator of the mitochondrial stress response in mammals. J Cell Biol 216(7):2027–2045
118. Han J, Back SH, Hur J, Lin YH, Gildersleeve R, Shan J et al (2013) ER-stress-induced transcriptional regulation increases protein synthesis leading to cell death. Nat Cell Biol 15(5):481–490
119. Ye J, Rawson RB, Komuro R, Chen X, Dave UP, Prywes R et al (2000) ER stress induces cleavage of membrane-bound ATF6 by the same proteases that process SREBPs. Mol Cell 6(6):1355–1364
120. Yamamoto K, Sato T, Matsui T, Sato M, Okada T, Yoshida H et al (2007) Transcriptional induction of mammalian ER quality control proteins is mediated by single or combined action of ATF6α and XBP1. Dev Cell 13(3):365–376
121. Yoshida H, Matsui T, Yamamoto A, Okada T, Mori K (2001) XBP1 mRNA is induced by ATF6 and spliced by IRE1 in response to ER stress to produce a highly active transcription factor. Cell 107(7):881–891
122. Liou HC, Boothby MR, Finn PW, Davidon R, Nabavi N, Zeleznik-Le NJ et al (1990) A new member of the leucine zipper class of proteins that binds to the HLA DR alpha promoter. Science 247(4950):1581–1584
123. Travers KJ, Patil CK, Wodicka L, Lockhart DJ, Weissman JS, Walter P (2000) Functional and genomic analyses reveal an essential coordination between the unfolded protein response and ER-associated degradation. Cell 101(3):249–258
124. Bonifacino JS, Weissman AM (1998) Ubiquitin and the control of protein fate in the secretory and endocytic pathways. Annu Rev Cell Dev Biol 14:19–57
125. Hwang J, Qi L (2018) Quality control in the endoplasmic reticulum: crosstalk between ERAD and UPR pathways. Trends Biochem Sci 43(8):593–605
126. Wu X, Rapoport TA (2018) Mechanistic insights into ER-associated protein degradation. Curr Opin Cell Biol 53:22–28
127. Hollien J, Lin JH, Li H, Stevens N, Walter P, Weissman JS (2009) Regulated Ire1-dependent decay of messenger RNAs in mammalian cells. J Cell Biol 186(3):323–331
128. Oikawa D, Tokuda M, Hosoda A, Iwawaki T (2010) Identification of a consensus element recognized and cleaved by IRE1α. Nucleic Acids Res 38(18):6265–6273
129. Hollien J, Weissman JS (2006) Decay of endoplasmic reticulum-localized mRNAs during the unfolded protein response. Science 313(5783):104–107
130. Maurel M, Chevet E, Tavernier J, Gerlo S (2014) Getting RIDD of RNA: IRE1 in cell fate regulation. Trends Biochem Sci 39(5):245–254
131. Verfaillie T, Rubio N, Garg AD, Bultynck G, Rizzuto R, Decuypere JP et al (2012) PERK is required at the ER-mitochondrial contact sites to convey apoptosis after ROS-based ER stress. Cell Death Differ 19(11):1880–1891

132. Lynes EM, Bui M, Yap MC, Benson MD, Schneider B, Ellgaard L et al (2012) Palmitoylated TMX and calnexin target to the mitochondria-associated membrane. EMBO J 31(2):457–470
133. Yang S, Bhardwaj A, Cheng J, Alkayed NJ, Hurn PD, Kirsch JR (2007) Sigma receptor agonists provide neuroprotection in vitro by preserving bcl-2. Anesth Analg 104(5):1179–1184
134. Penke B, Fulop L, Szucs M, Frecska E (2018) The role of sigma-1 receptor, an intracellular chaperone in neurodegenerative diseases. Curr Neuropharmacol 16(1):97–116
135. Ola MS, Moore P, El-Sherbeny A, Roon P, Agarwal N, Sarthy VP et al (2001) Expression pattern of sigma receptor 1 mRNA and protein in mammalian retina. Brain Res Mol Brain Res 95(1-2):86–95
136. Ha Y, Shanmugam AK, Markand S, Zorrilla E, Ganapathy V, Smith SB (2014) Sigma receptor 1 modulates ER stress and Bcl2 in murine retina. Cell Tissue Res 356(1):15–27
137. Wang J, Saul A, Cui X, Roon P, Smith SB (2017) Absence of sigma 1 receptor accelerates photoreceptor cell death in a Murine model of retinitis pigmentosa. Invest Ophthalmol Vis Sci 58(11):4545–4558
138. Ha Y, Dun Y, Thangaraju M, Duplantier J, Dong Z, Liu K et al (2011) Sigma receptor 1 modulates endoplasmic reticulum stress in retinal neurons. Invest Ophthalmol Vis Sci 52(1):527–540
139. Krishnamoorthy RR, Agarwal P, Prasanna G, Vopat K, Lambert W, Sheedlo HJ et al (2001) Characterization of a transformed rat retinal ganglion cell line. Brain Res Mol Brain Res 86(1-2):1–12
140. Wang L, Eldred JA, Sidaway P, Sanderson J, Smith AJ, Bowater RP et al (2012) Sigma 1 receptor stimulation protects against oxidative damage through suppression of the ER stress responses in the human lens. Mech Ageing Dev 133(11-12):665–674
141. Tanimukai H, Kudo T (2015) Fluvoxamine alleviates paclitaxel-induced neurotoxicity. Biochem Biophys Rep 4:202–206
142. Omi T, Tanimukai H, Kanayama D, Sakagami Y, Tagami S, Okochi M et al (2014) Fluvoxamine alleviates ER stress via induction of Sigma-1 receptor. Cell Death Dis 5:e1332
143. Morihara R, Yamashita T, Liu X, Nakano Y, Fukui Y, Sato K et al (2018) Protective effect of a novel sigma-1 receptor agonist is associated with reduced endoplasmic reticulum stress in stroke male mice. J Neurosci Res 96:1707–1716
144. Ono Y, Tanaka H, Tsuruma K, Shimazawa M, Hara H (2013) A sigma-1 receptor antagonist (NE-100) prevents tunicamycin-induced cell death via GRP78 induction in hippocampal cells. Biochem Biophys Res Commun 434(4):904–909
145. Hong J, Wang L, Zhang T, Zhang B, Chen L (2017) Sigma-1 receptor knockout increases alpha-synuclein aggregation and phosphorylation with loss of dopaminergic neurons in substantia nigra. Neurobiol Aging 59:171–183
146. Alam S, Abdullah CS, Aishwarya R, Orr AW, Traylor J, Miriyala S, et al. (2017) SigmaR1 regulates endoplasmic reticulum stress-induced C/EBP-homologous protein expression in cardiomyocytes. Biosci Rep 37(4)
147. Watanabe S, Ilieva H, Tamada H, Nomura H, Komine O, Endo F et al (2016) Mitochondria-associated membrane collapse is a common pathomechanism in SIGMAR1- and SOD1-linked ALS. EMBO Mol Med 8(12):1421–1437
148. Gregianin E, Pallafacchina G, Zanin S, Crippa V, Rusmini P, Poletti A et al (2016) Loss-of-function mutations in the SIGMAR1 gene cause distal hereditary motor neuropathy by impairing ER-mitochondria tethering and Ca^{2+} signalling. Hum Mol Genet 25(17):3741–3753
149. Horga A, Tomaselli PJ, Gonzalez MA, Laura M, Muntoni F, Manzur AY et al (2016) SIGMAR1 mutation associated with autosomal recessive Silver-like syndrome. Neurology 87(15):1607–1612
150. Li X, Hu Z, Liu L, Xie Y, Zhan Y, Zi X et al (2015) A SIGMAR1 splice-site mutation causes distal hereditary motor neuropathy. Neurology 84(24):2430–2437

151. Nandhagopal R, Meftah D, Al-Kalbani S, Scott P (2018) Recessive distal motor neuropathy with pyramidal signs in an Omani kindred: underlying novel mutation in the SIGMAR1 gene. Eur J Neurol 25(2):395–403
152. Fukunaga K, Shinoda Y, Tagashira H (2015) The role of SIGMAR1 gene mutation and mitochondrial dysfunction in amyotrophic lateral sclerosis. J Pharmacol Sci 127(1):36–41
153. Mavlyutov TA, Yang H, Epstein ML, Ruoho AE, Yang J, Guo LW (2017) APEX2-enhanced electron microscopy distinguishes sigma-1 receptor localization in the nucleoplasmic reticulum. Oncotarget 8(31):51317–51330
154. Miki Y, Mori F, Kon T, Tanji K, Toyoshima Y, Yoshida M et al (2014) Accumulation of the sigma-1 receptor is common to neuronal nuclear inclusions in various neurodegenerative diseases. Neuropathology 34(2):148–158
155. Natsvlishvili N, Goguadze N, Zhuravliova E, Mikeladze D (2015 Apr 30) Sigma-1 receptor directly interacts with Rac1-GTPase in the brain mitochondria. BMC Biochem 16:11
156. Mitsuda T, Omi T, Tanimukai H, Sakagami Y, Tagami S, Okochi M, Kudo T, Takeda M (2011 Nov 25) Sigma-1Rs are upregulated via PERK/eIF2α/ATF4 pathway and execute protective function in ER stress. Biochem Biophys Res Commun 415(3):519–525
157. Kourrich S (2017) Sigma-1 receptor and neuronal excitability. Handb Exp Pharmacol 244:109–130. https://doi.org/10.1007/164_2017_8

Chapter 29
ER-Mitochondria Calcium Transfer, Organelle Contacts and Neurodegenerative Diseases

Francesca Vallese, Lucia Barazzuol, Lorenzo Maso, Marisa Brini, and Tito Calì

Abstract It is generally accepted that interorganellar contacts are central to the control of cellular physiology. Virtually, any intracellular organelle can come into proximity with each other and, by establishing physical protein-mediated contacts within a selected fraction of the membrane surface, novel specific functions are acquired. Endoplasmic reticulum (ER) contacts with mitochondria are among the best studied and have a major role in Ca^{2+} and lipid transfer, signaling, and membrane dynamics.

Their functional (and structural) diversity, their dynamic nature as well as the growing number of new players involved in the tethering concurred to make their monitoring difficult especially in living cells. This review focuses on the most established examples of tethers/modulators of the ER-mitochondria interface and on the roles of these contacts in health and disease by specifically dissecting how Ca^{2+} transfer occurs and how mishandling eventually leads to disease. Additional functions of the ER-mitochondria interface and an overview of the currently available methods to measure/quantify the ER-mitochondria interface will also be discussed.

Keywords Calcium · Mitochondria · Endoplasmic reticulum · ER-mitochondria contact sites · Neurodegenerative diseases · Organelle communication · Signalling · MAM · Protein tethers · Bioenergetics

---

F. Vallese · L. Barazzuol
Department of Biomedical Sciences, University of Padua, Padua, Italy

L. Maso · M. Brini (✉)
Department of Biology, University of Padua, Padua, Italy
e-mail: marisa.brini@unipd.it

T. Calì (✉)
Department of Biomedical Sciences, University of Padua, Padua, Italy

Padua Neuroscience Center (PNC), Padua, Italy
e-mail: tito.cali@unipd.it

© Springer Nature Switzerland AG 2020

M. S. Islam (ed.), *Calcium Signaling*, Advances in Experimental Medicine and Biology 1131, https://doi.org/10.1007/978-3-030-12457-1_29

29.1 Introduction

Eukaryotic cells are characterized by the presence of defined intracellular membranous structures, the organelles that enable the compartmentalization of metabolites and the spatial separation of biological processes.

This chapter focuses on the most recent findings on the tethering complexes found between the endoplasmic reticulum (ER) and mitochondria, the so-called MAMs (mitochondria-associated ER membranes) region, the currently available methods for their monitoring, and their multiple physiological functions including Ca^{2+} signaling and lipid synthesis. How alterations in these connections are linked to the onset of neurodegenerative diseases such as Alzheimer's Disease (AD), Parkinson's Disease (PD) and Amyotrophic Lateral Sclerosis (ALS), will also be discussed.

29.2 Players at the ER-Mitochondria Interface

The maintenance of a correct tethering between the ER and mitochondria is guaranteed by several pairs of cytosolic proteins and integral membrane proteins, that keep the distance between the organelles in a proper range [1–4] (Fig. 29.1). The tethering complex composition and thickness, however, are not constant and

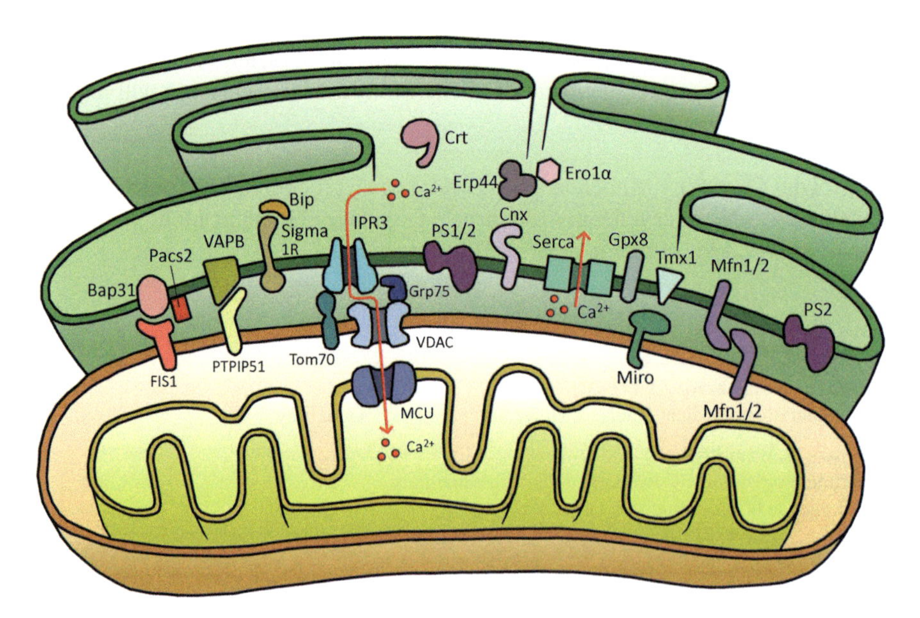

Fig. 29.1 ER-mitochondria contact sites and Ca^{2+} homeostasis. The figure depicts the main MAM proteins involved in Ca^{2+} exchange between ER and mitochondria

the distance between the ER and the outer mitochondrial membrane (OMM) can range from 10 nm up to 80–100 nm depending on the cell needs and response to different conditions [5]. In the following section the most established examples of tethers/modulators will be considered, for a general view of the growing number of proposed new players at this interface the reader is referred to [6–9].

29.2.1 MFN2

Mitofusin 2 (MFN2) is a large GTPase, it is involved in the fusion of mitochondrial outer membrane and it is enriched in MAMs. It is widely accepted that this protein is a major regulator of mitochondria-ER interface in different tissues [10, 11], however the exact role played at the mitochondria-ER contacts is still unclear. MFN2 located in the ER can form hetero- or homo-dimers in trans with mitochondrial mitofusins (i.e., MFN1 or MFN2) to fine tune the MAMs-dependent functions in a concerted manner. Although the precise mechanism by which the distance between the ER and mitochondria is adapted to cope with the cell needs must still be fully elucidated [12–15], the idea that the ER-mitochondria contacts can functionally and physically occur at different distances to perform specific functions is recently emerging [15, 16]. Accordingly, MFN2 has been shown to have opposite effects in the crosstalk between mitochondria and the rough or the smooth ER [15], a finding that was also confirmed by using a novel genetically encoded probe for the evaluation of narrow and wide organelle interactions [17].

29.2.2 VAPB-PTPIP51

Vesicle-associated membrane protein-associated protein B (VAPB) is an integral protein in the ER membrane that is involved in unfolded protein response (UPR) and in the regulation of cellular Ca^{2+} homeostasis [18]. VAPB interacts with a mitochondrial outer membrane protein, the tyrosine phosphatase-interacting protein-51 (PTPIP51) and its overexpression increases ER-mitochondria tethering, while knockdown of either protein decreases the tethering between the two organelles [19]. Multiple approaches demonstrated that VAPB and PTPIP51 are ER-mitochondria tethers: (i) VAPB is enriched in MAMs, and PTPIP51 is a known outer mitochondrial membrane protein [20, 21]. (ii) VAPB and PTPIP51 directly interact and their modulation affects Ca^{2+} exchange between ER and mitochondria [19, 20, 22, 23]. (iii) the manipulation of VAPB and/or PTPIP51 expression induces changes in ER-mitochondria contacts [19, 24].

29.2.3 Fis1-Bap31

Fis1-Bap31 interaction creates a preformed scaffold complex that is proven to tether ER and mitochondria together. B-cell receptor-associated protein 31 (Bap31) is an integral membrane chaperone protein of the ER that forms several protein complexes and controls the fate of newly synthesized integral membrane proteins. Recently, the mitochondrial fission protein 1 homologue (Fis1) has been shown to interact with Bap31 to form an ER-mitochondrial platform which is essential for either the recruitment and activation of procaspase 8 and for the conveyance of the apoptotic signal from mitochondria to the ER [25]. The tethering function of Bap31 may be modulated by phosphofurin acidic cluster sorting protein 2 (PACS-2), a multifunctional cytosolic protein. Downregulation of PACS-2 causes Bap31-dependent mitochondrial fragmentation and uncoupling from the ER along with inhibition of Ca^{2+} signal transmission [26, 27].

29.2.4 IP3R-Grp75-VDAC1

The ER Ca^{2+} channel inositol 1,4,5-trisphosphate receptor (IP3R) and the OMM protein VDAC1 (voltage-dependent anion channel 1) interact via the mitochondrial chaperone Grp75 (glucose-regulated protein 75) [28, 29]. This complex may not have a proper physical tethering role, but rather it is essential to functionally couple ER and mitochondria and favour Ca^{2+} exchanges. A role of TOM70 in the formation of this complex has also been recently demonstrated: it clusters at ER-mitochondria contacts, where it recruits IP3R and promotes inter-organelle Ca^{2+} transfer (likely through the IP3R-Grp75-VDAC1 complex), cell bioenergetics and proliferation [30].

29.3 Ca^{2+} Transfer at the ER-Mitochondria Interface

Ca^{2+} ion is one of the most important second messenger in eukaryotic cells, being a multitude of biological processes associated to the transient variation of its intracellular concentration. Basal levels of free cytoplasmic Ca^{2+} are maintained in the nM range by energy-dependent mechanisms regulated by proteins such as the plasma membrane Ca^{2+} ATPase (PMCA), the SR/ER Ca^{2+} ATPase (SERCA) and the secretory pathway Ca^{2+} ATPase (SPCA) of the Golgi apparatus. Ca^{2+} pumps, together with the plasma membrane Na^{+}/Ca^{2+} exchanger (NCX), promptly counteract cytosolic Ca^{2+} increases by extruding Ca^{2+} ions in the extracellular milieu or by pumping it in the intracellular stores, thus guaranteeing their refilling [31–33]. It is also well known that intracellular organelles, such as mitochondria, participate in the control of cytosolic Ca^{2+} signal. Evidence that Ca^{2+} is taken up by mitochondria has been clear since the 60s, when it was found that the respiratory chain generates, across the inner mitochondrial membrane (IMM), an electrochemical gradient responsible for the import of the cation in the mitochondrial matrix [34, 35]. Only 50 years later, the molecular identity of the channel that mediates this ion transfer was discovered. The Mitochondrial Ca^{2+} uniporter (MCU) was identified in 2011 [36, 37]. Molecularly, the fungal and metazoan MCU is assembled in a tetrameric structure to form the active Ca^{2+} channel [38–41] while MCU from *C. elegans* is a homo-oligomer with pentameric symmetry [42]. The complex also includes a dominant negative isoform called MCUb [43]. Their expression level also differs among the tissues, possibly representing a way to regulate MCU activity according to the different cells type demands [43, 44]. A regulatory dimer formed by the two EF-hand containing proteins MICU1 and MICU2 [45, 46] gates the opening of MCU and depends on the extra-mitochondrial Ca^{2+} concentration. Additional components have also been identified, among them MICU3 [47, 48], highly expressed in neuronal tissue, EMRE [49], an essential protein for the correct assembly of the complex and MCUR1 [50], whose role and the inclusion in the MCU complex is still debated. As already documented by numerous studies before its molecular identification, MCU is characterized by a very low affinity for Ca^{2+} (K_d 15–20 μM), amply far away from the physiological Ca^{2+} concentration values reached in the cytosol of living cells, which are in the range of 50–100 nM in resting conditions and of 1–3 μM upon stimulation. This problem was solved by Rizzuto and co-workers, who showed for the first time that Ca^{2+} uptake by mitochondria occurs preferentially at the sites of contact between ER and mitochondria [1, 51]. Indeed, they observed that the two organelles are strategically located in close proximity, allowing the formation of Ca^{2+} hotspots that overcome the low affinity threshold of mitochondrial Ca^{2+} uptake mechanism. We now know that alterations of the interaction between ER and mitochondria lead to a disruption of Ca^{2+} transfer between the organelles and to ER stress [26]. Ca^{2+} accumulation into mitochondria is crucial not only for the maintenance of cellular Ca^{2+} homeostasis [52] but also for mitochondrial metabolism and adenosine triphosphate (ATP) production, especially for neuronal cells which are highly energy demanding. Ca^{2+} transport proteins, especially those located in MAMs, play a key role in buffering and shaping cytosolic Ca^{2+} transients [53–55] (Fig. 29.1).

The multiprotein complex involved in ER-mitochondria Ca^{2+} transfer is the IP3R-VDAC-Grp75 complex [28]. When MAMs are disrupted, the release of Ca^{2+} from the ER mediated by IP3R is suppressed and ATP production and cell survival are impaired [9, 56]. At the opposite, massive and/or a prolonged accumulation of Ca^{2+} into mitochondria can lead to the opening of the permeability transition pore (PTP) in the IMM, the swelling of the organelles and the induction of apoptosis [57].

SERCA pump, in addition to play a key role in maintaining bulk cytosolic Ca^{2+} at basal level and in replenishing intracellular stores, controls local Ca^{2+} transfer at ER-mitochondrial contacts level [3]. Several mechanisms regulate SERCA activity at MAMs level. Phosphorylation of the transmembrane chaperone calnexin (CNX) has been proposed to inhibit SERCA activity and, on the other side, its dephosphorylation occurs upon IP3R-mediated Ca^{2+} release, suggesting a tight regulation via the phosphorylation status of the CNX cytosolic domain [58]. However CNX has also been suggested to act as a positive SERCA2b regulator [59, 60]. In conjunction with CNX, thioredoxin-related TMX1 has been shown to inhibit SERCA2b activity [8] and promote ER-mitochondria contact formation, thus suggesting that inhibition of SERCA2b might be a compensatory measure to reduce Ca^{2+} levels in the ER and to prevent possible mitochondrial Ca^{2+} overload [61]. Recently, the glutathione peroxidase (GPX8), another ER membrane and MAMs resident redox regulator, has been shown to decrease SERCA2b activity [62].

Ca^{2+} fluxes at the ER-mitochondria interface are thus tightly tuned, it is not surprising that many components acting at the MAM are indeed influenced by Ca^{2+} itself acting as a key regulator of many components required to sustain the plethora of physiologic processes occurring at this location [63]. On the mitochondrial side, there is the mitochondrial Rho GTPase (Miro), an essential protein for the regulation of mitochondrial movements. Interestingly, it responds to Ca^{2+} above a concentration range that is never attained in the bulk cytosol of living cells (10–100 μM) [64], but that might be easily reached at the ER-mitochondria contact sites. On the ER side, the above mentioned CNX, as well as calreticulin (CRT) and the Binding immunoglobulin protein BiP are MAMs-specific Ca^{2+} -binding and glucose regulated chaperones [65, 66] and serve as high-capacity Ca^{2+} pools in the ER [67, 68]. CRT has also a regulatory role on Ca^{2+} signaling since its overexpression has an inhibitory effect on the IP3R [69], while its downregulation impaired Ca^{2+} homeostasis [70]. The oxidoreductase Ero1a and its co-chaperone ERp44 [71] are enriched at the MAMs and regulate the activity of the IP3R via a direct interaction [72, 73]. Furthermore, Ero1a has been shown to directly regulate ER-mitochondria Ca^{2+} fluxes and to influence the activity of the mitochondrial transporters [74]. Analogously, the ER protein Sigma-1 receptor (Sig-1R), a Ca^{2+}-sensitive chaperone, operates at MAMs forming a complex with another chaperone, Bip. Upon ER Ca^{2+} depletion, Sig-1R dissociates from BiP, leading to a prolonged Ca^{2+} signaling into mitochondria via IP3Rs [65].

29.4 ER-Mitochondria Ca^{2+} Mishandling in Neurodegenerative Diseases

Neurodegenerative diseases, including PD, AD and ALS/FTD (fronto-temporal dementia), affect millions of people worldwide and are characterized by progressive nervous system dysfunction. Common pathogenic mechanisms have been described for these diseases, including abnormal proteostasis, often associated with the formation of intracellular and/or extracellular protein inclusions, ER and oxidative stress, mitochondrial dysfunction, and neuroinflammation [75, 76]. The existence of ER-mitochondria contacts also in neuronal cells, which are mainly dependent from Ca^{2+} influx from the extracellular ambient for the control of their function, suggests

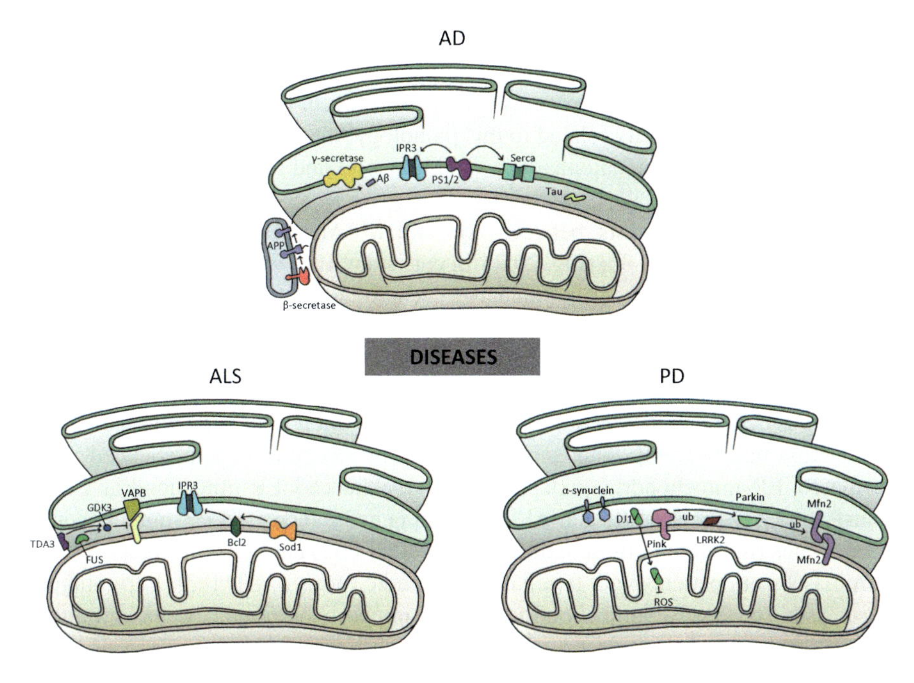

Fig. 29.2 ER-mitochondria contact sites in neurodegenerative diseases. The figure depicts the main players of MAMs involved in the pathogenesis of Alzheimer's disease (AD), Parkinson's disease (PD) and amyotrophic lateral sclerosis (ALS)

that MAMs may have a critical role also in regulating synaptic activity [77, 78]. Over the last decade, alterations in the contact sites between ER and mitochondria have been found as a common characteristic in many neurodegenerative diseases, consistently with the fact that several disease-related proteins are transiently or constitutively associated with the ER-mitochondria interface. The presence of mutations in these proteins has been shown to alter the structural and functional features of this subcellular compartment (Fig. 29.2). In the following sections, we will focus on representative examples of altered ER-mitochondria communication in the context of AD, PD and ALS.

29.4.1 Alzheimer's Disease (AD)

The first evidence on ER-mitochondria contacts involvement in neurodegeneration has been highlighted in the study of AD. AD is an irreversible neurodegenerative disorder characterized by progressive loss of episodic memory and by cognitive and behavioral impairment caused by degeneration of hippocampal and cortical neurons. All the familial forms of AD (FAD) so far identified are related to mutations in the genes coding for amyloid precursor protein (APP) or for presenilins (PS1 and

PS2), the major components of the γ-secretase complex that processes APP to release Amiloid β (Aβ) peptide. The generation or the accumulation of Aβ is tightly associated with the onset and progression of the disease, but this is not the unique phenotypic feature directly related to this pathology: alterations in Ca^{2+} and lipids homeostasis, ROS levels, autophagy, axonal transport and mitochondrial dynamics are commonly observed [79].

In the last years, a link between AD and MAMs dysfunction has become increasingly evident [21, 80] (Fig. 29.2). In particular, PSs, which play an important role in Ca^{2+} homeostasis [81, 82], have been found at the MAMs [6, 83] and proposed to form ER Ca^{2+} channels [84], as well as to act as modulators of the IP3R or ryanodine receptor (RyR) open probability [85, 86] or of the SERCA pump activity [87, 88]. Pizzo and coworkers have reported that PS2 action in Ca^{2+} signaling is dependent on the modulation of ER-mitochondria interactions and their Ca^{2+} cross talk [89, 90]. Pathogenic mutations of PS2 have been associated with increased ER-mitochondria juxtaposition and enhanced ER-mitochondria Ca^{2+} transfer in FAD-mutant PS1 and APP expressing cells. Increased ER-mitochondria connections (resembling those induced by mutated PS2) have also been reported in fibroblasts from patients with familial and sporadic forms of AD [83]. Intriguingly, the possibility that ER-mitochondria contact impairments might be the common feature in different neurodegenerative conditions becomes even more consistent considering that Parkin, a PD-related protein, has been recently shown to differently modulate PS1 and PS2 expression at the transcriptional level [91]. Interestingly, when primary hippocampal neurons are incubated with nM concentrations of oligomeric Aβ, ER-mitochondria proximity and Ca^{2+} exchanges between the two organelles increase [92].

Another typical feature of AD is the accumulation of intracellular neurofibrillary tangles (NFT) composed by tau and Aβ plaques. Both of them have also been shown to have a direct influence on several mitochondria-related activities: interestingly, a fraction of tau protein has been recently found at the OMM and within the inner mitochondrial space (IMS) [93], suggesting a potential tau-dependent regulation of mitochondrial functions and the possibility that alterations in its distribution may precede the appearance of the detrimental effects induced by its aggregation. Mis-handling and defects in the ER-mitochondria communications might be an important pathological event in tau-related dysfunction and thereby contributing to neurodegeneration [93, 94].

29.4.2 Parkinson's Disease (PD)

PD is the second most common neurodegenerative disease after AD. PD patients typically develop slowness of movements (bradykinesia), involuntary shaking (tremor), increased resistance to passive movement (rigidity) and postural instability [95, 96].

The phenotypes of both sporadic and familial forms of PD are essentially indistinguishable, implying that they might share common underlying etiological mechanisms. Dominant forms of the disease are related to mutations in three proteins, α-synuclein, LRRK2 (Leucine-rich repeat kinase 2), and VPS35 (Vacuolar protein sorting-associated protein 35), whereas mutations in Parkin, PINK1 (PTEN-induced putative kinase 1), and DJ-1 cause autosomal recessive-inherited forms [97]. Recently, many studies have found dysfunctions in ER-mitochondria communication in different cellular models for PD (Fig. 29.2) and consequently alterations in many different pathways depending on it [21, 94, 98].

α-synuclein is a protein with a central role in the modulation of synaptic integrity and function, its disease-related mutations alter a plethora of physiological processes that are regulated by the signaling between ER and mitochondria, and, consistently, α-synuclein has been shown to play a key role at the ER–mitochondria contacts by favouring Ca^{2+} transfer between the organelles and sustaining mitochondrial metabolism [99–102]. PINK1 and Parkin are associated with mitochondria and in addition to play an important role in the process of mitochondrial quality control (mitophagy) [103–106], they may be involved in the modulation of the ER-mitochondria apposition, thus suggesting that mutations in the two proteins could be linked to MAMs anomalies. Furthermore, contact regions between ER and mitochondria have been shown to be prime locations for Parkin-mediated mitophagy and local recruitment of autophagosome precursors. Parkin was also shown to be present at the ER and the mitochondrial membranes under basal conditions [107–112] and to have a role in the modulation of the mitochondrial-ER interactions [111] and the proteosome activity. Cumulating evidence strongly supports a role for Parkin in general protein quality control and ER stress pathways [113–118]. The absence of Parkin in mice fibroblast or the presence of Parkin mutations in PD-patients fibroblasts have been shown to decrease the ER-mitochondrial tethering and, interestingly, mutant MFN2 that cannot be ubiquitinated, failed to restore ER-mitochondria contacts, suggesting that Parkin mediated ubiquitination of MFN2 may be required for inter-organelle association [119].

Mutations in DJ-1 protein have been linked to autosomal recessive early-onset parkinsonism [120]. DJ-1 protein plays a protective role against oxidative stress and it is essential to maintain proper mitochondria dynamics. It is mainly localized in the cytosol and in the nucleus, but it has also been found in mitochondria [121] and at MAMs level [122]. DJ-1 overexpression in cell models increased the ER–mitochondria association, whereas its downregulation caused mitochondria fragmentation and decreased mitochondrial Ca^{2+} uptake, suggesting a direct role of this protein in MAMs functions [122].

29.4.3 Amyotrophic Lateral Sclerosis (ALS)

ALS is the most common form of motor neuron disease that leads to a progressive muscle paralysis caused by a prominent degeneration of upper and lower motor neurons that communicate with muscle cells, leading to death within few years of diagnosis [123]. Over 90% of ALS cases occur sporadically, while familial ALS (FALS) patients with known genetic mutations are relatively rare. Approximately 50% of the familial forms of the disease are associated with mutations in VAPB, TAR-DNA binding protein 43 (TDP43), fused in sarcoma (FUS) and superoxide dismutase 1 (SOD1). All the studies on these proteins strongly suggest that alterations of the ER-mitochondria platform could be an early event at the basis of ALS (Fig. 29.2). VAPB is an ER membrane anchored-protein enriched at the MAMs [20] and it is involved in the regulation of the ER stress response and consequent Ca^{2+}-mediated death in motor neurons [124]. VAPB interacts with PTPIP51 (a mitochondrial outer membrane protein) and this interaction is necessary to support ER-mitochondria Ca^{2+} transfer between the two organelles. Indeed, mutation of VAPB (P56S) induces impaired mitochondria Ca^{2+} uptake, increased cytosolic Ca^{2+} levels and mitochondria clustering [20, 125].

Concerning the involvement of TDP-43 in ALS, its overexpression in both wt and mutated form decreases ER-mitochondria physical and functional coupling by disrupting VAPB-PTPIP51 interaction through the activation of GSK-3 [19]. The disassembling of the VAPB-PTPIP51 tethering complex via GSK-3 is also induced upon FUS overexpression (both in the wt and ALS-linked mutant form) [23]. Impaired ER-mitochondria juxtaposition caused by the overexpression of these proteins results in all cases in defects in Ca^{2+} transfer and, consequently, in mitochondrial ATP production.

SOD1 is the most frequent mutated protein in ALS, accounting for the 20% of the familial cases [126]. SOD1 is a key enzyme in the defense against oxidative stress. At the moment, the best model for the study of ALS is represented by the transgenic mice expressing mutant SOD1. The overexpression of the ALS-related hSOD1 mutant G93A has been observed to alter ER and mitochondrial Ca^{2+} homeostasis in mouse embryonic motor neurons [127, 128]. The aberrant interactions of mutant SOD1 with Bcl-2 [129], a protein present both at mitochondrial and ER membranes, could in part define the mechanisms whereby mutant SOD1 affects Ca^{2+} regulation, as Bcl-2 has been proposed to modulate IP3R activity [130] and control ER Ca^{2+} levels [131].

29.5 Additional Cellular Functions Associated with the ER-Mitochondria Interface

Functional roles of ER-mitochondria contacts were first uncovered in phospholipid biosynthesis and transport [132, 133]. Subsequently, ER-mitochondria contacts were implicated in other different cellular processes (Ca^{2+} and ROS homeostasis, phosphoregulation, mitochondrial fission and fusion, autophagy), emerging as a complex hub fundamental for the correct integration of numerous signaling pathways [4, 134–136] (Fig. 29.3).

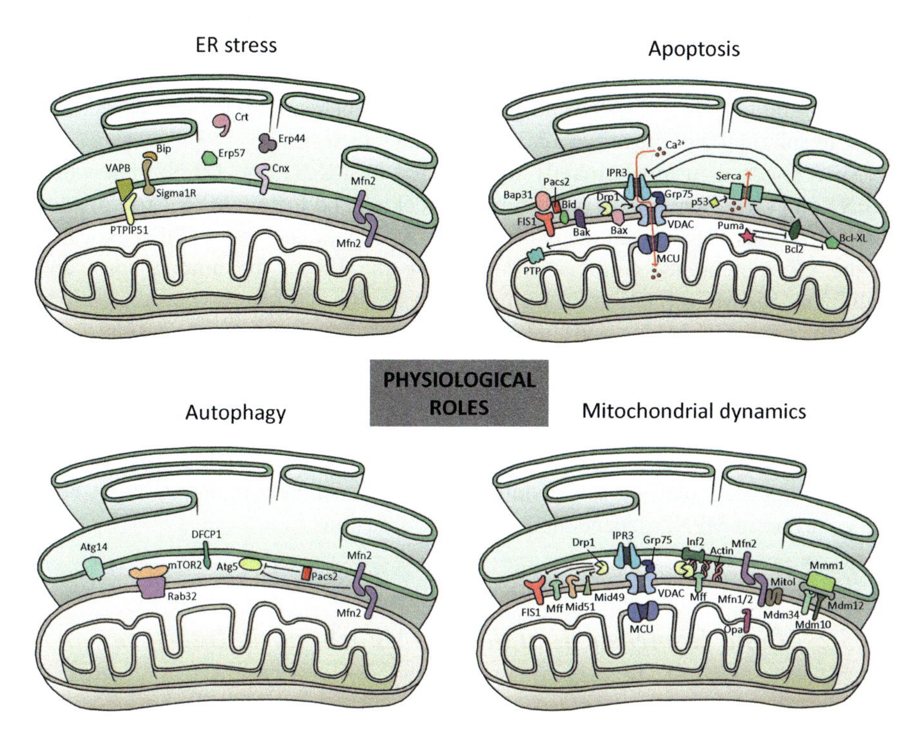

Fig. 29.3 Main physiological roles of MAM junctions. The figure summarizes the most important key components of MAMs involved in ER stress, apoptosis, autophagy and mitochondrial dynamics

29.5.1 Mithocondrial Dynamics

The best studied connection between ER-mitochondria contacts and mitochondrial dynamics involves mitochondrial fission and fusion machinery. Mitochondrial fission is mediated by human dynamin-related protein 1 (Drp1). It localizes primarily at the cytoplasm and is recruited to mitochondria by different proteins: Fis [137], the mitochondrial fission factor (Mff) [138] and the mitochondrial dynamics 51 and 49 kDa proteins (MiD51 and MiD49) [2] that regulate mitochondrial fission by anchoring Drp1 on the outer mitochondrial membrane [139]. Drp1 associates with the OMM, where, thanks to a conformational change due to its GTPase activity, forms oligomers that wrap around the constricted portions of the mitochondria membrane and lead to a fission event [140, 141]. However, it has been observed that mitochondrial fission may start even when Drp1 or Mff are down-regulated [2] since ER membranes were able to wrap around the mitochondrial tubules, indicating that ER-mitochondria contact represents a conserved platform for the regulation of mitochondrial division. Accordingly, recent findings from the Higgs group proposed a potential mechanism for ER association-induced mitochondrial fission involving actin polymerization and a ER-localized protein, i.e. the inverted formin 2 (INF2) [142]. INF2 is activated to polymerize actin, which in turn might generate the driving force for the initial constriction of mitochondria [136, 142]. Once assembled, the ER-associated constricted mitochondria enable polymerized Drp1 to spiral around the mitochondria to mediate their fission [143].

Mitochondrial fusion is mainly orchestrated by MFN1 and MFN2, but while MFN1 plays a critical role in mitochondrial docking and fusion, MFN2 coordinates the interactions between mitochondria [144]. As mentioned above, MFN2 located in the ER membrane is crucial for tethering the ER to the mitochondria and stabilizing MAMs formation by forming both homo- and heterotypic interactions with mitochondrial MFN1 and 2 [12]. The mitochondrial ubiquitin ligase protein, MITOL, regulates MFN2 activity at the ER–mitochondria interface interacting with mitochondrial MFN2 and mediating its ubiquitination and its subsequent oligomerization, a fundamental step in MFN2-induced ER-mitochondria tethering [145]. The dual role of MFN2 in both mitochondrial fusion and ER-mitochondria tethering suggests that the establishment of ER–mitochondria contact might be a critical event in MFN2-dependent mitochondrial fusion.

29.5.2 Lipid Homeostasis at the ER-Mitochondria Interface

The maintenance of a defined lipid composition in mitochondrial membranes depends on bidirectional lipid trafficking between the ER and mitochondria. ER is the predominant production site for lipids and makes close contacts with other organelles to ensure lipid exchange [146]. Enzymes involved in phospholipid biosynthesis are enriched at MAMs [147, 148]. Despite of the important role of MAMs in phospholipid exchange between ER and mitochondria is well recognized, the precise role of the proteins involved in the ER-mitochondria tether in mammalian cells remain to be defined [149]. In yeast, the ER-mitochondrial encounter structure (ERMES) that mediates ER-mitochondria contacts has been well characterized: it is composed by Mdm34 (mitochondrial distribution and morphology 34) and Mdm10 that are integral OMM proteins, Mmm1 (maintenance of mitochondrial morphology 1) that resides in the ER membrane and Mdm12 that is associated with OMM peripherally [150–152]. It is believed that ERMES complex may act as a lipid transferase, since Mmm1, Mdm12, and Mdm34 contain a synaptotagmin-like mitochondrial-lipid-binding protein (SMP) domain, which forms a hydrophobic cavity and likely binds to hydrophobic molecules such as phospholipids [153, 154].

29.5.3 Apoptosis

The role of ER-mitochondria communication during apoptosis has been extensively documented, in particular in relation with Ca^{2+} transfer from ER to mitochondria, that is essential to sustain mitochondrial function but can also be deleterious and promote cell death. This process is modulated by the presence of various Bcl-2 family proteins on the membranes of both organelles [155, 156], whose role has been discussed in several excellent reviews on the field [157–159]. In a simplified view, the anti-apoptotic members Bcl-2 and Bcl-XL inhibit ER Ca^{2+} release and apoptosis [160, 161], whereas the pro-apoptotic members Bax/Bak Puma and Bik acts as positive regulators [162–164].

Upon induction of apoptosis, mitochondrial permeability transition pore (MPTP) opening occurs, the mitochondrial network fragments and cytochrome c are released activating downstream caspases. Among the targets of caspases there is BAP31, an ER-membrane protein. Cleavage of BAP31 generates a pro-apoptotic p20 fragment that remains at the ER membrane and stimulates ER Ca^{2+} release that consequently induces mitochondrial uptake and the recruitment of the dynamin-related protein Drp1 that promotes mitochondrial fragmentation [165]. Exogenous inducers of apoptosis promote physical contacts between BAP31 and Fis1. This interaction is required for the activation of caspase-8 and the generation of the p20 fragment, demonstrating a bi-directional communication between the two organelles and a pattern of feedback amplification. Consistent with this, cell death caused by Fis1 over-expression is dependent on the ER-mitochondria Ca^{2+} transfer [166]. Mitochondrial dynamics are strictly related to apoptosis, indeed, during this process, Drp1 is massively recruited to the OMM, where it assembles into foci that mediate mitochondrial division, causing a dramatic fragmentation of the mitochondrial network. The pro-apoptotic protein Bax behaves similarly to Drp1 during apoptosis; it is recruited at the OMM, where it inserts and oligomerizes to form foci that are functionally associated with OMM permeabilization. Under apoptotic conditions, Drp1 and MFN2 were both found in foci with Bax [167] [168].

29.5.4 Autophagy

Autophagy is a regulated mechanism of the cell that controls both the specific disassembles of unnecessary or dysfunctional cellular components and the non-selective response to the deprivation of nutrients. Defects in autophagy may play a significant role in several human pathologies, including cancer and neurodegeneration [169]. The selective degradation of damaged mitochondria is called mitophagy. During this processes mitochondria are first excluded from the mitochondrial network, through fission events, and then delivered to the lysosomes by the autophagy machinery [170]. The recognition of mitochondria that need to be degraded takes place through specific molecular pathways which involve the OMM protein NIP3-like protein X (NIX; also known as BNIP3L) and PINK1/Parkin [106]. The generation of the autophagosome is central for the autophagic process and several observations suggest that phagophore membranes could be formed primarily from the ER. Moreover, the evidence for a connection between mitochondria and autophagosome is very strong: several autophagy-related proteins can localize to the mitochondria, different mitochondrial proteins can modulate the autophagic process, and the OMM participates in autophagosome biogenesis under starvation conditions [171]. In recent studies, it has been shown that dynamic crosstalk between the ER and mitochondria is critical for autophagosome formation [172]. For example, when autophagy is induced upon starvation, the pre-autophagosome marker ATG14 (present at cytosolic and ER sites under resting conditions) and the omegasome marker DFCP1 (present in the MAMs fractions during starvation) are markedly concentrated at the MAMs and serves as a platform for autophagosome formation [172]. Additionally, ATG5, which is critical for autophagosome formation, translocate to the MAMs compartment during phagophore biogenesis and then dissociates from MAMs upon completion of the autophagosome formation, thus establishing a stable interaction with the ER and transient associations with the mitochondria. Autophagosome formations is prevented when there is an interference in the formation of ER-mitochondria contacts. The ER-mitochondria interface and autophagy are thus tightly linked and, interestingly, the diverse MAMs related functions are also coordinated and fine-tuned to keep cell homeostasis. One such example is represented by the mTOR complex 2 (mTORC2) [173], a MAMs resident complex that regulates autophagosome formation and mitochondrial dynamics in a Rab32 dependent manner [174], additionally, mTORC2 has been shown to control Ca^{2+} uptake, mitochondrial bioenergetics and apoptosis via Akt-mediated regulation of IP3R, hexokinase2 and PACS2 [175] suggesting that the ER-mitochondria interface is not only important for sensing cellular stress and coordinating Ca^{2+} transfer and the apoptotic response, but it also represents the primary platform for autophagosome formation and integration of many fundamental cell processes.

29.5.5 The ER Stress Response at the MAMs

ER stress consists in a perturbation of ER homeostasis that alters protein folding process and activates the unfolded protein response (UPR), an intracellular signaling pathway that is activated by the accumulation of unfolded proteins in the lumen of the ER and stimulates transcriptional responses of the genes that encode ER chaperones (such as BiP, CNX or CTR) or ER biosynthetic machinery components to increase the protein-folding/degradation capacity of the ER [176, 177]. ER-mitochondria contacts have been linked to ER stress and UPR, in particular, ER stress leads to a redistribution of mitochondrial and reticular networks at the perinuclear zone, with increased contact points and Ca^{2+} transfer. Chaperone complexes containing Grp75, and other proteins such as Rab32 [178] or PACS-2 [26], are involved in the adaptive response to ER stress. A variety of ER chaperones involved in protein folding, such as BiP, CNX, CRT, ERp44, ERp57, and the Sigma-1 receptor are enriched at MAMs [63], in addition, some players whose roles are apparently unrelated to the ER folding/degradation processes have also been shown to play a role in the UPR by influencing the ER-stress response, for example Mfn2 and VAPB can modulate the UPR and mitochondrial function via PERK and ATF6, respectively [179, 180]. The observation that structural uncoupling of ER from mitochondria induces ER stress by impeding a correct unfolded protein response [26] suggests that the crosstalk between ER and mitochondria at MAMs level may have a role in facilitating UPR.

29.6 Methods to Measure ER-Mitochondria Contact Sites

Several approaches are currently available to study ER-mitochondria contact sites. The oldest and more consolidate approach to elucidate the structure of ER-mitochondria contacts and to calculate their distance is the electron microscopy (EM) [2, 12, 139, 181, 182].

Starting from 1958 EM studies produced a complete and detailed view of the ER-mitochondria contacts: different 3D reconstructions of these structures were generated by serial, tilt-angle tomography in yeast cells [139] as well as by focused ion beam scanning EM at 4-nm resolution in neurons from mice brain slices [78], and soft X-ray tomography with 50-nm resolution in mammalian lymphoblastoid cells [183]. However, the acquisition and reconstruction processes for these 3D approaches remain laborious and therefore they are not yet widely applicable. Furthermore, EM analysis is not able to generate sufficient data for statistical comparisons of organelle geometry. Historically, the existence of the MAMs fraction has been proven when it was found that, during cell-fractionation experiments, sub compartmentalized ER membranes co-sedimented with mitochondria [184]. To date, subcellular fractionation is still considered the standard method to prove presence/levels of specific players at the ER-mitochondria interface [185]. Other approaches, mainly based on the co-localization of ER and mitochondrial markers, are described in the literature and helped to describe ER-mitochondria contacts architecture [14, 51, 186, 187], but none of them has yet been applied to quantify the overall extent or geometry of the interface. The *in-situ* proximity ligation assay (PLA) also allows to visualize and quantify endogenous ER-mitochondria interactions in fixed cells by using pairs of primary antibodies against proteins on opposing membranes [188]. This technique is widely used [23, 189, 190] though it can only be applied to fixed cells and is limited by the availability as well as by the specificity of the antibodies. For the specific detection and quantification of the contact points it has been developed a tool based on the system of rapamycin-inducible linkers that were tagged with a pair of fluorophores capable of generating Förster resonance energy transfer (FRET) [191]. One half of the linker was targeted to the OMM and tagged with CFP, while the other half was targeted to the ER surface and tagged with YFP. A short treatment with rapamycin or its analogues causes the linkage of the two halves, whenever they are in sufficient close proximity. The linkage can be visualized as an increase in FRET signal. Major limitations in the use of this FRET-based probe are due to the fact that it requires equimolar expression of the two moieties [13] and that its in vivo applications are limited by the use of rapamycin, a potent inducer of autophagy [192]. Novel tools to measure inter organelle proximity have been recently developed [17, 193–197]. The sensor developed by our group, called SPLICS (split-GFP-based Contact site Sensor), specifically allows to detect ER-mitochondria contact sites in living cells and in vivo [17]. Two non-fluorescent portions of the superfolder GFP, the GFP_{1-10} moiety and the GFP β-strand 11 [198, 199], engineered to fluoresce when are in close proximity, were targeted to the ER and the OMM by the addition of targeting sequences. Two SPLICS versions were generated that efficiently measured narrow (8–10 nm) and wide (40–50 nm) juxtapositions between ER and mitochondria, and interestingly, they were able to document the existence of at least two types of contact sites in human cells that undergo to differential modulation upon different pharmacological treatments or conditions [17].

Acknowledgements This work was supported by grants from the Università degli Studi di Padova (Progetto di Ateneo 2015 n. CPDA 153402 to MB, Progetto Giovani 2012 n.GRIC128SP0 to TC and Progetto di Ateneo 2016 n. CALI_SID16_01 to TC) and from the Ministry of University and Research (Bando SIR 2014 n. RBSI14C65Z to TC).

References

1. Rizzuto R (1998) Close contacts with the endoplasmic reticulum as determinants of mitochondrial Ca^{2+} responses. Science 280(5370):1763–1766. https://doi.org/10.1126/science.280.5370.1763
2. Friedman JR, Lackner LL, West M, DiBenedetto JR, Nunnari J, Voeltz GK (2011) ER tubules mark sites of mitochondrial division. Science 334(6054):358–362. https://doi.org/10.1126/science.1207385
3. Csordas G, Hajnoczky G (2001) Sorting of calcium signals at the junctions of endoplasmic reticulum and mitochondria. Cell Calcium 29(4):249–262. https://doi.org/10.1054/ceca.2000.0191
4. Wu H, Carvalho P, Voeltz GK (2018) Here, there, and everywhere: the importance of ER membrane contact sites. Science 361(6401):eaan5835. https://doi.org/10.1126/ science.aan5835
5. Giacomello M, Pellegrini L (2016) The coming of age of the mitochondria-ER contact: a matter of thickness. Cell Death Differ 23(9):1417–1427. https://doi.org/10.1038/cdd.2016.52
6. Csordas G, Weaver D, Hajnoczky G (2018) Endoplasmic reticulum-mitochondrial contactology: structure and signaling functions. Trends Cell Biol 28(7):523–540. https://doi.org/10.1016/j.tcb.2018.02.009
7. Herrera-Cruz MS, Simmen T (2017) Over six decades of discovery and characterization of the architecture at Mitochondria-Associated Membranes (MAMs). Adv Exp Med Biol 997:13–31. https://doi.org/10.1007/978-981-10-4567-7_2
8. Raturi A, Gutierrez T, Ortiz-Sandoval C, Ruangkittisakul A, Herrera-Cruz MS, Rockley JP, Gesson K, Ourdev D, Lou PH, Lucchinetti E, Tahbaz N, Zaugg M, Baksh S, Ballanyi K, Simmen T (2016) TMX1 determines cancer cell metabolism as a thiol-based modulator of ER-mitochondria Ca^{2+} flux. J Cell Biol 214(4):433–444. https://doi.org/10.1083/jcb.201512077
9. Rowland AA, Voeltz GK (2012) Endoplasmic reticulum-mitochondria contacts: function of the junction. Nat Rev Mol Cell Biol 13(10):607–625. https://doi.org/10.1038/nrm3440
10. Naon D, Scorrano L (2014) At the right distance: ER-mitochondria juxtaposition in cell life and death. Biochim Biophys Acta 1843(10):2184–2194. https://doi.org/10.1016/j.bbamcr.2014.05.011

11. Filadi R, Pendin D, Pizzo P (2018) Mitofusin 2: from functions to disease. Cell Death Dis 9(3):330. https://doi.org/10.1038/s41419-017-0023-6
12. de Brito OM, Scorrano L (2008) Mitofusin 2 tethers endoplasmic reticulum to mitochondria. Nature 456(7222):605–610. https://doi.org/10.1038/nature07534
13. Naon D, Zaninello M, Giacomello M, Varanita T, Grespi F, Lakshminaranayan S, Serafini A, Semenzato M, Herkenne S, Hernandez-Alvarez MI, Zorzano A, De Stefani D, Dorn GW 2nd, Scorrano L (2016) Critical reappraisal confirms that Mitofusin 2 is an endoplasmic reticulum-mitochondria tether. Proc Natl Acad Sci U S A 113(40):11249–11254. https://doi.org/10.1073/pnas.1606786113
14. Filadi R, Greotti E, Turacchio G, Luini A, Pozzan T, Pizzo P (2015) Mitofusin 2 ablation increases endoplasmic reticulum-mitochondria coupling. Proc Natl Acad Sci U S A 112(17):E2174–E2181. https://doi.org/10.1073/pnas.1504880112
15. Wang PT, Garcin PO, Fu M, Masoudi M, St-Pierre P, Pante N, Nabi IR (2015) Distinct mechanisms controlling rough and smooth endoplasmic reticulum contacts with mitochondria. J Cell Sci 128(15):2759–2765. https://doi.org/10.1242/jcs.171132
16. Bravo-Sagua R, Lopez-Crisosto C, Parra V, Rodriguez-Pena M, Rothermel BA, Quest AF, Lavandero S (2016) mTORC1 inhibitor rapamycin and ER stressor tunicamycin induce differential patterns of ER-mitochondria coupling. Sci Rep 6:36394. https://doi.org/10.1038/srep36394
17. Cieri D, Vicario M, Giacomello M, Vallese F, Filadi R, Wagner T, Pozzan T, Pizzo P, Scorrano L, Brini M, Cali T (2017) SPLICS: a split green fluorescent protein-based contact site sensor for narrow and wide heterotypic organelle juxtaposition. Cell Death Differ 25(6):1131–1145. https://doi.org/10.1038/s41418-017-0033-z
18. Kanekura K, Nishimoto I, Aiso S, Matsuoka M (2006) Characterization of amyotrophic lateral sclerosis-linked P56S mutation of vesicle-associated membrane protein-associated protein B (VAPB/ALS8). J Biol Chem 281(40):30223–30233. https://doi.org/10.1074/jbc.M605049200
19. Stoica R, De Vos KJ, Paillusson S, Mueller S, Sancho RM, Lau KF, Vizcay-Barrena G, Lin WL, Xu YF, Lewis J, Dickson DW, Petrucelli L, Mitchell JC, Shaw CE, Miller CC (2014) ER-mitochondria associations are regulated by the VAPB-PTPIP51 interaction and are disrupted by ALS/FTD-associated TDP-43. Nat Commun 5:3996. https://doi.org/10.1038/ncomms4996
20. De Vos KJ, Morotz GM, Stoica R, Tudor EL, Lau KF, Ackerley S, Warley A, Shaw CE, Miller CC (2012) VAPB interacts with the mitochondrial protein PTPIP51 to regulate calcium homeostasis. Hum Mol Genet 21(6):1299–1311. https://doi.org/10.1093/hmg/ddr559
21. Paillusson S, Stoica R, Gomez-Suaga P, Lau DH, Mueller S, Miller T, Miller CC (2016) There's something wrong with my MAM; the ER-mitochondria axis and neurodegenerative diseases. Trends Neurosci 39(3):146–157. https://doi.org/10.1016/j.tins.2016.01.008
22. Huttlin EL, Ting L, Bruckner RJ, Gebreab F, Gygi MP, Szpyt J, Tam S, Zarraga G, Colby G, Baltier K, Dong R, Guarani V, Vaites LP, Ordureau A, Rad R, Erickson BK, Wuhr M, Chick J, Zhai B, Kolippakkam D, Mintseris J, Obar RA, Harris T, Artavanis-Tsakonas S, Sowa ME, De Camilli P, Paulo JA, Harper JW, Gygi SP (2015) The BioPlex network: a systematic exploration of the human interactome. Cell 162(2):425–440. https://doi.org/10.1016/j.cell.2015.06.043
23. Stoica R, Paillusson S, Gomez-Suaga P, Mitchell JC, Lau DH, Gray EH, Sancho RM, Vizcay-Barrena G, De Vos KJ, Shaw CE, Hanger DP, Noble W, Miller CC (2016) ALS/FTD-associated FUS activates GSK-3beta to disrupt the VAPB-PTPIP51 interaction and ER-mitochondria associations. EMBO Rep 17(9):1326–1342. https://doi.org/10.15252/embr.201541726
24. Galmes R, Houcine A, van Vliet AR, Agostinis P, Jackson CL, Giordano F (2016) ORP5/ORP8 localize to endoplasmic reticulum-mitochondria contacts and are involved in mitochondrial function. EMBO Rep 17(6):800–810. https://doi.org/10.15252/embr.201541108

25. Iwasawa R, Mahul-Mellier AL, Datler C, Pazarentzos E, Grimm S (2011) Fis1 and Bap31 bridge the mitochondria-ER interface to establish a platform for apoptosis induction. EMBO J 30(3):556–568. https://doi.org/10.1038/emboj.2010.346
26. Simmen T, Aslan JE, Blagoveshchenskaya AD, Thomas L, Wan L, Xiang Y, Feliciangeli SF, Hung CH, Crump CM, Thomas G (2005) PACS-2 controls endoplasmic reticulum-mitochondria communication and Bid-mediated apoptosis. EMBO J 24(4):717–729. https://doi.org/10.1038/sj.emboj.7600559
27. Betz C, Stracka D, Prescianotto-Baschong C, Frieden M, Demaurex N, Hall MN (2013) mTOR complex 2-Akt signaling at mitochondria-associated endoplasmic reticulum membranes (MAM) regulates mitochondrial physiology. Proc Natl Acad Sci U S A 110(31):12526–12534. https://doi.org/10.1073/pnas.1302455110
28. Szabadkai G, Bianchi K, Varnai P, De Stefani D, Wieckowski MR, Cavagna D, Nagy AI, Balla T, Rizzuto R (2006) Chaperone-mediated coupling of endoplasmic reticulum and mitochondrial Ca^{2+} channels. J Cell Biol 175(6):901–911. https://doi.org/10.1083/jcb.200608073
29. De Stefani D, Bononi A, Romagnoli A, Messina A, De Pinto V, Pinton P, Rizzuto R (2012) VDAC1 selectively transfers apoptotic Ca^{2+} signals to mitochondria. Cell Death Differ 19(2):267–273. https://doi.org/10.1038/cdd.2011.92
30. Filadi R, Leal NS, Schreiner B, Rossi A, Dentoni G, Pinho CM, Wiehager B, Cieri D, Calì T, Pizzo P, Ankarcrona M (2018) TOM70 sustains cell bioenergetics by promoting IP3R3-mediated ER to mitochondria Ca 2+ transfer. Curr Biol 28(3):369–382.e6. https://doi.org/10.1016/j.cub.2017.12.047
31. Berridge MJ, Bootman MD, Roderick HL (2003) Calcium signalling: dynamics, homeostasis and remodelling. Nat Rev Mol Cell Biol 4(7):517–529. https://doi.org/10.1038/nrm1155
32. Clapham DE (2007) Calcium signaling. Cell 131(6):1047–1058. https://doi.org/10.1016/j.cell.2007.11.028
33. Brini M, Carafoli E (2009) Calcium pumps in health and disease. Physiol Rev 89(4):1341–1378. https://doi.org/10.1152/physrev.00032.2008
34. Deluca HF, Engstrom GW (1961) Calcium uptake by rat kidney mitochondria. Proc Natl Acad Sci U S A 47:1744–1750
35. Mitchell P (1961) Coupling of phosphorilation to electron transfer by a chemi-osmotic type of mechanism. Nature 191(4784):144–148
36. De Stefani D, Raffaello A, Teardo E, Szabo I, Rizzuto R (2011) A forty-kilodalton protein of the inner membrane is the mitochondrial calcium uniporter. Nature 476(7360):336–340. https://doi.org/10.1038/nature10230
37. Baughman JM, Perocchi F, Girgis HS, Plovanich M, Belcher-Timme CA, Sancak Y, Bao XR, Strittmatter L, Goldberger O, Bogorad RL, Koteliansky V, Mootha VK (2011) Integrative genomics identifies MCU as an essential component of the mitochondrial calcium uniporter. Nature 476(7360):341–345. https://doi.org/10.1038/nature10234
38. Fan C, Fan M, Orlando BJ, Fastman NM, Zhang J, Xu Y, Chambers MG, Xu X, Perry K, Liao M, Feng L (2018) X-ray and cryo-EM structures of the mitochondrial calcium uniporter. Nature 559(7715):575–579. https://doi.org/10.1038/s41586-018-0330-9
39. Nguyen NX, Armache JP, Lee C, Yang Y, Zeng W, Mootha VK, Cheng Y, Bai XC, Jiang Y (2018) Cryo-EM structure of a fungal mitochondrial calcium uniporter. Nature 559(7715):570–574. https://doi.org/10.1038/s41586-018-0333-6
40. Yoo J, Wu M, Yin Y, Herzik MA Jr, Lander GC, Lee SY (2018) Cryo-EM structure of a mitochondrial calcium uniporter. Science 361(6401):506–511. https://doi.org/10.1126/science.aar4056
41. Baradaran R, Wang C, Siliciano AF, Long SB (2018) Cryo-EM structures of fungal and metazoan mitochondrial calcium uniporters. Nature 559(7715):580–584. https://doi.org/10.1038/s41586-018-0331-8
42. Oxenoid K, Dong Y, Cao C, Cui T, Sancak Y, Markhard AL, Grabarek Z, Kong L, Liu Z, Ouyang B, Cong Y, Mootha VK, Chou JJ (2016) Architecture of the mitochondrial calcium uniporter. Nature 533(7602):269–273. https://doi.org/10.1038/nature17656

43. Raffaello A, De Stefani D, Sabbadin D, Teardo E, Merli G, Picard A, Checchetto V, Moro S, Szabo I, Rizzuto R (2013) The mitochondrial calcium uniporter is a multimer that can include a dominant-negative pore-forming subunit. EMBO J 32(17):2362–2376. https://doi.org/10.1038/emboj.2013.157
44. De Stefani D, Rizzuto R, Pozzan T (2016) Enjoy the trip: calcium in mitochondria back and forth. Annu Rev Biochem 85:161–192. https://doi.org/10.1146/annurev-biochem-060614-034216
45. Patron M, Checchetto V, Raffaello A, Teardo E, Vecellio Reane D, Mantoan M, Granatiero V, Szabo I, De Stefani D, Rizzuto R (2014) MICU1 and MICU2 finely tune the mitochondrial Ca^{2+} uniporter by exerting opposite effects on MCU activity. Mol Cell 53(5):726–737. https://doi.org/10.1016/j.molcel.2014.01.013
46. Perocchi F, Gohil VM, Girgis HS, Bao XR, McCombs JE, Palmer AE, Mootha VK (2010) MICU1 encodes a mitochondrial EF hand protein required for Ca^{2+} uptake. Nature 467(7313):291–296. https://doi.org/10.1038/nature09358
47. Plovanich M, Bogorad RL, Sancak Y, Kamer KJ, Strittmatter L, Li AA, Girgis HS, Kuchimanchi S, De Groot J, Speciner L, Taneja N, Oshea J, Koteliansky V, Mootha VK (2013) MICU2, a paralog of MICU1, resides within the mitochondrial uniporter complex to regulate calcium handling. PLoS One 8(2):e55785. https://doi.org/10.1371/journal.pone.0055785
48. Patron M, Granatiero V, Espino J, Rizzuto R, De Stefani D (2018) MICU3 is a tissue-specific enhancer of mitochondrial calcium uptake. Cell Death Differ 26(1):179–195. https://doi.org/10.1038/s41418-018-0113-8
49. Sancak Y, Markhard AL, Kitami T, Kovacs-Bogdan E, Kamer KJ, Udeshi ND, Carr SA, Chaudhuri D, Clapham DE, Li AA, Calvo SE, Goldberger O, Mootha VK (2013) EMRE is an essential component of the mitochondrial calcium uniporter complex. Science 342(6164):1379–1382. https://doi.org/10.1126/science.1242993
50. Mallilankaraman K, Cardenas C, Doonan PJ, Chandramoorthy HC, Irrinki KM, Golenar T, Csordas G, Madireddi P, Yang J, Muller M, Miller R, Kolesar JE, Molgo J, Kaufman B, Hajnoczky G, Foskett JK, Madesh M (2012) MCUR1 is an essential component of mitochondrial Ca^{2+} uptake that regulates cellular metabolism. Nat Cell Biol 15(1):123. https://doi.org/10.1038/ncb2669
51. Rizzuto R, Brini M, Murgia M, Pozzan T (1993) Microdomains with high Ca^{2+} close to IP3-sensitive channels that are sensed by neighboring mitochondria. Science 262(5134):744–747
52. Pinton P, Giorgi C, Siviero R, Zecchini E, Rizzuto R (2008) Calcium and apoptosis: ER-mitochondria Ca^{2+} transfer in the control of apoptosis. Oncogene 27(50):6407–6418. https://doi.org/10.1038/onc.2008.308
53. Calì T, Ottolini D, Brini M (2012) Mitochondrial Ca^{2+} as a key regulator of mitochondrial activities. Adv Exp Med Biol 942:53–73. https://doi.org/10.1007/978-94-007-2869-1_3
54. Calì T, Ottolini D, Brini M (2012) Mitochondrial Ca^{2+} and neurodegeneration. Cell Calcium 52(1):73–85. https://doi.org/10.1016/j.ceca.2012.04.015
55. Rizzuto R, De Stefani D, Raffaello A, Mammucari C (2012) Mitochondria as sensors and regulators of calcium signalling. Nat Rev Mol Cell Biol 13(9):566–578. https://doi.org/10.1038/nrm3412
56. Cardenas C, Muller M, McNeal A, Lovy A, Jana F, Bustos G, Urra F, Smith N, Molgo J, Diehl JA, Ridky TW, Foskett JK (2016) Selective vulnerability of cancer cells by inhibition of Ca^{2+} transfer from endoplasmic reticulum to mitochondria. Cell Rep 14(10):2313–2324. https://doi.org/10.1016/j.celrep.2016.02.030
57. Bononi A, Bonora M, Marchi S, Missiroli S, Poletti F, Giorgi C, Pandolfi PP, Pinton P (2013) Identification of PTEN at the ER and MAMs and its regulation of Ca^{2+} signaling and apoptosis in a protein phosphatase-dependent manner. Cell Death Differ 20(12):1631–1643. https://doi.org/10.1038/cdd.2013.77
58. Roderick HL, Lechleiter JD, Camacho P (2000) Cytosolic phosphorylation of calnexin controls intracellular Ca^{2+} oscillations via an interaction with SERCA2b. J Cell Biol 149(6):1235–1248

59. Lynes EM, Raturi A, Shenkman M, Ortiz Sandoval C, Yap MC, Wu J, Janowicz A, Myhill N, Benson MD, Campbell RE, Berthiaume LG, Lederkremer GZ, Simmen T (2013) Palmitoylation is the switch that assigns calnexin to quality control or ER Ca^{2+} signaling. J Cell Sci 126(Pt 17):3893–3903. https://doi.org/10.1242/jcs.125856
60. Lynes EM, Bui M, Yap MC, Benson MD, Schneider B, Ellgaard L, Berthiaume LG, Simmen T (2012) Palmitoylated TMX and calnexin target to the mitochondria-associated membrane. EMBO J 31(2):457–470. https://doi.org/10.1038/emboj.2011.384
61. Krols M, Bultynck G, Janssens S (2016) ER-mitochondria contact sites: a new regulator of cellular calcium flux comes into play. J Cell Biol 214(4):367–370. https://doi.org/10.1083/jcb.201607124
62. Yoboue ED, Rimessi A, Anelli T, Pinton P, Sitia R (2017) Regulation of calcium fluxes by GPX8, a type-II transmembrane peroxidase enriched at the mitochondria-associated endoplasmic reticulum membrane. Antioxid Redox Signal 27(9):583–595. https://doi.org/10.1089/ars.2016.6866
63. Hayashi T, Rizzuto R, Hajnoczky G, Su TP (2009) MAM: more than just a housekeeper. Trends Cell Biol 19(2):81–88. https://doi.org/10.1016/j.tcb.2008.12.002
64. Wang X, Schwarz TL (2009) The mechanism of Ca^{2+} -dependent regulation of kinesin-mediated mitochondrial motility. Cell 136(1):163–174. https://doi.org/10.1016/j.cell.2008.11.046
65. Hayashi T, Su TP (2007) Sigma-1 receptor chaperones at the ER-mitochondrion interface regulate Ca^{2+} signaling and cell survival. Cell 131(3):596–610. https://doi.org/10.1016/j.cell.2007.08.036
66. Myhill N, Lynes EM, Nanji JA, Blagoveshchenskaya AD, Fei H, Carmine Simmen K, Cooper TJ, Thomas G, Simmen T (2008) The subcellular distribution of calnexin is mediated by PACS-2. Mol Biol Cell 19(7):2777–2788. https://doi.org/10.1091/mbc.E07-10-0995
67. Bastianutto C, Clementi E, Codazzi F, Podini P, De Giorgi F, Rizzuto R, Meldolesi J, Pozzan T (1995) Overexpression of calreticulin increases the Ca^{2+} capacity of rapidly exchanging Ca^{2+} stores and reveals aspects of their lumenal microenvironment and function. J Cell Biol 130(4):847–855
68. Hendershot LM (2004) The ER function BiP is a master regulator of ER function. Mt Sinai J Med 71(5):289–297
69. Camacho P, Lechleiter JD (1995) Calreticulin inhibits repetitive intracellular Ca^{2+} waves. Cell 82(5):765–771
70. Michalak M, Groenendyk J, Szabo E, Gold LI, Opas M (2009) Calreticulin, a multi-process calcium-buffering chaperone of the endoplasmic reticulum. Biochem J 417(3):651–666. https://doi.org/10.1042/BJ20081847
71. Anelli T, Alessio M, Mezghrani A, Simmen T, Talamo F, Bachi A, Sitia R (2002) ERp44, a novel endoplasmic reticulum folding assistant of the thioredoxin family. EMBO J 21(4):835–844
72. Higo T, Hattori M, Nakamura T, Natsume T, Michikawa T, Mikoshiba K (2005) Subtype-specific and ER lumenal environment-dependent regulation of inositol 1,4,5-trisphosphate receptor type 1 by ERp44. Cell 120(1):85–98. https://doi.org/10.1016/j.cell.2004.11.048
73. Li G, Mongillo M, Chin KT, Harding H, Ron D, Marks AR, Tabas I (2009) Role of ERO1-alpha-mediated stimulation of inositol 1,4,5-triphosphate receptor activity in endoplasmic reticulum stress-induced apoptosis. J Cell Biol 186(6):783–792. https://doi.org/10.1083/jcb.200904060
74. Anelli T, Bergamelli L, Margittai E, Rimessi A, Fagioli C, Malgaroli A, Pinton P, Ripamonti M, Rizzuto R, Sitia R (2012) Ero1alpha regulates Ca^{2+} fluxes at the endoplasmic reticulum-mitochondria interface (MAM). Antioxid Redox Signal 16(10):1077–1087. https://doi.org/10.1089/ars.2011.4004
75. Ghavami S, Shojaei S, Yeganeh B, Ande SR, Jangamreddy JR, Mehrpour M, Christoffersson J, Chaabane W, Moghadam AR, Kashani HH, Hashemi M, Owji AA, Los MJ (2014) Autophagy and apoptosis dysfunction in neurodegenerative disorders. Prog Neurobiol 112:24–49. https://doi.org/10.1016/j.pneurobio.2013.10.004

76. Millecamps S, Julien JP (2013) Axonal transport deficits and neurodegenerative diseases. Nat Rev Neurosci 14(3):161–176. https://doi.org/10.1038/nrn3380
77. Mironov SL, Symonchuk N (2006) ER vesicles and mitochondria move and communicate at synapses. J Cell Sci 119(Pt 23):4926–4934. https://doi.org/10.1242/jcs.03254
78. Wu Y, Whiteus C, Xu CS, Hayworth KJ, Weinberg RJ, Hess HF, De Camilli P (2017) Contacts between the endoplasmic reticulum and other membranes in neurons. Proc Natl Acad Sci U S A 114(24):E4859–E4867. https://doi.org/10.1073/pnas.1701078114
79. Goedert M (2015) NEURODEGENERATION. Alzheimer's and Parkinson's diseases: the prion concept in relation to assembled Abeta, tau, and alpha-synuclein. Science 349(6248):1255555. https://doi.org/10.1126/science.1255555
80. Area-Gomez E, de Groof A, Bonilla E, Montesinos J, Tanji K, Boldogh I, Pon L, Schon EA (2018) A key role for MAM in mediating mitochondrial dysfunction in Alzheimer disease. Cell Death Dis 9(3):335. https://doi.org/10.1038/s41419-017-0215-0
81. Bezprozvanny I (2013) Presenilins and calcium signaling-systems biology to the rescue. Sci Signal 6(283):pe24. https://doi.org/10.1126/scisignal.2004296
82. Popugaeva E, Pchitskaya E, Bezprozvanny I (2017) Dysregulation of neuronal calcium homeostasis in Alzheimer's disease – a therapeutic opportunity? Biochem Biophys Res Commun 483(4):998–1004. https://doi.org/10.1016/j.bbrc.2016.09.053
83. Area-Gomez E, de Groof AJ, Boldogh I, Bird TD, Gibson GE, Koehler CM, Yu WH, Duff KE, Yaffe MP, Pon LA, Schon EA (2009) Presenilins are enriched in endoplasmic reticulum membranes associated with mitochondria. Am J Pathol 175(5):1810–1816. https://doi.org/10.2353/ajpath.2009.090219
84. Tu H, Nelson O, Bezprozvanny A, Wang Z, Lee SF, Hao YH, Serneels L, De Strooper B, Yu G, Bezprozvanny I (2006) Presenilins form ER Ca^{2+} leak channels, a function disrupted by familial Alzheimer's disease-linked mutations. Cell 126(5):981–993. https://doi.org/10.1016/j.cell.2006.06.059
85. Cheung KH, Shineman D, Muller M, Cardenas C, Mei L, Yang J, Tomita T, Iwatsubo T, Lee VM, Foskett JK (2008) Mechanism of Ca^{2+} disruption in Alzheimer's disease by presenilin regulation of InsP3 receptor channel gating. Neuron 58(6):871–883. https://doi.org/10.1016/j.neuron.2008.04.015
86. D'Adamio L, Castillo PE (2013) Presenilin-ryanodine receptor connection. Proc Natl Acad Sci U S A 110(37):14825–14826. https://doi.org/10.1073/pnas.1313996110
87. Green KN, Demuro A, Akbari Y, Hitt BD, Smith IF, Parker I, LaFerla FM (2008) SERCA pump activity is physiologically regulated by presenilin and regulates amyloid beta production. J Cell Biol 181(7):1107–1116. https://doi.org/10.1083/jcb.200706171
88. Brunello L, Zampese E, Florean C, Pozzan T, Pizzo P, Fasolato C (2009) Presenilin-2 dampens intracellular Ca^{2+} stores by increasing Ca^{2+} leakage and reducing Ca^{2+} uptake. J Cell Mol Med 13(9B):3358–3369. https://doi.org/10.1111/j.1582-4934.2009.00755.x
89. Zampese E, Fasolato C, Kipanyula MJ, Bortolozzi M, Pozzan T, Pizzo P (2011) Presenilin 2 modulates endoplasmic reticulum (ER)-mitochondria interactions and Ca^{2+} cross-talk. Proc Natl Acad Sci U S A 108(7):2777–2782. https://doi.org/10.1073/pnas.1100735108
90. Kipanyula MJ, Contreras L, Zampese E, Lazzari C, Wong AK, Pizzo P, Fasolato C, Pozzan T (2012) Ca^{2+} dysregulation in neurons from transgenic mice expressing mutant presenilin 2. Aging Cell 11(5):885–893. https://doi.org/10.1111/j.1474-9726.2012.00858.x
91. Duplan E, Sevalle J, Viotti J, Goiran T, Druon C, Renbaum P, Levy-Lahad E, Gautier C, Corti O, Leroudier N, Checler F, da Costa CA (2013) Parkin differently regulates presenilin-1 and presenilin-2 function by direct control of their promoter transcription. J Mol Cell Biol 5(2):132–142. https://doi.org/10.1093/jmcb/mjt003
92. Hedskog L, Pinho CM, Filadi R, Ronnback A, Hertwig L, Wiehager B, Larssen P, Gellhaar S, Sandebring A, Westerlund M, Graff C, Winblad B, Galter D, Behbahani H, Pizzo P, Glaser E, Ankarcrona M (2013) Modulation of the endoplasmic reticulum-mitochondria interface in Alzheimer's disease and related models. Proc Natl Acad Sci U S A 110(19):7916–7921. https://doi.org/10.1073/pnas.1300677110

93. Cieri D, Vicario M, Vallese F, D'Orsi B, Berto P, Grinzato A, Catoni C, De Stefani D, Rizzuto R, Brini M, Cali T (2018) Tau localises within mitochondrial sub-compartments and its caspase cleavage affects ER-mitochondria interactions and cellular Ca^{2+} handling. Biochim Biophys Acta Mol basis Dis 1864(10):3247–3256. https://doi.org/10.1016/j.bbadis.2018.07.011
94. Cieri D, Brini M, Cali T (2016) Emerging (and converging) pathways in Parkinson's disease: keeping mitochondrial wellness. Biochem Biophys Res Commun 483(4):1020–1030. https://doi.org/10.1016/j.bbrc.2016.08.153
95. Poewe W, Seppi K, Tanner CM, Halliday GM, Brundin P, Volkmann J, Schrag AE, Lang AE (2017) Parkinson disease. Nat Rev Dis Primers 3:17013. https://doi.org/10.1038/nrdp.2017.13
96. Calì T, Ottolini D, Brini M (2014) Calcium signaling in Parkinson's disease. Cell Tissue Res 357:439–454. https://doi.org/10.1007/s00441-014-1866-0
97. Lin MK, Farrer MJ (2014) Genetics and genomics of Parkinson's disease. Genome Med 6(6):48. https://doi.org/10.1186/gm566
98. Rodriguez-Arribas M, SMS Y-D, JMB P, Gomez-Suaga P, Gomez-Sanchez R, Martinez-Chacon G, Fuentes JM, Gonzalez-Polo RA, Niso-Santano M (2017) Mitochondria-Associated Membranes (MAMs): overview and its role in Parkinson's disease. Mol Neurobiol 54(8):6287–6303. https://doi.org/10.1007/s12035-016-0140-8
99. Calì T, Ottolini D, Negro A, Brini M (2012) alpha-synuclein controls mitochondrial calcium homeostasis by enhancing endoplasmic reticulum-mitochondria interactions. J Biol Chem 287(22):17914–17929. https://doi.org/10.1074/jbc.M111.302794
100. Ottolini D, Cali T, Szabo I, Brini M (2017) Alpha-synuclein at the intracellular and the extracellular side: functional and dysfunctional implications. Biol Chem 398(1):77–100. https://doi.org/10.1515/hsz-2016-0201
101. Vicario M, Cieri D, Brini M, Calì T (2018) The close encounter between alpha-synuclein and mitochondria. Front Neurosci 12:388. https://doi.org/10.3389/fnins.2018.00388
102. Guardia-Laguarta C, Area-Gomez E, Schon EA, Przedborski S (2015) A new role for alpha-synuclein in Parkinson's disease: alteration of ER-mitochondrial communication. Mov Disord 30(8):1026–1033. https://doi.org/10.1002/mds.26239
103. Gan L, Cookson MR, Petrucelli L, La Spada AR (2018) Converging pathways in neurodegeneration, from genetics to mechanisms. Nat Neurosci 21(10):1300–1309. https://doi.org/10.1038/s41593-018-0237-7
104. Palikaras K, Lionaki E, Tavernarakis N (2018) Mechanisms of mitophagy in cellular homeostasis, physiology and pathology. Nat Cell Biol 20(9):1013–1022. https://doi.org/10.1038/s41556-018-0176-2
105. Pickles S, Vigie P, Youle RJ (2018) Mitophagy and quality control mechanisms in mitochondrial maintenance. Curr Biol 28(4):R170–R185. https://doi.org/10.1016/j.cub.2018.01.004
106. Youle RJ, Narendra DP (2011) Mechanisms of mitophagy. Nat Rev Mol Cell Biol 12(1):9–14. https://doi.org/10.1038/nrm3028
107. Van Laar VS, Roy N, Liu A, Rajprohat S, Arnold B, Dukes AA, Holbein CD, Berman SB (2014) Glutamate excitotoxicity in neurons triggers mitochondrial and endoplasmic reticulum accumulation of Parkin, and, in the presence of N-acetyl cysteine, mitophagy. Neurobiol Dis 74:180–193. https://doi.org/10.1016/j.nbd.2014.11.015
108. Darios F, Corti O, Lucking CB, Hampe C, Muriel MP, Abbas N, Gu WJ, Hirsch EC, Rooney T, Ruberg M, Brice A (2003) Parkin prevents mitochondrial swelling and cytochrome c release in mitochondria-dependent cell death. Hum Mol Genet 12(5):517–526
109. Shin JH, Ko HS, Kang H, Lee Y, Lee YI, Pletinkova O, Troconso JC, Dawson VL, Dawson TM (2011) PARIS (ZNF746) repression of PGC-1alpha contributes to neurodegeneration in Parkinson's disease. Cell 144(5):689–702. https://doi.org/10.1016/j.cell.2011.02.010
110. Bertolin G, Ferrando-Miguel R, Jacoupy M, Traver S, Grenier K, Greene AW, Dauphin A, Waharte F, Bayot A, Salamero J, Lombes A, Bulteau AL, Fon EA, Brice A, Corti O (2013) The TOMM machinery is a molecular switch in PINK1 and PARK2/PARKIN-dependent mitochondrial clearance. Autophagy 9(11):1801–1817. https://doi.org/10.4161/auto.25884

111. Calì T, Ottolini D, Negro A, Brini M (2013) Enhanced parkin levels favour ER-mitochondria crosstalk and guarantee Ca^{2+} transfer to sustain cell bioenergetics. BBA-Mol Basis Dis 1832(4):495–508. https://doi.org/10.1016/j.bbadis.2013.01.004
112. Imai Y, Soda M, Inoue H, Hattori N, Mizuno Y, Takahashi R (2001) An unfolded putative transmembrane polypeptide, which can lead to endoplasmic reticulum stress, is a substrate of Parkin. Cell 105(7):891–902
113. Wang HQ, Imai Y, Kataoka A, Takahashi R (2007) Cell type-specific upregulation of Parkin in response to ER stress. Antioxid Redox Signal 9(5):533–542. https://doi.org/10.1089/ars.2006.1522
114. Bouman L, Schlierf A, Lutz AK, Shan J, Deinlein A, Kast J, Galehdar Z, Palmisano V, Patenge N, Berg D, Gasser T, Augustin R, Trumbach D, Irrcher I, Park DS, Wurst W, Kilberg MS, Tatzelt J, Winklhofer KF (2011) Parkin is transcriptionally regulated by ATF4: evidence for an interconnection between mitochondrial stress and ER stress. Cell Death Differ 18(5):769–782. https://doi.org/10.1038/cdd.2010.142
115. Imai Y, Soda M, Hatakeyama S, Akagi T, Hashikawa T, Nakayama KI, Takahashi R (2002) CHIP is associated with Parkin, a gene responsible for familial Parkinson's disease, and enhances its ubiquitin ligase activity. Mol Cell 10(1):55–67
116. Avraham E, Rott R, Liani E, Szargel R, Engelender S (2007) Phosphorylation of Parkin by the cyclin-dependent kinase 5 at the linker region modulates its ubiquitin-ligase activity and aggregation. J Biol Chem 282(17):12842–12850. https://doi.org/10.1074/jbc.M608243200
117. Yao D, Gu Z, Nakamura T, Shi ZQ, Ma Y, Gaston B, Palmer LA, Rockenstein EM, Zhang Z, Masliah E, Uehara T, Lipton SA (2004) Nitrosative stress linked to sporadic Parkinson's disease: S-nitrosylation of parkin regulates its E3 ubiquitin ligase activity. Proc Natl Acad Sci U S A 101(29):10810–10814. https://doi.org/10.1073/pnas.0404161101
118. Han K, Hassanzadeh S, Singh K, Menazza S, Nguyen TT, Stevens MV, Nguyen A, San H, Anderson SA, Lin Y, Zou J, Murphy E, Sack MN (2017) Parkin regulation of CHOP modulates susceptibility to cardiac endoplasmic reticulum stress. Sci Rep 7(1):2093. https://doi.org/10.1038/s41598-017-02339-2
119. Basso V, Marchesan E, Peggion C, Chakraborty J, von Stockum S, Giacomello M, Ottolini D, Debattisti V, Caicci F, Tasca E, Pegoraro V, Angelini C, Antonini A, Bertoli A, Brini M, Ziviani E (2018) Regulation of endoplasmic reticulum-mitochondria contacts by Parkin via Mfn2. Pharmacol Res 138:43–56. https://doi.org/10.1016/j.phrs.2018.09.006
120. van der Merwe C, Jalali Sefid Dashti Z, Christoffels A, Loos B, Bardien S (2015) Evidence for a common biological pathway linking three Parkinson's disease-causing genes: parkin, PINK1 and DJ-1. Eur J Neurosci 41(9):1113–1125. https://doi.org/10.1111/ejn.12872
121. Cali T, Ottolini D, Soriano ME, Brini M (2014) A new split-GFP-based probe reveals DJ-1 translocation into the mitochondrial matrix to sustain ATP synthesis upon nutrient deprivation. Hum Mol Genet 24(4):1045–1160. https://doi.org/10.1093/hmg/ddu519
122. Ottolini D, Calì T, Negro A, Brini M (2013) The Parkinson disease-related protein DJ-1 counteracts mitochondrial impairment induced by the tumour suppressor protein p53 by enhancing endoplasmic reticulum-mitochondria tethering. Hum Mol Genet 22(11):2152–2168. https://doi.org/10.1093/hmg/ddt068
123. Hardiman O, Al-Chalabi A, Chio A, Corr EM, Logroscino G, Robberecht W, Shaw PJ, Simmons Z, van den Berg LH (2017) Amyotrophic lateral sclerosis. Nat Rev Dis Primers 3:17071. https://doi.org/10.1038/nrdp.2017.71
124. Langou K, Moumen A, Pellegrino C, Aebischer J, Medina I, Aebischer P, Raoul C (2010) AAV-mediated expression of wild-type and ALS-linked mutant VAPB selectively triggers death of motoneurons through a Ca^{2+}-dependent ER-associated pathway. J Neurochem 114(3):795–809. https://doi.org/10.1111/j.1471-4159.2010.06806.x
125. Morotz GM, De Vos KJ, Vagnoni A, Ackerley S, Shaw CE, Miller CC (2012) Amyotrophic lateral sclerosis-associated mutant VAPBP56S perturbs calcium homeostasis to disrupt axonal transport of mitochondria. Hum Mol Genet 21(9):1979–1988. https://doi.org/10.1093/hmg/dds011

126. Rosen DR (1993) Mutations in Cu/Zn superoxide dismutase gene are associated with familial amyotrophic lateral sclerosis. Nature 364(6435):362. https://doi.org/10.1038/364362c0
127. Damiano M, Starkov AA, Petri S, Kipiani K, Kiaei M, Mattiazzi M, Flint Beal M, Manfredi G (2006) Neural mitochondrial Ca^{2+} capacity impairment precedes the onset of motor symptoms in G93A Cu/Zn-superoxide dismutase mutant mice. J Neurochem 96(5):1349–1361. https://doi.org/10.1111/j.1471-4159.2006.03619.x
128. Parone PA, Da Cruz S, Han JS, McAlonis-Downes M, Vetto AP, Lee SK, Tseng E, Cleveland DW (2013) Enhancing mitochondrial calcium buffering capacity reduces aggregation of misfolded SOD1 and motor neuron cell death without extending survival in mouse models of inherited amyotrophic lateral sclerosis. J Neurosci 33(11):4657–4671. https://doi.org/10.1523/JNEUROSCI.1119-12.2013
129. Pedrini S, Sau D, Guareschi S, Bogush M, Brown RH Jr, Naniche N, Kia A, Trotti D, Pasinelli P (2010) ALS-linked mutant SOD1 damages mitochondria by promoting conformational changes in Bcl-2. Hum Mol Genet 19(15):2974–2986. https://doi.org/10.1093/hmg/ddq202
130. Eckenrode EF, Yang J, Velmurugan GV, Foskett JK, White C (2010) Apoptosis protection by Mcl-1 and Bcl-2 modulation of inositol 1,4,5-trisphosphate receptor-dependent Ca^{2+} signaling. J Biol Chem 285(18):13678–13684. https://doi.org/10.1074/jbc.M109.096040
131. Pinton P, Ferrari D, Magalhaes P, Schulze-Osthoff K, Di Virgilio F, Pozzan T, Rizzuto R (2000) Reduced loading of intracellular Ca^{2+} stores and downregulation of capacitative Ca^{2+} influx in Bcl-2-overexpressing cells. J Cell Biol 148(5):857–862
132. Dimmer KS, Rapaport D (2017) Mitochondrial contact sites as platforms for phospholipid exchange. Biochim Biophys Acta Mol Cell Biol Lipids 1862(1):69–80. https://doi.org/10.1016/j.bbalip.2016.07.010
133. Vance JE (2015) Phospholipid synthesis and transport in mammalian cells. Traffic 16(1):1–18. https://doi.org/10.1111/tra.12230
134. Csordas G, Hajnoczky G (2009) SR/ER-mitochondrial local communication: calcium and ROS. Biochim Biophys Acta 1787(11):1352–1362. https://doi.org/10.1016/j.bbabio.2009.06.004
135. Raffaello A, Mammucari C, Gherardi G, Rizzuto R (2016) Calcium at the center of cell signaling: interplay between endoplasmic reticulum, mitochondria, and lysosomes. Trends Biochem Sci 41(12):1035–1049. https://doi.org/10.1016/j.tibs.2016.09.001
136. Phillips MJ, Voeltz GK (2016) Structure and function of ER membrane contact sites with other organelles. Nat Rev Mol Cell Biol 17(2):69–82. https://doi.org/10.1038/nrm.2015.8
137. Yoon Y, Krueger EW, Oswald BJ, McNiven MA (2003) The mitochondrial protein hFis1 regulates mitochondrial fission in mammalian cells through an interaction with the dynamin-like protein DLP1. Mol Cell Biol 23(15):5409–5420
138. Otera H, Wang C, Cleland MM, Setoguchi K, Yokota S, Youle RJ, Mihara K (2010) Mff is an essential factor for mitochondrial recruitment of Drp1 during mitochondrial fission in mammalian cells. J Cell Biol 191(6):1141–1158. https://doi.org/10.1083/jcb.201007152
139. Murley A, Lackner LL, Osman C, West M, Voeltz GK, Walter P, Nunnari J (2013) ER-associated mitochondrial division links the distribution of mitochondria and mitochondrial DNA in yeast. elife 2:e00422. https://doi.org/10.7554/eLife.00422
140. Mears JA, Lackner LL, Fang S, Ingerman E, Nunnari J, Hinshaw JE (2011) Conformational changes in Dnm1 support a contractile mechanism for mitochondrial fission. Nat Struct Mol Biol 18(1):20–26. https://doi.org/10.1038/nsmb.1949
141. Zhao J, Liu T, Jin S, Wang X, Qu M, Uhlen P, Tomilin N, Shupliakov O, Lendahl U, Nister M (2011) Human MIEF1 recruits Drp1 to mitochondrial outer membranes and promotes mitochondrial fusion rather than fission. EMBO J 30(14):2762–2778. https://doi.org/10.1038/emboj.2011.198
142. Korobova F, Ramabhadran V, Higgs HN (2013) An actin-dependent step in mitochondrial fission mediated by the ER-associated formin INF2. Science 339(6118):464–467. https://doi.org/10.1126/science.1228360
143. Yoon Y, Pitts KR, McNiven MA (2001) Mammalian dynamin-like protein DLP1 tubulates membranes. Mol Biol Cell 12(9):2894–2905. https://doi.org/10.1091/mbc.12.9.2894

144. Koshiba T, Detmer SA, Kaiser JT, Chen H, McCaffery JM, Chan DC (2004) Structural basis of mitochondrial tethering by mitofusin complexes. Science 305(5685):858–862. https://doi.org/10.1126/science.1099793
145. Sugiura A, Nagashima S, Tokuyama T, Amo T, Matsuki Y, Ishido S, Kudo Y, McBride HM, Fukuda T, Matsushita N, Inatome R, Yanagi S (2013) MITOL regulates endoplasmic reticulum-mitochondria contacts via Mitofusin2. Mol Cell 51(1):20–34. https://doi.org/10.1016/j.molcel.2013.04.023
146. Lev S (2012) Nonvesicular lipid transfer from the endoplasmic reticulum. Cold Spring Harb Perspect Biol 4(10):a013300. https://doi.org/10.1101/cshperspect.a013300
147. Achleitner G, Zweytick D, Trotter PJ, Voelker DR, Daum G (1995) Synthesis and intracellular transport of aminoglycerophospholipids in permeabilized cells of the yeast, Saccharomyces cerevisiae. J Biol Chem 270(50):29836–29842
148. Vance JE (1990) Phospholipid synthesis in a membrane fraction associated with mitochondria. J Biol Chem 265(13):7248–7256
149. Kornmann B (2013) The molecular hug between the ER and the mitochondria. Curr Opin Cell Biol 25(4):443–448. https://doi.org/10.1016/j.ceb.2013.02.010
150. Kornmann B, Currie E, Collins SR, Schuldiner M, Nunnari J, Weissman JS, Walter P (2009) An ER-mitochondria tethering complex revealed by a synthetic biology screen. Science 325(5939):477–481. https://doi.org/10.1126/science.1175088
151. Kornmann B, Walter P (2010) ERMES-mediated ER-mitochondria contacts: molecular hubs for the regulation of mitochondrial biology. J Cell Sci 123(Pt 9):1389–1393. https://doi.org/10.1242/jcs.058636
152. Hirabayashi Y, Kwon SK, Paek H, Pernice WM, Paul MA, Lee J, Erfani P, Raczkowski A, Petrey DS, Pon LA, Polleux F (2017) ER-mitochondria tethering by PDZD8 regulates Ca^{2+} dynamics in mammalian neurons. Science 358(6363):623–630. https://doi.org/10.1126/science.aan6009
153. Kawano S, Tamura Y, Kojima R, Bala S, Asai E, Michel AH, Kornmann B, Riezman I, Riezman H, Sakae Y, Okamoto Y, Endo T (2018) Structure-function insights into direct lipid transfer between membranes by Mmm1-Mdm12 of ERMES. J Cell Biol 217(3):959–974. https://doi.org/10.1083/jcb.201704119
154. AhYoung AP, Jiang J, Zhang J, Khoi Dang X, Loo JA, Zhou ZH, Egea PF (2015) Conserved SMP domains of the ERMES complex bind phospholipids and mediate tether assembly. Proc Natl Acad Sci U S A 112(25):E3179–E3188. https://doi.org/10.1073/pnas.1422363112
155. Crompton M, Costi A, Hayat L (1987) Evidence for the presence of a reversible Ca^{2+}-dependent pore activated by oxidative stress in heart mitochondria. Biochem J 245(3):915–918
156. Deniaud A, Sharaf el dein O, Maillier E, Poncet D, Kroemer G, Lemaire C, Brenner C (2008) Endoplasmic reticulum stress induces calcium-dependent permeability transition, mitochondrial outer membrane permeabilization and apoptosis. Oncogene 27(3):285–299. https://doi.org/10.1038/sj.onc.1210638
157. Adams JM, Cory S (2018) The BCL-2 arbiters of apoptosis and their growing role as cancer targets. Cell Death Differ 25(1):27–36. https://doi.org/10.1038/cdd.2017.161
158. Strzyz P (2017) Cell death: BCL-2 proteins feed their own expression. Nat Rev Mol Cell Biol 18(11):652–653. https://doi.org/10.1038/nrm.2017.106
159. Giorgi C, Danese A, Missiroli S, Patergnani S, Pinton P (2018) Calcium dynamics as a machine for decoding signals. Trends Cell Biol 28(4):258–273. https://doi.org/10.1016/j.tcb.2018.01.002
160. Rong YP, Bultynck G, Aromolaran AS, Zhong F, Parys JB, De Smedt H, Mignery GA, Roderick HL, Bootman MD, Distelhorst CW (2009) The BH4 domain of Bcl-2 inhibits ER calcium release and apoptosis by binding the regulatory and coupling domain of the IP3 receptor. Proc Natl Acad Sci U S A 106(34):14397–14402. https://doi.org/10.1073/pnas.0907555106
161. White C, Li C, Yang J, Petrenko NB, Madesh M, Thompson CB, Foskett JK (2005) The endoplasmic reticulum gateway to apoptosis by Bcl-X(L) modulation of the InsP3R. Nat Cell Biol 7(10):1021–1028. https://doi.org/10.1038/ncb1302

162. Mathai JP, Germain M, Shore GC (2005) BH3-only BIK regulates BAX,BAK-dependent release of Ca^{2+} from endoplasmic reticulum stores and mitochondrial apoptosis during stress-induced cell death. J Biol Chem 280(25):23829–23836. https://doi.org/10.1074/jbc.M500800200
163. Nutt LK, Pataer A, Pahler J, Fang B, Roth J, McConkey DJ, Swisher SG (2002) Bax and Bak promote apoptosis by modulating endoplasmic reticular and mitochondrial Ca^{2+} stores. J Biol Chem 277(11):9219–9225. https://doi.org/10.1074/jbc.M106817200
164. Shibue T, Suzuki S, Okamoto H, Yoshida H, Ohba Y, Takaoka A, Taniguchi T (2006) Differential contribution of Puma and Noxa in dual regulation of p53-mediated apoptotic pathways. EMBO J 25(20):4952–4962. https://doi.org/10.1038/sj.emboj.7601359
165. Breckenridge DG, Stojanovic M, Marcellus RC, Shore GC (2003) Caspase cleavage product of BAP31 induces mitochondrial fission through endoplasmic reticulum calcium signals, enhancing cytochrome c release to the cytosol. J Cell Biol 160(7):1115–1127. https://doi.org/10.1083/jcb.200212059
166. Alirol E, James D, Huber D, Marchetto A, Vergani L, Martinou JC, Scorrano L (2006) The mitochondrial fission protein hFis1 requires the endoplasmic reticulum gateway to induce apoptosis. Mol Biol Cell 17(11):4593–4605. https://doi.org/10.1091/mbc.e06-05-0377
167. Karbowski M, Lee YJ, Gaume B, Jeong SY, Frank S, Nechushtan A, Santel A, Fuller M, Smith CL, Youle RJ (2002) Spatial and temporal association of Bax with mitochondrial fission sites, Drp1, and Mfn2 during apoptosis. J Cell Biol 159(6):931–938. https://doi.org/10.1083/jcb.200209124
168. Cleland MM, Norris KL, Karbowski M, Wang C, Suen DF, Jiao S, George NM, Luo X, Li Z, Youle RJ (2011) Bcl-2 family interaction with the mitochondrial morphogenesis machinery. Cell Death Differ 18(2):235–247. https://doi.org/10.1038/cdd.2010.89
169. He C, Klionsky DJ (2009) Regulation mechanisms and signaling pathways of autophagy. Annu Rev Genet 43:67–93. https://doi.org/10.1146/annurev-genet-102808-114910
170. MacVicar T (2013) Mitophagy. Essays Biochem 55:93–104. https://doi.org/10.1042/bse0550093
171. Hailey DW, Rambold AS, Satpute-Krishnan P, Mitra K, Sougrat R, Kim PK, Lippincott-Schwartz J (2010) Mitochondria supply membranes for autophagosome biogenesis during starvation. Cell 141(4):656–667. https://doi.org/10.1016/j.cell.2010.04.009
172. Hamasaki M, Furuta N, Matsuda A, Nezu A, Yamamoto A, Fujita N, Oomori H, Noda T, Haraguchi T, Hiraoka Y, Amano A, Yoshimori T (2013) Autophagosomes form at ER-mitochondria contact sites. Nature 495(7441):389–393. https://doi.org/10.1038/nature11910
173. Sarbassov DD, Guertin DA, Ali SM, Sabatini DM (2005) Phosphorylation and regulation of Akt/PKB by the rictor-mTOR complex. Science 307(5712):1098–1101. https://doi.org/10.1126/science.1106148
174. Hirota Y, Tanaka Y (2009) A small GTPase, human Rab32, is required for the formation of autophagic vacuoles under basal conditions. Cell Mol Life Sci 66(17):2913–2932. https://doi.org/10.1007/s00018-009-0080-9
175. Betz C, Stracka D, Prescianotto-Baschong C, Frieden M, Demaurex N, Hall MN (2013) Feature Article: mTOR complex 2-Akt signaling at mitochondria-associated endoplasmic reticulum membranes (MAM) regulates mitochondrial physiology. Proc Natl Acad Sci U S A 110(31):12526–12534. https://doi.org/10.1073/pnas.1302455110
176. Bernales S, Soto MM, McCullagh E (2012) Unfolded protein stress in the endoplasmic reticulum and mitochondria: a role in neurodegeneration. Front Aging Neurosci 4:5. https://doi.org/10.3389/fnagi.2012.00005
177. Wang M, Kaufman RJ (2016) Protein misfolding in the endoplasmic reticulum as a conduit to human disease. Nature 529(7586):326–335. https://doi.org/10.1038/nature17041
178. Bui M, Gilady SY, Fitzsimmons RE, Benson MD, Lynes EM, Gesson K, Alto NM, Strack S, Scott JD, Simmen T (2010) Rab32 modulates apoptosis onset and mitochondria-associated membrane (MAM) properties. J Biol Chem 285(41):31590–31602. https://doi.org/10.1074/jbc.M110.101584

179. Munoz JP, Ivanova S, Sanchez-Wandelmer J, Martinez-Cristobal P, Noguera E, Sancho A, Diaz-Ramos A, Hernandez-Alvarez MI, Sebastian D, Mauvezin C, Palacin M, Zorzano A (2013) Mfn2 modulates the UPR and mitochondrial function via repression of PERK. EMBO J 32(17):2348–2361. https://doi.org/10.1038/emboj.2013.168
180. Gkogkas C, Middleton S, Kremer AM, Wardrope C, Hannah M, Gillingwater TH, Skehel P (2008) VAPB interacts with and modulates the activity of ATF6. Hum Mol Genet 17(11):1517–1526. https://doi.org/10.1093/hmg/ddn040
181. Mannella CA, Buttle K, Rath BK, Marko M (1998) Electron microscopic tomography of rat-liver mitochondria and their interaction with the endoplasmic reticulum. Biofactors 8(3–4):225–228
182. Csordas G, Renken C, Varnai P, Walter L, Weaver D, Buttle KF, Balla T, Mannella CA, Hajnoczky G (2006) Structural and functional features and significance of the physical linkage between ER and mitochondria. J Cell Biol 174(7):915–921. https://doi.org/10.1083/jcb.200604016
183. Elgass KD, Smith EA, LeGros MA, Larabell CA, Ryan MT (2015) Analysis of ER-mitochondria contacts using correlative fluorescence microscopy and soft X-ray tomography of mammalian cells. J Cell Sci 128(15):2795–2804. https://doi.org/10.1242/jcs.169136
184. Rusinol AE, Cui Z, Chen MH, Vance JE (1994) A unique mitochondria-associated membrane fraction from rat liver has a high capacity for lipid synthesis and contains pre-Golgi secretory proteins including nascent lipoproteins. J Biol Chem 269(44):27494–27502
185. Wieckowski MR, Giorgi C, Lebiedzinska M, Duszynski J, Pinton P (2009) Isolation of mitochondria-associated membranes and mitochondria from animal tissues and cells. Nat Protoc 4(11):1582–1590. https://doi.org/10.1038/nprot.2009.151
186. Brunstein M, Wicker K, Herault K, Heintzmann R, Oheim M (2013) Full-field dual-color 100-nm super-resolution imaging reveals organization and dynamics of mitochondrial and ER networks. Opt Express 21(22):26162–26173. https://doi.org/10.1364/OE.21.026162
187. Bottanelli F, Kromann EB, Allgeyer ES, Erdmann RS, Wood Baguley S, Sirinakis G, Schepartz A, Baddeley D, Toomre DK, Rothman JE, Bewersdorf J (2016) Two-colour live-cell nanoscale imaging of intracellular targets. Nat Commun 7:10778. https://doi.org/10.1038/ncomms10778
188. Tubbs E, Rieusset J (2016) Study of endoplasmic reticulum and mitochondria interactions by in situ proximity ligation assay in fixed cells. J Vis Exp: JoVE (118). https://doi.org/10.3791/54899
189. Gomez-Suaga P, Paillusson S, Stoica R, Noble W, Hanger DP, Miller CC (2017) The ER-mitochondria tethering complex VAPB-PTPIP51 regulates autophagy. Curr Biol 27(3):371–385. https://doi.org/10.1016/j.cub.2016.12.038
190. Paillusson S, Gomez-Suaga P, Stoica R, Little D, Gissen P, Devine MJ, Noble W, Hanger DP, Miller CCJ (2017) alpha-Synuclein binds to the ER-mitochondria tethering protein VAPB to disrupt Ca^{2+} homeostasis and mitochondrial ATP production. Acta Neuropathol 134(1):129–149. https://doi.org/10.1007/s00401-017-1704-z
191. Csordas G, Varnai P, Golenar T, Roy S, Purkins G, Schneider TG, Balla T, Hajnoczky G (2010) Imaging interorganelle contacts and local calcium dynamics at the ER-mitochondrial interface. Mol Cell 39(1):121–132. https://doi.org/10.1016/j.molcel.2010.06.029
192. Ravikumar B, Duden R, Rubinsztein DC (2002) Aggregate-prone proteins with polyglutamine and polyalanine expansions are degraded by autophagy. Hum Mol Genet 11(9):1107–1117
193. Shi F, Kawano F, Park SE, Komazaki S, Hirabayashi Y, Polleux F, Yazawa M (2018) Optogenetic control of endoplasmic reticulum-mitochondria tethering. ACS Synth Biol 7(1):2–9. https://doi.org/10.1021/acssynbio.7b00248
194. Yang Z, Zhao X, Xu J, Shang W, Tong C (2018) A novel fluorescent reporter detects plastic remodeling of mitochondria-ER contact sites. J Cell Sci 131(1):jcs208686. https://doi.org/10.1242/jcs.208686

195. Harmon M, Larkman P, Hardingham G, Jackson M, Skehel P (2017) A Bi-fluorescence complementation system to detect associations between the endoplasmic reticulum and mitochondria. Sci Rep 7(1):17467. https://doi.org/10.1038/s41598-017-17278-1
196. Kakimoto Y, Tashiro S, Kojima R, Morozumi Y, Endo T, Tamura Y (2018) Visualizing multiple inter-organelle contact sites using the organelle-targeted split-GFP system. Sci Rep 8(1):6175. https://doi.org/10.1038/s41598-018-24466-0
197. Shai N, Yifrach E, van Roermund CWT, Cohen N, Bibi C, IJ L, Cavellini L, Meurisse J, Schuster R, Zada L, Mari MC, Reggiori FM, Hughes AL, Escobar-Henriques M, Cohen MM, Waterham HR, Wanders RJA, Schuldiner M, Zalckvar E (2018) Systematic mapping of contact sites reveals tethers and a function for the peroxisome-mitochondria contact. Nat Commun 9(1):1761. https://doi.org/10.1038/s41467-018-03957-8
198. Cabantous S, Terwilliger TC, Waldo GS (2005) Protein tagging and detection with engineered self-assembling fragments of green fluorescent protein. Nat Biotechnol 23(1):102–107. https://doi.org/10.1038/nbt1044
199. Pedelacq JD, Cabantous S, Tran T, Terwilliger TC, Waldo GS (2006) Engineering and characterization of a superfolder green fluorescent protein. Nat Biotechnol 24(1):79–88. https://doi.org/10.1038/nbt1172

Chapter 30
The Role of Mitochondrial Calcium Signaling in the Pathophysiology of Cancer Cells

Andra M. Sterea and Yassine El Hiani

Abstract The pioneering work of Richard Altman on the presence of mitochondria in cells set in motion a field of research dedicated to uncovering the secrets of the mitochondria. Despite limitations in studying the structure and function of the mitochondria, advances in our understanding of this organelle prompted the development of potential treatments for various diseases, from neurodegenerative conditions to muscular dystrophy and cancer. As the powerhouses of the cell, the mitochondria represent the essence of cellular life and as such, a selective advantage for cancer cells. Much of the function of the mitochondria relies on Ca^{2+} homeostasis and the presence of effective Ca^{2+} signaling to maintain the balance between mitochondrial function and dysfunction and subsequently, cell survival. Ca^{2+} regulates the mitochondrial respiration rate which in turn increases ATP synthesis, but too much Ca^{2+} can also trigger the mitochondrial apoptosis pathway; however, cancer cells have evolved mechanisms to modulate mitochondrial Ca^{2+} influx and efflux in order to sustain their metabolic demand and ensure their survival. Therefore, targeting the mitochondrial Ca^{2+} signaling involved in the bioenergetic and apoptotic pathways could serve as potential approaches to treat cancer patients. This chapter will review the role of Ca^{2+} signaling in mediating the function of the mitochondria and its involvement in health and disease with special focus on the pathophysiology of cancer.

Keywords Mitochondria · Calcium signaling · Cancer · Calcium uptake · ROS · Mitochondrial dysfunction · Cancer treatment

---

A. M. Sterea · Y. El Hiani (✉)
Departments of Physiology and Biophysics, Dalhousie University, Halifax, NS, Canada
e-mail: yassine.elhiani@dal.ca

© Springer Nature Switzerland AG 2020

M. S. Islam (ed.), *Calcium Signaling*, Advances in Experimental Medicine and Biology 1131, https://doi.org/10.1007/978-3-030-12457-1_30

30.1 An Introduction to Calcium Signaling

Calcium signaling is an important process in all aspects of cellular function and at its center lies the calcium atom, an element first isolated in 1808 [1, 2]. This method of communication within cells evolved as a means of adaptation to a changing environment, a survival strategy involving a multitude of complex and dynamic signaling cascades. Initially, low amounts of calcium where present on Earth and as a result, primitive cells contained very little calcium in their cytoplasm [3, 4]. Because of the low environmental calcium, the cell machinery evolved to tolerate nanomolar concentrations of this element. However, as the Earth's crust began to release more calcium, the accumulation of this element within cells became toxic; thus, initiating the evolution of calcium removal systems. Interestingly, unlike its ability to alter proteins and modulate cellular processes, calcium itself cannot be chemically modified [5, 6]. This property of the calcium atom along with the increasing calcium concentration in the atmosphere prompted the cell to establish methods of control over the levels of calcium in the cytosol through chelation, sequestration within organelles and extrusion [7, 8]. These processes require numerous calcium-binding proteins, sensors, pumps and ion channels. In its ionic form, Ca^{2+} concentrations vary depending on its location within the cell. For example, at rest, the cytoplasmic Ca^{2+} concentration is around 100 nM while the extracellular and endoplasmic reticulum Ca^{2+} reaches concentrations of approximately 1 mM and 0.5 mM, respectively [9]. These values are subject to change, in particular upon cell stimulation when calcium-selective ion channels open and allow for Ca^{2+} influx; thus, setting the basis for calcium signaling as an intricate signal transduction network. Furthermore, calcium signaling is involved in muscle contraction, cell growth, cellular motility, synaptic plasticity, but can also impact apoptosis, the permeability of ion channels and the cytoskeleton [10–18].

The concept of Ca^{2+} signaling and the importance of Ca^{2+} as a ubiquitous second messenger became apparent more than 100 years ago (circa 1883) when studies on heart cells demonstrated that the presence of Ca^{2+} was necessary for the contraction of cardiomyocytes [18]. These experiments set the stage for Ca^{2+} as an important intracellular regulator of muscle contraction. However, during the 1960s, studies moved beyond muscle research and established a pivotal role for Ca^{2+} as a modulator of cellular processes and identified the presence of buffering systems that can accommodate for the change in Ca^{2+} concentration. Ca^{2+} is naturally present in the extracellular environment and it enters cells via two types of proteins, channels and pumps which are gated by external messengers (receptor-operated receptors, ROCs) or voltage (voltage-gated receptors, VOCs) [19]. These proteins are present within the plasma membrane and upon stimulation (e.g. stretch, agonists, depletion of intracellular stores, etc.), they allow the inflow of Ca^{2+} and the initiation of Ca^{2+} signaling pathways [20]. Furthermore, the cell can also generate Ca^{2+} signals internally through the activation of Phospholipase C (PLC) found in the plasma membrane. Upon activation of PLC, phosphatidylinositol 4,5-biphosphate gets hydrolyzed to IP3 and DAG [21]. IP3 then binds to its receptors on the surface of the endoplasmic reticulum (ER) and stimulates the release of Ca^{2+} from the ER [20]. The mitochondria and the ER have a web-like distribution throughout the cell which facilitates the uptake of Ca^{2+}, but also its distribution. Ca^{2+} is transported within the mitochondria via a uniporter (MCU) which allows for the rapid inflow of Ca^{2+} from the cytosol or the ER [22–25]. Once inside the matrix of the mitochondria, Ca^{2+} can alter the mitochondrial function, especially their ability to produce ATP [26]. Studies have shown that an elevation in the Ca^{2+} concentration inside the matrix can increase mitochondrial respiration and ATP synthesis [27–29]. This bioenergetic dependence of the mitochondria on the presence of Ca^{2+} allows these organelles to coordinate ATP synthesis with the needs of the cell while maintaining Ca^{2+} homeostasis. Nonetheless, excessive buildup of Ca^{2+} can lead to mitochondrial swelling and cell death, a feature that has been exploited in cancer in efforts to eliminate aberrant cells [30, 31]. In addition, while Ca^{2+} can modulate the function of the mitochondria, the organelle itself can in turn affect Ca^{2+} signaling. Numerous research groups have shifted their focus on the involvement of mitochondria Ca^{2+} signaling and its role in disease, especially in cancer cells which are metabolically distinct from normal cells as evidenced by their dependence on mitochondrial ATP to sustain cell proliferation. This chapter will focus on mitochondrial Ca^{2+} signaling, its impact on the pathophysiology of cancer and current mitochondria-based therapies for the treatment of cancer patients.

30.2 Mitochondria, a Historical Overview and Functional Analysis

Commonly referred to as the powerhouses of the cell, the mitochondria have been the subject of extensive scientific interest, from cytology and biochemistry to molecular biology. Early records of mitochondria-like features date back to the 1840s, a time when these structures were yet to be identified as the double membraned organelles we know today. However, in 1890, Richard Altmann, a German pathologist was the first to report the existence of mitochondria within cells, describing them as “living, elementary organisms” or “bioblasts” [32, 33]. Altmann believed that the presence of “bioblasts” was essential for cell metabolism and various genetic functions; thus, making them a vital component of the cell’s physiology, a belief that would soon dominate the research world, giving rise to theories on the origins of the mitochondria and their place as the driving force behind the evolution of eukaryotes [32]. Two main evolutionary scenarios stemming from the same endosymbiosis theory have attempted to explain the origin of the mitochondria and the basis behind the cell’s energetic dependence on these organelles [34, 35]. The symbiogenesis scenario provides evidence supporting the existence of mitochondria as free-living α-proteobacteria that were engulfed by a prokaryotic cell forming a symbiotic relationship between the mitochondria and the host cell [36–38]. With time, the symbiont reduced its genome size by transferring its genetic material to the host cell and eventually becoming an organelle. This theory maintains the idea that the complexity of the modern eukaryotic cell evolved after this symbiotic event [39]. In contrast, in the archezoan scenario, the host cell was an early eukaryotic cell as opposed to a prokaryotic cell [39]. This model suggests that primitive eukaryotic cells became what we now refer to as eukaryotic cells before the mitochondria was integrated within the host cell. Nonetheless, despite the accumulating body of evidence supporting the symbiogenesis theory, the exact origin of the mitochondria and their place within the evolutionary timeframe has proven to be much more ambiguous and complicated to pinpoint. Elements from both scenarios can be used to explain the evolution of the mitochondria, but neither possibility can be rejected with unwavering certainty at this time.

The morphological heterogeneity of the mitochondria was confirmed by various studies demonstrating the existence of the mitochondria as small spheres, but also tubular structures as a consequence of the balance between fusion and fission, more commonly referred to as mitochondrial dynamics [40–42]. True to their name, the mitochondria have become the epitope of structural and functional complexity that extends far beyond their double membranes and serves as an example of the intimate relationship between morphology and functionality. The most distinct structural characteristic of the mitochondrion is the presence of a smooth outer membrane and a folded inner membrane surrounding the mitochondrial matrix [41, 43]. Each fold in the inner membrane creates cristae which act to increase the surface area of the mitochondria to allow for greater processing efficiency [44]. Interestingly, the outer and inner membranes are compositionally different and functionally independent from one another, with each membrane performing distinct roles necessary to maintain the viability of the mitochondrion and subsequently, the cell [45–47]. The outer mitochondrial membrane contains many porins or pore-forming membrane proteins which allow for the passage of ions and other uncharged molecules resulting in the lack of a membrane potential [48]. In contrast, molecules and ions can only cross the inner mitochondrial membrane when bound to specific transporter proteins or by passing through ion channels (e.g. MCU, discussed later); thus, creating an electrochemical gradient across the inner membrane which in turn, dictates its ion selectivity [48]. The presence of an electrochemical gradient is also indicative of the function of the inner mitochondrial membrane as the center for oxidative phosphorylation and the electron transport chain. Within the inner membrane there are four complexes (I: NADPH dehydrogenase, II: Succinate dehydrogenase, III: Cytochrome c reductase and IV: Cytochrome c oxidase) which facilitate the synthesis of ATP through ion trafficking across the membrane [49, 50]. However, despite the difference in the composition of their membranes, one of the most striking features of the mitochondrion is the presence of mitochondrial DNA (mtDNA) within the matrix compartment, a remnant of their bacterial ancestry [51]. Unlike other organelles, the mitochondrion harbors its own circular DNA which is transcribed and replicated in the matrix and can be inherited (in mammals it occurs only through maternal inheritance) [51]. This mitochondrial genome is packed into nucleoids which contain DNA binding proteins for DNA repair and signaling [52, 53]. These nucleoids also facilitate the development of a signaling network between the mitochondria and the nucleus [54]. During nutrient starvations or mitochondrial damage, signals are sent from the mitochondria to the nucleus where they can induce mitochondrial gene transcription [54, 55]. While it is fascinating that a single organelle can influence nuclear gene expression, the nucleus is not the only organelle that shares a functional relationship with the mitochondria. The "social organelle network" of the mitochondria also includes the endoplasmic reticulum (ER) and lysosomes, two organelles essential for the maintenance of calcium homeostasis in the cell [9, 56, 57].

The mitochondria perform a myriad of functions, each orchestrated in a way as to accommodate the needs of the cell. Aside from generating energy for the cell through oxidative phosphorylation and the electron transport chain, the mitochondria are also responsible for Ca^{2+} signaling, apoptosis, fatty acid and amino acid metabolism [58–62]. Changes in mitochondrial metabolism often leads to the production of reactive oxygen species (ROS), hence the classification of the mitochondrion as a major site of ROS output [63, 64]. The presence of ROS is often regarded as Pandora's box where failure to maintain optimal levels of ROS can be detrimental to the cell and contribute to the development of diseases such as cancer [65–67]. However, although the accumulation of ROS can promote the onset of different pathologies, sustained ROS can damage the mitochondria and ultimately lead to cell death [68]. Normally, ROS act as signaling molecules that can modulate various intracellular pathways that influence cellular function. The levels of ROS are maintained by the cell's antioxidant system through the activity of catalases and peroxiredoxins [69]. As with many of the functions of the mitochondria, one of the factors that can affect the production of ROS is Ca^{2+} [70, 71]. Many studies have shown that the presence of Ca^{2+} promotes the production of ATP inside the mitochondria while excess Ca^{2+} increases ROS production [72, 73]. The interplay between Ca^{2+} and the mitochondria is well studied, but a few pieces of the puzzle are still missing. Research efforts are still focusing on mitochondrial Ca^{2+} signaling and how these Ca^{2+}-dependent pathways help shape the fate of the cell.

30.3 Mitochondria and Calcium Signaling

The mitochondria control many facets of the cell's normal physiology by acting as a Ca^{2+} buffer and sensor. In response to various stimuli, the mitochondria can modulate Ca^{2+} signaling directly by importing it via specific transporters or indirectly by releasing metabolites that can act on the Ca^{2+} signaling machinery and in turn, Ca^{2+} regulates mitochondrial function [74]. This dynamic relationship between the mitochondria and Ca^{2+} signaling affects cell metabolism and survival. However, it was only in the 1960s when scientists discovered that mitochondria can uptake and accumulate Ca^{2+}, but it took another 50 years to identify the mechanism behind this process [75–77]. Most discoveries regarding the mitochondrial Ca^{2+}

uptake mechanism and the proteins involved were made within the last 10 years due to advancements in molecular techniques to study the structure of the mitochondria. The paragraphs below will discuss the mitochondrial Ca^{2+} uptake and release system, modulation of mitochondrial functions by Ca^{2+} and the impact of mitochondrial Ca^{2+} signaling on cellular function (Fig. 30.1).

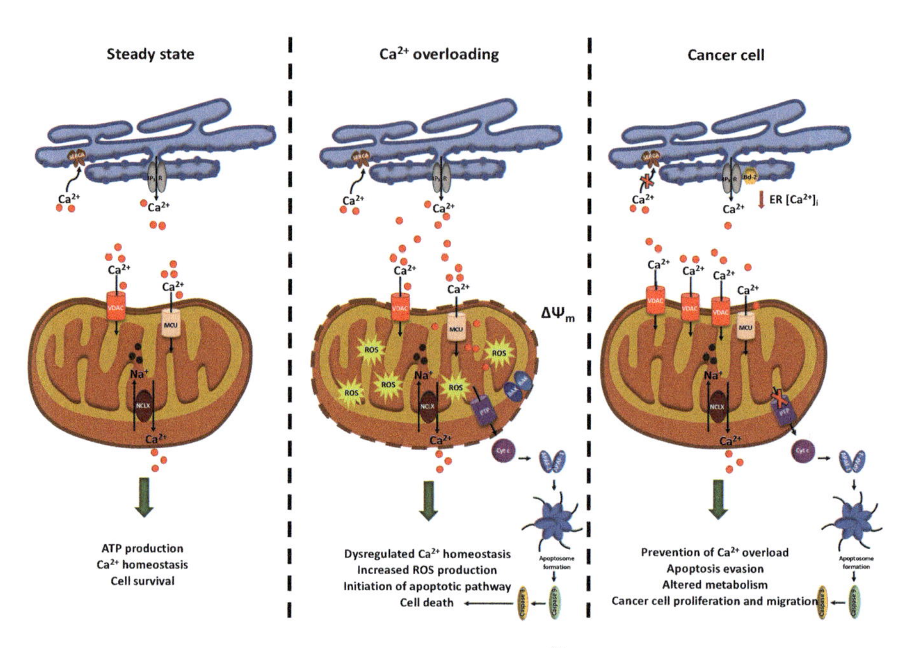

Fig. 30.1 The mitochondrion during steady state, Ca^{2+}overloading and cancer. As the powerhouses of the cell, the mitochondria utilize Ca^{2+} to make ATP in order to supply the cell's metabolic demands. The mitochondria uptake calcium mainly via two Ca^{2+} channels: VDAC (outer mitochondrial membrane) and MCU (inner mitochondrial membrane). In addition, to prevent overloading and to maintain cellular Ca^{2+} homeostasis, the Ca^{2+} must be removed from the mitochondria, a process mediated by the NCLX exchanger. However, if Ca^{2+} accumulates within the mitochondria, PTP opens, the outer membrane depolarizes and cytochrome c is release; thus, initiating the apoptosis pathway. Ca^{2+} overloading can also increase ROS production through the electron transport (located in the inner mitochondrial membrane) which ultimately leads to DNA damage and cell death. Given the importance of the mitochondria in normal cell physiology, these organelles represent a selective advantage for cancer cells. One of the many hallmarks of cancer is their ability to alter their metabolism to sustain proliferation. Unlike normal cells, cancer cells adopt a aerobic glycolysis phenotype whereas normal cells rely mainly on oxidative phosphorylation. Furthermore, cancer cells have evolved mechanisms to hijack the cell's machinery to prevent mitochondrial Ca^{2+} overloading and the activation of the apoptosis pathway. By depleted the ER stores and/or decreasing Ca^{2+} uptake into the ER., cancer cells are able to bypass the calcium induced mitochondrial apoptotic pathway. In addition, the presence of Ca^{2+} influences the expression of VDAC which in turn increases the uptake of Ca^{2+} and promotes cancer cell migration

30.3.1 Mitochondrial Calcium Uptake and Release

As mentioned in the introduction section, when the cytosolic Ca^{2+} concentration is high, Ca^{2+} is passively transported across the inner mitochondrial membrane via the Mitochondrial Calcium Uniporter (MCU), an inward rectifying Ca^{2+} tetrameric channel [25, 77, 78]. The movement of Ca^{2+} through MCU is driven by the negative mitochondrial membrane potential, unlike the outer membrane where the Ca^{2+} passes through porins [79]. MCU requires the binding of the Ca^{2+} to its cytosolic domain in order to be activated. However, MCU appears to work in a biphasic manner where extremely high cytosolic Ca^{2+} inhibit the channel; thus, preventing the accumulation of Ca^{2+} inside the mitochondria and potentially acting as a regulatory mechanism for the uptake of Ca^{2+}. Furthermore, due to its low affinity for Ca^{2+}, MCU allows for micromolar concentrations of Ca^{2+} ($\sim$10–20 μM) to pass through it [80]. With advances in molecular proteomic techniques, some of the architectural components of MCU were identified which enabled us to better understand the mechanism behind MCU-mediated Ca^{2+} uptake in the mitochondria. Studies have shown that the MCU channel pore is formed by three subunits: Mitochondrial Calcium Uniporter (MCU, found in plants and most vertebrates), Mitochondrial Ca^{2+} Uniporter b (MCUb, present in most vertebrates) and Essential MCU Regulator (EMRE, not found in plants, fungi and protozoa) [81]. While MCU and MUCb share 50% sequence similarity, MUCb is believed to function as a negative subunit that inhibits Ca^{2+} entry and reduces the activity of the MCU uniporter complex [82]. Furthermore, EMRE was found as necessary for the transport of Ca^{2+} into the mitochondria by maintaining MCU in an open confirmation while recruiting Mitochondrial Calcium Uptake 1 (MICU1) and 2 (MICU2), two of the three regulatory proteins forming the MCU complex [83, 84]. Experimental manipulation demonstrated that the knockdown of EMRE prevented the influx of Ca^{2+} into the mitochondria. Interestingly, the loss of EMRE leads to a decrease in the size of the MCU complex which suggests a possible role for EMRE as a regulator of MCU complex assembly. As mentioned in previous sentences, the MCU uniporter complex also consists of three regulatory proteins MICU1, 2 and 3. These proteins form a heterodimer and act as gatekeepers of the MCU channel. MCU2 was demonstrated as the inhibitory subunit whose inhibition is released upon MCU channel activation while MCU1 was found to control the activation of MCU and Ca^{2+} uptake [85–89]. Of note, because of the interaction of MCU1 with EMRE and their involvement in promoting the opening/activation of the MCU channel, the loss of either of these subunits leads to a decrease in Ca^{2+} influx into the mitochondria [83]. In addition, in 2018, MICU3 was described as an enhancer of Ca^{2+} uptake upon binding to MICU1, but not MICU2 [90]. Furthermore, the loss of MICU3 was shown to impair Ca^{2+} influx in cortical neurons; thus, suggesting an important role of MICU3 in neuronal function [90]. Moreover, recent studies have identified another protein responsible for calcium uptake through MCU and mitochondrial bioenergetics. Scientists have found that MCU regulator 1 (MCUR1) functions as a scaffold protein necessary for MCU complex formation and subsequently, the loss of MCUR1 prevents the formation of the MCU complex leading to impaired mitochondrial metabolism [91, 92].

Aside from the Ca^{2+} influx through the MCU uniporter, mitochondria have developed contact sites with the ER, the cell's main Ca^{2+} stores [93, 94]. These contact sites are composed of several proteins (e.g. Mitofusin1 and 2) that form tethers between the mitochondria and the ER known as mitochondria-associated membranes (MAM) [95–97]. The MAMs express IP_3 receptors (IP_3Rs, responsible for the Ca^{2+} release from the ER) in close proximity to voltage-dependent anion channels type 1 (VDAC1, a ruthenium red- sensitive Ca^{2+} channel) present on the outer mitochondrial membrane [98]. These two channels form a stable physical connection via Grp75, a chaperon part of the heat shock protein family [99]. Other proteins have been proposed as potential mediators of Ca^{2+} influx including the leucine zipper-EF-hand-containing transmembrane protein 1 (Letm1) and ryanodine receptor (RyR); however, the experimental evidence supporting these suggestions is insufficient as proof of concept [100]. Nonetheless, to prevent Ca^{2+} overloading within the mitochondrial matrix and damage to the mitochondria, efflux mechanisms are set in place to release the excess Ca^{2+}. The efflux of Ca^{2+} is mediated by the presence of the Na^+/Ca^{2+} exchanger, NCLX within the inner mitochondrial membrane [101]. Numerous research groups have demonstrated that the loss of NCLX abolished Ca^{2+} efflux while overexpression of this exchanger enhanced the removal of Ca^{2+} from the mitochondria [101]. The mechanism behind the NCLX-mediated Ca^{2+} efflux is largely unknown, but studies suggest that the Na^+/Ca^{2+} is powered by an electrochemical gradient generated by the presence of high levels of Na^+ in the cytosol [102, 103]. Subsequently, the change in the mitochondrial inner membrane potential facilitates the exchange of three Na^+ for every one Ca^{2+} extruded from the mitochondrial matrix; however, the exchange stoichiometry of NCLX is controversial with some scientists supporting the 3:1 ion ratio while others suggesting a 2:1 ratio of Na^+ to Ca^{2+} [104, 105]. Interestingly, the NCLX exchanger is also involved in the transport of Li^+, a characteristic unique to NCLX [106]. Nonetheless, our knowledge regarding the exact structure and mode of action of NCLX is limited, partially due to the difficulty in isolating the inner mitochondrial membrane. Scientists speculate that the structure of NCLX is similar to that of NCX, a plasma membrane transporter [107]. Furthermore, in order to maintain a steady-state, the influx and efflux mechanisms of the mitochondria must perform in a synchronized manner as to allow the appropriate amount of Ca^{2+} in and out of the mitochondria. Once the Ca^{2+} is released from the mitochondria it is redistributed throughout the cell where it is likely taken up by various organelles including the ER and the lysosomes.

30.3.2 Calcium Modulation of Mitochondrial Function

The main function of the mitochondria is the maintenance of cell metabolism, mainly through oxidative phosphorylation, a characteristic that is heavily influenced by the presence of Ca^{2+} [27]. The passage of calcium across the inner mitochondrial membrane is driven by a large, negative mitochondrial membrane potential which favors passive Ca^{2+} entry, but also the transport of Ca^{2+} down its electrochemical gradient [108]. Generated by mitochondrial respiration, the negative membrane potential (150–200 mV) coupled with the low Ca^{2+} concentration inside the matrix facilitates the transport of Ca^{2+} across the inner mitochondrial membrane [109, 110]. In turn, Ca^{2+} entry causes the depolarization of the mitochondrial membrane [78]. Upon entering the mitochondrial matrix, the Ca^{2+} concentration rises and activates various enzymes necessary for the initiation of the Krebs cycle, and the synthesis of ATP [111, 112]. In particular, the Ca^{2+}-binding α-ketoglutarate dehydrogenase and isocitrate dehydrogenase [113, 114]. Additionally, pyruvate dehydrogenase, an enzyme associated with the conversion of pyruvate to acetyl-CoA becomes activated by a Ca^{2+}-dependent phosphatase [114]. The activation of these rate-limiting enzymes increases the mitochondrial respiration rate and subsequently, enhancing the ATP production rate [115, 116]. The elegant orchestration of the Ca^{2+} signal and the output of energy further emphasizes the dynamic relationship between Ca^{2+} and the mitochondria.

30.3.3 Mitochondrial Calcium Overloading

While Ca^{2+} uptake is a normal, regulatory process for the mitochondria, excessive accumulation of this element can have detrimental consequences, not only for the organelle, but for the whole cell. The Permeability Transition Pore (PTP) is a large conductance channel found in the mitochondrial membrane and its activation represents the pathological effect of Ca^{2+} overloading (Fig. 30.1) [117, 118]. PTP is activated by elevated Ca^{2+} levels inside the mitochondria and results in the permeabilization of the outer mitochondrial membrane (MOMP, mitochondrial outer membrane permeabilization) [119, 120]. The PTP channel is believed to be regulated by several proteins including VDAC, cyclophilin D and Adenine nucleotide translocator (ANT); however, the exact structure of this channel remains unknown [121–123]. When the channel is activated, PTP opens and triggers the release of cytochrome c into the cytosol, an initiating event preceding the beginning of the apoptotic signaling cascade [124–126]. Normally, cytochrome c is involved in the electron transport chain and the generation of ATP, but once released from the confinements of the mitochondria, cytochrome c binds to apoptotic peptidase activating factor 1 (APAF-1) leading to the formation of the apoptosome [127, 128]. The apoptosome then activates procaspase-9 and subsequently caspase 3, ultimately resulting in the death of the cell [128, 129]. The opening of the PTP disrupts the mitochondrial membrane potential and Ca^{2+} leaks out, keeping the channel open. Due to the significance of MOMP in determining whether the cell lives or dies, the process of permeabilizing the outer mitochondrial membrane is tightly regulated by pro- and anti-apoptotic proteins. The anti-apoptotic proteins are part of the Bcl-2 (Bcl-2, Bcl-xL, MCL-1, A1, Bcl-B and Bcl-w) family while the pro-apoptotic proteins are BAX and BAK along with other proteins belonging to the BH3 family (BID, BIM, BAD, PUMA, etc.) [130, 131]. The activation of BAX and BAK removes the apoptotic suppression set by the Bcl-2 proteins and initiates MOMP, and the release of cytochrome c [132–136]. Moreover, the initiation of the apoptotic pathway increases the production of reactive oxygen species (ROS) and decreases ATP synthesis; thus, altering the Ca^{2+} homeostasis within the mitochondria. Additionally, ROS generation maintains the PTP channel in its open conformation [137].

30.4 Defective Mitochondrial Calcium Signaling and Disease

Although the mitochondria play an integral role in normal cell physiology, these organelles proved to be of significance in the development and progression of various pathologies. The seemingly paradoxical relationship between the mitochondria and Ca^{2+} acts as a double edge sword where optimal Ca^{2+} levels benefit the cell while too much Ca^{2+} serves as the basis of many diseases including neurodegenerative and muscular diseases. For example, many research groups have identified that mitochondrial Ca^{2+} overloading contributes to the pathogenesis of Huntington's disease (HD), a neurodegenerative condition characterized by emotional instability and motor impairment [138]. In HD humans and mouse models, neuronal mitochondria are particularly susceptible to MOMP due to a decreased Ca^{2+} holding capacity [139–141]. Furthermore, Ca^{2+} overloading during diastole causes mitochondrial dysfunction and the overproduction of ROS [142]. The excessive ROS generation is thought to impair cardiac function after infarction and as such, contributing to heart failure [142]. Mitochondrial Ca^{2+} signaling has also been implicated in Alzheimer's disease (AD), Parkinson's disease (PD) and Amyotrophic lateral sclerosis (ALS); however, these topics are beyond the scope of this chapter and will not be discussed further [143–149].

30.5 Mitochondrial Calcium Handling in the Pathophysiology of Cancer

Cancer cells are characterized by uncontrolled proliferation and the ability to invade distant tissues; however, one of the emerging hallmarks of cancer cells comes in the form of metabolic reprogramming [150, 151]. Normally, cells generate energy through oxidative phosphorylation to meet their metabolic demand. But unlike normal cells, cancer cells undergo a metabolic shift that switches their cellular machinery from oxidative phosphorylation to aerobic glycolysis [152]. This peculiar event was first described by Otto Warburg who hypothesized that the metabolic change is due defects in the mitochondria, a theory that was later challenged, but never completely disproved [153, 154]. Regardless of the state of the mitochondria, the driver behind ATP generation is Ca^{2+}; hence, for the cancer cell, Ca^{2+} represents the ability to produce enough energy to sustain their aberrant growth (Fig. 30.1). For the scientist, Ca^{2+}-driven metabolism represents a potential target for the development of new anti-cancer drugs.

30.5.1 IP_3R

As stated in previous paragraphs, Ca^{2+} overloading triggers the mitochondrial apoptotic pathway. Nonetheless, insufficient Ca^{2+} transfer from the ER to the mitochondria leads to decreased mitochondrial Ca^{2+} uptake and the activation of autophagy, a well-known degradation process. A paper published in 2010 demonstrated the role of Ca^{2+} transfer through ER IP_3Rs as a major determinant of normal cell bioenergetics [155]. In their paper, Cardenas et al. provided evidence that Ca^{2+} released specifically through IP_3Rs activates AMP-activated kinase (AMPK) which then activates the mechanistic target of Rapamycin (mTOR) and initiates the autophagy pathway [155]. AMPK becomes activated in the absence of adequate levels of ATP from the mitochondria and signals to the cell to start breaking down biomolecules to supply the cell with energy for survival. In terms of a cancer cell, this increase in autophagy is beneficial as it allows the cell to bypass the apoptotic pathway normally triggered by increased ER Ca^{2+} transfer to the mitochondria. This is of particular benefit to cancer cells when exposed to chemotherapeutics targeted at increasing ER Ca^{2+} release. This seemingly paradoxical alteration in function was shown in acute promyelocytic leukemia (APL) where the loss-of-function in the promyelocytic leukemia (PML) tumor suppressor prevented the release of Ca^{2+} from the ER and induced autophagy, promoting cancer cell survival as a result [156, 157]. To further support their work, the same research group conducted experiments on breast, prostate and cervix cancer cells where they found that inhibition of IP_3R induced a “bioenergetic crisis” and ultimately, cell death [158]. The underlying mechanism behind the diminished viability of IP_3R inhibited cancer cells was attributed to the attempt of cancer cells to proliferate while energetically-compromised [158]. Because cancer cells are able to bypass the intrinsic mitochondrial apoptotic pathway by reducing either Ca^{2+} uptake or release, they induce autophagy to supply their metabolic needs. However, in this case, when IP_3R was inhibited, autophagy activation was insufficient to support cell division resulting in cell cycle arrest and necrosis [158]. Another facet of the involvement of IP_3R in cancer was found in glioblastoma were the overexpression IP_3R subtype 3 (IP_3R3) promoted migration. Interestingly, caffeine inhibition of IP_3R3 prevented Ca^{2+} release and hampered the ability of glioblastoma cells to migrate and invade [159]. The caffeine-induced inhibition also decreased glioblastoma cell viability [159].

30.5.2 MCU

Studies performed in human colon cancer have identified a MCU-targeting microRNA (miR-25) able to downregulate the expression of MCU in the mitochondria [160, 161]. The research conducted by Marchi et al. found that miR-25 is overexpressed in colon cancer where it correlates with a decrease in Ca^{2+} uptake by the mitochondria and favors survival of cancer cells by enhancing proliferation [160]. In addition, re-expression of MCU sensitized colon cancer cells to apoptotic signals from the mitochondria [160]. However, in contrast to the results presented in the Marchi suty, a recent paper published in 2018 revealed that the interaction between the receptor-interacting protein kinase 1 (RIPK1) and MCU promotes colon cancer cell proliferation [162]. The study found that overexpression of RIPK1 in colon cancer cell lines increased mitochondrial Ca^{2+} uptake through MCU which resulted in an increase in cell proliferation [162]. In addition, upon analysis of the protein and mRNA expression level of RIPK1, the authors observed a significant increase in the expression of RIPK1 in colorectal cancer patient tissues as compared to normal tissues suggesting a potential for the RIPK1:MCU pathway as a target for colorectal cancer [162]. However, despite the evidence provided by the two studies described above, the role of MCU in colon cancer remains controversial.

The involvement of MCU in the carcinogenesis has also been described in breast cancer models where the expression of MCU increased the migration and invasion abilities of breast cancer cells under the regulation of the microRNA miR-340 [163]. This microRNA was found to be suppressed in breast cancer cells; thus, allowing for the expression of MCU. This concept was further confirmed by knocking down MCU and observing a decrease in migration [163]. Additional research on MCU and breast cancer uncovered the underlying mechanism behind its effect on migration. In 2016, work performed by Tosatto and colleagues found that MCU expression caused an increase in Ca^{2+} uptake by breast cancer cells leading to the production of ROS [164]. Moreover, the ROS released from the mitochondria induced the transcription of Hypoxia-inducible factor 1-alpha ($HIF1\alpha$) [164]. The downstream pathways affected by $HIF1\alpha$ are responsible for migration, glycolytic protein expression and invasion. These effects were lost upon silencing of MCU expression. Aside from the mechanistic involvement of MCU in breast cancer, scientists have found a negative correlation between breast cancer patient survival and the expression of MCU [165].

The impact of mitochondrial Ca^{2+} signaling extends into hepatocellular carcinomas where the expression of mitochondrial MCUR1 was found to be enhanced [166]. In accordance with previous studies showing similar mechanisms, the overexpression of MCUR1 increased the uptake of Ca^{2+} into the mitochondria allowing the cancer cells to overcome any pro-apoptotic challenges. Subsequently, the loss of MCURI restored the mitochondrial Ca^{2+} signaling pathway. The authors concluded that the effects seen in hepatocellular carcinoma cells were in part due to the production of ROS and the activation of the AKT/MDM2 pathway [166]. Activation of this pathway triggers the ubiquitination and degradation of p53, a major regulator of p53-dependent apoptosis [166]. The degradation of p53 then promotes the survival of cancer cells by preventing cell death.

30.5.3 VDAC

VDAC is an essential mediator in the cross talk between the mitochondria and the cell [167]. In addition to facilitating the flow of Ca^{2+} into and out of the mitochondria, VDAC also interacts with hexokinases (HK) to allow for the bridge between oxidative phosphorylation and glycolysis [168–170]. As mentioned in the paragraphs above, cancer cells undergo a metabolic switch to sustain cell division making VDAC an important determinant of the metabolic phenotype of a cancer cell. In addition, VDAC is upregulated in many cancer and its expression is influenced by the level of Ca^{2+} in the cell [171]. As such, it is unsurprising that the loss of VDAC1 (three VDAC isoforms exist) resulted in decreased cell growth and migration in lung, pancreatic and colon cancer cell lines in vitro, but also in vivo [172]. This decrease was concomitant with diminished ATP levels, suggesting that VDAC1 downregulation interferes with cancer cell metabolism; thus creating energetically unfavorable conditions for the cancer cell [172]. These studies were further confirmed in breast cancer where VDAC was demonstrated to interact with Bcl-xL to promote migration. The underlying mechanism behind the enhanced cancer cell migration was mediated by the interaction of Bcl-xL, CD95 and VDAC [173]. Together, these proteins facilitated the influx of Ca^{2+} from the ER to the mitochondria and the subsequent production of ATP which supplied the cancer cells with enough energy to migrate [173]. Similar studies were conducted in other cancers [174–176].

30.5.4 Bcl-2

Besides the involvement of MCU and IP_3R in cancer bioenergetics, the Bcl-2 protein family has also been demonstrated to affect Ca^{2+} flux into the mitochondria through an ER-mediated manner. In fact, one mechanism through which Bcl-2 proteins can affect cancer cell survival is by decreasing the Ca^{2+} concentration in the ER and preventing mitochondrial Ca^{2+} overloading [177]. Bcl-2 has been shown to inhibit the sarco/endoplasmic reticulum calcium ATPase (SERCA) to decrease Ca^{2+} levels within the ER and as such, protecting the mitochondria against Ca^{2+} overloading [178]. Several other modes of actions have been proposed to explain the impact of Bcl-2 on mitochondrial Ca^{2+} uptake such as the potential of Bcl-2 existing as an IP_3R sensitizer to reduce the level of Ca^{2+} in the ER. In addition, Bcl-2 is believed to be able to bind to IP_3R directly and inhibit its function, limiting Ca^{2+} release from the ER as a result. Furthermore, Bcl-2 was also found to be overexpressed in a number of cancers including non-small cell lung carcinoma, breast, neuroblastoma, B-cell lymphoma, chronic lymphocytic leukemia and others [179–183]. Interestingly, experimental evidence has demonstrated that cells overexpressing Bcl-2 experience an increase in Ca^{2+} leakage from the ER which limits the level of Ca^{2+} available for release and mitochondrial uptake [184–186]. Although Bcl-2 is able to affect intracellular Ca^{2+} in various ways, the common goal is to protect the cancer cell against mitochondrial-induced cell death.

30.6 Closing Remarks

The availability of cancer treatments along with resistance against current therapeutics represent factors that limit a patient's chance at survival and their quality of life. Research efforts have been aimed at discovering novel targets to advance cancer therapy and improve patient outcome. These targets include mitochondrial Ca^{2+} signaling and metabolism. One of the major hallmarks of cancer is their ability to divide uncontrollability, a quality that requires a constant supply of energy. Therefore, as the main sources of ATP in the cell, the mitochondria hold tremendous potential as therapeutic targets for the treatment of cancer. In addition, the mitochondria are involved in mediating apoptosis making them an attractive option in the quest for an effective tool to eliminate cancer cells. However, although the role of the mitochondria in cancer is undeniable, our knowledge regarding the mechanisms behind the cancer cell's energetic dependence on this organelle are not well understood; a limitation which is currently putting a damper on the development of mitochondria-targeted therapies. Part of the reason behind our information gaps stems from the unavailability of proper tools to study the mitochondria, in particular the inner membrane of the mitochondria. But despite the difficulty in studying the mitochondrial structure and function, progress has been made in terms of identifying potential therapeutic targets [187, 188]. As of date, numerous compounds targeting the mitochondria are being tested and some even undergoing clinical trials. For example, paclitaxel, a Taxol-based drug used to treat breast and ovarian cancer, has been shown to modulate the cytosolic Ca^{2+} signal by acting on the mitochondria and inducing the opening of the permeability transition pore in pancreatic acinar cells [189]. The opening of this channel disturbs the mitochondrial membrane potential and leads to the loss of Ca^{2+} from the mitochondria [189]. Another example of drug-induced modulation of mitochondrial Ca^{2+} currently under testing is arsenic trioxide [190]. This drug was demonstrated to induce apoptosis in multiple myeloma cells by promoting the release of cytochrome c from the mitochondria through the upregulation of VDAC [190]. Furthermore, aspirin, a common drug for the treatment of fevers and pain, has also been shown to induce apoptosis in cervical cancer cells through the modulation of VDAC1 [191]. In addition, Ca^{2+} influx modulation through MCU has been shown to sensitize pancreatic cancer cells to gemcitabine-induced apoptosis [192]. Other methods of targeting the mitochondrial function in cancer cells include the synthesis of peptides designed to act on the Bcl-2 proteins to induce apoptosis

[193, 194]. Although targeting the mitochondria poses some challenges, advances in our knowledge of the mitochondrial structure and function are bringing researchers closer to developing effective cancer treatments.

Acknowledgements AS is supported through the cancer research training program (CRTP) administered by the Beatrice Hunter Cancer Research Institute (BHCRI) and funded by The Canadian Institute of Health Research (CIHR), Terry Fox Research Institute (TFRI), Cancer Care Nova Scotia, Dalhousie Medical Research Foundation (DMRF) and the Canadian Cancer Society Nova Scotia Division.

Conflict of Interest The authors declare no conflict of interest.

References

1. E.S.T. Sir Humphry Davy (1873) Notes queries s4-XI(276):304. https://doi.org/10.1093/nq/s4-XI.276.304-i
2. Brini M, Carafoli E (2000) Calcium signalling: a historical account, recent developments and future perspectives. Cell Mol Life Sci 57(3):354–370
3. Case RM, Eisner D, Gurney A, Jones O, Muallem S, Verkhratsky A (2007) Evolution of calcium homeostasis: from birth of the first cell to an omnipresent signalling system. Cell Calcium 42(4–5):345–350. https://doi.org/10.1016/j.ceca.2007.05.001
4. Verkhratsky A, Parpura V (2014) Calcium signalling and calcium channels: evolution and general principles. Eur J Pharmacol 739(C):1–3. https://doi.org/10.1016/j.ejphar.2013.11.013
5. Kazmierczak J, Kempe S, Kremer B (2013) Calcium in the early evolution of living systems: a biohistorical approach. Curr Org Chem 17(16):1738–1750. https://doi.org/10.2174/13852728113179990081
6. Cai X, Wang X, Patel S, Clapham DE (2015) Insights into the early evolution of animal calcium signaling machinery: a unicellular point of view. Cell Calcium 57(3):166–173. https://doi.org/10.1016/j.ceca.2014.11.007
7. Kass GEN, Orrenius S (1999) Calcium signaling and cytotoxicity. Environ Health Perspect 107(SUPPL. 1):25–35. https://doi.org/10.1289/ehp.99107s125
8. Montero M, Brini M, Marsault R et al (1995) Monitoring dynamic changes in free Ca^{2+} concentration in the endoplasmic reticulum of intact cells. EMBO J 14:5467
9. Raffaello A, Mammucari C, Gherardi G, Rizzuto R (2016) Calcium at the center of cell signaling: interplay between endoplasmic reticulum, mitochondria, and lysosomes. Trends Biochem Sci 41(12):1035–1049. https://doi.org/10.1016/j.tibs.2016.09.001
10. Hepler PK (1994) The role of calcium in cell division. Cell Calcium 16(4):322–330. https://doi.org/10.1016/0143-4160(94)90096-5
11. Mattson MP, Chan SL (2003) Calcium orchestrates apoptosis. Nat Cell Biol 5(12):1041–1043. https://doi.org/10.1038/ncb1203-1041
12. McConkey DJ, Orrenius S (1997) The role of calcium in the regulation of apoptosis. Biochem Biophys Res Commun 59(239):775–783. https://doi.org/10.1006/bbrc.1997.7409
13. Pinto MCX, Kihara AH, Goulart VAM et al (2015) Calcium signaling and cell proliferation. Cell Signal 27(11):2139–2149. https://doi.org/10.1016/j.cellsig.2015.08.006
14. Berridge MJ (1995) Calcium signalling and cell proliferation. BioEssays 17(6):491–500. https://doi.org/10.1002/bies.950170605
15. Hepler PK (2016) The cytoskeleton and its regulation by calcium and protons. Plant Physiol 170(1):3–22. https://doi.org/10.1104/pp.15.01506
16. Fitzjohn SM, Collingridge GL (2002) Calcium stores and synaptic plasticity. Cell Calcium 32(5–6):405–411. https://doi.org/10.1016/S0143-4160(02)00199-9

17. Lamont MG, Weber JT (2012) The role of calcium in synaptic plasticity and motor learning in the cerebellar cortex. Neurosci Biobehav Rev 36(4):1153–1162. https://doi.org/10.1016/j.neubiorev.2012.01.005
18. Ringer S (1883) A further contribution regarding the influence of the different constituents of the blood on the contraction of the heart. J Physiol 4(1):29–42. https://doi.org/10.1113/jphysiol.1883.sp000120
19. McFadzean I, Gibson A (2002) The developing relationship between receptor-operated and store-operated calcium channels in smooth muscle. Br J Pharmacol 135(1):1–13. https://doi.org/10.1038/sj.bjp.0704468
20. Berridge MJ, Bootman MD, Roderick HL (2003) Calcium signalling: dynamics, homeostasis and remodelling. Nat Rev Mol Cell Biol. https://doi.org/10.1038/nrm1155
21. Putney JW, Tomita T (2012) Phospholipase C signaling and calcium influx. Adv Biol Regul. https://doi.org/10.1016/j.advenzreg.2011.09.005
22. Carafoli E, Lehninger AL (1971) A survey of the interaction of calcium ions with mitochondria from different tissues and species. Biochem J. https://doi.org/10.1042/bj1220681
23. Gunter KK, Gunter TE (1994) Transport of calcium by mitochondria. J Bioenerg Biomembr. https://doi.org/10.1007/BF00762732
24. Rizzuto R, Simpson AWM, Brini M, Pozzan T (1992) Rapid changes of mitochondrial Ca^{2+} revealed by specifically targeted recombinant aequorin. Nature. https://doi.org/10.1038/358325a0
25. De Stefani D, Raffaello A, Teardo E, Szabó I, Rizzuto R (2011) A forty-kilodalton protein of the inner membrane is the mitochondrial calcium uniporter. Nature. https://doi.org/10.1038/nature10230
26. Denton RM, McCormack JG (1980) The role of calcium in the regulation of mitochondrial metabolism. Biochem Soc Trans. https://doi.org/10.1042/bst0080266
27. Denton RM, McCormack JG (1980) On the role of the calcium transport cycle in heart and other mammalian mitochondria. FEBS Lett. https://doi.org/10.1016/0014-5793(80)80986-0
28. Glancy B, Willis WT, Chess DJ, Balaban RS (2013) Effect of calcium on the oxidative phosphorylation cascade in skeletal muscle mitochondria. Biochemistry. https://doi.org/10.1021/bi3015983
29. Territo PR, Mootha VK, French SA, Balaban RS (2000) Ca^{2+} activation of heart mitochondrial oxidative phosphorylation: role of the F(0)/F(1)-ATPase. Am J Physiol Cell Physiol. https://doi.org/10.1152/ajpcell.2000.278.2.C423
30. Huang X, Zhai D, Huang Y (2000) Study on the relationship between calcium-induced calcium release from mitochondria and PTP opening. Mol Cell Biochem. https://doi.org/10.1023/A:1007138818124
31. Ichas F, Mazat JP (1998) From calcium signaling to cell death: two conformations for the mitochondrial permeability transition pore. Switching from low- to high-conductance state. Biochim Biophys Acta Bioenerg. https://doi.org/10.1016/S0005-2728(98)00119-4
32. O'Rourke B (2010) From bioblasts to mitochondria: ever expanding roles of mitochondria in cell physiology. Front Physiol. https://doi.org/10.3389/fphys.2010.00007
33. Ernster L, Schatz G (1981) Mitochondria: a historical review. J Cell Biol. https://doi.org/10.1083/jcb.91.3.227s
34. Check C (2012, Jan 1) Mitochondrial evolution 2–4
35. Gray MW, Burger G, Lang BF (1999) Mitochondrial evolution. Science (80-). https://doi.org/10.1126/science.283.5407.1476
36. Martin WF, Müller M (2007) Origin of mitochondria and hydrogenosomes. https://doi.org/10.1007/978-3-540-38502-8
37. Dyall SD, Brown MT, Johnson PJ (2004) Ancient invasions: from endosymbionts to organelles. Science (80-). https://doi.org/10.1126/science.1094884
38. Embley TM, Martin W (2006) Eukaryotic evolution, changes and challenges. Nature. https://doi.org/10.1038/nature04546
39. Gray MW (2014) The pre-endosymbiont hypothesis: a new perspective on the origin and evolution of mitochondria. Cold Spring Harb Perspect Biol. https://doi.org/10.1101/cshperspect.a016097

40. Westermann B (2012) Bioenergetic role of mitochondrial fusion and fission. Biochim Biophys Acta Bioenerg. https://doi.org/10.1016/j.bbabio.2012.02.033
41. McCarron JG, Wilson C, Sandison ME et al (2013) From structure to function: mitochondrial morphology, motion and shaping in vascular smooth muscle. J Vasc Res. https://doi.org/10.1159/000353883
42. Rafelski SM (2013) Mitochondrial network morphology: building an integrative, geometrical view. BMC Biol. https://doi.org/10.1186/1741-7007-11-71
43. Sjöstrand FS (1953) Electron microscopy of mitochondria and cytoplasmic double membranes: ultra-structure of rod-shaped mitochondria. Nature. https://doi.org/10.1038/171030a0
44. PALADE GE (1953) An electron microscope study of the mitochondrial structure. J Histochem Cytochem. https://doi.org/10.1177/1.4.188
45. Comte J, Maisterrena B, Gautheron DC, Mary SM (1976) Lipid composition and protein profiles of outer and inner membranes from pig heart mitochondria: comparison with microsomes. Biochim Biophys Acta. https://doi.org/10.1016/0005-2736(76)90353-9
46. Gohil VM, Greenberg ML (2009) Mitochondrial membrane biogenesis: phospholipids and proteins go hand in hand. J Cell Biol. https://doi.org/10.1083/jcb.200901127
47. Prasai K (2017) Regulation of mitochondrial structure and function by protein import: a current review. Pathophysiology. https://doi.org/10.1016/j.pathophys.2017.03.001
48. K?hlbrandt W (2015) Structure and function of mitochondrial membrane protein complexes. BMC Biol. https://doi.org/10.1186/s12915-015-0201-x
49. Davies KM, Strauss M, Daum B et al (2011) Macromolecular organization of ATP synthase and complex I in whole mitochondria. Proc Natl Acad Sci. https://doi.org/10.1073/pnas.1103621108
50. Lodish H, Berk A, Zipursky SL, Matsudaira P, Baltimore D, Darnell J (2000) Molecular cell biology. 4th ed. https://doi.org/10.1017/CBO9781107415324.004
51. Taanman J-W (1999) The mitochondrial genome: structure, transcription, translation and replication. Biochim Biophys Acta Bioenerg. https://doi.org/10.1016/S0005-2728(98)00161-3
52. Satoh M, Kuroiwa T (1991) Organization of multiple nucleoids and DNA molecules in mitochondria of a human cell. Exp Cell Res. https://doi.org/10.1016/0014-4827(91)90467-9
53. Bogenhagen DF (2012) Mitochondrial DNA nucleoid structure. Biochim Biophys Acta Gene Regul Mech. https://doi.org/10.1016/j.bbagrm.2011.11.005
54. Iborra FJ, Kimura H, Cook PR (2004) The functional organization of mitochondrial genomes in human cells. BMC Biol. https://doi.org/10.1186/1741-7007-2-9
55. Shaw AK, Kalem MC, Zimmer SL (2016) Mitochondrial gene expression is responsive to starvation stress and developmental transition in *Trypanosoma cruzi*. mSphere. https://doi.org/10.1128/mSphere.00051-16
56. Diogo CV, Yambire KF, Fernández Mosquera L, Branco FT, Raimundo N (2017) Mitochondrial adventures at the organelle society. Biochem Biophys Res Commun. https://doi.org/10.1016/j.bbrc.2017.04.124
57. Murley A, Nunnari J (2016) The emerging network of mitochondria-organelle contacts. Mol Cell. https://doi.org/10.1016/j.molcel.2016.01.031
58. Kunau WH, Dommes V, Schulz H (1995) β-oxidation of fatty acids in mitochondria, peroxisomes, and bacteria: a century of continued progress. Prog Lipid ResProg Lipid Res. https://doi.org/10.1016/0163-7827(95)00011-9
59. Bartlett K, Eaton S (2004) Mitochondrial beta-oxidation. Eur J Biochem. https://doi.org/10.1046/j.1432-1033.2003.03947.x
60. Guda P, Guda C, Subramaniam S (2007) Reconstruction of pathways associated with amino acid metabolism in human mitochondria. Genomics Proteomics Bioinformatics. https://doi.org/10.1016/S1672-0229(08)60004-2
61. Hutson SM, Fenstermacher D, Mahar C (1988) Role of mitochondrial transamination in branched chain amino acid metabolism. J Biol Chem
62. Liu X, Kim CN, Yang J, Jemmerson R, Wang X (1996) Induction of apoptotic program in cell-free extracts: requirement for dATP and cytochrome c. Cell. https://doi.org/10.1016/S0092-8674(00)80085-9

63. Chandel NS, Maltepe E, Goldwasser E, Mathieu CE, Simon MC, Schumacker PT (1998) Mitochondrial reactive oxygen species trigger hypoxia-induced transcription. Proc Natl Acad Sci U S A. https://doi.org/10.1073/pnas.95.20.11715
64. Finkel T (1998) Oxygen radicals and signaling. Curr Opin Cell Biol. https://doi.org/10.1016/S0955-0674(98)80147-6
65. Sena LA, Chandel NS (2012) Physiological roles of mitochondrial reactive oxygen species. Mol Cell. https://doi.org/10.1016/j.molcel.2012.09.025
66. De Sá Junior PL, Câmara DAD, Porcacchia AS et al (2017) The roles of ROS in cancer heterogeneity and therapy. Oxidative Med Cell Longev. https://doi.org/10.1155/2017/2467940
67. Liou M-Y, Storz P (2010) Reactive oxygen species in cancer. Free Radic Res 44:479–496. https://doi.org/10.3109/10715761003667554
68. Fleury C, Mignotte B, Vayssière J-L (2002) Mitochondrial reactive oxygen species in cell death signaling. Biochimie. https://doi.org/10.1016/S0300-9084(02)01369-X
69. Kotiadis VN, Duchen MR, Osellame LD (2014) Mitochondrial quality control and communications with the nucleus are important in maintaining mitochondrial function and cell health. Biochim Biophys Acta Gen Subj. https://doi.org/10.1016/j.bbagen.2013.10.041
70. Sohal RS, Allen RG (1985) Relationship between metabolic rate, free radicals, differentiation and aging: a unified theory. In: Molecular biology of aging. https://doi.org/10.1007/978-1-4899-2218-2_4
71. Görlach A, Bertram K, Hudecova S, Krizanova O (2015) Calcium and ROS: a mutual interplay. Redox Biol. https://doi.org/10.1016/j.redox.2015.08.010
72. Li X, Fang P, Mai J, Choi ET, Wang H, Yang X (2013) Targeting mitochondrial reactive oxygen species as novel therapy for inflammatory diseases and cancers. J Hematol Oncol. https://doi.org/10.1186/1756-8722-6-19
73. Brookes PS (2004) Calcium, ATP, and ROS: a mitochondrial love-hate triangle. AJP Cell Physiol. https://doi.org/10.1152/ajpcell.00139.2004
74. Walsh C, Barrow S, Voronina S, Chvanov M, Petersen OH, Tepikin A (2009) Modulation of calcium signalling by mitochondria. Biochim Biophys Acta Bioenerg. https://doi.org/10.1016/j.bbabio.2009.01.007
75. Deluca HF, Engstrom GW (1961) Calcium uptake by rat kidney mitochondria. Proc Natl Acad Sci U S A. https://doi.org/10.1073/pnas.47.11.1744
76. Vasington FD, Murphy J (1962) Ca++ ion uptake by rat kidney mitochondria and its dependence on respiration and phosphorylation. J Biol Chem. https://doi.org/10.1016/J.CELL.2008.08.006
77. Baughman JM, Perocchi F, Girgis HS et al (2011) Integrative genomics identifies MCU as an essential component of the mitochondrial calcium uniporter. Nature. https://doi.org/10.1038/nature10234
78. Kirichok Y, Krapivinsky G, Clapham DE (2004) The mitochondrial calcium uniporter is a highly selective ion channel. Nature. https://doi.org/10.1038/nature02246
79. Bernardi P (1999) Mitochondrial transport of cations: channels, exchangers, and permeability transition. Physiol Rev. https://doi.org/10.1152/physrev.1999.79.4.1127
80. Marchi S, Pinton P (2014) The mitochondrial calcium uniporter complex: molecular components, structure and physiopathological implications. J Physiol. https://doi.org/10.1113/jphysiol.2013.268235
81. De Stefani D, Rizzuto R, Pozzan T (2016) Enjoy the trip: calcium in mitochondria back and forth. Annu Rev Biochem. https://doi.org/10.1146/annurev-biochem-060614-034216
82. Raffaello A, De Stefani D, Sabbadin D et al (2013) The mitochondrial calcium uniporter is a multimer that can include a dominant-negative pore-forming subunit. EMBO J. https://doi.org/10.1038/emboj.2013.157
83. Sancak Y, Markhard AL, Kitami T, et al (2013) EMRE is an essential component of the mitochondrial calcium uniporter complex. Science (80-). https://doi.org/10.1126/science.1242993
84. Baradaran R, Wang C, Siliciano AF, Long SB (2018) Cryo-EM structures of fungal and metazoan mitochondrial calcium uniporters. Nature. https://doi.org/10.1038/s41586-018-0331-8

85. Csordás G, Golenár T, Seifert EL et al (2013) MICU1 controls both the threshold and cooperative activation of the mitochondrial Ca^{2+}uniporter. Cell Metab. https://doi.org/10.1016/j.cmet.2013.04.020
86. Hoffman NE, Chandramoorthy HC, Shamugapriya S et al (2013) MICU1 motifs define mitochondrial calcium uniporter binding and activity. Cell Rep. https://doi.org/10.1016/j.celrep.2013.11.026
87. Kamer KJ, Mootha VK (2014) MICU1 and MICU2 play nonredundant roles in the regulation of the mitochondrial calcium uniporter. EMBO Rep. https://doi.org/10.1002/embr.201337946
88. Perocchi F, Gohil VM, Girgis HS et al (2010) MICU1 encodes a mitochondrial EF hand protein required for Ca^{2+}uptake. Nature. https://doi.org/10.1038/nature09358
89. Mallilankaraman K, Doonan P, Cárdenas C et al (2012) MICU1 is an essential gatekeeper for mcu-mediated mitochondrial Ca 2+ uptake that regulates cell survival. Cell. https://doi.org/10.1016/j.cell.2012.10.011
90. Patron M, Granatiero V, Espino J (2018) MICU3 is a tissue-speci fi c enhancer of mitochondrial calcium uptake. Cell Death Differ. https://doi.org/10.1038/s41418-018-0113-8
91. Tomar D, Dong Z, Shanmughapriya S et al (2016) MCUR1 is a scaffold factor for the MCU complex function and promotes mitochondrial bioenergetics. Cell Rep. https://doi.org/10.1016/j.celrep.2016.04.050
92. Mallilankaraman K, Cardenas C, Doonan PJ et al (2012) MCUR1 is an essential component of mitochondrial Ca^{2+} uptake that regulates cellular metabolism. Nat Cell Biol. https://doi.org/10.1038/ncb2622
93. Rizzuto R, Pinton P, Carrington W, et al (1998) Close contacts with the endoplasmic reticulum as determinants of mitochondrial Ca^{2+} responses. Science (80-). https://doi.org/10.1126/science.280.5370.1763
94. Csordás G, Renken C, Várnai P et al (2006) Structural and functional features and significance of the physical linkage between ER and mitochondria. J Cell Biol. https://doi.org/10.1083/jcb.200604016
95. Wieckowski MRMR, Giorgi C, Lebiedzinska M, Duszynski J, Pinton P (2009) Isolation of mitochondria-associated membranes and mitochondria from animal tissues and cells. Nat Protoc. https://doi.org/10.1038/nprot.2009.151
96. Giorgi C, Missiroli S, Patergnani S, Duszynski J, Wieckowski MR, Pinton P (2015) Mitochondria-associated membranes: composition, molecular mechanisms, and physiopathological implications. Antioxid Redox Signal. https://doi.org/10.1089/ars.2014.6223
97. De Brito OM, Scorrano L (2008) Mitofusin 2 tethers endoplasmic reticulum to mitochondria. Nature. https://doi.org/10.1038/nature07534
98. Rowland AA, Voeltz GK (2012) Endoplasmic reticulum-mitochondria contacts: function of the junction. Nat Rev Mol Cell Biol. https://doi.org/10.1038/nrm3440
99. Szabadkai G, Bianchi K, Várnai P et al (2006) Chaperone-mediated coupling of endoplasmic reticulum and mitochondrial Ca^{2+} channels. J Cell Biol. https://doi.org/10.1083/jcb.200608073
100. Jiang D, Zhao L, Clapham DE (2009) Genome-wide RNAi screen identifies Letm1 as a mitochondrial Ca^{2+}/H^{+} antiporter. Science (80-). https://doi.org/10.1126/science.1175145
101. Palty R, Silverman WF, Hershfinkel M et al (2010) NCLX is an essential component of mitochondrial Na^{+}/Ca^{2+} exchange. Proc Natl Acad Sci. https://doi.org/10.1073/pnas.0908099107
102. CROMPTON M, KÜNZI M, CARAFOLI E (1977) The calcium-induced and sodium-induced effluxes of calcium from heart mitochondria: evidence for a sodium-calcium carrier. Eur J Biochem. https://doi.org/10.1111/j.1432-1033.1977.tb11839.x
103. Carafoli E, Tiozzo R, Lugli G, Crovetti F, Kratzing C (1974) The release of calcium from heart mitochondria by sodium. J Mol Cell Cardiol. https://doi.org/10.1016/0022-2828(74)90077-7
104. Brand MD (1985) The stoichiometry of the exchange catalysed by the mitochondrial calcium/sodium antiporter. Biochem J. https://doi.org/10.1042/bj2290161

105. Baysal K, Jung DW, Gunter KK, Gunter TE, Brierley GP (1994) Na(+)-dependent Ca^{2+} efflux mechanism of heart mitochondria is not a passive $Ca^{2+}/2Na^{+}$ exchanger. Am J Phys
106. Palty R, Ohana E, Hershfinkel M et al (2004) Lithium-calcium exchange is mediated by a distinct potassium-independent sodium-calcium exchanger. J Biol Chem. https://doi.org/10.1074/jbc.M401229200
107. Palty R, Hershfinkel M, Sekler I (2012) Molecular identity and functional properties of the mitochondrial Na $+/Ca^{2+}$ exchanger. J Biol Chem. https://doi.org/10.1074/jbc.R112.355867
108. Duchen M (2000) Mitochondria and Ca^{2+} in cell physiology and pathophysiology. Cell Calcium
109. Brand MD, Chen CH, Lehninger AL (1976) Stoichiometry of H^{+} ejection during respiration dependent accumulation of Ca^{2+} by rat liver mitochoneria. J Biol Chem
110. Pozzan T, Magalhães P, Rizzuto R (2000) The comeback of mitochondria to calcium signalling. Cell Calcium. https://doi.org/10.1054/ceca.2000.0166
111. Hansford RG, Zorov D (1998) Role of mitochondrial calcium transport in the control of substrate oxidation. Mol Cell Biochem
112. McCormack JG, Halestrap AP, Denton RM (1990) Role of calcium ions in regulation of mammalian intramitochondrial metabolism. Physiol Rev. https://doi.org/10.1152/physrev.1990.70.2.391
113. Traaseth N, Elfering S, Solien J, Haynes V, Giulivi C (2004) Role of calcium signaling in the activation of mitochondrial nitric oxide synthase and citric acid cycle. Biochim Biophys Acta. https://doi.org/10.1016/j.bbabio.2004.04.015
114. Wan B, LaNoue KF, Cheung JY, Scaduto RC (1989) Regulation of citric acid cycle by calcium. J Biol Chem
115. Jouaville LS, Pinton P, Bastianutto C, G a R, Rizzuto R (1999) Regulation of mitochondrial ATP synthesis by calcium: evidence for a long-term metabolic priming. Proc Natl Acad Sci U S A. https://doi.org/10.1073/pnas.96.24.13807
116. Fink BD, Bai F, Yu L, Sivitz WI (2017) Regulation of ATP production: dependence on calcium concentration and respiratory state. Am J Phys Cell Phys. https://doi.org/10.1152/ajpcell.00086.2017
117. Giorgi C, Romagnoli A, Pinton P, Rizzuto R (2008) Ca^{2+} signaling, mitochondria and cell death. Curr Mol Med. https://doi.org/10.2174/156652408783769571
118. Hunter DR, Haworth RA, Southard JH (1976) Relationship between configuration, function, and permeability in calcium treated mitochondria. J Biol Chem. https://doi.org/10.1016/0304-4157(95)00003-A
119. Chalmers S, Nicholls DG (2003) The relationship between free and total calcium concentrations in the matrix of liver and brain mitochondria. J Biol Chem. https://doi.org/10.1074/jbc.M212661200
120. Basso E, Fante L, Fowlkes J, Petronilli V, Forte MA, Bernardi P (2005) Properties of the permeability transition pore in mitochondria devoid of cyclophilin D. J Biol Chem. https://doi.org/10.1074/jbc.C500089200
121. Crompton M, Ellinger H, Costi A (1988) Inhibition by cyclosporin A of a Ca^{2+}-dependent pore in heart mitochondria activated by inorganic phosphate and oxidative stress. Biochem J
122. Griffiths EJ, Halestrap AP (1991) Further evidence that cyclosporin A protects mitochondria from calcium overload by inhibiting a matrix peptidyl-prolyl cis-trans isomerase. Implications for the immunosuppressive and toxic effects of cyclosporin. Biochem J
123. Belzacq AS, Vieira HLA, Kroemer G, Brenner C (2002) The adenine nucleotide translocator in apoptosis. Biochimie. https://doi.org/10.1016/S0300-9084(02)01366-4
124. Green DR, Kroemer G (2004) The pathophysiology of mitochondrial cell death. Science (80-). https://doi.org/10.1126/science.1099320
125. Crompton M (1999) The mitochondrial permeability transition pore and its role in cell death. Biochem J. https://doi.org/10.1042/BJ3410233
126. Orrenius S, Zhivotovsky B, Nicotera P (2003) Regulation of cell death: the calcium-apoptosis link. Nat Rev Mol Cell Biol. https://doi.org/10.1038/nrm1150

127. Von Ahsen O, Waterhouse N, Kuwana T, Newmeyer D, Green D (2000) The "harmless" release of cytochrome C. Cell Death Differ. https://doi.org/10.1038/sj.cdd.4400782
128. Zou H, Li Y, Liu X, Wang X (1999) An APAf-1 · cytochrome C multimeric complex is a functional apoptosome that activates procaspase-9. J Biol Chem. https://doi.org/10.1074/jbc.274.17.11549
129. Thornberry NA (1998) Caspases: enemies within. Science (80-). https://doi.org/10.1126/science.281.5381.1312
130. Wei MC, Zong WX, Cheng EHY et al (2001) Proapoptotic BAX and BAK: a requisite gateway to mitochondrial dysfunction and death. Science (80-). https://doi.org/10.1126/science1059108
131. Kuwana T, Newmeyer DD (2003) Bcl-2-family proteins and the role of mitochondria in apoptosis. Curr Opin Cell Biol. https://doi.org/10.1016/j.ceb.2003.10.004
132. Jurgensmeier JM, Xie Z, Deveraux Q, Ellerby L, Bredesen D, Reed JC (1998) Bax directly induces release of cytochrome c from isolated mitochondria. Proc Natl Acad Sci. https://doi.org/10.1073/pnas.95.9.4997
133. Finucane DM, Bossy-Wetzel E, Waterhouse NJ, Cotter TG, Green DR (1999) Bax-induced caspase activation and apoptosis via cytochrome c release from mitochondria is inhibitable by Bcl-xL. J Biol Chem. https://doi.org/10.1074/jbc.274.4.2225
134. Luo X, Budihardjo I, Zou H, Slaughter C, Wang X (1998) Bid, a Bcl2 interacting protein, mediates cytochrome c release from mitochondria in response to activation of cell surface death receptors. Cell. https://doi.org/10.1016/S0092-8674(00)81589-5
135. Kluck RM (1997) The release of cytochrome c from mitochondria: a primary site for Bcl-2 regulation of apoptosis. Science (80-). https://doi.org/10.1126/science.275.5303.1132
136. Kuwana T, Bouchier-Hayes L, Chipuk JE et al (2005) BH3 domains of BH3-only proteins differentially regulate Bax-mediated mitochondrial membrane permeabilization both directly and indirectly. Mol Cell. https://doi.org/10.1016/j.molcel.2005.02.003
137. Vercesi AE, Kowaltowski AJ, Grijalba MT, Meinicke AR, Castilho RF (1997) The role of reactive oxygen species in mitochondrial permeability transition. Biosci Rep
138. Walker FO (2007) Huntington's disease. Semin Neurol. https://doi.org/10.1055/s-2007-971176
139. Choo YS, Johnson GVW, MacDonald M, Detloff PJ, Lesort M (2004) Mutant huntingtin directly increases susceptibility of mitochondria to the calcium-induced permeability transition and cytochrome c release. Hum Mol Genet. https://doi.org/10.1093/hmg/ddh162
140. Panov AV, Gutekunst CA, Leavitt BR et al (2002) Early mitochondrial calcium defects in Huntington's disease are a direct effect of polyglutamines. Nat Neurosci. https://doi.org/10.1038/nn884
141. Lim D, Fedrizzi L, Tartari M et al (2008) Calcium homeostasis and mitochondrial dysfunction in striatal neurons of Huntington disease. J Biol Chem. https://doi.org/10.1074/jbc.M704704200
142. Santulli G, Xie W, Reiken SR, Marks AR (2015) Mitochondrial calcium overload is a key determinant in heart failure. Proc Natl Acad Sci U S A. https://doi.org/10.1073/pnas.1513047112
143. Luth ES, Stavrovskaya IG, Bartels T, Kristal BS, Selkoe DJ (2014) Soluble, prefibrillar α-synuclein oligomers promote complex I-dependent, Ca^{2+}-induced mitochondrial dysfunction. J Biol Chem. https://doi.org/10.1074/jbc.M113.545749
144. Gómez-Suaga P, Bravo-San Pedro JM, González-Polo RA, Fuentes JM, Niso-Santano M (2018) ER-mitochondria signaling in Parkinson's disease review-article. Cell Death Dis. https://doi.org/10.1038/s41419-017-0079-3
145. Guardia-Laguarta C, Area-Gomez E, Rub C et al (2014) α-synuclein is localized to mitochondria-associated ER membranes. J Neurosci. https://doi.org/10.1523/JNEUROSCI.2507-13.2014
146. Paillusson S, Gomez-Suaga P, Stoica R et al (2017) α-synuclein binds to the ER–mitochondria tethering protein VAPB to disrupt Ca^{2+} homeostasis and mitochondrial ATP production. Acta Neuropathol. https://doi.org/10.1007/s00401-017-1704-z

147. Picone P, Nuzzo D, Caruana L, Scafidi V, Di Carlo MD (2014) Mitochondrial dysfunction: different routes to Alzheimer's disease therapy. Oxidative Med Cell Longev. https://doi.org/10.1155/2014/780179
148. Magi S, Castaldo P, MacRi ML et al (2016) Intracellular calcium dysregulation: implications for Alzheimer's disease. Biomed Res Int 2016. https://doi.org/10.1155/2016/6701324
149. Yi J, Ma C, Li Y et al (2011) Mitochondrial calcium uptake regulates rapid calcium transients in skeletal muscle during excitation-contraction (E-C) coupling. J Biol Chem. https://doi.org/10.1074/jbc.M110.217711
150. Oermann EK, Wu J, Guan KL, Xiong Y (2012) Alterations of metabolic genes and metabolites in cancer. Semin Cell Dev Biol. https://doi.org/10.1016/j.semcdb.2012.01.013
151. Hanahan D, Weinberg RA (2011) Hallmarks of cancer: the next generation. Cell. https://doi.org/10.1016/j.cell.2011.02.013
152. Vander Heiden MG, Cantley LC, Thompson CB (2009) Understanding the Warburg effect: cell proliferation. Science (80-). https://doi.org/10.1126/science.1160809
153. Warburg O (1956) On the origin of cancer cells on the origin of cance. Source Sci New Ser. https://doi.org/10.1126/science.123.3191.309
154. Koppenol WH, Bounds PL, Dang CV (2011) Otto Warburg's contributions to current concepts of cancer metabolism. Nat Rev Cancer. https://doi.org/10.1038/nrc3038
155. Cárdenas C, Miller RA, Smith I et al (2010) Essential regulation of cell bioenergetics by constitutive InsP3 receptor Ca^{2+} transfer to mitochondria. Cell. https://doi.org/10.1016/j.cell.2010.06.007
156. Giorgi C, Ito K, Lin HK, et al (2010) PML regulates apoptosis at endoplasmic reticulum by modulating calcium release. Science (80-). https://doi.org/10.1126/science.1189157
157. Missiroli S, Bonora M, Patergnani S et al (2016) PML at mitochondria-associated membranes is critical for the repression of autophagy and cancer development. Cell Rep. https://doi.org/10.1016/j.celrep.2016.07.082
158. Cárdenas C, Müller M, McNeal A et al (2016) Selective vulnerability of cancer cells by inhibition of Ca^{2+} transfer from endoplasmic reticulum to mitochondria. Cell Rep. https://doi.org/10.1016/j.celrep.2016.02.030
159. Kang SS, Han KS, Ku BM et al (2010) Caffeine-mediated inhibition of calcium release channel inositol 1,4,5-trisphosphate receptor subtype 3 blocks glioblastoma invasion and extends survival. Cancer Res. https://doi.org/10.1158/0008-5472.CAN-09-2886
160. Marchi S, Lupini L, Patergnani S et al (2013) Downregulation of the mitochondrial calcium uniporter by cancer-related miR-25. Curr Biol. https://doi.org/10.1016/j.cub.2012.11.026
161. Marchi S, Pinton P (2013) Mitochondrial calcium uniporter, MiRNA and cancer. Commun Integr Biol. https://doi.org/10.4161/cib.23818
162. Zeng F, Chen X, Cui W et al (2018) RIPK1 binds MCU to mediate induction of mitochondrial Ca2þuptake and promotes colorectal oncogenesis. Cancer Res. https://doi.org/10.1158/0008-5472.CAN-17-3082
163. Yu C, Wang Y, Peng J et al (2017) Mitochondrial calcium uniporter as a target of microRNA-340 and promoter of metastasis via enhancing the Warburg effect. Oncotarget. https://doi.org/10.18632/oncotarget.19747
164. Tosatto A, Sommaggio R, Kummerow C et al (2016) The mitochondrial calcium uniporter regulates breast cancer progression via HIF-1α. EMBO Mol Med. https://doi.org/10.15252/emmm.201606255
165. Mammucari C, Gherardi G, Rizzuto R (2017) Structure, activity regulation, and role of the mitochondrial calcium uniporter in health and disease. Front Oncol. https://doi.org/10.3389/fonc.2017.00139
166. Xing J, Ren T, Wang J et al (2017) MCUR1-mediated mitochondrial calcium signaling facilitates cell survival of hepatocellular carcinoma via ROS-dependent P53 degradation. Antioxid Redox Signal. https://doi.org/10.1089/ars.2017.6990
167. Shoshan-Barmatz V, De Pinto V, Zweckstetter M, Raviv Z, Keinan N, Arbel N (2010) VDAC, a multi-functional mitochondrial protein regulating cell life and death. Mol Asp Med. https://doi.org/10.1016/j.mam.2010.03.002

168. Wu W, Zhao S (2013) Metabolic changes in cancer: beyond the Warburg effect. Acta Biochim Biophys Sin Shanghai. https://doi.org/10.1093/abbs/gms104
169. Pastorino JG, Hoek JB (2008) Regulation of hexokinase binding to VDAC. J Bioenerg Biomembr. https://doi.org/10.1007/s10863-008-9148-8
170. Mathupala SP, Ko YH, Pedersen PL (2009) Hexokinase-2 bound to mitochondria: cancer's stygian link to the "Warburg effect" and a pivotal target for effective therapy. Semin Cancer Biol. https://doi.org/10.1016/j.semcancer.2008.11.006
171. Weisthal S, Keinan N, Ben-Hail D, Arif T, Shoshan-Barmatz V (2014) Ca^{2+}-mediated regulation of VDAC1 expression levels is associated with cell death induction. Biochim Biophys Acta Mol Cell Res. https://doi.org/10.1016/j.bbamcr.2014.03.021
172. Arif T, Vasilkovsky L, Refaely Y, Konson A, Shoshan-Barmatz V (2014) Silencing VDAC1 expression by siRNA inhibits cancer cell proliferation and tumor growth in vivo. Mol Ther Nucleic Acids. https://doi.org/10.1038/mtna.2014.9
173. Fouqué A, Lepvrier E, Debure L et al (2016) The apoptotic members CD95, BclxL, and Bcl-2 cooperate to promote cell migration by inducing Ca^{2+} flux from the endoplasmic reticulum to mitochondria. Cell Death Differ. https://doi.org/10.1038/cdd.2016.61
174. Arif T, Krelin Y, Nakdimon I et al (2017) VDAC1 is a molecular target in glioblastoma, with its depletion leading to reprogrammed metabolism and reversed oncogenic properties. Neuro-Oncology. https://doi.org/10.1093/neuonc/now297
175. Abu-Hamad S, Sivan S, Shoshan-Barmatz V (2006) The expression level of the voltage-dependent anion channel controls life and death of the cell. Proc Natl Acad Sci U S A. https://doi.org/10.1073/pnas.0600103103
176. Shoshan-Barmatz V, Ben-Hail D, Admoni L, Krelin Y, Tripathi SS (2014) The mitochondrial voltage-dependent anion channel 1 in tumor cells. Biochim Biophys Acta. https://doi.org/10.1016/j.bbamem.2014.10.040
177. Sammels E, Parys JB, Missiaen L, De Smedt H, Bultynck G (2010) Intracellular Ca^{2+} storage in health and disease: a dynamic equilibrium. Cell Calcium. https://doi.org/10.1016/j.ceca.2010.02.001
178. MacLennan DH, Rice WJ, Green NM (1997) The mechanism of Ca^{2+} transport by sarco(endo)plasmic reticulum Ca^{2+}-ATPases. J Biol Chem. https://doi.org/10.1074/jbc.272.46.28815
179. Coustan-Smith E, Kitanaka A, Pui CH et al (1996) Clinical relevance of BCL-2 overexpression in childhood acute lymphoblastic leukemia. Blood
180. Noujaim D, van Golen CM, van Golen KL, Grauman A, Feldman EL (2002) N-Myc and Bcl-2 coexpression induces MMP-2 secretion and activation in human neuroblastoma cells. Oncogene. https://doi.org/10.1038/sj.onc.1205552
181. Del Bufalo D, Biroccio a LC, Zupi G (1997) Bcl-2 overexpression enhances the metastatic potential of a human breast cancer line. FASEB J
182. Choi J, Choi K, Benveniste EN et al (2005) Bcl-2 promotes invasion and lung metastasis by inducing matrix metalloproteinase-2. Cancer Res. https://doi.org/10.1158/0008-5472.CAN-04-4570
183. Sánchez-Beato M, Sánchez-Aguilera A, Piris MA (2003) Cell cycle deregulation in B-cell lymphomas. Blood. https://doi.org/10.1182/blood-2002-07-2009
184. Foyouzi-Youssefi R, Arnaudeau S, Borner C et al (2000) Bcl-2 decreases the free Ca^{2+} concentration within the endoplasmic reticulum. Proc Natl Acad Sci. https://doi.org/10.1073/pnas.97.11.5723
185. Pinton P, Ferrari D, Magalhaes P et al (2000) Reduced loading of intracellular Ca^{2+} stores and downregulation of capacitative Ca^{2+} influx in Bcl-2-overexpressing cells. J Cell Biol. https://doi.org/10.1083/jcb.148.5.857
186. Chiu W-T, Chang H-A, Lin Y-H et al (2018) Bcl-2 regulates store-operated Ca^{2+} entry to modulate ER stress-induced apoptosis. Cell Death Dis. https://doi.org/10.1038/s41420-018-0039-4
187. Hockenbery DM (2010) Targeting mitochondria for cancer therapy. Environ Mol Mutagen. https://doi.org/10.1002/em.20552

188. Weinberg SE, Chandel NS (2015) Targeting mitochondria metabolism for cancer therapy. Nat Chem Biol. https://doi.org/10.1038/nchembio.1712
189. Kidd JF, Pilkington MF, Schell MJ et al (2002) Paclitaxel affects cytosolic calcium signals by opening the mitochondrial permeability transition pore. J Biol Chem. https://doi.org/10.1074/jbc.M106802200
190. Zheng Y, Shi Y, Tian C et al (2004) Essential role of the voltage-dependent anion channel (VDAC) in mitochondrial permeability transition pore opening and cytochrome c release induced by arsenic trioxide. Oncogene. https://doi.org/10.1038/sj.onc.1207205
191. Tewari D, Majumdar D, Vallabhaneni S, Bera AK (2017) Aspirin induces cell death by directly modulating mitochondrial voltage-dependent anion channel (VDAC). Sci Rep. https://doi.org/10.1038/srep45184
192. Chen L, Sun Q, Zhou D et al (2017) HINT2 triggers mitochondrial Ca^{2+} influx by regulating the mitochondrial Ca^{2+} uniporter (MCU) complex and enhances gemcitabine apoptotic effect in pancreatic cancer. Cancer Lett. https://doi.org/10.1016/j.canlet.2017.09.020
193. Wang J-L, Liu D, Zhang Z-J, et al (2000) Structure-based discovery of an organic compound that binds Bcl-2 protein and induces apoptosis of tumor cells. Proc Natl Acad Sci U S A. 97/13/7124 [pii]
194. Inoue-Yamauchi A, Jeng PS, Kim K et al (2017) Targeting the differential addiction to anti-apoptotic BCL-2 family for cancer therapy. Nat Commun. https://doi.org/10.1038/ncomms16078

Chapter 31
Simulation Strategies for Calcium Microdomains and Calcium Noise

Nicolas Wieder, Rainer H. A. Fink, and Frederic von Wegner

Abstract In this article, we present an overview of simulation strategies in the context of subcellular domains where calcium-dependent signaling plays an important role. The presentation follows the spatial and temporal scales involved and represented by each algorithm. As an exemplary cell type, we will mainly cite work done on striated muscle cells, i.e. skeletal and cardiac muscle. For these cells, a wealth of ultrastructural, biophysical and electrophysiological data is at hand. Moreover, these cells also express ubiquitous signaling pathways as they are found in many other cell types and thus, the generalization of the methods and results presented here is straightforward.

The models considered comprise the basic calcium signaling machinery as found in most excitable cell types including Ca^{2+} ions, diffusible and stationary buffer systems, and calcium regulated calcium release channels. Simulation strategies can be differentiated in stochastic and deterministic algorithms. Historically, deterministic approaches based on the macroscopic reaction rate equations were the first models considered. As experimental methods elucidated highly localized Ca^{2+} signaling events occurring in femtoliter volumes, stochastic methods were

N. Wieder
Broad Institute of MIT and Harvard, Cambridge, MA, USA

Department of Medicine, Brigham and Women's Hospital and Harvard Medical School, Boston, MA, USA

Medical Biophysics Group, Institute of Physiology and Pathophysiology, University of Heidelberg, Heidelberg, Germany

R. H. A. Fink
Medical Biophysics Group, Institute of Physiology and Pathophysiology, University of Heidelberg, Heidelberg, Germany

F. von Wegner (✉)
Department of Neurology and Brain Imaging Center, Goethe University Frankfurt, Frankfurt am Main, Germany

Medical Biophysics Group, Institute of Physiology and Pathophysiology, University of Heidelberg, Heidelberg, Germany
e-mail: vonWegner@med.uni-frankfurt.de

© Springer Nature Switzerland AG 2020

M. S. Islam (ed.), *Calcium Signaling*, Advances in Experimental Medicine and Biology 1131, https://doi.org/10.1007/978-3-030-12457-1_31

increasingly considered. However, detailed simulations of single molecule trajectories are rarely performed as the computational cost implied is too large. On the mesoscopic level, Gillespie's algorithm is extensively used in the systems biology community and with increasing frequency also in models of microdomain calcium signaling. To increase computational speed, fast approximations were derived from Gillespie's exact algorithm, most notably the chemical Langevin equation and the τ-leap algorithm. Finally, in order to integrate deterministic and stochastic effects in multiscale simulations, hybrid algorithms are increasingly used. These include stochastic models of ion channels combined with deterministic descriptions of the calcium buffering and diffusion system on the one hand, and algorithms that switch between deterministic and stochastic simulation steps in a context-dependent manner on the other. The basic assumptions of the listed methods as well as implementation schemes are given in the text. We conclude with a perspective on possible future developments of the field.

Keywords Calcium · Calcium signaling · Microdomains · Stochastic modeling · Calcium noise · Colored noise · Chemical master equation · Gillespie's algorithm · Langevin equation · IP3R

31.1 Biological Relevance of Calcium Microdomains

Ca^{2+} is an important second messenger in virtually every cell type. It acts as a signaling molecule in a variety of different cellular processes, ranging from synaptic neurotransmitter release and exocytosis in general to muscle contraction, the regulation of gene networks and many others [1]. Over the last decades, it became clear that the distribution of Ca^{2+} inside a cell cannot be considered to be homogenous. Quite the contrary, it emerged that Ca^{2+} signaling has a strong spatial component [2]. Ca^{2+} release sites were found to be highly organized entities that allow for cytosolic Ca^{2+} elevations in specific signaling relevant locations. The extent of these release events is controlled by a large cytosolic Ca^{2+} buffer capacity, Ca^{2+} pumps and transporters that transfer Ca^{2+} against a steep concentration gradient out of the cytosol and predefined morphological substructures such as neuronal synapses or the dyadic cleft in skeletal muscle cells. These transient, subcellular elevations of Ca^{2+} create temporary signaling entities that are termed Ca^{2+} microdomains [3].

From a systemic perspective, Ca^{2+} microdomains constitute the most fundamental building blocks of Ca^{2+} signaling networks and their biological relevance has been studied extensively. Ca^{2+} microdomains play an important role in the generation of complex spatio-temporal signaling patterns such as waves and oscillations. They are based on synchronized openings of endoplasmic reticulum (ER) Ca^{2+} release channels which are limited by negative feedback mechanisms such as membranous Ca^{2+} pumps. Well studied examples are insulin release from pancreatic beta cells [4], fertilization, muscle contraction and the regulation of gene networks [5]. Another prominent example where Ca^{2+} microdomains play a crucial role in fundamental cellular processes are contact sites between the endoplasmic reticulum and mitochondria, known as mitochondria-associated membranes (MAMs). An interplay between clusters of endoplasmic Ca^{2+} release channels (e.g. IP_3R receptors) and the mitochondrial Ca^{2+} uniporter (MCU) allows to shuttle Ca^{2+} ions from the ER to the mitochondria via localized Ca^{2+} release events at MAMs. The impairment of this mechanism has been associated with different disease relevant cellular states such as inflammation [6], metabolism [7] and apoptosis [8]. A well understood example of structurally confined Ca^{2+} microdomains are neuronal synapses where highly spatially organized Ca^{2+} microdomains play an important role in synaptic plasticity and the generation of long term potentation (LTPs) that is involved in learning and memory [9]. The list of relevant examples goes on, but a comprehensive review of Ca^{2+} microdomains lies beyond the scope of this book chapter. For a more detailed discussion the reader is referred to [1–3, 10].

31.2 Canonical Models of Calcium Microdomains

Identical subcellular compartments and systems have been modeled using different strategies. Synaptic activation including vesicle release at the neuronal synapse, for instance, has been modeled with different approaches [11, 12] ranging from simulations at the scale of individual molecules to deterministic reaction rate equations. Similarly, different model classes of subcellular calcium dynamics in cardiomyocytes and skeletal muscle fibers are still being evaluated for their explanatory and predictive power. The simulated volume and average reactant concentrations can help to decide which simulation strategy is the most adequate for a given problem.

Virtually all cell types share a common family of molecules involved in the regulation of the local, subcellular calcium concentration and many cell lines even share common modules, i.e. small signaling networks that regulate key processes such as cell cycle regulation, adaptation of the metabolic rate, vesicle secretion, motility and excitability [13]. In order to keep the presentation of simulation strategies compact, we here focus on some key molecular species, in particular calcium buffers and calcium channels, from where the transfer to more specific problems should be easy. For a more general introduction to modeling and the simulation of chemical reactions, the reader is referred to [14–17].

The presentation and the implementation of the models used here is further facilitated by the fact that it is sufficient to take into account chemical reactions of second order, at most. In this context, zero-order reactions model the constant generation or degradation of a molecular species at a fixed rate, for instance, when a molecule is assumed to be unstable on time scales relevant for the simulation. Zero-order reactions also come into play when certain reaction networks are not simulated in detail but one of their products (resp. educts) appears as a reaction partner in the system that is simulated in detail. Also, ion channel currents can be modeled as zero-order reactions occurring in time intervals when the channel is in an open state. Modeling ion channel currents as chemical reaction events has the advantage that the inclusion of channel currents does not require major modifications of the computational model but only the addition of another reaction type. First-order reactions are mainly concerned when a calcium-bound molecular complex dissociates and thereby liberates a calcium ion. Second-order reactions describe the corresponding association reaction of a calcium ion and a calcium-binding molecule such as buffer proteins, calcium-sensitive enzymes, membrane constituents, ion channels or fluorescent dye molecules. In principle, any of these calcium-binding molecules can be considered a calcium buffer.

The molecular species used in this presentation are calcium ions (Ca^{2+}), a set of calcium buffers (B_i, $I = 1, \ldots, N_B$) that can be either diffusible or immobile and calcium-regulated ion channels (Ch). In the case of ion channels, those regulated by calcium ions but conducting another ionic species (e.g. calcium regulated chloride or potassium channels) must be distinguished from calcium-regulated calcium channels. The latter, e.g. IP_3- and RyR-channels, provide a highly localized, nonlinear calcium-sensitive feedback system since these channels have activating as well as inhibitory calcium binding sites [18, 19].

31.3 Microscopic Simulation

To achieve a maximum of spatial detail, one has to simulate the Brownian motion of all molecules and introduce reaction events whenever a molecular collision of sufficiently high energy (the activation energy of the reaction) occurs [20]. The price of this level of detail is paid in computation time. Runtime scales unfavorably with increasing simulation volume and increasing number of reactants as diffusion events occur at much higher rates than chemical reactions and the number of trajectories grows linearly with the number of reactants. In the context of calcium dynamics, simulations of individual molecular trajectories are used less frequently than the approaches explained below. They can be found mainly in the context of calcium-regulated synaptic signaling [11, 21, 22] as well as Ca^{2+} dynamics in cardiac myocytes [23–26]. Practically, these systems can be implemented with the freely available software package MCell developed by the Salk Institute (www.mcell.cnl.salk.edu).

31.4 Mesoscopic Stochastic Simulation

Mesoscopic simulation approaches have been a growing area in computational cell biology in the past 15 years [15, 27]. Using the mesoscopic perspective, the fluctuating number of molecules of each type is tracked by one of several stochastic algorithms. To better understand different mesoscopic algorithms, it is necessary to introduce some notation and a few definitions. The state of a given model system of N molecular species $S_1. \ldots, S_N$ will from here on be described by the time-dependent ***state vector*** $X_t = (x_t^{(1)}, \ldots, x_t^{(N)})$ that contains the ***copy number*** $x_t^{(i)}$ of each species (index i) at time t. By copy number, we understand the number of molecules of a certain molecular species in a defined volume at a given time, e.g. a Ca^{2+} concentration of 100 nM in a 1 fl volume yields a copy number of 60. The term copy number is mostly used in the context of small reaction volumes and low molar concentrations resulting in relatively few molecules of each type. The N chemical species in our model system interact through M types of chemical reactions R_j, $j = 1, \ldots, M$. Next, a ***state change vector*** V_j is introduced for each reaction R_j. The component v_{ji} describes the change in the copy number of molecular species S_i caused by a single occurrence of reaction type R_j. After a reaction R_j occurs, the state vector is updated according to $X_t \leftarrow X_t + V_j$. Furthermore, we assume that given a state X_t, there is a defined probability for each reaction R_j to occur within a small (infinitesimal) time interval $[t, t + dt]$. In the context of chemical reactions, this probability equals $a_j(X_t)dt$, and $a_j(X_t)$ is called the ***reaction propensity*** of reaction R_j. As the state vector X_t is time-dependent, the reaction propensity $a_j(X_t)$ is also time-dependent, however, this notation is often suppressed for the sake of simplicity. The following stochastic algorithms can all be derived from the chemical master equation (CME). The CME is a deterministic differential (or difference) equation that describes the temporal evolution of the probability density $p(X_t)$, i.e. the multivariate distribution of the state vector X_t. Using the notation introduced above, the CME reads:

$$\partial_t p(X_t) = \sum_{j=1}^{M} \left[a_j (X_t - V_j) p (X_t - V_j) - a_j (X_t) p (X_t) \right] \tag{31.1}$$

In words, the CME states that the change in $p(X_t)$ is calculated as the net probability flow conveyed by flows from state $X_t - V_j$ into state X_t (via reaction R_j) and the reverse flows out of state X_t. A closed-form, analytical solution for $p(X_t)$ is only accessible for very simple systems (8). As for many other stochastic systems, this is where Monte Carlo sampling schemes come into play. Numerical solutions are also difficult or impossible to obtain due to the large state space that arises when all possible combinations of molecule counts are considered. An approach to circumvent or approximate the CME problem is given by the Finite State projection algorithm [28].

31.5 Gillespie's Algorithm

Gillespie's algorithm or Stochastic Simulation Algorithm (SSA), provides a Monte Carlo simulation scheme that samples the time-dependent probability density $p(X_t)$ exactly [29]. The crucial point in Gillespie's algorithm is the insight that the evolution of the state vector X_t follows a multivariate Markov process on the N-dimensional integer lattice Z_N. The transition rate between the lattice points X_t and $X_t + V_j$ is given by the reaction propensity $a_j(X_t)$. As Markov processes are characterized by exponential waiting time distributions, the waiting time until the next reaction event can be calculated in a single step. The waiting time distribution is parametrized by the cumulative reaction rate

$$a_0 (X_t) = \sum_{j=1}^{M} a_j (X_t) \tag{31.2}$$

The rate $a_0(X_t)$ determines the probability $p(\tau, j|X_t)$ that, given state X_t at time t, the next reaction event will be of type R_j and will occur in the small time interval $[t + \tau, t + \tau + dt]$:

$$p (\tau, j \mid X_t) = a_j (X_t) e^{-a_0(X_t)\tau} \tag{31.3}$$

In the actual algorithm, the probability is factored in two parts. First, the waiting times τ for the next event are obtained as samples from an exponential distribution with parameter $a_0(X_t)$ according to

$$\tau = \frac{-\ln (r_1)}{a_0 (X_t)} \tag{31.4}$$

where $r_1 \sim U_{[0,\, 1]}$ is a uniformly distributed random variable. Here it is assumed that a_j remains constant during the next time step τ. Next, a second uniformly distributed random variable $r_2 \sim U_{[0,\, 1]}$ is drawn which determines the next reaction type R_j according to the ratios of $\frac{a_j(X_t)}{a_0(X_t)}$.

Thus, the complete Gillespie algorithm reads:

1. ***Initialization***: initialize the state vector components $X_{t=0}$ with the copy number of chemical species S_i expected in equilibrium or any other desired initial condition. Calculate all individual reaction propensities $a_j(X_t)$ and the cumulative propensity $a_0(X_t)$.
2. ***Waiting time and reaction selection***: sample a waiting time τ and a reaction index j from the distribution (Eq. 31.3).
3. ***Update state and time***: $X_t \leftarrow X_t + V_j$ and $t \leftarrow t + \tau$.
4. ***Exit condition***: if $t \geq T_{max}$ where T_{max} is the desired length of the simulation.
5. ***Update propensities***: recalculate the propensities of all reactions affected by the last state change.
6. go to step 2.

An efficient implementation based on the dependency structure between reactions and the reuse of random numbers was developed later [30]. The result of Gillespie's algorithm is a set of N time series reflecting the fluctuating copy number of each molecular species S_i. The advantage of the algorithm becomes most clear when compared to earlier approaches. An alternative approach to simulate the reaction process relies on the fact that the probability of reaction R_j to occur in a small interval $[t, t + dt]$ equals $a_j(X_t)dt$. If dt is chosen small enough to ensure that $a_j(X_t)dt \ll 1$, assuming that only one reaction occurs during dt, one can iteratively advance the simulation time by dt, accepting or rejecting reaction events by comparison with the value of a uniformly distributed random variable r. This strategy is not recommended however, as one will find that most time steps dt will pass without any reaction event happening. The problem becomes even worse when the desired precision demands a small time step dt. Gillespie's algorithm overcomes the problem by sampling the exponential waiting time distribution directly and by jumping to the next event time without further checks. Moreover, no fixed integration time step dt has to be chosen and therefore, the algorithm is stochastically 'exact'. Though widely used in systems biology, there are comparatively few examples for the use of Gillespie's algorithm in the modeling of intracellular calcium dynamics [31]. Gillespie's algorithm has been used to model stochastic resonance effects in whole cells [32, 33], calmodulin-dependent synaptic plasticity in dendritic spine microdomains [12] and to model calcium microdomains in the vicinity of individual L-type calcium channels [34]. Although the algorithm can be used for arbitrary simulation volumes, computation time quickly increases for larger volumes because the copy numbers enter the propensity terms in a combinatorial way. Therefore, several approximations of the exact simulation algorithm have been developed.

31.6 The Binomial and the τ-leap Methods

For small time steps dt, the probability for a reaction R_j to occur in the next time interval $[t, t + dt]$ is given by $a_j(X_t)dt$. Assuming that the propensity $a_j(X_t)$ remains approximately constant during the time span τ, the number of times the reaction R_j occurs during the span τ is a Poisson-distributed random variable with expected value $a_j(X_t)\tau$ [35]. Now, if τ is small enough to guarantee approximately constant reaction propensities $a_j(X_t)$ and at the same time large enough so that $a_j(X_t)\tau \gg 1$, i.e. a significant number of reactions will occur during the time interval of length τ, then Gillespie's algorithm can be accelerated by using the larger time step τ and by sampling the number of reactions from Poissonian distributions with parameters $a_j(X_t)\tau$ directly. This procedure is called the ***τ-leap method*** [15]. Obviously, the main task is to set the right time step τ adaptively throughout the simulation. Small time steps will produce low numbers of reactions (even zero), and in this situation the exact Gillespie algorithm could be used instead. Too large time steps however, lead to a larger error compared to the exact solution and furthermore, can lead to negative molecule numbers. A detailed discussion of τ-selection procedures can be found in the literature [36]. Using Poisson random numbers, there is always a risk of producing negative entries in the state vector as Poisson variables range from zero to infinity. As an alternative, the ***binomial leap method*** has been proposed.

Instead of Poisson variables, appropriately parametrized binomial random variables are used, the main advantage being that the range of the variable can be bounded by the current copy number of each molecule [37]. Irrespective of the method used, the update rule can be written as

$$x_{t+\tau}^{(i)} = x_t^{(i)} + \sum_{j=1}^{M} v_{ji} \xi_j \left(a_j \left(X_t \right) \right) \tag{31.5}$$

where ξ_j represents either a Poissonian or a binomial random variable. Applications and systematic evaluations of these methods for the simulation of calcium microdomains are still scarce, a comparative study can be found in [37].

31.7 The Chemical Langevin and Fokker-Planck Equations

The approximation of the number of reactions that occur in a given time interval by samples from a defined probability density can be taken a step further. In the case of the chemical Langevin equation (CLE), the number of reactions R_j occurring in a small time interval of length dt is sampled from a normal distribution with mean and variance equal to $a_j(X_t)dt$. From a statistical point of view, this approach is justified as the Poisson random variable introduced in the last paragraph converges to a normal distribution in the case of a large mean value. When the expected number of reactions is too small, the symmetric normal distribution does not yield a good approximation of the corresponding right-skewed Poisson distribution and negative molecule counts can be obtained. This becomes clear when considering models with a single ion channel [34, 38]. When each state of the ion channel is modeled as a distinct chemical species, e.g. open and closed states, the count of each state is either 0 or 1. Chemical reactions including the ion channel can therefore not be modeled adequately by the CLE approach. For these cases, alternative schemes have been proposed in the literature [39, 40]. In our code, we set $X_t = 0$ whenever the integration yields $X_t < 0$, as proposed for the Cox-Ingersoll-Ross process in [41]. The most important differences of the CLE approach are (i) that the time interval dt is fixed, and (ii) that, due to the normally distributed random variable, the copy numbers are real numbers rather than integers. The updating rule is given by

$$x_{t+dt}^{(i)} = x_t^{(i)} + \sum_{j=1}^{M} v_{ji} a_j \left(X_t \right) dt + \sum_{j=1}^{M} v_{ji} \sqrt{a_j \left(X_t \right)}\, dW_t \tag{31.6}$$

where dW_t are the increments of M mutually independent standard Wiener processes [14]. Using Eq. 31.6 in simulations directly corresponds to the Euler-Maruyama integration scheme for stochastic differential equations [17]. The first sum in Eq. (31.6) contains the deterministic dynamics that will be discussed in the following paragraph while the second sum adds appropriately scaled stochastic fluctuations. Compared to the algorithms discussed so far, the CLE provides an extremely fast way to obtain an approximate solution of the CME. A basic result in the theory of stochastic processes allows a transformation of the Langevin equation to an associated Fokker-Planck equation (FPE) that describes the temporal evolution of the probability density $p(X_t)$ [14]. For discrete variables, the dynamics are captured completely by the CME, a family of first-order differential equations for the probabilities over all possible system states. For a Markov process, the CME is identical to the Chapman-Kolmogorov equation describing the dynamics of the transition probabilities $P(X_{t+n} = y|X_t = x)$. The CME's Taylor expansion in the spatial step size dx is called the Kramers-Moyal expansion. It generates a deterministic PDE from the integro-differential CME, both describing the dynamics of the system's underlying probability distributions. Using a second-order truncation of the infinite Kramers-Moyal expansion yields the Fokker-Planck equation [14]. Applications can focus on individual sample paths, in analogy to a single run of an experiment, or on distributions, summarizing the full stochastic structure of the system at hand. Ideally, all solutions would provide the complete distributions of all variables, but in practice, computation times often force the user to analyze a few representative sample paths, as produced by the CLE, for instance. Implementations of the CLE approach have been presented for calcium-dependent signaling pathways in neurons [42], IP_3-mediated calcium sensitive pathways in non-excitable cells [32, 43, 44] and Ca^{2+} release events in general [45]. An example of the FPE method to calcium dynamics in the dyadic cleft of cardiomyocytes is given in [46].

31.8 Deterministic Simulation

When stochastic effects in the dynamics of the modeled system are ignored, the classical deterministic reaction-rate equations are recovered. These correspond to the deterministic terms found in the chemical Langevin equation [35]. From a statistical physics point of view, the deterministic dynamics reflect the limit of an infinitely large simulation volume while keeping all reactant concentrations constant:

$$x_{t+dt}^{(i)} = x_t^{(i)} + \sum_{j=1}^{M} v_{ji} a_j (X_t)\, dt \tag{31.7}$$

Simulating the deterministic system using Eq. 31.7 directly represents a first-order Euler integration scheme. As stochastic effects are no longer present, a set of ordinary differential equations is obtained that can be conveniently integrated using standard schemes as implemented in most numerical software packages. If diffusion is included in the model, an additional diffusion term arises for each diffusible species. In deterministic simulations, diffusion would be modeled by a discrete Laplace operator applied to the spatial concentration profile of the diffusing species. In the stochastic setting, diffusion is modeled as a pair of additional chemical reactions, as detailed further below. The model is now described by set of partial differential equations. Deterministic methods represent the major part of the literature on calcium dynamics in excitable and non-excitable cells. In the context of microdomain calcium dynamics in cardiac and skeletal muscle cells, we list only a small selection of representative and landmark studies [47–51].

31.9 Hybrid Simulation

Complex subcellular systems often generate patterns across several temporal scales, especially when the participating reactions have rate constants that span several orders of magnitude. In these cases, it may happen that Gillespie's exact algorithm performs quite slowly due to the exact tracking of effectively irrelevant fluctuations while the system's dynamics may perform in an almost deterministic way. Moreover, models of subcellular calcium dynamics often contain plasma membrane or endoplasmic reticulum ion channels that modulate the local calcium concentration or are modulated by Ca^{2+} ions. As ion channel gating is almost exclusively described with stochastic methods, mainly with Markov chain models, the introduction of ion channels in deterministic models is not obvious. For these cases, hybrid models combining several of the aforementioned techniques have been developed. For instance, hybrid models using deterministic reaction dynamics for the subset of calcium diffusion, permeation and buffering reactions and stochastic models of ion channel gating have successfully been used for the modeling of skeletal and cardiac muscle cells [46, 52, 53] as well as for generic cell types with an IP_3 regulated calcium signaling system [54]. Other approaches switch adaptively between deterministic dynamics and the CLE-approximation [55], or between the Gillespie algorithm, the CLE-approximation and the τ-leap method [56].

31.10 Reaction Propensities for Calcium Microdomains

In the preceding sections, reaction propensities were introduced in an abstract way to calculate certain probability densities. Now, we will give some explicit expressions for reaction types essential to microdomain calcium signaling.

1. ***Chemical reactions***. $a_j(X_t) = c_j h_j(X_t)$, where c_j is the stochastic reaction rate and $h_j(X_t)$ denotes the number of possible molecular combinations available for reaction R_j at time t. The stochastic rates are calculated from the macroscopic reaction rate constants k_j and the system volume V, for monomolecular reactions we get $c_j = k_j$, for bimolecular reactions $c_j = \frac{k_j}{V}$. Calcium buffering reactions

consist of a pair of reactions, a second-order association reaction and a first-order dissociation reaction. Calcium pumps can be modeled in the same way.

2. ***Ion channel gating***. Given a gating scheme and the associated transition rates, transitions between channel substates can be treated as first-order chemical reactions where the different channel states are seen as different chemical species that are transformed into each other as specified by the gating scheme. Calcium-sensitive gating steps are modeled as a second-order association reaction between a calcium ion and the unbound channel state. The propensity terms are identical to the preceding case.
3. ***Ion channel currents***. If the mean channel current amplitude is known, the process of ion conduction and release from the channel pore can be simplified as a zero-order reaction, i.e. a Poisson process with a fixed rate [34]. Even though a constant release rate may only be a rough approximation of the real permeation process [57], the model yields the correct mean current and furthermore takes into account the quantal nature (release of a single ion at a time) and stochasticity inherent to ion channel currents. Conversion of the current amplitude I_{ch} to a stochastic rate constant is achieved by

$$c_j = \left| \frac{I_{ch}}{e \times z_{ion}} \right|$$

where e is the elementary charge and z_{ion} is the charge of the permeating ion. A calcium current of 1 pA, for instance, has an 'event rate' of $c_j \approx 3000/ms$. The factor $h_j(X_t)$ is equal to the number of *open* channels at time t, which is easily implemented as each channel state is regarded as a separate chemical species.
4. ***Diffusion***. Assuming a multivoxel simulation geometry, diffusion at the mesoscopic level can be implemented using the concept of stochastic event rates [58]. In analogy to the previously introduced techniques, the diffusion event rate of species S_i is calculated from the macroscopic diffusion constant D_i in order to yield the correct behavior in the deterministic limit. At the microscopic level, the diffusion event constant controls the speed of the molecular random walk. Assuming a voxel side length δx, the rate is given by

$$a_j = \frac{D_i}{(\delta x)^2} x_t^{(i)}$$

where $x_t^{(i)}$ is the copy number of species S_i at time t. In practice, diffusion rates of small molecules often lie several orders of magnitude above the common reaction rates. This can lead to a significant increase in computation time. However, even in single voxel simulations the implementation of diffusion may be necessary to avoid accumulation of e.g. calcium ions released from calcium channels [34]. In the single voxel case, this reduces to a fixed efflux rate of the ion.

31.11 The Colored Noise Interpretation

The chemical Langevin equation transforms the discrete picture of individual molecules, diffusing and reacting within the reaction volume, into a continuous stochastic process. While the copy number of any molecular species can only change by an integer value in the discrete models, the mesoscopic description provided by the chemical Langevin equation allows arbitrarily small changes in all variables. The mesoscopic CLE model represents an intermediate scale between discrete molecule counts and continuous chemical concentrations.

Conceptually, the CLE describes chemical reaction dynamics within the mathematical framework of stochastic differential equations, for which a large body of theoretical and numerical results exist [14]. Though Eq. (31.6) is a generic way to describe chemical reactions stochastically, the application to microdomain calcium signaling may not be straightforward. Moreover, we are interested in approximations that simplify the expression in Eq. (31.6), while still representing physical reality to a reasonable extent. The theory for calcium ions binding to a single or multiple buffers, with and without diffusion, has been developed in Ref. [10, 59]. In order to illustrate the procedure, we here present the elemental case of calcium ions interacting with a single buffer species. We consider the reversible bimolecular reaction of Ca^{2+} with a single buffer species B:

$$Ca^{2+} + B \rightleftharpoons CaB$$

In terms of the macroscopic, deterministic reaction rates k^+ and k^-, the stochastic association and dissociation rates for the reaction volume V are $c^+ = \frac{k^+}{V}$ and $c^- = k^-$, respectively.

Next, let x_t denote the copy number of free calcium ions, Ca^{2+}, y_t the copy number of free buffer molecules B, and z_t the number of calcium bound buffer molecules CaB. The total amount of calcium in the system is $x_T = x_t + z_t$, and the total number of buffer molecules is $y_T = y_t + z_t$.

As we model two reactions, the full chemical Langevin equation of the system contains two independent Wiener processes $dW_t^{(1)}$ and $dW_t^{(2)}$, that can be combined into a single process dW_t, as detailed in [10]. In order to simplify the CLE, we will apply the excess buffer approximation [60], which assumes that the fluctuations in the number of free calcium ions is small compared to the number of free buffer molecules. Therefore, the number of free buffer molecules y_t can be approximated by its equilibrium concentration, denoted y_{eq} [10]. Though the assumption may fail in case of a massive intracellular calcium increase, it is appropriate for resting conditions. The simplified CLE then reads:

$$dx_t = -\left(c^+ y_{eq} + c^-\right) x_t dt + c^- x_T + \sqrt{c^+ x_t y_{eq} + c^- z_t}\, dW_t$$

Defining the constants $C_1 = c^+ y_{eq} + c^-$ and $C_2 = c^- x_T$, we obtain

$$dx_t = -C_1 \left(x_t - \frac{C_2}{C_1} \right) dt + \sqrt{c^+ x_t y_{eq} + c^- z_t} dW_t$$

The equilibrium calcium concentration is given by $[Ca^{2+}]_{\mathrm{eq}} V = x_{eq} = \frac{C_2}{C_1}$, and defining the relaxation time $\tau = \frac{1}{C_1}$ and the time-dependent variance $\sigma_t = \sqrt{c^+ x_t y_{eq} + c^- z_t}$, the expression is

$$dx_t = -\frac{1}{\tau} \left(x_t - x_{eq} \right) dt + \sigma_t dW_t$$

A further approximation can be introduced by considering a stationary variance term.

$\sigma_t = \sqrt{c^+ x_{eq} y_{eq} + c^- z_{eq}}$, using the equilibrium calcium concentration x_{eq} rather than the time-dependent exact calcium concentration x_t. The resulting expression

$$dx_t = -\frac{1}{\tau} \left(x_t - x_{eq} \right) dt + \sigma dW_t$$

is the so-called Ornstein-Uhlenbeck process, a generic stochastic process with well known analytical properties [14]. The Ornstein-Uhlenbeck process is a Markovian process often used to represent and to generate so-called colored noise samples. To summarize linear noise properties, the time autocorrelation function, or equivalently, the power spectral density of a noisy time series can be used. Formally, the time series has to be time-stationary for these methods to work. For an oscillatory system, for instance, an oscillation frequency of $f = 1/T$ leads to a spectral peak at frequency f in the power spectrum, and to a dominant time lag of T in the autocorrelation function. Temporally uncorrelated noise yields a flat power spectrum, hence the name 'white' noise, in correspondence to the physical spectrum of white light. In contrast, a non-flat electromagnetic spectrum yields colored light, therefore the name colored noise for stochastic processes with non-flat spectrum. The Ornstein-Uhlenbeck process is the minimal representation of an exponentially autocorrelated, Gaussian distributed noise. Thus, realistic calcium noise samples with realistic first and second-order statistical properties, i.e. identical mean, variance and auto-correlation function, can be generated very quickly using a simple Ornstein-Uhlenbeck process. Moreover, the wealth of exact stochastic properties derived for the Ornstein-Uhlenbeck process can be transferred to the environment of microdomain calcium signaling. Using the calculations presented in this paragraph, direct algebraic relations between physico-chemical constants and stochastic properties like the relaxation time and the time-dependent variance can be derived in many cases, including multiple buffers and calcium diffusion [10, 59].

31.12 The Impact of Microdomain Calcium Noise on IP_3R Gating

The relevance of noise in non-linear systems in nature has received increasing attention over the last years [38, 61–63]. Considering the non-linear dynamics of many downstream Ca^{2+} effectors, one could hypothesize about a physiological relevant information content of Ca^{2+} fluctuations [64]. A suiting model system for the investigation of such an effect is the endoplasmic, homotetrameric inositol 1,4,5-trisphosphate receptor (IP_3R), a fundamental part of the intracellular Ca^{2+} signaling apparatus [65]. It is crucially involved in the generation and regulation of complex spatiotemporal Ca^{2+} signals [66], which have been observed in a variety of different cell types [67]. Each of the four channel subunits has a binding site for the second messenger IP_3 and two Ca^{2+} binding sites that give rise to the typical bell-shaped Ca^{2+} response curve of IP_3Rs [18]. For low Ca^{2+} concentrations a high-affinity Ca^{2+} binding site induces a positive feedback known as calcium-induced calcium release, while a low-affinity Ca^{2+} binding site is responsible for the negative feedback mechanism which eventually induces the inactivation of the IP_3 receptor at higher $[Ca^{2+}]$ [19]. The DeYoung-Keizer model is a well-established mathematical representation of the IP_3R gating dynamics [68] and was therefore used in the here presented study. The investigation of Ca^{2+} noise induced effects to more recent mathematical models of the IP_3R [69, 70] lies beyond the scope of the here presented example but will be subject of future studies. As discussed above, Ca^{2+} noise is shaped by diffusion and the presence of Ca^{2+} binding molecules, such as endogenous Ca^{2+} buffer proteins [59]. For a stochastically exact representation of those effects, one has to model the system at least at the mesoscopic level [31]. Considering the timescales of channel gating events (up to seconds), Gillespie's algorithm constitutes the only feasible approach to capture the exact noise phenomena of interest. A minimal system capable of exhibiting noise induced effects on the non-linear gating dynamics of the IP_3R should contain at least a single IP_3R in the DeYoung-Keizer representation, a constant IP_3 concentration and Ca^{2+} diffusion. The variation of Ca^{2+} diffusion coefficients allows us to tune the autocorrelation of Ca^{2+} noise in this model. The IP_3R can be functionally characterized by its open probability, its mean open time and its mean channel close time. A C++ implementation of Gillespie's algorithm, developed for the simulation of the here introduced model, is available on github (https://github.com/nwieder/SSTS).

In order to investigate the potential effects of Ca^{2+} noise on downstream signaling pathways, while maintaining computational feasibility, it is necessary to reduce a biological model system to its core elements. Hence, following an example detailed in [38], we assume an in-silico experiment where we first model a single non-conducting IP_3R in the absence of diffusion and buffer dynamics with a constant number of Ca^{2+} ions. In a second step, we introduce Ca^{2+} diffusion, hence Ca^{2+} fluctuations to the system. For each of these conditions we generate trajectories for different mean Ca^{2+} concentrations, essentially generating the characteristic bell-shaped open probability curve of IP_3Rs for each of the different conditions as a read out. Interestingly, we find that the modulations of noise autocorrelation time by diffusion, indeed influences the shape of the IP_3R response curves in a [Ca^{2+}] dependent manner. Figure 31.1 summarizes IP_3R gating properties for different theoretical Ca^{2+} diffusion constants. The dashed line (empty squares) represents a model in the absence of Ca^{2+} diffusion and thus the absence of Ca^{2+} fluctuations. This condition mimics a deterministic setting with a constant Ca^{2+} concentration. With the introduction of Ca^{2+} diffusion (solid lines), noise is added to the system. Considering a physiological Ca^{2+} diffusion coefficient $D_{Ca} = 200\ \mu m^2 s^{-1}$ (solid triangles), for [Ca^{2+}] > 10 μM the presence of Ca^{2+} fluctuations does induce a significant increase in open probability, mean open times and a decrease in mean closed times. For increasing diffusion constants, the noise autocorrelation time decreases and the IP_3R open probability for [Ca^{2+}] > 1 μM increases as a function of the diffusion constant and hence the autocorrelation time τ. In summary, this example demonstrates how the introduction of Ca^{2+} fluctuations alters IP_3R gating characteristics, which in turn has the potential to affect the properties of more complex spatiotemporal Ca^{2+} signals.

We are well aware of the fact that this example is based on an oversimplified model, accounting only for a fraction of known biologically relevant interactions between Ca^{2+} signaling molecules and endoplasmic Ca^{2+} release channels. Most importantly, the IP_3R was modeled as a non-conducting ion channel. This has two reasons: (a) Experimental quantification of Ca^{2+} dependent IP_3R gating dynamics has been conducted in a setting where Ca^{2+} was replaced by K^+ as primary charge carrier to extract information about the Ca^{2+} dependence of the IP_3R [19]. This experimental dataset was used to verify the here presented modeling approach. (b) The introduction of Ca^{2+} to the modeled system changes the mean number of Ca^{2+} ions in the system and hence alters the amplitude of Ca^{2+} fluctuations. Both these quantities need to be kept constant to isolate the effect of colored Ca^{2+} noise on the IP_3R. Hence the simplification is necessary to carefully extract a causal link between Ca^{2+} fluctuations (here induced by diffusion or Ca^{2+} buffers) and the non-linear dynamics of the down-stream effector (here the IP_3R). In [38], it has been shown that the presence of noise, as well its stochastic characteristics, has the potential to affect the dynamics of key components of biological signaling networks. Based on those observations, we argue that the relevance of stochastic fluctuations will most likely scale up to more complex systems found in living systems.

Over the last decades, Ca^{2+} microdomains emerged as the most fundamental entity of Ca^{2+} signaling [2]. In neurons and muscle cells, structurally distinct microdomains, such as dendritic spines or the dyadic cleft of cardiac myocytes, are recognized for their functional relevance in learning, memory, and heart- beat generation [71–73]. Even though the importance of stochastic Ca^{2+} fluctuations in such microdomains received increasing attention over the last years [66], the question to what extent such fluctuations influence global Ca^{2+} signals in general remains yet unanswered. The difficulties in painting a full picture is largely due to the immense complexity of Ca^{2+} signaling networks as well as the limited experimental approaches at hand to investigate noise induced effects. Nevertheless,

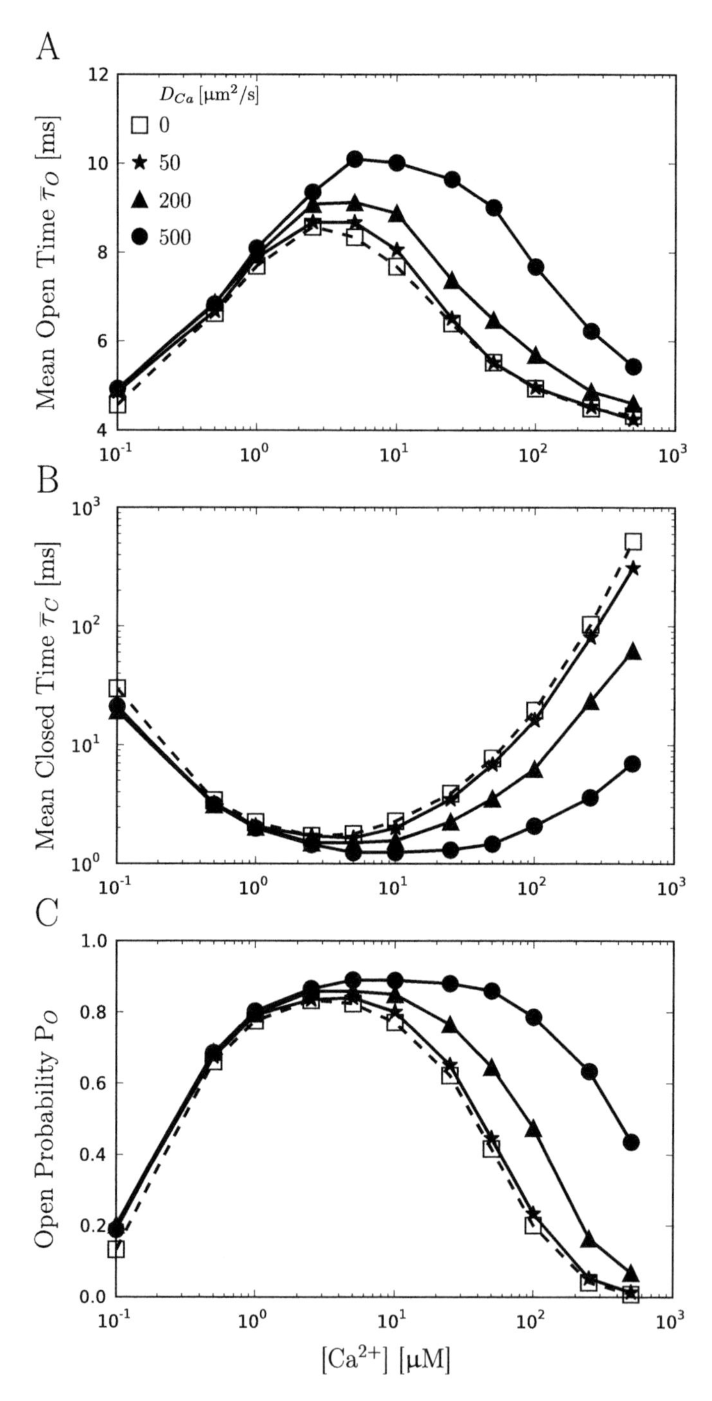

Fig. 31.1 Summary of the characteristic, bell shaped gating properties of the IP_3R for increasing diffusion coefficients D_{Ca}. Empty squares (dashed curve) show results for the zero noise model and represent expected results from deterministic algorithms. Solid curves show results for increasing diffusion coefficients $D_{Ca} = 50\ \mu m^2\ s^{-1}$ (stars), $200\mu m^2\ s^{-1}$ (triangles) and $500\mu m^2\ s^{-1}$ (circles). For low $[Ca^{2+}]$, Ca^{2+} noise has no obvious influence on the IP_3R. Starting at $[Ca^{2+}] = 1\ \mu M$, (**a**) mean open times τ_O increase with increasing D_{Ca}, whereas (**b**) mean close times τ_c decrease. (**c**) consequently, the channel open probability P_O increases with increasing D_{Ca}. (Reprint from [38])

this example emphasizes the importance of thermodynamic noise and its potential to scale up to the magnitude of intracellular signaling networks. We therefore have to keep in mind that, whenever low molecular copy numbers of key molecules meet nonlinear signaling effectors, deterministic assumptions might be limited and molecular noise has to be considered.

31.13 Working Examples

Example 1: We consider a simple system that contains Ca^{2+} ions and two buffers, called $B1$ and $B2$. The model geometry is given by a cube with a side length of 1 μm, i.e. the system volume is 1 *fl*. The copy numbers of the five molecular species, in the order $[Ca^{2+}, B_1, B_2, CaB_1, CaB_2]$ are stored in the state vector $X_t = [x_t^{(1)}, \ldots, x_t^{(5)}]$. We assume an equilibrium concentration of $[Ca^{2+}]_{eq} = 0.1\ \mu M$ (free Ca^{2+}), and total buffer concentrations of $[B_1]_T = [B_2]_T = 100\ \mu M$. To convert deterministic reaction rate constants into stochastic rate constants, we use the reaction propensities defined above (Sect. 31.10). Starting with the simulation box length l given in nm, we have $V = l^3 10^{-24}$ liters. Using Avogadro's number N_A, we get $c_{on} = k_{on}/(V^*N_A{}^*10^{-6})$ and $c_{off} = k_{off}$. The rate constants are set here to $k_1^+ = 0.1\ \mu M^{-1} ms^{-1}$, $k_1^- = 0.1 ms^{-1}$ for the reaction of Ca^{2+} and the first buffer B_1, and to $k_2^+ = 0.01\ \mu M^{-1} ms^{-1}$, $k_2^- = 0.01 ms^{-1}$ for the reaction of Ca^{2+} with the second buffer B_2. Thus, the four reactions read as follows.

Buffer 1:

$$Ca^{2+} + B_1 \rightarrow CaB_1$$

$$CaB_1 \rightarrow Ca^{2+} + B_1$$

Buffer 2:

$$Ca^{2+} + B_2 \rightarrow CaB_2$$

$$CaB_2 \rightarrow Ca^{2+} + B_2$$

The corresponding state change matrix $V = (v_{ji})$ contains the state change vector V_j at row index j, from which the stoichiometric coefficients can be read off directly:

$$V = \begin{pmatrix} -1 & -1 & 0 & 1 & 0 \\ 1 & 1 & 0 & -1 & 0 \\ -1 & 0 & -1 & 0 & 1 \\ 1 & 0 & 1 & 0 & -1 \end{pmatrix}$$

If we want to start the system in equilibrium, we obtain the initial state vector (after rounding) $X_0 = [60, 54755, 54755, 5476, 5476]$. The stochastic rate constants c_j are calculated as explained in the preceding paragraph and yield $c_1^+ = 1.66 * 10^{-4} ms^{-1}$, $c_1^- = 0.1 ms^{-1}$, $c_2^+ = 1.66 * 10^{-5} ms^{-1}$ and $c_2^- = 0.01 ms^{-1}$. For convenience, we adapted the notation using superscripts in order to better recognize the relation with macroscopic rate constants, i.e. c_1^+ is the stochastic rate constant associated with the macroscopic rate k_1^+. The time and state dependent propensities $a_j(X_t)$ are given by:

$$a_1(X_t) = c_1^+ x_t^{(1)} x_t^{(2)}$$

$$a_2(X_t) = c_1^- x_t^{(4)}$$

$$a_3(X_t) = c_2^+ x_t^{(1)} x_t^{(3)}$$

$$a_4(X_t) = c_2^- x_t^{(5)}$$

Finally, using the cumulative propensity $a_0(X_t) = a_1(X_t) + \cdots + a_4(X_t)$ and the initial state vector X_0, all parameters to run the Gillespie algorithm as defined above are given. Using the state change matrix and the propensities defined above, we obtain the corresponding chemical Langevin Eq. (31.6):

$$\begin{aligned} dx_t^{(1)} = & -a_1(X_t)\,dt - \sqrt{a_1(X_t)}\mathrm{d}W_t^{(1)} + a_2(X_t)\,dt + \sqrt{a_2(X_t)}dW_t^{(2)} \\ & -a_3(X_t)\,dt - \sqrt{a_3(X_t)}dW_t^{(3)} + a_4(X_t)\,dt + \sqrt{a_4(X_t)}dW_t^{(4)} \end{aligned}$$

$$dx_t^{(2)} = -a_1(X_t)\,dt - \sqrt{a_1(X_t)}dW_t^{(1)} + a_2(X_t)\,dt + \sqrt{a_2(X_t)}dW_t^{(2)}$$

$$dx_t^{(3)} = -a_3(X_t)\,dt - \sqrt{a_3(X_t)}dW_t^{(3)} + a_4(X_t)\,dt + \sqrt{a_4(X_t)}dW_t^{(4)}$$

$$dx_t^{(4)} = a_1(X_t)\,dt + \sqrt{a_1(X_t)}dW_t^{(1)} - a_2(X_t)\,dt - \sqrt{a_2(X_t)}dW_t^{(2)}$$

$$dx_t^{(5)} = a_3(X_t)\,dt + \sqrt{a_3(X_t)}dW_t^{(3)} - a_4(X_t)\,dt - \sqrt{a_4(X_t)}dW_t^{(4)}$$

Here, $dW_t^{(j)}$ represents the increments of four independent Brownian motions, i.e. each $dW_t^{(j)}$ is an independent and identically distributed Gaussian random process with mean zero and variance *dt*. A first-order, explicit Euler integration scheme is often satisfactory and allows a straightforward implementation of the above formulas in virtually any programming language. Sample code (Octave, Python) can be obtained from the authors.

Example 2: We consider a 1 *fl* single voxel system of diffusible Ca^{2+} ions and a single, immobile buffer B_1. The simulation volume is surrounded by a 'constant pool', i.e. an infinite volume with constant (equilibrium) calcium concentration. The state vector $X_t = [x_t^{(1)}, x_t^{(2)}, x_t^{(3)}]$ represents the copy numbers of $[Ca^{2+}, B_1, CaB_1]$. We again set $[Ca^{2+}]_{eq} = 0.1\ \mu M$, $[B_1]_T = 100\ \mu M$ and $k_1^+ = 0.1 \mu M^{-1} ms^{-1}$, $k_1^- = 0.1 ms^{-1}$ and set the diffusion coefficient of free calcium ions to $D_{Ca} = 200\ \mu m^2 s^{-1}$. Including the diffusion reactions of Ca^{2+} ions out of the simulation volume and influx of Ca^{2+} ions from the constant pool into the simulation volume, we get the following set of reactions.

Buffer 1:

$$Ca^{2+} + B_1 \rightarrow CaB_1$$

$$CaB_1 \rightarrow Ca^{2+} + B_1$$

Diffusion:

$$Ca^{2+} \rightarrow \varnothing$$

$$\varnothing \rightarrow Ca^{2+}$$

The state change matrix V now reads:

$$V = \begin{pmatrix} -1 & -1 & 1 \\ 1 & 1 & -1 \\ -1 & 0 & 0 \\ 1 & 0 & 0 \end{pmatrix}$$

Under equilibrium conditions, we obtain the initial state vector $X_0 = [60, 54755, 5476]$. The stochastic rate constants c_1^+ and c_1^- have the same magnitude as in the preceding example, and the stochastic diffusion rate c_D, calculated as explained in the preceding paragraph, evaluates to $c_D = 0.2\ ms^{-1}$. The diffusion propensities a_j are

$$a_3(X_t) = c_D x_t^{(1)}$$

$$a_4(X_t) = c_D x_0^{(1)}$$

Note that the rate $a_4(X_t)$ is constant as the surrounding calcium concentration is assumed to be fixed. The according chemical Langevin equation is

$$dx_t^{(1)} = -a_1(X_t)\,dt - \sqrt{a_1(X_t)}dW_t^{(1)} + a_2(X_t)\,dt + \sqrt{a_2(X_t)}dW_t^{(2)} \\ - a_3(X_t)\,dt - \sqrt{a_3(X_t)}dW_t^{(3)} + a_4(X_t)\,dt + \sqrt{a_4(X_t)}dW_t^{(4)}$$

$$dx_t^{(2)} = -a_1(X_t)\,dt - \sqrt{a_1(X_t)}dW_t^{(1)} + a_2(X_t)\,dt + \sqrt{a_2(X_t)}dW_t^{(2)}$$

$$dx_t^{(3)} = a_1(X_t)\,dt + \sqrt{a_1(X_t)}dW_t^{(1)} - a_2(X_t)\,dt - \sqrt{a_2(X_t)}dW_t^{(2)}$$

Example 3: In the last example, we extend the system considered in Example 2 by a single ion channel that has a closed (C) and an open open substate (O) and conducts Ca^{2+} ions. It is convenient to model the two sub-states as different molecular species that can be transformed into each other. Obviously, the channel substate species are not diffusible. The state vector $X_t = [x_t^{(1)}, \ldots, x_t^{(5)}]$ represents the copy numbers of $[Ca^{2+}, B_1, CaB_1, C, O]$. As we consider a single ion channel, both $x_t^{(4)}$ and $x_t^{(5)}$ must be equal to either zero or one and $x_t^{(4)} + x_t^{(5)} = 1$ must be fulfilled at all times t. Introducing an extra reaction that models Ca^{2+} ion release from the O state (reaction 5), we extend the set of reactions introduced in the Example 2 by three ion channel related reactions.

Channel gating:

$$C \to O$$

$$O \to C$$

Calcium permeation:

$$O \to O + Ca^{2+}$$

The state change matrix V now reads:

$$\begin{pmatrix} -1 & -1 & 1 & 0 & 0 \\ 1 & 1 & -1 & 0 & 0 \\ -1 & 0 & 0 & 0 & 0 \\ 1 & 0 & 0 & 0 & 0 \\ 0 & 0 & 0 & -1 & 1 \\ 0 & 0 & 0 & 1 & -1 \\ 1 & 0 & 0 & 0 & 0 \end{pmatrix}$$

Reaction propensities for the new reactions are derived from the assumed channel kinetics and the channel permeability, respectively. As channel substate transitions are first-order reactions, we write $a_5(X_t) = c_{ch}^{+} x_t^{(4)}$ and $a_6(X_t) = c_{ch}^{-} x_t^{(5)}$ where $c_{ch}^{+} = k_{ch}^{+}$ and $c_{ch}^{-} = k_{ch}^{-}$ and k_{ch}^{+}, k_{ch}^{-} are the macroscopic rate constants for channel opening and closing as determined from electrophysiological measurements for instance. Finally, Ca^{2+} ion release is modeled by $a_7(X_t) = c_P x_t^{(5)}$ with $c_P = I_{ch}/(2e)$.

In this case, it is important to note that a chemical Langevin equation approach cannot be applied because the system contains only a single ion channel and due to the use of normally distributed random numbers, negative values for $x_t^{(4)}$ or $x_t^{(5)}$ are highly probable. Negative copy numbers in turn lead to negative reaction propensities and render the square root terms in the CLE undefined.

31.14 Comparision of SSA and CLE Solutions

We ran $N = 20$ simulations of $T = 100$ *ms* length each, using an equilibrium calcium concentration of 0.1 $\mu mol/l$, a total buffer concentration of 10 $\mu mol/l$ and reaction rate constants of $k^+ = 0.5$ $\mu mol/ms$ and $k^- = 0.25$ */ms*. To study the dependency of simulation accuracy and computation times on the simulation volume, we considered cubes of side length $L = 100$, 250, 500, 1000, 2500 and 5000 *nm*, corresponding to volumes of $V = 0.001$, 0.016, 0.125, 1.0, 15.625 and 125 *fl*, respectively. The CLE was numerically solved using a fixed integration time step of $dt = 0.01$ *ms* and an Euler-Maruyama integration scheme. As the basis of the SSA are randomly sampled time intervals between individual reactions, the generated time series of molecular counts are irregularly spaced. Before further processing, SSA-generated time series are linearly interpolated to the same sampling interval as the CLE solution, i.e. $dt = 0.01$ *ms*.

Figure 31.2 shows four exemplary samples. The first two panels (A, B) show simulations of a small reaction volume ($L = 250$ *nm*, $V = 1/64$ *fl*). The upper panel

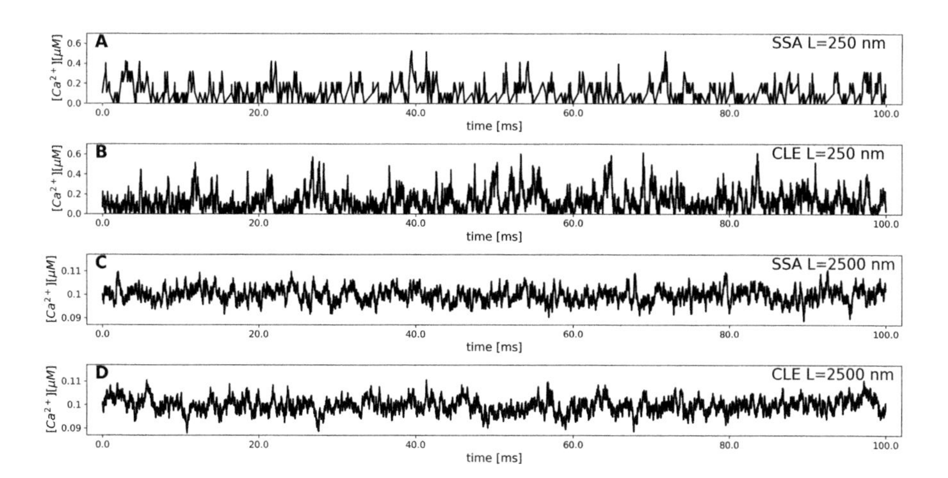

Fig. 31.2 Comparison between simulated 100 ms trajectories generated with either the SSA (A,C) or the CLE (B,D) approach for a small volume V = 1/64 fl and a large volume V = 15.625 fl

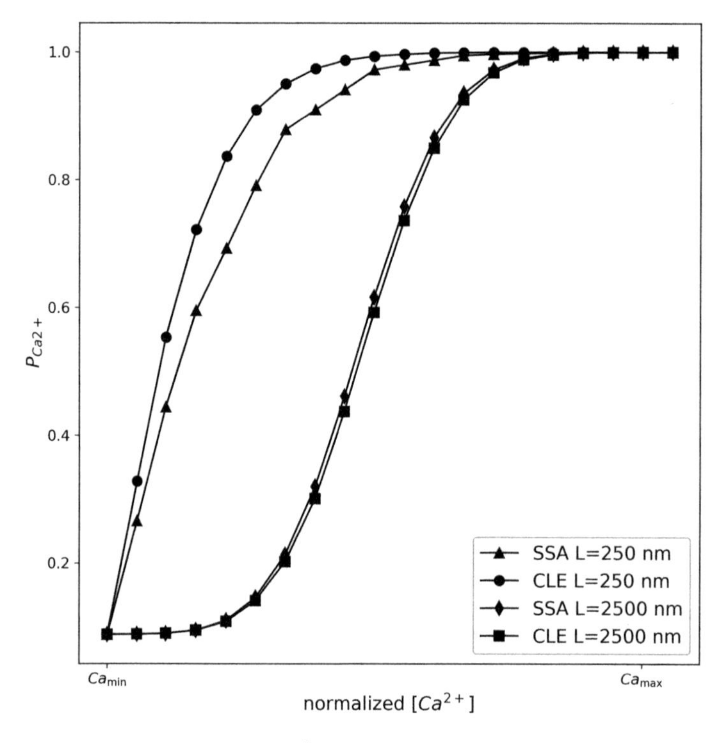

Fig. 31.3 Comparison of cumulative Ca^{2+} distributions from N = 20 simulation runs of the models presented in Fig. 31.2

shows the SSA solution, that is recognized by the discrete steps in the Ca^{2+} ion count. The corresponding CLE solution looks more continuous due to the small increments generated by the Brownian motion. We also observe that the Ca^{2+} concentration in both cases hits zero, indicating that the formal requirements of the CLE are not met for this small volume. Panels C (SSA) and D (CLE) show the case of a larger simulation volume ($L = 2500$ *nm*, $V = 15.625$ *fl*). Both solutions, SSA and CLE, are visually very similar and far from zero counts.

The accuracy of the two algorithms for the same volumes as shown in Fig. 31.2 is further analyzed in Fig. 31.3. To compare the algorithms, we plot the cumulative probability distribution of the calcium concentration (P_{Ca}), pooling the results of $N = 20$ simulations. To show the results of both simulation volumes in the same plot, we use a normalized calcium concentration on the abscissa. The values increase linearly from the minimum to the maximum calcium concentration obtained in each case. This visualization step is necessary as the width (or variance) of the calcium distribution for the larger volume is much smaller as that of the small reaction volume. We observe that the calcium distribution in the SSA and CLE solutions for the small volume (triangles and circles, respectively) differ to a larger degree than those in the large volume case (diamonds and squares, respectively).

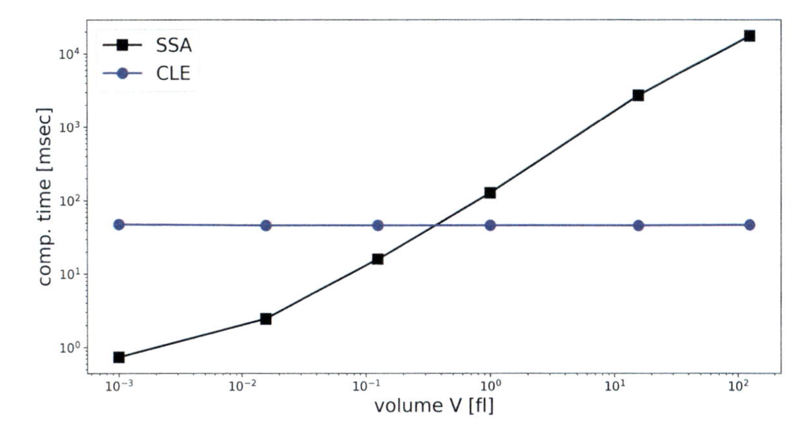

Fig. 31.4 Comparison of computation times between SSA (blue) and CLE (black) for increasing simulation volumes V

Taking the SSA solution as a reference, the difference of the CLE derived calcium distribution quantifies the accuracy of the CLE for a given volume. Even though a Kolmogorov-Smirnov test distinguishes the two solutions for both volumes, the better approximation for the larger simulation volume is clearly observed.

Finally, we address the computation times associated with both algorithms, SSA and CLE. Figure 31.4 shows the mean computation times of $N = 20$ simulations, as a function of the simulation volume, using the same parameters as given above. For the SSA algorithm, we expect the computation time to be approximately proportional to the number of molecules present in the simulation volume. For fixed concentrations, the computation time should therefore increase with the first power of the volume V, or with the third power of the simulation box length L, i.e. $T_c \sim L^3$. This increase is shown in log-log coordinates in Fig. 31.4 (black line and squares). A least-squares fit of L vs. T_c gives an empirical exponent of 2.7, comparable to the expected value of 3. The computation time of the CLE algorithm does not vary with the simulation volume, as expected for the Euler-Maruyama integration scheme with a fixed time interval dt (blue line and circles).

31.15 Outlook

Stochastic methods have received an increasing amount of attention in the modeling of subcellular signaling systems. Given the central role of calcium ions in the regulation of many cellular functions and the very low number of calcium ions present in the relevant volumes, e.g. a calcium concentration of 100 nM translates to some 60 calcium ions in a 1 μm^3 volume, surprisingly few studies model the stochastic effects of calcium diffusion and buffering in small volumes [34, 46]. Here, it is important to note that possibly important effects such as bistability of a signaling pathway can even be missed completely when only deterministic dynamics are modeled [74]. The rising interest in simulating large systems with several interacting subsystems in some of which deterministic dynamics may dominate while in others stochastic effects may be important will lead to novel approaches combining deterministic and stochastic simulations. Most importantly, the validity of different approaches on a given scale will have to be evaluated for numerous experimental systems. At the same time, stochastic models of subcellular reactions will help to estimate the parameters of experimentally recorded signals, in which stochastic effects are uncovered thanks to technical advances such as high-resolution laser microscopy.

References

1. Clapham DE (2007) Calcium Signaling. Cell 131:1047–1058
2. Berridge MJ (2006) Calcium microdomains: organization and function. Cell Calcium 40:405–412. https://doi.org/10.1016/j.ceca.2006.09.002
3. Rizzuto R, Pozzan T (2006) Microdomains of intracellular ca 2+ : molecular determinants and functional consequences. Physiol Rev 86:369–408. https://doi.org/10.1152/physrev.00004.2005
4. Martín F, Soria B (1996) Glucose-induced [Ca^{2+}]i oscillations in single human pancreatic islets. Cell Calcium 20:409–414. https://doi.org/10.1016/S0143-4160(96)90003-2
5. Jaffe LF (1999) Organization of early development by calcium patterns. BioEssays 21:657–667
6. Giorgi C, Missiroli S, Patergnani S et al (2015) Mitochondria-associated membranes: composition, molecular mechanisms, and physiopathological implications. Antioxid Redox Signal 22(12):995–1019. https://doi.org/10.1089/ars.2014.6223
7. Rieusset J, Fauconnier J, Paillard M et al (2016) Disruption of calcium transfer from ER to mitochondria links alterations of mitochondria-associated ER membrane integrity to hepatic insulin resistance. Diabetologia 59:614–623. https://doi.org/10.1007/s00125-015-3829-8
8. Orrenius S, Zhivotovsky B, Nicotera P (2003) Regulation of cell death: the calcium-apoptosis link. Nat Rev Mol Cell Biol 4:552–565
9. Myoga MH, Regehr WG (2011) Calcium microdomains near R-type calcium channels control the induction of presynaptic long-term potentiation at parallel Fiber to Purkinje cell synapses. J Neurosci 31:5235–5243. https://doi.org/10.1523/JNEUROSCI.5252-10.2011
10. von Wegner F, Wieder N, Fink RHA (2014) Microdomain calcium fluctuations as a colored noise process. Front Genet 5:376. https://doi.org/10.3389/fgene.2014.00376
11. Keller DX, Franks KM, Bartol TM, Sejnowski TJ (2008) Calmodulin activation by calcium transients in the postsynaptic density of dendritic spines. PLoS One 3:e2045. https://doi.org/10.1371/journal.pone.0002045
12. Zeng S, Holmes WR (2010) The effect of noise on CaMKII activation in a dendritic spine during LTP induction. J Neurophysiol 103:1798–1808. https://doi.org/10.1152/jn.91235.2008
13. Berridge MJ (2009) Inositol trisphosphate and calcium signalling mechanisms. Biochim Biophys Acta, Mol Cell Res 1793:933–940
14. Gardiner CW (1996) Handbook of stochastic methods: for physics, chemistry and the natural sciences, Springer series in synergetics. Springer, Berlin
15. Gillespie DT (2007) Stochastic simulation of chemical kinetics. Annu Rev Phys Chem 58:35–55. https://doi.org/10.1146/annurev.physchem.58.032806.104637
16. Higham DJ (2008) Modeling and simulating chemical reactions. SIAM Rev 50:347–368. https://doi.org/10.1137/060666457

17. Higham DJ, Higham DJ (2001) An algorithmic introduction to numerical simulation of stochastic differential equations. SIAM Rev 43:525–546. https://doi.org/10.1137/S0036144500378302
18. Bezprozvanny I, Watras J, Ehrlich BE (1991) Bell-shaped calcium-response curves of lns(l,4,5)P3- and calcium-gated channels from endoplasmic reticulum of cerebellum. Nature 351:751–754. https://doi.org/10.1038/351751a0
19. Mak D-OD, McBride S, Foskett JK (1998) Inositol 1,4,5-tris-phosphate activation of inositol tris-phosphate receptor Ca^{2+} channel by ligand tuning of Ca^{2+} inhibition. Proc Natl Acad Sci 95:15821–15825. https://doi.org/10.1073/pnas.95.26.15821
20. Andrews SS, Bray D (2004) Stochastic simulation of chemical reactions with spatial resolution and single molecule detail. Phys Biol 1:137–151. https://doi.org/10.1088/1478-3967/1/3/001
21. Franks KM, Bartol TM, Sejnowski TJ (2002) A Monte Carlo model reveals independent signaling at central glutamatergic synapses. Biophys J 83:2333–2348. https://doi.org/10.1016/S0006-3495(02)75248-X
22. Shahrezaei V, Delaney KR (2004) Consequences of molecular-level Ca^{2+} channel and synaptic vesicle colocalization for the Ca^{2+} microdomain and neurotransmitter exocytosis: a Monte Carlo study. Biophys J 87:2352–2364. https://doi.org/10.1529/biophysj.104.043380
23. Tanskanen AJ, Greenstein JL, O'Rourke B, Winslow RL (2005) The role of stochastic and modal gating of cardiac L-type Ca^{2+} channels on early after-depolarizations. Biophys J 88:85–95. https://doi.org/10.1529/biophysj.104.051508
24. Hake J, Lines GT (2008) Stochastic binding of Ca^{2+} ions in the dyadic cleft; continuous versus random walk description of diffusion. Biophys J 94:4184–4201. https://doi.org/10.1529/biophysj.106.103523
25. Flegg MB, Rüdiger S, Erban R (2013) Diffusive spatio-temporal noise in a first-passage time model for intracellular calcium release. J Chem Phys 138:154103. https://doi.org/10.1063/1.4796417
26. Dobramysl U, Rüdiger S, Erban R (2016) Particle-based multiscale modeling of calcium puff dynamics. Multiscale Model Simul 14:997–1016. https://doi.org/10.1137/15M1015030
27. Nguyen V, Mathias R, Smith GD (2005) A stochastic automata network descriptor for Markov chain models of instantaneously coupled intracellular Ca^{2+}channels. Bull Math Biol 67:393–432. https://doi.org/10.1016/j.bulm.2004.08.010
28. Munsky B, Khammash M (2006) The finite state projection algorithm for the solution of the chemical master equation. J Chem Phys 124:044104. https://doi.org/10.1063/1.2145882
29. Gillespie DT (1977) Exact stochastic simulation of coupled chemical reactions. J Phys Chem 81:2340–2361. https://doi.org/10.1021/j100540a008
30. Gibson MA, Bruck J (2000) Efficient exact stochastic simulation of chemical systems with many species and many channels. J Phys Chem A 104:1876–1889. https://doi.org/10.1021/jp993732q
31. Weinberg SH (2016) Microdomain [ca 2+] fluctuations Alter temporal dynamics in models of ca 2+ −dependent signaling cascades and synaptic vesicle release. Neural Comput 28:493–524. https://doi.org/10.1162/NECO_a_00811
32. Li H, Hou Z, Xin H (2005) Internal noise stochastic resonance for intracellular calcium oscillations in a cell system. Phys Rev E Stat Nonlin Soft Matter Phys 71:061916. https://doi.org/10.1103/PhysRevE.71.061916
33. Kummer U, Krajnc B, Pahle J et al (2005) Transition from stochastic to deterministic behavior in calcium oscillations. Biophys J 89:1603–1611. https://doi.org/10.1529/biophysj.104.057216
34. Von Wegner F, Fink RHA (2010) Stochastic simulation of calcium microdomains in the vicinity of an L-type calcium channel. Eur Biophys J 39:1079–1088
35. Gillespie DT (2000) Chemical Langevin equation. J Chem Phys 113:297–306. https://doi.org/10.1063/1.481811
36. Cao Y, Gillespie DT, Petzold LR (2006) Efficient step size selection for the tau-leaping simulation method. J Chem Phys 124:044109. https://doi.org/10.1063/1.2159468

37. Tian T, Burrage K (2004) Binomial leap methods for simulating stochastic chemical kinetics. J Chem Phys 121:10356–10364. https://doi.org/10.1063/1.1810475
38. Wieder N, Fink R, Von Wegner F (2015) Exact stochastic simulation of a calcium microdomain reveals the impact of Ca^{2+} fluctuations on IP3R gating. Biophys J 108:557–567. https://doi.org/10.1016/j.bpj.2014.11.3458
39. Goldwyn JH, Imennov NS, Famulare M, Shea-Brown E (2011) Stochastic differential equation models for ion channel noise in Hodgkin-Huxley neurons. Phys Rev E Stat Nonlin Soft Matter Phys 83:041908. https://doi.org/10.1103/PhysRevE.83.041908
40. Dangerfield CE, Kay D, Burrage K (2012) Modeling ion channel dynamics through reflected stochastic differential equations. Phys Rev E Stat Nonlin Soft Matter Phys 85:051907. https://doi.org/10.1103/PhysRevE.85.051907
41. Alfonsi A (2005) On the discretization schemes for the CIR (and Bessel squared) processes. Monte Carlo Methods Appl 11:355–384. https://doi.org/10.1163/156939605777438569
42. Manninen T, Linne M-L, Ruohonen K (2006) Developing Itô stochastic differential equation models for neuronal signal transduction pathways. Comput Biol Chem 30:280–291. https://doi.org/10.1016/j.compbiolchem.2006.04.002
43. Zhang J, Hou Z, Xin H (2004) System-size biresonance for intracellular calcium signaling. ChemPhysChem 5:1041–1045. https://doi.org/10.1002/cphc.200400089
44. lian ZC, Jia Y, Liu Q et al (2007) A mesoscopic stochastic mechanism of cytosolic calcium oscillations. Biophys Chem 125:201–212. https://doi.org/10.1016/j.bpc.2006.08.001
45. Wang X, Hao Y, Weinberg SH, Smith GD (2015) Ca^{2+}−activation kinetics modulate successive puff/spark amplitude, duration and inter-event-interval correlations in a Langevin model of stochastic Ca^{2+}release. Math Biosci 264:101–107. https://doi.org/10.1016/j.mbs.2015.03.012
46. Winslow RL, Tanskanen A, Chen M, Greenstein JL (2006) Multiscale modeling of calcium signaling in the cardiac dyad. Ann NY Acad Sci 1080:362–375
47. Soeller C, Cannell MB (1997) Numerical simulation of local calcium movements during L-type calcium channel gating in the cardiac diad. Biophys J 73:97–111. https://doi.org/10.1016/S0006-3495(97)78051-2
48. Cannell MB, Soeller C (1997) Numerical analysis of ryanodine receptor activation by L-type channel activity in the cardiac muscle diad. Biophys J. doi: S0006-3495(97)78052-4 [pii]\r10.1016/S0006-3495(97)78052-4, vol 73, pp 112–122
49. Smith GD, Keizer JE, Stern MD et al (1998) A simple numerical model of calcium spark formation and detection in cardiac myocytes. Biophys J 75:15–32. https://doi.org/10.1016/S0006-3495(98)77491-0
50. Jiang YH, Klein MG, Schneider MF (1999) Numerical simulation of Ca^{2+} "sparks" in skeletal muscle. Biophys J 92:308–332. https://doi.org/10.1016/j.pbiomolbio.2005.05.016
51. Baylor SM, Hollingworth S (2007) Simulation of ca $^{2+}$ movements within the sarcomere of fast-twitch mouse fibers stimulated by action potentials. J Gen Physiol 130:283–302. https://doi.org/10.1085/jgp.200709827
52. Stern MD, Pizarro G, Ríos E (1997) Local control model of excitation-contraction coupling in skeletal muscle. J Gen Physiol 110:415–440. https://doi.org/10.1085/jgp.110.4.415
53. Greenstein JL, Winslow RL (2002) An integrative model of the cardiac ventricular myocyte incorporating local control of Ca^{2+} release. Biophys J 83:2918–2945. https://doi.org/10.1016/S0006-3495(02)75301-0
54. Rüdiger S, Shuai JW, Huisinga W et al (2007) Hybrid stochastic and deterministic simulations of calcium blips. Biophys J 93:1847–1857. https://doi.org/10.1529/biophysj.106.099879
55. Kalantzis G (2009) Hybrid stochastic simulations of intracellular reaction-diffusion systems. Comput Biol Chem 33:205–215. https://doi.org/10.1016/j.compbiolchem.2009.03.002
56. Choi T, Maurya MR, Tartakovsky DM, Subramaniam S (2010) Stochastic hybrid modeling of intracellular calcium dynamics. J Chem Phys 133:165101. https://doi.org/10.1063/1.3496996
57. Krishnamurthy V, Chung SH (2007) Large-scale dynamical models and estimation for permeation in biological membrane ion channels. Proc IEEE 95:853–880. https://doi.org/10.1109/JPROC.2007.893246

58. Elf J, Doncic A, Ehrenberg M (2003) Mesoscopic reaction-diffusion in intracellular signaling. In: Bezrukov S, Frauenfelder H, Moss F (eds) Fluctuations and noise in biological, biophysical, and biomedical systems. Proceedings of the SPIE, pp 114–124
59. Weinberg SH, Smith GD (2014) The influence of Ca^{2+} buffers on free [Ca^{2+}] fluctuations and the effective volume of Ca^{2+} microdomains. Biophys J 106:2693–2709. https://doi.org/10.1016/j.bpj.2014.04.045
60. Sherman A, Smith GD, Dai L, Miura RM (2001) Asymptotic analysis of buffered calcium diffusion near a point source. SIAM J Appl Math 61:1816–1838. https://doi.org/10.1137/S0036139900368996
61. Li QS, Wang P (2004) Internal signal stochastic resonance induced by colored noise in an intracellular calcium oscillations model. Chem Phys Lett 387:383–387. https://doi.org/10.1016/j.cplett.2004.02.042
62. Blomberg C (2006) Fluctuations for good and bad: the role of noise in living systems. Phys Life Rev 3:133–161
63. Faisal AA, Selen LPJ, Wolpert DM (2008) Noise in the nervous system. Nat Rev Neurosci 9:292–303
64. Zhong S, Qi F, Xin H (2001) Internal stochastic resonance in a model system for intracellular calcium oscillations. Chem Phys Lett 342:583–586. https://doi.org/10.1016/S0009-2614(01)00625-X
65. Thul R, Falcke M (2004) Release currents of IP3 Receptor Channel clusters and concentration profiles. Biophys J 86:2660–2673. https://doi.org/10.1016/S0006-3495(04)74322-2
66. Skupin A, Falcke M (2009) From puffs to global Ca^{2+} signals: how molecular properties shape global signals. Chaos 19:037111. https://doi.org/10.1063/1.3184537
67. Marchant JS, Parker I (2001) Role of elementary Ca^{2+} puffs in generating repetitive Ca^{2+} oscillations. EMBO J 20:65–76. https://doi.org/10.1093/emboj/20.1.65
68. De Young GW, Keizer J (1992) A single-pool inositol 1,4,5-trisphosphate-receptor-based model for agonist-stimulated oscillations in Ca^{2+} concentration. Proc Natl Acad Sci 89:9895–9899. https://doi.org/10.1073/pnas.89.20.9895
69. Ullah G, Daniel Mak D-O, Pearson JE (2012) A data-driven model of a modal gated ion channel: the inositol 1,4,5-trisphosphate receptor in insect Sf9 cells. J Gen Physiol 140:159–173. https://doi.org/10.1085/jgp.201110753
70. Siekmann I, Wagner LE, Yule D et al (2012) A kinetic model for type i and II IP3R accounting for mode changes. Biophys J 103:658–668. https://doi.org/10.1016/j.bpj.2012.07.016
71. Xu T, Yu X, Perlik AJ et al (2009) Rapid formation and selective stabilization of synapses for enduring motor memories. Nature 462:915–919. https://doi.org/10.1038/nature08389
72. Bers DM (2002) Cardiac excitation contraction coupling. Nature 415:198–215. https://doi.org/10.1016/B978-0-12-378630-2.00221-8
73. Bers DM, Despa S (2013) Cardiac excitation-contraction coupling. In: Lennarz WJ, Lane MD (eds) The Encyclopedia of biological chemistry, 2nd edn. Academic Press, Waltham, MA, pp 379–383
74. Artyomov MN, Das J, Kardar M, Chakraborty AK (2007) Purely stochastic binary decisions in cell signaling models without underlying deterministic bistabilities. Proc Natl Acad Sci U S A 104:18958–18963. https://doi.org/10.1073/pnas.0706110104

Chapter 32
A Statistical View on Calcium Oscillations

Jake Powell, Martin Falcke, Alexander Skupin, Tomas C. Bellamy, Theodore Kypraios, and Rüdiger Thul

Abstract Transient rises and falls of the intracellular calcium concentration have been observed in numerous cell types and under a plethora of conditions. There is now a growing body of evidence that these whole-cell calcium oscillations are stochastic, which poses a significant challenge for modelling. In this review, we take a closer look at recently developed statistical approaches to calcium oscillations. These models describe the timing of whole-cell calcium spikes, yet their parametrisations reflect subcellular processes. We show how non-stationary calcium spike sequences, which e.g. occur during slow depletion of intracellular calcium stores or in the presence of time-dependent stimulation, can be analysed with the help of so-called intensity functions. By utilising Bayesian concepts, we demonstrate how values of key parameters of the statistical model can be inferred from single cell calcium spike sequences and illustrate what information whole-cell statistical models can provide about the subcellular mechanistic processes that drive calcium oscillations. In particular, we find that the interspike interval distribution of HEK293 cells under constant stimulation is captured by a Gamma distribution.

J. Powell · T. Kypraios · R. Thul (✉)
Centre for Mathematical Medicine and Biology, School of Mathematical Sciences, University of Nottingham, Nottingham, UK
e-mail: pmxjp8@exmail.nottingham.ac.uk; theodore.kypraios@nottingham.ac.uk; ruediger.thul@nottingham.ac.uk

M. Falcke
Max Delbrück Centre for Molecular Medicine, Berlin, Germany

Department of Physics, Humboldt University, Berlin, Germany
e-mail: martin.falcke@mdc-berlin.de

A. Skupin
Luxembourg Centre for Systems Biomedicine, University of Luxembourg, Belval, Luxembourg

National Biomedical Computation Resource, University California San Diego, La Jolla, CA, USA
e-mail: alexander.skupin@uni.lu

T. C. Bellamy
School of Life Sciences, University of Nottingham, Nottingham, UK
e-mail: tomas.bellamy@nottingham.ac.uk

© Springer Nature Switzerland AG 2020

M. S. Islam (ed.), *Calcium Signaling*, Advances in Experimental Medicine and Biology 1131, https://doi.org/10.1007/978-3-030-12457-1_32

Keywords Calcium spikes · Bayesian inference · Intensity functions · Heterogeneous cell populations

32.1 Introduction

Calcium (Ca^{2+}) oscillations have long been recognised as a centrepiece in the world of intracellular Ca^{2+} signals [1–9]. Acting as a ubiquitous and versatile signalling mechanism, Ca^{2+} oscillations are responsible for inducing gene expression [10–12], controlling hormone secretion [13–17], orchestrating fertilisation [18–20] and steering bacterial invasion [21], to name but a few cellular functions. The notion of Ca^{2+} oscillations usually refers to transient increases in the whole-cell Ca^{2+} concentration that present themselves as a series of Ca^{2+} spikes. Since whole-cell calcium recordings yield averaged concentration values, it has often been assumed that mathematical models of intracellular Ca^{2+} oscillations can be directly based on the averaged Ca^{2+} concentration. To illustrate this concept, consider Ca^{2+} oscillations driven by Ca^{2+} release from the endoplasmic reticulum (ER) through inositol-1,4,5-trisphosphate ($InsP_3$) receptors ($InsP_3Rs$). In its simplest incarnation, these mathematical models assume that Ca^{2+} transport through all open $InsP_3Rs$ and the activity of all sarco-endoplasmic Ca^{2+} ATP (SERCA) pumps can be averaged across the cell to yield averaged Ca^{2+} release and resequestration, respectively. Since the activity of both $InsP_3Rs$ and SERCA pumps depends on the cytosolic Ca^{2+} concentration, these models implicitly assume that the gating of $InsP_3Rs$ and SERCA pumps is controlled by averaged Ca^{2+} concentration values.

This assumption may serve as a starting point to explore Ca^{2+} dynamics in systems for which detailed Ca^{2+} measurements are missing, and models based on averaged Ca^{2+} concentrations have been instrumental in furthering our understanding of Ca^{2+} oscillations [13, 16, 22–48]. However, the notion of mean Ca^{2+} values generally falls short of capturing the biology that underlies Ca^{2+} oscillations. The main reason for this is that $InsP_3Rs$ form clusters that are distributed throughout the cell at distances of 2–7μm [49–53]. This entails that the dynamics of $InsP_3Rs$ is controlled by the *local* Ca^{2+} concentration, not a global average. In other words, measuring the Ca^{2+} concentration across a cell, taking the spatial average and determining the gating of all $InsP_3Rs$ subject to the averaged Ca^{2+} concentration misrepresents the actual $InsP_3R$ dynamics. In addition, there are only a few tens of $InsP_3Rs$ per cluster [54–56]. Since binding of Ca^{2+} and $InsP_3$ to $InsP_3Rs$ is random and hence transitions between different states of the $InsP_3R$ occur stochastically, the relative fluctuation in the number of open $InsP_3Rs$ is considerable. This stochasticity might even be enhanced by the fact that at basal Ca^{2+} concentration, the actual numbers of Ca^{2+} ions in the vicinity of an $InsP_3$ is small [57–60]. Taken together, these observations strongly suggest that intracellular Ca^{2+} is a spatially extended stochastic medium, which prompts the question on how to best describe $InsP_3$ mediated Ca^{2+} oscillations mathematically.

One approach starts with the dynamics of single $InsP_3Rs$, groups them into clusters and then places the clusters into a three-dimensional representation of the cytosol—see [61] for a recent perspective. In these models, $InsP_3Rs$ are described by stochastic models known as Markov chains, which consist of different states of the $InsP_3R$ such as open, closed and inhibited and contain rules for stochastic transitions between different states. Clusters of $InsP_3Rs$ communicate with each other through Ca^{2+} diffusion. One advantage of such hierarchical modelling lies in its mechanistic interpretation. It allows questions to be answered about how Ca^{2+} oscillations are shaped by e.g. the distance between $InsP_3R$ clusters, single channel current and Ca^{2+} buffers. However, these models require as input a significant number of parameters, such as gating constants for the $InsP_3R$, and are computationally expensive. In order to reduce the computational load, Langevin-type models have been put forward. In essence, they approximate the exact stochastic dynamics of the Markov chains.

In terms of modelling philosophy, the above approaches fall into the category of bottom-up techniques. At the other end of the spectrum lie so-called top-down methods. Here, we construct models that directly describe key properties of Ca^{2+} spikes such as amplitude and frequency without explicitly incorporated mechanistic details as e.g. the possible states of an $InsP_3R$. At first sight, this might appear less advantageous as different model behaviours cannot immediately be linked to specific molecular processes. However, there are distinct advantages. Firstly, the computational demand is significantly lower than with bottom-up approaches. This puts us in an ideal position to generate large numbers of realistic Ca^{2+} spike sequences, which in turn can serve as input to signalling cascades that decode Ca^{2+} spikes. Secondly, top-down models provide a powerful framework for fitting data and testing hypotheses on Ca^{2+} spike generation. Consequentially, we can use the knowledge gained from top-down models to improve bottom-up approaches, which in turn will advance our mechanistic understanding of Ca^{2+} oscillations.

In this review, we present the current state of statistical modelling of Ca^{2+} oscillations. The techniques that we employ are well established amongst statisticians, but are less familiar to modellers and experimentalists in the field of Ca^{2+} signalling. We therefore mainly focus on describing the underlying concepts and how they are related to the physiology of Ca^{2+} signalling. We discuss practical approaches for how to ascertain whether our statistical assumptions are consistent with measured Ca^{2+} spike sequences and what we can learn from our statistical analysis regarding the mechanisms that underlie Ca^{2+} spike generation.

32.2 Interspike Interval Statistics

We outlined in the introduction the mechanistic reasons for why Ca^{2+} oscillations are stochastic. At this point, one might argue—as is often done—that the molecular fluctuations present at $InsP_3R$ clusters average out at the whole-cell level. In other words, since a cell can contain a large number of $InsP_3R$ clusters, the stochastic contributions cancel. To test this hypothesis, Skupin et al [62] measured spontaneous Ca^{2+} oscillations in microglia, astrocytes and PLA cells as well as carbachol-induced oscillations in HEK293 cells. They found that their data is consistent with stochastic whole-cell Ca^{2+} oscillations, which was also confirmed in later experiments [63]. This conclusion rests on results as shown in Fig. 32.1. We plot representative fluorescence traces for carbachol stimulated HEK293 cells in Fig. 32.1a, b. Cells were initially stimulated with 20 μM carbachol before the solution was switched to 50 μM carbachol. Figure 32.1a illustrates the well-known phenomenon of frequency encoding, by which the frequency of Ca^{2+} oscillations increases with an increase in stimulation strength. In Fig. 32.1c we plot the Ca^{2+} spike times for a larger number of cells. If Ca^{2+} oscillations were deterministic and governed by averaged Ca^{2+} concentrations, we would expect an almost constant spread of Ca^{2+} spike times, i.e. an almost constant value for the interspike interval (ISI), not the observed large variability, which is present at both stimulation strengths. Our argument for stochastic Ca^{2+} oscillations is further strengthened by the results shown in Fig. 32.1d. Here, each triangle corresponds to a sequence of Ca^{2+} spikes from one cell and denotes its mean μ and its standard deviation σ. We observe that the standard deviation is of the same magnitude as the mean, which is another strong indicator of stochastic behaviour. Importantly, similar results have been obtained for a number of additional cell types and under different conditions [61, 62], which lends even more support for the stochasticity of Ca^{2+} oscillations. Given the insights that $\sigma - \mu$ plots can provide into the nature of Ca^{2+} oscillations, we have recently released CaSiAn [64], a user friendly tool that allows for automatic ISI detection from fluorescence time course data and interactive investigation of the relationship between μ and σ.

To appreciate the fact that μ and σ are of the same order of magnitude, we introduce a key concept for this review: the conditional Ca^{2+} spike intensity $q(t|s)$, $t > s$. Based on it, we obtain the conditional Ca^{2+} spike probability $q(t|s)\mathrm{d}t$, which represents the probability to observe a Ca^{2+} spike in the time interval $[t, t + \mathrm{d}t]$ given a Ca^{2+} spike at time s. In [62], the following ansatz was made:

$$q(t|s) = \begin{cases} 0\,, & s \le t \le T_r + s\,, \\ \lambda\left[1 - \mathrm{e}^{-\xi(t-s-T_r)}\right], & T_r + s \le t\,. \end{cases} \tag{32.1}$$

Here, T_r denotes the cellular refractory period. Numerous experiments have shown that there exists a minimal amount of time T_r after a Ca^{2+} spike before another Ca^{2+} spike can be triggered [62, 63, 66]. Therefore, the conditional intensity vanishes, i.e. $q = 0$, for a time T_r after the last Ca^{2+} spike. It is important to note that T_r is significantly longer than the recovery time of $InsP_3Rs$ [63]. Once the refractory period has passed, the conditional intensity for a Ca^{2+} spike starts to increase at a rate ξ and eventually approaches an equilibrium value λ. This reflects the notion that a cell has to recover from the last Ca^{2+} spike. While ξ is a single number, it subsumes numerous recovery processes such as refilling of the ER or replenishment of $InsP_3$ following degradation by $InsP_3$-3-kinase and $InsP_3$-5-phosphatase. The values of T_r, ξ and λ can be directly inferred from Fig. 32.1d as outlined below. Due to the strong linear relationship between the mean and the standard deviation, we posit that

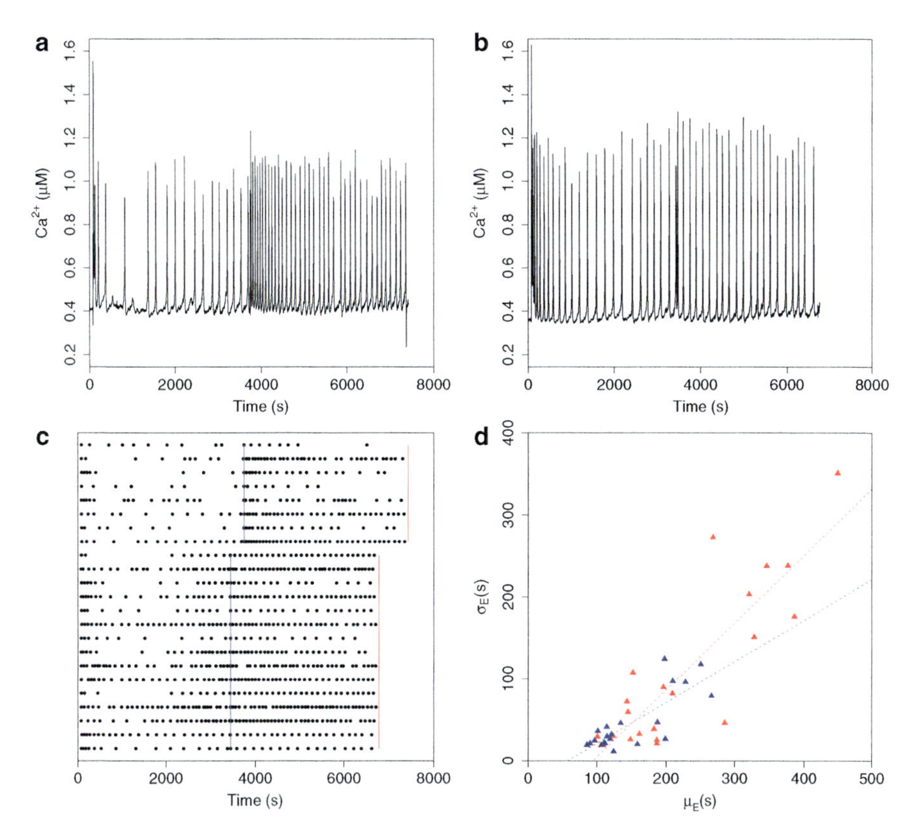

Fig. 32.1 (**a**, **b**) Fura-2 fluorescence intensity traces of two HEK293 cells stimulated first with 20 μM carbachol and then with 50 μM carbachol. The solution was exchanged at 3738s in (**a**) and at 3444s in (**b**). (**c**) Raster plot of Ca^{2+} spike times for the same stimulus protocol as in (**a**, **b**). The blue line indicates solution exchange and the red line denotes the end of the experiment. (**d**) Relationship between the standard deviation σ_E and the mean μ_E for the data shown in (**c**). Each triangle corresponds to data from one cell, and the line is the best linear fit. Red refers to 20 μM carbachol, and blue to 50 μM carbachol. (For details of the experiments see [65])

$$\sigma = \alpha(\mu - T_r), \tag{32.2}$$

a relationship that has been shown to hold true for another 8 cell types and 10 conditions (see [61] for further discussion). When the standard deviation equals zero, successive Ca^{2+} spikes are separated by a constant period. Such Ca^{2+} spike sequences appear deterministic since there is no variation in the ISI, but the interpretation is different. The lack of ISI variability results from the fact that when the Ca^{2+} spike generation probability is high, i.e. λ is large, a Ca^{2+} spike is initiated as soon as the cell exits its refractory period. Therefore, the mean of the ISI distribution at a vanishing standard deviation equals T_r. This corresponds to the intersections of the red and blue lines with the x-axis in Fig. 32.1d, respectively. To determine ξ and λ, we start from Eq. (32.1) and derive the ISI probability density $f(t, s)$, i.e. the probability density for Ca^{2+} spikes to occur at times t and s. This is equivalent to the probability of a Ca^{2+} spike at t given that the last spike occurred at s and no Ca^{2+} spike during the time $(s - t)$. Based on this interpretation of the ISI probability, we obtain

$$f(t, s) = q(t|s) \exp\left\{-\int_s^t q(u|s)\mathrm{d}u\right\} , \tag{32.3}$$

where the exponential term corresponds to the absence of Ca^{2+} spikes between s and t. The mean μ and the standard deviation σ of the ISI distribution then follow from Eq. (32.3) as

$$\mu = \int_0^\infty t f(t, 0)\mathrm{d}t , \qquad \sigma^2 = \int_0^\infty t^2 f(t, 0)\mathrm{d}t - \mu^2 . \tag{32.4}$$

For practical purposes, we can set $T_r = 0$ in the computation of μ and σ, since a constant T_r only shifts the mean and does not affect the standard deviation. To put it another way, we evaluate Eq. (32.4) for $T_r = 0$ and then add T_r to obtain the mean ISI μ. Next, we fit the equations in (32.4) to data such as shown in Fig. 32.1d to obtain cell specific values for ξ and λ. This is achieved in a two-step process. Firstly, we determine the experimental mean μ_E and standard deviation σ_E from individual Ca^{2+} spike sequences as shown in Fig. 32.1c. This gives one data point in Fig. 32.1d. Since μ and σ^2 in Eq. (32.4) depend on ξ and λ through $f(t, 0)$ via $q(t|s)$, we can perform a least square fit of Eq. (32.4) to the experimental data μ_E and σ_E to obtain single cell estimates for ξ and λ. Figure 32.2a, b display results for HEK293 cells stimulated with 30 μM carbachol. While the distribution for ξ exhibits a localised peak, the distribution of λ is much broader. A similar behaviour is observed for spontaneously spiking astrocytes as seen in Fig. 32.2c, d. A comparison of the Ca^{2+} spike rate λ reveals that it is almost an order of magnitude larger for HEK293 cells than for astrocytes, which might be attributed to the fact that the former is stimulated, but the latter is not. Intriguingly, the time scale for recovery ξ is almost 10-fold larger for HEK293 cells than for astrocytes, indicating that HEK293 cells recover more slowly than astrocytes after a Ca^{2+} spike. The existence of wide distributions for ξ and λ also points towards significant cell-to-cell variability, which provides another argument in favour of a statistical description of Ca^{2+} spikes.

It is now instructive to evaluate Eq. (32.4) for a constant conditional intensity function $q = r > 0$, which corresponds to a homogenous Poisson process. It emerges from the general form of the conditional intensity function in Eq. (32.1) in the limit of fast recovery, i.e. a large value of ξ. In this case, the integrals can be computed analytically and we obtain $\mu = \sigma = r$, which is consistent with the scaling in Eq. (32.2). This provides further intuition for the statement made above that stochastic effects need to be taken into account when the mean and the standard deviation are of similar magnitude.

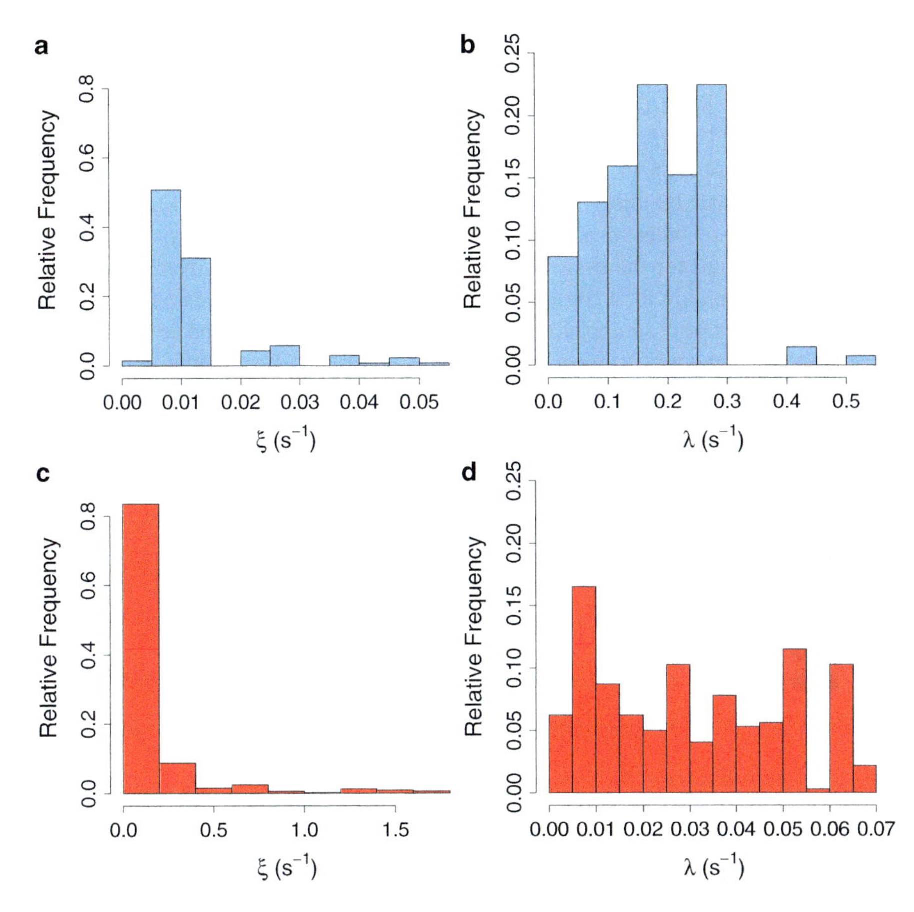

Fig. 32.2 Relative frequency for ξ (**a**, **c**) and λ (**b**, **d**) for HEK293 cells (top, blue) and astrocytes (bottom, red). HEK293 cells were stimulated with 30 μM carbachol, while Ca^{2+} spikes in astrocytes were spontaneous. $N = 138$ for HEK293 cells and $N = 321$ astrocytes. For experimental details, see [62]

Equation (32.3) expresses the ISI distribution in terms of the conditional intensity function. It is often convenient to reverse the approach and start from an ISI distribution. Firstly, we obtain ISIs directly from experimental recordings, which inform us about the possible shapes of ISI distributions. Secondly, some ISI distributions that have been shown to capture experimental data cannot be derived from closed form intensity functions as e.g. in Eq. (32.1). A point in case is the Gamma distribution, which is consistent with Ca^{2+} oscillations in HEK293 cells [67] and also with voltage spikes in neurons [68, 69]. One common representation for the density of the Gamma distribution reads as

$$f_G(t, s) = \frac{\beta^{\alpha}}{\Gamma(\alpha)}(t - s)^{\alpha-1}e^{-\beta(t-s)}, \qquad (32.5)$$

where α and β are called the shape parameter and rate, respectively, and Γ denotes the standard Gamma function. Suppose for a moment that the time between successive Ca^{2+} puffs follows a Poisson distribution with rate β. In contrast to Ca^{2+} spikes, Ca^{2+} puffs correspond to localised Ca^{2+} liberation through a cluster of $InsP_3Rs$. In addition to Ca^{2+} release through single $InsP_3Rs$, Ca^{2+} puffs are considered the basic building blocks in the hierarchy of Ca^{2+} signals [2, 61, 70]. A Gamma distribution where α is a positive integer returns the probability that α Ca^{2+} puffs have occurred for the first time. In other words, the Gamma distribution is a probability distribution for a combination of events to happen for the first time. This interpretation makes it an appealing candidate for Ca^{2+} spikes. The reason is that Ca^{2+} spikes are thought to form when a small number of Ca^{2+} puffs generates a region of elevated Ca^{2+} in the cell, which then initiates Ca^{2+} release throughout the cell. Alternatively, recent experiments in astrocytes suggest that the co-occurrence of a certain number of Ca^{2+} puffs is sufficient to trigger a Ca^{2+} spike [71]. This also fits well with a body of research that shows that Ca^{2+} puffs and Ca^{2+} spikes can be described as first-passage time problems [63, 72–76]. As an interesting observation, note that the mean ISI for Eq. (32.5) is α/β, so that the mean interpuff interval for α puffs is $1/\beta$, which is consistent with the mean interpuff time when puffs are described by a Poisson process with rate β. To relate a given ISI distribution to the conditional intensity function, we find that

$$q(t|s) = \frac{f(t,s)}{1 - \int_s^t f(u,s)\mathrm{d}u}, \tag{32.6}$$

which is equivalent to Eq. (32.3) as shown in Appendix 1. Equations (32.3) and (32.6) allow us to switch between conditional intensity functions and ISI distributions depending on what our modelling question requires.

32.3 Beyond Stationary Ca^{2+} Spike Sequences

The discussion so far assumed that successive Ca^{2+} spikes are independent and are described by the same statistics. The conditional intensity function $q(t|s)$ only depends on the time since the last spike $(t - s)$, but not on the absolute Ca^{2+} spike times t and s. Hence, the probability for two spikes to be separated by say $80s$ is the same irrespective of whether the first spike occurs 10s into the experiment or 1000s. The same holds true for the ISI density in Eq. (32.5) which only depends on the time difference $(t - s)$ between successive Ca^{2+} spikes. A consequence of the independence of Ca^{2+} spikes is that we can immediately write down the probability density for n Ca^{2+} spikes occurring at times $y_1, y_2, \ldots, y_n$. If we collect the Ca^{2+} spike times in a set $\mathbf{y} = \{y_1, \ldots y_n\}$ the probability density for the entire Ca^{2+} spike sequence is given by

$$p(\mathbf{y}) = f_1(y_1, 0) f(y_2, y_1) \cdots f(y_n, y_{n-1}) f_n(T, y_n), \tag{32.7}$$

where $f_1(y_1, 0)$ denotes the probability density for the first spike to occur at y_1 and $f_n(T, y_n)$ is the probability that no spike happens after y_n until the end of the experiment at time T. The probability for a Ca^{2+} spike sequence factorises in the probabilities of individual and identical ISIs, which are properties often referred to as independence and stationarity, respectively. We separate out the contributions from f_1 and f_n since they do not correspond to ISI probabilities and hence are often modelled by different probability distributions, e.g. a Poisson distribution.

However, there are numerous reasons for why ISI probabilities do not remain constant over time and hence ISIs at different times of the experiment follow different probability distributions. For example, while the ER refills between Ca^{2+} spikes, the level of refilling can decrease as Ca^{2+} leaves the cell across the plasma membrane. In most experiments, $InsP_3$ is formed in response to activation of cell surface receptors, but the efficiency of this pathway may decrease over time. Both factors lower the propensity for the generation of Ca^{2+} spikes as the experiment progresses and introduces trends when plotting ISIs. When analysing Ca^{2+} spikes, we can remove trends and only consider Ca^{2+} spikes after initial transients. This presents a sensible approach when cells experience constant stimulation such as in step change experiments. However, under physiological conditions, hormones arrive in a time-dependent manner, so do neurotransmitters and paracrine signals. To mimic such an in vivo environment, cells need to be challenged with time-varying stimuli. As soon as we introduce an explicit time-dependence, ISI distributions are no longer stationary, but depend on the absolute time of the experiment.

This raises the question on how to mathematically describe the non-stationarity of Ca^{2+} spike sequences. One approach is to introduce an explicit time-dependence into the ISI distribution by making the parameters change over time. While conceptually appealing, the practicalities of this approach are limited. For instance, if we believe that the parameters change continuously over time, it is not apparent how to constrain the model best given that we sample the values of the parameters at only a few discrete time points, viz. the times of Ca^{2+} spikes. Another issue arises from the fact that the probability of a Ca^{2+} spike sequence does not necessarily factorise any more as in Eq. (32.7), but we need to consider the full multivariate probability $p(\mathbf{y}) = p(y_1, \ldots, y_n)$, which can pose significant challenges.

A more practical approach was put forward in [68]. At the heart of it lies a time transformation that maps the time of the original experiment, denoted by t, to a new time u via

$$u(t) = \int_0^t x(v)\mathrm{d}v\,, \tag{32.8}$$

where x is called the intensity function and relates to the level of Ca^{2+} spiking as we will illustrate below. As such, x is always strictly positive and hence associates each value of t with a unique values of u through Eq. (32.8). A consequence of this mapping is that in the new time u, ISIs become independent [68]. This means that the probability density for a Ca^{2+} spike sequence factorises again and we have

$$p(\mathbf{y}|x) = g_1(u_1, 0|x)g(u_2, u_1|x)\cdots g(u_n, u_{n-1}|x)g_n(U, u_n|x), \tag{32.9}$$

where $u_i = u(y_i)$, $U = u(T)$ and the dependence of $\mathbf{y}$ on the left hand side enters on the right hand side through u being a function of t. We explicitly include x to emphasise that the transformation depends on the intensity function. What makes Eq. (32.9) particularly useful is that the probability density g is related to the original ISI probability density f via

$$g(u_i, u_{i-1}|x) = x(y_i)f(u_i, u_{i-1}), \tag{32.10}$$

which follows from the conservation of probability [67, 77]. We illustrate a practical calculation for Eq. (32.10) in Appendix 2.

At this point, it might appear that the intensity function is mathematically convenient, but detached from the actual biology. As it turns out, the contrary holds true. For the models of Ca^{2+} spiking considered here, $x(t)$ corresponds to the probability of Ca^{2+} spiking *independent* of the history of the Ca^{2+} spike sequence. Put differently, if there are N identical Ca^{2+} spiking cells, $Nx(t)$ is the expected number of Ca^{2+} spikes at time t. To illustrate this concept, we chose an intensity function (red line in Fig. 32.3), generated 10,000 Ca^{2+} spike sequences from it and binned them (light blue histogram). By using a large number of Ca^{2+} spike sequences, binning is equivalent to taking the average across all possible histories that led to a Ca^{2+} spike in the respective bin. The excellent agreement between the intensity function and the histogram confirms the above interpretation of $x(t)$. For the practicalities of generating the Ca^{2+} spike sequences, we refer the reader to Appendix 3.

32.4 Bayesian Computation

A main motivation for pursuing a statistical approach is to fit models of Ca^{2+} spike generation more easily to experimental data and hence learn more about the nature of Ca^{2+} oscillations. This requires us to infer the parameters of the ISI distribution, e.g. α and β for the Gamma distribution, and the time course of $x(t)$ from measured Ca^{2+} spike sequences. For ease of reference, we will call all unknowns of the model, i.e. the intensity function and the parameters of the ISI distribution, hyperparameters and denote them by θ.

There are a number of ways for achieving this goal. Here, we will make use of Bayesian inference, that addresses the following question: what does the data tell us about the parameters of the model? Expressed more formally, we are interested in $p(\theta|\mathbf{y})$, i.e. the probability distribution of the hyperparameters given a Ca^{2+} spike sequence. This probability is called the *posterior* distribution. The advantage of this

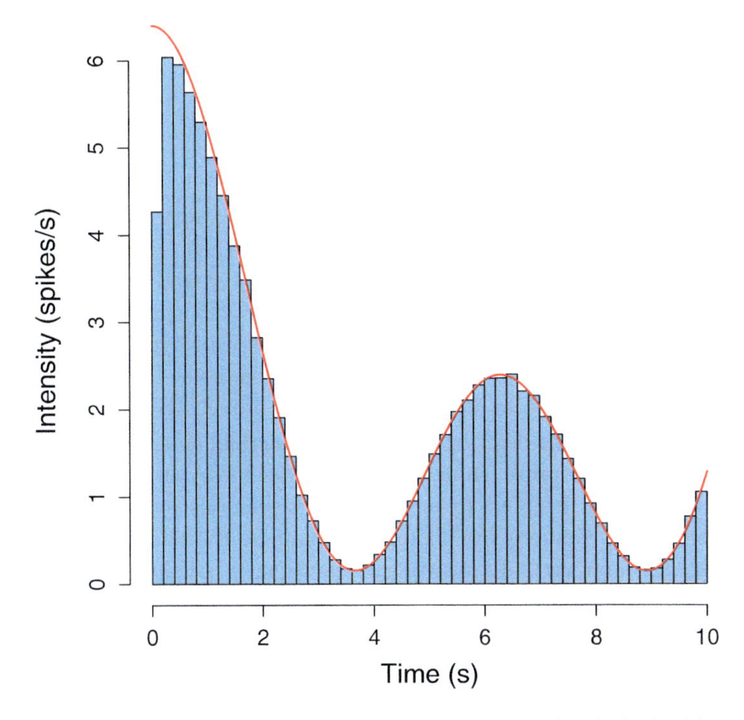

Fig. 32.3 Intensity function (red) and peristimulus-time histogram (blue) obtained from 10,000 Ca^{2+} spikes when the ISI distribution is given by a Gamma distribution. Parameter values are $x(t) = 2\cos(t) + 2\cos(0.5t) + 2.4$ and $\alpha = 6.2$ and $\beta = 6.2$s

approach is that we do not merely obtain a single value, but the full probability distribution for the parameters that are consistent with the data. This allows us to judge how well the model captures the data and what parameter values to use to describe the underlying biology. For instance, consider one of the hyperparameters, say θ_1. If the distribution for θ_1 is sharply peaked around a value θ_1^*, we can be confident that θ_1^* is a sensible estimate for θ_1. On the other hand, if the probability distribution is broad or exhibits multiple maxima, we are pressed much harder to interpret the results. It might also indicate that we based our original model on incorrect assumptions. In addition to these conceptual benefits, the posterior distribution is all we need to answer any question we have about the experiment. For instance, we can determine summary statistics such as mean and variance as well as the behaviour of functions that depend on hyperparameters.

To compute the posterior probability, we make use of Bayes' theorem, which states that

$$p(\theta|\mathbf{y}) = \frac{p(\mathbf{y}|\theta)p(\theta)}{p(\mathbf{y})}. \tag{32.11}$$

The right hand side contains the likelihood function $p(\mathbf{y}|\theta)$, the so-called prior $p(\theta)$ and the normalisation

$$p(\mathbf{y}) = \int p(\mathbf{y}|\theta)p(\theta)\mathrm{d}\theta\,. \tag{32.12}$$

The conceptual appeal of Eq. (32.11) stems from its numerator. We already encountered an example for a likelihood function in Eq. (32.9). It represents how likely it is to observe what we have measured for a given θ and hence reflects our believes about the potential mechanisms that drive Ca^{2+} spike generation. The prior distribution allows us to provide sensible input for the hyperparameter values before we see the data. For instance, if we believe that some hyperparameter, say θ_1 again, has a value close to some θ_1^0, we pick a probability distribution that is centred around θ_1^0. On the other hand, if we are uncertain about possible values of θ_1, we choose a wider prior. Thinking about priors for hyperparameters that are numbers immediately leads us to probability distributions in the classical sense such as Poisson distributions or Gamma distributions. But what about a prior for the intensity function $x(t)$? To answer this question, it is helpful to return to the biological interpretation of $x(t)$, viz. the probability density for a Ca^{2+} spike at time t irrespective of the history of the Ca^{2+} spike sequence. If we challenge cells with a constant stimulus as in e.g. a step-change experiment, a reasonable assumption is that $x(t)$ remains constant over longer periods of time, but not necessarily at the same value for the entire experiment. For instance, as the experiment continues, Ca^{2+} spikes may become less frequent compared to the beginning of the experiment due to receptor desensitisation or changes to ER Ca^{2+} load. We can mimic this biological response by assuming an $x(t)$ that has a large constant value when the experiment is started and smaller constant value towards the end. In this particular illustration, we assumed that there are two different levels. To allow more flexibility, suppose that there are k levels and that the probability for having k levels is Poissonian with rate γ. If we further assume that each level h_i is drawn from a Gamma distribution with parameters a and b, we find that the prior for the intensity function is

$$p(x) = \mathrm{e}^{-\gamma}\frac{\gamma^k}{k!}\prod_{i=1}^{k}\frac{a^b}{\Gamma(a)}h_i^{a-1}\mathrm{e}^{-bh_i}\,. \tag{32.13}$$

Because the number of changepoints is independent from the levels h_i, which again are independent from each other, $p(x)$ factorises into individual contributions [78]. In Fig. 32.4a we illustrate three candidates for such piecewise constant priors with different numbers of changepoints and different level values.

While piecewise linear intensity functions possess computational advantages—e.g. the integral in Eq. (32.22) in Appendix 2 can be computed analytically—one issue with them is that they are discontinuous, i.e. they have jumps. This might be undesirable in some situation, which leads us to priors for continuous functions. An example for this is when cells are challenged with a time-varying stimulus as e.g.

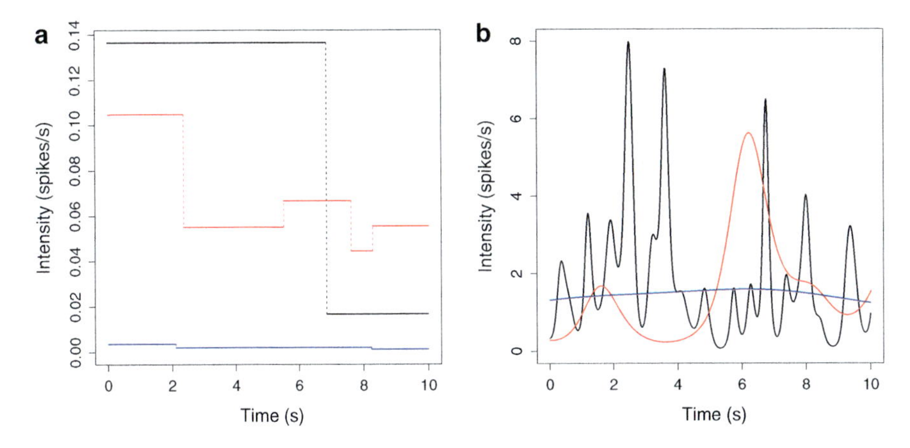

Fig. 32.4 Candidate intensity functions for (**a**) a piecewise linear prior and (**b**) a GP prior. The different colours indicate (**a**) different numbers of change points and different levels and (**b**) different values of κ. Here, blue corresponds to $\kappa = 5$s, red to $\kappa = 1$ s and black to $\kappa = 0.2$ s

in [67]. Since the stimulus changes smoothly over time, a reasonable assumption is that the intensity function inherits this smoothness. Among the different choices that can be made for continuous intensity functions, we here focus on so-called Gaussian processes (GPs). Consider the intensity function at some time point t. Instead of fixing a unique value $x = x(t)$, we prescribe a probability distribution $g(x)$. In other words, for a fixed time t there is a probability $g(x(t))\mathrm{d}x$ that the value of the intensity function lies in the interval $[x(t), x(t) + \mathrm{d}x]$. We here assume that the logarithm of the intensity function follows a Gaussian distribution of the form

$$f_{\mathrm{GP}}(x) = \frac{1}{\sqrt{2\pi\sigma^2}} \exp\left\{\frac{(\mu - x)^2}{2\sigma^2}\right\}, \tag{32.14}$$

where μ denotes the mean and σ the standard deviation, respectively. GPs derive their name from the fact that we employ a Gaussian distribution. The reason for assuming that $\log x(t)$ rather than $x(t)$ itself follows a Gaussian distribution is that $x(t)$ is always positive, but a Gaussian distribution can yield negative values. By using the logarithm, we enforce the positiveness of the intensity function. To ensure that the intensity function is continuous, we need to guarantee that the values of x at two close-by time points t and s are not too far apart. This is achieved by imposing a correlation function

$$\Sigma(s, t) = \sigma^2 \exp\left\{\frac{(s - t)^2}{\kappa^2}\right\}, \tag{32.15}$$

which we have chosen to be a squared exponential. Here, σ is the same as in Eq. (32.14) and κ measures how smooth the GP is. The larger κ the smoother the intensity function. Figure 32.4b shows three different realisations of a GP for varying values of κ. Observe that all three functions are smooth, and that there are less wiggles for larger values of κ, which is consistent with the interpretation above. Since we have to discretise time for any practical computation, suppose that there are n time points, i.e. we represent the time of the experiments at n discrete time points ranging from $t_1 = 0$ to $t_n = T$, where T is the duration of the experiment. A practical representation for these time points are the values at which the Ca^{2+} concentration is measured and is determined by e.g. the frame rate of the microscope cameras. The prior for a GP is a multivariate Gaussian distribution and reads as

$$p(x) = \frac{1}{\sqrt{(2\pi)^n \det \Sigma}} \exp\left\{\frac{1}{2}(\mathbf{t}-\boldsymbol{\mu})\Sigma^{-1}(\mathbf{t}-\boldsymbol{\mu})\right\}, \tag{32.16}$$

where $\mathbf{t}$ is a vector of length n representing the discretised time of the experiment, $\boldsymbol{\mu}$ is a vector of length n denoting the mean of the GP at each time point t_n, and Σ is given by Eq. (32.15).

Having introduced different priors for the hyperparameters of the model including the intensity function x, we can return to Eq. (32.11). While it is conceptually appealing and offers us a full picture of the parameters of the model as constrained by the experiment, it is computationally challenging. The reason is the integral in the denominator, which runs over the entire hyperparameter space. Since this can be high-dimensional, we require computationally efficient methods as a direct integration is often prohibitively expensive if not impossible. There are various methods for tackling this problem. For instance, instead of computing $p(\theta|\mathbf{y})$ directly including the integration of the denominator, we can determine the maximum of the distribution and its variance [79, 80]. This will provide us e.g. with the most likely intensity function that is consistent with the data as well as confidence intervals, see [67]. A different approach is to try and sample from $p(\theta|\mathbf{y})$ without having to explicitly compute it. The main idea is that if we can sample from a probability distribution, we know the possible values and the associated probabilities (values that are more likely than others are sampled more frequently) without having to determine a closed form solution. This often suffices for practical purposes. A technique that does this is known as Markov chain Monte Carlo [81].

32.5 Analysing Ca^{2+} Spike Sequences

Having introduced key concepts for a statistical analysis of Ca^{2+} spike sequences in the previous sections, we now apply them to different experiments. A crucial input to our model is the ISI probability density, see Eq. (32.9). However, we do not know *a priori* which distribution is most consistent with the data. To establish this, we can make use of the following transformation. Let the measured Ca^{2+} spike times be given by $y_1, \ldots y_n$ again. We define transformed ISIs by

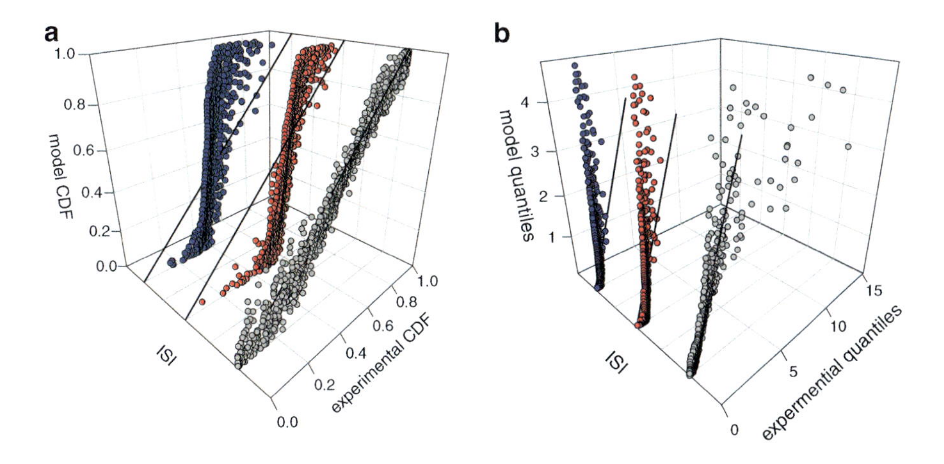

Fig. 32.5 K-S (**a**) and Q-Q (**b**) plots for data from 23 individual cells stimulated with carbachol cells. The initial concentration was 20 μM, which was increased to 50 μM. The ISI distributions are inverse Gaussian (blue), Poisson (red) and Gamma (grey). We used piecewise linear functions as prior for the intensity function

$$\tau_k = \int_{y_{k-1}}^{y_k} q(s|y_{k-1})\mathrm{d}s\,, \tag{32.17}$$

with q given by Eq. (32.6). It can now be shown that if the mechanisms that generate the observed Ca^{2+} spikes are consistent with the ISI distribution that we use in Eq. (32.6), the transformed ISIs τ_k are exponentially distributed with unit rate [82]. This leaves us with the task of comparing two probability distributions: the standard exponential distribution and the distribution of the transformed ISIs. The quantile-quantile (Q-Q) plot and the Kolmogorov Smirnov (K-S) plot are two powerful graphical approaches to examine differences between probability distributions. In Fig. 32.5 we show Q-Q and K-S plots for HEK293 cells in a multistep experiment. We tested three different ISI distributions: a Poisson, an inverse Gaussian and a Gamma distribution. Each cell was analysed individually and gave rise to a separate sequence of dots; no data assimilation was performed. For the Q-Q plot, we determine the quantiles of the transformed ISIs and plot them against the quantiles of the exponential distributions. For the K-S plot, we compute the cumulative distributions of the transformed ISIs and the exponential distribution, respectively, and plot them against each other. Identical probability distributions possess identical quantiles and identical cumulative distributions, respectively. Hence any deviation from a 45° straight line in the Q-Q and K-S plot points towards differences between the distributions and hence indicates that we need to improve our assumptions about the ISI distributions.

For both, the K-S and the Q-Q plot, the data points deviate significantly from a straight line with slope 1 for the Poisson and the inverse Gaussian distribution. On the other hand, we observe a strong correlation between the 45° line and the data points for the Gamma distribution. This visual inspection is corroborated by

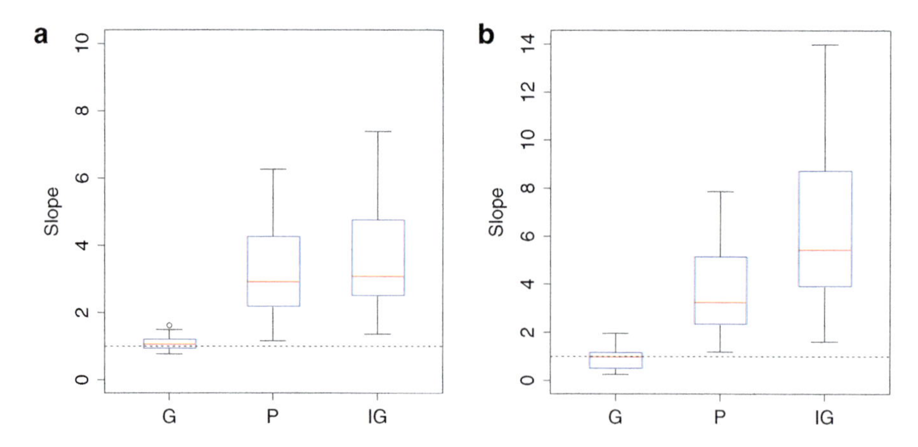

Fig. 32.6 Box and whisker plots for the data presented in Fig. 32.5 showing results for the (**a**) K-S plot and (**b**) Q-Q plot. The red line represents the median of the distribution, and the box delineates the range of data from the first to the third quartile. The whiskers indicate the spread of the data

the box-and-whisker plots in Fig. 32.6. Because we treated cells individually, we can determine the slope of a linear fit for each cell. The plots in Fig. 32.6 show the statistics for these slopes. The box represents the spread of data within the second and third quartile, and the red line indicates the median. The whiskers provide a measure for the overall spread of the data. Generally speaking, the smaller the box and the closer the whiskers to the box, the less spread is in the data. The Poisson and the inverse Gaussian distribution generally exhibit a larger spread than the Gamma distribution. Moreover, the median of the Gamma distribution is closer to one. In [67], we found that for HEK293 cells stimulated with 10 μM and 100 μM charbachol, respectively, the Gamma distribution worked best. These results and the new analysis presented here strongly suggest that the ISI statistics for Ca^{2+} spike sequences are captured by a Gamma distribution. A further argument to support this conclusion is that the data in [67] and [65] were acquired independently in different labs with different setups.

In order to obtain the results in Figs. 32.5 and 32.6 we had to estimate the intensity function $x(t)$ for each cell. Figure 32.7 displays $x(t)$ for two different cells. Since we analyse step change experiments, we first chose piecewise constant functions as a prior for x as discussed in Sect. 32.4. The mean intensity function is shown as a solid blue line, while the 95% confidence interval is represented by the shaded blue area. We clearly see an increase in the intensity function as the stimulus strength is stepped up. Moreover, during extended periods of time, the intensity function is almost constant, which has significant consequences for the interpretation of the mechanisms that drive Ca^{2+} spike generation as discussed below.

As pointed out in Sect. 32.4, piecewise constant functions are not the only possible prior. GPs constitute another possible class, and corresponding results are

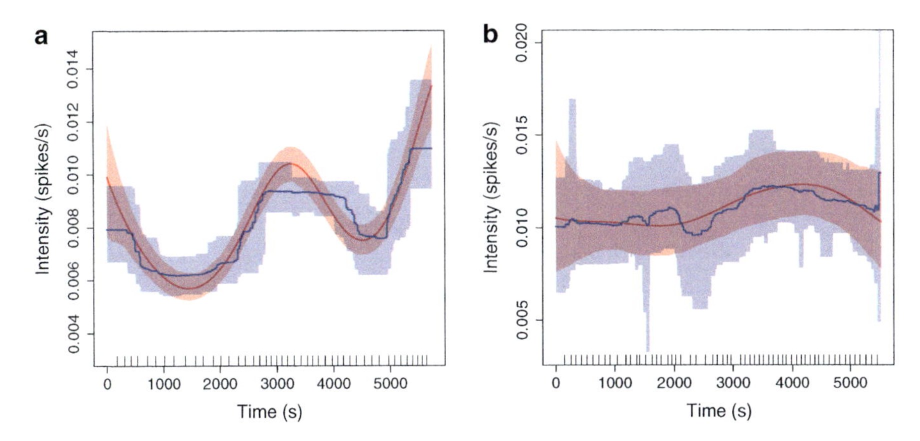

Fig. 32.7 Intensity functions (solid lines) and 95% confidence interval (shaded regions) for two cells in a multistep experiment. The initial stimulus was 20 μM carbachol and changed to 50 μM carbachol at $t = 2581$s (**a**) and $t = 2524$s (**b**). The prior for the intensity function is a GP (red) or piecewise constant (blue). The ticks along the x-axes indicate the Ca^{2+} spike times

shown in red in Fig. 32.7. Vitally, the intensity function obtained with a GP prior closely follows that derived from a piecewise constant prior. Given that the two priors represent significantly different functional forms of the intensity function, the consistency between the two approaches lends strong support for the validity of the derived intensity functions. Moreover, if we were to only use GPs as priors, a valid step is to interpolate the smooth prior with a piecewise linear function, which allows us to compute the mean ISI.

Having identified intensity functions that are consistent with measured Ca^{2+} spike sequences, we can now determine the conditional intensity functions $q(t|s) = q(y_i|y_{i-1})$. We start from Eq. (32.6), set $t = y_i$, $s = y_{i-1}$ and then replace $f(y_i, y_{i-1})$ with $g(u_i, u_{i-1}|x)$ from Eq. (32.10) by using Eq. (32.8) (see also Appendix 2). In other words, the conditional intensity function $q(t|s)$ is a highly nonlinear transformation of the estimated intensity function $x(t)$ given the Ca^{2+} spike times. In Fig. 32.8 we plot $q(t|s)$ for the data shown in Fig. 32.7. We notice that immediately after a Ca^{2+} spike $q(t|s)$ remains almost zero for some time before it increases. This indicates that during this period, Ca^{2+} spikes cannot occur, which is equivalent to saying that there is a refractory period. Importantly, we did not include an explicit refractory period in our model, i.e. we did not choose an ISI probability distribution that vanishes for a certain amount after the last Ca^{2+} spike. For instance, the Gamma distribution in Eq. (32.5) does not *per se* stay close to zero for small values of $(t - s)$. It only does so for certain values of α. Since the value of α is part of estimating $x(t)$, the vanishing of the conditional intensity function is an emergent result of the model. These findings are consistent with the presence of a refractory period T_r in the ansatz in Eq. (32.1). There, we chose the conditional intensity function and derived the ISI distribution, while for Fig. 32.8, we decided upon a certain ISI distribution and derived the conditional intensity function. The

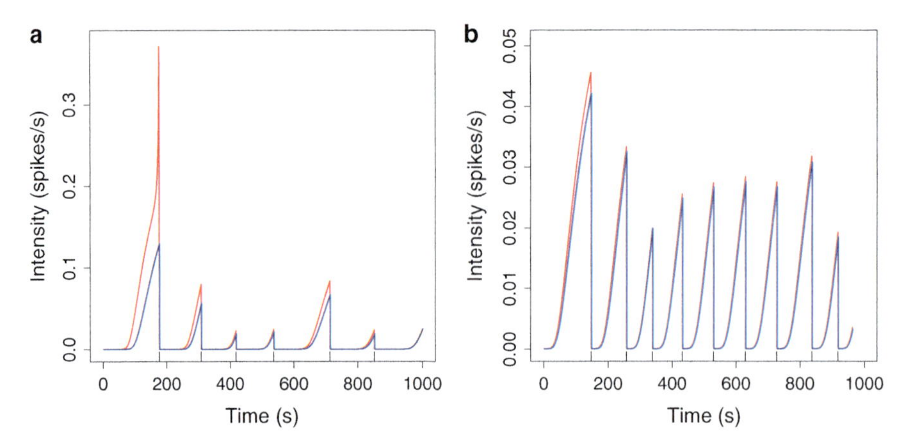

Fig. 32.8 Conditional intensity functions corresponding to the data in Fig. 32.7. The ticks along the x-axes indicate the Ca^{2+} spike times

agreement between the ansatz in Eq. (32.1) and the estimated conditional intensity function in Fig. 32.8 lends strong support to the former.

In addition to the conditional intensity function, we can also interrogate the ISI distribution. For a time-dependent intensity function $x(t)$, the corresponding ISI distribution is time-dependent as well, see Eqs. (32.21) and (32.22) in Appendix 2. However, when the intensity function is constant, this time-dependence is lost, and we can use the *same* ISI distribution for the entire period that x does not change. Inspecting Fig. 32.7a, we observe that the intensity function obtained with a PWL prior (blue line) is almost constant between 600s and 2000s, while a similar behaviour is seen in Fig. 32.7b during the first 2000s for the GP prior (red curve). Taking these values for x together with the inferred parameter values for the Gamma distribution, we now plot the corresponding ISI distributions in Fig. 32.9 based on Eq. (32.25). For the stronger stimulus (50 μM, blue line), the ISI distribution is shifted towards the left compared to the weaker stimulus (20 μM, red curve). In addition, the variance is more pronounced in the former than in the latter. To quantify this, we compute the mean μ and the standard deviation σ using the inferred intensity function x and the associated values of the Gamma distribution. We obtain $\mu = 159.07$s and $\sigma = 22.50$s for the small stimulus and $\mu = 96.73$s and $\sigma = 32.24$s for the stronger stimulus, respectively. We can compare this with the mean and standard deviation determined directly from the Ca^{2+} spike sequences shown in Fig. 32.7. We find $\mu_E = 166.57$s and $\sigma_E = 24.416$s at 20 μM and $\mu_E = 96.31$s and $\sigma_E = 27.03$s at 50 μM, respectively. The good agreement between the experimentally determined statistics (μ_E, σ_E) and the estimated quantities (μ, σ) demonstrates the usefulness of the Bayesian inference approach that we have employed here.

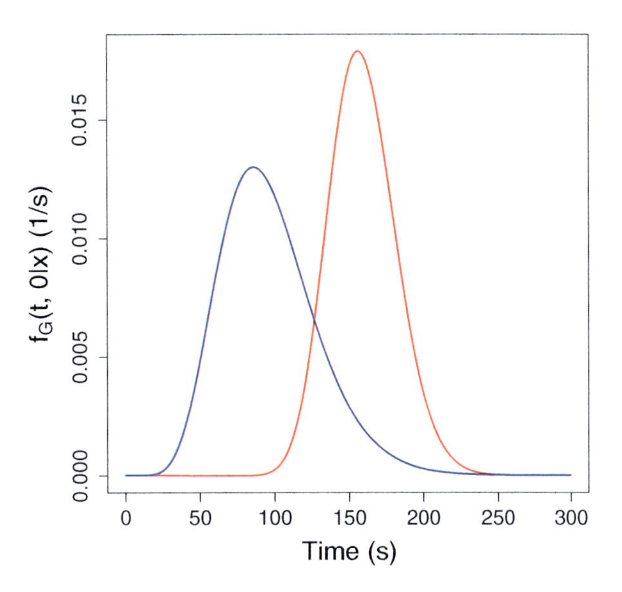

Fig. 32.9 ISI probability density $f_G(t, 0|x)$ for the data shown in Fig. 32.7a (red) and Fig. 32.7b (blue) for t between approximately 600s and 2000s (red) and t between 1s and 2000s (blue)

32.6 Concluding Discussion

Ca^{2+} spikes constitute a well-established mode of intracellular Ca^{2+} signalling across a large number of cell types. We can now draw on a substantial body of experimental measurements that have identified and characterised the cellular components that drive Ca^{2+} oscillations. Despite these successes, central questions remain open. Amongst them is a seemingly innocuous query: given a stimulus, can we predict the sequence of Ca^{2+} spikes? Since there is wide-ranging consensus that information about the stimulus is encoded in the properties of Ca^{2+} spike sequences, answering this question is critical for our understanding of intracellular Ca^{2+} signalling.

Addressing this issue from a modelling perspective is challenging for two reasons. The generation of Ca^{2+} spikes is firstly stochastic and secondly driven by the interaction of spatially distributed clusters of $InsP_3$ receptors. One avenue to make progress is to simulate partial differential equations for the intracellular calcium concentration (see [61] for a recent perspective). Here, we have reviewed a different framework that is a conceptual antipode to the first approach. While partial differential equations rely on mechanistic details and build oscillations from the bottom-up, the statistical ideas in this review aim at describing Ca^{2+} spikes directly at the cell level.

One advantage of a statistical model lies in its computational demands. It is considerably cheaper to generate Ca^{2+} spike sequences from a statistical model than to solve partial differential equations. This is particularly useful when studying cell populations, where intercellular heterogeneity calls for multiple Ca^{2+} spike sequences with different parameter values. But statistical models may also help conceptually. Ultimately, Ca^{2+} dependent signalling is driven in many instances by the sequence of Ca^{2+} spikes. Hence, the properties of Ca^{2+} spikes such as their ISI distribution are of central interest. Since it is conceivable that different microscopic models based on detailed molecular processes all lead to the same cellular behaviour, statistical frameworks are ideally placed to capture this common identity of Ca^{2+} spiking.

The first step in our statistical analysis is to derive a model for whole-cell Ca^{2+} spiking. Since Ca^{2+} spikes are stochastic, we can express their occurrence most naturally in the language of probabilities. A core ingredient is the ISI distribution $f(t, s)$, or equivalently the conditional intensity function $q(t|s)$. It is worth noting that both depend on only two times, i.e. we assume that the generation of a Ca^{2+} spike only depends on the history since the last Ca^{2+} spike. This independence of successive Ca^{2+} spikes has been shown for astrocytes, PLA cells and HEK293 cells in [83] for constant stimulation. In general, however, this might be too strong an assumption. In particular, when cells are challenged with continuously changing stimuli—in order to mimic a more realistic cellular environment—correlations within the signal might be inherited by the Ca^{2+} spikes. One possibility is to generalise the conditional intensity function. Instead of depending only on the last time t, it now relies on the entire Ca^{2+} spike history H_t, i.e. $q = q(t|H_t)$. This, however, does not necessarily lead to a mathematically tractable problem. A more practical approach is the introduction of an intensity function. Essentially, it transforms the original Ca^{2+} spike times in such a way that they become independent. As a consequence, we can use the original ISI distributions $f(t, s)$ or conditional intensity functions $q(t|s)$, which entails that the parameters of the model are those of the ISI distribution and the intensity function, respectively.

This leaves us with the task of finding the parameter values given the Ca^{2+} spike sequences. The last condition is of particular relevance to the current approach. Our goal is to derive a model that is constrained by experimental data and whose parameter values can be sensibly estimated. We have achieved this by employing Bayesian ideas, which allow us to determine the probability distribution of the parameters given the Ca^{2+} spike sequences, i.e. $p(\theta|\mathbf{y})$. This is a distinct advantage of the Bayesian framework. Other methods, such as maximum likelihood estimators, also provide information about the parameters of the model. However, they only deliver *one* set of parameter values associated with a standard error, not entire probability distributions. Moreover, these approaches come with numerical challenges and are hard to pursue in higher dimensions.

Since the ISI distribution is a core component of the model, we first ascertained if our choices are consistent with the recorded Ca^{2+} spike sequences. As Figs. 32.5 and 32.6 illustrate, the Gamma distribution captures the data well, while the inverse Gaussian and the Poisson distribution fail to do so. It is worth noting that the data analysed here were obtained in different experiments than those used in [67], yet both data sets lead to the same conclusion: Ca^{2+} spikes are well described by a Gamma distribution. This might point to the mechanisms that generate Ca^{2+} spikes. Since the Gamma distribution returns the probability of the first time that

α events have occurred (see Eq. (32.5)), it is consistent with the interpretation that the formation of a critical nucleus of elevated Ca^{2+}, driven by the occurrence of a certain number of Ca^{2+} puffs, underlies the generation of a Ca^{2+} spike. At the moment, we cannot rule out that other probability distributions that we have not tested yet describe Ca^{2+} spikes equally well or even better. The advantage of our Bayesian modelling framework is that it works for *any* probability distribution, which allows us to test more candidate distributions in the future. Moreover, we have two complementary tests at our disposal, the Q-Q and the K-S plot. While both approaches check whether two probability distributions coincide, the K-S plot is more sensitive towards the centre of the distribution, while the Q-Q plot focusses on the tails.

When testing for the most likely ISI distribution, we had to estimate the intensity function $x(t)$ at the same time, since the ISI distribution explicitly depends on $x(t)$ (see Eqs. (32.21) and (32.22)). Following on from our results so far, we focussed on intensity functions obtained for the Gamma distribution. The intensity function is central to our understanding of Ca^{2+} spike generation. An almost constant intensity function indicates that Ca^{2+} spikes originate from *stationary* dynamics. This means that the ISI distribution is identical for each recorded Ca^{2+} spike time, which allows us to compute the mean and the variance of a Ca^{2+} spike sequence (see Fig. 32.9). From a biological perspective, this corresponds to a cell with no explicitly time-dependent processes such as a continuous depletion of the ER or an accumulating degree of receptor desensitisation. In the latter, this does not mean that receptor desensitisation does not occur, but that the fraction of desensitised receptors across the cell stays constant. A change of experimental conditions is directly translated into changes of the intensity functions. For instance, recent experiments with sinusoidal stimuli led to intensity functions that reflect the rises and falls of the applied agonist [67]. Moreover, since intensity functions are estimated from Ca^{2+} spike sequences of individual cells, they mirror the variability of observed responses. A key line of research is therefore to quantify and classify such diverse intensity functions.

Using both the ISI distribution and the intensity function we can compute the conditional intensity function $q(t|s)$, which corresponds to the probability of a Ca^{2+} spike at time t given that a Ca^{2+} spike occurred at time s. The shape of q allows us to discuss potential mechanisms that are involved in Ca^{2+} spike generation. For instance, as Fig. 32.8 illustrates, the conditional intensity function remains small after a Ca^{2+} spike before it increases. This period of low Ca^{2+} spiking probability is consistent with the observations of refractoriness. Plots like Fig. 32.8 also allow us to estimate the range of refractory periods, which we can then compare to the refractory period obtained from plots of the mean ISI against the ISI standard deviation as seen in Fig. 32.1d. In addition, we can compare the rise time of the conditional probability function with known timescales of e.g. ER refilling or $InsP_3$ receptor recovery to ascertain whether any of the molecular timescales match the cellular time scale, or whether we deal with an emergent timescale.

One motivation for pursuing a statistical approach is to obtain *distributions* for the parameter values that govern Ca^{2+} spike generation. The reason for why distributions exist in the first place—and not a single parameter value only—lies in the inherent single-cell variability. Consider e.g. the variation of ξ and λ shown in Fig. 32.2. The recovery from global cellular inhibition is controlled by ξ. This involves inter alia resequestration of Ca^{2+} from the cytosol to the ER via SERCA pumps or recovery of $InsP_3$ levels. Since expression levels of SERCA pumps can vary amongst cells, recovery proceeds at different speed in different cells, which is captured by the variability of ξ. As for λ it controls the asymptotic Ca^{2+} spike rate. As Ca^{2+} spikes are believed to occur via the formation of a critical nucleus of elevated Ca^{2+} and subsequent propagation of a Ca^{2+} wave, the spatial distributions of $InsP_3$Rs and SERCA pumps are crucial. These distributions vary significantly between cells, which directly impacts on the spread of λ.

As with all modelling approaches, the methodology presented here is not without its caveats. In its current form, we only consider Ca^{2+} spike times and leave aside other Ca^{2+} spike properties such as amplitude, duration, shape, or baseline Ca^{2+} concentration levels. However, these characteristics have been shown to impact on a number of Ca^{2+} dependent processes [4]. An interesting point in this respect is a potential interplay between release amplitude, release duration and the absolute refractory period. Our results in [65] suggest that the absolute refractory period is not controlled by cell variability and hence that there is only one value for all cells. We can further test this hypothesis by extending the model for a Ca^{2+} spike as in Eq. (32.25) in Appendix 3 to explicitly include a distributed refractory period. The advantage of the Bayesian approach is that the estimation process remains conceptually the same, but we need to estimate additional parameters. As stated above, one incentive for the current work is to relate the estimated parameter values to biophysical processes. Care needs to be taken here as different processes can potentially give rise to the same whole cell parameter values that we infer. Hence further tests are needed to discriminate between different alternatives. A consequence of this consideration is that cells might employ a number of different strategies to generate the same whole cell signal, and it will be fascinating to tease apart the advantages and disadvantages of specific routes to global Ca^{2+} signals.

The preceding discussion illustrates that advanced statistical modelling can provide valuable insights into the dynamics of Ca^{2+} spiking. Vitally, our approach works for cells that are dynamically stimulated with agonist time courses that mimic physiological conditions in vivo, which allows us to model cellular Ca^{2+} spiking in a realistic environment. By inferring parameter values from single cell measurements, we can determine their ISI distribution, which is a central ingredient to modelling Ca^{2+} spikes. Moreover, it allows us to compute statistical properties such as means and variances, which in turn quantify stochastic Ca^{2+} spikes. In addition, we showed how the statistical model allows us to infer potential mechanisms of Ca^{2+} spike generation. This connects the statistical approach with the mechanistic framework of simulating partial differential equations for Ca^{2+} signalling. In the future, it will be desirable to see these two complementary techniques working hand in hand, which has the potential to significantly enhance our understanding of the Ca^{2+} signalling toolkit.

Appendix 1

We here show the equivalence of Eqs. (32.3) and (32.6). For this, it is convenient to introduce

$$F(t,s) = 1 - \int_s^t f(u,s)\mathrm{d}u \,. \tag{32.18}$$

The right hand side of Eq. (32.6) can be written as a full derivative in the form

$$q(t|s) = -\frac{\mathrm{d}}{\mathrm{d}t} \ln F(t,s) \tag{32.19}$$

Multiplying through by (-1) and integrating both sides with respect to t yields

$$-\int_s^t q(u|s)\mathrm{d}u = \ln F(t,s)\,, \tag{32.20}$$

noting that $\ln F(s,s) = 0$. When we exponentiate both sides and use the fact that $F(t,s) = f(t,s)/q(t|s)$ as per Eq. (32.6) we arrive at Eq. (32.3).

Appendix 2

Here, we demonstrate how to practically apply Eq. (32.10) when f is given by the density for the Gamma distribution as in Eq. (32.5). We obtain

$$g_{\mathrm{G}}(u_i, u_{i-1}|x) = x(y_i)\frac{\beta^\alpha}{\Gamma(\alpha)} X_{i,i-1}^{\alpha-1} \mathrm{e}^{-\beta X_{i,i-1}}\,, \tag{32.21}$$

where

$$X_{i,i-1} = \int_{y_{i-1}}^{y_i} x(v)\mathrm{d}v\,, \tag{32.22}$$

since the difference $(t - s)$ in the transformed time u is given by

$$u(t) - u(s) = u(y_i) - u(y_{i-1}) = \int_0^{y_i} x(v)\mathrm{d}v - \int_0^{y_{i-1}} x(v)\mathrm{d}v = \int_{y_{i-1}}^{y_i} x(v)\mathrm{d}v\,. \tag{32.23}$$

A common choice for f_1 and f_n is a Poisson distribution, which leads to

$$g_1(u_1, 0|x) = x(y_1)\mathrm{e}^{-X_{1,0}}\,, \qquad g_n(U, u_n|x) = e^{-X_{n+1,n}}\,, \tag{32.24}$$

in the transformed time, where we have set $y_0 = 0$ and $y_{n+1} = T$.

Appendix 3

To generate the Ca^{2+} spikes that underlie the histogram in Fig. 32.3 we use inverse sampling [84]. Since $x(t)$ is non-constant, we need to use the time-dependent ISI density $g(u_i, u_{i-1}|x)$ from Eq. (32.10). For ease of presentation, we rewrite Eq. (32.10) in terms of the non-transformed Ca^{2+} spike times y_i using the results from Appendix 2 as

$$f_{\mathrm{G}}(y_i, y_{i-1}|x) = x(y_i)\frac{\beta^\alpha}{\Gamma(\alpha)}\left[\int_{y_{i-1}}^{y_i} x(v)\mathrm{d}v\right]^{\alpha-1} \exp\left\{-\beta\int_{y_{i-1}}^{y_i} x(v)\mathrm{d}v\right\}, \tag{32.25}$$

which we use in the definition of the cumulative probability function

$$F_{\mathrm{G}}(t, y_{i-1}|x) = \int_{y_{i-1}}^{t} f_{\mathrm{G}}(s, y_{i-1}|x)\mathrm{d}s\,. \tag{32.26}$$

Suppose now that the last Ca^{2+} spike occurred at time y_{i-1}. We find the next Ca^{2+} spike time as $y_i = y_{i-1} + \Delta$, where

$$\Delta = \inf\{t|F_{\mathrm{G}}(t, y_{i-1}|x) > \omega\}\,, \tag{32.27}$$

and ω is a random number that is uniformly distributed between 0 and 1. To put it another way, we need to integrate the ISI probability density f_G from y_{i-1} until we obtain a value of ω for the integral and then add the corresponding upper bound of the integral to the previous Ca^{2+} spike time y_{i-1}.

References

1. Berridge MJ, Galione A (1988) Cytosolic calcium oscillators. FASEB J 2:3074–3082
2. Berridge MJ, Lipp P, Bootman MD (2000) The versatility and universality of calcium signalling. Nat Rev Mol Cell Biol 1:11–21
3. Parekh AB (2011) Decoding cytosolic Ca^{2+} oscillations. Trends Biochem Sci 36:78–87
4. Dupont G, Combettes L (2016) Fine tuning of cytosolic Ca^{2+} oscillations. F1000Research 5. https://f1000research.com/articles/5-2036/v1
5. Thul R, Bellamy TC, Roderick HL, Bootman MD, Coombes S (2008) Calcium oscillations. Adv Exp Med Biol 641:1–27
6. Dupont G, Combettes L, Bird GS, Putney JW (2011) Calcium oscillations. Cold Spring Harb Perspect Biol 3:pii:a004226
7. Dupont G, Falcke M, Kirk V, Sneyd J (2016) Models of calcium signalling. Interdisciplinary Applied Mathematics, vol 43. Springer, Cham
8. Schuster S, Marhl M, Höfer T (2002) Modelling of simple and complex calcium oscillations. From single-cell responses to intercellular signalling. Eur J Biochem 269:1333–1355
9. Uhlén P, Fritz N (2010) Biochemistry of calcium oscillations. Biochem Biophys Res Commun 396:28–32

10. Dolmetsch RE, Xu K, Lewis RS (1998) Calcium oscillations increase the efficiency and specificity of gene expression. Nature 392:933–936
11. Noren DP, Chou WH, Lee SH, Qutub AA, Warmflash A, Wagner DS et al (2016) Endothelial cells decode VEGF-mediated Ca^{2+} signaling patterns to produce distinct functional responses. Sci Signal 9:ra20–ra20
12. Di Capite J, Ng SW, Parekh AB (2009) Decoding of cytoplasmic Ca^{2+} oscillations through the spatial signature drives gene expression. Curr Biol 19:853–858
13. Nunemaker CS, Bertram R, Sherman A, Tsaneva-Atanasova K, Daniel CR, Satin LS (2006) Glucose modulates $[Ca^{2+}]_i$ oscillations in pancreatic islets via ionic and glycolytic mechanisms. Biophys J 91:2082–2096
14. Tse A, Tse FW, Almers W, Hille B (1993) Rhythmic exocytosis stimulated by GnRH-induced calcium oscillations in rat gonadotropes. Science (New York, NY). 260:82–84
15. Colsoul B, Schraenen A, Lemaire K, Quintens R, Van Lommel L, Segal A et al (2010) Loss of high-frequency glucose-induced Ca^{2+} oscillations in pancreatic islets correlates with impaired glucose tolerance in $Trpm5^{-/-}$ mice. Proc Natl Acad Sci U S A 107:5208–5213
16. Tsaneva-Atanasova K, Yule DI, Sneyd J (2005) Calcium oscillations in a triplet of pancreatic acinar cells. Biophys J 88:1535–1551
17. Tse A, Lee AK, Tse FW (2012) Ca^{2+} signaling and exocytosis in pituitary corticotropes. Cell Calcium 51:253–259
18. Santella L, Lim D, Moccia F (2004) Calcium and fertilization: the beginning of life. Trends Biochem Sci 29:400–408
19. Whitaker M (2006) Calcium at fertilization and in early development. Physiol Rev 86:25–88
20. Denninger P, Bleckmann A, Lausser A, Vogler F, Ott T, Ehrhardt DW et al (2014) Male-female communication triggers calcium signatures during fertilization in arabidopsis. Nat Commun 5:4645
21. Van Nhieu GT, Clair C, Bruzzone R, Mesnil M, Sansonetti P, Combettes L (2003) Connexin-dependent inter-cellular communication increases invasion and dissemination of Shigella in epithelial cells. Nat Cell Biol 5:720–726
22. Ullah G, Jung P, Machaca K (2007) Modeling Ca^{2+} signaling differentiation during oocyte maturation. Cell Calcium 42:556–564
23. Ullah G, Jung P (2006) Modeling the statistics of elementary calcium release events. Biophys J 90:3485–3495
24. Shuai JW, Jung P (2003) Optimal ion channel clustering for intracellular calcium signaling. Proc Natl Acad Sci U S A 100:506–510
25. Shuai JW, Jung P (2002) Stochastic properties of Ca^{2+} release of inositol 1,4,5-trisphosphate receptor clusters. Biophys J 83:87–97
26. Shuai J, Jung P (2002) Optimal intracellular calcium signaling. Phys Rev Lett 88:068102
27. De Young GW, Keizer J (1992) A single-pool inositol 1,4,5-trisphosphate-receptor-based model for agonist-stimulated oscillations in Ca^{2+} concentration. Proc Natl Acad Sci U S A 89:9895–9899
28. Li Y, Rinzel J (1994) Equations for $InsP_3$ receptor-mediated $[Ca^{2+}]_i$ oscillations derived from a detailed kinetic model: a Hodgkin-Huxley like formalism. J Theor Biol 166:461–473
29. Tang Y, Stephenson J, Othmer H (1996) Simplification and analysis of models of calcium dynamics based on IP_3-sensitive calcium channel kinetics. Biophys J 70:246–263
30. Gaspers LD, Bartlett PJ, Politi A, Burnett P, Metzger W, Johnston J et al (2014) Hormone-induced calcium oscillations depend on cross-coupling with inositol 1,4,5-trisphosphate oscillations. Cell Rep 9:1209–1218
31. Politi A, Gaspers LD, Thomas J, Höfer T (2006) Models of IP_3 and Ca^{2+} oscillations: frequency encoding and identification of underlying feedbacks. Biophys J 90:3120–3133
32. Sun CH, Wacquier B, Aguilar DI, Carayol N, Denis K, Boucherie S et al (2017) The Shigella type III effector IpgD recodes Ca^{2+} signals during invasion of epithelial cells. EMBO J 36(17):2567–2580
33. Bessonnard S, De Mot L, Gonze D, Barriol M, Dennis C, Goldbeter A et al (2014) Gata6, Nanog and Erk signaling control cell fate in the inner cell mass through a tristable regulatory network. Development (Cambridge, England) 141:3637–3648

34. De Caluwé J, Dupont G (2013) The progression towards Alzheimer's disease described as a bistable switch arising from the positive loop between amyloids and Ca^{2+}. J Theor Biol 331:12–18
35. Dupont G, Lokenye EFL, Challiss RAJ (2011) A model for Ca^{2+} oscillations stimulated by the type 5 metabotropic glutamate receptor: an unusual mechanism based on repetitive, reversible phosphorylation of the receptor. Biochimie 93:2132–2138
36. Dupont G, Abou-Lovergne A, Combettes L (2008) Stochastic aspects of oscillatory Ca^{2+} dynamics in hepatocytes. Biophys J 95:2193–2202
37. Swillens S, Champeil P, Combettes L, Dupont G (1998) Stochastic simulation of a single inositol 1,4,5-trisphosphate-sensitive Ca^{2+} channel reveals repetitive openings during 'blip-like' Ca^{2+} transients. Cell Calcium. 23:291–302
38. Tran Van Nhieu G, Kai Liu B, Zhang J, Pierre F, Prigent S, Sansonetti P et al (2013) Actin-based confinement of calcium responses during Shigella invasion. Nat Commun 4:1567
39. Salazar C, Zaccaria Politi A, Höfer T (2008) Decoding of calcium oscillations by phosphorylation cycles: analytic results. Biophys J 94:1203–1215
40. Kummer U, Olsen LF, Dixon CJ, Green AK, Bornberg-Bauer E, Baier G (2000) Switching from simple to complex oscillations in calcium signaling. Biophys J 79:1188–1195
41. Kummer U, Krajnc B, Pahle J, Green AK, Dixon CJ, Marhl M (2005) Transition from stochastic to deterministic behavior in calcium oscillations. Biophys J 89:1603–1611
42. Sneyd J, Han JM, Wang L, Chen J, Yang X, Tanimura A et al (2017) On the dynamical structure of calcium oscillations. Proc Natl Acad Sci U S A 114:1456–1461
43. Cao P, Tan X, Donovan G, Sanderson MJ, Sneyd J (2014) A deterministic model predicts the properties of stochastic calcium oscillations in airway smooth muscle cells. PLoS Comput Biol 10:e1003783
44. Wang IY, Bai Y, Sanderson MJ, Sneyd J (2010) A mathematical analysis of agonist- and KCl-induced Ca^{2+} oscillations in mouse airway smooth muscle cells. Biophys J 98:1170–1181
45. Higgins ER, Cannell MB, Sneyd J (2006) A buffering SERCA pump in models of calcium dynamics. Biophys J 91:151–163
46. Sneyd J, Tsaneva-Atanasova K, Yule DI, Thompson JL, Shuttleworth TJ (2004) Control of calcium oscillations by membrane fluxes. Proc Natl Acad Sci U S A 101:1392–1396
47. Sneyd J, Tsaneva-Atanasova K, Reznikov V, Bai Y, Sanderson MJ, Yule DI (2006) A method for determining the dependence of calcium oscillations on inositol trisphosphate oscillations. Proc Natl Acad Sci USA 103:1675–1680
48. Nowacki J, Mazlan S, Osinga HM, Tsaneva-Atanasova K (2010) The role of large-conductance Calcium-activated K^+ (BK) channels in shaping bursting oscillations of a somatotroph cell model. Physica D Nonlinear Phenomena 239:485–493
49. Bootman M, Niggli E, Berridge M, Lipp P (1997) Imaging the hierarchical Ca^{2+} signalling system in HeLa cells. J Physiol 499 (Pt 2):307–314
50. Smith IF, Wiltgen SM, Parker I (2009) Localization of puff sites adjacent to the plasma membrane: functional and spatial characterization of Ca^{2+} signaling in SH-SY5Y cells utilizing membrane-permeant caged IP_3. Cell Calcium 45:65–76
51. Taufiq-Ur-Rahman, Skupin A, Falcke M, Taylor CW (2009) Clustering of $InsP_3$receptors by $InsP_3$retunes their regulation by $InsP_3$ and Ca^{2+}. Nature 458:655–659
52. Keebler MV, Taylor CW (2017) Endogenous signalling pathways and caged IP_3 evoke Ca^{2+} puffs at the same abundant immobile intracellular sites. J Cell Sci 130:3728–3739
53. Konieczny V, Keebler MV, Taylor CW (2012) Spatial organization of intracellular Ca^{2+} signals. Semin Cell Dev Biol 23:172–180
54. Bruno L, Solovey G, Ventura AC, Dargan S, Dawson SP (2010) Quantifying calcium fluxes underlying calcium puffs in Xenopus laevis oocytes. Cell Calcium 47:273–286
55. Shuai J, Pearson JE, Foskett JK, Mak DOD, Parker I (2007) A kinetic model of single and clustered IP_3 receptors in the absence of Ca^{2+} feedback. Biophys J 93:1151–1162
56. Smith IF, Parker I (2009) Imaging the quantal substructure of single IP_3R channel activity during Ca^{2+} puffs in intact mammalian cells. Proc Natl Acad Sci U S A 106:6404–6409

57. Flegg MB, Rüdiger S, Erban R (2013) Diffusive spatio-temporal noise in a first-passage time model for intracellular calcium release. J Chem Phys 138:154103
58. Dobramysl U, Rüdiger S, Erban R (2016) Particle-based multiscale modeling of calcium puff dynamics. Multiscale Model Simul 14:997–1016
59. Weinberg SH, Smith GD (2014) The influence of Ca^{2+} buffers on free [Ca^{2+}] fluctuations and the effective volume of Ca^{2+} microdomains. Biophys J 106:2693–2709
60. Wieder N, Fink R, von Wegner F (2015) Exact stochastic simulation of a calcium microdomain reveals the impact of Ca^{2+} fluctuations on IP_3R gating. Biophys J 108:557–567
61. Falcke M, Moein M, Tilunaite A, Thul R, Skupin A (2018) On the phase space structure of IP_3 induced Ca^{2+} signalling and concepts for predictive modeling. Chaos (Woodbury, NY) 28:045115
62. Skupin A, Kettenmann H, Winkler U, Wartenberg M, Sauer H, Tovey SC et al (2008) How does intracellular Ca^{2+} oscillate: by chance or by the clock? Biophys J 94:2404–2411
63. Thurley K, Smith IF, Tovey SC, Taylor CW, Parker I, Falcke M (2011) Timescales of IP_3-evoked Ca^{2+} spikes emerge from Ca^{2+} puffs only at the cellular level. Biophys J 101:2638–2644
64. Moein M, Grzyb K, Gonçalves Martins T, Komoto S, Peri F, Crawford A et al (2018) CaSiAn: a calcium signaling analyzer tool. Bioinformatics (Oxford, England) 1:11
65. Thurley K, Tovey SC, Moenke G, Prince VL, Meena A, Thomas J et al (2014) Reliable encoding of stimulus intensities within random sequences of intracellular Ca^{2+} spikes. Sci Signal 7:ra59
66. Rooney TA, Sass EJ, Thomas AP (1990) Agonist-induced cytosolic calcium oscillations originate from a specific locus in single hepatocytes. J Biol Chem 265:10792–10796
67. Tilunaite A, Croft W, Russell N, Bellamy TC, Thul R (2017) A Bayesian approach to modelling heterogeneous calcium responses in cell populations. PLoS Comput Biol 13:e1005794
68. Barbieri R, Quirk MC, Frank LM, Wilson MA, Brown EN (2001) Construction and analysis of non-Poisson stimulus-response models of neural spiking activity. J Neurosci Methods 105:25–37
69. Miura K, Tsubo Y, Okada M, Fukai T (2007) Balanced excitatory and inhibitory inputs to cortical neurons decouple firing irregularity from rate modulations. J Neurosci 27:13802–13812
70. Thurley K, Skupin A, Thul R, Falcke M (2012) Fundamental properties of Ca^{2+} signals. Biochim Biophys Acta 1820:1185–1194
71. Croft W, Reusch K, Tilunaite A, Russel N, Thul R, Bellamy TC (2016) Probabilistic encoding of stimulus strength in astrocyte global calcium signals. Glia 64:537–552
72. Thul R, Falcke M (2006) Frequency of elemental events of intracellular Ca^{2+} dynamics. Phys Rev E 73(6 Pt 1):061923
73. Thul R, Falcke M (2007) Waiting time distributions for clusters of complex molecules. Europhys Lett 79:38003
74. Thul R, Thurley K, Falcke M (2009) Toward a predictive model of Ca^{2+} puffs. Chaos (Woodbury, NY) 19:037108
75. Thurley K, Falcke M (2011) Derivation of Ca^{2+} signals from puff properties reveals that pathway function is robust against cell variability but sensitive for control. Proc Natl Acad Sci U S A 108:427–432
76. Moenke G, Falcke M, Thurley K (2012) Hierarchic stochastic modelling applied to intracellular Ca^{2+} signals. PLoS One 7:e51178
77. Papoulis A, Pillai SU (2002) Probability, random variables and stochastic processes, 4th edn. McGraw Hill, Boston
78. Ross SM (2003) Introduction to probability models, 8th edn. Academic, Amsterdam
79. Cunningham JP, Shenoy KV, Sahani M (2008) Fast gaussian process methods for point process intensity estimation. In: Proceedings of the 25th International Conference on Machine Learning, Stanford University, Palo Alto, pp 192–199

80. Cunningham JP, Yu B, Shenoy K, Sahani M (2008) Inferring neural firing rates from spike trains using gaussian processes. In: Platt JC, Koller D, Singer Y, Roweis S (eds) Advances in neural information processing systems, vol 20. MIT Press. Cambridge, MA, pp 329–336
81. Robert CP, Casella G (2005) Monte Carlo statistical methods. Springer Texts in Statistics. Springer, New York
82. Brown EN, Barbieri R, Ventura V, Kass RE, Frank LM (2002) The time-rescaling theorem and its application to neural spike train data analysis. Neural Comput 14:325–346
83. Skupin A, Falcke M (2010) Statistical analysis of calcium oscillations. Eu Phys J Spec Top 187:231–240
84. Devroye L (1986) Non-uniform random variate generation. Springer, New York

Chapter 33
Calcium Regulation of Bacterial Virulence

Michelle M. King, Biraj B. Kayastha, Michael J. Franklin, and Marianna A. Patrauchan

Abstract Calcium (Ca^{2+}) is a universal signaling ion, whose major informational role shaped the evolution of signaling pathways, enabling cellular communications and responsiveness to both the intracellular and extracellular environments. Elaborate Ca^{2+} regulatory networks have been well characterized in eukaryotic cells, where Ca^{2+} regulates a number of essential cellular processes, ranging from cell division, transport and motility, to apoptosis and pathogenesis. However, in bacteria, the knowledge on Ca^{2+} signaling is still fragmentary. This is complicated by the large variability of environments that bacteria inhabit with diverse levels of Ca^{2+}. Yet another complication arises when bacterial pathogens invade a host and become exposed to different levels of Ca^{2+} that (1) are tightly regulated by the host, (2) control host defenses including immune responses to bacterial infections, and (3) become impaired during diseases. The invading pathogens evolved to recognize and respond to the host Ca^{2+}, triggering the molecular mechanisms of adhesion, biofilm formation, host cellular damage, and host-defense resistance, processes enabling the development of persistent infections. In this review, we discuss: (1) Ca^{2+} as a determinant of a host environment for invading bacterial pathogens, (2) the role of Ca^{2+} in regulating main events of host colonization and bacterial virulence, and (3) the molecular mechanisms of Ca^{2+} signaling in bacterial pathogens.

Keywords Calcium signaling · Calcium channels · Calcium sensors · Toxins · Adhesins · Biofilm · Attachment · Two component regulatory systems · Secretion · Bacterial pathogens

M. M. King · B. B. Kayastha · M. A. Patrauchan (✉)
Department of Microbiology and Molecular Genetics, Oklahoma State University, Stillwater, OK, USA
e-mail: m.patrauchan@okstate.edu

M. J. Franklin
Department of Microbiology and Center for Biofilm Engineering, Montana State University, Bozeman, MT, USA

© Springer Nature Switzerland AG 2020

M. S. Islam (ed.), *Calcium Signaling*, Advances in Experimental Medicine and Biology 1131, https://doi.org/10.1007/978-3-030-12457-1_33

33.1 Elevated External Calcium (Ca^{2+}) Regulates Adaptation of Bacterial Pathogens to Their Host Environment

33.1.1 Host-Associated Ca^{2+}

In order to survive, bacteria must sense the chemical landscape of their environment and respond to it by adjusting their biological activities. Bacterial pathogens have an additional challenge of recognizing the transition between outside and inside the host and efficiently rearranging their gene expression to enable survival in the hostile host. The environment inside the host has a drastically different chemistry regulated by complex signaling systems, including one of the most versatile intracellular messengers, calcium (Ca^{2+}).

Ca^{2+} signaling has been widely studied in eukaryotes [1, 2]. Ca^{2+} signaling is based on tightly regulated fluctuations in the levels of the ion in different cellular compartments, that trigger multiple molecular pathways. Whereas the cytoplasmic concentration of free Ca^{2+} is maintained at high nanomolar level, the extracellular concentration of the ion reaches millimolar levels [3–5] differing between different body fluids, tissues, and organs. Several examples are summarized in Table 33.1.

Since Ca^{2+} signaling regulates most essential cellular processes, slight abnormalities in Ca^{2+} homeostasis cause diseases or are a result of certain pathologies. For example, in cystic fibrosis (CF) [6, 7], different types of cells, including skin fibroblasts and bronchial epithelium cells, show elevated intracellular Ca^{2+} pools [8, 9]. In addition, abnormally elevated levels of Ca^{2+} were registered in multiple body fluids of CF patients (Table 33.1). Further, the elevation of cytosolic Ca^{2+} concentration was shown to trigger host immune responses against invading pathogens. For example, intestinal epithelial cells infected with *Salmonella* serotype Typhimurium require an increased cytosolic Ca^{2+} to express pro-inflammatory chemokine IL-8 [10]. Elevated Ca^{2+} in CF sputum positively correlates with the release of IL-8 in the necrotic immune cells [11]. As a part of the innate immunity defense, production of antimicrobial peptides (AMPs) by epidermal keratinocytes in response to infection by *Pseudomonas aeruginosa, Staphylococcus aureus* and

Table 33.1 Examples of free Ca^{2+} levels in human body fluids

Body fluid	$[Ca^{2+}]$	References
Joint fluids	4 mM	[26]
Plasma	1.3–1.5 mM	[23, 27–29]
Serum	0.7 to 1.4 mM.	[23, 30–34]
Saliva (in CF patients)	0.3 mM (4.8 ± 0.7 mM)	[35–41]
Nasal secretions (in CF patients)	3.1 ± 1.6 mM (4.7 ± 2.2 mM)	[42–44]
Sputum (in CF patients)	1.1 mM (2.5 mM)	[11]
Urine	1.6–5 mM	[45]

other pathogens is induced by elevated levels of Ca^{2+} [12]. Some of the AMPs, including a family of Ca^{2+} binding EF-hand S100 family, require Ca^{2+} for their interactions with targets [13].

Some bacterial pathogens are able to alter the hosts $[Ca^{2+}]_{in}$ levels through activating Ca^{2+} flux across the plasma membrane and, releasing Ca^{2+} from the intracellular stores into the cytosol [10, 14–17]. These interactions can be mediated by bacterial surface associated proteins such as PilC of *Neisseria meningitidis* [17], FliC of *P. aeruginosa* and *Salmonella* [18], and FimH of *Escherichia coli* [19] or by secreted effectors, such as hemolysin A from *S. aureus* [20], pyocyanin and homoserine lactones from *P. aeruginosa* and *Serratia liquefaciens* [21–25]. Such alterations in the host Ca^{2+} have been shown to facilitate bacterial adherence and subsequent internalization into the host cells.

In plants, Ca^{2+} is one of the earliest signaling elements that coordinate adaptive immune responses to invading pathogenic bacteria. Cytoplasmic Ca^{2+} ($[Ca^{2+}]_{cyt}$) increases in response to infecting pathogens, such as *P. syringae* [46]. A sustained elevation of $[Ca^{2+}]_{cyt}$ serves as an important early signal, which links the recognition of infection to downstream defenses including generation of reactive oxygen species (ROS) and oxidative burst [47, 48]. The ROS burst may lead to cell death preventing the pathogen establishment inside the plant [49].

Overall, Ca^{2+} is an essential component of the host environment that both responds to the presence of bacterial pathogens, and regulates specific defense mechanisms. Ca^{2+} levels in a host may signal to the invading pathogens that they are entering a host and also indicate the status of immune protection in the host. Therefore, recognizing the host Ca^{2+} level can be beneficial to the invaders and trigger their adaptation to the host environment, and lead to their increased virulence and survival of the pathogen.

33.1.2 Ca^{2+} Triggers Life Style Switches in Bacterial Pathogens

Bacteria possess efficient regulatory systems that enable their adaptation to continuously changing environments. Regulation of gene expression is key for bacterial survival in a variety of environments. One particularly efficient and complex mechanism of surviving hostile environments is a switch between free-swimming or planktonic lifestyle to sessile life as surface-associated community, called biofilm. This transition is enabled by major molecular rearrangements ultimately enabling increased resistance, cell-cell communication and efficient metabolism [50, 51]. This mechanism is of particularly high importance to extracellular pathogenic bacteria colonizing host surfaces and surviving both host defenses and antimicrobial treatments.

There is a growing body of evidence that Ca^{2+} plays both a structural and a regulatory role in the transition to surface-associated biofilm lifestyle. Bacterial adhesion is the first step in biofilm formation, and itself is a survival mechanism, as nutrients, for example, tend to accumulate at surfaces [52]. The effect of Ca^{2+} on adhesion is partially due to electrostatic interactions, but also due to strong interactions of the surfaces with the cell structures, such as pili and fimbriae [53–55], and other macromolecules including teichoic acids, adhesins, lipopolysaccharide (LPS), and extracellular polysaccharides (EPS). It was shown that cell surface properties and their electrostatic interactions with the substratum contribute to Ca^{2+}-enhanced adhesion of non-motile and motile *P. aeruginosa* [56]. Ca^{2+}-enhanced cell adhesion to diverse host molecules and in vitro substrates, as well as cell-cell aggregation, relies on the presence of type I and type IV pili in a number of pathogens, including *Xylella fastidiosa,* [53], *P. aeruginosa* [57], *Vibrio vulnificus* [58], and *N. gonorrhoeae* [59]. The Ca^{2+} regulation of the type IV pilus is determined by its binding to pilus-biogenesis factor, PilY1, enabling pilus extension and retraction [60]. This interaction is also required for the bacterium twitching motility. By interacting with type I pili and fimbriae, Ca^{2+} modulates invasion of bacterial pathogens, such as *E. coli,* into host cells [19, 61].

Ca^{2+} also enhances bacterial adhesion via large cell surface Ca^{2+}-binding adhesins, such as SdrC and SdrD in *S. aureus* [62, 63] and BapA in *Paracoccus denitrificans* [64]. The former contain EF hand-like motifs that bind Ca^{2+} required for protein folding. The latter belongs to repeats-in-toxin (RTX) family, containing multiple nonapeptide Ca^{2+}-binding domains, secreted via Type I Secretion System (TISS), and serving a variety of functions, including cell-cell or cell-surface interactions or contributing to protection against hostile environments by forming bacterial surface (S)-layers (reviewed in [65]). In *Listeria monocytogenes,* elevated Ca^{2+} has been reported to stabilize the complex between the adhesin InlA (Internalin A) and the human extracellular E-cadherin domain 1 (hEC1). Once inside the host cell, where Ca^{2+} concentrations are lower, the InlA-hEC1 complex dissociates, which facilitates the liberation of the bacteria from the host cell membrane into the cytosol [66]. Ca^{2+} is required for multimerization of large adhesin LapF involved in colonization, microcolony formation, and biofilm maturation of *P. putida* [67–69].

Due to its interactions with surface proteins and by forming ionic bridges between negatively charged macromolecules, Ca^{2+} enhances cell aggregation and strengthens biofilm matrixes, including cell aggregation in oral *Streptococci* [70] and alginate cross-linking of *P. aeruginosa* biofilm matrix [71]. Ca^{2+} was also shown to bind extracellular DNA (eDNA), another important component of biofilm matrix, and this thermodynamically favorable binding increases bacterial aggregation in several Gram-positive and Gram-negative species, including *S. aureus* and *P. aeruginosa*. The authors concluded that Ca^{2+} did not affect DNA release [72]. However, this observation is likely species- and strain-specific [73], as Ca^{2+} was shown to induce production of *P. aeruginosa* pyocyanin, which promotes DNA release [74]. Furthermore, the presence of Ca^{2+} increased eDNA release, contributing to biofilm formation in *Streptococcus mutans* [75]. Ca^{2+} was also shown to increase the adhesive nature of *P. fluorescence* biofilm, but reduced its elastic properties [76].

In different bacterial species, elevated Ca^{2+} either stimulates or reduces biofilm formation. Positive regulation was observed in response to 1–10 mM Ca^{2+} in *Pectobacterium carotovorum* [77], *Rhizobium leguminosarum* [78], *Pseudoalteromonas* sp. [79], *Shewanella oneidensis* [80], *P. aeruginosa* [73], *X. fastidiosa* [81], *V. cholerae* [82], and *V. fischeri* [83]. This regulation was shown to be mediated by diverse mechanisms. For example, elevated Ca^{2+} activates the transcription of genes responsible for production of surface adhesins and EPS: alginate in *P. aeruginosa* [73] and *P. syringae* [84]; symbiosis polysaccharide (*syp*) or cellulose in *V. fisheri*. Ca^{2+}-dependent hemophilic interactions of surface-associated adhesin SdrC promotes biofilm formation in *S. aureus* [62].

Negative regulation of biofilm by elevated Ca^{2+} was reported in *S. aureus* [85] and *V. cholerae* [86]. In *V. cholerae,* this regulation is mediated by negatively regulated two-component system CarSR and *vps* (Vibrio polysaccharide) genes. However, *V. cholerae* also produces Vps-independent biofilm, which is preferred under high Ca^{2+} sea water conditions, where Ca^{2+} interacts directly with the O-antigen polysaccharide [87]. *S. aureus* produces several surface adhesins, such as clumping factors A and B (ClfA and ClfB) [88, 89] and biofilm-associated protein (Bap) [85]. These proteins contain Ca^{2+}-binding EF-hand-like motifs, and binding the ion inhibits their role in cell adhesion and biofilm formation. A point mutation in protease aureolysin (*aur)* gene in one of *S. aureus* strains led to increased activity of ClfB, required for biofilm growth under Ca^{2+}-depleted conditions [90].

Some factors contributing to biofilm formation are known to be regulated by cyclic-di-GMP (c-di-GMP) (reviewed in [91]) and quorum sensing (QS) (reviewed in [92]). This raises the possibility of interconnections between c-di-GMP, QS and Ca^{2+} regulatory networks that warrant further studies.

33.1.3 Virulence Factors Regulated by Ca^{2+}

Factors that enable pathogenic bacteria to cause diseases can be broadly grouped into several categories, such as penetration, colonization, damage of host cells, evasion of host defenses, and proliferation, all ultimately contributing to the developing infections. Colonization requires pathogens to establish interactions with host tissues by producing extracellular or cell-associated molecules. It may also involve communication between invaders themselves or those with commensals. The relationship between some of these factors and Ca^{2+} is discussed above. Here we outline virulence factors attributed to more invasive host-pathogen interactions that are directly or indirectly regulated by Ca^{2+}.

Bacterial invasion is commonly facilitated by the production and secretion of molecules that cause enzymatic or non-enzymatic damage to the host cells [93, 94]. A number of secreted enzymes are known to be regulated by Ca^{2+} in bacterial pathogens. For example, in *P. aeruginosa,* Ca^{2+} promotes the production of extracellular proteases LasB, LasA, PrpL (protease IV), and AprA [73, 95–97].

In the case of elastase LasB, Ca^{2+} not only increases the production of the protein, but modulates its processing, export, stability and enzymatic activity [95, 98, 99]. The enzymatic activity covers a wide repertoire of substrates, including elastin, collagen, and human immunoglobulins, underlining the significance of the protein and its Ca^{2+} regulation in *P. aeruginosa* pathogenicity. The alkaline protease A (AprA) binds Ca^{2+} through its C-terminal RTX domain, enabling folding of both C- and N-terminal proteolytic domains, which is required for stable conformation and enzymatic activity of the protease [73, 80, 96]. Both AprA and LasB are capable of degrading exogenous flagellin monomers under Ca^{2+}-replete condition, which prevents flagellin-mediated immune recognition and killing of *P. aeruginosa via* complement-mediated phagocytosis [99, 100]. The Ca^{2+}-enhanced production of the two proteases with anti-flagellin activity provides a robust strategy for *P. aeruginosa* to escape the detection by the complement system.

Our earlier studies showed that production of pyocyanin, the extracellular redox cycling compound and a virulence factor of *P. aeruginosa* [101, 102] is up-regulated in response to elevated Ca^{2+} [73]. Pyocyanin is found in pulmonary fluids of CF patients and shown to disrupt Ca^{2+} homeostasis of the host epithelial cells [21, 103], potentially contributing to a further increase of extracellular host Ca^{2+} and therefore induction of Ca^{2+}-regulated virulence.

Toxins represent one of the most powerful strategies of bacterial pathogens to conquer a host. Ca^{2+} modulates the production, secretion, and function of several toxins in a number of pathogens. In *E. coli,* Ca^{2+} is required for Hemolysin A (HlyA) binding to erythrocytes [104]. Binding Ca^{2+} causes conformational change in the toxin increasing its surface hydrophobicity and promoting the irreversible binding to the lipid bilayer of erythrocytes. This interaction preludes and ensures the lytic effect [105]. In *V. cholerae*, Ca^{2+} enhances bile salt-dependent activation of virulence. The mechanism relies on Ca^{2+} promoting the bile salt-induced activation of transmembrane virulence regulator TcpP, which then induces the production of major virulence factors, including toxin-coregulated pilus (TCP) [106]. The presence of Ca^{2+} has been reported to be essential for the toxicity of anthrax-edema toxin (composed of protective antigen and edema factor) produced by *Bacillus anthracis* [107]. The edema factor has adenylate cyclase activity synthesizing cAMP. Once in the host cytosol, the edema factor produces cAMP, which causes a rapid influx of Ca^{2+}. The accumulation of cAMP in the cytosol requires the presence of extracellular Ca^{2+}. As a potent inhibitor of immune response, accumulated cAMP leads to suppression of proinflammatory cytokines, phagocytosis and bactericidal activity of leukocytes thereby facilitating the survival of bacteria in the host [107, 108].

On the other hand, elevated host Ca^{2+} may have a negative regulatory effect on virulence and thus contribute to host defenses. One example is a cell wall degrading enzyme endopolygalacturonase (PehA) that is down-regulated by high (10–30 mM) levels of Ca^{2+} in a plant pathogen *Pectobacterium carotovorum*. This prevents the pathogen from infecting the plant [109].

Overall, Ca^{2+} regulates many virulence factors of invading bacterial pathogens, which stresses the importance of a detailed understanding of Ca^{2+} regulatory pathways in these pathogens.

33.1.4 Ca^{2+}-Regulated Secretion Systems

Most bacteria can respond to and manipulate their environment through the secretion of extracellular proteins. Bacterial secreted proteins are often involved in breakdown of macromolecules, such as polysaccharides or polypeptides to simple sugars or amino acids that the bacteria can take up and utilize as carbon, nitrogen, and energy sources. Secreted proteins may also act as virulence factors, as in the case for the proteases described above, LasA, LasB, PrpL, and AprA, which modulate immune effectors and degrade elastic tissues [73, 95, 96, 99, 100]. Pathogenic bacteria also use protein secretion to kill other cells, including eukaryotic cells [110] and, in some cases, competing bacteria within biofilm communities [111]. Extracellular Ca^{2+} concentration plays a direct or indirect signaling role in many of the bacterial protein secretion systems.

Bacteria use at least six different strategies to secrete proteins (termed: T1SS to T6SS) reviewed in [112]. The T1SS transports specific proteins directly from the cytoplasm to the extracellular medium, with no apparent periplasmic intermediate. The Ca^{2+} requiring protease, AprA [113], is secreted by the T1SS, composed of three components, AprDEF, which include cytoplasmic ATPase, an inner membrane protein component, and an outer membrane protein [114]. These proteins form a molecular complex, dedicated to the export of AprA [113]. AprA accumulates in the biofilms of *P. aeruginosa* under Ca^{2+}-replete conditions, but not under Ca^{2+} – limiting conditions [73]. The other protease virulence factors described above that require Ca^{2+} for activity or structural integrity, LasA, LasB, and the PrpL [98], are secreted by the T2SS [115–117]. The T2SS is a general pathway for secretion of a variety of extracellular proteins. In the T2SS, proteins with N-terminal export signal peptides, are first exported to the periplasm by either the Sec export machinery or the twin-arginine translocation (TAT) system [118, 119]. Sec, exports proteins in an unfolded state, then folds the proteins into their three dimensional confirmation in the periplasm, with the help of proteins such as disulfide bond isomerase, DsbA. The TAT system exports folded proteins into the periplasm. Once in the periplasm, proteins are secreted across the outer member *via* the secretion apparatus. For example, in *P. aeruginosa* the secretion apparatus is composed of the Xcp proteins (XcpA and XcpP-Z) or the homologous system, Hxc (composed of HxcP-Z) [112]. In addition to secretion of enzymes into the extracellular medium, the T2SS also plays a role in generation of certain types of pili, the Type IV pilus, which plays role in bacterial attachment and movement along surface. In twitching motility, bacteria move along surfaces by extension and retraction of the pili, through polymerization and depolymerization of the pilin subunits. Some bacteria, including *E. coli* and several *Vibrio* spps, requires Ca^{2+} for structural integrity of the major pseudopilin subunit, GspG [120].

The type III secretion system (T3SS), encoded on pathogenicity islands of many pathogenic bacteria, delivers effector protein toxins directly into the cytosol of eukaryotic cells during infection. The toxins, including enzymes such as ADP-ribosyltransferases, phospholipases, or adenylate cyclases, disrupt such host cell activities as actin remodeling, and gene regulation [112]. Perhaps the most interesting role of Ca^{2+} in secretion of bacterial virulence factors is its direct role in expression regulation (activation or repression) of the genes encoding the secretion apparatuses. It has been known for many years that expression of the T3SS genes is induced by either host-cell contact or by chelation of Ca^{2+} from the medium (low $[Ca^{2+}]$) [121, 122]. The T3SS forms a complex needle-like structure that is related to the bacterial flagella basal body [123]. The T3SS includes inner and outer membrane ring structures, and cytoplasmic protein components that dock to the inner membrane ring. The needle-like structure protrudes from the basal body, punctures the host cell membrane, and secretes toxins directly into the host cells. For this reason, the T3SS has also been termed the injectosome.

Regulation of expression of the T3SS gene clusters by host cell contact has been well characterized in *P. aeruginosa* [124, 125]. Transcription of the T3SS in *P. aeruginosa* is controlled by the transcriptional activator, ExsA, an AraC/XylS-type regulator. ExsA is inhibited by a cascade of protein-protein interactions that prevent ExsA binding to the DNA. The cascade involves interactions of ExsA, ExsD, ExsC, and ExsE. Expression of the T3SS genes is induced when ExsE is translocated from the cell through the T3SS, ultimately titrating the anti-activator, ExsD away from the ExsA and allowing transcription. If translocation of ExsE through the T3SS is functional (e.g. during host cell contact or at low Ca^{2+}), transcription of the TS33 genes is activated. If ExsE builds up in the cell due to lack of host cell contact, then further transcription of the TS33 genes is inhibited.

Another role of Ca^{2+} in regulation of T3SS was recently shown to involve the Ca^{2+}-sensor protein, LadS [126]. LadS is a hybrid membrane-bound sensor, containing both a histidine kinase domain and a periplasmic Ca^{2+}-binding DISMED2 domain. Broder et al. [126] mutated potential Ca^{2+}-binding residues in *P. aeruginosa* and found that the resulted T3SS gene expression became insensitive to Ca^{2+} conditions. Ca^{2+} binding to the LadS DISMED2 domain is the first step in a regulatory cascade that responds to external Ca^{2+}. The cascade involves two component system GacCS, two small regulatory RNAs, RsmY and RsmZ, and the RNA binding protein, RsmA. Ultimately, binding of RsmA to specific mRNA sequences results in gene regulation at the translational level [126].

Assembly of the T3SS is a dynamic process that responds to external Ca^{2+}. Using *Yersinia entercolitica*, Diepold et al. [127] tagged components of the T3SS with fluorescent reporter proteins, so that they could image the membrane and cytosolic components. Using Fluorescence correlation spectroscopy, they calculated the diffusion rates of the T3SS cytoplasmic components under inducing (low Ca^{2+}) and non-inducing (high Ca^{2+}) conditions, as they assembled. They found that the rate of diffusion of the cytoplasmic components changed with the external Ca^{2+} concentration and proposed this as a novel mechanism for the role of Ca^{2+} in regulation of T3SS assembly.

The switch between expression of T3SS (low Ca^{2+}) to more recently discovered type T6SS [128] may correlate with the switch from acute to chronic infections of *P. aeruginosa* [126]. The T6SS also uses direct injection of effector proteins into other cells, which may be host cells, or other competing bacteria [111]. However, rather than being evolutionarily related to the flagella basal body, the T6SS appears related to bacteriophage tail-associated proteins [129]. The tail-spike is used to puncture the membranes of cells, and the effector molecules at the tip of the spike are injected directly into the cytosol. Regulation of expression of T6SS is not well characterized and the role of Ca^{2+} in regulation of T6SS is not well known. However, Allsopp et al. [130] using a transposon mutagenesis screening approach identified the RNA binding protein, RsmA as a primary component involved in the regulation of all the three T6SS gene operons of *P. aeruginosa*. Therefore, a common link in the inverse regulation of T3SS and T6SS involves the RNA binding protein RsmA.

33.2 Molecular Mechanisms of Ca^{2+} Responses in Pathogens

33.2.1 Two Component Systems

As discussed above, Ca^{2+} levels differ within a host, fluctuate during disease progression, and thus form a complex signaling landscape for invading pathogens. Therefore, sensing host Ca^{2+} is an important task enabling pathogens to efficiently adjust to the host environment by triggering the expression of genes responsible for virulence and resistance. Bacteria accomplish this in part by employing two-component regulatory systems (TCS). A traditional TCS consists of a sensor kinase and a response regulator. The sensor kinase is usually embedded into the inner membrane with the sensor region often facing the periplasm. Upon signal binding, the sensor kinase auto-phosphorylates followed by phosphorylating the response regulator, typically regulating transcription of a set of target genes [131–133].

Several Ca^{2+} sensing TCSs have been identified. Some of them are positively and some are negatively regulated by elevated Ca^{2+} (Table 33.2). To test whether the relationship with the ion is defined by the recognition sequence within the TCS sensor, we aligned the sensor sequences of the TCS experimentally shown to be regulated by Ca^{2+}. Based on the Clustal Omega alignment, a maximum likelihood tree was constructed using MEGA4 algorithm [134, 135] (Fig.33.1). The distinct grouping of positively and negatively regulated sensors supports the idea of sequence-dependent relationship with Ca^{2+}.

The TCSs that are negatively regulated by Ca^{2+} have been studied in more details. PhoPQ is a well-characterized TCS, found in a variety of Gram-negative pathogens, such as *P. aeruginosa*, *E. coli*, *Yersinia pestis, Shigella flexneri* and *S.*

Table 33.2 Sensors from two-component regulatory systems (TCS) regulated by Ca^{2+}

Name	GenBank ID	Stimuli	Ca^{2+}-dependent regulatory targets	Regulated Phenotype	Organism	References
Positively Regulated by Elevated Ca^{2+}						
CarS	AAG06044.1	Ca^{2+}	*carP, carO*	Virulence	*P. aeruginosa*	[153]
CvsS	AAO56858.1	Ca^{2+}	*hspR, algU*	Biofilm formation, cellulose production, virulence, T3SS	*P. syringae*	[84]
AtoS	AEH68827.1	Ca^{2+} SpermidineHis-tamineAcetoacetate	*Ato operon*	Synthesis of PHB-PP, motility	*E. coli*	[154–157]
VicK	PNM00564.1	Ca^{2+}, Mn^{2+}, sucrose	*atlA* operon	eDNA release, response to host immunity, attachment, and biofilm formation	*S. mutans*	[91, 158, 159]
Negatively Regulated by Elevated Ca^{2+}						
PhoQ	AIB53821.1	Mg^{2+}, Ca^{2+}, acetate		LPS modification	*E. coli*	[137, 147]
PhoQ	P0DM80.1	Mg^{2+}, Ca^{2+}		LPS modification, SOS response	*S. enterica*	[136, 145, 160]
PhoQ	AEU17992.1	Mg^{2+}, Ca^{2+}		LPS modification	*P. aeruginosa*	[137, 141, 143]
PehS	BAJ11971.1	Ca^{2+}	*pehA*	Virulence	*P. carotovorum*	[148]
CiaH	P0A4I6.1	Ca^{2+}	*ciaX*	Ca^{2+} mediated cell functions and biofilm production	*S. pneumoniae*	[161]
CarS	2,614,773	Ca^{2+}	*vps*	Vps-dependent biofilms, antibiotic resistance	*V. cholerae*	[86, 162]

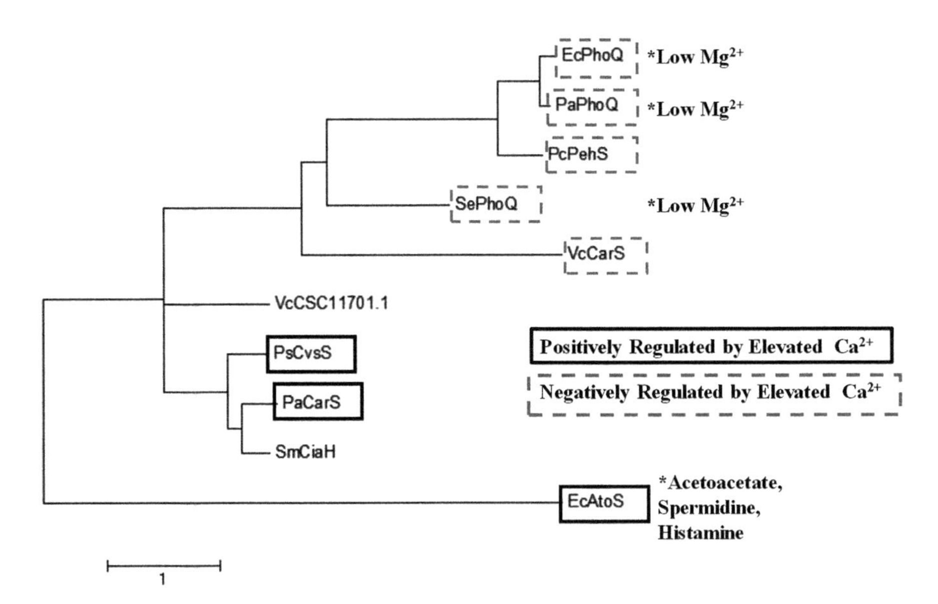

Fig. 33.1 Molecular Phylogenetic analysis of Ca^{2+} regulated TCS. To analyze the sequences of Ca^{2+} regulated TCS sensors, we first determined the sensor regions by selecting the periplasmic loop of the proteins based on TMHMM analyses. After the 11 sequences were aligned, all positions containing gaps and missing data were eliminated. The phylogenetic relationship of the sensor regions was inferred by using the Maximum Likelihood algorithm based on the JTT matrix-based model [1] in MEGA7 [2]. The tree with the highest log likelihood (−2667.30) is shown. The tree is drawn to scale, with branch lengths measured in the number of substitutions per site

enterica [82, 131, 136–138]. PhoPQ systems regulate multiple virulence factors, resistance and motility. The PhoPQ regulon includes the SOS response of *S. enterica* [139], and the *arn* operon of *P. aeruginosa*, which is responsible for lipid A modifications and increased resistance to antimicrobial peptides [138, 140–144]. Interestingly, PhoPQ systems are also repressed by elevated Mg^{2+}, and two-distinct Mg^{2+} and Ca^{2+} binding regions were identified in PhoQ sensor of *S. enterica* [136]. These regions are conserved in other PhoQ homologs [145, 146]. While the PhoPQ systems share overall sequence similarly, they differ in their responses. Thus, mutating Ca^{2+}-binding residues in PhoQ have less of an impact on transcriptional regulation in *E. coli* than in *S. enterica* [146]. PhoQ interactions with ligands also differ in different species: Mg^{2+} binds to PhoQ dimer causing destabilization and preventing signal transduction in *P. aeruginosa* PhoQ (*Pa*PhoQ), but not in *E. coli* PhoQ (*Ec*PhoQ) [137]. Finally, PhoQ may respond to additional ligands such as acetate in the *E. coli* PhoQ [147]. The TCS most closely grouped with PhoPQ is PehSR from a plant pathogen *Pectobacterium carotovorum* (Fig. 33.1). This system regulates the production of a secreted endopolygalacturonase, PehA which is required for initial invasion of the pathogen into a host [109, 148–150]. Homologs of PehA, although not yet characterized, have been identified in another plant pathogen *Erwinia chrysanthemi* [151, 152].

Another TCS negatively regulated by elevated Ca^{2+} is CarSR from *V. cholera* [86]. CarSR regulates *Vibrio* polysaccharide (*vps*), the main matrix component of *vps*-dependent biofilms. Although elevated Ca^{2+} decreases the formation of *vps*-dependent biofilms, it increases the formation of *vps*-independent biofilm in *V. cholerae,* as well as production of bile-salt-dependent virulence factors required for colonization of the gut [82, 163, 164]. The mechanisms for this positive Ca^{2+} regulation of *V. cholerae* virulence are not known and potentially involve an alternative TCS that is positively regulated by Ca^{2+}. To predict such a TCS, we performed BLASTP alignment of the sensor region from the positively regulated by Ca^{2+} *P. aeruginosa* CarS against the *V. cholerae* genome and identified a putative Ca^{2+} responsive sensor protein CSC11701.1. The close grouping of its sensor sequence with *P. aeruginosa* CarS and *P. syringae* CvsS supports its potentially positive regulation by Ca^{2+} (Fig. 33.1).

An atypical TCS, CiaHR, was identified in *Streptococcus mutans.* CiaHR contains a third component, CiaX, a small protein, which upon binding Ca^{2+}, interacts with the sensor portion of CiaH and prevents the phosphor-relay. The system was shown to regulate antibiotic resistance, biofilm formation, eDNA uptake, as well as stress response [161, 165–167]. Interestingly, CiaH grouped closely with the positively regulated sensors, indicating a possibility that CiaH itself may be regulated by Ca^{2+}. In addition, *S. mutans* biofilm formation and eDNA release can be positively regulated in response to Ca^{2+} *via* the TCS VicKR [91, 159, 168, 169]. However, VicK is more distantly related to the positively regulated Ca^{2+} sensors, which could be attributed to its potential to respond to other stimuli (Fig. 33.1).

The TCS that are positively regulated by elevated Ca^{2+} were shown to be involved in regulating virulence and resistance factors in response to Ca^{2+}. Earlier, our group identified a TCS that is positively regulated by Ca^{2+} in *P. aeruginosa.* It was named it Calcium Regulated Senor/ Regulator, CarSR [153]. This system regulates at least two identified targets, CarO and CarP, involved in Ca^{2+}–induced production of virulence factors pyocyanin and pyoverdine and contributes to the regulation of the intracellular Ca^{2+} homeostasis and tolerance to elevated Ca^{2+}. Recently, another Ca^{2+}-induced TCS was found in plant pathogen *P. syringae,* named Calcium, Virulence, and Swarming Sensor and Regulator, CvsSR. CvsSR is required for *P. syringae* pathogenicity in plants by enhancing transcription of genes for T3SS and small RNAs while repressing alginate and flagella biosynthesis [84]. In *E. coli,* synthesis of the polyhydoxybutyrate polyphosphate (PHB-PP) Ca^{2+} channels is positively regulated by the AtoSC TCS. These non-proteinaceous Ca^{2+} channels play a role in eDNA uptake and folding of the outer membrane protein OmpA [154, 170, 171]. In addition, AtoSC regulates other processes, such as motility in response to acetoacetate, histamine, and spermidine [155–157]. This may be reflected in its distant grouping from CarSR and CvsSR (Fig. 33.1).

In summary, bacterial pathogens utilize multiple TCS to recognize changes in the environmental Ca^{2+} and adjust their transcriptional activity. These systems are versatile as they evolved to enable bacterial adaptations to multi-variant environments. Understanding the regulation by TCS is challenged by several factors: the presence of multiple systems in one organism, sensors responding to different stimuli, and additional components involved in signal recognition [172–175]. Different TCS may also cross-talk enabling multiple signals to control similar responses, as in the case of SypFG, proposed to mediate Ca^{2+} induction of biofilm formation in *V. fischeri* [83]. The sensing portion of SypF, however, was not required for Ca^{2+} induction, and involved an alternative sensor kinase RscS phosphorylating SypF in response to Ca^{2+} [83]. A much better understanding of the TCS signaling networks triggered by Ca^{2+} is needed to fully appreciate their role in Ca^{2+}-mediated communication between invading pathogens and their hosts.

33.2.2 Ca^{2+} Sensors

In eukaryotes, members of the calmodulin superfamily are the best studied Ca^{2+} sensors [176, 177]. Calmodulin (CaM) contains two canonical EF-hand motifs coordinating Ca^{2+} binding [178, 179]. Upon binding Ca^{2+}, CaM undergoes conformational changes enabling binding and activating diverse target proteins [180]. Therefore, searches for components of Ca^{2+} signaling networks in bacteria have focused on proteins with EF hands [181]. In addition, proteins that play roles in translocating or buffering Ca^{2+} have been studied [182–188]. A number of calmodulin-like proteins have been reported based on their sequence and structure similarity to CaM and binding to anti-calmodulin antibodies. Several of these Ca^{2+} sensors are summarized in Table 33.3. The first proposed bacterial Ca^{2+} sensor was CasA, from *Rhizobium etli.* CasA has three pairs of EF hand domains, similarly to eukaryotic calbindin and calretinin [189]. CasA mediates Ca^{2+}-dependent symbiotic relationship between *R. etli* and its plant host. Our group identified a homolog of CasA in *P. aeruginosa* and named it EfhP (EF-hand protein) [190]. EfhP is required for maintaining $Ca^{2+}{}_{in}$ homeostasis and involved in Ca^{2+} regulation of *P. aeruginosa* virulence. Our current studies verified the ability of this protein to selectively bind Ca^{2+} and undergo Ca^{2+}–dependent conformational changes supporting its role as a Ca^{2+} sensor (Kayastha et al. in preparation). Another EF hand protein, proposed to function as a Ca^{2+}-sensor, is CabD from *Streptomyces coelicolor*. CabD contributes to Ca^{2+} regulation of aerial mycelium formation. CabD has high- and low-affinity Ca^{2+}-binding sites, suggesting their distinct roles in mediating Ca^{2+}-regulatory and buffering roles, respectively [177].

CaM-like proteins have been reported in *Mycobacteria* and are suggested to play a role in sensing Ca^{2+} [199]. The CaM-like protein, Rv1211 of *M. tuberculosis* binds Ca^{2+} through its single EF hand domain and stimulates the activities of NAD kinase and phosphodiesterase (PDE), targets that are similar to those of eukaryotic CaM. Reduced expression of this protein has been shown to impair *M. tuberculosis* growth and survival in macrophages, suggesting its importance during infection [192, 193]. A homologous CaM-like protein from *M. smegmatis* has been shown to stimulate phosphodiesterase activity and regulate the metabolism of phospholipids [200] supporting the role of this protein as a Ca^{2+} sensor.

Table 33.3 Ca^{2+} Sensors

Protein name/GenBank ID	Organism	Function/Properties	Domain	References
EfhP AAG07494.1	*P. aeruginosa*	Regulates Ca^{2+} induced virulence and intracellular Ca^{2+} homeostasis	One pair of EF hands	Kayastha et al. (in preparation)
CasA AF288533	*R. etli*	Mediates Ca^{2+} dependent symbiosis with leguminous host	Three pairs of EF hands	[189]
CabD 3AKA_A	*S. coelicolor*	Affects formation of aerial mycelium	Two pairs of EF hands	[177]
Protein S WP_020477824	*M. xanthus*	Required for assembly of spore coat	Beta-gamma crystalline fold	[191]
CAMLP NP_215727	*M. tuberculosis*	Activates NAD kinase and phosphodiesterase upon Ca^{2+} binding	Single EF hand motif	[192, 193]
CAMLP AY319523.1	*M. smegmatis*	Activates phosphodiesterase	Single EF hand motif	[194]
CALP YP_004243569	*B. subtilis*	Activates phosphodiesterase in Ca^{2+} dependent manner		[195]
CLP	*B. pertussis*	Activates adenylate cyclase in Ca^{2+} dependent manner		[196]
LadS AAG07361	*P. aeruginosa*	Ca^{2+} dependent phosphor-relay to GacSA	Histidine kinase, 7 transmembrane, DISMED2	[197, 198]

Protein S from *Myxococcus xanthus* is a member of ßγ-crystalline family. This protein shares structural similarity to CaM and binds Ca^{2+}, which is required for the protein assembly on the surface of the spores [201]. A more recent report showed that the protein's Ca^{2+}-binding site forms a high charge density pocket similar to that in calsequestrin and human Hsp70. However it is still not clear whether the Ca^{2+}-induced conformational changes in protein S play roles in signaling events [191].

Recently, a hybrid histidine kinase LadS was shown to trigger a Ca^{2+}-induced switching between acute and chronic type of virulence in *P. aeruginosa* [198]. As discussed above, LadS belongs to a unique class of bacterial sensors that possess histidine kinase, seven transmembrane, and the periplasmic DISMED2 domain. The latter was shown to bind Ca^{2+} *via* Asp80 and Asp90 residues and activate the kinase activity leading to phosphorelay cascade [197, 198]. In contrast to typical sensors of TCS that phosphorylate partnered response regulators, LadS phosphorylates the TCS GacSA and thus activates GacS/Rsm pathway responsible for the global regulation of chronic infection-type virulence and tolerance [88]. *In silico* analysis of the sequence conservation of LadS showed that this protein is unique to *Pseudomonas*. Interestingly, during searches for homologous DISMED2 domains among selected bacterial pathogens, we identified several proteins including putative alkaline phosphatase synthesis sensors CJK91172.1 and CJK40304.1, and a membrane protein with GGDEF domain, CJK88170.1, in *S. pneumoniae*. The GGDEF domain is known to act as a diguanylate cyclase responsible for synthesis of cyclic-di-GMP, a second messenger mainly regulating biofilm formation [202]. In addition to DISMED2 domain, these proteins contain the two residues (Asp80 and Asp90) required for Ca^{2+} recognition, suggesting they may sense and respond to Ca^{2+}.

Pathogenic bacteria possess Ca^{2+}-sensing proteins that play an important role in modulating their pathogenicity. However, despite evidence of the significance of Ca^{2+} regulation in bacterial physiology and virulence, the knowledge on Ca^{2+} sensors and regulators in bacteria is still limited. Further studies are needed to determine whether these proteins sense the changes in the intracellular Ca^{2+} and thus enable Ca^{2+} to serve as second messenger.

33.3 Components of Intracellular Ca^{2+} Signaling in Pathogenic Bacteria Mediating Ca^{2+} Regulation of Virulence

A number of studies have shown that bacterial metabolic processes respond to elevated Ca^{2+} (reviewed in [203–205]). Some of these processes are regulated by extracellular Ca^{2+} levels and are mediated by TCS. However, other processes may respond to the transient changes in the intracellular Ca^{2+} ($Ca^{2+}{}_{in}$), thus implicating Ca^{2+} as a second messenger. Although the latter still requires experimental proof, here we discuss the components of $Ca^{2+}{}_{in}$ signaling network that have been shown to play a role in Ca^{2+} regulation, supporting the idea of a regulatory role of $Ca^{2+}{}_{in}$ in bacteria.

Similarly to eukaryotic cells, bacteria maintain their basal $[Ca^{2+}]_{in}$ at high nanomolar level and generate transient changes in $[Ca^{2+}]_{in}$ in response to diverse stimuli [206–210]. These stimuli include a variety of extracellular factors, such as $Ca^{2+}{}_{ex}$, pH, mechanical stimulation, intermediates of carbohydrate metabolism, and oxidative stress; all factors potentially to be encountered upon entering a host [206, 208, 210–213]. A number of proteins have been shown to contribute to the maintenance of $Ca^{2+}{}_{in}$ homeostasis and to the generation of $Ca^{2+}{}_{in}$ fluctuations [203, 206]. However, it is still not clear whether bacteria have an intracellular source of Ca^{2+} for these fluctuations (e.g. a compartment for storing and releasing Ca^{2+} into the cytoplasm) or if they rely on influx of extracellular Ca^{2+}. *E. coli* accumulates Ca^{2+} in the periplasm to millimolar levels when grown in the presence of millimolar extracellular Ca^{2+} [207]. It is possible that the periplasmic Ca^{2+} may be released into the cytoplasm and used for generating intracellular Ca^{2+} transients. However, further mechanistic studies are imperative.

33.3.1 Ca^{2+} Channels

Several types of Ca^{2+} transporters have been identified in bacteria. The transporters contribute to the regulation of virulence and host-pathogen interactions. First, the poly-β-hydroxybutyrate polyphosphate (PHB-PP) in *E. coli* forms non-proteinaceous Ca^{2+} channels and translocates Ca^{2+} into the cytoplasm [214]. In addition, PHB-PP channels are required for Ca^{2+}-dependent genetic competence, which plays a key role in uptake of foreign DNA, eventually enhancing bacterial adaptation to the host environment and resistance [215]. The production of the corresponding PHB-PP synthases was shown to be induced by elevated extracellular Ca^{2+} and several other stimuli *via* Ca^{2+}-dependent TCS AtoSC [154].

Another type of Ca^{2+} channels, a pH-dependent Ca^{2+} leak channel, YetJ was identified in *B. subtilis* [216]. This protein contains a BAX-1 inhibitor domain homologous to the one in a Ca^{2+} leak channel found in the endoplasmic reticulum (ER) membrane. Eukaryotic channels containing transmembrane BAX inhibitor-1 motif (TMBIM) mediate $Ca^{2+}{}_{in}$ homeostasis and apoptosis [217]. Interestingly, this highly conserved domain has been identified in a number of bacterial proteins. Initially, this domain was identified in *E. coli* protein YccA, and was shown to play a role in biofilm maturation [99]. While more research is needed to determine the roles of YetJ and YccA in *B. subtilis* and *E. coli* physiology, our group recently identified another homolog of the channel in *P. aeruginosa* [Guragain et al. in preparation]. We named it CalC for <u>Ca</u>$^{2+}$ <u>L</u>eak <u>C</u>hannel and determined that the protein is responsible for generating transient changes in $[Ca^{2+}]$ in the *P. aeruginosa* cytoplasm in response to extracellular Ca^{2+}. Transcriptional profiling of the mutant strain with disrupted *calC* revealed that the responses to elevated Ca^{2+} were impaired, particularly for genes encoding virulence factors and biofilm determinants [Guragain et al. in preparation]. This work provides experimental proof of the regulatory role of $Ca^{2+}{}_{in}$ transients in bacterial responses to Ca^{2+}. Furthermore, homology searches identified a number of putative BAX-1 Ca^{2+} leak channels in bacterial pathogens including *S. pneumoniae*, *P. carotovorum*, *Coexiella burnetti*, *S. enterica*, and *H. pylori*, indicating the conserved nature of the Ca^{2+} leak channel in bacterial pathogens, and possibly suggesting a role in the pathogenic life style.

Mechanosensitive channels (MSC) are large, non-selective channels that usually allow the passage of ions in response to mechanical or osmotic stress (reviewed in [205, 218, 219]). MSC are found in a variety of different bacteria including human and plant pathogens [220–222]. In *B. subtilis*, MSC SpoVAC releases Ca-dipicolinic acid complex, which is required for spore formation [223]. Mechanical stress in *E. coli* was shown to cause an increase in $[Ca^{2+}]_{in}$ leading to altered gene expression [212]. However, the MSC, MscL in this organism did not impact $Ca^{2+}{}_{in}$ homeostasis [224], raising a possibility of an alternative Ca^{2+} channel responding to mechanical stress. Our studies with *P. aeruginosa* identified several Ca^{2+} transporters including a putative MSC encoded by PA4614, which contributed to the restoration of the $[Ca^{2+}]_{in}$ basal level and the regulation of Ca^{2+}-induced swarming motility [206].

33.3.2 Ca^{2+} Pumps

Elevated levels of free Ca^{2+}_{in} can be toxic to bacterial cells, and the recovery to the basal $[Ca^{2+}]_{in}$ is critical for re-sensitizing cells for the next wave of $[Ca^{2+}]_{in}$ signaling. Therefore, the mechanisms of Ca^{2+} efflux are of high importance for cellular survival and for Ca^{2+}_{in} regulation. Underlining their physiological significance, multiple families of efflux transporters have evolved and been shown to play a role in bacterial physiology and virulence (reviewed in [203]). The first group includes two types (P and F) of ATPases that couple Ca^{2+} export to ATP hydrolysis [60, 225, 226]. These proteins are highly conserved and were identified in diverse bacterial pathogens. In addition to translocating Ca^{2+}, or likely because of it, some of these proteins are important in diverse bacterial processes related to survival in a host. For example, CaxP plays a role in host colonization by *S. pneumoniae* [227], CtpE of *M. smegmatis* contributes to cell surface integrity [228], and PA3920 and PA2435 of *P. aeruginosa* mediate Ca^{2+} regulation of swarming motility [206]. The second group includes ion exchange transporters coupling Ca^{2+} export to co-transport of other ions. Although many of these transporters have been identified in bacterial pathogens (reviewed in [203]), there is little evidence yet about their role in virulence.

33.3.3 Predicting Novel Components of Ca^{2+} Signaling Network

To expand our knowledge on the components of Ca^{2+}_{in} signaling network in pathogenic bacteria, we aimed to predict novel Ca^{2+}-recognizing or translocating proteins based on their homology to well-characterized components of eukaryotic Ca^{2+} network. For this, we selected well-characterized eukaryotic Ca^{2+}-binding proteins, whose homologs in bacteria have not been reported. All sequence alignments were carried out using NCBI BLASTP [76, 229]. CarR, a Ca^{2+} sensor that is a G protein-coupled receptor, plays an essential role in fluctuating intracellular Ca^{2+} homeostasis in response to minute changes in $[Ca^{2+}]_{ex}$ and other stimuli [230–234]. The protein contains three cooperative Ca^{2+} binding sites. To our surprise, four (underlined) out of five (in bold) residues required for Ca^{2+} binding (**R**XX**E**XX**EE**A**E**ERD) were found in a large number of bacterial proteins involved in a variety of life-sustaining functions, including recruitment of replisome in *S. aureus* [235], cell division protein FtsA in *E. persicina* [236], putative Rhs toxin and DNA recombination regulation system in *E. coli* [237–239], and putative pili assembly gene in *Enterobacter cloacae*. Another eukaryotic Ca^{2+} signaling protein is RyR1, which is a Ca^{2+} channel known to release Ca^{2+} stored in the sarcoplasmic reticulum into the cytoplasm [240]. Sequence homology searches against bacterial genomes identified only a small fragment of the protein as aligning with bacterial proteins. This region happened to be located within the lining of the pore required for Ca^{2+} sensitivity [240–242]. Four (underlined) out of five residues E3893, H3895, E3967, Q3970, T5001 required for Ca^{2+} binding were found with similar spacing in several putative ABC transporters of pathogenic bacteria, including *S. aureus* and *S. pneumoniae*. We also detected these residues in putative bestrophin transporters of many bacterial species including *P. aeruginosa* and *E. coli*, in an orphan transcriptional regulator unique to a *P. aeruginosa* clinical isolate, and in several enzymes including a putative purine phosphatase of *P. aeruginosa*. The discovery of a putative Ca^{2+} binding site in bacterial bestrophin channels is particularly interesting, since human bestrophin is a Ca^{2+}-gated potassium channel [243]. However, the only characterized bacterial bestrophin channel (in *K. pneumoniae*) was shown to not require Ca^{2+} for its function [244] nor did it contain the Ca^{2+}-binding site found in the human bestrophin. This raises a possibility of two types of bestrophin channels in bacteria, Ca^{2+}-dependent and Ca^{2+}-independent. Interestingly, the eukaryotic bestrophin has been demonstrated to possess multiple splicing variants: with and without Ca^{2+} binding region [245]. Overall, these findings suggest that a significantly greater number of bacterial processes are likely regulated by Ca^{2+} than already known. The fact that most of these predicted Ca^{2+}-binding proteins were detected in bacterial pathogens suggests the importance of Ca^{2+} regulation in their physiology and, possibly, interactions with hosts, whose vital processes are regulated by Ca^{2+}.

33.4 Concluding Remarks

Bacteria have very dynamic and complex responses to Ca^{2+}. Over the past 10 years, the evidence that bacteria utilize Ca^{2+} for signaling has grown, yet important pieces are still missing. An area that needs study is on intracellular Ca^{2+} signaling. While a number of Ca^{2+} sensors and Ca^{2+}-dependent regulatory systems have been shown to regulate essential functions, most of the findings are of correlative nature with no direct experimental evidence linking the changes in the intracellular Ca^{2+} to the regulation of transcription or translation. Even less is known about how the amplitude and the frequency of intracellular Ca^{2+} signals modulate the response. An interesting aspect is the conservation of many Ca^{2+}-binding domains in eukaryotes and bacteria indicating an evolutionary lineage between Ca^{2+} signaling networks in these domains of life. One technical problem, that is worth mentioning, is the disregard for the presence of Ca^{2+} in commonly used rich bacteriological growth media, such as LB or BHI. Consequently, Ca^{2+} regulation of the resultant bacterial phenotypes may be underestimated. Overall, Ca^{2+} signaling in bacteria is an exciting and quickly developing field, which is providing not only the fundamental understanding of bacterial life and evolution but also generating insights into the regulation of bacterial pathogenicity.

References

1. Edel KH, Kudla J (2015) Increasing complexity and versatility: how the calcium signaling toolkit was shaped during plant land colonization. Cell Calcium 57(3):231–246
2. Permyakov EA, Kretsinger RH (2009) Cell signaling, beyond cytosolic calcium in eukaryotes. J Inorg Biochem 103(1):77–86
3. Clapham DE (1995) Calcium signaling. Cell 80(2):259–268
4. Bush DS, Jones RL (1990) Measuring intracellular Ca^{2+} levels in plant cells using the fluorescent probes, Indo-1 and Fura-2: progress and prospects. Plant Physiol 93(3):841–845
5. Bose J et al (2011) Calcium efflux systems in stress signaling and adaptation in plants. Front Plant Sci 2:85
6. Von Ruecker AA, Bertele R, Harms HK (1984) Calcium metabolism and cystic fibrosis: mitochondrial abnormalities suggest a modification of the mitochondrial membrane. Pediatr Res 18(7):594
7. Aris RM et al (1999) Altered calcium homeostasis in adults with cystic fibrosis. Osteoporos Int 10(2):102–108
8. Ceder O, Roomans G, Hösli P (1982) Increased calcium content in cultured fibroblasts from trisomy patients: comparison with cystic fibrosis fibroblasts. Scan Electron Microsc 1982(Pt 2):723–730
9. Roomans GM (1986) Calcium and cystic fibrosis. Scan Electron Microsc 1986(Pt 1):165–178
10. Gewirtz AT et al (2000) Salmonella typhimurium induces epithelial IL-8 expression via Ca^{2+}−mediated activation of the NF-κB pathway. J Clin Invest 105(1):79–92
11. Smith DJ et al (2014) Elevated metal concentrations in the CF airway correlate with cellular injury and disease severity. J Cyst Fibros 13(3):289–295
12. Büchau AS, Gallo RL (2007) Innate immunity and antimicrobial defense systems in psoriasis. Clin Dermatol 25(6):616–624
13. Donato R (2001) S100: a multigenic family of calcium-modulated proteins of the EF-hand type with intracellular and extracellular functional roles. Int J Biochem Cell Biol 33(7):637–668
14. Pace J, Hayman MJ, Galán JE (1993) Signal transduction and invasion of epithelial cells by S. Typhimurium. Cell 72(4):505–514
15. Gekara NO et al (2007) The multiple mechanisms of Ca^{2+} signalling by listeriolysin O, the cholesterol-dependent cytolysin of Listeria monocytogenes. Cell Microbiol 9(8):2008–2021
16. Hu Y et al (2011) Structures of Anabaena calcium-binding protein CcbP INSIGHTS INTO CA^{2+} SIGNALING DURING HETEROCYST DIFFERENTIATION. J Biol Chem 286(14):12381–12388
17. Asmat TM et al (2011) Streptococcus pneumoniae infection of host epithelial cells via polymeric immunoglobulin receptor transiently induces calcium release from intracellular stores. J Biol Chem 286(20):17861–17869
18. Nhieu GT et al (2004) Calcium signalling during cell interactions with bacterial pathogens. Biol Cell 96(1):93–101
19. Khan NA et al (2007) FimH-mediated Escherichia coli K1 invasion of human brain microvascular endothelial cells. Cell Microbiol 9(1):169–178
20. Eichstaedt S et al (2009) Effects of Staphylococcus aureus-hemolysin a on calcium signalling in immortalized human airway epithelial cells. Cell Calcium 45(2):165–176
21. Denning GM et al (1998) Pseudomonas pyocyanine alters calcium signaling in human airway epithelial cells. Am J Phys Lung Cell Mol Phys 274(6):L893–L900
22. Schwarzer C et al (2010) Pseudomonas aeruginosa homoserine lactone activates store-operated cAMP and cystic fibrosis transmembrane regulator-dependent cl− secretion by human airway epithelia. J Biol Chem 285(45):34850–34863
23. Forsen S, Kordel J (1994) Calcium in biological systems. University Science Books, Mill Valley, CA, p 107

24. Vikström E et al (2010) Role of calcium signalling and phosphorylations in disruption of the epithelial junctions by Pseudomonas aeruginosa quorum sensing molecule. Eur J Cell Biol 89(8):584–597
25. Werthén M, Lundgren T (2001) Intracellular Ca^{2+} mobilization and kinase activity during acylated homoserine lactone-dependent quorum sensing in Serratia liquefaciens. J Biol Chem 276(9):6468–6472
26. Maroudas A (1979) Physicochemical properties of articular cartilage. In: Freeman MAS (ed) Adult articular cartilage. Pitman Medical, Tunbridge Wells, pp 215–290
27. Prohaska C, Pomazal K, Steffan I (2000) Determination of ca, mg, Fe, cu, and Zn in blood fractions and whole blood of humans by ICP-OES. Fresenius J Anal Chem 367(5):479–484
28. Baker S, Worthley L (2002) *The essentials of calcium, magnesium and phosphate metabolism: part I. Physiology.* Critical care and. Resuscitation 4:301–306
29. Oreskes I et al (1968) Measurement of ionized calcium in human plasma with a calcium selective electrode. Clin Chim Acta 21(3):303–313
30. Sava L et al (2005) Serum calcium measurement: total versus free (ionized) calcium. Indian J Clin Biochem 20(2):158–161
31. Moore EW (1970) Ionized calcium in normal serum, ultrafiltrates, and whole blood determined by ion-exchange electrodes. J Clin Invest 49(2):318–334
32. Schwartz H, McConville B, Christopherson E (1971) Serum ionized calcium by specific ion electrode. Clin Chim Acta 31(1):97–107
33. Reinhart RA (1988) Magnesium metabolism: a review with special reference to the relationship between intracellular content and serum levels. Arch Intern Med 148(11):2415–2420
34. Fanconi A, Rose GA (1958) The ionized, complexed, and protein-bound fractions of calcium in plasma: an investigation of patients with various diseases which affect calcium metabolism, with an additional study of the role of calcium ions in the prevention of tetany. Q J Med 27:463–494
35. Fiyaz M et al (2013) Association of salivary calcium, phosphate, pH and flow rate on oral health: a study on 90 subjects. J Ind Soc Periodontol 17(4):454
36. Blomfield J, Warton KL, Brown J (1973) Flow rate and inorganic components of submandibular saliva in cystic fibrosis. Arch Dis Child 48(4):267–274
37. Chernick WS, Barbero GJ, Parkins FM (1961) Studies on submaxillary saliva in cystic fibrosis. J Pediatr 59(6):890–898
38. Marmar J, Barbero GJ, Sibinga MS (1966) The pattern of parotid gland secretion in cystic fibrosis of the pancreas. Gastroenterology 50(4):551–556
39. Blomfield J et al (1976) Parotid gland function in children with cystic fibrosis and child control subjects. Pediatr Res 10(6):574
40. Moreira A et al (2009) Flow rate, pH and calcium concentration of saliva of children and adolescents with type 1 diabetes mellitus. Braz J Med Biol Res 42(8):707–711
41. Agha-Hosseini F, Dizgah IM, Amirkhani S (2006) The composition of unstimulated whole saliva of healthy dental students. J Contemp Dent Pract 7(2):104–111
42. Halmerbauer G et al (2000) The relationship of eosinophil granule proteins to ions in the sputum of patients with cystic fibrosis. Clin Exp Allergy 30(12):1771–1776
43. Lorin MI, Gaerlan PF, Mandel ID (1972) Quantitative composition of nasal secretions in normal subjects. J Lab Clin Med 80(2):275–281
44. Lorin M et al (1976) Composition of nasal secretion in patients with cystic fibrosis. J Lab Clin Med 88(1):114–117
45. Taylor EN, Curhan GC (2007) Differences in 24-hour urine composition between black and white women. J Am Soc Nephrol 18(2):654–659
46. Nemchinov LG, Shabala L, Shabala S (2008) Calcium efflux as a component of the hypersensitive response of Nicotiana benthamiana to Pseudomonas syringae. Plant Cell Physiol 49(1):40–46
47. Grant M et al (2000) The RPM1 plant disease resistance gene facilitates a rapid and sustained increase in cytosolic calcium that is necessary for the oxidative burst and hypersensitive cell death. Plant J 23(4):441–450

48. Zhang L, Du L, Poovaiah B (2014) Calcium signaling and biotic defense responses in plants. Plant Signal Behav 9(11):e973818
49. Ma W, Berkowitz GA (2007) The grateful dead: calcium and cell death in plant innate immunity. Cell Microbiol 9(11):2571–2585
50. Hall-Stoodley L, Costerton JW, Stoodley P (2004) Bacterial biofilms: from the natural environment to infectious diseases. Nat Rev Microbiol 2(2):95–108
51. Sauer K (2003) The genomics and proteomics of biofilm formation. Genome Biol 4(6):219
52. van Loosdrecht MC et al (1990) Influence of interfaces on microbial activity. Microbiol Rev 54(1):75–87
53. Cruz LF, Cobine PA, De La Fuente L (2012) Calcium increases Xylella fastidiosa surface attachment, biofilm formation, and twitching motility. Appl Environ Microbiol 78(5):1321–1331
54. Romantschuk M (1992) Attachment of plant pathogenic bacteria to plant surfaces. Annu Rev Phytopathol 30(1):225–243
55. Yamaguchi T et al (2009) Gene cloning and characterization of Streptococcus intermedius fimbriae involved in saliva-mediated aggregation. Res Microbiol 160(10):809–816
56. Kerchove AJ, Elimelech M (2008) Calcium and magnesium cations enhance the adhesion of motile and nonmotile pseudomonas aeruginosa on alginate films. Langmuir 24(7):3392–3399
57. Johnson MD et al (2011) Pseudomonas aeruginosa PilY1 binds integrin in an RGD-and calcium-dependent manner. PLoS One 6(12):e29629
58. Williams TC, Ayrapetyan M, Oliver JD (2015) Molecular and physical factors that influence attachment of Vibrio vulnificus to chitin. Appl Environ Microbiol 81(18):6158–6165
59. Cheng Y et al (2013) Mutation of the conserved calcium-binding motif in Neisseria gonorrhoeae PilC1 impacts adhesion but not piliation. Infect Immun 81(11):4280–4289
60. Orans J et al (2010) Crystal structure analysis reveals Pseudomonas PilY1 as an essential calcium-dependent regulator of bacterial surface motility. Proc Natl Acad Sci U S A 107(3):1065–1070
61. Eto DS et al (2008) Clathrin, AP-2, and the NPXY-binding subset of alternate endocytic adaptors facilitate FimH-mediated bacterial invasion of host cells. Cell Microbiol 10(12):2553–2567
62. Barbu EM et al (2014) SdrC induces staphylococcal biofilm formation through a homophilic interaction. Mol Microbiol 94(1):172–185
63. Josefsson E et al (1998) The binding of calcium to the B-repeat segment of SdrD, a cell surface protein of Staphylococcus aureus. J Biol Chem 273(47):31145–31152
64. Kumar S, Spiro S (2017) Environmental and genetic determinants of biofilm formation in Paracoccus denitrificans. mSphere 2(5):e00350-17
65. Linhartova I et al (2010) RTX proteins: a highly diverse family secreted by a common mechanism. FEMS Microbiol Rev 34(6):1076–1112
66. Niemann HH, Schubert W-D, Heinz DW (2004) Adhesins and invasins of pathogenic bacteria: a structural view. Microbes Infect 6(1):101–112
67. Martínez-Gil M et al (2012) Calcium causes multimerization of the large adhesin LapF and modulates biofilm formation by Pseudomonas putida. J Bacteriol 194(24):6782–6789
68. Martínez-Gil M, Yousef-Coronado F, Espinosa-Urgel M (2010) LapF, the second largest Pseudomonas putida protein, contributes to plant root colonization and determines biofilm architecture. Mol Microbiol 77(3):549–561
69. Espinosa-Urgel M, Salido A, Ramos J-L (2000) Genetic analysis of functions involved in adhesion of Pseudomonas putida to seeds. J Bacteriol 182(9):2363–2369
70. Rose RK (2000) The role of calcium in oral streptococcal aggregation and the implications for biofilm formation and retention. BBA-Gen Subjects 1475(1):76–82
71. Korstgens V et al (2001) Influence of calcium ions on the mechanical properties of a model biofilm of mucoid Pseudomonas aeruginosa. Water Sci Technol 43(6):49–57
72. Das T et al (2014) Influence of calcium in extracellular DNA mediated bacterial aggregation and biofilm formation. PLoS One 9(3):e91935

73. Sarkisova S et al (2005) Calcium-induced virulence factors associated with the extracellular matrix of mucoid Pseudomonas aeruginosa biofilms. J Bacteriol 187(13):4327–4337
74. Das T, Manefield M (2012) Pyocyanin promotes extracellular DNA release in Pseudomonas aeruginosa. PLoS One 7(10):e46718
75. Jung CJ et al (2017) AtlA mediates extracellular DNA release, which contributes to Streptococcus mutans biofilm formation in an experimental rat model of infective endocarditis. Infect Immun 85(9):pii: e00252-17
76. Safari A et al (2014) The significance of calcium ions on Pseudomonas fluorescens biofilms - a structural and mechanical study. Biofouling 30(7):859–869
77. Haque MM et al (2017) CytR homolog of Pectobacterium carotovorum subsp. carotovorum controls air-liquid biofilm formation by regulating multiple genes involved in cellulose production, c-di-GMP signaling, motility, and type III secretion system in response to nutritional and environmental signals. Front Microbiol 8:972
78. Vozza NF et al (2016) A rhizobium leguminosarum CHDL- (cadherin-like-) Lectin participates in assembly and remodeling of the biofilm matrix. Front Microbiol 7:1608
79. Patrauchan MA et al (2005) Calcium influences cellular and extracellular product formation during biofilm-associated growth of a marine Pseudoalteromonas sp. Microbiology-Sgm 151:2885–2897
80. Theunissen S et al (2010) The 285 kDa bap/RTX hybrid cell surface protein (SO4317) of Shewanella oneidensis MR-1 is a key mediator of biofilm formation. Res Microbiol 161(2):144–152
81. Parker JK et al (2016) Calcium transcriptionally regulates the biofilm machinery of Xylella fastidiosa to promote continued biofilm development in batch cultures. Environ Microbiol 18(5):1620–1634
82. Hay AJ et al (2017) Calcium enhances bile salt-dependent virulence activation in Vibrio cholerae. Infect Immun 85(1):pii: e00707-16
83. Tischler AH et al (2018) Discovery of calcium as a biofilm-promoting signal for Vibrio fischeri reveals new phenotypes and underlying regulatory complexity. J Bacteriol 200(15):pii: e00016-18
84. Fishman MR et al (2018) Ca^{2+}-induced two-component system CvsSR regulates the type III secretion system and the Extracytoplasmic function sigma factor AlgU in Pseudomonas syringae pv. Tomato DC3000. J Bacteriol 200(5):e00538-17
85. Arrizubieta MJ et al (2004) Calcium inhibits bap-dependent multicellular behavior in Staphylococcus aureus. J Bacteriol 186(22):7490–7498
86. Bilecen K, Yildiz FH (2009) Identification of a calcium-controlled negative regulatory system affecting Vibrio cholerae biofilm formation. Environ Microbiol 11(8):2015–2029
87. Watnick PI et al (2001) The absence of a flagellum leads to altered colony morphology, biofilm development and virulence in Vibrio cholerae O139. Mol Microbiol 39(2):223–235
88. O'Connell DP et al (1998) The fibrinogen-binding MSCRAMM (clumping factor) of Staphylococcus aureus has a Ca^{2+}–dependent inhibitory site. J Biol Chem 273(12):6821–6829
89. Eidhin DN et al (1998) Clumping factor B (ClfB), a new surface-located fibrinogen-binding adhesin of Staphylococcus aureus. Mol Microbiol 30(2):245–257
90. Abraham NM, Jefferson KK (2012) Staphylococcus aureus clumping factor B mediates biofilm formation in the absence of calcium. Microbiology 158(Pt 6):1504–1512
91. Romling U, Galperin MY, Gomelsky M (2013) Cyclic di-GMP: the first 25 years of a universal bacterial second messenger. Microbiol Mol Biol Rev 77(1):1–52
92. Whiteley M, Diggle SP, Greenberg EP (2017) Progress in and promise of bacterial quorum sensing research. Nature 551(7680):313–320
93. Wilson J, Schurr M, LeBlanc C (2002) Mechanisms of bacterial pathogenicity. Postgrad Med J 78:216–224
94. Ribet D, Cossart P (2015) How bacterial pathogens colonize their hosts and invade deeper tissues. Microbes Infect 17(3):173–183

95. Olson JC, Ohman DE (1992) 1992, *Efficient production and processing of elastase and LasA by Pseudomonas aeruginosa require zinc and calcium ions*. J Bacteriol 174:4140–4147
96. Zhang L, Conway JF, Thibodeau PH (2012) Calcium-induced folding and stabilization of the Pseudomonas aeruginosa alkaline protease. J Biol Chem 287(6):4311–4322
97. Marquart ME et al (2005) Calcium and magnesium enhance the production of Pseudomonas aeruginosa protease IV, a corneal virulence factor. Med Microbiol Immunol 194(1-2):39–45
98. Thayer M, Flaherty KM, McKay DB (1991) Three-dimensional structure of the elastase of Pseudomonas aeruginosa at 1.5-a resolution. J Biol Chem 266(5):2864–2871
99. Casilag F et al (2016) The LasB elastase of Pseudomonas aeruginosa acts in concert with alkaline protease AprA to prevent flagellin-mediated immune recognition. Infect Immun 84(1):162–171
100. Laarman AJ et al (2012) Pseudomonas aeruginosa alkaline protease blocks complement activation via the classical and lectin pathways. J Immunol 188(1):386–393
101. Hall S et al (2016) Cellular effects of pyocyanin, a secreted virulence factor of Pseudomonas aeruginosa. Toxins 8(8):236
102. Rada B, Leto TL (2011) The redox-active Pseudomonas virulence factor pyocyanin induces formation of neutrophil extracellular traps. FASEB J 25(1 Supplement):360.1–360.1
103. Ran H, Hassett DJ, Lau GW (2003) Human targets of Pseudomonas aeruginosa pyocyanin. Proc Natl Acad Sci 100(24):14315–14320
104. Boehm DF (1990) R a Welch, and I S Snyder., *Calcium Is Required for Binding of Escherichia Coli Hemolysin (HlyA) to Erythrocyte Membranes*. Infect Immun 58(6):1951–1958
105. Bakas L et al (1998) Calcium-dependent conformation of E. coli α-haemolysin. Implications for the mechanism of membrane insertion and lysis. Biochim Biophys Acta 1368(2):225–234
106. Hay AJ et al (2017) Calcium enhances bile salt-dependent virulence activation in Vibrio cholerae. Infect Immun 85(1):e00707–e00716
107. Kumar P, Ahuja N, Bhatnagar R (2002) Anthrax edema toxin requires influx of calcium for inducing cyclic AMP toxicity in target cells. Infect Immun 70(9):4997–5007
108. Serezani CH et al (2008) Cyclic AMP: master regulator of innate immune cell function. Am J Respir Cell Mol Biol 39(2):127–132
109. Flego D et al (1997) Control of virulence gene expression by plant calcium in the phytopathogen Erwinia carotovora. Mol Microbiol 25(5):831–838
110. Deng W et al (2017) Assembly, structure, function and regulation of type III secretion systems. Nat Rev Microbiol 15(6):323–337
111. Ho BT, Dong TG, Mekalanos JJ (2014) A view to a kill: the bacterial type VI secretion system. Cell Host Microbe 15(1):9–21
112. Filloux A (2011) Protein secretion Systems in Pseudomonas aeruginosa: an essay on diversity, evolution, and function. Front Microbiol 2:155
113. Baumann U et al (1993) Three-dimensional structure of the alkaline protease of Pseudomonas aeruginosa: a two-domain protein with a calcium binding parallel beta roll motif. EMBO J 12(9):3357–3364
114. Duong F et al (1992) Sequence of a cluster of genes controlling synthesis and secretion of alkaline protease in Pseudomonas aeruginosa: relationships to other secretory pathways. Gene 121:47–54
115. Kessler E et al (1993) Secreted LasA of Pseudomonas aeruginosa is a staphylolytic protease. J Biol Chem 268(10):7503–7508
116. Thayer MM, Flaherty KM, McKay DB (1991) Three-dimensional structure of the elastase of Pseudomonas aeruginosa at 1.5-Å resolution. J Biol Chem 266(5):2864–2871
117. Wilderman PJ et al (2001) Characterization of an endoprotease (PrpL) encoded by a PvdS-regulated gene in Pseudomonas aeruginosa. Infect Immun 69(9):5385–5394
118. Palmer T, Berks BC (2012) The twin-arginine translocation (tat) protein export pathway. Nat Rev Microbiol 10(7):483–496
119. Papanikou E, Karamanou S, Economou A (2007) Bacterial protein secretion through the translocase nanomachine. Nat Rev Microbiol 5(11):839–851

120. Korotkov KV et al (2009) Calcium is essential for the major pseudopilin in the type 2 secretion system. J Biol Chem 284(38):25466–25470
121. Dasgupta N et al (2006) Transcriptional induction of the Pseudomonas aeruginosa type III secretion system by low Ca^{2+} and host cell contact proceeds through two distinct signaling pathways. Infect Immun 74(6):3334–3341
122. Vallis AJ et al (1999) Regulation of ExoS production and secretion by Pseudomonas aeruginosa in response to tissue culture conditions. Infect Immun 67(2):914–920
123. Schraidt O, Marlovits TC (2011) Three-dimensional model of Salmonella's needle complex at subnanometer resolution. Science 331(6021):1192–1195
124. Vakulskas CA, Brutinel ED, Yahr TL (2010) ExsA recruits RNA polymerase to an extended −10 promoter by contacting region 4.2 of sigma-70. J Bacteriol 192(14):3597–3607
125. Brutinel ED, Vakulskas CA, Yahr TL (2010) ExsD inhibits expression of the Pseudomonas aeruginosa type III secretion system by disrupting ExsA self-association and DNA binding activity. J Bacteriol 192(6):1479–1486
126. Broder UN, Jaeger T, Jenal U (2016) LadS is a calcium-responsive kinase that induces acute-to-chronic virulence switch in Pseudomonas aeruginosa. Nat Microbiol 2:16184
127. Diepold A et al (2017) A dynamic and adaptive network of cytosolic interactions governs protein export by the T3SS injectisome. Nat Commun 8:15940
128. Mougous JD et al (2006) A virulence locus of Pseudomonas aeruginosa encodes a protein secretion apparatus. Science 312(5779):1526–1530
129. Leiman PG et al (2009) Type VI secretion apparatus and phage tail-associated protein complexes share a common evolutionary origin. Proc Natl Acad Sci U S A 106(11):4154–4159
130. Allsopp LP et al (2017) RsmA and AmrZ orchestrate the assembly of all three type VI secretion systems in Pseudomonas aeruginosa. Proc Natl Acad Sci U S A 114(29):7707–7712
131. He X, Wang S (2014) DNA consensus sequence motif for binding response regulator PhoP, a virulence regulator of Mycobacterium tuberculosis. Biochemistry 53(51):8008–8020
132. Kreamer NN, Costa F, Newman DK (2015) The ferrous iron-responsive BqsRS two-component system activates genes that promote cationic stress tolerance. MBio 6(2):e02549
133. Schaaf S, Bott M (2007) Target genes and DNA-binding sites of the response regulator PhoR from Corynebacterium glutamicum. J Bacteriol 189(14):5002–5011
134. Sievers F et al (2011) Fast, scalable generation of high-quality protein multiple sequence alignments using Clustal omega. Mol Syst Biol 7:539–539
135. Tamura K et al (2007) MEGA4: molecular evolutionary genetics analysis (MEGA) software version 4.0. Mol Biol Evol 24(8):1596–1599
136. Vescovi EG et al (1997) Characterization of the bacterial sensor protein PhoQ. Evidence for distinct binding sites for Mg^{2+} and Ca^{2+}. J Biol Chem 272(3):1440–1443
137. Lesley JA, Waldburger CD (2001) Comparison of the Pseudomonas aeruginosa and Escherichia coli PhoQ sensor domains: evidence for distinct mechanisms of signal detection. J Biol Chem 276(33):30827–30833
138. Cai X et al (2011) The effect of the potential PhoQ histidine kinase inhibitors on Shigella flexneri virulence. PLoS One 6(8):e23100
139. Tu X et al (2006) The PhoP/PhoQ two-component system stabilizes the alternative sigma factor RpoS in <em>Salmonella enterica</em>. Proc Natl Acad Sci 103(36):13503–13508
140. Bishop RE et al (2004) Enzymology of lipid a palmitoylation in bacterial outer membranes. J Endotoxin Res 10(2):107–112
141. Macfarlane EL et al (1999) PhoP–PhoQ homologues in Pseudomonas aeruginosa regulate expression of the outer-membrane protein OprH and polymyxin B resistance. Mol Microbiol 34(2):305–316
142. Shin D et al (2006) A positive feedback loop promotes transcription surge that jump-starts <em>Salmonella</em> virulence circuit. Science 314(5805):1607–1609
143. Macfarlane EL, Kwasnicka A, Hancock RE (2000) Role of Pseudomonas aeruginosa PhoP-PhoQ in resistance to antimicrobial cationic peptides and aminoglycosides. Microbiology 146(10):2543–2554

144. Gooderham WJ et al (2009) The sensor kinase PhoQ mediates virulence in Pseudomonas aeruginosa. Microbiology 155(3):699–711
145. Chamnongpol S, Cromie M, Groisman EA (2003) Mg^{2+} sensing by the Mg^{2+} sensor PhoQ of Salmonella enterica. J Mol Biol 325(4):795–807
146. Regelmann AG et al (2002) Mutational analysis of the Escherichia coli PhoQ sensor kinase: differences with the Salmonella enterica serovar typhimurium PhoQ protein and in the mechanism of Mg^{2+} and Ca^{2+} sensing. J Bacteriol 184(19):5468–5478
147. Lesley JA, Waldburger CD (2003) Repression of Escherichia coli PhoP-PhoQ signaling by acetate reveals a regulatory role for acetyl coenzyme a. J Bacteriol 185(8):2563–2570
148. Flego D et al (2000) A two-component regulatory system, pehR-pehS, controls Endopolygalacturonase production and virulence in the plant pathogen Erwinia carotovora subsp. carotovora. Mol Plant-Microbe Interact 13(4):447–455
149. Kariola T et al (2003) Erwinia carotovora subsp carotovora and Erwinia-derived elicitors HrpN and PehA trigger distinct but interacting defense responses and cell death in Arabidopsis. Mol Plant-Microbe Interact 16(3):179–187
150. Palomaki T et al (2002) A putative three-dimensional targeting motif of polygalacturonase (PehA), a protein secreted through the type II (GSP) pathway in Erwinia carotovora. Mol Microbiol 43(3):585–596
151. Kaneshiro WS et al (2008) Characterization of Erwinia chrysanthemi from a bacterial heart rot of pineapple outbreak in Hawaii. Plant Dis 92(10):1444–1450
152. Hugouvieux-Cotte-Pattat N, Shevchik VE, Nasser W (2002) PehN, a Polygalacturonase homologue with a low hydrolase activity, is Coregulated with the other Erwinia chrysanthemi Polygalacturonases. J Bacteriol 184(10):2664–2673
153. Guragain M et al (2016) The Pseudomonas aeruginosa PAO1 two-component regulator CarSR regulates calcium homeostasis and calcium-induced virulence factor production through its regulatory targets CarO and CarP. J Bacteriol 198(6):951–963
154. Theodorou MC et al (2006) Involvement of the AtoS-AtoC signal transduction system in poly-(R)-3-hydroxybutyrate biosynthesis in Escherichia coli. Biochim Biophys Acta 1760(6):896–906
155. Theodorou MC, Theodorou EC, Kyriakidis DA (2012) Involvement of AtoSC two-component system in Escherichia coli flagellar regulon. Amino Acids 43(2):833–844
156. Theodorou MC et al (2007) Spermidine triggering effect to the signal transduction through the AtoS-AtoC/Az two-component system in Escherichia coli. Biochim Biophys Acta 1770(8):1104–1114
157. Theodorou EC, Theodorou MC, Kyriakidis DA (2013) Regulation of poly-(R)-(3-hydroxybutyrate-co-3-hydroxyvalerate) biosynthesis by the AtoSCDAEB regulon in phaCAB(+) Escherichia coli. Appl Microbiol Biotechnol 97(12):5259–5274
158. Jung C-J et al (2017) AtlA mediates extracellular DNA release, which contributes to Streptococcus mutans biofilm formation in an experimental rat model of infective endocarditis. Infect Immun 85(9):e00252–e00217
159. Downey JS et al (2014) In vitro manganese-dependent cross-talk between Streptococcus mutans VicK and GcrR: implications for overlapping stress response pathways. PLoS One 9(12):e115975
160. Kawasaki K (2012) Complexity of lipopolysaccharide modifications in Salmonella enterica: its effects on endotoxin activity, membrane permeability, and resistance to antimicrobial peptides. Food Res Int 45(2):493–501
161. He X et al (2008) The cia operon of Streptococcus mutans encodes a unique component required for calcium-mediated autoregulation. Mol Microbiol 70(1):112–126
162. Bilecen K et al (2015) Polymyxin B resistance and biofilm formation in Vibrio cholerae are controlled by the response regulator CarR. Infect Immun 83(3):1199–1209
163. Kierek K, Watnick PI (2003) The <em>Vibrio cholerae</em> O139 O-antigen polysaccharide is essential for ca²⁺–dependent biofilm development in sea water. Proc Natl Acad Sci 100(24):14357–14362

164. Kierek K, Watnick PI (2003) Environmental determinants of Vibrio cholerae biofilm development. Appl Environ Microbiol 69(9):5079–5088
165. Giammarinaro P, Sicard M, Gasc AM (1999) Genetic and physiological studies of the CiaH-CiaR two-component signal-transducing system involved in cefotaxime resistance and competence of Streptococcus pneumoniae. Microbiology 145(Pt 8):1859–1869
166. Ibrahim YM et al (2004) Control of virulence by the two-component system CiaR/H is mediated via HtrA, a major virulence factor of Streptococcus pneumoniae. J Bacteriol 186(16):5258–5266
167. Wu C et al (2010) Regulation of ciaXRH operon expression and identification of the CiaR regulon in Streptococcus mutans. J Bacteriol 192(18):4669–4679
168. Jung CJ et al (2017) AtlA mediates extracellular DNA release, which contributes to Streptococcus mutans biofilm formation in an experimental rat model of infective endocarditis. Infect Immun 85(9):10
169. Senadheera MD et al (2005) A VicRK signal transduction system in Streptococcus mutans affects gtfBCD, gbpB, and ftf expression, biofilm formation, and genetic competence development. J Bacteriol 187(12):4064–4076
170. Reusch R (2013) The role of short-chain conjugated poly-(R)-3-Hydroxybutyrate (cPHB) in protein folding. Int J Mol Sci 14(6):10727
171. Theodorou MC, Tiligada E, Kyriakidis DA (2009) Extracellular Ca^{2+} transients affect poly-(R)-3-hydroxybutyrate regulation by the AtoS-AtoC system in Escherichia coli. Biochem J 417(3):667–672
172. Yamamoto K et al (2005) Functional characterization in vitro of all two-component signal transduction systems from Escherichia coli. J Biol Chem 280(2):1448–1456
173. Rodrigue A et al (2000) Cell signalling by oligosaccharides. Two-component systems in Pseudomonas aeruginosa: why so many? Trends Microbiol 8(11):498–504
174. Siryaporn A, Goulian M (2008) Cross-talk suppression between the CpxA–CpxR and EnvZ–OmpR two-component systems in E. Coli. Mol Microbiol 70(2):494–506
175. Wanner BL (1992) Is cross regulation by phosphorylation of two-component response regulator proteins important in bacteria? J Bacteriol 174(7):2053
176. Chin D, Means AR (2000) Calmodulin: a prototypical calcium sensor. Trends Cell Biol 10(8):322–328
177. Zhao X et al (2010) Structural basis for prokaryotic calciummediated regulation by a Streptomyces coelicolor calcium binding protein. Protein Cell 1(8):771–779
178. Zhang M et al (2012) Structural basis for calmodulin as a dynamic calcium sensor. Structure 20(5):911–923
179. Ikura M (1996) Calcium binding and conformational response in EF-hand proteins. Trends Biochem Sci 21(1):14–17
180. Crivici A, Ikura M (1995) Molecular and structural basis of target recognition by calmodulin. Annu Rev Biophys Biomol Struct 24(1):85–116
181. Zhou Y et al (2006) Prediction of EF-hand calcium-binding proteins and analysis of bacterial EF-hand proteins. Proteins 65(3):643–655
182. Inouye S, Franceschini T, Inouye M (1983) Structural similarities between the development-specific protein S from a gram-negative bacterium, Myxococcus xanthus, and calmodulin. Proc Natl Acad Sci 80(22):6829–6833
183. Iwasa Y et al (1981) Calmodulin-like activity in the soluble fraction of Escherichia coli. Biochem Biophys Res Commun 98(3):656–660
184. Kerson GW, Miernyk JA, Budd K (1984) Evidence for the occurrence of, and possible physiological role for, cyanobacterial calmodulin. Plant Physiol 75(1):222–224
185. Leadlay PF, Roberts G, Walker JE (1984) Isolation of a novel calcium-binding protein from streptomyces erythreus. FEBS Lett 178(1):157–160
186. Swan D et al (1989) Cloning, characterization, and heterologous expression of the Saccharopolyspora erythraea (Streptomyces erythraeus) gene encoding an EF-hand calcium-binding protein. J Bacteriol 171(10):5614–5619

187. Swan DG et al (1987) A bacterial calcium-binding protein homologous to calmodulin. Nature 329(6134):84
188. Yonekawa T, Ohnishi Y, Horinouchi S (2005) A calmodulin-like protein in the bacterial genus Streptomyces. FEMS Microbiol Lett 244(2):315–321
189. Xi C et al (2000) Symbiosis-specific expression of rhizobium etli casA encoding a secreted calmodulin-related protein. Proc Natl Acad Sci 97(20):11114–11119
190. Sarkisova SA et al (2014) A Pseudomonas aeruginosa EF-hand protein, EfhP (PA4107), modulates stress responses and virulence at high calcium concentration. PLoS One 9(6):e98985
191. Scholl ZN et al (2016) Single-molecule force spectroscopy reveals the calcium dependence of the alternative conformations in the native state of a βγ-Crystallin protein. J Biol Chem 291(35):18263–18275
192. Koul, S., et al., A novel calcium binding protein in Mycobacterium tuberculosis—potential target for trifluoperazine. 2009
193. Advani MJ, Rajagopalan M, Reddy PH (2014) Calmodulin-like protein from M. Tuberculosis H37Rv is required during infection. Sci Rep 4:6861
194. Reddy PT et al (2003) Cloning and expression of the gene for a novel protein from Mycobacterium smegmatis with functional similarity to eukaryotic calmodulin. J Bacteriol 185(17):5263–5268
195. Fry I, Becker-Hapak M, Hageman J (1991) Purification and properties of an intracellular calmodulinlike protein from Bacillus subtilis cells. J Bacteriol 173(8):2506–2513
196. Nagai M et al (1994) Purification and characterization of Bordetella calmodulin-like protein. FEMS Microbiol Lett 116(2):169–174
197. Vincent F et al (2010) Distinct oligomeric forms of the Pseudomonas aeruginosa RetS sensor domain modulate accessibility to the ligand binding site. Environ Microbiol 12(6):1775–1786
198. Broder UN, Jaeger T, Jenal U (2017) LadS is a calcium-responsive kinase that induces acute-to-chronic virulence switch in Pseudomonas aeruginosa. Nat Microbiol 2(1):16184
199. Falah A et al (1988) On the presence of calmodulin-like protein in mycobacteria. FEMS Microbiol Lett 56(1):89–93
200. Burra SS et al (1991) Calmodulin-like protein and the phospholipids of Mycobacterium smegmatis. FEMS Microbiol Lett 80(2-3):189–194
201. Teintze M, Inouye M, Inouye S (1988) Characterization of calcium-binding sites in development-specific protein S of Myxococcus xanthus using site-specific mutagenesis. J Biol Chem 263(3):1199–1203
202. Ryjenkov DA et al (2005) Cyclic diguanylate is a ubiquitous signaling molecule in bacteria: insights into biochemistry of the GGDEF protein domain. J Bacteriol 187(5):1792–1798
203. Dominguez DC, Guragain M, Patrauchan M (2015) Calcium binding proteins and calcium signaling in prokaryotes. Cell Calcium 57(3):151–165
204. Dominguez DC (2004) Calcium signalling in bacteria. Mol Microbiol 54(2):291–297
205. Booth IR et al (2015) The evolution of bacterial mechanosensitive channels. Cell Calcium 57(3):140–150
206. Guragain M et al (2013) Calcium homeostasis in Pseudomonas aeruginosa requires multiple transporters and modulates swarming motility. Cell Calcium 54(5):350–361
207. Jones HE, Holland IB, Campbell AK (2002) Direct measurement of free Ca^{2+} shows different regulation of Ca^{2+} between the periplasm and the cytosol of Escherichia coli. Cell Calcium 32(4):183–192
208. Naseem R et al (2008) pH and monovalent cations regulate cytosolic free Ca^{2+} in E. Coli. Biochim Biophys Acta Biomembr 1778(6):1415–1422
209. Knight MR et al (1991) Recombinant aequorin as a probe for cytosolic free Ca^{2+} in Escherichia coli. FEBS Lett 282(2):405–408
210. Jones HE et al (1999) Slow changes in cytosolic free Ca^{2+} inEscherichia colihighlight two putative influx mechanisms in response to changes in extracellular calcium. Cell Calcium 25(3):265–274
211. Campbell AK et al (2007) Fermentation product butane 2, 3-diol induces Ca^{2+} transients in E. Coli through activation of lanthanum-sensitive Ca^{2+} channels. Cell Calcium 41(2):97–106

212. Bruni GN et al (2017) Voltage-gated calcium flux mediates <em>Escherichia coli</em> mechanosensation. Proc Natl Acad Sci 114(35):9445–9450
213. Herbaud M-L et al (1998) Calcium signalling in Bacillus subtilis. Biochim Biophys Acta 1448(2):212–226
214. Pavlov E et al (2005) A high-conductance mode of a poly-3-hydroxybutyrate/calcium/polyphosphate channel isolated from competent Escherichia coli cells. FEBS Lett 579(23):5187–5192
215. Huang R, Reusch RN (1995) Genetic competence in Escherichia coli requires poly-beta-hydroxybutyrate/calcium polyphosphate membrane complexes and certain divalent cations. J Bacteriol 177(2):486–490
216. Chang Y et al (2014) Structural basis for a pH-sensitive calcium leak across membranes. Science 344(6188):1131–1135
217. Liu Q (2017) TMBIM-mediated Ca^{2+} homeostasis and cell death. BBA-Mol Cell Res 1864(6):850–857
218. Haswell ES, Phillips R, Rees DC (2011) Mechanosensitive channels: what can they do and how do they do it? Structure 19(10):1356–1369
219. Martinac B (2004) Mechanosensitive ion channels: molecules of mechanotransduction. J Cell Sci 117(12):2449–2460
220. Moe PC, Blount P, Kung C (1998) Functional and structural conservation in the mechanosensitive channel MscL implicates elements crucial for mechanosensation. Mol Microbiol 28(3):583–592
221. Chang G et al (1998) Structure of the MscL homolog from Mycobacterium tuberculosis: a gated mechanosensitive ion channel. Science 282(5397):2220–2226
222. Blount P et al (1996) Membrane topology and multimeric structure of a mechanosensitive channel protein of Escherichia coli. EMBO J 15(18):4798–4805
223. Velasquez J et al (2014) Bacillus subtilis spore protein SpoVAC functions as a mechanosensitive channel. Mol Microbiol 92(4):813–823
224. Cox CD et al (2013) Selectivity mechanism of the mechanosensitive channel MscS revealed by probing channel subconducting states. Nat Commun 4:2137
225. Kühlbrandt W (2004) Biology, structure and mechanism of P-type ATPases. Nat Rev Mol Cell Biol 5:282
226. Rensing C et al (2000) CopA: an <em>Escherichia coli</em> cu(I)-translocating P-type ATPase. Proc Natl Acad Sci 97(2):652–656
227. Rosch JW et al (2008) Calcium efflux is essential for bacterial survival in the eukaryotic host. Mol Microbiol 70(2):435–444
228. Gupta HK, Shrivastava S, Sharma R (2017) A novel calcium uptake transporter of uncharacterized P-type ATPase family supplies calcium for cell surface integrity in Mycobacterium smegmatis. MBio 8(5):e01388-17
229. States DJ, Gish W (1994) Combined use of sequence similarity and codon bias for coding region identification. J Comput Biol 1(1):39–50
230. Huang Y et al (2009) Multiple Ca^{2+} binding sites in the extracellular domain of Ca^{2+}-sensing receptor corresponding to cooperative Ca^{2+} response. Biochemistry 48(2):388–398
231. Hendy GN et al (2000) Mutations of the calcium-sensing receptor (CASR) in familial hypocalciuric hypercalcemia, neonatal severe hyperparathyroidism, and autosomal dominant hypocalcemia. Hum Mutat 16(4):281–296
232. Holstein DM et al (2004) Calcium-sensing receptor-mediated ERK1/2 activation requires $G\alpha i2$ coupling and dynamin-independent receptor internalization. J Biol Chem 279(11):10060–10069
233. Di Mise A et al (2018) Activation of calcium-sensing receptor increases intracellular calcium and decreases cAMP and mTOR in PKD1 deficient cells. Sci Rep 8(1):5704
234. Smith KA et al (2016) Calcium-sensing receptor regulates cytosolic $[Ca^{2+}]$ and plays a major role in the development of pulmonary hypertension. Front Physiol 7:517

235. Huang Y-H et al (2016) Characterization of Staphylococcus aureus Primosomal DnaD protein: highly conserved C-terminal region is crucial for ssDNA and PriA helicase binding but not for DnaA protein-binding and self-Tetramerization. PLoS One 11(6):e0157593
236. Mosyak L et al (2000) The bacterial cell-division protein ZipA and its interaction with an FtsZ fragment revealed by X-ray crystallography. EMBO J 19(13):3179–3191
237. Koskiniemi S et al (2013) Rhs proteins from diverse bacteria mediate intercellular competition. Proc Natl Acad Sci U S A 110(17):7032–7037
238. Drees JC et al (2006) Inhibition of RecA protein function by the RdgC protein from Escherichia coli. J Biol Chem 281(8):4708–4717
239. Briggs GS et al (2007) Ring structure of the Escherichia coli DNA-binding protein RdgC associated with recombination and replication fork repair. J Biol Chem 282(17):12353–12357
240. Hernández-Ochoa EO et al (2015) Critical role of intracellular RyR1 calcium release channels in skeletal muscle function and disease. Front Physiol 6:420
241. Xu L et al (2018) G4941K substitution in the pore-lining S6 helix of the skeletal muscle ryanodine receptor increases RyR1 sensitivity to cytosolic and luminal Ca^{2+}. J Biol Chem 293(6):2015–2028
242. Clarke OB, Hendrickson WA (2016) Structures of the colossal RyR1 calcium Release Channel. Curr Opin Struct Biol 39:144–152
243. Kane Dickson V, Pedi L, Long SB (2014) Structure and insights into the function of a Ca^{2+}-activated Cl(−) channel. Nature 516(7530):213–218
244. Yang T et al (2014) Structure and selectivity in bestrophin ion channels. Science 346(6207):355–359
245. Kuo YH et al (2014) Effects of alternative splicing on the function of bestrophin-1 calcium-activated chloride channels. Biochem J 458:575–583

Chapter 34
Ca^{2+} Signaling in *Drosophila* Photoreceptor Cells

Olaf Voolstra and Armin Huber

Abstract In *Drosophila* photoreceptor cells, Ca^{2+} exerts regulatory functions that control the shape, duration, and amplitude of the light response. Ca^{2+} also orchestrates light adaptation allowing *Drosophila* to see in light intensity regimes that span several orders of magnitude ranging from single photons to bright sunlight. The prime source for Ca^{2+} elevation in the cytosol is Ca^{2+} influx from the extracellular space through light-activated TRP channels. This Ca^{2+} influx is counterbalanced by constitutive Ca^{2+} extrusion via the Na^{+}/Ca^{2+} exchanger, CalX. The light-triggered rise in intracellular Ca^{2+} exerts its regulatory functions through interaction with about a dozen well-characterized Ca^{2+} and Ca^{2+}/CaM binding proteins. In this review we will discuss the dynamic changes in Ca^{2+} concentration upon illumination of photoreceptor cells. We will present the proteins that are known to interact with Ca^{2+} (/CaM) and elucidate the physiological functions of these interactions.

Keywords Arrestin · Calcium signaling · *Drosophila* · Light adaptation · Phospholipase C · Phototransduction · Protein kinase C · Rhodopsin · TRP channel · Vision

34.1 Introduction

The signaling cascade in photoreceptor cells of the *Drosophila* compound eye is among the best studied G protein-coupled, phosphoinositide-mediated signal transduction pathways. Identification of the components of this signaling cascade began in the 1970s when Bill Pak and others performed genetic screens directed to identify genes encoding proteins of the visual signaling pathway [1]. Now, a long list of proteins were found to be involved in fly vision and *Drosophila* mutants for

O. Voolstra · A. Huber (✉)
Department of Biochemistry, Institute of Physiology, University of Hohenheim, Stuttgart, Germany
e-mail: armin.huber@uni-hohenheim.de

© Springer Nature Switzerland AG 2020

M. S. Islam (ed.), *Calcium Signaling*, Advances in Experimental Medicine and Biology 1131, https://doi.org/10.1007/978-3-030-12457-1_34

each of these proteins are available (see e.g [2–6]). This list includes, for example, the first cloned invertebrate rhodopsin (RH1) and, importantly, the first ion channels of the TRP class of ion channels to be discovered. Because the light-activated TRP channels are Ca^{2+} permeable, activation of the visual cascade results in a profound increase in the intracellular Ca^{2+} concentration. Ca^{2+} in turn feeds back on several components of the signaling cascade, thereby ensuring fast photoresponses and adaptation to variable light intensities. Many of these Ca^{2+} feedback mechanisms have been clarified only recently and some are still elusive. The role of Ca^{2+} in fly photoreceptors has been dealt with in other reviews e.g [4, 7–9]. In this review, we focus mainly on more recent results elucidating new aspects of Ca^{2+} signaling in fly photoreceptor cells.

34.2 Regulation of Ca^{2+} Levels in *Drosophila* Photoreceptor Cells

34.2.1 Activation of the Phototransduction Cascade

The compound eye of *Drosophila* is composed of ~800 unit eyes, called ommatidia. Each ommatidium consists of a lens to gather the light, eight photoreceptor cells, and auxiliary cells. The elongated photoreceptor cells harbor a light-guiding structure, the rhabdomere that is built from its apical plasma membrane. ~40,000 finger-like evaginations of this membrane, called microvilli, form the rhabdomere. In the rhabdomere, the components of the phototransduction cascade, including the visual pigment rhodopsin, are located. Absorption of a photon triggers the phototransduction cascade by converting rhodopsin to its active state, metarhodopsin. Metarhodopsin activates a heterotrimeric G protein causing the exchange of GDP bound to the α subunit with cytosolic GTP and the release of the α subunit from the βγ subunits. The activated α subunit diffuses to its effector protein, phospholipase Cβ (PLCβ), and activates it (Fig. 34.1). PLCβ in turn cleaves the membrane component phosphatidylinositol-4,5-bisphosphate (PIP_2), to yield water-soluble inositol-1,4,5-trisphosphate ($InsP_3$), membrane-bound diacyl glycerol (DAG), and a proton leading to the opening of the ion channels TRP and TRPL. Currently, the exact activation mechanism of the plasma membrane channels is still a matter of debate. It has been shown that the decrease in PIP_2 together with the hitherto neglected generation of protons causes the opening of the channels [10]. Additionally, the removal of the bulky hydrophilic $InsP_3$ group from PIP_2, which leaves the much smaller lipid DAG in the microvillar membrane results in a change of lipid packing and generation of mechanical forces that might contribute to the gating of the TRP and TRPL ion channels [11]. The mechanical forces generated by PIP_2 cleavage are revealed by a contraction of the entire rhabdomere upon bright light illumination that was quantified by atomic force microscopy and shown to strictly depend on activation of the phototransduction cascade [11]. On the other

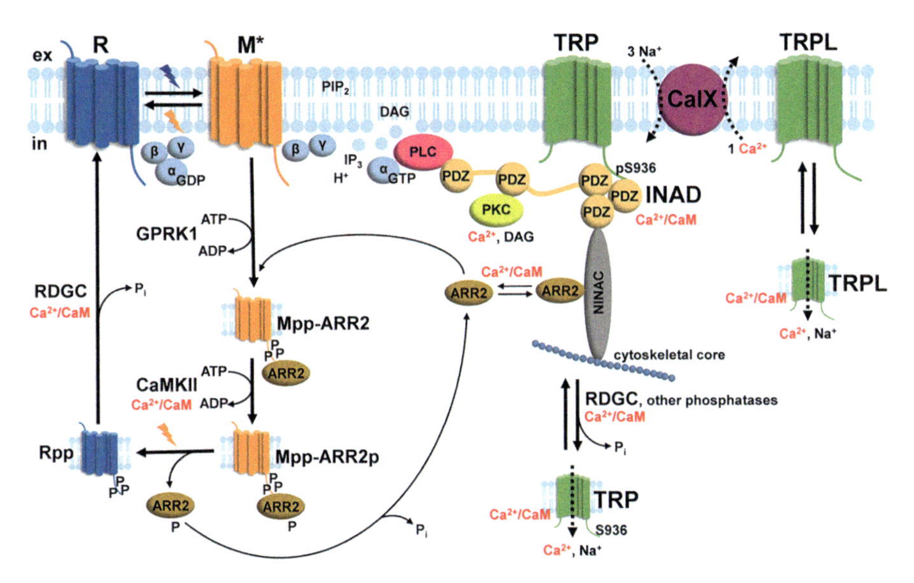

Fig. 34.1 The rhodopsin cycle and the phototransduction cascade. Absorption of a photon converts rhodopsin (R) to metarhodopsin (M*). M* can be directly converted back to R by absorption of another photon. If no photon is absorbed by M* within ~5 ms, the phototransduction cascade is activated [68]. The α subunit of the heterotrimeric G protein exchanges GDP for GTP and diffuses from the βγ subunits. Then it activates phospholipase Cβ that in turn cleaves phosphatidylinositol 4,5-bisphosphate (PIP_2) to yield diacylglycerol (DAG), inositol 1,4,5-trisphosphate (IP_3), and a proton (H^+). This reduction of PIP_2 levels and buildup of protons, together with mechanical forces that result from the removal of the bulky IP_3 head groups trigger the opening of the TRP channels [10, 11]. Ca^{2+} and Na^+ ions enter the photoreceptor cell generating the photoreceptor potential. Ca^{2+} then interacts with numerous targets. Activated metarhodopsin (M*) constitutes a substrate for the G protein-coupled receptor kinase 1 (GPRK1) that phosphorylates metarhodopsin at its C-terminus [54, 56]. Ca^{2+}/CaM releases ARR2 from NINAC and ARR2 binds to M* to inactivate it. CaMKII is activated by Ca^{2+}/CaM and phosphorylates ARR2 that is bound to Mpp. Phosphorylation of ARR2 facilitates its release from Mpp [71]. Absorption of another photon converts Mpp to Rpp and triggers the release of ARR2p [58]. RDGC that is activated by Ca^{2+}/CaM dephosphorylates Rpp or Mpp to yield R. Ca^{2+} together with DAG activate eye-specific protein kinase C (PKC) that in turn phosphorylates INAD and TRP [89, 93–97]. The TRP ion channel is phosphorylated in the dark at S936 by a yet unknown kinase and becomes RDGC-dependently dephosphorylated in the light. The phosphorylation state of S936 influences the kinetics of the channel opening [75, 102]. Ca^{2+} alone or together with CaM has strong regulatory effects on the TRP and TRPL ion channels. Low Ca^{2+} concentrations of ~300 nM facilitate channel opening whereas higher concentrations of ~1 μM promote closing of the channel. Concentrations of ~50 μM that are only reached transiently in vivo, inhibit PLC activity in a PKC-dependent manner. To reestablish resting Ca^{2+} concentrations, the Na^+/Ca^{2+} exchanger CalX extrudes 1 Ca^{2+} for every 3 Na^+ that are taken up. (Parts of the Figure were modified from [68])

hand, Delgado and co-workers reported on evidence that points to DAG as the endogenous TRP agonist [12]. It is also important to mention that TRP can be activated by polyunsaturated fatty acids (PUFAs) which could be generated by hydrolysis of PUFAs from DAG [13]. However, to date, only a DAG lipase, INAE, that cleaves off the saturated but not the unsaturated fatty acid from DAG has been discovered [14]. Thus, the role of PUFAs in the activation of TRP ion channels in vivo remains questionable.

34.2.2 Generation of the Photoreceptor Potential by Na^+ and Ca^{2+} Influx

The light-triggered opening of the ion channels TRP and TRPL results in an influx of Ca^{2+} and Na^+ ions into the photoreceptor cell. The influx of cations depolarizes the photoreceptor cell and generates the photoreceptor potential. Thus, the light stimulus is converted into an electrical signal in the photoreceptor cell. The resting Ca^{2+} concentration in dark-adapted photoreceptors was measured using INDO-1 calcium indicator dye [15] and the genetically-encoded calcium indicator GCaMP6f [16]. Usage of INDO-1 resulted in an intracellular Ca^{2+} concentration of ~160 nM and usage of GCaMP6f in ~80 nM. These results are not too far from each other and provide a rough estimate of the intracellular resting Ca^{2+} concentration. Since these measurements were carried out on isolated ommatidia and depend on some calibration and assumptions, it is not clear if they reliably represent the in *vivo dark* level of photoreceptor Ca^{2+} concentration, which can be much lower. During a photoreceptor potential, the Ca^{2+} concentration can transiently reach almost millimolar concentrations within the small volume of the rhabdomeral microvilli where Ca^{2+} enters the cell mainly through TRP and to a smaller extent through TRPL channels [17–20]. Ca^{2+} diffuses from the rhabdomere into the cell body where it is diluted and buffered [17, 21] so that the cellular Ca^{2+} concentration reaches maximally 10 μM [8, 17, 18, 22]. Resting concentrations of Ca^{2+} are restored mainly by CalX, a Na^+/Ca^{2+} exchanger that extrudes one Ca^{2+} ion from the photoreceptor cell in exchange for the uptake of three Na^+ ions [23]. Loss of CalX function resulted in a transient light response to sustained light, a decrease in signal amplification, and abnormally rapid light adaptation [24]. Conversely, overexpression of CalX led to the opposite effects and rescued retinal degeneration caused by a constitutively active TRP variant [24]. Interestingly, the Na^+/Ca^{2+} exchange equilibrium can be manipulated to experimentally control intracellular Ca^{2+} concentrations [25] (see Chap. 3.3). Besides Ca^{2+} extrusion by CalX, uptake of Ca^{2+} into internal stores might constitute an additional mechanism to restore resting concentrations of Ca^{2+}. The sarcoendoplasmic reticulum Ca^{2+} ATPase (SERCA) mediates the uptake of Ca^{2+} into the endoplasmic reticulum (ER). SERCA is expressed in *Drosophila* photoreceptor cells [26] and pharmacological inhibition of SERCA resulted in elevated intracellular Ca^{2+} concentrations [20, 27]. However, the relative contribution of SERCA to the re-establishment of resting Ca^{2+} concentrations is elusive.

The source of the Ca^{2+} ions entering the cytosol of the photoreceptor cell upon light stimulation is somewhat controversial. It is general consensus that the extracellular Ca^{2+} that enters the cell through TRP and TRPL channels is the major source for light-triggered Ca^{2+} elevation in the cytosol. However, $InsP_3$ that is generated by PLC action in *Drosophila* photoreceptor cells could activate an $InsP_3$ receptor in the endoplasmatic reticulum leading to Ca^{2+} release from intracellular stores. In photoreceptor cells of other invertebrates, for example in compound eyes of bees or in the ventral photoreceptor of *Limulus*, release of Ca^{2+} from internal stores is the major source of cellular Ca^{2+} elevation, which is well documented [28, 29]. In the *Drosophila* genome, a single *insP*$_3$ receptor gene exists. Two groups investigated *insP*$_3$ receptor mutant flies but did not find impairments in phototransduction [30, 31]. Thereafter, the idea of light-dependent Ca^{2+} release from intracellular stores in *Drosophila* photoreceptor cells was abandoned until Kohn and co-workers took up the topic again [32]. The authors used a *gmr-gal4* construct to drive $InsP_3$ receptor-RNAi transcription. Using RNAi to reduce $InsP_3$ receptor levels, they were able to circumvent structural eye damage that was found in $InsP_3$ receptor mutants. Additionally, Kohn and co-workers used a genetically-encoded calcium indicator, GCaMP6F, to monitor intracellular Ca^{2+} concentrations. They found that reduced $insP_3$ receptor levels or Ca^{2+} store depletion resulted in a reduction in light sensitivity and concluded that release of Ca^{2+} from intracellular stores plays an essential role in light excitation. The findings by Kohn and colleagues were challenged by Bollepalli and co-workers who found no effect of $insP_3$ receptor RNAi or $insP_3$ receptor mutation on photoresponses or Ca^{2+} influx, but observed that Gal4 expression under control of the *gmr* promotor resulted in altered photoreceptor physiology [33]. To avoid problems with *gmr*-driven GAL4 expression, Asteriti and colleagues used a fly expressing GCaMP6f under direct control of the *rh1* promoter [16]. Using dissociated photoreceptor cells from this fly, they were able to manipulate the ionic composition of the extracellular buffer. Under physiological conditions, rises in GCaMP6f fluorescence were observed 10–25 ms after a light stimulus and exhibited increases of up to 20fold. In contrast, in Ca^{2+}-free bath, latencies increased to 200 ms and fluorescence increases dropped to 4fold. Mutation of the $insP_3$ receptor had no effect showing that the residual signal was not generated by Ca^{2+} release from internal stores. Fluorescence increases were abolished in a *trpl;trp* double mutant, in a *calX* mutant, or when Na^+ was omitted from the bath, suggesting dependence on Na^+ and activity of the Na^+/Ca^{2+} exchanger. Thus, light-dependent rises in the fluorescence of calcium indicators under Ca^{2+}-free conditions can most probably be explained by the reversal of the Na^+/Ca^{2+} exchange [16]. It was proposed that extracellular Ca^{2+} that is resistant to buffering probably enters the photoreceptor cell through the reversed Na^+/Ca^{2+} exchange and leads to generation of the residual fluorescence. A question remains about the nature and location of this proposed pool of buffering-resistant extracellular Ca^{2+}. Alternatively, a fraction of CalX could be located in the membrane of intracellular Ca^{2+} stores and pump Ca^{2+} from these stores into the cytosol, when operating in reverse mode under high cytosolic Na^+ conditions [16].

34.2.3 Contribution of Ca^{2+} to the Photoreceptor Potential

Because Ca^{2+} is a charge carrier in the generation of the photoreceptor potential, but also serves a multitude of regulatory functions as we will discuss in Chap. 3, the determination of the fractional contribution of Ca^{2+} to the light-induced current (LIC) has been of general interest. The fractional current carried by Ca^{2+} influx can theoretically be estimated by the Goldman-Hodgkin-Katz (GHK) theory [34, 35]. However, Chu and colleagues doubted that the assumptions of this theory were met by the *Drosophila* photoreceptor cell. The authors therefore exploited the existence of a tail current that is caused by the CalX-dependent Ca^{2+}/Na^{+} exchange to experimentally determine the fractional current that is evoked by influx of Ca^{2+} into the photoreceptor cell [36]. While the GHK theory predicts the fractional Ca^{2+} current to be 42%, Chu and co-workers experimentally determined the fractional Ca^{2+} current to be 26%. The remaining fractional current can largely be attributed to Na^{+}. Collectively, Ca^{2+} significantly contributes to the LIC, but does not represent the largest fraction of the LIC. While the permeability of the TRP ion channel for Ca^{2+} is about 50 times higher than for Na^{+}, the permeability of the TRPL ion channel for Ca^{2+} is only about 4 times higher than for Na^{+} [37, 38]. The fact that Ca^{2+} does not generate the largest fractional current is due to the extracellular ion composition where the concentration of Na^{+} is 120 mM and that of Ca^{2+} is 1.5 mM [39].

34.3 Effects that Ca^{2+} Exerts in the Photoreceptor Cell

Through interaction with a variety of target proteins, Ca^{2+} exerts regulatory functions that control underlying processes to influence the shape, duration, and amplitude of the light response. Ca^{2+} mediates positive as well as negative feedback, resulting in a fast rising electrophysiological response to a light stimulus [40, 41]. Furthermore, Ca^{2+} orchestrates light adaptation allowing *Drosophila* to see in light intensity regimes that span several orders of magnitude ranging from single photons to bright sunlight ($\sim 10^6$ photons/photoreceptor/s). Proteins of fly photoreceptor cells that interact with Ca^{2+} are listed in Table 34.1.

34.3.1 General Ca^{2+} Binding Proteins

34.3.1.1 Ca^{2+} Buffering by Calphotin

Calphotin is a 85 kDa Ca^{2+} binding protein that is located in a defined cytoplasmic region in *Drosophila* photoreceptor cells adjacent to the base of the rhabdomere [42, 43]. This cellular region is virtually free of electron dense structures and

Table 34.1 Ca^{2+}-interacting proteins in *Drosophila* photoreceptor cells

Protein	Function of this protein	Loss-of-function phenotype	Effect of Ca^{2+}
Calmodulin	Ca^{2+} sensor/adaptor to convey Ca^{2+} regulation	Prolonged deactivation of the phototransduction cascade [53]	Triggers binding of CaM to target proteins
Calphotin	Ca^{2+} buffer	Light-dependent retinal degeneration [21]	
CaMKII	Phosphorylation of ARR2 [48]	ARR2 hypophosphorylation and strong binding to rhodopsin [71]	Activation (together with CaM)
INAD	Scaffold for TRP, NINAC, PKC, and PLC	Mislocalization of signaling complex members, retinal degeneration	Ca^{2+}/CaM binding has unknown consequences
Myosin V	Translocation of pigment granula		Movement along RTW filament towards the rhabdomere base
Neurocalcin		Rhodopsin hyperphosphorylation? [116]	Inhibition of rhodopsin phosphorylation [116]
NINAC	Binding of ARR2	Light-dependent retinal degeneration, electrophysiological defect	Ca^{2+}/CaM triggers release of ARR2
PKC	Phosphorylation of INAD, TRP, and possibly other substrates	Prolonged deactivation of the photoresponse, transient receptor potential	Activation (together with DAG)
PKC53E	Phosphorylation of INAD, TRP, and possibly other substrates	No apparent phenotype	Activation (together with DAG)
PLC	Cleavage of PIP_2	No receptor potential	Activation at $\leq 1\ \mu M$
			Inhibition at 50 μM
RDGC	Dephosphorylation of rhodopsin and TRP	Light-dependent retinal degeneration, prolonged deactivation of the phototransduction cascade, premature entrance into PDA	Activation (together with CaM)
TRP	Na^{+}/Ca^{2+} channel	Transient receptor potential, light-dependent retinal degeneration	Activation at ~300 nM
			Inhibition at 0.1–10 μM
TRPL	Na^{+}/Ca^{2+} channel	Minor electrophysiological defects	Inhibition at 0.1–10 μM

separates the rhabdomeres from organelles in the cell body. Upon illumination, Ca^{2+} concentrations reach high values in the near millimolar range within the microvilli of the rhabdomere, but Ca^{2+} concentrations in the photoreceptor cell body do not rise above 10 μM [8, 17, 18, 22]. In part, this can simply be attributed to the higher volume of the cell body compared to the very small volume of the rhabdomeral microvilli. In addition, calphotin seems to act as a Ca^{2+} buffer that binds Ca^{2+} ions that diffuse out of the rhabdomeral compartment and thereby prevents high Ca^{2+} concentrations in the cell body [21]. Using calcium imaging, Weiss and colleagues observed an abnormally fast rise in intracellular Ca^{2+} levels upon illumination of photoreceptors that expressed reduced amounts of calphotin [21]. The attenuation of high Ca^{2+} concentrations in the cell body by calphotin may help to avoid toxic effects of high cellular calcium. Indeed, mutations in the calphotin gene result in retinal degeneration and in rough eyes [21, 44]. Interestingly, retinal degeneration in calphotin-defective flies was rescued by overexpression of CalX resulting in enhanced extrusion of Ca^{2+} ions [21]. These results indicate that calphotin acts as an immobile Ca^{2+} buffer that attenuates free Ca^{2+} concentrations that spread from the rhabdomeric compartment to the cell body of light-activated photoreceptor cells.

34.3.1.2 Calmodulin

Calmodulin is a ubiquitous, Ca^{2+}-binding, regulatory protein. Calmodulin can bind four Ca^{2+} ions and, upon binding, changes its conformation. Ca^{2+}/Calmodulin in turn exerts its regulatory roles upon binding to target proteins. Among the major signal transduction proteins in the *Drosophila* eye, calmodulin can bind to the cation channels TRP [45] and TRPL [46], NINAC [47], Ca^{2+}/calmodulin-dependent kinase II (CamKII) [48], RDGC [49], and INAD [50]. Additional calmodulin binding proteins comprise the ryanodine receptor [51] as well as additional Ca^{2+}/calmodulin-dependent protein kinases and phosphatases [52]. The physiological role of calmodulin binding to these target proteins will be discussed in detail below. Target proteins harbor calmodulin binding sites that are hard to predict from the amino acid sequence.

cam mutant flies exhibit a prolonged deactivation of the photoresponse that is also reflected in the elementary responses of the photoreceptor cells, called quantum bumps. In contrast to wild type photoreceptors, *cam* mutant photoreceptors produce a train of quantum bumps upon stimulation with a single photon [53]. These findings suggest an involvement of calmodulin in the termination of the photoresponse.

34.3.2 Ca^{2+}-Interacting Proteins and the Rhodopsin Cycle

When rhodopsin (R) absorbs a photon, the 11-*cis* 3-OH retinal chromophore is isomerized into its all-*trans* form and rhodopsin is converted to its active form, metarhodopsin (M*) that triggers the phototransduction cascade (Fig. 34.1). M* can be reconverted to R by absorption of a second photon and thereby becomes inactivated. An alternative way for M* inactivation is binding to arrestin 2 that sterically hinders activation of the Gq protein. At least a fraction of arrestin 2 is bound to an unconventional myosin, NINAC, in the dark, from which it has to be released for binding to M*. M* becomes phosphorylated by a G protein-coupled receptor kinase, GPRK1 [54], although this phosphorylation is not required for arrestin 2 binding and M* inactivation [55–57]. Arrestin 2 becomes phosphorylated by CamKII [48]. After reconversion of M to R by absorption of a photon, arrestin 2 dissociates from the receptor and rhodopsin becomes dephosphorylated by RDGC and returns to its ground state [58, 59]. Three events in this circle involve Ca^{2+}: (1) The release of arrestin 2 from NINAC, (2) the phosphorylation of arrestin 2 by CamKII, and (3) the dephosphorylation of rhodopsin by RDGC.

34.3.2.1 NINAC

Like the other *nina* and *ina* mutants, *ninaC* was originally isolated in a mutagenesis screen in which mutations were identified by their electroretinogram phenotype [1]. The *ninaC* locus encodes a 132 kDa (p132) and a 174 kDa (p174) protein variant. The two variants share an N-terminal protein kinase domain, a domain that is homologous to the head region of the myosin heavy chain, and a calmodulin binding site. p174 harbors an additional calmodulin binding site and extends farther in the C-terminal direction [47, 60]. p132 is localized to the cytosol of the photoreceptor cell and p174 is localized to the rhabdomeres [61, 62]. Mutation of *ninaC* results in light- and age-dependent retinal degeneration and in an electrophysiological defect that is independent from this retinal degeneration [61]. While disruption of p132 alone had no effect, elimination of p174 alone resulted in a phenotype reminiscent of the *ninaC* null mutant [61]. To assign roles to the kinase and myosin domains, Porter and Montell mutated these domains in p174. Upon mutation of the kinase domain, they observed an ERG defect but no retinal degeneration. Deletion of the myosin domain resulted in an altered subcellular localization, an ERG defect, and retinal degeneration [63]. To investigate the roles of the CaM binding domains, Porter and colleagues deleted each of these or both in the p174 variant [47]. All these mutations resulted in a higher susceptibility to enter the prolonged depolarizing afterpotential (PDA) state. A PDA manifests in a persisting depolarization after cessation of the light stimulus [64]. The authors concluded that the CaM binding sites in NINAC function in termination of the light response. It has then been proposed that NINAC actively transports arrestin 2-loaded vesicles into the rhabdomeres, interacting with PIP_3 in the vesicle membrane [65, 66]. However, Satoh and Ready did not observe a requirement of NINAC for ARR2 translocation [67]. In an elegant set of experiments, Liu and co-workers used a fly expressing the UV-sensitive opsin, RH3, under control of the *ninaE* promoter in a *ninaE* null background [68]. The benefit of using RH3 was that absorption maxima of the inactivated (R) and the activated (M*) form are better separated than for RH1. This enabled the authors to convert a large portion of R to M* and M* to R by using the respective light qualities. They observed a rapid, Ca^{2+}-dependent inactivation of M* in wild type flies. However, the Ca^{2+} dependence was abolished in *cam* and *ninaC* mutant flies. Therefore, the authors proposed that NINAC is an arrestin 2 binding protein that releases arrestin 2 upon Ca^{2+}/CaM binding so that arrestin 2 can inactivate M*. Further evidence for a diffusion-based translocation mechanism of arrestin 2 is the finding that arrestin 2 translocation to the rhabdomeres is stoichiometric to R to M* isomerization [69]. Light-dependent arrestin 2 translocation into the rhabdomere constitutes a light adaptation mechanism (see 3.4).

34.3.2.2 CamKII

Phosphorylation of arrestin 2 at serine 366 is the earliest light-induced phosphorylation step in *Drosophila* photoreceptors [70]. This phosphorylation is catalyzed by the Ca^{2+}/calmodulin-dependent kinase II (CaMKII) [48]. Alloway and Dolph showed that mutation of serine 366 to alanine had no effect on the binding of arrestin 2 to rhodopsin, but it hindered release of arrestin 2 from rhodopsin [71]. It was also known that hyperphosphorylation of rhodopsin results in a strong interaction of arrestin and rhodopsin and that the resulting complexes are internalized into the photoreceptor cell and trigger retinal degeneration (see RDGC). Therefore, Kristaponyte and colleagues tried to dissect the consequences of rhodopsin hyperphosphorylation and arrestin 2 hypophosphorylation [72]. They used phospholipase C null ($norpA^{P24}$) flies since in these flies, no Ca^{2+} enters the photoreceptor cells to activate RDGC and CaMKII. Thus, in $norpA^{P24}$ flies, arrestin 2 should remain dephosphorylated and rhodopsin should become hyperphosphorylated upon illumination. To test the consequences of permanent arrestin 2 dephosphorylation alone, the authors used a fly that expresses an inhibitory peptide of the CaMKII [73]. As a result, inhibition of CaMKII by inhibitory peptides did not result in photoreceptor degeneration while induction of rhodopsin hyperphosphorylation did. These results indicate that photoreceptor degeneration due to reduced Ca^{2+} influx after light activation, as in the $norpA^{P24}$ mutant, is a consequence of rhodopsin hyperphosphorylation rather than a lack of arrestin 2 phosphorylation.

34.3.2.3 RDGC

Drosophila retinal degeneration C (RDGC) is the founding member of the protein phosphatases with EF hands (PPEF) family. RDGC is expressed in the retina, in ocelli, in the mushroom bodies of the brain, and to a lower extend, in the lamina and medulla [74]. The *rdgC* locus encodes three different RDGC protein isoforms [49]. RDGC isoforms harbor an N-terminal IQ domain that interacts with Ca^{2+}/Calmodulin [49], a catalytic domain, and C-terminal EF hand domains that probably directly interact with Ca^{2+}. Upon Ca^{2+} activation, RDGC mediates dephosphorylation of rhodopsin [56] and the TRP ion channel at S936 [75]. *rdgC* null mutant flies exhibit rhodopsin and TRP hyperphosphorylation [56, 75]. Hyperphosphorylation of rhodopsin results in an abnormally strong interaction between rhodopsin and arrestin. As an immediate consequence, arrestin 2 molecules are no longer available to deactivate metarhodopsin, resulting in prolonged deactivation of the phototransduction cascade upon orange light stimulation and higher susceptibility for persistent activation upon blue light stimulation (prolonged depolarizing afterpotential) [56]. Additionally, rhodopsin/arrestin complexes are internalized from the rhabdomeric membrane into the photoreceptor cell triggering apoptosis and ultimately resulting in photoreceptor degeneration. Collectively, Ca^{2+}-mediated regulation of RDGC ensures proper photoreceptor function and prevents retinal degeneration.

34.3.3 *Ca^{2+}-Mediated Positive and Negative Feedback for Generating Highly Time-Resolved Photoreceptor Responses*

Absorption of a single photon both in *Drosophila* and in vertebrate photoreceptors activates the phototransduction cascade and results in a distinct electrical response, termed quantum bump [76]. In *Drosophila* a quantum bump probably corresponds to the opening of TRP and TRPL channels within a single rhabdomeral microvillus [8]. The quantum bumps of *Drosophila* have an average amplitude of 12 pA at −70 mV membrane voltage and a duration of less than 50 ms. A macroscopic response to brighter illumination represents the summation of generated quantum bumps. The *Drosophila* quantum bump is much shorter than vertebrate quantum bumps resulting in better time resolved visual responses and a better time resolution of the fly visual system as compared to vertebrate eyes [76]. The sharpness of *Drosophila* quantum bumps is largely due to Ca^{2+}-mediated positive and negative feedback. In addition, it has been suggested that the assembly of main signaling components of the phototransduction cascade into a signaling complex by the scaffolding protein INAD may ensure specific and fast signal transduction [77, 78].

34.3.3.1 Positive and Negative Feedback on the TRP Channels

In 1991, Hardie as well as Ranganathan and co-workers observed that the time to peak of the photoresponse was shortened when the concentration of Ca^{2+} in the bath was elevated in dissociated ommatidia preparations [40, 79]. Release of caged Ca^{2+} during the rising phase of the photoresponse facilitated the LIC. This facilitation was mediated through the TRP but not the TRPL channel [80]. The EC_{50} of this facilitation was later measured to be ∼300 nM [81]. Besides positive feedback, Ca^{2+} apparently also promotes negative feedback on the TRP channels [79, 80]. Release of caged Ca^{2+} during the plateau phase of the light response resulted in an accelerated inactivation of the photoresponse [80]. Through manipulation of the Na^{+}/Ca^{2+} exchange equilibrium, Gu and colleagues varied the intracellular Ca^{2+} concentration [25]. They observed inhibition of the LIC within a range of 0.1–10 μM Ca^{2+} (IC50 ~1 μM). At these concentrations, PLC activity was not affected, leading Gu and colleagues to the assumption that inhibition takes place at the level of the ion channels. In line with this suggestion is the presence of CaM binding sites within both the TRP and TRPL channels [45, 46]. TRP harbors CaM binding site within its C-terminus that bind CaM in a Ca^{2+}-independent manner [45, 82–84]. TRPL harbors two calmodulin binding sites in its C-terminus, one that exhibits Ca^{2+}-dependent CaM binding and one that shows Ca^{2+}-independent CaM binding [46]. Mutation of the Ca^{2+}-dependent CaM binding site resulted in prolonged inactivation of the light response [53]. Hypomorphic cam^{352}/cam^{339} mutants expressing less than 10% CaM of the wild type showed a similar phenotype [53].

It was proposed that the facilitation of rising and falling phase of the photoresponse that accompanies the rise of the intracellular Ca^{2+} concentration is caused by sensitization and desensitization of TRP channels and the following model was suggested [76]. Upon illumination, PLC action leads to accumulation of second messengers and an increase in membrane tension until a certain threshold is reached and the first TRP channel opens. Influx of Ca^{2+} ions through the first channel into the photoreceptor cell results in an almost instantaneous rise of the Ca^{2+} concentration in the respective microvillus because of its restricted volume. The other TRP channels present in the microvillus are rendered more sensitive towards second messengers and also open. When Ca^{2+} reaches a yet higher concentration, it inhibits the TRP and TRPL channels resulting in fast deactivation of the photoresponse.

34.3.3.2 Ca^{2+}-Mediated Feedback on PLC

In vitro studies showed that *Drosophila* PLC is stimulated at Ca^{2+} concentrations up to 1 μM while higher Ca^{2+} concentrations (5 μM) inhibited its activity [85]. A requirement of Ca^{2+} for PLC activity was also shown in intact photoreceptor cells [86, 87]. On the other hand and in line with positive as well as negative feedback of Ca^{2+} on PLC, Gu and colleagues observed inhibition of the PLC at Ca^{2+} concentrations of 50 μM or higher [25]. Such concentrations are only reached transiently in vivo [17, 18, 22]. The authors proposed that Ca^{2+}-mediated inhibition of PLC is a mechanism to avoid depletion of the PIP_2 pool in vivo. This observation might also provide an explanation for the *trp* phenotype. In a *trp* null mutant fly, due to the lack of the major Ca^{2+}-selective channel, Ca^{2+} influx into the photoreceptor cells is drastically reduced. Reduction of Ca^{2+} influx results in a transient receptor potential, meaning that the photoreceptor potential repolarizes back to resting voltages during a light stimulus. This transient receptor potential correlates with a depletion of the membrane component PIP_2. Due to the drastic reduction in Ca^{2+} influx, Ca^{2+}-dependent negative feedback mechanisms are not activated and PLC is not inhibited so that PLC cleaves all available PIP_2 molecules to IP_3 and DAG. *trp* null mutant flies undergo light dependent retinal degeneration that has been shown to be caused by PIP_2 depletion [88]. Ca^{2+}-mediated inhibition of the PLC depended on eye-PKC [25]. Thus, a possible mechanism is that eye-PKC is activated by high Ca^{2+} concentrations and phosphorylates a target that inactivates PLC. This phosphorylation target does not seem to be PLC itself as phosphorylation of PLC so far has neither been observed in vitro [89] nor in vivo (Voolstra and Huber, unpublished results). In accordance with a role of ePKC in PLC inactivation, *inaC* null mutant flies exhibit quantum bumps that do not terminate properly. Additionally, these flies exhibit an ERG phenotype that can be described as a transient receptor potential combined with a prolonged deactivation. Like in the *trp* null mutant, the transient receptor potential phenotype of the *inaC* mutant might be attributed to PIP_2 depletion that results from the inability to phosphorylate target proteins to switch off phototransduction. In contrast to the *trp* null mutant, the $inaC^{P209}$ fly still expresses a functional TRP channel. It has been demonstrated that the TRP channel is less sensitive towards PIP_2 depletion than the TRPL channel [90]. This might explain why in the *inaC* mutant, transient receptor potential phenotype is less pronounced than in the *trp* mutant. Alternatively, a second protein kinase C, PKC53E, that probably resulted from a gene duplication event of the *inaC* gene and is also expressed in photoreceptor cells, may partially substitute for eye-PKC function in the *inaC* null mutant [91]. It has to be noted, however, that it is not understood how PIP_2 depletion is a prerequisite for TRP channel activation [10] but at the same time causes channel inhibition. Additionally, the transient receptor potential phenotype and early retinal degeneration is rescued in an *rdgA;;trp* double mutant [92]. This is astonishing since in this fly, due to the *trp* mutation, the Ca^{2+} influx is drastically lowered and solely the PIP_2 depletion-sensitive TRPL channel is expressed. It would thus be interesting to investigate PIP_2 levels in an *rdgA;;trp* double mutant. However, DAG-dependent activation of the TRP channels has been reported [12] and therefore, excessive DAG levels resulting from the *rdgA* mutation might mask effects of PIP_2 depletion. The TRP ion channel and the INAD scaffolding protein constitute targets of eye-PKC [89, 93–97]. However, mutation of putative eye-PKC phosphorylation sites in TRP or INAD did not evoke an electrophysiological phenotype that resembled the *inaC* mutant phenotype (Voolstra and Huber, unpublished results). It can therefore be concluded that either the relevant phosphorylation sites on TRP and INAD have not yet been identified or that other eye-PKC target proteins exist. In future work, the identification of these target proteins would contribute significantly to the understanding of the regulation of the visual response in *Drosophila*.

After PLC-mediated cleavage of PIP_2 to DAG and IP_3, PIP_2 is regenerated from DAG (for review, see [98]). Enzymes that are involved in the regeneration of PIP_2 represent potential targets for Ca^{2+} regulation. However, Ca^{2+}-mediated regulation of phospholipid cycle enzymes other than PLC is not likely. Using a fluorescently-tagged PIP_2 probe in intact flies, wild type and *trp* null mutant flies showed similar time courses of PIP_2 resynthesis and various background illumination intensities that were applied to trigger Ca^{2+} influx failed to accelerate PIP_2 synthesis in the wild type [90].

34.3.4 Ca^{2+}-Mediated Light Adaptation

34.3.4.1 General Aspects of Light Adaptation

Ambient light intensities span eleven orders of magnitude during a typical cycle of day and night [99]. Therefore, animals had to develop strategies to broaden the dynamic range of their visual systems in order to obtain information of relative brightness (contrast) of different objects in their environment at different levels of ambient light intensities. This is called light adaptation and it is achieved by shifting the relatively steep operational range (V-log I curve) along the 11 orders of magnitude of ambient light intensities. A second aspect of light adaptation affects the temporal resolution of a visual system, which preserves the contrast sensitivity when photoreceptors become light-adapted. Light adaptation in *Drosophila* is Ca^{2+}-dependent and takes place at different time scales by different mechanisms. First, *Drosophila* uses a pupil mechanism achieved by pigment granula migration [100, 101]. Second, during illumination, excitatory components of the phototransduction cascade are translocated out of the rhabdomere while inhibitory components are translocated into the rhabdomere [3]. Pigment granula migration and translocation of inhibitory and excitatory components constitute long-term light adaptation mechanisms. Third, short-term light adaptation is achieved by Ca^{2+} (/CaM) binding directly to the ion channels TRP and TRPL [25]. This mechanism has the strongest effect on light adaptation. Recently, an additional light adaptation mechanism was unraveled. The phosphorylation state of serine 936 of the TRP ion channel influences kinetic features of the photoresponse [75, 102]. We will discuss these aspects of Ca^{2+}-mediated light adaptation in detail below.

34.3.4.2 Short Term Light Adaptation

By manipulating cytosolic Ca^{2+} via the Na^{+}/Ca^{2+} exchange equilibrium, Gu and colleagues found that Ca^{2+} inhibited the light-induced current (LIC) over a range corresponding to steady-state light-adapted Ca^{2+} levels (0.1–10 μM) and accurately mimicked light adaptation [25]. In contrast, PLC activity was unaffected by the steady-state Ca^{2+} concentrations reached during light adaptation, but it was inhibited over the range of concentrations experienced during the Ca^{2+} transients and this effect may be involved in contrast sensitivity.

28 phosphorylation sites have been identified within the TRP ion channel [97]. 15 of these sites are phosphorylated in the light and become dephosphorylated in the dark. The phosphorylation of most of these sites depends on the phototransduction cascade and on the activity of the TRP ion channel. Most probably, phosphorylation thus depends on Ca^{2+} ion influx into the photoreceptor cell. The physiological function of these phosphorylation sites is still elusive. Besides these sites that show increased phosphorylation in the light, a single site, S936, exhibits elevated phosphorylation in the dark and becomes dephosphorylated in the light. The

dephosphorylation of this site directly depends on Ca^{2+} and at least in part is mediated by the RDGC phosphatase [75]. The S936 phosphorylation site is involved in aspects of light adaptation related to temporal resolution of a visual system. In the light, when S936 is dephosphorylated, photoreceptors are able to immediately follow a light stimulus flickering at high frequency. In the dark, when S936 is phosphorylated, it takes several seconds for photoreceptors to follow the same light stimulus. Thus, phosphorylation at S936 is regulated by Ca^{2+} and is involved in one aspect of light adaptation.

34.3.4.3 Long Term Light Adaptation by Subcellular Translocation of Signaling Components

The properties of neurons depend on the abundance of signaling components in the plasma membrane. Hence, removal from or reinsertion into the plasma membrane of receptors, ion channels, or other signaling components can function as a mechanism for light adaptation. At least three *Drosophila* photoreceptor proteins undergo a light-triggered translocation between the rhabdomeral plasma membrane and the cell body: the TRPL ion channel, ARR2 and the $G\alpha q$ subunit of the visual G-protein [66, 103, 104]. The TRPL ion channel is located to the rhabdomeres in the dark. Upon illumination, it translocates into the photoreceptor cell body [103]. Light-dependent TRPL translocation occurs in two stages which comprise transport of TRPL to the base of the rhabdomere and the adjacent stalk membrane in the first stage and a vesicle-mediated transport into the cell body within several hours [105, 106]. TRPL translocation depends on the phototransduction cascade and on Ca^{2+} influx [107].

The $G\alpha q$ subunit is also partially removed from the rhabdomere upon illumination, thus reducing the amount of available G-protein for phototransduction [103]. This mechanism has been shown to significantly reduce the sensitivity of fly photoreceptors [107]. The association of $G\alpha q$ to the membrane-attached $G\beta\gamma$ in the dark after illumination requires NINAC [108, 109]. Translocation of $G\alpha q$ may be mediated by dynamic palmitoylation/depalmitoylation of the subunit but it is not known whether it depends on the light-induced Ca^{2+} influx [104].

Arrestin 2 is translocated in the opposite direction, i.e. it is highly abundant in the rhabdomere in the light but diffuses out of the rhabdomere in the dark [66]. As mentioned in Chap. 3.2, Ca^{2+} affects translocation of arrestin 2 into the rhabdomere as it is required to release arrestin 2 from NINAC [68].

34.3.4.4 Pupil Mechanism Mediated by Myosin V

Rhabdomeres act as optical light guides comparable to glass fibers. Therefore, photons that pass the dioptric apparatus enter the rhabdomere at its distal end and are guided through the rhabdomere towards the proximal end until they are eventually absorbed by a rhodopsin molecule or leave the rhabdomere at its proximal end. Upon light-induced Ca^{2+} influx, pigment granules within the *Drosophila* photoreceptor cell migrate from the cytosol towards the base of the rhabdomere where they can absorb photons travelling through the rhabdomere [100, 101]. This mechanism in effect progressively reduces the light intensity in the rhabdomeres and has been compared to the effect of a closing pupil in vertebrate eyes. This "pupil mechanism" accounts for an attenuation of 0.8 log units and constitutes an important part of light adaptation [110, 111]. Satoh and coworkers showed that the migration of pigment molecules is mediated by myosin V. Upon illumination, myosin V that is loaded with a pigment granule migrates along a rhabdomere terminal web (RTW) filament towards the rhabdomere base in a Ca^{2+}- and calmodulin-dependent manner [112].

34.3.5 Ca^{2+} Interactions with a Still Elusive Physiological Role

34.3.5.1 INAD

INAD is a scaffolding protein that tethers some of the components of the phototransduction cascade resulting in short diffusion times and high fidelity of the system [113]. INAD binds CaM in a Ca^{2+}-dependent manner [45]. Using calmodulin overlay assays, Xu and colleagues showed direct interaction of calmodulin with an INAD protein fragment spanning amino acids 146 to 235 [50]. The exact consequences of calmodulin binding to INAD are elusive. However, recently, it has become increasingly clear that INAD has to be regarded as a dynamic machine rather than a rigid scaffold. This notion is corroborated by the finding that INAD undergoes a light-dependent conformational change to transiently release one of its binding partners, probably TRP, to ensure proper termination of visual response [95, 114].

34.3.5.2 Regulation of Neurocalcin

Drosophila neurocalcin was first identified in a polymerase chain reaction approach searching for recoverin-like proteins [115]. Neurocalcin is expressed in the central nervous system and in the eye [115] (flyatlas.org). The deduced amino acid sequence revealed three putative EF hands and an N-terminal myristoylation site. Indeed, recombinantly-expressed neurocalcin bound $^{45}Ca^{2+}$ and displayed a Ca^{2+}-dependent mobility shift in electrophoresis assays. Coexpression of neurocalcin with N-myristoyl transferase in bacteria in the presence of [^{3}H]myristic acid resulted in radioactive labeling of neurocalcin [115]. These results showed that *Drosophila* neurocalcin is myristoylated and can bind Ca^{2+}. Myristoylated neurocalcin exhibited a Ca^{2+}-dependent translocation to membranes [116]. Since it had been reported that vertebrate recoverin inhibits phosphorylation of rhodopsin [117], Faurobert and colleagues tested this for *Drosophila* neurocalcin [116]. Indeed, the authors observed a Ca^{2+}- and myristoylation-dependent inhibition of phosphorylation of bovine rhodopsin by neurocalcin in vitro. However, the exact physiological role of neurocalcin in flies is elusive. It might be interesting to analyze neurocalcin-defective flies for rhodopsin hyperphosphorylation and concomitant.

34.4 Concluding Remarks

After ~50 years of *Drosophila* vision research, many interesting aspects of the *Drosophila* phototransduction cascade have been elucidated. We are now able to describe quantitatively how absorption of a photon ultimately leads to the generation of the photoreceptor potential [118]. As elaborated above, Ca^{2+} plays a major role in the regulation of the underlying processes. It controls positive and negative feedback mechanisms to increase the fidelity of the light response and regulates the sensitivity of the whole visual system to increase the dynamic range.

However, several questions have not been answered so far: How does the Ca^{2+}- and eye-PKC-dependent inactivation of the photoresponse operate? There is evidence that eye-PKC is mandatory for Ca^{2+}-dependent inhibition of PLCβ, but PLCβ does not seem to be a direct substrate of this protein kinase C. How does Ca^{2+} interact with the TRP and TRPL ion channels to provide positive and negative feedback? How can low Ca^{2+} concentrations promote channel opening while high Ca^{2+} concentrations inhibit channel opening? Finally, Ca^{2+}-mediated light adaptation is not well understood. Although it is likely that light adaptation occurs at the level of the TRP ion channels, it is not clear how this mechanism operates. Future studies will continue to increase our understanding of the role of Ca^{2+} in biological signaling pathways.

Acknowledgements Work in the laboratory of the authors has been supported by the German Research Foundation (DFG Hu839/7-1, Vo1741/1-1).

References

1. Pak WL, Grossfield J, Arnold KS (1970) Mutants of the visual pathway of *Drosophila melanogaster*. Nature 227(257):518–520
2. Smith DP, Stamnes MA, Zuker CS (1991) Signal transduction in the visual system of Drosophila. AnnuRevCell Biol 7:161–90: 161–190
3. Wang T, Montell C (2007) Phototransduction and retinal degeneration in *Drosophila*. Pflugers Arch 454(5):821–847
4. Minke B, Hardie RC (2000) Genetic dissection of Drosophila phototransduction. In: Stavenga D, DeGrip WJ, Pugh EN Jr (eds) Handbook of biological physics, Molecular mechanisms in visual transduction, vol 3. Elsevier, Amsterdam/London/New York/Oxford/Paris/Shannon/Tokyo, pp 449–525

5. Huber A (2004) Invertebrate phototransduction: multimolecular signaling complexes and the role of TRP and TRPL channels. In: Frings S, Bradley J (eds) Transduction channels in sensory cells. WILEY-VCH, Weinheim, pp 179–206
6. Tian Y, Hu W, Tong H et al (2012) Phototransduction in *Drosophila*. Sci China Life Sci 55(1):27–34. https://doi.org/10.1007/s11427-012-4272-4
7. O'Tousa JE (2002) $Ca2^{+}$ regulation of Drosophila phototransduction. Adv Exp Med Biol 514:493–505
8. Hardie RC, Juusola M (2015) Phototransduction in Drosophila. Curr Opin Neurobiol 34:37–45. https://doi.org/10.1016/j.conb.2015.01.008
9. Katz B, Minke B (2018) The *Drosophila* light-activated TRP and TRPL channels - targets of the phosphoinositide signaling cascade. Prog Retin Eye Res 66:200–219. https://doi.org/10.1016/j.preteyeres.2018.05.001
10. Huang J, Liu C-S, Hughes SA et al (2010) Activation of TRP channels by protons and Phosphoinositide depletion in *Drosophila* photoreceptors. Curr.Biol. 20:189–197
11. Hardie RC, Franze K (2012) Photomechanical responses in *Drosophila* photoreceptors. Science 338(6104):260–263
12. Delgado R, Muñoz Y, Peña-Cortés H et al (2014) Diacylglycerol activates the light-dependent channel TRP in the photosensitive microvilli of Drosophila melanogaster photoreceptors. J Neurosci 34(19):6679–6686. https://doi.org/10.1523/JNEUROSCI.0513-14.2014
13. Chyb S, Raghu P, Hardie RC (1999) Polyunsaturated fatty acids activate the Drosophila light-sensitive channels TRP and TRPL. Nature 397(6716):255–259
14. Leung HT, Tseng-Crank J, Kim M et al (2008) DAG lipase activity is necessary for TRP channel regulation in *Drosophila* photoreceptors. Neuron 58(6):825–827
15. Hardie RC (1996) INDO-1 measurements of absolute resting and light-induced $Ca2^{+}$ concentration in Drosophila photoreceptors. J.Neurosci. 16(9):2924–2933
16. Asteriti S, Liu C-H, Hardie RC (2017) Calcium signalling in Drosophila photoreceptors measured with GCaMP6f. Cell Calcium 65:40–51. https://doi.org/10.1016/j.ceca.2017.02.006
17. Oberwinkler J, Stavenga DG (2000) Calcium transients in the rhabdomeres of dark- and light-adapted fly photoreceptor cells. J.Neurosci. 20(5):1701–1709
18. Postma M, Oberwinkler J, Stavenga DG (1999) Does Ca^{2+} reach millimolar concentrations after single photon absorption in *Drosophila* photoreceptor microvilli? Biophys.J. 77(4):1811–1823
19. Peretz A, Suss-Toby E, Rom-Glas A et al (1994) The light response of *Drosophila* photoreceptors is accompanied by an increase in cellular calcium: effects of specific mutations. Neuron 12(6):1257–1267
20. Ranganathan R, Bacskai BJ, Tsien RY et al (1994) Cytosolic calcium transients: spatial localization and role in Drosophila photoreceptor cell function. Neuron 13(4):837–848
21. Weiss S, Kohn E, Dadon D et al (2012) Compartmentalization and $Ca2^{+}$ buffering are essential for prevention of light-induced retinal degeneration. J Neurosci 32(42):14696–14708. https://doi.org/10.1523/JNEUROSCI.2456-12.2012
22. Hardie RC (1996) A quantitative estimate of the maximum amount of light-induced Ca^{2+} release in *Drosophila* photoreceptors. JPhotochemPhotobiolB 35(1–2):83–89
23. Schwarz EM, Benzer S (1997) Calx, a Na-ca exchanger gene of *Drosophila melanogaster*. Proc Natl Acad Sci U S A 94(19):10249–10254
24. Wang T, Xu H, Oberwinkler J et al (2005) Light activation, adaptation, and cell survival functions of the Na^{+}/Ca^{2+} exchanger CalX. Neuron 45(3):367–378. https://doi.org/10.1016/j.neuron.2004.12.046
25. Gu Y, Oberwinkler J, Postma M et al (2005) Mechanisms of light adaptation in *Drosophila* photoreceptors. CurrBiol 15(13):1228–1234
26. Magyar A, Bakos E, Váradi A (1995) Structure and tissue-specific expression of the *Drosophila* melanogaster organellar-type Ca^{2+}-ATPase gene. Biochem J 310(Pt 3):757–763
27. Hardie RC (1996) Excitation of Drosophila photoreceptors by BAPTA and ionomycin: evidence for capacitative $Ca2^{+}$ entry? Cell Calcium 20(4):315–327

28. Walz B, Baumann O (1995) Structure and cellular physiology of Ca^{2+} Stores in Invertebrate Photoreceptors. Cell Calcium 18(4):342–351
29. Brown JE, Blinks JR (1974) Changes in intracellular free calcium concentration during illumination of invertebrate photoreceptors. Detection with aequorin. J Gen Physiol 64(6):643–665
30. Acharya JK, Jalink K, Hardy RW et al (1997) InsP3 receptor is essential for growth and differentiation but not for vision in *Drosophila*. Neuron 18(6):881–887
31. Raghu P, Colley NJ, Webel R et al (2000) Normal phototransduction in *Drosophila* photoreceptors lacking an InsP(3) receptor gene. MolCell Neurosci 15(5):429–445
32. Kohn E, Katz B, Yasin B et al (2015) Functional cooperation between the IP3 receptor and phospholipase C secures the high sensitivity to light of *Drosophila* photoreceptors in vivo. J Neurosci 35(6):2530–2546. https://doi.org/10.1523/JNEUROSCI.3933-14.2015
33. Bollepalli MK, Kuipers ME, Liu C-H et al (2017) Phototransduction in Drosophila is compromised by Gal4 expression but not by InsP3 receptor knockdown or mutation. eNeuro 4(3):ENEURO.0143–ENEU17.2017. https://doi.org/10.1523/ENEURO.0143-17.2017
34. Schneggenburger R, Zhou Z, Konnerth A et al (1993) Fractional contribution of calcium to the cation current through glutamate receptor channels. Neuron 11(1):133–143
35. Hille B (2001) Ionic channels of excitable membranes, 3rd edn. Sinauer Associates, Sunderland
36. Chu B, Postma M, Hardie RC (2013) Fractional Ca(2^+) currents through TRP and TRPL channels in Drosophila photoreceptors. BiophysJ 104(9):1905–1916
37. Reuss H, Mojet MH, Chyb S et al (1997) In vivo analysis of the drosophila light-sensitive channels, TRP and TRPL. Neuron 19(6):1249–1259
38. Liu CH, Wang T, Postma M et al (2007) *In vivo* identification and manipulation of the Ca^{2+} selectivity filter in the *Drosophila* transient receptor potential channel. J Neurosci 27(3):604–615. https://doi.org/10.1523/JNEUROSCI.4099-06.2007
39. Hofstee CA, Stavenga DG (1996) Calcium homeostasis in photoreceptor cells of Drosophila mutants inaC and trp studied with the pupil mechanism. VisNeurosci 13(2):257–263
40. Hardie RC (1991) Whole-cell recordings of the light induced current in dissociated *Drosophila* photoreceptors: evidence for feedback by calcium permeating the light-sensitive channels. Proc R Soc Lond Ser B Biol Sci 245(1314):203–210. https://doi.org/10.1098/rspb.1991.0110
41. Hardie RC, Minke B (1994) Calcium-dependent inactivation of light-sensitive channels in *Drosophila* photoreceptors. JGenPhysiol 103(3):409–427
42. Ballinger DG, Xue N, Harshman KD (1993) A Drosophila photoreceptor cell-specific protein, calphotin, binds calcium and contains a leucine zipper. Proc Natl Acad Sci U S A 90(4):1536–1540
43. Martin JH, Benzer S, Rudnicka M et al (1993) Calphotin: a Drosophila photoreceptor cell calcium-binding protein. Proc Natl Acad Sci U S A 90(4):1531–1535
44. Yang Y, Ballinger D (1994) Mutations in calphotin, the gene encoding a Drosophila photoreceptor cell-specific calcium-binding protein, reveal roles in cellular morphogenesis and survival. Genetics 138(2):413–421
45. Chevesich J, Kreuz AJ, Montell C (1997) Requirement for the PDZ domain protein, INAD, for localization of the TRP store-operated channel to a signaling complex. Neuron 18(1):95–105
46. Warr CG, Kelly LE (1996) Identification and characterization of two distinct calmodulin-binding sites in the Trpl ion-channel protein of *Drosophila* melanogaster. Biochem J 314(Pt 2):497–503
47. Porter JA, Minke B, Montell C (1995) Calmodulin binding to *Drosophila* NinaC required for termination of phototransduction. EMBO J 14(18):4450–4459
48. Kahn ES, Matsumoto H (1997) Calcium/calmodulin-dependent kinase II phosphorylates Drosophila visual arrestin. JNeurochem 68(1):169–175
49. Lee SJ, Montell C (2001) Regulation of the rhodopsin protein phosphatase, RDGC, through interaction with calmodulin. Neuron 32(6):1097–1106

50. Xu XZ, Choudhury A, Li X et al (1998) Coordination of an array of signaling proteins through homo- and heteromeric interactions between PDZ domains and target proteins. J.Cell Biol. 142(2):545–555
51. Hasan G, Rosbash M (1992) *Drosophila* homologs of two mammalian intracellular Ca(2^+)-release channels: identification and expression patterns of the inositol 1,4,5-triphosphate and the ryanodine receptor genes. Development 116(4):967–975
52. Xu XZS, Wes PD, Chen H et al (1998) Retinal targets for calmodulin include proteins implicated in synaptic transmission. In: The journal of biological chemistry (USA), vol 273, pp 31297–31307
53. Scott K, Sun Y, Beckingham K et al (1997) Calmodulin regulation of *Drosophila* light-activated channels and receptor function mediates termination of the light response in vivo. Cell 91(3):375–383
54. Lee SJ, Xu H, Montell C (2004) Rhodopsin kinase activity modulates the amplitude of the visual response in Drosophila. ProcNatlAcadSciUSA 101(32):11874–11879
55. Plangger A, Malicki D, Whitney M et al (1994) Mechanism of arrestin 2 function in rhabdomeric photoreceptors. J.Biol.Chem. 269(43):26969–26975
56. Vinos J, Jalink K, Hardy RW et al (1997) A G protein-coupled receptor phosphatase required for rhodopsin function. Science 277(5326):687–690
57. Kiselev A, Socolich M, Vinos J et al (2000) A molecular pathway for light-dependent photoreceptor apoptosis in *Drosophila*. Neuron 28(1):139–152
58. Byk T, Bar-Yaacov M, Doza YN et al (1993) Regulatory arrestin cycle secures the fidelity and maintenance of the fly photoreceptor cell. ProcNatlAcadSciUSA 90(5):1907–1911
59. Selinger Z, Doza YN, Minke B (1993) Mechanisms and genetics of photoreceptors desensitization in *Drosophila* flies. BiochimBiophysActa 1179(3):283–299
60. Montell C, Rubin GM (1988) The Drosophila ninaC locus encodes two photoreceptor cell specific proteins with domains homologous to protein kinases and the myosin heavy chain head. Cell 52(5):757–772
61. Porter JA, Hicks JL, Williams DS et al (1992) Differential localizations of and requirements for the two Drosophila ninaC kinase/myosins in photoreceptor cells. J.Cell Biol. 116(3):683–693
62. Hicks JL, Williams DS (1992) Distribution of the myosin I-like ninaC proteins in the *Drosophila* retina and ultrastructural analysis of mutant phenotypes. J Cell Sci 101(Pt 1):247–254
63. Porter JA, Montell C (1993) Distinct roles of the Drosophila ninaC kinase and myosin domains revealed by systematic mutagenesis. JCell Biol 122(3):601–612
64. Minke B (2012) The history of the prolonged depolarizing afterpotential (PDA) and its role in genetic dissection of Drosophila phototransduction. J Neurogenet 26(2):106–117. https://doi.org/10.3109/01677063.2012.666299
65. Lee S-J, Montell C (2004) Light-dependent translocation of visual arrestin regulated by the NINAC myosin III. Neuron 43(1):95–103. https://doi.org/10.1016/j.neuron.2004.06.014
66. Lee S-J, Xu H, Kang L-W et al (2003) Light adaptation through phosphoinositide-regulated translocation of Drosophila visual arrestin. Neuron 39(1):121–132
67. Satoh AK, Ready DF (2005) Arrestin1 mediates light-dependent rhodopsin endocytosis and cell survival. Curr.Biol. 15(19):1722–1733
68. Liu CH, Satoh AK, Postma M et al (2008) $Ca2^+$–dependent metarhodopsin inactivation mediated by calmodulin and NINAC myosin III. Neuron 59(5):778–789. https://doi.org/10.1016/j.neuron.2008.07.007
69. Satoh AK, Xia H, Yan L et al (2010) Arrestin translocation is stoichiometric to rhodopsin isomerization and accelerated by phototransduction in Drosophila photoreceptors. Neuron 67(6):997–1008. https://doi.org/10.1016/j.neuron.2010.08.024
70. Komori N, Usukura J, Kurien B et al (1994) Phosrestin I, an arrestin homolog that undergoes light-induced phosphorylation in dipteran photoreceptors. Insect Biochem Mol Biol 24(6):607–617

71. Alloway PG, Dolph PJ (1999) A role for the light-dependent phosphorylation of visual arrestin. Proc.Natl.Acad.Sci.U.S.A 96(11):6072–6077
72. Kristaponyte I, Hong Y, Lu H et al (2012) Role of rhodopsin and arrestin phosphorylation in retinal degeneration of Drosophila. J.Neurosci. 32(31):10758–10766
73. Griffith LC, Verselis LM, Aitken KM et al (1993) Inhibition of calcium/calmodulin-dependent protein kinase in Drosophila disrupts behavioral plasticity. Neuron 10(3):501–509
74. Steele FR, Washburn T, Rieger R et al (1992) *Drosophila* retinal-degeneration-C (Rdgc) encodes a novel serine threonine protein phosphatase. Cell 69(4):669–676
75. Voolstra O, Rhodes-Mordov E, Katz B et al (2017) The phosphorylation state of the *Drosophila* TRP channel modulates the frequency response to oscillating light in vivo. J Neurosci 37(15):4213–4224. https://doi.org/10.1523/JNEUROSCI.3670-16.2017
76. Hardie RC, Raghu P (2001) Visual transduction in *Drosophila*. Nature 413(6852):186–193
77. Huber A, Sander P, Gobert A et al (1996) The transient receptor potential protein (Trp), a putative store-operated Ca^{2+} channel essential for phosphoinositide-mediated photoreception, forms a signaling complex with NorpA, InaC and InaD. EMBO J 15(24):7036–7045
78. Scott K, Zuker CS (1998) Assembly of the Drosophila phototransduction cascade into a signalling complex shapes elementary responses. Nature 395(6704):805–808
79. Ranganathan R, Harris GL, Stevens CF et al (1991) A *Drosophila* mutant defective in extracellular calcium-dependent photoreceptor deactivation and rapid desensitization. Nature 354(6350):230–232
80. Hardie RC (1995) Photolysis of caged Ca2^{+} facilitates and inactivates but does not directly excite light-sensitive channels in Drosophila photoreceptors. J Neurosci 15(1 Pt 2):889–902
81. Chu B, Liu C-H, Sengupta S et al (2013) Common mechanisms regulating dark noise and quantum bump amplification in Drosophila photoreceptors. J Neurophysiol 109(8):2044–2055. https://doi.org/10.1152/jn.00001.2013
82. Tang J, Lin Y, Zhang Z et al (2001) Identification of common binding sites for calmodulin and inositol 1,4,5-trisphosphate receptors on the carboxyl termini of trp channels. J Biol Chem 276(24):21303–21310. https://doi.org/10.1074/jbc.M102316200
83. Zheng Y-H, Liu W (2016) Identification and characterization of a new Calmodulin binding site at the C-terminus of Drosophila TRP Channel. Chinese J Biochem Mol Biol 32(7):790–797. https://doi.org/10.13865/j.cnki.cjbmb.2016.07.09
84. Sun Z, Zheng Y, Liu W (2018) Identification and characterization of a novel calmodulin binding site in Drosophila TRP C-terminus. Biochem Biophys Res Commun 501:434–439. https://doi.org/10.1016/j.bbrc.2018.05.007
85. Running Deer JL, Hurley JB, Yarfitz SL (1995) G protein control of *Drosophila* photoreceptor phospholipase C. J.Biol.Chem. 270(21):12623–12628
86. Katz B, Minke B (2012) Phospholipase C-mediated suppression of dark noise enables single-photon detection in Drosophila photoreceptors. J Neurosci 32(8):2722–2733. https://doi.org/10.1523/JNEUROSCI.5221-11.2012
87. Hardie RC (2005) Inhibition of phospholipase C activity in *Drosophila* photoreceptors by 1,2-bis(2-aminophenoxy)ethane N,N,N',N'-tetraacetic acid (BAPTA) and di-bromo BAPTA. Cell Calcium 38(6):547–556. https://doi.org/10.1016/j.ceca.2005.07.005
88. Sengupta S, Barber TR, Xia H et al (2013) Depletion of PtdIns(4,5)P(2) underlies retinal degeneration in Drosophila trp mutants. J Cell Sci 126(Pt 5):1247–1259. https://doi.org/10.1242/jcs.120592
89. Huber A, Sander P, Bahner M et al (1998) The TRP Ca^{2+} channel assembled in a signaling complex by the PDZ domain protein INAD is phosphorylated through the interaction with protein kinase C (ePKC). FEBS Lett 425(2):317–322
90. Hardie RC, Liu C-H, Randall AS et al (2015) In vivo tracking of phosphoinositides in Drosophila photoreceptors. J Cell Sci 128(23):4328–4340. https://doi.org/10.1242/jcs.180364
91. Schaeffer E, Smith D, Mardon G et al (1989) Isolation and characterization of two new Drosophila protein kinase C genes, including one specifically expressed in photoreceptor cells. Cell 57(3):403–412

92. Raghu P, Usher K, Jonas S et al (2000.Apr.;26.(1.):169.-79) Constitutive activity of the light-sensitive channels TRP and TRPL in the Drosophila diacylglycerol kinase mutant, rdgA. Neuron 26(1):169–179
93. Huber A, Sander P, Paulsen R (1996) Phosphorylation of the InaD gene product, a photoreceptor membrane protein required for recovery of visual excitation. JBiolChem 271(20):11710–11717
94. Liu M, Parker LL, Wadzinski BE et al (2000) Reversible phosphorylation of the signal transduction complex in *Drosophila* photoreceptors. J.Biol.Chem. 275(16):12194–12199
95. Mishra P, Socolich M, Wall MA et al (2007) Dynamic scaffolding in a G protein-coupled signaling system. Cell 131(1):80–92
96. Voolstra O, Spät P, Oberegelsbacher C et al (2015) Light-dependent phosphorylation of the Drosophila inactivation no afterpotential D (INAD) scaffolding protein at Thr170 and Ser174 by eye-specific protein kinase C. PLoS One 10(3):e0122039. https://doi.org/10.1371/journal.pone.0122039
97. Voolstra O, Bartels J-P, Oberegelsbacher C et al (2013) Phosphorylation of the *Drosophila* transient receptor potential ion channel is regulated by the phototransduction cascade and involves several protein kinases and phosphatases. PLoS One 8(9):e73787. https://doi.org/10.1371/journal.pone.0073787
98. Balakrishnan SS, Basu U, Raghu P (2015) Phosphoinositide signalling in *Drosophila*. Biochim Biophys Acta 1851(6):770–784. https://doi.org/10.1016/j.bbalip.2014.10.010
99. Rodieck RW (1998) The first steps in seeing. Sinauer Associates, Inc, Sunderland
100. Kirschfeld K, Franceschini N (1969) Ein Mechanismus zur Steuerung des Lichtflusses in den Rhabdomeren des Komplexauges von Musca (a mechanism for the control of the light flow in the rhabdomeres of the complex eye of Musca). Kybernetik 6(1):13–22
101. Kirschfeld K, Vogt K (1980) Calcium ions and pigment migration in fly photoreceptors. Naturwissenschaften 67(10):516–517. https://doi.org/10.1007/BF01047639
102. Katz B, Voolstra O, Tzadok H et al (2017) The latency of the light response is modulated by the phosphorylation state of Drosophila TRP at a specific site. Channels (Austin) 11(6):678–685. https://doi.org/10.1080/19336950.2017.1361073
103. Bähner M, Frechter S, Da Silva N et al (2002) Light-regulated subcellular translocation of *Drosophila* TRPL channels induces long-term adaptation and modifies the light-induced current. Neuron 34:83–93
104. Kosloff M, Elia N, Joel-Almagor T et al (2003) Regulation of light-dependent Gq alpha translocation and morphological changes in fly photoreceptors. EMBO J 22(3):459–468
105. Cronin MA, Lieu MH, Tsunoda S (2006) Two stages of light-dependent TRPL-channel translocation in Drosophila photoreceptors. JCell Sci 119:2935–2944
106. Oberegelsbacher C, Schneidler C, Voolstra O et al (2011) The Drosophila TRPL ion channel shares a Rab-dependent translocation pathway with rhodopsin. EurJCell Biol 90(8):620–630
107. Meyer NE, Joel-Almagor T, Frechter S et al (2006) Subcellular translocation of the eGFP-tagged TRPL channel in Drosophila photoreceptors requires activation of the phototransduction cascade. J.Cell Sci. 119:2592–2603
108. Frechter S, Elia N, Tzarfaty V et al (2007) Translocation of Gq alpha mediates long-term adaptation in *Drosophila* photoreceptors. J.Neurosci. 27(21):5571–5583
109. Cronin MA, Diao F, Tsunoda S (2004) Light-dependent subcellular translocation of Gqalpha in Drosophila photoreceptors is facilitated by the photoreceptor-specific myosin III NINAC. J.Cell Sci. 117(Pt 20):4797–4806
110. Stavenga DG (2004) Angular and spectral sensitivity of fly photoreceptors. III. Dependence on the pupil mechanism in the blowfly Calliphora. J Comp Physiol A Neuroethol Sens Neural Behav Physiol 190(2):115–129. https://doi.org/10.1007/s00359-003-0477-0
111. Franceschini N (1972) Pupil and Pseudopupil in the compound eye of Drosophila. In: Wehner R (ed) Information processing in the visual Systems of Anthropods: symposium held at the Department of Zoology, University of Zurich, March 6–9, 1972. Springer, Berlin/Heidelberg, pp 75–82

112. Satoh AK, Li BX, Xia H et al (2008) Calcium-activated myosin V closes the Drosophila pupil. Curr Biol 18(13):951–955. https://doi.org/10.1016/j.cub.2008.05.046
113. Tsunoda S, Zuker CS (1999) The organization of INAD-signaling complexes by a multivalent PDZ domain protein in Drosophila photoreceptor cells ensures sensitivity and speed of signaling. Cell Calcium 26(5):165–171
114. Liu W, Wen W, Wei Z et al (2011) The INAD scaffold is a dynamic, redox-regulated modulator of signaling in the *Drosophila* eye. Cell 145(7):1088–1101. https://doi.org/10.1016/j.cell.2011.05.015
115. Teng DH, Chen CK, Hurley JB (1994) A highly conserved homologue of bovine neurocalcin in Drosophila melanogaster is a Ca(2^+)-binding protein expressed in neuronal tissues. J Biol Chem 269(50):31900–31907
116. Faurobert E, Chen CK, Hurley JB et al (1996) Drosophila neurocalcin, a fatty acylated, $Ca2^+$−binding protein that associates with membranes and inhibits in vitro phosphorylation of bovine rhodopsin. J Biol Chem 271(17):10256–10262
117. Kawamura S, Hisatomi O, Kayada S et al (1993) Recoverin has S-modulin activity in frog rods. J Biol Chem 268(20):14579–14582
118. Song Z, Postma M, Billings SA et al (2012) Stochastic, adaptive sampling of information by microvilli in fly photoreceptors. Curr Biol 22(15):1371–1380. https://doi.org/10.1016/j.cub.2012.05.047

Chapter 35
Calcium Imaging in *Drosophila melanogaster*

Nicola Vajente, Rosa Norante, Paola Pizzo, and Diana Pendin

Abstract *Drosophila melanogaster*, colloquially known as the fruit fly, is one of the most commonly used model organisms in scientific research. Although the final architecture of a fly and a human differs greatly, most of the fundamental biological mechanisms and pathways controlling development and survival are conserved through evolution between the two species. For this reason, *Drosophila* has been productively used as a model organism for over a century, to study a diverse range of biological processes, including development, learning, behavior and aging. Ca^{2+} signaling comprises complex pathways that impact on virtually every aspect of cellular physiology. Within such a complex field of study, *Drosophila* offers the advantages of consolidated molecular and genetic techniques, lack of genetic redundancy and a completely annotated genome since 2000. These and other characteristics provided the basis for the identification of many genes encoding Ca^{2+} signaling molecules and the disclosure of conserved Ca^{2+} signaling pathways. In this review, we will analyze the applications of Ca^{2+} imaging in the fruit fly model, highlighting in particular their impact on the study of normal brain function and pathogenesis of neurodegenerative diseases.

Keywords Calcium imaging · *Drosophila* · Calcium indicators · GECI · Calcium signaling · Neurodegenerative diseases

Nicola Vajente and Rosa Norante have contributed equally.

N. Vajente · R. Norante
Department of Biomedical Sciences, University of Padova, Padova, Italy

P. Pizzo · D. Pendin (✉)
Department of Biomedical Sciences, University of Padova, Padova, Italy

Neuroscience Institute, National Research Council (CNR), Padova, Italy
e-mail: diana.pendin@unipd.it

© Springer Nature Switzerland AG 2020

M. S. Islam (ed.), *Calcium Signaling*, Advances in Experimental Medicine and Biology 1131, https://doi.org/10.1007/978-3-030-12457-1_35

35.1 A Brief History

The path of *Drosophila* as a research model is a history of groundbreaking achievements, underpinned by 6 Nobel Prizes since 1933. The first went to Thomas Hunt Morgan, who delineated the theory of inheritance by using *Drosophila* to define genes location on chromosomes [1]. Some years later, Hermann Muller defined the effects of X-rays on mutation rate in fruit flies [2], opening the field to modern genetics. These seminal discoveries allowed the generation of genetic tools that still prosper, *e.g.*, balancer chromosomes, special chromosomes that, preventing meiotic crossing-over, are used to maintain complex stocks with multiple mutations on single chromosomes over generations [3]. New genetic tools developed over the years allowed the fruit fly to move with times. As a significant example, CRISPR/Cas9 genome editing strategies allow simple and rapid engineering of the fly genome [4].

What makes *Drosophila* the model organism of choice of many researchers is the observation that relevant genes, cellular processes and basic building blocks in cellular and animal biology are conserved between flies and mammals [5]. Moreover, compared to vertebrate models, *Drosophila* has considerably less genetic redundancy, making the characterization of protein function less complicated. The function of a gene product can be inferred by generating fly lines for its up- or down-regulation and then analyzing the resulting phenotypes. The fruit fly represents also an ideal model organism to study human diseases. Remarkably, over 60% of known human disease-causing genes have a fly orthologue [6]. Most of the cellular processes known to be involved in human disorders pathogenesis, including apoptosis signaling cascades, intracellular calcium (Ca^{2+}) homeostasis, as well as oxidative stress, are conserved in flies. Of note, the high accessibility of the nervous system at different developmental stages, makes also neuroscience experiments feasible in the fly model. Moreover, flies exhibit complex behaviors and, like in humans, many of these behaviors, including learning, memory and motor ability, deteriorate with age [7, 8].

Box 35.1 Advantages of Using *Drosophila* as a Research Animal Model
Beside genetics, the strongest selling point of using *Drosophila* as an animal model are: (i) *Drosophila* are relatively inexpensive and easy to keep, as they are raised in bottles or vials containing cheap jelly-like food. (ii) Generally, there are very few restrictions, minimal ethical and safety issues on their laboratory use. (iii) Flies life cycle is very fast, lasting about 10–12 days at 25 °C. Newly laid eggs take 24 h to undergo embryogenesis before hatching into first instar larvae, which develop into second, and then third instar larvae. The duration of these stages varies with the temperature: at 20 °C, the average length of the egg-larval period is 8 days; at 25 °C it is reduced to 5 days. Larvae transform into immobile pupa, undergo metamorphosis and eclose in the adult form 5–7 days later. (iv) A single fly can produce hundreds

(continued)

Box 35.1 (continued)
of offspring within days, thus it is relatively easy to quickly generate large numbers of embryos, larvae or flies of a given genotype. Individual flies are easily manipulated when anaesthetized with carbon dioxide, allowing identification of selectable phenotypic features under a stereomicroscope [9].

35.2 *Drosophila* Ca^{2+} Toolkit

The Ca^{2+} ion is the major intracellular messenger, mediating a variety of physiological responses to chemical and electrical stimulations. Therefore, cell Ca^{2+} concentration $[Ca^{2+}]$ must be tightly controlled in terms of both space and time, a task that is accomplished by several Ca^{2+} transporting and buffering systems. Easily accessible knock-down and knock-out strategies, applicable to cell lines, as well as to living animals, have helped discovering in flies a number of molecules involved in Ca^{2+} signaling. As in mammals, basal cytosolic Ca^{2+} levels are controlled in flies by the interplay of Ca^{2+} transport systems, localized in the plasma membrane (PM) and the membranes of intracellular organelles that function as internal Ca^{2+} stores (Fig. 35.1). This toolkit, together with a number of Ca^{2+}-binding proteins, concurs in creating and regulating the dynamics and spatial localization of Ca^{2+} signals. The major players in Ca^{2+} signaling in *Drosophila* are briefly described below; the interested reader is referred to a more extensive review on the topic [10].

Ca^{2+} enters the PM through Ca^{2+} channels, *e.g.*, voltage- and ligand- gated. As for voltage-gated Ca^{2+} channels, the fly genome encodes three $\alpha 1$ subunits (Ca-α1D, cacophony, Ca-α1T) forming Ca_v1, Ca_v2, and Ca_v3 type channels, respectively, mainly expressed in the nervous system and muscles [10]. Among ligand-gated channels, glutamate-gated ionotropic receptors (iGluRs) are represented in *Drosophila* by 15 genes encoding different subunits. As in other animal species, *Drosophila* uses glutamate as a fast neurotransmitter in neuromuscular junctions (NMJs), and highly Ca^{2+} permeable iGluRs are clustered in active zones in postsynaptic motor neuron terminals [11]. Cations enter sensory neurons through Transient Receptor Potential (TRP) channels. The gene encoding the first member of the trp superfamily was identified in *Drosophila* photoreceptors as a PM Ca^{2+} permeable channel, required for mediating the light response [12]. A vast number of trp homologs were found in vertebrates that have been classified in seven major subfamilies in metazoans.

Cytosolic Ca^{2+} increase can be also triggered by the activation of phospholipase C (PLC), which produces inositol 1,4,5-trisphosphate (IP_3) that interacts with Ca^{2+} channels located in the Endoplasmic Reticulum (ER) and Golgi apparatus (GA), causing their opening. Three IP_3 receptor (IP_3R) isoforms are expressed in mammals, while a single IP_3R is present in *Drosophila* (itpr) [13]. The channel shares the highest level of similarity as well as functional properties (channel conductance, gating properties, IP_3- and Ca^{2+}-dependence) with the mouse IP_3R1. The release of Ca^{2+} from intracellular stores occurs also through Ryanodine

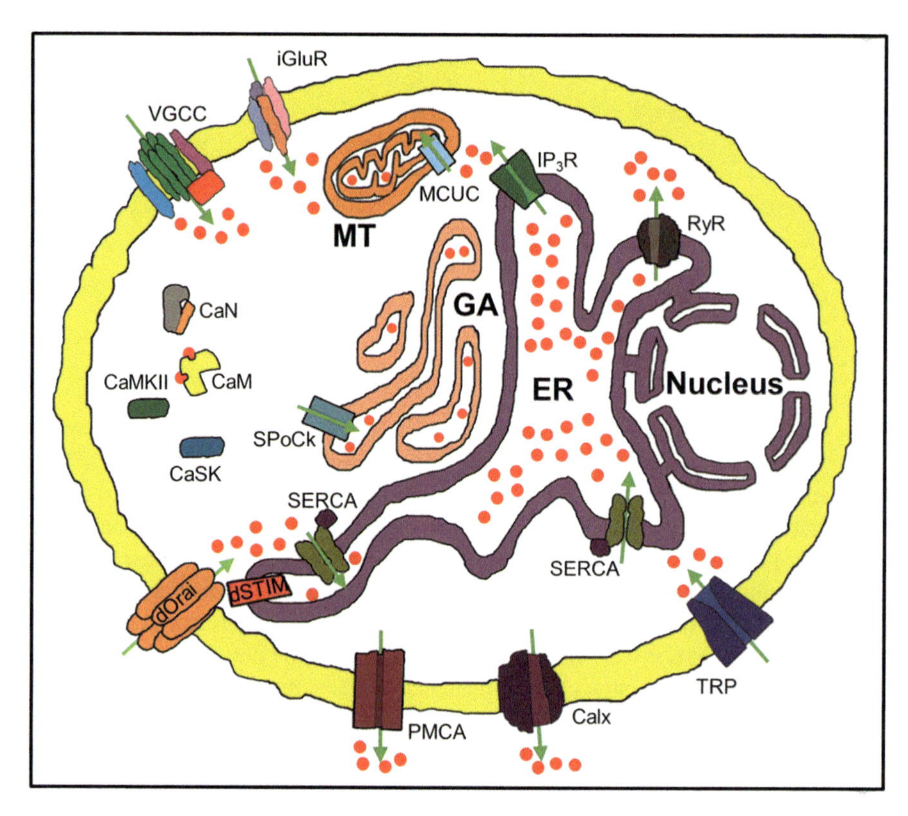

Fig. 35.1 A *Drosophila* cell with its Ca^{2+}toolkit. The movement of Ca^{2+} ions (red spots) are indicated as green arrows. The Ca^{2+} handling proteins inserted in the PM, from the upper left corner, are: Voltage-Gated Ca^{2+} channels (VGCC), glutamate-gated ionotropic receptors (iGluRs), Transient Receptor Potential (TRP) channels, Na^{+}/Ca^{2+} exchanger (Calx), PM Ca^{2+} ATPase (PMCA) and ORAI1 oligomers forming a channel. In the ER membrane from the upper left side are present: inositol 1,4,5-trisphosphate receptor (IP_3R), Ryanodine Receptor (RyR), Sarco-Endoplasmic Reticulum Ca^{2+} ATPase (SERCA) and STIM1. In the GA membrane is present the Secretory Pathway Ca^{2+} ATPase (SPoCk). The inner mitochondrial membrane hosts the mitochondrial calcium uniporter complex (MCUC). Different Ca^{2+} interacting proteins are resident in the cytosol: Calcineurin (CaN), Calmodulin (Cam), Ca^{2+}/Calmodulin-dependent protein kinase II (CaMKII), and Ca^{2+}/Calmodulin-dependent serine protein kinase (CaSK)

Receptor (RyR), located in the sarco/endoplasmic reticulum (SER) membrane. In vertebrates, three isoforms are described (RyR 1–3), while the *Drosophila* genome contains a single RyR gene that encodes a protein with approximately 45% identity with the vertebrate family members [14, 15].

Ca^{2+} release from intracellular stores is most often accompanied by Ca^{2+} influx through PM channels in the regulated process of Store-Operated Ca^{2+} Entry (SOCE). The molecular basis of SOCE, whereby Ca^{2+} influx across the PM is activated in response to depletion of ER Ca^{2+} stores, has been under investigation for more than 20 years and was finally revealed thanks to the identification of the two molecular key players in RNAi screens performed in *Drosophila* S2 cells [16, 17]. The presence of a single fly homologue for stromal interacting molecule 1 (STIM1) and ORAI1, whereas mammals have two and three copies respectively, offered a flying start for the identification of the proteins.

The main route for Ca^{2+} uptake into mitochondria is through the mitochondrial calcium uniporter (MCU) complex, a Ca^{2+}-selective ion channel located at the inner mitochondrial membrane. The channel subunit MCU is regulated through other regulatory components, i.e. MICU1/2/3 and EMRE [18]. A MCU homologue has been identified and characterized in *Drosophila* [19, 20], along with its regulatory subunits MICU1 [19, 20] and EMRE [21].

Ca^{2+} signals are terminated by the combined activities of Ca^{2+} ATPases, located on PM, ER, GA membranes and the Na^{+}/Ca^{2+} exchanger, NCX. PM Ca^{2+} ATPase (PMCA) is a protein present in all animals, characterized by a high Ca^{2+} affinity and a low-transport capacity that extrudes Ca^{2+} from the cytosol to maintain $[Ca^{2+}]$ at the basal value of about 100 nM. In humans and other mammals, four major PMCA isoforms are encoded by separate genes, while the *Drosophila* genome encodes a single, ubiquitously expressed PMCA [22]. The ATPases located on ER and GA membranes acts to re-accumulate the cation in the organelles' lumen. The SER Ca^{2+} ATPase (SERCA), transports inside ER/SR two Ca^{2+} ions per ATP hydrolyzed. In vertebrates, three SERCA protein isoforms are encoded by three distinct genes, while in *Drosophila* a single gene was identified [23]. Fly SERCA has a higher identity with mammalian SERCA1 and SERCA2 (71–73%) and is expressed at a very high level in the central nervous system (CNS) and muscles. A single homolog of the Secretory Pathway Ca^{2+} ATPase (SPCA) is present in *Drosophila*, named SPoCk. The gene results in three isoforms, but only one (SPoCk-A) has been reported to localize in GA membranes, as its mammalian counterpart [24]. The other two variants have been reported to localize in the ER and peroxisomal membranes. The NCX is a non-ATP-dependent antiporter that mediates the efflux of Ca^{2+} ions in exchange for Na^{2+} import. The *Drosophila* NCX, named Calx, is highly expressed in brain and muscle and has 55% identity with the three mammalian isoforms NCX1, NCX2 and NCX3, which are differentially expressed mainly in the heart, brain and skeletal muscles, respectively [25].

35.3 Experimental Set Up for Ca^{2+} Imaging in *Drosophila*

The conserved Ca^{2+} molecular toolkit, together with the advantages of the model depicted above (Box 35.1), set the basis for the fruit fly to be a major model organism for Ca^{2+} signaling research. Thanks to the development of a broad range of Ca^{2+} indicators (Box 35.2), Ca^{2+} imaging procedures have been specifically designed for their application in flies.

Advancements in Ca^{2+} imaging techniques have proceeded through two distinct although interconnected avenues: the improvement of Ca^{2+} indicators and the development of appropriate instrumentations. In the field of Ca^{2+} imaging of live tissues/animals, the application of wide-field microscopy is limited by light scattering across the z-axis of extended pieces of tissue, thus the use of two-photon

(2P) microscopy is usually preferred. 2P microscopy have allowed measurements in the intact brain of an entire transgenic animal, improving spatial resolution by restricting the excitation of chromophores to defined focal planes.

In order to perform optical Ca^{2+} imaging experiments in *Drosophila*, the access of light for excitation to the structures of interest must be assured. The simplest possibility is to excite the genetically encoded Ca^{2+} indicator (GECI, see Box 35.2, Fig. 35.2) directly through the animal's cuticle without any surgical manipulation. When baseline fluorescence of the GECI used is strong enough, the partial transparency of third instar larvae allows for optical access of brain, dorsal sensory neurons and muscles. Since imaging is hurdled by continuous larval movements, a few experimental tips have been developed, *e.g.*, immobilization of the larva on the coverslip with a transparent sticky tape [26] or in microfluidic clamps [27]. Despite the immobilization, slight contractions and movements cannot be completely eliminated, possibly leading to shifts in the focal plane and thus alterations in the fluorescence intensity. The problem can be overcome by using ratiometric GECIs (Box 35.2).

Box 35.2 Ca^{2+} Indicators for Imaging in Flies

Approximately 30 years ago, scientists started to design and engineer organic fluorescent Ca^{2+} indicators, opening the door for cellular Ca^{2+} imaging. Since then, a variety of probes have been developed, which differ in their mode of action, Ca^{2+} affinities, intrinsic baseline fluorescence and kinetic properties. Two major classes of Ca^{2+} indicators have been developed, *i.e.*, chemical probes and genetically encoded Ca^{2+} indicators (GECIs).

Chemical indicators (*e.g.*, fura-2, indo-1, fluo-4) are small fluorescent molecules that are able to chelate Ca^{2+} ions. These molecules are based on BAPTA, an EGTA homologue with high selectivity for Ca^{2+}. The Ca^{2+} chelating carboxyl groups are usually masked as acetoxymethyl esters, making the molecule more lipophilic and allowing an easy entrance into cells. Once the molecule is inside the cell, the Ca^{2+} binding domains are freed by cellular esterases. Binding of a Ca^{2+} ion to the molecule leads to either an increase in quantum yield of fluorescence or an emission/excitation wavelength shift. Chemical indicators are mostly used to measure cytosolic $[Ca^{2+}]$. Early attempts to measure presynaptic $[Ca^{2+}]$ in *Drosophila* employed membrane permeant chemical Ca^{2+} indicators [30]. However, this technique is hardly reliable due to uneven dye loading, high background fluorescence and lack of cell type selectivity. To overcome current limitations, dextran-conjugated fluorescent Ca^{2+} indicators have been loaded in cut axons. The approach allowed to measure resting $[Ca^{2+}]$ and nerve-evoked Ca^{2+} signals during high-frequency activity [31].

GECIs include different types of engineered proteins, such as single fluorescent protein-based indicators (*e.g.*, GCaMP), bioluminescent probes (*e.g.*, aequorin) and fluorescence (or Förster) resonance energy transfer (FRET)-based indicators (*e.g.*, cameleons) [32].

(continued)

Box 35.2 (continued)

GCaMP is one of the most used GECIs, based on a circularly-permuted variant of Green Fluorescent Protein (cpGFP). The N-terminus of the cpGFP is connected to the M13 fragment of the myosin light chain kinase, while the C-terminus ends with the Ca^{2+}-binding region of calmodulin (CaM). In the presence of Ca^{2+}, M13 wraps around Ca^{2+}-bound CaM, leading to a conformational change that increases the fluorescence protein (FP) fluorescence intensity [33] (Fig. 35.2, panel a). During recent years, different variants of GCaMP indicators have been developed, with improved characteristics in terms of Ca^{2+} affinity, brightness, dynamic range. Other variants of FPs, *e.g.*, red-coloured, have been used to develop sensors allowing simultaneous measurement in different organelles, making GCaMPs a whole family of great tools to follow Ca^{2+} dynamics.

One limitation in the use of GCaMPs, and in general of single protein-based GECIs, is the sensitivity to movement artifacts as well as focal plane shifts, which can be mistaken for $[Ca^{2+}]$ changes. A method used to correct for this type of artifacts is to co-express a FP together with the GECI [34]. Alternatively, the limit can be overcome by using ratiometric indicators, such as cameleon. This molecule uses the same Ca^{2+} binding domains of the GCaMP (M13 and CaM), that are bound to two different variants of the GFP: a cyan (CFP) and a yellow (YFP) variant. In the absence of Ca^{2+}, the excited CFP emits at 480 nm, while in the presence of Ca^{2+} the interaction between Ca^{2+}, CaM and M13 brings the two FPs at a closer distance (2–6 nm), and the energy released from the CFP is absorbed by the YFP, that emits at a different wavelength (535 nm) (Fig. 35.2, panel b). By calculating the ratio of EYFP/ECFP emissions, one obtains a clear indication of intracellular $[Ca^{2+}]$ variations, excluding changes of fluorescence caused by artefactual movements of the sample.

GECIs allow the monitoring of Ca^{2+} not only in the cytosol, but also in organelles (*e.g.*, ER, mitochondria, GA, etc.) thanks to the addition of specific targeting sequences. GECIs have demonstrated valuable in the measurement of $[Ca^{2+}]$ in cells and within organelles in several in vivo models. Notably, in flies, the ease of transgenesis allowed the generation of several lines for the expression of GECIs, both cytosolic and organelle-targeted. Moreover, the Gal4-UAS expression system [35, 36] allows the targeting of the probes to specific tissues or even cell subtypes. In this two-part approach, one fly strain carries the Ca^{2+} sensor cDNA under the control of an upstream activator sequence (UAS), so that the gene is silent in the absence of the transcription factor Gal4. A second fly strain expressing Gal4 in a cell type-specific manner is mated to the UAS strain, resulting in progeny expressing the probe in a transcriptional pattern that reflects the expression pattern of the Gal4 line promoter.

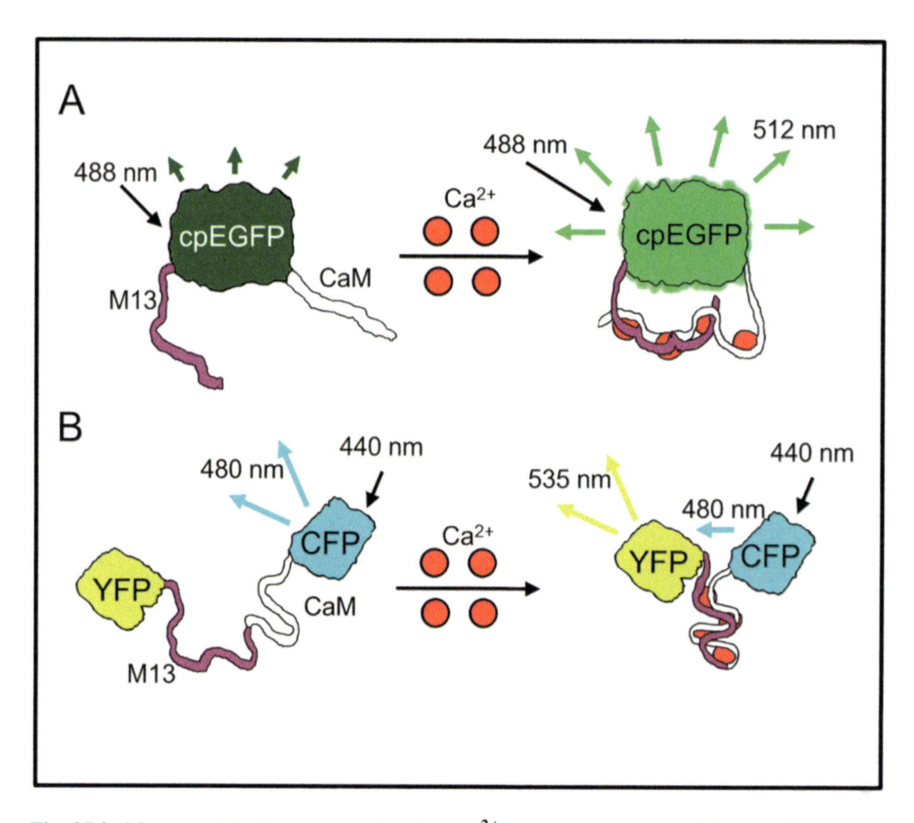

Fig. 35.2 Most used indicators for in vivo Ca^{2+} imaging in *Drosophila*. (**a**) The single-wavelength indicator GCaMP is composed by: the M13 fragment of the myosin light chain kinase domain (M13, pink), the circularly-permutated enhanced Green Fluorescent Protein (cpEGFP, green) and Ca^{2+}-binding region of calmodulin (CaM, white). The FP is excited at 488 nm and the emission is detected at 512 nm. Upon Ca^{2+} binding, a conformational change increases the emitted fluorescence intensity. (**b**) The FRET-based Cameleon probe, composed by: Yellow Fluorescent Protein (YFP, yellow), the M13 domain (pink), the CaM domain (white) and the Cyan Fluorescent Protein (CFP, cyan). The protein is excited at 440 nm and in absence of Ca^{2+} the emission is detected at 480 nm; upon Ca^{2+} binding, conformational changes provide the optimal distance to get Forster Resonance Energy Transfer (FRET), and the YFP emission is detected at 535 nm

Imaging Ca^{2+} activities in the CNS of adult flies usually requires a surgical intervention to achieve optical access to the brain. However, trans-cuticular imaging have also been applied to intact adult brains using 3P microscopy [28]. We report, as an example, a protocol applied to monitor odor-evoked Ca^{2+} dynamics. Anesthetized flies are restrained between a sticky tape and a fine-meshed metal grid which enables air exchange around the abdomen. A small hole cut through the sticky tape into the head capsule allows the exposure of the brain and the antennal lobes expressing the GECI. The odors are then applied to the fly's antennae and the temporal dynamics and spatial distribution of Ca^{2+} activities are monitored using an imaging microscope [26] (Fig. 35.3). Since movements are reduced in these

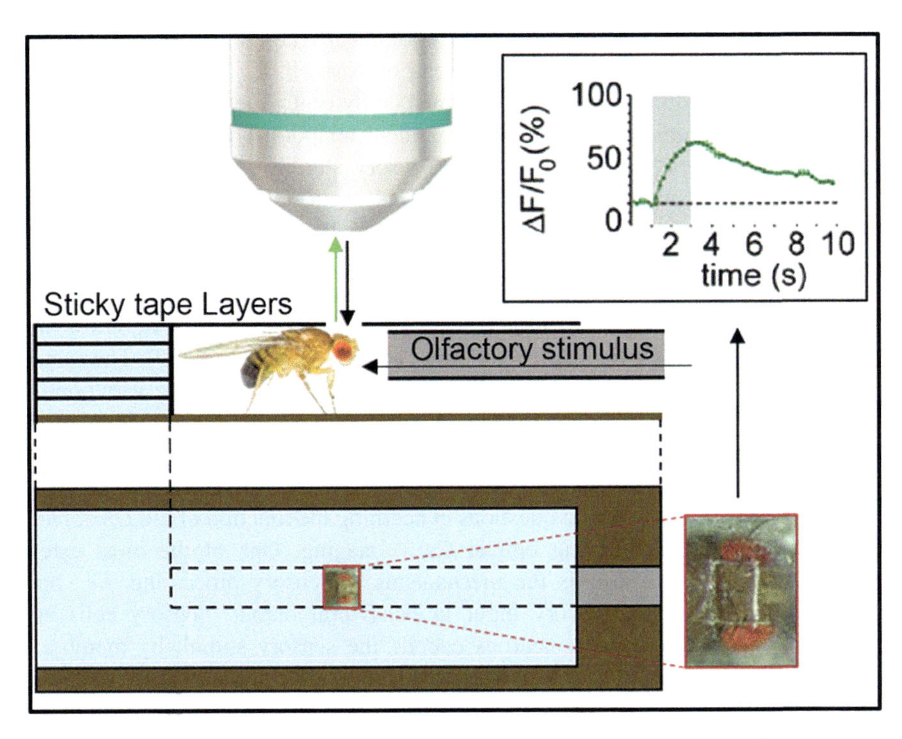

Fig. 35.3 Schematic illustration of a set up for in vivo odor-stimulated Ca^{2+}imaging in Drosophila. A grid is placed on the microscope slide and fixed with several layers of sticky tape. A small passage is cut into the layers of tape, fitting a single fly and a small tube for the delivery of the odor. The chamber is sealed with another layer of adhesive tape where a small window is cut, providing access to the fly's head. The dynamics of intracellular Ca^{2+} in the olfactory sensory neurons of the antennal lobe has been detected upon application of 3-octanol using the sensor G-CaMP 3.0. The duration of the odor stimulus is indicated as a grey bar. (Adapted from: Ref. [29])

preparations, single-wavelength GECIs, such as GCaMPs, are usually preferred, due to their higher dynamic range and because they permit the use of simpler imaging systems.

Thanks to the parallel development of indicators and imaging systems, Ca^{2+} imaging has matured over the years into a powerful tool for imaging of cellular activity in living animals. We propose now an overview of Ca^{2+} imaging experiments that can be performed in *Drosophila*, aware that by far this is not all encompassing, and the list of interesting investigations could certainly be extended.

35.4 Ca^{2+} Imaging: Sensory Neuroscience and Beyond

Drosophila melanogaster contributed to many aspects of neuroscience. In the past, the analysis of fly brain function has been challenging due to the small size of neurons that initially restricted electrophysiological recordings to specific highly accessible regions, *e.g.*, larval preparations of NMJs [37]. Recently, whole-cell patch-clamp recordings have been performed on fly central neurons [38], providing insights into neuronal activity with an excellent temporal precision. However, as in the brains of vertebrates, *Drosophila* sensory stimuli, motor outputs as well as central processing events, are encoded as spatio-temporal activity patterns that require the concerted activity of many neurons. As a consequence, besides recordings from individual neurons, monitoring the activity across many cells is mandatory to explore complex circuits. In neurons, membrane depolarization is accompanied by fast Ca^{2+} influx via voltage-gated Ca^{2+} channels, as well as slower Ca^{2+} signals deriving from the ER and mitochondrial Ca^{2+} pools [39]. The easiest way to indirectly measure membrane depolarization is measuring the variations in $[Ca^{2+}]$ inside the cells. The development of GECIs allowed for these measurements in vivo in multi-cellular animals, by targeting the probes to specific cells and sub-cellular compartments.

Different types of scientific questions concerning the function of the *Drosophila* brain can be addressed using optical Ca^{2+} imaging. One of the most extensively explored fields regards the mechanisms of sensory processing, *i.e.*, how neural activity encodes sensory input in behavioral output. Sensory cells and directly coupled downstream neurons encode the sensory stimuli by membrane depolarization-induced action potential frequencies. High-intensity stimuli result in high-firing frequencies, leading to strong intracellular Ca^{2+} transients, allowing to fully exploit the potential of GECIs. A number of studies in this field have helped identify and measure the response of specific brain regions to various sensory stimuli including olfaction, taste, and thermosensation [26]. We present here some significant examples.

Optical Ca^{2+} imaging has been successfully applied to study neuronal activity in the olfactory system of the fly's brain. Flies display robust odor-evoked behaviors in response to cues from plants or other flies. More than 1000 olfactory sensory neurons located in the olfactory sensory organs of the head (*i.e.*, the third segment of the antenna and the maxillary palps) project with their axons to the antennal lobe, the primary olfactory center of the fly's brain. The neuronal terminal arborizations are organized into spherical structures called glomeruli that contact projection neurons and local interneurons. Projection neurons signal to higher brain centers, such as the mushroom body and the lateral horn. Since each sensory neuron expresses a limited number of olfactory receptors with a specific ligand-binding profile, each odor information is represented as a specific spatiotemporal code before it is sent to higher brain centers. Optical Ca^{2+} imaging has been performed in each order of olfactory neurons, including the antennal lobe (example in: [40]; protocol in [41]) and the neurons of the mushroom bodies (Kenyon cells) [42]. The mushroom body has been shown over many years to be a brain region necessary and sufficient for the association of odor stimuli through learning with rewarding or punishing cues [43, 44]. Electrophysiological studies on individual cells allowed to propose a model for odors encoding in the mushroom bodies. The model proposed that only very few out of a large array of Kenyon cells are selectively responding to any given odor stimulus, due to the convergence of several projection neurons onto a given Kenyon cell, combined with the high firing thresholds of Kenyon cells. The use of GECIs recently allowed to confirm the proposed model: the activity of >100 mushroom body neurons was simultaneously monitored in vivo by two-photon imaging of the Ca^{2+} indicator GCaMP3, allowing the visualization of the distinct patterns of sparse mushroom body neurons activated upon different odors stimulation [45].

Ca^{2+} imaging has also proven to be of enormous value for the analysis of how auditory stimuli are encoded in flies. The structure that has been primarily associated with hearing is the Johnston's organ, located on the second antennal segment. Interestingly, Ca^{2+} imaging experiments performed in intact animals demonstrated that Johnston's organ contains also wind-sensitive neurons. GCaMP-1.3 was expressed under the control of different Gal4 enhancer trap lines in distinct groups of neurons in Johnston's organ. Live flies were mounted in an inverted orientation under a two-photon microscope, and an air flow or a near-field sound, were delivered while recording Ca^{2+} dynamics. Optical Ca^{2+} imaging represented here a powerful tool to dissect this novel circuit, providing evidence that a common sensory organ is used to encode sound-evoked stimuli and air movements [46].

Other relevant aspects for which optical Ca^{2+} imaging using GECIs has been successfully applied to sensory neuroscience comprise: propagation of fly taste perception [47], neuronal plasticity underlying associative learning and memory formation [42], visual circuits dissection [48], mapping of mechanosensory circuits [49].

An interesting recent work, exploiting whole-brain Ca^{2+} imaging in adult flies, aimed at assessing intrinsic functional connectivity. Ca^{2+} signals were acquired from the central brain and functional data were assigned to atlas regions. This allowed to correlate activity between distinct brain regions, providing a framework for using *Drosophila* to study functional large-scale brain networks [50]. Whole-brain imaging has also been attempted during open field behavior in adult flies [51], allowing functional imaging of brain activity of untethered, freely walking flies during sensorial and socially evoked behaviors. Of note, despite imaging over extended periods in live animals is critical to dissect the mechanisms of plasticity, neural development, degeneration and aging, chronic preparations for long-term (>24 h) microscopy have been difficult in *Drosophila*, due to the fly's fragility and opaque exoskeleton. Only recently, laser microsurgery has been employed to create a chronic fly preparation for repeated imaging of neural dynamics for up to 50 days. Ca^{2+} and voltage imaging was performed in fly mushroom body neurons, in particular odor-evoked Ca^{2+} transients were recorded over 7 weeks [52]. This chronic preparation is compatible with a broad range of optical techniques to address in live flies biological questions previously unanswerable.

An interesting approach developed to evaluate functional connections is the combination of Ca^{2+} imaging with genetically encoded optogenetic tools. Optogenetic activation in presynaptic neurons and Ca^{2+} imaging in postsynaptic neurons have been used to map circuits governing different aspects of fly behavior, *e.g.*, antennal grooming behavior [53], courtship [54] and aggression [55]. Some critical aspects need to be taken into account when setting up this kind of experiments, *i.e.*, minimize spectral overlaps and use independent gene expression systems for the two transgenes [56].

Another example of the power of combined functional imaging in the fly takes advantage of both GECIs and Genetically Encoded Voltage Indicators (GEVIs) (reviewed in [57]). Yang and colleagues compared voltage and Ca^{2+} responses within compartments of the same neuron, using the ultrafast GCaMP6f sensor together with a newly developed GFP-based voltage sensor, named Asap2f. *In vivo* two-photon imaging of the two indicators was performed in the *Drosophila* visual system. Remarkably, intracellular [Ca^{2+}] do not simply follow the decay of voltage signals. Instead, Ca^{2+} responses appear compartmentalized, *i.e.*, they are different in their amplitude and kinetics among distinct regions of the same cell [58]. Voltage and Ca^{2+} signals appear distinct and neurons may convey varying information to their postsynaptic partners in different synaptic layers. The unprecedented resolution afforded by both indicators allowed to shed light on local neural computations during visual information processing.

In addition to adult flies, also other developmental stages can be subjected to Ca^{2+} imaging. In a recent paper, insight in the molecular pathway underlying network refinement was obtained by performing Ca^{2+} imaging at the embryonic NMJs. The authors demonstrated that oscillatory Ca^{2+} signals via voltage-gated Ca^{2+} channels orchestrate the activity of several kinases and phosphatases, key components in pruning aberrant synapses during embryonic synaptic refinement [59].

Another developmental stage much studied and appreciated for its accessibility and well-established organization is the larval stage. Ca^{2+} imaging has been performed in intact larvae, as well as in isolated larval CNS. An interesting recent example that underpins the power of the fly model is represented by a screen of unknown compounds for their potential to function as anticonvulsants [60]. In this work, GCaMP was expressed in motoneurons and the isolated CNS of third instar larvae was imaged to evaluate the effectiveness of novel anticonvulsive compounds to reduce seizure-like CNS activity.

Whole-brain imaging has also been performed in larvae (*e.g.*, imaging of ventral nerve cord during motor programs execution [61]; imaging of isolated CNS during coordinated motor pattern generation [62]).

Photoactivatable GECIs have been exploited for targeted neuronal imaging in cultured neurons and in fruit fly larvae [63]. Light-induced photoactivation allows single cells to be selected out of dense populations, for visualization of morphology and high signal-to-noise measurements of activity, synaptic transmission and connectivity. This tool combines the reporting Ca^{2+} activity with the selective highlighting of individual cells *in situ* in live tissues, facilitating tracking fine neuronal processes with a clarity that cannot be achieved with dense expression of standard FPs.

Besides the obvious importance in neurotransmission, Ca^{2+} signaling is fundamental for the survival and welfare of all cell types. Indeed, Ca^{2+} imaging experiments have been performed in other tissues, most relevantly in fly muscles. As an interesting recent example, an attempt to image all flight muscles together has been tried in intact flying animals. In flies, wing motion is adjusted for both quick voluntary maneuvers and slow compensatory reflexes using only a dozen pairs of muscles. By applying visual motion stimuli while recording the pattern of activity across the complete set of steering muscles using GCaMP6f, the authors propose a model whereby the motor array regulates aerodynamically functional features of wing motion [64].

35.4.1 Ca^{2+} Imaging Inside Organelles

Organelle Ca^{2+} handling plays a fundamental role in cell Ca^{2+} homeostasis. At the cellular level, techniques to measure intra-organelle Ca^{2+} are nowadays available and routinely used. However, this type of measurements is still poorly exploited in living animals and few examples are available in *Drosophila*. Among them, the most commonly measured is mitochondrial Ca^{2+} [20, 65] although some attempts have also been done in other organelles.

As a recent example, Drago and Davis revealed a developmental role for the MCUC in memory formation in adult flies. The authors generated a transgenic line for a mitochondria-targeted GCaMP (named 4mtGCaMP3) that was expressed in MB neurons to measure mitochondria Ca^{2+} uptake upon downregulation of the MCUC components MCU or MICU1. Ca^{2+} imaging experiments have been associated to behavioral studies, demonstrating that the inhibition of mitochondrial Ca^{2+} entry in the developing fly MB neurons causes memory impairment [20].

A sensor of the GFP-aequorin protein (GAP) family, optimized for measurements in high-[Ca^{2+}] environments have been also developed and used in drosophila [66]. Among other applications, the authors propose the imaging of SR Ca^{2+} dynamics in the muscle of transgenic flies in vivo, providing evidence for a valuable tool to explore subcellular complex Ca^{2+} signaling in flies.

35.5 *Drosophila* Models of Neurodegenerative Diseases

Changes in intracellular [Ca^{2+}] mediate a wide range of cellular processes that are relevant to neurodegenerative disorder etiology, including learning and memory, as well as cell death. Indeed, a close link between the pathogenesis of different neurodegenerative disorders and Ca^{2+} regulating systems, the so called "Ca^{2+} hypothesis of neurodegenerative diseases", has been convincingly corroborated by several experimental findings [67, 68]. The potential of the approaches described above makes flies a powerful model system to elucidate pathogenic processes in neurobiology. Indeed, a wide collection of fly models of neurodegenerative disorders have been developed. Often, the model consists of targeted expression of human disease-associated protein. In the case of loss of function pathologic mutations, also knock out/knock down approaches have proved successful in mimicking the pathology. In many cases, robust neurodegeneration is observed in these models.

Ca^{2+} imaging experiments performed in fly models of neurodegenerative disorders have helped unravel the pathogenesis of diseases, including Alzheimer's disease (AD), Parkinson's Disease (PD), Huntington Disease (HD). AD is a neurodegenerative disorder characterized by deposition of amyloid β (Aβ) in extracellular neuritic plaques, formation of intracellular neurofibrillary tangles and neuronal cell death. Among familial (FAD) cases, approximately 50% have been attributed to mutations in three genes: amyloid β precursor protein (*APP*) [69], presenilin 1 (*PSEN1*) [70] and presenilin 2 (*PSEN2*) [71]. The fly genome encodes a single Presenilin gene (*Psn*) [72] and a single APP orthologue (*Appl*) which encodes for β-amyloid precursor-like protein. *Drosophila* models of AD have been developed, mostly expressing wild type and FAD-mutant forms of human APP and presenilins, reproducing AD phenotypes, such as Aβ deposition, progressive learning defects, extensive neurodegeneration and ultimately a shortened lifespan [73]. It is now accepted that FAD-linked presenilins mutants are responsible for a dysregulation of cellular Ca^{2+} homeostasis. An imbalance of Ca^{2+} homeostasis is supposed to represent an early event in the pathogenesis of FAD, but the mechanisms through which FAD-linked mutants affect Ca^{2+} homeostasis are controversial [74]. Using a fly model of FAD, Michno et al. [75], showed that expression of wild type or FAD-mutant Psn in *Drosophila* cholinergic neurons results in cell-autonomous deficits in Ca^{2+} stores, highlighted using the chemical Ca^{2+} probe Fura-2. Importantly, these deficits occur independently of Aβ generation. They also describe a novel genetic, physiological and physical interaction between Psn and Calmodulin, a key regulator of intracellular Ca^{2+} homeostasis. More recently, a study conducted by Li et al. [76] exploited Ca^{2+} probes and confocal imaging to demonstrate that Imidazole, by decreasing the level of intracellular Ca^{2+}, can rescue the mental defect in Aβ42-expressing flies.

PD is the most common movement disorder, typically affecting people between 50 and 60 years of age. The disease is mostly sporadic, only a small fraction of PD cases have been linked to mutations in specific genes, including a-synuclein [77], parkin [78], PINK1 [79]. Cytoplasmic aggregates mainly formed by α-synuclein protein, called Lewy bodies [80], are usually found in the *substantia nigra* of brain tissue. Dopaminergic neurons are the most susceptible to degeneration in PD. Fruit flies are largely used as model for PD. Expression of human α-synuclein in flies leads to selective loss of dopaminergic neurons in the adult brain over time and accumulation of protein in cytoplasmic inclusions [81]. Pan-neural expression of α-synuclein either wild type or carrying PD-linked mutations results in premature loss of climbing ability. Noteworthy, the first animal models revealing an interaction between the two PD genes homologues Pink1 and parkin have been developed in *Drosophila* [82, 83]. Recently, Ca^{2+}-induced neurotoxicity have been explored in a *Drosophila* model of retinal degeneration. The authors found that increasing the autophagic flux prevented cell death in mutant flies, and this depended on the Pink1/parkin pathway [84]. The results indicated that maintaining mitochondrial homeostasis via Pink1/parkin-dependent mitochondrial quality control could potentially alleviate cell death in a wide range of neurodegenerative diseases.

HD is a progressive brain disorder characterized by uncontrolled movements, emotional problems, and loss of cognition. The disease is caused by autosomal dominant mutations in the gene encoding for huntingtin (*HTT*), resulting in an abnormal expansion of the number of CAG triplets, encoding for Glutamine. A *Drosophila* model of HD have helped investigating the mechanisms by which expanded full-length huntingtin (htt) impairs synaptic transmission [85]. The authors showed that expression of expanded full-length htt led to increased neurotransmitter release and increased resting intracellular Ca^{2+} levels, compared to controls. Moreover, mutations in voltage-gated Ca^{2+} channels restored the elevated $[Ca^{2+}]$ and improved neurotransmitter release efficiency, as well as neurodegenerative phenotypes. This suggests that a defect in Ca^{2+} homeostasis contributes to the pathogenesis of the disease, which is in agreement with observations in mammalian systems [86–89].

Drosophila models have been created recapitulating many other diseases affecting the neuronal system, e.g. Hereditary Spastic Paraplegias (HSPs), are a group of inherited neurodegenerative disorders characterized by retrograde degeneration of corticospinal neurons, leading to muscle weakness and spasticity of the lower limbs. HSPs are highly genetically heterogeneous, with over 70 spastic paraplegia gene (SPG) loci associated [90]. Despite this diversity, it is now clear that one of the most common causes of HSPs are mutations in genes encoding proteins that, directly or indirectly, regulate ER morphology and/or distribution. Proper shape is necessary for the ER diverse functions, that are crucial for neuronal welfare [91]. Among these functions, Ca^{2+} sequestration and release play a fundamental role in shaping cytosolic signals [92]. *Drosophila* models have been generated for many HSP-related genes, among them the homologues of Atlastin-1 (SPG3A) [93], Spastin (SPG4) [94], Reticulon-2 (SPG12), ARL6IP1 (SPG61) [95]. Available tools for Ca^{2+} imaging applied to these models would provide valuable insights in the role of Ca^{2+} in the pathogenesis of HSPs.

35.6 Conclusions and Future Perspectives

Ca^{2+} signaling plays a critical role in cellular physiology and, in particular, in fundamental neuronal functions, such as synaptic transmission, synaptogenesis, neuronal plasticity, memory and cell survival. Understanding how the concerted action of neurons, synapses and circuits underlie brain function is a core challenge for neuroscience. The examples we described in this review underscore the contribution of *Drosophila* as a model system to explore cellular and circuits neurophysiology, highlighting potential future directions in the field. Given its genetic accessibility, complex behavioral repertoire and functional similarities with mammalian brain, the fruit fly represents an attractive model organism to approach relevant physiological and pathological questions. The combination of Ca^{2+} imaging with other tools, such as voltage indicators or optogenetics, provides a valuable developing strategy for investigating neural function and dysfunction. New variants of both green and red Ca^{2+} indicators are continually developed, offering improved sensitivity, brightness, photostability and kinetics and can be fruitfully applied to the *Drosophila* model.

Acknowledgments The authors thank the CARIPARO Foundation for Starting Grant 2015 to DP and University of Padova for a fellowship to RN.

References

1. Hunt TM, Bridges CB (1916) Sex linked inheritance in Drosophila, vol 352. Carnegie Institution of Washington, Washington, DC
2. Muller HJ (1928 Sep) The production of mutations by X-rays. Proc Natl Acad Sci U S A 14(9):714–726
3. Lindsley DL, Zimm GG (1992) The genome of Drosophila melanogaster. Annu Rev Genomics Hum Genet 4:89–117
4. Ewen-Campen B, Yang-Zhou D, Fernandes VR, González DP, Liu L-P, Tao R et al (2017) Optimized strategy for in vivo Cas9-activation in *Drosophila*. Proc Natl Acad Sci 114:9409–9414
5. Yoshihara M, Ensminger AW, Littleton JT (2001) Neurobiology and the Drosophila genome. Funct Integr Genomics 1:235–240
6. Wangler MF, Yamamoto S, Bellen HJ (2015) Fruit flies in biomedical research. Genetics 199(3):639–653
7. Mockett RJ, Bayne ACV, Kwong LK, Orr WC, Sohal RS (2003) Ectopic expression of catalase in Drosophila mitochondria increases stress resistance but not longevity. Free Radic Biol Med 34(2):207–217
8. Simon AF, Liang DT, Krantz DE (2006) Differential decline in behavioral performance of Drosophila melanogaster with age. Mech Ageing Dev 127(7):647–651
9. Stocker H, Gallant P (2008) Getting started : an overview on raising and handling Drosophila. Methods Mol Biol 420:27–44
10. Chorna T, Hasan G (2012) The genetics of calcium signaling in Drosophila melanogaster. Biochim Biophys Acta 1820(8):1269–1282
11. DiAntonio A (2006) Glutamate receptors at the Drosophila neuromuscular junction. Int Rev Neurobiol 75:165–179
12. Hardie RC, Minke B (1992) The trp gene is essential for a light-activated Ca^{2+} channel in Drosophila photoreceptors. Neuron 8(4):643–651
13. Yoshikawa S, Tanimura T, Miyawaki A, Nakamura M, Yuzaki M, Furuichi T et al (1992) Molecular cloning and characterization of the inositol 1,4,5-trisphosphate receptor in Drosophila melanogaster. J Biol Chem 267(23):16613–16619
14. Takeshima H, Nishi M, Iwabe N, Miyata T, Hosoya T, Masai I et al (1994) Isolation and characterization of a gene for a ryanodine receptor/calcium release channel in Drosophila melanogaster. FEBS Lett 337(1):81–87
15. Hasan G, Rosbash M (1992) Drosophila homologs of two mammalian intracellular ca(2+)-release channels: identification and expression patterns of the inositol 1,4,5-triphosphate and the ryanodine receptor genes. Development 116(4):967–975
16. Zhang SL, Yeromin AV, Zhang XH-F, Yu Y, Safrina O, Penna A et al (2006) Genome-wide RNAi screen of Ca^{2+} influx identifies genes that regulate Ca^{2+} release-activated Ca^{2+} channel activity. Proc Natl Acad Sci 103(24):9357–9362
17. Roos J, DiGregorio PJ, Yeromin AV, Ohlsen K, Lioudyno M, Zhang S et al (2005) STIM1, an essential and conserved component of store-operated Ca^{2+} channel function. J Cell Biol 169(3):435–445
18. Pendin D, Greotti E, Pozzan T (2014) The elusive importance of being a mitochondrial Ca^{2+} uniporter. Cell Calcium 55(3):139–145
19. Walkinshaw E, Gai Y, Farkas C, Richter D, Nicholas E, Keleman K et al (2015) Identification of genes that promote or inhibit olfactory memory formation in Drosophila. Genetics 199(4):1173–1182

20. Drago I, Davis RL (2016) Inhibiting the mitochondrial calcium uniporter during development impairs memory in adult Drosophila. Cell Rep 16(10):2763–2776
21. Choi S, Quan X, Bang S, Yoo H, Kim J, Park J et al (2017) Mitochondrial calcium uniporter in Drosophila transfers calcium between the endoplasmic reticulum and mitochondria in oxidative stress-induced cell death. J Biol Chem 292(35):14473–14485
22. Bai J, Binari R, Ni J-Q, Vijayakanthan M, Li H-S, Perrimon N (2008) RNA interference screening in Drosophila primary cells for genes involved in muscle assembly and maintenance. Development 135(8):1439–1449
23. Magyar A, Váradi A (1990) Molecular cloning and chromosomal localization of a sarco/endoplasmic reticulum-type Ca^{2+}-ATPase of drosophila melanogaster. Biochem Biophys Res Commun 173(3):872–877
24. Southall TD (2006) Novel subcellular locations and functions for secretory pathway Ca^{2+}/Mn^{2+}-ATPases. Physiol Genomics 26(1):35–45
25. Schwarz EM, Benzer S (1997) Calx, a Na-Ca exchanger gene of Drosophila melanogaster. Proc Natl Acad Sci U S A 94(19):10249–10254
26. Riemensperger T, Pech U, Dipt S, Fiala A (2012) Optical calcium imaging in the nervous system of Drosophila melanogaster. Biochim Biophys Acta Gen Subj 1820:1169–1178
27. Ghaemi R, Rezai P, Nejad FR, Selvaganapathy PR (2017) Characterization of microfluidic clamps for immobilizing and imaging of Drosophila melanogaster larva's central nervous system. Biomicrofluidics 11(3):034113
28. Tao X, Lin H-H, Lam T, Rodriguez R, Wang JW, Kubby J (2017) Transcutical imaging with cellular and subcellular resolution. Biomed Opt Express 8(3):1277–1289
29. Optical calcium imaging using DNA-encoded fluorescence sensors in transgenic fruit flies, Drosophila melanogaster. Dipt S, Riemensperger T, Fiala A. Methods Mol Biol. 2014;1071:195–206
30. Karunanithi S, Georgiou J, Charlton MP, Atwood HL (1997) Imaging of calcium in Drosophila larval motor nerve terminals. J Neurophysiol 78(6):3465–3467
31. Macleod GT (2002) Fast calcium signals in Drosophila motor neuron terminals. J Neurophysiol 88(5):2659–2663
32. Pendin D, Greotti E, Lefkimmiatis K, Pozzan T (2016) Exploring cells with targeted biosensors. J Gen Physiol 149(1):1–36
33. Nakai J, Ohkura M, Imoto K (2001) A high signal-to-noise Ca^{2+} probe composed of a single green fluorescent protein. Nat Biotechnol 19(2):137–141
34. Berry JA, Cervantes-Sandoval I, Chakraborty M, Davis RL (2015) Sleep facilitates memory by blocking dopamine neuron-mediated forgetting. Cell 161(7):1656–1667
35. Brand AH, Perrimon N (1993) Targeted gene expression as a means of altering cell fates and generating dominant phenotypes. Development 118(2):401–415
36. Klueg KM, Alvarado D, Muskavitch MAT, Duffy JB (2002) Creation of a GAL4/UAS-coupled inducible gene expression system for use in Drosophila cultured cell lines. Genesis 34(1–2):119–122
37. Broadie KS (1994) Synaptogenesis in Drosophila: coupling genetics and electrophysiology. J Physiol Paris 88(2):123–139
38. Wilson RI, Turner GC, Laurent G (2004) Transformation of olfactory representations in the Drosophila antennal lobe. Science 303(5656):366–370
39. Berridge MJ (1998) Neuronal calcium signaling. Neuron 21(1):13–26
40. Strube-Bloss MF, Grabe V, Hansson BS, Sachse S (2017) Calcium imaging revealed no modulatory effect on odor-evoked responses of the Drosophila antennal lobe by two populations of inhibitory local interneurons. Sci Rep 7(1):7854
41. Silbering AF, Bell R, Galizia CG, Benton R (2012) Calcium imaging of odor-evoked responses in the Drosophila antennal lobe. J Vis Exp 60:1–10
42. Barnstedt O, Owald D, Felsenberg J, Brain R, Moszynski JP, Talbot CB et al (2016) Memory-relevant mushroom body output synapses are cholinergic. Neuron 89(6):1237–1247
43. Heisenberg M (2003) Mushroom body memoir: from maps to models. Nat Rev Neurosci 4(4):266–275

44. Menzel R (2014) The insect mushroom body, an experience-dependent recoding device. J Physiol Paris 108(2–3):84–95
45. Honegger KS, Campbell RAA, Turner GC (2011) Cellular-resolution population imaging reveals robust sparse coding in the Drosophila mushroom body. J Neurosci 31(33):11772–11785
46. Yorozu S, Wong A, Fischer BJ, Dankert H, Kernan MJ, Kamikouchi A et al (2009) Distinct sensory representations of wind and near-field sound in the Drosophila brain. Nature 458(7235):201–205
47. Harris DT, Kallman BR, Mullaney BC, Scott K (2015) Representations of taste modality in the Drosophila Brain. Neuron 86(6):1449–1460
48. Schnaitmann C, Haikala V, Abraham E, Oberhauser V, Thestrup T, Griesbeck O et al (2018) Color processing in the early visual system of Drosophila. Cell 172(1–2):318–318.e18
49. Patella P, Wilson RI (2018) Functional maps of Mechanosensory features in the Drosophila Brain. Curr Biol 28:1189–1203.e5
50. Mann K, Gallen CL, Clandinin TR (2017) Whole-Brain calcium imaging reveals an intrinsic functional network in Drosophila. Curr Biol 27(15):2389–2396.e4
51. Grover D, Katsuki T, Greenspan RJ (2016) Flyception: imaging brain activity in freely walking fruit flies. Nat Methods 13(7):569–572
52. Huang C, Maxey JR, Sinha S, Savall J, Gong Y, Schnitzer MJ (2018) Long-term optical brain imaging in live adult fruit flies. Nat Commun 9(1):872
53. Hampel S, Franconville R, Simpson JH, Seeds AM (2015) A neural command circuit for grooming movement control. elife 4(9):e08758
54. Shirangi TR, Wong AM, Truman JW, Stern DL (2016) *Doublesex* regulates the connectivity of a neural circuit controlling *Drosophila* male courtship song. Dev Cell 37(6):533–544
55. Hoopfer ED, Jung Y, Inagaki HK, Rubin GM, Anderson DJ (2015) P1 interneurons promote a persistent internal state that enhances inter-male aggression in Drosophila. elife 4(12):pii: e11346
56. Simpson JH, Looger LL (2018) Functional imaging and optogenetics in drosophila. Genetics 208(4):1291–1309
57. Kaschula R, Salecker I (2016) Neuronal computations made visible with subcellular resolution. Cell 166:18–20
58. Yang HHH, St-Pierre F, Sun X, Ding X, Lin MZZ, Clandinin TRR (2016) Subcellular imaging of voltage and calcium signals reveals neural processing in vivo. Cell 166(1):245–257
59. Vonhoff F, Keshishian H (2017) In vivo calcium signaling during synaptic refinement at the Drosophila neuromuscular junction. J Neurosci 37(22):2922–2916
60. Streit AK, Fan YN, Masullo L, Baines RA (2016) Calcium imaging of neuronal activity in Drosophila can identify anticonvulsive compounds. PLoS One 11(2):e0148461
61. Lemon WC, Pulver SR, Höckendorf B, McDole K, Branson K, Freeman J et al (2015) Whole-central nervous system functional imaging in larval Drosophila. Nat Commun 6:7924
62. Pulver SR, Bayley TG, Taylor AL, Berni J, Bate M, Hedwig B (2015) Imaging fictive locomotor patterns in larval *Drosophila*. J Neurophysiol 114(5):2564–2577
63. Berlin S, Carroll EC, Newman ZL, Okada HO, Quinn CM, Kallman B et al (2015) Photoactivatable genetically encoded calcium indicators for targeted neuronal imaging. Nat Methods 12(9):852–858
64. Lindsay T, Sustar A, Dickinson M (2017) The function and organization of the motor system controlling flight maneuvers in flies. Curr Biol 27(3):345–358
65. Ivannikov MV, Macleod GT (2013) Mitochondrial free Ca^{2+} levels and their effects on energy metabolism in drosophila motor nerve terminals. Biophys J 104(11):2353–2361
66. Navas-Navarro P, Rojo-Ruiz J, Rodriguez-Prados M, Ganfornina MD, Looger LL, Alonso MT et al (2016) GFP-Aequorin protein sensor for Ex vivo and in vivo imaging of Ca 2+ dynamics in high-Ca 2+ organelles. Cell Chem Biol 23(6):738–745
67. Mattson MR (2007) Calcium and neurodegeneration. Aging Cell 6:337–350
68. Mattson MP (2004) Pathways towards and away from Alzheimer's disease. Nature 430:631–639

69. Goate A, Chartier-Harlin MC, Mullan M, Brown J, Crawford F, Fidani L et al (1991) Segregation of a missense mutation in the amyloid precursor protein gene with familial Alzheimer's disease. Nature 349(6311):704–706
70. Sherrington R, Rogaev EI, Liang Y, Rogaeva EA, Levesque G, Ikeda M et al (1995) Cloning of a gene bearing missense mutations in early-onset familial Alzheimer's disease. Nature 375(6534):754–760
71. Rogaev EI, Sherrington R, Rogaeva EA, Levesque G, Ikeda M, Liang Y et al (1995) Familial Alzheimer's disease in kindreds with missense mutations in a gene on chromosome 1 related to the Alzheimer's disease type 3 gene. Nature 376(6543):775–778
72. Boulianne GL, Livne-Bar I, Humphreys JM, Liang Y, Lin C, Rogaev E et al (1997) Cloning and characterization of the Drosophila presenilin homologue. Neuroreport 8(4):1025–1029
73. Iijima K, Liu H-P, Chiang A-S, S a H, Konsolaki M, Zhong Y (2004) Dissecting the pathological effects of human Abeta40 and Abeta42 in Drosophila: a potential model for Alzheimer's disease. Proc Natl Acad Sci U S A 101(17):6623–6628
74. Agostini M, Fasolato C (2016) When, where and how? Focus on neuronal calcium dysfunctions in Alzheimer's disease. Cell Calcium 60:289–298
75. Michno K, Knight D, Campussano JM, van de Hoef D, Boulianne GL (2009) Intracellular calcium deficits in Drosophila cholinergic neurons expressing wild type or FAD-mutant presenilin. PLoS One 4(9):e6904
76. Li M, Zhang W, Wang W, He Q, Yin M, Qin X et al (2018 Apr) Imidazole improves cognition and balances Alzheimer's-like intracellular calcium homeostasis in transgenic Drosophila model. Neurourol Urodyn 37(4):1250–1257
77. Polymeropoulos MH, Lavedan C, Leroy E, Ide SE, Dehejia A, Dutra A et al (1997) Mutation in the alpha-synuclein gene identified in families with Parkinson's disease. Science 276(5321):2045–2047
78. Kitada T, Asakawa S, Hattori N, Matsumine H, Yamamura Y, Minoshima S et al (1998) Mutations in the parkin gene cause autosomal recessive juvenile parkinsonism. Nature 392(6676):605–608
79. Valente EM, Abou-Sleiman PM, Caputo V, Muqit MMK, Harvey K, Gispert S et al (2004) Hereditary early-onset Parkinson's disease caused by mutations in PINK1. Science 304(5674):1158–1160
80. Spillantini MG, Schmidt ML, Lee VM-Y, Trojanowski JQ, Jakes R, Goedert M (1997) Alpha-synuclein in Lewy bodies. Nature 388(6645):839–840
81. Feany MB, Bender WW (2000) A Drosophila model of Parkinson's disease. Nature 404(6776):394–398
82. Park J, Lee SB, Lee S, Kim Y, Song S, Kim S et al (2006) Mitochondrial dysfunction in Drosophila PINK1 mutants is complemented by parkin. Nature 441(7097):1157–1161
83. Clark IE, Dodson MW, Jiang C, Cao JH, Huh JR, Seol JH et al (2006) Drosophila pink1 is required for mitochondrial function and interacts genetically with parkin. Nature 441(7097):1162–1166
84. Huang Z, Ren S, Jiang Y, Wang T (2016) PINK1 and Parkin cooperatively protect neurons against constitutively active TRP channel-induced retinal degeneration in Drosophila. Cell Death Dis 7:e2179
85. Romero E, Cha GH, Verstreken P, Ly CV, Hughes RE, Bellen HJ et al (2008) Suppression of neurodegeneration and increased neurotransmission caused by expanded full-length huntingtin accumulating in the cytoplasm. Neuron 57(1):27–40
86. Bezprozvanny I, Hayden MR (2004) Deranged neuronal calcium signaling and Huntington disease. Biochem Biophys Res Commun 322:1310–1317
87. Cepeda C, Ariano MA, Calvert CR, Flores-Hernández J, Chandler SH, Leavitt BR et al (2001) NMDA receptor function in mouse models of Huntington disease. J Neurosci Res 66(4):525–539
88. Hodgson JG, Agopyan N, Gutekunst CA, Leavitt BR, Lepiane F, Singaraja R et al (1999) A YAC mouse model for Huntington's disease with full-length mutant huntingtin, cytoplasmic toxicity, and selective striatal neurodegeneration. Neuron 23(1):181–192

89. Tang T-S, Slow E, Lupu V, Stavrovskaya IG, Sugimori M, Llinas R et al (2005) Disturbed Ca^{2+} signaling and apoptosis of medium spiny neurons in Huntington's disease. Proc Natl Acad Sci 102(7):2602–2607
90. Blackstone C (2012) Cellular pathways of hereditary spastic paraplegia. Annu Rev Neurosci 35(1):25–47
91. Renvoisé B, Blackstone C (2010) Emerging themes of ER organization in the development and maintenance of axons. Curr Opin Neurobiol 20:531–537
92. Verkhratsky A (2005) Physiology and pathophysiology of the calcium store in the endoplasmic reticulum of neurons. Physiol Rev 85(1):201–279
93. Orso G, Pendin D, Liu S, Tosetto J, Moss TJ, Faust JE et al (2009) Homotypic fusion of ER membranes requires the dynamin-like GTPase Atlastin. Nature 460(7258):978–983
94. Trotta N, Orso G, Rossetto MG, Daga A, Broadie K (2004) The hereditary spastic paraplegia gene, spastin, regulates microtubule stability to modulate synaptic structure and function. Curr Biol 14(13):1135–1147
95. Fowler PC, O'Sullivan NC (2016) ER-shaping proteins are required for ER and mitochondrial network organization in motor neurons. Hum Mol Genet 25(13):2827–2837

Chapter 36
Calcium Imaging in the Zebrafish

Petronella Kettunen

Abstract The zebrafish (*Danio rerio*) has emerged as a widely used model system during the last four decades. The fact that the zebrafish larva is transparent enables sophisticated in vivo imaging, including calcium imaging of intracellular transients in many different tissues. While being a vertebrate, the reduced complexity of its nervous system and small size make it possible to follow large-scale activity in the whole brain. Its genome is sequenced and many genetic and molecular tools have been developed that simplify the study of gene function in health and disease. Since the mid 90's, the development and neuronal function of the embryonic, larval, and later, adult zebrafish have been studied using calcium imaging methods. This updated chapter is reviewing the advances in methods and research findings of zebrafish calcium imaging during the last decade. The choice of calcium indicator depends on the desired number of cells to study and cell accessibility. Synthetic calcium indicators, conjugated to dextrans and acetoxymethyl (AM) esters, are still used to label specific neuronal cell types in the hindbrain and the olfactory system. However, genetically encoded calcium indicators, such as aequorin and the GCaMP family of indicators, expressed in various tissues by the use of cell-specific promoters, are now the choice for most applications, including brain-wide imaging. Calcium imaging in the zebrafish has contributed greatly to our understanding of basic biological principles during development and adulthood, and the function of disease-related genes in a vertebrate system.

Keywords Calcium · Development · Genetically encoded calcium indicator · Transgenic · Zebrafish · Embryo · Olfaction · Mauthner · Circuit · Tectum

---

P. Kettunen (✉)
Institute of Neuroscience and Physiology, Sahlgrenska Academy at the University of Gothenburg, Gothenburg, Sweden
e-mail: petronella.kettunen@neuro.gu.se

© Springer Nature Switzerland AG 2020

M. S. Islam (ed.), *Calcium Signaling*, Advances in Experimental Medicine and Biology 1131, https://doi.org/10.1007/978-3-030-12457-1_36

36.1 Introduction

The zebrafish (*Danio rerio*) (Fig. 36.1) has emerged as a widely used model system during the last four decades. Its attractiveness as an experimental system stems from a number of factors that are pushing the zebrafish to the forefront as a model system for biomedical research. The fact that the zebrafish embryo is transparent during its first days, and can stay transparent longer with chemical treatment or as non-pigmented strains (Fig. 36.1b) [1] enables sophisticated in vivo imaging. While being a vertebrate, the reduced complexity of its nervous system and small size make it possible to follow large-scale activity in the whole brain. Its genome is sequenced and many genetic and molecular tools have been developed that simplify the study of gene function. For example, a variety of mutants strains have been isolated in large mutagenesis screens and identification of cell-specific enhancers and promoters has been used for the development of transgenic animals. Moreover, proteins can be easily overexpressed transiently or as stable lines using RNA and DNA injections into the fertilized egg. Similarly, genes can be knocked out using the CRISPR/Cas9 technology, and proteins can be transiently down-regulated during the first 5 days by the use of morpholino oligonucleotides, resulting in so-called morphants [2].

Due to its external and fast embryonic development, the zebrafish initially attracted developmental scientists. With time, the zebrafish has showed usefulness in different areas, including cardiovascular research [3], neurodegenerative diseases [4], psychiatry [5] and cancer research [6]. One quickly growing field using the zebrafish is high-throughput chemical and toxicological screening benefitting from the small size and simplicity of drug delivery through the skin [7]. In addition, various techniques such as electrophysiological recordings, calcium imaging and behavioral tests have demonstrated the convenience of the zebrafish as a model system. Particularly, the combination of in vivo physiological and behavioral techniques in wild-type, mutant or transgenic zebrafish has made it possible to perform studies that would be hard or impossible in other preparations.

The aim of this updated chapter is to give a broad summary of the calcium indicators and labeling techniques used to record intracellular calcium in various zebrafish preparations and review new findings from projects using calcium imaging in the zebrafish since the last edition of this chapter [8].

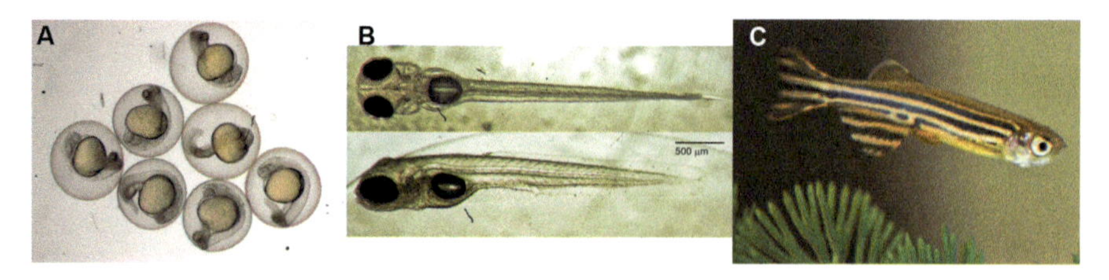

Fig. 36.1 The zebrafish can be studied at various ages. (**a**) Photo of wild-type zebrafish embryos, ~24 h post-fertilization. (Reprinted from Kaufman et al. [169] by permission from Springer Nature, Nature Protocols, Copyright 2009). (**b**) Zebrafish *nacre* mutant larva at ~72 h post-fertilization. (Photo: Todd O. Anderson. (**c**) Adult male zebrafish. Photo: Todd O. Anderson)

36.2 Calcium Indicators

The main advantages of using the zebrafish for calcium imaging are the variety of labeling methods and the small size, enabling in vivo imaging of many different structures in anesthetized or awake animals. Since the mid 90's, the embryonic development and neuronal function of the larval, and later, adult zebrafish have been studied using calcium imaging methods. Calcium indicators can be divided according to their chemical structure, optical properties and means of delivery [9]. Dextran-conjugated indicators, membrane-permeable acetoxymethyl (AM) ester dyes and genetically encoded calcium indicators (GECIs) have all been used in the zebrafish, studying phenomena ranging from the fertilization of the oocyte to neuronal activation in the adult animal. The choice of calcium indicator depends on the desired number of cells to study and cell accessibility. For example, dextran indicators can label a limited number of cells of a cell type that lacks appropriate cell specific promoters, or label cells in a cell lineage during development [10]. In contrast to dextrans, the perfusion of AM esters permits labeling of larger areas of tissue using multi-cell bolus loading [11] or can be used to label isolated organs or cells in culture [12]. During the last decade, expression of GECIs has been replacing the use of AM esters for general labeling of large volumes of cells, since general promoters, such as the neuronal promoter *HuC* (referred to as *elavl3* (ELAV like neuron-specific RNA binding protein 3) throughout the text) can label the whole nervous system [13, 14]. In contrast, the use of cell-specific gene enhancers or promoters can also give expression that is limited to a set of specific neurons at a specific time during development.

36.2.1 Synthetic Calcium Indicators

36.2.1.1 Dextrans

The water-soluble dextran-conjugated calcium indicators have a wide variety of applications, ranging from injection into developing eggs, dialysis into cells or injection into axonal pathways, which leads to local dye uptake with subsequent anterograde and retrograde transport over hours and days. Dextran dyes can fill structures far away, meaning that imaging can be done at a distance from the injection site where the tissue might be damaged and nonspecifically stained, increasing the background fluorescence. In comparison with the AM ester dyes, they show no compartmentalization and have low toxicity.

Calcium green-1 dextran is the first and most common visible light-excitable calcium imaging dye used in the embryonic and larval zebrafish [15]. When injected into the spinal cord it can backfill neurons in the spinal cord, hindbrain and midbrain by severing axonal tracts [16]. Injections of calcium green-1 dextran into eggs at the one-cell stage have helped showing calcium fluctuations during fertilization, cleavage and blastula period [17–21]. If injected embryos are let to develop, the dye is retained in developing cells allowing imaging of those at later developmental stages.

The related dye Oregon green 488 BAPTA-1 has low calcium-binding affinity, high fluorescence yield and great resistance to photobleaching. It has been used similarly to retrograde label neurons and to stain cells following egg injections. It has been used together with Texas red dextran for ratiometric calcium measurements during gastrulation [22]. Both dextrans of calcium green-1 and Oregon green BAPTA-1 has been implemented in in vivo calcium imaging of the adult zebrafish tectum, via delivery to tectal cells via electroporation or tungsten pin injections of dye crystals [23].

36.2.1.2 AM Esters

Cell-permeable AM esters diffuse into cells where endogenous esterases will cleave the ester groups, trapping the indicators inside the cells. AM esters can simultaneously and nonselectively label numerous cells in a tissue. AM esters also serve as a faster means of labeling cells (30 min) than dextran dye backfilling (12–24 h) [24]. AM esters are often mixed with pluronic acid, increasing the solubilization of water-insoluble dyes.

The first attempt to label larger portions of the larval zebrafish spinal cord was done by Brustein and colleagues in 2003, using bolus injections of AM ester calcium dyes [24]. However, this method has since then not been generally used, perhaps due to the parallel development of GECIs at the same time. Oregon green 488 BAPTA-1 AM labeling has on the other hand been very useful when labeling large structures in the tectum [25]. In addition, embryonic hearts and isolated myocytes have been successfully labeled with fluo-4 AM before imaging [26–28]. Fluo-5N AM, an analog of fluo-4 with lower calcium-binding affinity which prevents saturation of signal in cells with high intracellular calcium levels, has been used to image calcium fluctuations in isolated muscle fibers [29]. In the red spectra, rhod-2 AM is often used to label the adult zebrafish olfactory bulb neurons in explant preparations [30].

36.2.2 Genetically Encoded Calcium Indicators

36.2.2.1 Bioluminescent Aequorins

Protein-based calcium indicators can be divided into being fluorescent or bioluminescent, i.e. either emitting light when excited by light, or emitting visible light following a chemical reaction in a living organism. The first bioluminescent calcium-sensitive photoprotein used in the zebrafish was aequorin, derived from the jellyfish *Aequorea victoria* [31, 32]. The binding of calcium ions to the photoprotein starts an enzymatic reaction leading to an emission of blue light. In contrast to fluorescent reporters, aequorin has no background light emission at basal calcium levels and does not require excitation light. Since no input of radiation energy is required, problems with photobleaching, phototoxicity and autofluorescence are avoided.

Recombinant aequorin has been directly injected into the zebrafish egg before fertilization to study calcium patterns during the whole embryogenesis up to ~24 h post-fertilization (hpf), from fertilization to the formation of somites [19, 33]. To allow calcium imaging at later stages, mRNA coding for the native apoaequorin has been injected into the fertilized egg, expressing the fluorescent protein in somites and the trunk as late as 48 hpf [34].

Despite a good signal to noise ratio, the low quantum yield of aequorin has limited its use. However, aequorin naturally exists in complex with green fluorescent protein (GFP), and energy from the chemical reaction of aequorin is transferred to GFP, leading to an emission of green light. This association with GFP is increasing the efficiency of calcium-dependent photoemission from aequorin from 10% to 90%, which inspired the development of a GFP-aequorin fusion protein [35]. Injection of chimeric GFP-aequorin mRNA gives protein expression as early as blastula stage [36] until at least 48 hpf [34]. Zebrafish with GFP-apoaequorin expression driven by the *neuro-β-tubulin* promoter in hypocretin neurons as late as 7 days post-fertilization (dpf) have been used to study neuronal activity during natural behaviors [37]. More recently, a zebrafish line with selective expression of an optimized version of aequorin in motor neurons was combined with treatment of the GFP-aequorin substrate coelenterazine, allowing imaging of calcium transients in freely moving larvae [38]. Finally, injections of mRNA coding for a red form of aequorin, Redquorin, have been used to image calcium transients associated with twitching behavior in developing zebrafish embryos during segmentation [39].

36.2.2.2 GFP-Derived Fluorescent Indicators

Cameleon

A majority of the fluorescent GECIs are derivatives of GFP [40], emitting light in the green spectra. The first GEICs used in zebrafish, cameleon and pericams (a detailed review of the use of pericams in zebrafish is found here [8]), have now mostly been replaced with the GCaMP family of indicators. Cameleon is a hybrid protein in which cyan fluorescent protein (CFP) and yellow fluorescent protein (YFP) are linked by calmodulin and a calmodulin-binding peptide of myosin light-chain kinase (M13) [41]. When calcium levels are increasing, calcium binds to calmodulin, resulting in fluorescence resonance energy transfer (FRET) from CFP to YFP [42]. When excited by a wavelength appropriate for CFP excitation, an increase in calcium concentration causes an increase in the YFP/CFP fluorescence intensity ratio. Thus, cameleon serves as a ratiometric calcium indicator.

Yellow cameleon 2.1 (YC2.1), YC2.12 and YC2.6 have been generally expressed in the zebrafish brain using the *elavl3* promoter, in myocytes using heat shock protein 8A (*hsp8A*) [43] and in spinal neurons using the ISL LIM homeobox 1 (*isl1*) promoter [44]. Injection of YC2.12 mRNA into the fertilized egg gives YC expression between 3 and 33 hpf [42, 45] making it possible to follow calcium changes from gastrulation to the early pharyngula period.

GCaMP Family

In 2001, Nakai and coworkers developed a calcium probe based on a single GFP molecule with high calcium affinity named GCaMP [46]. GCaMP, 1.6, 2, 3, 4.1, 6, 6 s and 7 have all been expressed in zebrafish and the latest studies will be described further on.

GCaMP-HS (GCaMP-*h*yper *s*ensitive) [47] is brighter at the resting level and more sensitive to the change of intracellular calcium concentrations than the previous GCaMP indicators. It has been used in larval zebrafish to record spontaneous activity in motoneurons [47], retinal bipolar cells [48], Mauthner cells and spiral fiber neurons [49], and visual responses in tectum and dorsal telencephalon [50].

SyGCaMP2 is a fusion of GCaMP2 to the cytoplasmic side of synaptophysin, a transmembrane protein in synaptic vesicles [51]. By imaging zebrafish in vivo, it has been demonstrated that SyGCaMP2 can be used to monitor visual activity in synapses of spiking neurons in the optic tectum and neurons in the retina, sampling hundreds of terminals simultaneously [51, 52]. Imaging of SyGCaMP2 in bipolar cell terminals in the retina has also revealed modulation visual processing by olfactory signals [53] and crossover inhibition in the inner retina [54].

To this date, few red-emitting GECIs have been used in the zebrafish, although red indicators allow for deeper imaging into tissue, reduced cytotoxicity, and can be used in combination with the opsins that are activated by blue light. R-GECO-1 and the synaptic form SyRGECO, which is fused to the protein synaptophysin, were tested in the retinotectal system of zebrafish larvae [55]. A group of red GECIs, including RCaMP [56], jRCaMP1a, jRCaMP1b and jRGECO1a [57], has been evaluated after expression in trigeminal neurons. K-GECO1, which is based on a circularly permutated RFP derived from the sea anemone *Entacmaea quadricolor*, has been transiently expressed in spinal Rohon-Beard cells visualizing calcium fluctuations in response to skin stimulation [58].

An alternative way to monitor changes in intracellular calcium levels in freely moving animals has been to make use of the calcium indicator calcium-modulated photoactivatable ratiometric integrator (CaMPARI), which undergoes irreversible green-to-red conversion only when elevated intracellular calcium and UV illumination coincide. In this way, individual cells and brain areas, with elevated calcium during a specific time period, can be identified [59].

In conclusion, depending on the cell-specificity and temporal resolution required, calcium photoproteins can be delivered and used in different ways. Injecting a purified photoprotein, such as aequorin, enables detection of calcium signals already at fertilization, while injection of mRNA or use of some promoters delay the time when the photoprotein is available, but gives a prolonged window of protein expression. Transgenic animals expressing GECIs driven by cell-specific promoters can give precise spatiotemporal patterns. Due to the external development of the zebrafish embryo, the ease of genetic manipulations and the possibility of in vivo imaging, the zebrafish has been a valuable model system for the development of new calcium probes.

36.3 Calcium Imaging Studies in the Zebrafish

36.3.1 Development

The popularity of the zebrafish initially started among the developmental biologists, hence leading to the outcome that the first zebrafish calcium imaging experiments were to study calcium signaling in the zebrafish egg and early embryo [60]. Embryological studies of the zebrafish are particularly rewarding since the developmental process is done *ex utero*. Fertilized eggs are collected from breeding couples in the morning and can then be studied throughout the day, reaching the segmentation period within the first 24 h. The early development of the zebrafish embryo is comparably fast, and the speed of development can be manipulated by increasing or decreasing the temperature of the eggs/embryos. The precise developmental periods have been described by Kimmel and colleagues [61]. When incubated at 28.5 °C the times for the developmental stages are the following: zygote (0–0.75 hpf), cleavage period (0.75–2.25 hpf), blastula period (2.25–5.25 hpf), gastrula period (5.25–10.33 hpf), segmentation period (10.33–24 hpf), pharyngula period (24–48 hpf) and hatching period (48–72 hpf).

Calcium signaling has been studied throughout the whole development of the zebrafish embryo [19, 43, 62, 63]. The zygote and cleavage periods are dominated by intracellular calcium signals, but as the embryonic cell number increases, there is an appearance of localized intercellular signals along with the intracellular ones [18]. This transition proceeds through the blastula and early gastrula periods. Then, as global patterning processes start during the rest of the gastrula period, pan-embryonic intercellular signals associated with the dramatic morphological events of gastrulation can be observed [64]. Once the germ layers and major body axes are established, there is a return to localized intercellular signals associated with the generation of specific structures, for example somite formation, brain partitioning, eye development, and heart formation [33].

Fluctuations in intracellular calcium during **fertilization** and later developmental periods has been studied using injections of recombinant aequorin [19, 65], mRNA of YC2.12 and dextran dyes [8, 17, 20, 66, 67], since imaging would require very early labeling of the egg and embryo. However, the identification of the constitutively active promoter *hsp8A* has made it possible to express YC2.60 and follow calcium signals from fertilization to segmentation [43]. Similarly, the early and ubiquitous activity of the *βactin2* promoter has allowed researchers to drive GCaMP6s expression as early as 30 min post-fertilization (mpf), throughout cleavage and the blastula stage [68].

When monitoring the calcium patterns during fertilization, the oocytes are held in place using an egg injection chamber when carefully injected with the calcium indicator without affecting the normal development of the oocyte [65]. In the zebrafish oocyte, the blastodisc (animal pole) is located on top of the yolk cell (vegetal pole). Activation and fertilization of the oocyte are then done under the confocal microscope to monitor the calcium changes during these processes. The unfertilized zebrafish oocyte exhibits little evidence of calcium signaling, however both activation with water and fertilization trigger rapid increases in intracellular calcium in the oocyte cortex. This fluorescence continues to increase during the first 12–15 min post-insemination (mpi), particularly at the animal pole [17]. The initial transient is followed by a more prolonged transient reaching a higher amplitude with a maximum at 8–10 mpi [17]. These two initial transients are followed by later calcium oscillations (30–60 mpi) associated with blastodisc expansion and cytokinesis [19, 65, 69].

Protein-tyrosine kinase 2-beta (PTK2B/PYK2), a calcium-sensitive protein tyrosine kinase was found to be activated in response to fertilization of the zebrafish oocyte [20]. Oocytes were injected with calcium green-1 dextran and then monitored by confocal microscopy. Fertilization-induced Ptk2b activation could be blocked by suppressing calcium transients via injection of BAPTA as a calcium chelator. Suppression of Ptk2b activity by chemical inhibition or by injection of a dominant-negative construct encoding the N-terminal ERM domain of Ptk2b inhibited formation of an organized fertilization cone and reduced the frequency of successful sperm incorporation.

The zebrafish **cleavage period** is represented by six rapid and synchronical cell cleavages, with one blastomere division approximately every 15 min (Fig. 36.2). The zebrafish embryo undergoes discoidal cleavage, meaning that the blastodisc is located at the animal pole and is divided with cleavage furrows that do not penetrate or divide the yolk. The cells remain interconnected by cytoplasmic bridges. Elevation of intracellular calcium can be seen at the cleavage furrow before the first cell cleavage, i.e. the furrow positioning signal, then during the furrow formation and deepening of the furrow [70]. Next, calcium elevations can be observed along the equators of the dividing cells, where the second cleavage furrow emerges (Fig. 36.2c, d). Thus, it appears that there is a close spatial correlation between elevated calcium and the formation of a cell cleavage furrow [68, 69, 71]. The calcium signals at cleavage furrows [72] and subsequent cell divisions [69] have indeed been prevented by injection of the calcium chelator BAPTA into the embryo, indicating the importance of calcium for cell division [73].

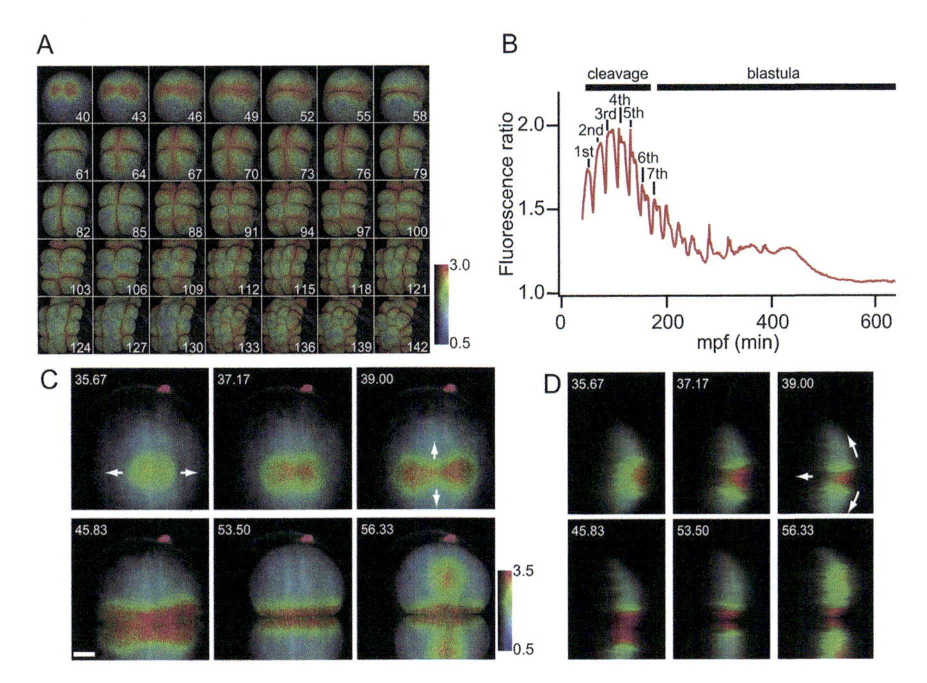

Fig. 36.2 Imaging of localized calcium propagations along the cleavage furrow. Zebrafish embryos are expressing yellow cameleon (YC) 2.60 driven by the promoter for shock protein 8A (*hsp8A*). The numbers on the images indicate minutes post fertilization (mpf). Images are pseudo-color ratiometric z-projections in intensity modulated display (IMD). (**a**) Montage of two-photon images of the calcium responses during the cleavage period. (**b**) Intensity trace of the fluorescence ratio from the whole embryo during two-photon imaging. The time for the 1st to 7th cleavage events are indicated. Bars on the top indicate the cleavage and blastula periods. (**c**) Images from line-scanning microscopy of calcium fluctuations during cleavage. (**d**) Cross sections performed using line-scanning microscopy showing calcium propagations along the cleavage furrow. Arrows in (**c**) and (**d**) indicate directions of calcium propagation. Scale bars are 100 μm. (Reprinted from Mizuno et al. [43])

Recently, Eno et al. showed that *nebel* mutants exhibit reduced furrow-associated slow calcium waves, caused by defective enrichment of calcium stores [66]. The imaging made use of labeling with the ratiometric indicator fura-2 dextran, Oregon green 488 BAPTA-1 dextran, and the transgenic line *Tg(βactin2:GCaMP6s)*.

At the 256-cell stage, the **blastula period** begins with the start of asynchronous cell divisions and zygotic gene transcription. During this period, the yolk syncytial layer (YSL) forms, and epiboly begins. This means spreading and flattening of the blastula to finally cover the whole yolk at the end of epiboly. During this period, there is a transition from intracellular signals to intercellular signals within the blastula. Fast, short-range, and slow, long-range calcium waves propagate exclusively through the external YSL (E-YSL), mainly initiated from the dorsal side (Figs. 36.3, 36.4) [74]. Bonneau and coauthors injected calcium green-1 dextran at the 128-cell stage to study the calcium waves in the YSL. They could conclude that

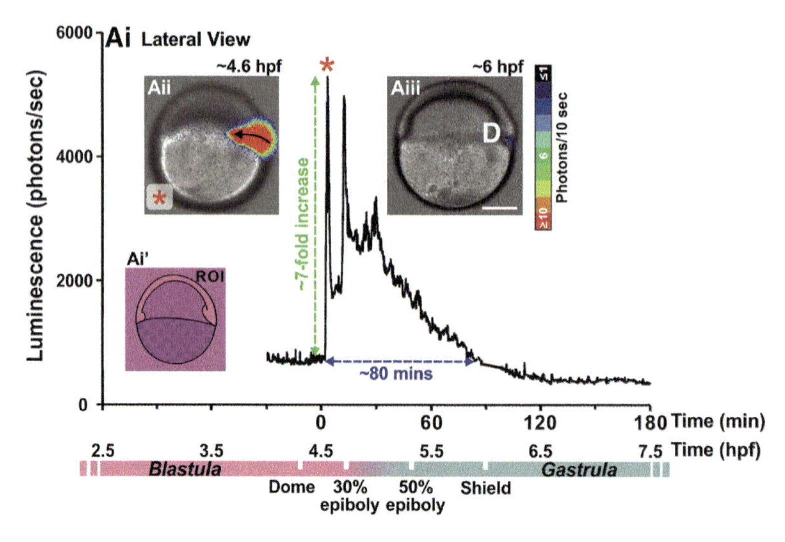

Fig. 36.3 Characterization of calcium signaling in the external yolk syncytial layer during the late blastula and early gastrula periods. (A) An embryo, injected with f-aequorin at the 128-cell stage, was imaged from a lateral view, where the location of the dorsal side remained constant throughout the experimental period. (Ai) Temporal profile of the aequorin-generated luminescence from ~ 4 h post-fertilization (hpf) to 7.5 hpf, collected from the region of interest (ROI) in panel Ai′. The duration of and maximum increase in luminescence of the calcium signals are indicated by the blue and green dashed arrows, respectively. (Aii, Aiii) Luminescence images were superimposed on to the corresponding bright-field images to show the orientation, morphology and calcium signals generated at: (Aii) the beginning of the signaling period (the calcium signal corresponds to peak marked by a red asterisk in panel Ai), and (Aiii) at the shield stage (D is dorsal). Scale bar is 200 μm. (Reprinted from Yuen et al. [74], Copyright 2013, with permission from Elsevier)

phosphorylation of the protein Bcl-2–like 10 (Nrz) enables the generation of YSL calcium waves at the beginning of epiboly [21].

Moreover, localized elevations of calcium, so called calcium spikes, are generated in individual cells or small groups of cells in the blastoderm. These signals are restricted to the enveloping layer (EVL) cells and appear to propagate as calcium waves [18, 43, 68].

During the **gastrula period**, each germ layer (endoderm, mesoderm, and ectoderm) is spatially organized so that organs and tissues can form in the correct location. The morphogenetic movements of involution, convergence, and extension form the epiblast, hypoblast, and embryonic axis through the end of epiboly [22]. The level of intracellular calcium reaches a maximum during early gastrulation (6.5 hpf), when epiboly resumes and the embryonic shield starts to extend towards the animal pole [19]. Fast intracellular calcium waves have been observed around the blastoderm margin during late epiboly, moving up the trunk in an anterior direction [62, 64]. The function of these axial waves are unknown but are thought to contribute to the calcium waves that underlie the large-scale migrations of cells during this period.

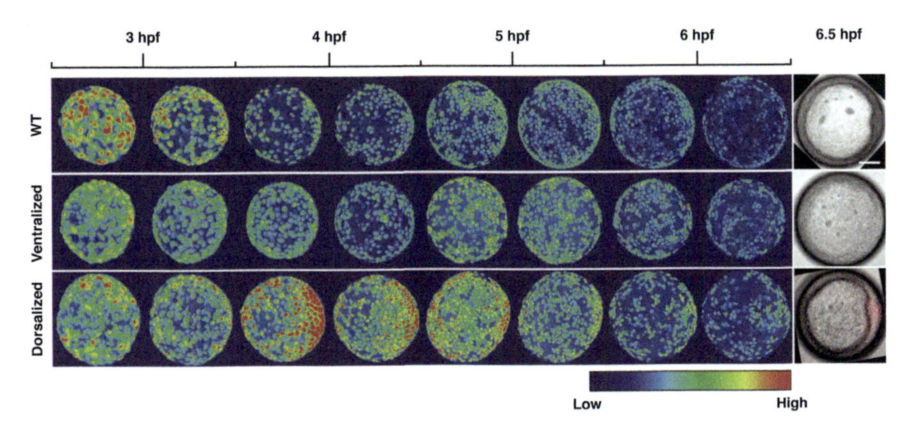

Fig. 36.4 External yolk syncytial layer calcium signaling pattern in ventralized and dorsalized embryos. Recordings of 30-min time-lapse overlay of calcium signals in wild-type, ventralized, and dorsalized *Tg(βactin2:GCaMP6s)* embryos from 2.5 h post-fertilization (hpf) to 6.5 hpf. Images were captured using a spinning-disk confocal microscope, showing that dorsal-biased calcium signaling was diminished in ventralized embryos and was ectopically induced in dorsalized ones. The ventralized phenotype was acquired using *ichabod/β-catenin2* mutant embryos and the dorsalized phenotype by injection of *β-catenin1* mRNA. Scale bar is 150 μm. (Reprinted from Chen et al. [68], Copyright 2017, with permission from Elsevier)

Other calcium release events that are seen during gastrulation are aperiodic transient fluxes found mainly in the EVL and dorsal forerunner cells becoming the Kupffer's vesicle, a structure implicated in laterality [75]. To answer whether an intraciliary calcium signal at the Kupffer's vesicle causes left-right development, Yuan et al. targeted GECIs into cilia via fusion to the ciliary protein Arl13b, a small GTPase [76]. *arl13b-GCaMP6* mRNA was injected into embryos at the one-cell stage and animals were imaged at the 1–4 somite stage. Confocal microscopy of the Kupffer's vesicle revealed the presence of highly dynamic intracellular calcium oscillations. The oscillations depended on polycystic kidney disease 2 (Pkd2) and were left-biased at the Kupffer's vesicle in response to ciliary motility. Suppression of oscillations using a cilia-targeted calcium sink revealed that they were essential for left-right development.

Following the dramatic global rearrangements during gastrulation, the embryo now undergoes a series of more localized morphogenetic movements that make up the **segmentation period**. Somites and the neural cord develop during this period, as well as the primary organs. The earliest body movements can be seen at this time. At 10–11 hpf, distinct calcium patterns can be recognized along the antero-posterior axis of the embryo [19]. High calcium levels can be observed in the presumptive mid- and forebrain in contrast to low calcium in the presumptive hindbrain (Fig. 36.5). This specific calcium pattern in the brain remains clearly visible for several hours and precedes morphological patterning of the brain [19]. Apart from the calcium patterns in the head, various calcium waves, gradients, and spikes can be observed in the trunk and tail region. The most pronounced is an ultraslow calcium

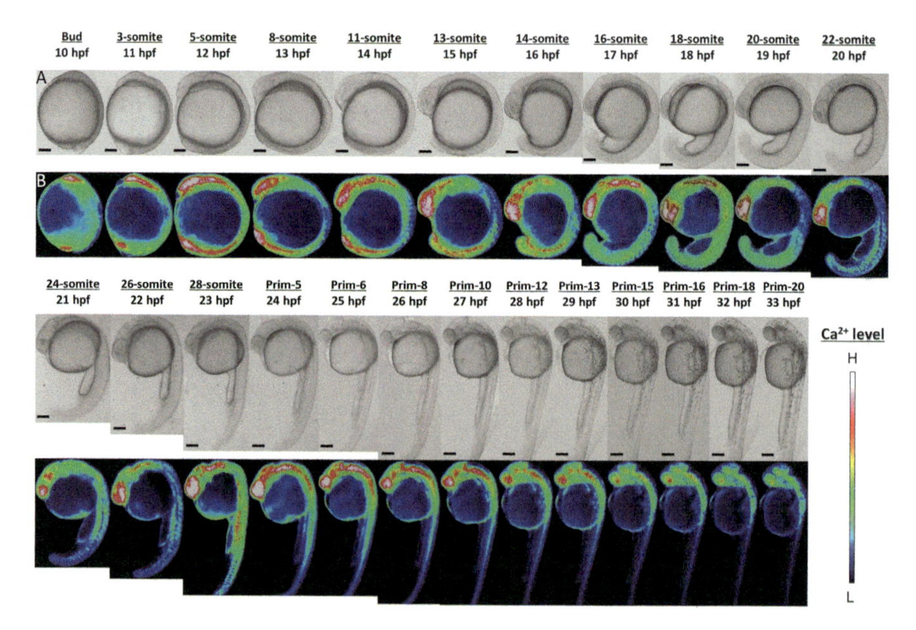

Fig. 36.5 Calcium dynamics in the late gastrula, segmentation, and early pharyngula periods. The genetically encoded calcium indicator yellow cameleon, YC2.12, was expressed in developing zebrafish embryos after mRNA injection at the one-cell stage. Imaging was performed on a fluorescent microscope. (**a**) Bright field image. (**b**) Color-coded image. Scale bar is 200 μm. The color-coded images show high calcium levels as white and low calcium levels as blue. High intracellular calcium levels could be observed in the anterior and posterior body regions from bud to 16-somite (10–17 h post-fertilization) stages. In the anterior trunk, the calcium levels reached a peak at 18-somite stage, whereas in the posterior trunk the calcium peak was present at 28-somite stage. (Reprinted from Tsuruwaka et al. [45])

wave moving posteriorly along with the formation of the somites and neural keel at 10–14 hpf [19].

In one experiment, YC2.12, injected as mRNA at the one-cell stage, was used to study calcium patterns up to the pharyngula period (Fig. 36.5). High levels of intracellular calcium were observed in the anterior and posterior body regions from bud to 16-somite (10–17 hpf) stages. In the anterior trunk, the calcium level reached a peak at 18-somite stage, whereas in the posterior trunk the calcium peak was shown at 28-somite stage [45].

The zebrafish embryo enters the **pharyngula period** at 24 hpf when the body axis begins to straighten. The circulation system develops and the heart starts to beat. The brain has now five distinct lobes and the pharyngula shows tactile sensitivity and uncoordinated movement. The **hatching** embryo (48–72 hpf) continues to form the primary organ systems and the sensory systems are complemented by hair cells and olfactory placodes [61]. The zebrafish is considered a **larva** from 72 hpf to 30 days (Fig. 36.1B) when it grows in size to become an adult animal (Fig. 36.1C). Pigmentation of the zebrafish skin starts around 24 hpf which requires 1-phenyl-2-thiourea treatment to keep embryos transparent, or the use of non-pigmented strains

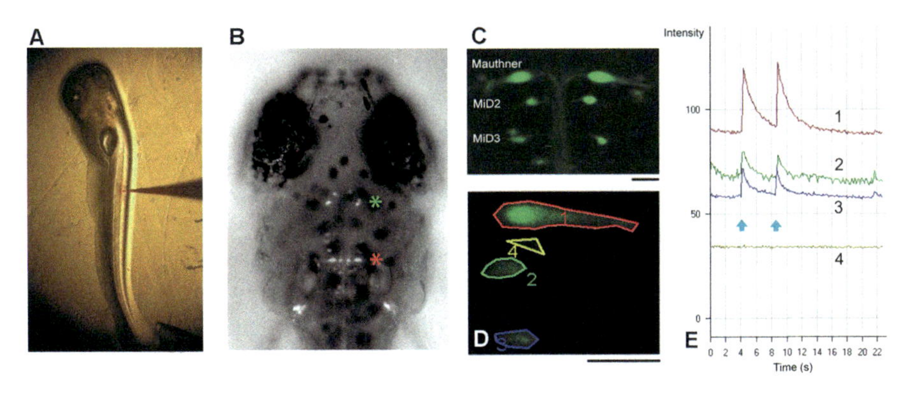

Fig. 36.6 Labeling of reticulospinal neurons in the larval zebrafish and recording of changes in intracellular calcium during the startle response. (**a**) Injection of a dextran dye (here calcium-insensitive rhodamine dextran) into the spinal cord of a larval zebrafish. (**b**) Descending projection neurons labeled with calcium green-1 dextran (10 kDa) in a wild-type larva with normal pigmentation 24 h after injection. The pair of Mauthner cells is marked with a red asterisk and the midbrain neurons with a green asterisk. (**c**) Confocal microscopy image showing Mauthner cells and their homologs labeled with calcium green-1 dextran in a 5-day-old larva. Scale bar is 25 μm. (**d**) The region of interest (ROI) 1 marks one Mauthner cell, ROI 2 marks the closest segmental homolog, MiD2cm, and ROI 3 marks the homolog MiD3cm. Background activity from non-labeled cells is recorded from ROI 4. Scale bar is 25 μm. (**e**) Intensity traces of the changes in intracellular calcium levels from ROIs 1–4 during weak electrical stimulation of the skin. The x-axis indicates time. Two skin shocks to the head are marked by the blue arrows. The y-axis indicates the fluorescence intensity in arbitrary units. The value 0 represents no fluorescence; the value of maximal fluorescence is 256. The numbering of traces corresponds to the ROIs in (**d**). The two skin shocks produced two fast peaks in intracellular calcium in the Mauthner cell and the segmental homologs. Each of the calcium transient lasted about 5 s. (Images: Petronella Kettunen)

like *nacre* [1], *casper* [77] or *crystal* [78]. Calcium imaging of different parts of the nervous system and muscle cells during these later stages will be described in the next sections of this chapter.

36.3.2 Locomotor Circuits

36.3.2.1 Reticulospinal Neurons

At the time when developmental biologists started to monitor calcium signaling events during development, physiologists started to inject dextran-conjugated calcium green-1 into the spinal cord to investigate firing properties of neurons in the spinal circuits and hindbrain [15, 79]. Since then, calcium imaging has been an important noninvasive tool in larval zebrafish to study neuronal activity and connectivity, simultaneously from a population of neurons.

The Mauthner cell network is one of the first circuits in the zebrafish brain that was studied using calcium imaging [15, 79]. The Mauthner cell is a large reticulospinal neuron located in the hindbrain (Fig. 36.6b, c) and responsible for the startle response, elicited by various sensory stimuli such as touch, sound, and visual input. It has two prominent dendrites and its axon crosses over to the contralateral side where it activates motoneurons, contracting the muscles to propel the fish trough the water. Two homologous reticulospinal neurons, MiD2cm and MiD3cm, are located in adjacent segments (Fig. 36.6c) and can elicit a startle response if the Mauthner cell is deleted [80].

Our knowledge of the structure and function of reticulospinal neurons has grown during the last three decades. As these neurons are large and their morphology easily distinguishable, and the fact that their activation leads to a distinct response make this system very useful to functional investigations of genetic modifications. With the discovery of a variety of mutant strains, generated in large-scale genetic screens, the possibilities to learn about the formation and function of the nervous system are unlimited. For example, in the large-scale screen for behavioral locomotive defects performed in 1996, over 150 motility mutants were identified [81]. Interestingly, many of these mutants had defects in the Mauthner cell networks, and calcium imaging of these mutants has taught us much about this circuit [8].

Labeling of reticulospinal neurons via pressure injection of calcium sensitive dextran dyes is still a used method. However, with the discovery of new cell-specific promoters for cells in the Mauthner cell network, this technique will probably be less used with time. For example, the enhancer *Gal4FF-62a* that selectively labels Mauthner cells [82] has been used to drive expression of GCaMP6s in experiments investigating calcium signaling during startle habituation [83] and regulation of the innate startle threshold in 5 dpf zebrafish larvae [84].

Soon, the field of reticulospinal cell imaging expanded to involve networks feeding into the Mauthner cell, regulating its function. For example, investigations were made of the three different inhibitory connections of the Mauthner cell network; recurrent inhibition mediated by an ipsilateral collateral of the Mauthner cell axon, feed-forward inhibition driven by sensory afferents, and reciprocal inhibition between bilaterally opposed Mauthner cells [85]. This inhibition could be confirmed by confocal recordings of calcium signals in Mauthner cells when these inhibitory connections were stimulated.

Spiral fiber neurons are excitatory interneurons, which project to the Mauthner cell axon hillock. Researchers took advantage of the *Tg(–6.7FRhcrtR:gal4VP16; UAS:GCaMP5)* line that has labeled spiral fiber neurons [49]. In these fish, Mauthner cells could be laser-ablated after backfilling using dextran dyes, which revealed that spiral fiber neurons are active in response to aversive stimuli capable of eliciting escapes.

The nucleus of the medial longitudinal fasciculus (nMLF) consists of a small group of reticulospinal neurons in the zebrafish midbrain that is the most rostral of the descending projecting neurons. They have previously been implicated in escape, swimming and prey capture behaviors [16, 86]. One common way to label these cells has been to perform spinal injections of dextran dyes. Sankrithi and O'Malley performed calcium imaging in Oregon green 488 BAPTA-1-labeled 3–5 dpf animals [87]. Calcium responses were monitored simultaneously with recording swimming, turning and struggling movements, elicited by head taps and abrupt illumination.

This type of preparation shows the advantage of using the transparent zebrafish larvae to study which cells are active during certain behaviors, which would be more complicated in other vertebrate systems.

Dye injection was also used by Wang and McLean labeling motoneurons and descending neurons from the nMLF with calcium greeen-1 dextran [88]. Locomotor responses to light via an LED were monitored in both cell groups using calcium imaging and simultaneous motor nerve recordings. Both light onset and offset activated nMLF neurons while mainly ventral (rather than dorsal) motoneurons were consistently activated by this stimulus.

Thiele et al. identified a Gal4 driver fish line, *Tg(Gal4^{s1171t})*, allowing targeted expression of proteins in nMLF neurons [89]. By crossing this fish line with one carrying *UAS:GCaMP6*, it was possible to monitor calcium fluctuations in 6 dpf fish using two-photon imaging while simultaneously monitoring tail movements during spontaneous behaviors with a high-speed camera. The calcium signals in different nMLF neurons were correlated with swimming, flips and struggles, showing that most nMLF cells were active during swimming.

36.3.2.2 Spinal Cord

In parallel with the start of labeling and recording activity in reticulospinal neurons, imaging of the downstream spinal cord neurons started. Each hemisegment of the zebrafish spinal cord consists of three primary motoneurons (PMNs), the caudal primary (CaP), middle primary (MiP), and rostral primary (RoP), and around 30 secondary motoneurons (SMNs). The development of spinal cord networks is partly regulated by spontaneous activity (Fig. 36.7). The first motor behavior performed by zebrafish embryos is the gap junction-driven coiling behavior present at 17–25 hpf. Researchers expressed the constructs *UAS:GCaMP3* and *Gal4^{s1020t}* in the embryonic spinal cord to follow the development of this spontaneous activity using a spinning disc confocal microscope (Fig. 36.7) [90]. *Gal4^{s1020t}* drives expression of the target protein in PMNs and SMNs, as well as in both ascending and descending interneurons. Spinal cord activity imaged at 18 hpf was sporadic, while 2 h later, at 20 hpf, the activity was organized in bursts of alternation between the left and right sides. Optogenetic inhibition revealed changes in functional connectivity between ipsilateral neurons during development.

Using the same transgenic GECI line, Plazas and colleagues followed calcium activity in individual PMNs during axonal extension and pathfinding between 17 and 24 hpf [91]. By suppressing calcium spiking activity selectively in single PMNs, they could observe errors in axon pathfinding of MiP and RoP axons, but not of CaP axons. This activity-dependent competition was mediated by modulation of the chemotropic receptor PlexinA3.

An alternative way to specifically silence synaptic activity was developed by Sternberg and colleagues [92]. A zebrafish-optimized botulinum neurotoxin light chain fused to GFP (*BoTxBLC-GFP*) was expressed in zebrafish embryonic and lar-

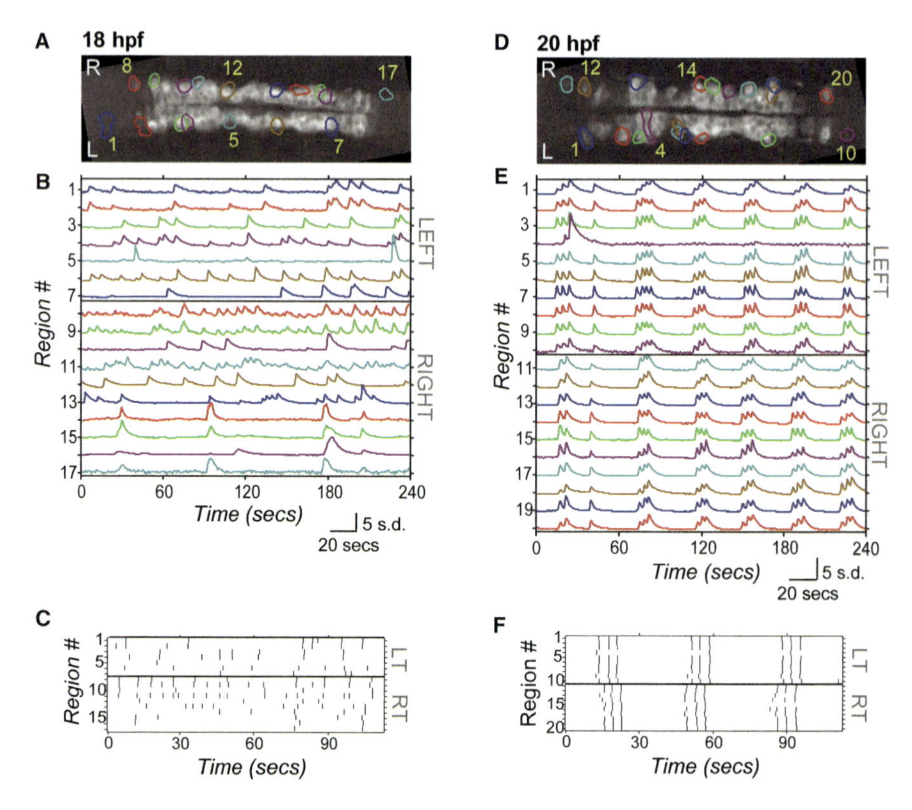

Fig. 36.7 Imaging of spontaneous calcium activity in spinal cord neurons during segmentation. Zebrafish *Tg(Gal4*^{s1020t}*;UAS:GCaMP3)* embryos were imaged at 18 and 20 h post-fertilization (hpf) using a spinning disc confocal microscope. (**a** and **d**) Dorsal views of GCaMP3 baseline fluorescence with circled regions of interest (rostral to the left). (**b** and **e**) Normalized intensity traces for active regions (identified on y-axis) for the left and right sides of the spinal cord at 18 hpf and 20 hpf. (**c** and **f**) Raster plots of detected events for a subsection of the data in (**b**) and (**e**), showing how population activity went from uncoordinated (**b**) to synchronized (**e**) bursting. (Reprinted from Warp et al. [90], Copyright 2012, with permission from Elsevier)

val spinal cord using *Gal4*^{s1020t}. Fish expressing this construct failed at performing the coiling behavior. Interestingly, imaging revealed that calcium signals were still present, showing that the effects of BoTxBLC-GFP were limited to abolishment of synaptic release.

Gal4^{s1020t} was also used to drive expression of GCaMP5 in the spinal cord, in experiments investigating the effect of nonvisual light stimulation of the locomotor circuits via the vertebrate ancient long opsin A (VALopA) [93]. Using illumination at 488 nm, the coiling behavior as well as the calcium activity, monitored via two-photon imaging, could be reduced in 24 hpf embryos after knockdown of *valopa*.

Other promoters/enhancers have been used to drive expression of GECIs in specific subsets of neurons in the spinal cord of zebrafish. For example, the gene trap driver *SAIGFF213A* gives expression in a subset of spinal neurons including the

CaP motoneurons [47]. *Tg(SAIGFF213A;UAS:GCaMP7a)* embryos were used for tracking spontaneous calcium signals using a confocal microscope in CaP neurons at 18 or 24 hpf [94]. In this study, it was found that inhibition of two-pore channels (TPCs) that mediate calcium release from intracellular acidic compartments, prevent the high-frequency calcium transients in CaPs corresponding to network maturation.

It has also been possible to record calcium fluctuations in presynaptic boutons, by the use of the synaptophysin (*syp*) promoter as a driver for GECIs. In the spinal cord, V2a excitatory premotor interneurons are recruited during increase in swimming speed. In a study by Menelaou et al., spinning disc confocal imaging with parallel motor nerve recordings was used to track calcium transients in GCaMP3-labeled synaptic terminals of V2a neurons following electrical skin stimulation [95].

Cerebrospinal fluid-contacting neurons (CSF-cNs) are GABAergic ciliated cells surrounding the central canal in the ventral spinal cord, specifically expressing the polycystic kidney disease 2-like 1 (PKD2L1) channel. Böhm and coauthors investigated the role of CSF-cNs in locomotion by using the zebrafish lines *Tg(pkd2l1:GCaMP5G)icm07* and *Tg(UAS:GCaMP6f;cryaa:mCherry)icm06*, allowing expression of GECIs in this cell type [96]. Two-photon imaging of 5–6 dpf larvae revealed that the CSF-cNs were not activated during fictive escapes elicited by a water jet, which activated motoneurons (labeled with the *mnx1:Gal4, UAS:GCaMP6f;cryaa:mCherry* constructs). On the other hand, CSF-cNs showed a strong activation of in response to muscle contractions during tail bends, which could be abolished in mutants for *pkd2l1*. In the subsequent paper from the group, calcium imaging confirmed that ventral CSF-cNs are recruited during spontaneous longitudinal contractions, allowing the fish to maintain balance during locomotion [97].

Recently, the use of codon-optimized aequorin has simplified the monitoring of spinal calcium signaling [98]. Global bioluminescence from the motoneurons was recorded from *Tg(mnx1:gal4;UAS:GFP-aequorin-opt)* 4 dpf zebrafish larvae during escape behaviors and swimming. Calcium signals from individual neurons were monitored with the *Tg(mnx1:gal4;UAS:GCaMP6f,cryaa:mCherry)* line. The authors showed that mechanosensory feedback enhances the recruitment of motor pools during active locomotion.

36.3.3 Visual System

36.3.3.1 Retina

The zebrafish retina is organized into three nuclear layers: outer nuclear layer (ONL), inner nuclear layer (INL) and ganglion cell layer (GCL) that are separated by synaptic/plexiform layers. The ONL harbors rod and cone photoreceptors. The INL harbors horizontal, bipolar, amacrine and Müller glial cells. The GCL harbors retinal ganglion cells whose axons make up the optic nerve. After being fairly inaccessible to calcium indicator labeling in vivo, the zebrafish retina research has benefitted highly from the development of constructs expressing GECIs in specific retinal cells.

One of the first zebrafish lines used to investigate retinal responses to light onset and offset expressed *SyGCaMP2* by the ribeye a (*ctbp2*) promoter, targeting the expression to ribbon synapses of bipolar cells within the inner plexiform layer. SyGCaMP2 is a fusion protein of synaptophysin and GCaMP2, allowing for calcium measurements at presynaptic synaptic boutons [51]. In a study by Odermatt and coauthors, SyGCaMP2 fluorescence was imaged in retinas in 9–12 dpf zebrafish larvae in response to on- and offset of a LED, investigating the luminance sensitivities of the bipolar cell synapses [99].

The same setup was used for investigation of crossover inhibition between bipolar cells and amacrine cells; the latter labeled by the *ptf1a:gal4;UAS:SyGCaMP3* constructs [54]. In OFF bipolar cells, sustained transmission was found to depend on crossover inhibition from the ON pathway through GABAergic amacrine cells. In ON bipolar cells, the amplitude of low-frequency signals was regulated by glycinergic amacrine cells, while GABAergic inhibition regulated the gain of bandpass signals.

The same group has also investigated the regulation of bipolar cell function by olfactory stimuli in 8–11 dpf larvae [53]. Zebrafish expressing SyGCaMP2 in bipolar cell terminals and GCaMP3.5 in ganglion cells, under the enolase 2 (*eno2*) promoter, were exposed to light flashes and oscillations during a bath application of the amino acid (AA) methionine. Two-photon calcium imaging revealed that olfactory stimulation reduced the gain and increased the sensitivity of responses to luminance and contrast transmitted through OFF bipolar cells. Activation of dopamine receptors increased the gain of synaptic transmission in vivo and potentiated synaptic calcium currents in isolated bipolar cells. These results indicate that olfactory stimuli alter the sensitivity of the retina through the dopaminergic regulation of presynaptic calcium channels that control the gain of synaptic transmission through OFF bipolar cells.

Tg(Isl2b:Gal4;UAS:SyGCaMP3) transgenic larvae were used to investigate the function of the cell-adhesive transmembrane protein Teneurin-3 (Tenm3) in structural and functional connectivity of retinal ganglion cells [100]. *tenm3* knockout animals showed mistargeting of dendritic processes and laminar arborization defects of these cells. Functional analysis of retinal ganglion cells targeting the tectum revealed a selective deficit in the development of orientation selectivity after *tenm3* knockdown.

The ribeye promoter was also used in an experiment, to express GCaMPHS in bipolar cell of zebrafish larvae (Fig. 36.8) [48]. Via optogenetic activation of the bipolar cell axon terminals, the authors could elicit calcium waves traveling over the retina.

Plasticity of the excitatory synapses formed by bipolar cells on retinal ganglion cells has been investigated by the combination of perforated whole-cell recordings and two-photon calcium imaging of 4 dpf transgenic *Tg(Gal4-VP16^{xfz43},UAS:GCaMP1.6)* larvae [101]. The driver *Gal4-VP16^{xfz43}* allows for labeling of a subset of bipolar cells in the retina [102]. Both repeated electrical

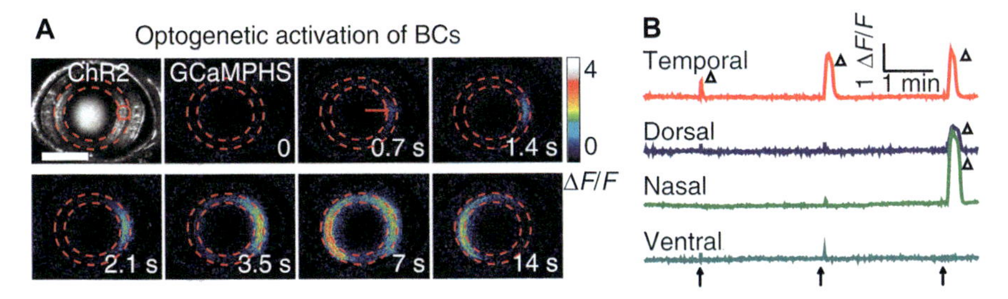

Fig. 36.8 Recording of retinal calcium waves induced by optogenetic activation of retinal bipolar cells. (**a**) Time-lapse images showing a retinal calcium wave evoked by optogenetic stimulation of a cluster of Channelrhodopsin-2 (ChR2) expressing bipolar cell terminals (red circle in first picture) in a transgenic *Tg(ctbp2:ChR2-CFP,Gal4-VP16xfz43,UAS:GCaMPHS)* larva. For optogenetic stimulation during calcium imaging, a 0.5 s pulse of 440 nm laser was activating the axon terminals, while calcium imaging data was simultaneously collected with a 900 nm or 488 nm laser using a two-photon microscope. Scale is 100 μm. (**b**) Calcium intensity traces during optogenetic stimulations of retinal bipolar cells in four different locations of the retina. Arrows mark repetitive optogenetic stimulations that induced calcium waves (triangles). (Reprinted from Zhang et al. [48])

and visual stimulations induced long-term potentiation (LTP) of bipolar cell synapses in larval but not juvenile zebrafish. LTP induction required the activation of postsynaptic N-methyl-D-aspartate (NMDA) receptors, and its expression involved arachidonic acid-dependent presynaptic changes in calcium dynamics and neurotransmitter release [101].

Finally, GCaMP3 has been selectively expressed in cone photoreceptors, by the use of the alpha subunit of cone transducin promoter (*TαCP*) [103]. Using this line, the authors analyzed whether photoreceptor degeneration induced by mutations in the phosphodiesterase-6 (*pde6*) gene was driven by excessive intracellular calcium levels. GCaMP3-expressing 5–6 dpf larvae were analyzed using multiphoton laser scanning. The time-lapse imaging monitored both changes in calcium transients and in cell shape every 20 min in a session that extended up to 9 h. In contrast to laser-killed cones, the ones in *pde6* mutants did not show elevated levels of intracellular calcium normally present at cell death [104].

36.3.3.2 Tectum

The zebrafish retinotectal system is responsible for converting moving visual inputs to the appropriate motor outputs, thus analogous to the mammalian superior colliculus [105]. The embryonic/larval tectum is located just beneath the skin and is therefore easily accessible to imaging and manipulations (Fig. 36.9a). The topographic map of the visual field present in the retina is conveyed into the tectum. Each retinal ganglion cell axon is targeted to a single lamina, and arborizes exclusively in this lamina. Information flows primarily from the superficial to the deeper layers in the tectum (Fig. 36.9b). In the deeper neuropil layers, information

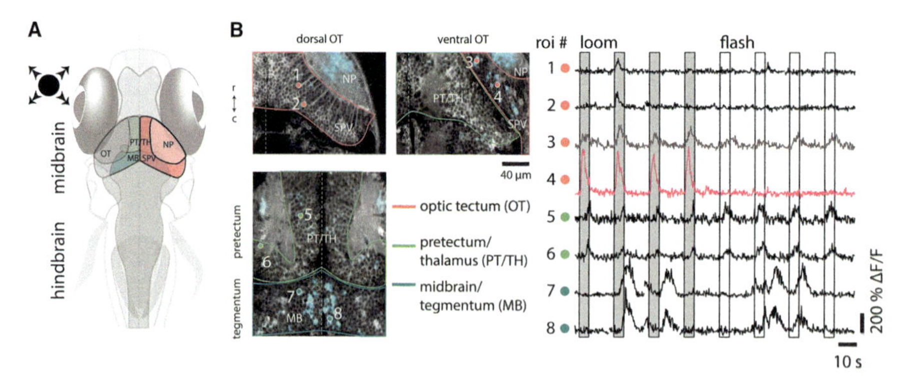

Fig. 36.9 Imaging calcium responses to visual stimuli from the tectum. (**a**) Schematic image of the brain regions processing visual information in larval zebrafish: the optic tectum (OT) with its neuropil (NP) and cell body layers (stratum periventriculare, SPV), the pretectum/thalamus (PT/TH), and the midbrain tegmental region (MB). (**b**) Neurons that responded to a looming visual stimuli, as well as to light flashes, were investigated in 5–6 days post-fertilization *Tg(elval3:GCaMP5G)* larvae using two-photon microscopy. Stimuli were presented onto a screen underneath the fish using a digital light processing (DLP) projector. The left picture shows the three brain regions imaged, i.e. the dorsal and ventral OT, the PT/TH, and the MB, where the recorded calcium activity is marked with blue. The regions of interest (ROIs) are marked as numbered circles, and they correspond to the time traces in the right diagram. Both neurons in the dorsal OT (1 and 2) and ventral OT (3 and 4) responded to looming stimuli. Neurons in the PT/TH (5 and 6) responded to both looming and flashed stimuli. Neurons in MB (7 and 8) acted spontaneously and not to the visual stimuli. (Reprinted from Dunn et al. [170], Copyright 2016, with permission from Elsevier)

is transmitted from the axons of interneurons to tectal projection neurons reaching premotor areas in the midbrain and hindbrain [106].

The first recordings of calcium activity in the larval zebrafish tectum were done after bolus labeling of Oregon green 488 BAPTA-1 AM into the tectal neuropil at 60 hpf to 9 dpf prior to visual stimulation using a miniature LCD screen [25]. Larvae were not anesthetized during imaging, as the agarose was sufficient to restrain them and prevent eye movements. A two-photon microscope collected images of visually evoked calcium fluctuation in tectal neurons, showing that zebrafish receptive field properties could be determined, such as visual topography, receptive field width, and direction and size selectivity [25].

Since then, this type of setup has been used in numerous studies investigating visual processing of diverse visual inputs, including moving dots, shadows, bars and the on- and offset of LEDs. However, today, the use of GECIs has almost entirely replaced the use of synthetic dyes for imaging of the tectum and therefore, the rest of this section will focus on research using GECIs. Due to the possibilities to combine calcium imaging of either individual cell types or entire brains in larval zebrafish in response to visual information, and to track the subsequent behavior, the studies of the visual system in zebrafish has grown exponentially during the last couple of years.

Researchers have investigated how visual information is encoded in the population activity of retinal ganglion cells. To address this, they used a genetically encoded reporter of presynaptic function (SyGCaMP3) to record visually evoked activity in the population of retinal ganglion cell axons innervating the zebrafish tectum [107]. Calcium imaging was performed using 7 dpf *Tg(isl2b:Gal4;UAS:SyGCaMP3)* zebrafish larvae with SyGCaMP3 expression in retinal ganglion cell axons within the tectal neuropil. Fish were presented light or dark drifting bars that were moving in 12 directions to one eye. Visually evoked SyGCaMP3 responses were recorded in the contralateral tectum. Analysis of SyGCaMP3 signals identified three subtypes of direction-selective and two subtypes of orientation-selective retinal input.

The same group then further studied the direction-selective and orientation-selective cells in the tectum in response to moving dark bars [108]. Imaging was performed using Oregon green 488 BAPTA-1 AM as well as the transgenic lines *Tg(Isl2b:Gal4; UAS:SyGCaMP3)* and *Tg(elavl3:GCaMP5)*, confirming that direction-selectivity is established in both the retina and tectum.

Gabriel et al. also used the *Tg(elavl3:Gal4;UAS:GFP)* line to better understand the direction-selective neurons in the tectum [109]. The team also developed new driver lines for tectal labeling with reduced labeling of retinal afferents. The two lines, *Tg(Oh:G-3)* and *Tg(Oh:G-4)*, are based on the gene orthopedia a (*otpa*) and a heat shock basal promoter fused to *Gal4VP16*. The driver lines were crossed with the *Tg(UAS:GCaMP3)* line to monitor calcium spikes in cell types of opposite directional tuning. A few years later, the group produced the *Tg(Oh:GCaMP6s)* transgene, used to map the activity of a group of superficial interneurons (SINs), showing exhibited distinct size-tuning properties [110].

Visual reflexes, such as the optomotor response, which is an orienting behavior evoked by visual motion, have been studied using calcium imaging in the zebrafish larva. In one study, the whole brains of 5–7 dpf *Tg(elavl3:GCaMP5G)* and *Tg(elavl3:GCaMP2)* zebrafish were scanned when the fish were exposed to moving gratings [111]. The fictive motor behavior was mapped simultaneously using motor nerve recordings. Interesting, the optomotor response was found to be processed by diverse neural response types distributed across multiple brain regions.

Prey capture behavior serves as a meaningful behavior to study in larval fish, as the visual stimulus and subsequent behavior are possible to induce in the lab by showing small moving objects to the test fish. A number of studies have investigated various aspects of prey capture behavior. In one of them, the *Tg(isl2b:Gal4, UAS:GCaMP6s)* line was used to label tectal cells in 8 dpf larvae [112]. The fish were restrained in agarose and were shown white dots on a black background. Two-photon calcium imaging revealed that a small visual area, AF7, was activated specifically by the optimal prey stimulus. This pretectal region was found to be innervated by two types of retinal ganglion cells, which also send collaterals to the optic tectum. The behavioral test was also performed with real paramecia, and laser ablations of AF7 neuropil confirmed the involvement of the brain region in prey capture behavior.

36.3.4 Olfaction

One of the few neural systems that has regularly involved calcium imaging of adult zebrafish is the olfactory system, due to the small size, transparent nature, and accessibility of the adult olfactory bulb for labeling, imaging and neuronal stimulation. In the nasal cavity, olfactory receptor neurons (ORNs) are stimulated by odors, and send a single unbranched axon to the first relay station in the central nervous system, the olfactory bulb (OB). ORN axons synapse onto mitral cells (MCs) and local interneurons in the olfactory glomeruli [113]. Each glomerulus receives convergent input from sensory neurons expressing the same odorant receptor [114]. MC axons exit the OB and project to several higher brain areas, such as the dorsal telencephalic area Dp, the homolog of olfactory cortex [115].

Initially, dextran and AM dyes were used to label neurons in the olfactory system, but with the development of transgenic lines expressing GECIs, these have been combined with synthetic dyes to allow for cell-specific labeling, activation and monitoring of cellular activity. Electrophysiology, as well as optogenetics [116] have successfully been used to map the functional connectivities in the networks processing olfactory information.

An explant preparation of the intact zebrafish brain and nose has been used extensively [117], allowing for two-photon imaging of different cell types in response to olfactory activation [118]. Olfactory stimuli such as AAs, bile acids or food extracts are applied to the naris of one olfactory epithelium. In this preparation, it is possible to monitor patterns of presynaptic activity in the OB and glomeruli induced by repeated applications in the same animal. In preparations where the telencephalon is spared [118], it is possible to record odor-generated activity in the telencephalic targets.

Calcium imaging from the zebrafish OB has shown that AAs and bile acids stimulate different parts of the OB [119]. AAs induce complex patterns of active glomerular modules that are unique for different stimuli and concentrations. Interestingly, the similarity of odorant-induced activity patterns is the highest for AAs that are chemically close.

Researchers used the explant preparation of the nose and brain from adult zebrafish to investigate the role of dopamine for olfactory signaling. The olfactory sensory neurons were loaded with Oregon green 488 BAPTA-1 dextran, and the preparation was exposed to amino acids. Here, dopamine had no effect on odorant-evoked calcium signals in sensory axon terminals. Next, neurons in the OB were loaded with rhod-2 AM by bolus injection in *Tg(elavl3:YC2.1)* zebrafish, a transgene where MCs are labeled with YC2.1. Electrophysiology showed that direct exposure by dopamine caused MCs cells to hyperpolarize. However, multiphoton calcium imaging of MCs showed that the mean response to odorants increased slightly, but not significantly, in the presence of DA [120]. In the following publication from the group, the role of dopaminergic modulation of the telencephalic area

Dp was investigated using bolus loading of rhod-2 AM into the tissue. Multiphoton calcium imaging showed that population responses of Dp neurons to olfactory tract stimulation or odor application were enhanced by dopamine [121].

Calcium activity patterns evoked by repeated odor stimulation were measured in the adult OB and Dp. Recordings of odor-evoked activity patterns of MCs in the OB were performed using the *Tg(elavl3:GCaMP5)* line while calcium responses in Dp was measured by bolus loading of rhod-2 AM. Whereas odor responses in the OB were highly reproducible, responses of Dp neurons were variable and reduced over trials. This data indicates that odor representations in higher brain areas are continuously modified by experience [122].

A group of GABAergic interneurons in the OB expresses the homeobox genes *dlx4* and *dlx6*. Zhu et al. combined electrophysiology with imaging of the dlx4/6 neurons by using the zebrafish line *Tg(dlx-itTA; Ptet-GCaMP2)*. In the ex vivo preparation, AA odorants or food extracts evoked widespread calcium changes throughout the glomerular layer. Individual dlx4/6 neurons responded to subsets of odorants and different odorants stimulated distributed, partially overlapping combinations of dlx4/6 neurons. Increasing odorant concentration enhanced responses and recruited additional dlx4/6 neurons. To examine how dlx4/6 neurons shape OB output during an odor response, the authors used optogenetics to silence dlx4/6 neurons while recording AA-stimulated MCs, labeled with rhod-2 AM. The inhibition of dlx4/6 cells suppressed the response of MCs to low odor concentrations, but enhanced their response to high concentrations [123].

Prostaglandin $F_{2\alpha}$ ($PGF_{2\alpha}$) is involved in reproductive behavior in fish, and it was found that it specifically activates two odorant receptors, or114-1 and or114-2 [124]. Imaging experiments were performed in the ex vivo preparation of the intact brain and nose from adult male zebrafish that were carrying the promoter for the olfactory marker protein (*omp*) which drove expression of *GCaMP7* or *GCaMP-HS* in ciliated olfactory sensory neurons. $PGF_{2\alpha}$ exposure gave rise to calcium signals in two ventromedular glomeruli in the OB. A low concentration of $PGF_{2\alpha}$ induced a calcium increase in one of the two glomeruli, while higher concentrations activated both glomeruli. Mutant male zebrafish, deficient in the high-affinity receptor or114-1, exhibited loss of attractive response to $PGF_{2\alpha}$ and impaired courtship behaviors toward female fish [124].

Labeling of ciliated olfactory sensory neurons by *Tg(omp:Gal4FF;UAS:GCaMP7)* has also been used in experiments aiming at understanding the signaling of the adenosine receptor A2c in the OB. Calcium imaging experiments of the nose-attached intact brain from adult male fish confirmed that adenine nucleotides (ATP, ADP and AMP) and adenosine, but not other nucleotides, nucleosides, AAs, or bile acids, specifically activated a single, large, identifiable glomerulus named lG2. The ATP- and alanine-induced neural activation in higher brain centers was investigated by c-Fos *in situ* hybridization. ATP and alanine applications resulted in significant increase in the number of c-Fos-positive neurons in multiple brain areas [125].

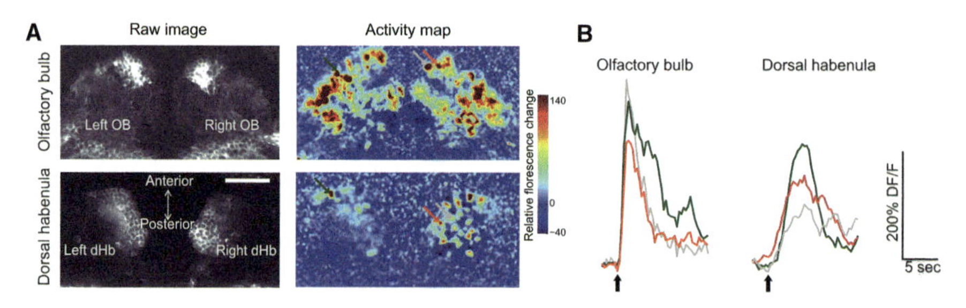

Fig. 36.10 Calcium signals in the olfactory bulb and dorsal habenula in response to odor stimulation. (**a**) Raw images and pseudo-colored food odor-evoked activity maps of the olfactory bulb and dorsal habenula recorded in 25-day-old *Tg(elavl3:GCaMP5)* zebrafish using two-photon microscopy. (**b**) Intensity traces of calcium signals in the two regions in response to food odor (black arrows). Colors of the traces correspond to the color of the arrows in (**a**). (Reprinted from Jetti et al. [126], Copyright 2014, with permission from Elsevier)

Due to its small size, the olfactory networks of larval and juvenile zebrafish can be imaged in the intact animal (Fig. 36.10). In most of these experiments, different versions of GCaMP have been expressed using the *elavl3* promoter, as it labels the MCs. In order to better understand the function of the habenular circuits and how habenulae process olfactory information, Jetti et al. measured and compared odor responses in the OB and the dorsal habenula (dHb) using two-photon calcium imaging, in 3- to 4-week-old *Tg(elavl3:GCaMP5)* zebrafish (Fig. 36.10) [126]. The fish were exposed to a large panel of odors, i.e. skin extract, urea, bile acid mixture, nucleotide mixture, chondroitin-6-sulfate, trimethylamine-n-oxide (TNO), putrescine, AA mixture, and food odor. Odor responses in the dHb were asymmetric and were primarily in the right dHb, unlike the symmetric odor responses in the OB. Interestingly, the odors evoked responses with different amplitudes in the OB, but amplitudes and distribution of dHb odor responses were indifferent to the stimulus strength as measured by OB activity.

Tree- to five-week-old larvae carrying the *Tα1:GCaMP2* construct were used to monitor odor-evoked neuronal activity in the OB related to fear responses. Exposure of odorants to the olfactory pits revealed that purified skin extracts and purified shark chondroitin sulfate activated one region of the OB, while bile acids and AA activated other regions [127].

To test the hypothesis that that zebrafish perceive MHC peptides as odorants, 6 dpf *Tg(elavl3:GCaMP2)* zebrafish larvae were exposed to a mixture of different MHC peptides. Multiphoton imaging was used to detect the odor-evoked calcium signals in the OB. Basal indicator fluorescence was higher in the mitral cell layer than in deeper, interneuron layers. Upon stimulation with the MHC peptides, fluorescence changes were observed in small, scattered populations of neurons and neuropil regions in the MC and interneuron layers [128].

OlfCc1 is a C family G-protein-coupled receptor that is expressed in a large population of microvillous sensory neurons in the zebrafish olfactory epithelium. It is a calcium-dependent, low-sensitivity receptor specific for the hydrophobic AAs isoleucine, leucine, and valine. DeMaria used three different promoters to drive expression of *GCaMP1.6* in the olfactory system of zebrafish larvae: *omp* for expression in ciliated olfactory sensory neurons, transient receptor potential cation channel C2 (*trpc2*) for expression in microvillous olfactory neurons, and *elavl3* for pan-neuronal expression in excitatory neurons throughout the CNS, including the OB. A confocal microscope was used to monitor activity to food extract, AA mix; bile acid mix and amine mix in 4.5 dpf animals. Morpholino knockdown of *olfcc1* demonstrated its importance in detecting various types of AAs [129].

Recently, researchers investigated the role of beating cilia in the nose of zebrafish in modulating the olfactory response. They made use of the *schmalhans* (*smh*) mutant line that displays ciliary motility defects. Fluorescently labeled food odor was delivered by a capillary, and odor flow dynamics and neural responses to odors were imaged using two-photon microscopy in 4 dpf wild-type and mutant *Tg(elavl3:GCaMP6s)* zebrafish. The experiments showed that beating cilia in the nose of zebrafish attract odors to the nose pit and facilitate detection of odors by the olfactory system. Flow fields from cilia facilitate odor detection and elicit odor responses at the ORNs and the OB. On the contrary, *smh-/-* zebrafish failed to attract odors to the nose and showed no detectable odor responses in ORNs or in the OB [130].

36.3.5 Brain-Wide Imaging

The small size and the transparency of the larval zebrafish allows for imaging experiments that are generally not feasible in many other vertebrates, i.e. the possibility to image activity of the entire nervous system in an awake animal while it is responding to sensory stimuli. The development of GECIs, in particular the GCaMP family of indicators, expressed by the generally expressed *elavl3* promoter has given rise to an exponentially increasing number of complex studies, often combined with new imaging techniques, such as light-sheet microscopy (Fig. 36.11) [131–136]. Just a few examples of studies performed using different forms of brain-wide imaging in transgenic zebrafish lines include monitoring of visually-induced brain activity during the optomotor [111] and optokinetic [137] responses, as well as brain activity during pharmacologically induced seizures [138, 139]. Selective plane illumination microscope (SPIM) imaging of *Tg(elavl3:H2B:GCaMP6f)* 6 dpf larvae has revealed brain regions and individual cells responsive for auditory processing [140]. Moreover, calcium imaging has successfully been performed in freely moving animals undergoing swimming [141] and prey capture behavior (Fig. 36.12) [142–144].

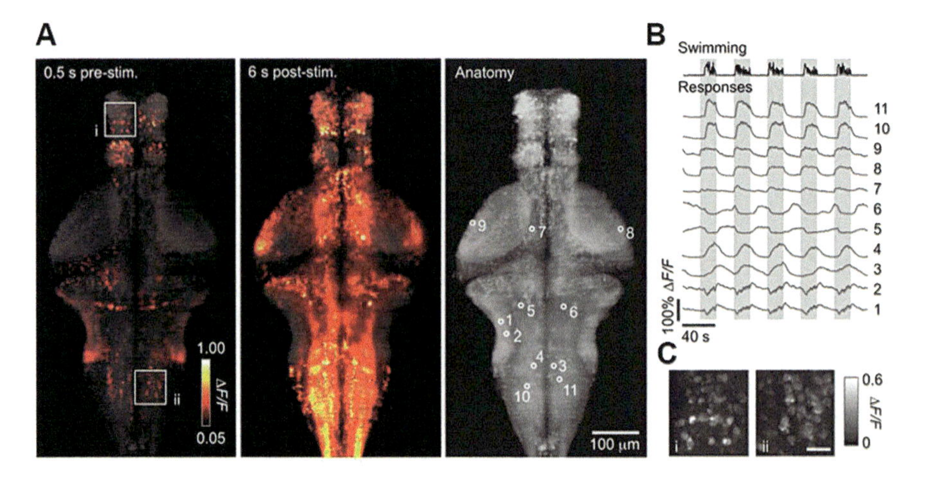

Fig. 36.11 Brain-wide calcium imaging of the optomotor response using light-sheet microscopy. A custom-built light-sheet microscope was used to image the full brains of 5–7 days post-fertilization *Tg(elavl3:GCaMP6s)* zebrafish larvae. The authors induced the optomotor response (OMR) in which swimming was elicited by visual gratings moving in the tail-to-head direction. Two extracellular suction electrodes recorded fictive swimming from the tail. (**a**) Maximum-intensity projections of ΔF/F over the entire volume before and after onset of grating movement averaged over 24 trials. The localizations of regions of interest (ROIs) are seen to the right. (**b**) Intensity traces of single neurons or patches of neuropil ROIs during fictive behavior. (**c**) Magnification of regions boxed in (**a**), showing single-neuron resolution. Scale bar is 20 μm. (Reprinted from Vladimirov et al. [136], Copyright 2014)

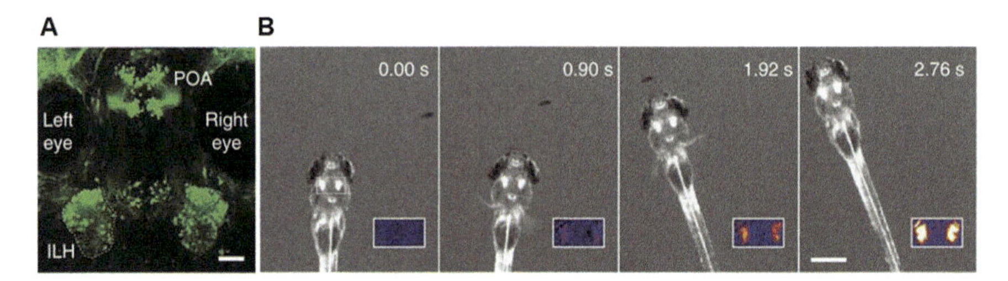

Fig. 36.12 Calcium imaging of freely swimming zebrafish larvae during prey capture behavior. (**a**) Labeling of the inferior lobe of the hypothalamus (ILH) of a 5 days post-fertilization (dpf) larva resulting from crossing the *Gal4* driver line *Tg(hspGFFDMC76A)* with *Tg(UAShspzGCaMP6s)* fish. POA: preoptic area. Scale bar is 50 μm. (**b**) ILH activation during prey capture behavior in a previously unfed 4 dpf larva. The imaging was performed using an epifluorescent microscope equipped with a complementary metal–oxide–semiconductor (CMOS) camera. Images show frames from a time-lapse recording during the behavior. Insets: Pseudo-color images of the frame divided by an averaged frame. Scale bar is 500 μm. (Reprinted from Muto et al. [144])

36.3.6 Muscle

Excitation–contraction coupling is the process that regulates contractions by skeletal muscle. A change in membrane voltage activates release of calcium from the sarcoplasmic reticulum (SR) via activation of ryanodine receptors (RyRs) type 1 or 3, leading to the initiation of muscle contraction via shortening of the myocyte myofilaments [145]. Although it has technically been possible to perform in vivo imaging of muscle fiber in the zebrafish to study this process, it has not commonly been done. One reason could be the ease of performing patch clamp recordings of muscle fibers in vivo [146]. In the cases when calcium signals have been recorded from myocytes, incubation with AM esters has been used, as well as injections of dextran dyes at the one-cell stage, or labeling using GECIs.

Lin et al. investigated the role of *microRNA 3906* (*miR-3906*) and *homer 1b* for muscle fiber differentiation (Fig. 36.13) [147]. They injected one-cell zebrafish embryos with a solution containing calcium green-1 and rhodamine dextran that was

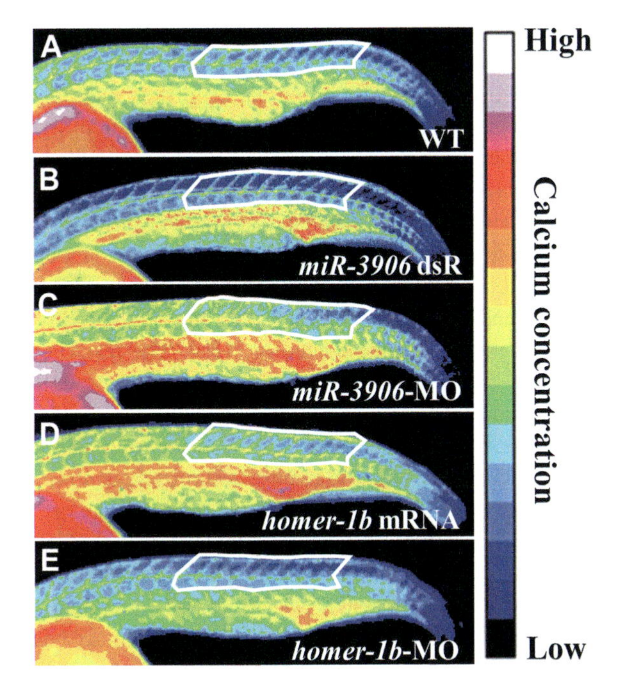

Fig. 36.13 Calcium imaging of myocytes in vivo. The contributions of the gene *homer1b* and microRNA *miR-3906* to development of calcium dynamics in zebrafish myocytes were investigated. A solution containing calcium green-1 and tetramethylrhodamine dextran was mixed with either: (**a**) buffer, (**b**) double-stranded *miR-3906* dsR, (**c**) *miR-3906* morpholino, (**d**) *homer1b* mRNA or (**e**) *homer1b* morpholino and was microinjected individually into one-cell zebrafish embryos. The fluorescent image of each embryo was captured by a fluorescent microscope at 24 h post-fertilization. The region of interest consisting of 11–20 trunk somites from where the calcium levels were recorded is marked with a white line. (Reprinted from Lin et al. [147])

mixed with morpholinos or RNA for either *miR-3906* or *homer 1b* to knock down or overexpress these in vivo. The fluorescent image of each embryo was captured at 24 hpf revealed that either *homer1b* overexpression or *miR-3906* knockdown increased the calcium concentration within muscle cells.

The role of RyRs has been investigated in zebrafish muscle. Klatt Shaw and coauthors made use of the RyR inhibitor azumolene and the RyR1 agonist 4-chloro-m-cresol (4-CmC) to manipulate RyR1 signaling [148]. Developing *Tg(act2b:GCaMP6f)* embryos were incubated in vehicle or drug from the gastrula stage (6 hpf) and onward. At 24 hpf, muscle contractions were electrically stimulated and calcium fluxes associated with muscle contraction were recorded. Under these conditions, azumolene acted in a dose-dependent manner to reduce calcium fluxes associated with muscle stimulation, whereas 4-CmC potentiated RyR activity, increasing both the duration and maximum levels of released calcium.

In another experiment, the role of RyR3 in calcium signaling and muscle function was investigated using morpholino knockdown [149]. Calcium sparks from the SR were monitored using confocal imaging of fluo-4 AM-loaded isolated tail myocytes from 72 hpf larvae. The bathing solution included caffeine to stimulate SR calcium release. It was found that the number of calcium sparks was reduced in the morphants, while the muscle function remained normal.

Two-pore channel type 2 (TPC2) is regulating differentiation of slow muscle fibers, and TPC2 has been implicated in intracellular calcium signaling as it has been suggested that the channel can mediate calcium release from internal stores. Researchers approached this theory by pharmacologically inhibiting, knocking down and overexpressing *tpc2* in zebrafish *Tg(α-actin-apoaequorin-IRES-EGFP)* [150]. Active aequorin was then reconstituted in vivo, by incubating transgenic embryos from the 4- to 64-cell stage to the ~16-somite stage with f-coelenterazine. Embryos were then transferred to a photomultiplier tube (PMT)-based luminescence detection systems for luminescence detection [150]. The resulting data indicated that localized calcium release via Tpc2 might trigger the generation of more global calcium release from the SR via calcium-induced calcium release.

Xiyuan and coauthors investigated the role of the nitric oxide-soluble guanylyl cyclase (NO-sGC) pathway on calcium release and muscle contraction in zebrafish [151]. Isolated skeletal muscle cells from 5–7 dpf zebrafish larvae were incubated with fluo-4 AM and were then electrically activated using platinum electrodes. Myocytes were treated with S-Nitroso-N-acetyl-DL-penicillamin (SNAP), a NO-donor, or N(G)-Nitro-L-arginine methyl ester (L-NAME), an unspecific blocker of nitric oxide synthase. The authors concluded that endogenous NO reduced force production through negative modulation of calcium transients via the NO-sGC pathway.

The cytosolic protein SH3 and cysteine rich domain 3 (Stac3) has been identified as an essential component for skeletal muscle excitation–contraction coupling and a mutation in human *STAC3* causes the debilitating Native American myopathy, a severe disorder characterized by altered skeletal muscle structure and function. The group of John Kuwada has investigated the role of Stac3 in zebrafish by either loading dissociated muscle fibers with fluo-4 AM [152] or by injecting the one-cell embryo with a plasmid carrying *α-actin:GCaMP3*, resulting in embryos that mosaically expressed GCaMP3 within skeletal muscle [153]. By performing calcium imaging of electrically stimulated isolated muscle fibers carrying the human *STAC3* mutation, it was found that Stac3 participates in excitation–contraction coupling in muscles [152]. Introduction of the mutated *stac3* in fish lacking the wild-type gene resulted in decreased calcium transients in fast-twitch muscles at 48 hpf and defective swimming [153].

Another myopathy, the autosomal recessive centronuclear myopathy, is caused by mutations in bridging integrator 1 (*BIN1*). To understand the role of Bin1 in muscle contraction, zebrafish embryos were injected at the one-cell stage with a plasmid encoding *GCaMP3–EGFP* under control of an *α-actin* promoter, alone or in combination with a *bin1* morpholino [154]. Calcium imaging of wild-type and *bin1* morphant embryos was done at 3 dpf using a spinning disk confocal microscope. KCl was used to initiate muscle contraction and relative fluorescence intensity changes of induced calcium transients in individual myofibers were recorded. Recordings confirmed that muscle calcium signaling was impaired in *bin1* morphants.

Another cause of centronuclear myopathy is mutations in the gene of dynamin 2 (*DNM2*), a ubiquitously expressed GTPase that regulates multiple subcellular processes. Expression of mutated human *DNM2-S619L* in zebrafish led to the accumulation of aberrant vesicular structures and to defective excitation-contraction coupling [155]. Embryos were co-injected with cDNA for *GCaMP3* and RNA for wildtype or mutant *DNM2* and calcium measurements of spontaneous muscle contractions were performed at 24 hpf. Although *DNM2-S619L* embryos displayed spontaneous muscle contractions with normal timing, there was no increase in intracellular calcium corresponding to calcium sparks.

The development of lines that selectively express GECIs in either fast or slow muscle fibers has big potential for the field. In a study by Jackson et al., the promoter for the slow myosin heavy chain 1 (*smyhc1*) was used to drive *GCaMP3* expression in slow-twitch muscle fibers [156]. Fast-twitch-specific transients could be monitored using the *Tg(mylz2:GCaMP3)* transgene, making use of the myosin light chain 2 promoter (*mylz2*).

Calcium imaging of 2 dpf *Tg(smyhc1:GCaMP3)* embryos was done with a confocal microscope, while muscle contractions were induced by pentylenetetrazol (PTZ) [156]. In wild-type embryos, the response time of slow-twitch fibers was shorter than in fast-twitch fibers, while the change in amplitude was smaller. The response time of *sox6* mutant fast-twitch fibers was significantly shorter than in wild types, resembling more that of wild-type slow-twitch fibers, whereas the amplitude change was unaffected. These findings indicate that the transcription factor Sox6 controls the calcium response of fast-twitch fibers.

Shahid et al. established an in vivo biosensor system to quantify potential toxicant effects on motor function, by generating a GFP line based on the regulatory element of the heat shock protein family B (small) member 11 (*hspb11*), i.e. *TgBAC(hspb11:GFP)* [157]. Exposure to substances that interfered with motor function induced a dose-dependent increase of GFP intensity, while also inducing muscle hyperactivity with increased calcium spike height and frequency. This was monitored by developing the muscle specific calcium biosensor line Tg(*[-505/-310]unc45b:GCaMP5A*), where the promoter for unc-45 myosin chaperone B is driving the *GCaMP5A* expression.

Finally, spontaneous calcium activity in embryonic myocytes has been observed in a zebrafish line containing the constitutively active promoter *hspa8* and the gene for YC2.60 [43]. At 24 hpf, twitching behavior only caused calcium signals in slow muscles while the calcium concentration in the fast muscles stayed at basal level. During stronger contraction, when left and right side myocytes contracted alternately and repeatedly, rises in calcium were observed both in the slow and fast muscles. Alternating contractions of left- and right-side myocytes were also observed at 51 hpf.

36.3.7 Heart

The development of the heart and cardiovascular system has been extensively studied in the zebrafish [158, 159]. Still, there is information lacking regarding the intracellular calcium handling of the zebrafish heart. The zebrafish heart has a simple structure, composed of the sinus venosus, the atrium, the ventricle, and the bulbus arteriosus connected in series. The heart is the first organ to form and function during development, and the heart of the embryonic and larval zebrafish are particularly easily imaged during this time. The first calcium imaging studies of the zebrafish heart made use of labeling of embryonic and larval hearts with dextran and AM ester indicators [8], while later studies have benefitted from transgenic zebrafish lines expressing GECIs via the heart-specific cardiac myosin light chain 2 (*cmlc2*)/myosin light chain 7 (*myl7*) promoter (referred to as *myl7* here) [160]. By crossing this line with mutant lines displaying heart defects such as contraction problems, researchers have gained significant molecular knowledge underlying the development and function of the heart as well as dysfunction related to cardiomyopathies.

Isolated zebrafish myocytes or intact hearts can be loaded with AM esters. Bove et al. used confocal imaging to investigate the role of SR in calcium signaling of isolated cardiac myocytes loaded with fluo-4 AM [26]. They found that contrary to mammalian myocytes, only 20% of the action potential-induced calcium transients in zebrafish are mediated via calcium release from the SR, and the majority was originating from calcium influx via L-type calcium channels.

Embryonic zebrafish hearts can be isolated and incubated with calcium indicators despite their small size. Researchers were interested in understanding the function of *pkd2* mutations that are causing autosomal dominant polycystic kidney disease in humans, with comorbidity of cardiovascular disease [27]. Hearts from 3 dpf wild-type and *pkd2* mutant embryos were isolated and labeled with fluo-4 AM. Cardiac contractions were stopped with blebbistatin to eliminate motion artifacts. Isolated *pkd2* mutant hearts displayed impaired intracellular calcium cycling and calcium alternans.

Moreover, recordings from embryonic hearts were performed both after isolation and incubation with fluo-4 AM and in vivo in the *Tg(cmlc2:gCaMP)*^{s878} line [28]. The goal was to understand how 3-O-sulfotransferase (3-OST) 7 affects cardiac calcium processing since *3-ost-7* knockdown results in a noncontracting cardiac ventricle at 48 hpf. Interestingly, the noncontracting ventricle in *3-ost-7* morphants generated normal action potentials and calcium transients. In addition, calcium activation and conduction velocity proceeded normally in *3-ost-7* morphants in vivo. It was found that elevated bone morphogenetic protein (BMP) signaling in the morphants leads to disruption of sarcomere assembly, leading to the defect in myocyte contraction.

Light-sheet microscopy has been used for detailed in vivo cardiac imaging in zebrafish embryos. *Tg(myl7:GCaMP5G-Arch)D95N* embryos between 36 and 52 hpf were imaged with a custom-built light-sheet microscope [161]. Using this technique, the authors could map the three-dimensional calcium signals of all cells in the entire electro-mechanically uncoupled heart during the looping stage.

Fukuda et al. were interested in better understanding the formation of heart tissue during development [162]. They made use of a mutant fish line for ankyrin repeat and SOCS box containing 2 (*asb2b*), an E3 ubiquitin ligase that functions to specify target proteins for proteasome-dependent degradation. Although *asb2b* mutants displayed cardiac dysfunction, imaging with a spinning disc confocal microscope revealed no calcium signaling dysfunction in 50 hpf mutant *Tg(myl7:gCaMP)*^{s878} zebrafish, when compared to wild-type fish. It was found that *asb2b* is required for myocyte maturation, causing structural defects in the mutants.

Cardiac trabeculation is a crucial morphogenetic process by which clusters of ventricular cardiomyocytes extrude and expand to form sheet-like projections. Liu et al. found that in zebrafish Erb-B2 receptor tyrosine kinase 2 (*erbb2*) mutants lack cardiac trabeculae and display a dysfunctional cardiac function [163]. Optical mapping of 3 and 10 dpf wild-type and *erbb2* mutants crossed to *Tg(cmlc2:gCaMP)s878* were embedded and cardiac contraction was suppressed with 2,3-butanedione monoxime before imaging using a high-speed camera. At 3 dpf, when trabeculae start to form in wild type fish, optical mapping revealed that the activation patterns in wild-type and *erbb2* mutant ventricles were indistinguishable. By contrast, following the formation of trabeculae, the ventricular calcium wave in 10 dpf wild-type hearts was different from that of *erbb2* mutant hearts. These findings indicate an important role for cardiac trabeculae and ErbB2 signaling in the maturation of the ventricular conduction system.

The zebrafish mutant line *tremblor* (*tre*^{tc318}) lacks the gene for the Na^+/Ca^{2+} exchanger 1, *slc8a1a*, giving it an irregular heart rhythm since its myocytes are unable to efficiently remove calcium ions from the cytoplasm. Via drug screening, it was possible to identify a compound, efsevin, that could restore the heartbeat in the *tre*^{tc318} mutants [164]. The *Tg(myl7:GCaMP4.1)* zebrafish line was used to record calcium waves in vivo at 36 dpf, confirming that efsevin could normalize the calcium signals in the hearts of mutants.

BIN1 is not only involved in centronuclear myopathy, but is also involved in intracellular calcium processing in cardiomyocytes, as it allows the correct processing of voltage-dependent calcium channels Cav1.2 to the cardiac T-tubules. *bin1* was knocked down in *Tg(cmlc2:gCaMP)*^{s878} zebrafish embryos where calcium transients and cardiac contractility were analyzed in vivo [165]. The analysis confirmed that calcium transients in 70 hpf *bin1* morphants was reduced, and the ventricular contractions were impaired.

The Popeye domain containing gene 2 (*popdc2*) encodes a transmembrane protein that expresses in the myocardium and transiently in the craniofacial and tail musculature. When *popdc2* was knocked down, the animals showed defects in skeletal muscle and the heart (Fig. 36.14) [166]. SPIM was used to image calcium oscillations in 5–6 dpf wild-type and *popdc2 Tg(cmlc2:gCaMP)*^{s878} embryos confirming the presence of a severe arrhythmia phenotype.

Desminopathies belong to a family of muscle disorders called myofibrillar myopathies that are caused by desmin mutations and lead to protein aggregates in muscle fibers. Desmin is involved in the formation of sarcomere architecture. The *desmin a* (*desma*) gene is homologous to human desmin and was further studied in the *desma*$^{sa5-/-}$ knock-out line and the *desma*^{Ct122aGt} gene trap mutant where citrine was inserted between AAs 460 and 461 in the C terminus of desmin a [167]. These lines were crossed with *Tg(myl7:galFF; UAS:GCaMP3)* animals to allow for confocal imaging of calcium transients in the heart. Calcium signals were measured from different regions of the heart, revealing reduced calcium signals at the tip of the ventricle in the mutant lines compared to controls.

Finally, calcium imaging has been performed in the endocardial cells that are sensing the flow forces of the heart at the time of the heart valve formation [168]. This was studied at the early stages of heart development, at 48 hpf, in the *Tg(fli1a:gal4ff; UAS:GCaMP3)* line that labels endothelial cells. It was observed that calcium levels were increased in the atrio-ventricular canal and was lower in the atrium and ventricle. The endocardial calcium response and the flow-responsive Kruppel-like factor 2a (*klf2a*) promoter were modulated by the oscillatory flow-through transient receptor potential cation channel subfamily V member 4 (Trpv4), a mechanosensitive ion channel specifically expressed in the endocardium during heart valve development.

36.4 Conclusion

The described preparations and experiments show the diversity of studies that have benefitted from calcium imaging in zebrafish cells, tissues and intact animals. These studies have contributed greatly to our understanding of basic biological principles during development and adulthood, as well as the function of disease-related genes in a vertebrate system. However, it is evident that we still lack knowledge in certain biomedical areas, where new preparations of the zebrafish would be valuable. Nevertheless, the technical advances in biomedical research promise exciting possibilities. For example, the discovery of new cell-specific promoters in combination with GECIs will undoubtedly continue to be of great importance for

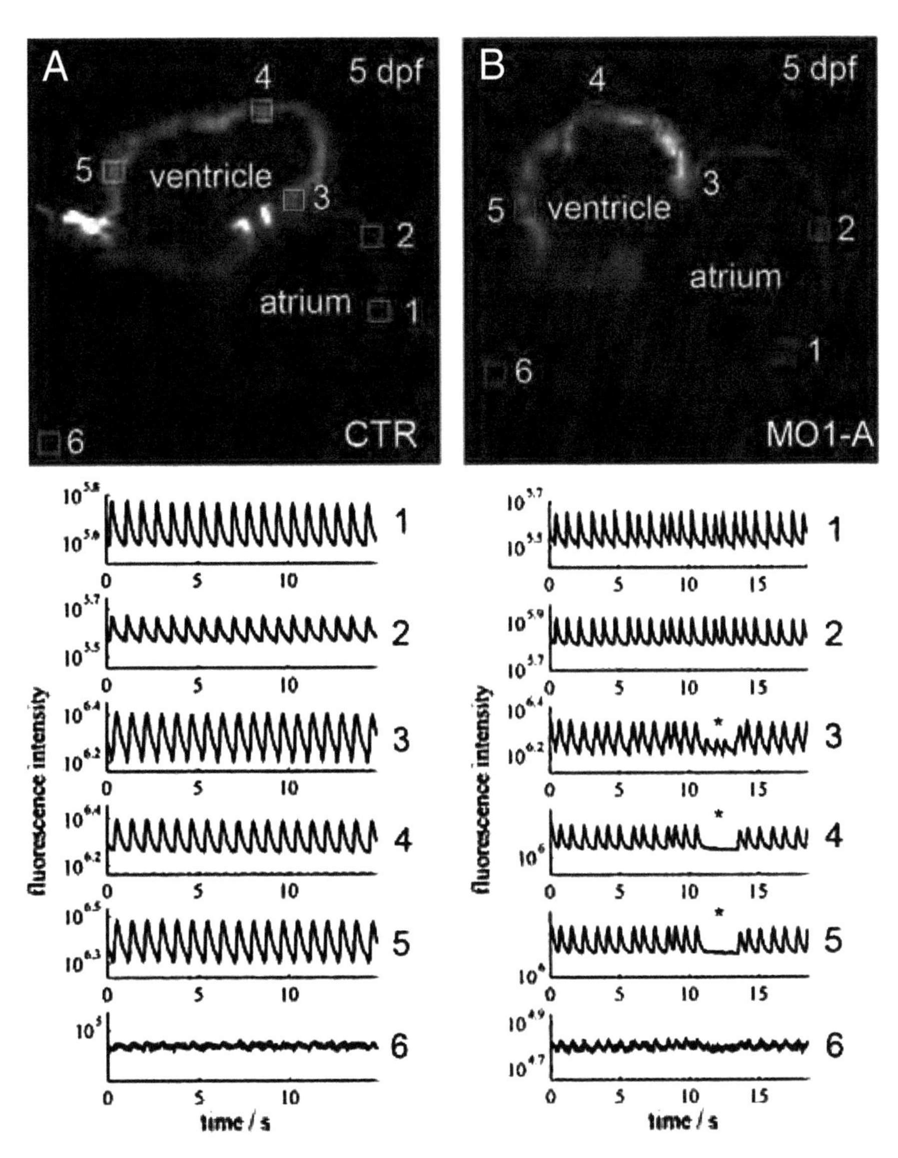

Fig. 36.14 Imaging of cardiomyocyte calcium signaling in vivo. The hearts of 5-day-old Popeye domain containing 2 (*popdc2*) morphants displayed cardiac conduction defects. A single plane illumination microscope (SPIM) was used to record videos of the heart of *Tg(cmlc2:gCaMP)s878* zebrafish embryos injected with (**a**) control (CTR) morpholino and (**b**) *MO1-popdc2* morpholino. The individual movies were processed to determine the fluorescence intensity of selected regions of the heart over time. Each selected region has a corresponding number plotted below. (**a**) In a control heart, fluorescence intensity varied with time in atrial and ventricular regions as the depolarization wave propagated through the heart. (**b**) The morphant heart displayed an atrioventricular (AV) block (asterisks). (Reprinted from Kirchmaier et al. [166], Copyright 2012, with permission from Elsevier)

future analyses of neuronal cell types and regions that so far have been unreachable by conventional labeling methods. Moreover, the combination of optogenetics with calcium imaging within the brain will mean valuable dissections of network function and connectivity. The development of new microscopy techniques with increased resolution, speed and imaging depth, such as light-sheet microscopy, will also be critical for the advancements of calcium imaging in the zebrafish, allowing the monitoring of processes in older animals and larger brain areas. New large-scale imaging methods and sophisticated analysis of activity from neuronal ensembles will bring our understanding of neuronal processing even further. The emerging use of the zebrafish as a model system for diseases, e.g. in the field of psychiatry, will mean exciting findings in the years to come. My hope is that this chapter will serve as an inspiration to continue the development of the zebrafish as a model organism to study the physiological processes and the complex biomedical mechanisms that are vital for biological beings.

References

1. Lister JA, Robertson CP, Lepage T, Johnson SL, Raible DW (1999) nacre encodes a zebrafish microphthalmia-related protein that regulates neural-crest-derived pigment cell fate. Development 126:3757–3767
2. Nasevicius A, Ekker SC (2000) Effective targeted gene 'knockdown' in zebrafish. Nat Genet 26:216–220
3. Asnani A, Peterson RT (2014) The zebrafish as a tool to identify novel therapies for human cardiovascular disease. Dis Model Mech 7:763–767
4. Xi Y, Noble S, Ekker M (2011) Modeling neurodegeneration in zebrafish. Curr Neurol Neurosci Rep 11:274–282
5. Norton WH (2013) Toward developmental models of psychiatric disorders in zebrafish. Front Neural Circuits 7:79
6. White R, Rose K, Zon L (2013) Zebrafish cancer: the state of the art and the path forward. Nat Rev Cancer 13:624–636
7. MacRae CA, Peterson RT (2015) Zebrafish as tools for drug discovery. Nat Rev Drug Discov 14:721–731
8. Kettunen P (2012) Calcium imaging in the zebrafish. Adv Exp Med Biol 740:1039–1071
9. Gobel W, Helmchen F (2007) In vivo calcium imaging of neural network function. Physiology (Bethesda) 22:358–365
10. O'Donovan MJ, Ho S, Sholomenko G, Yee W (1993) Real-time imaging of neurons retrogradely and anterogradely labelled with calcium-sensitive dyes. J Neurosci Methods 46:91–106
11. Stosiek C, Garaschuk O, Holthoff K, Konnerth A (2003) In vivo two-photon calcium imaging of neuronal networks. Proc Natl Acad Sci U S A 100:7319–7324
12. Tsien RY (1981) A non-disruptive technique for loading calcium buffers and indicators into cells. Nature 290:527–528
13. Park HC, Kim CH, Bae YK, Yeo SY, Kim SH, Hong SK, Shin J, Yoo KW, Hibi M, Hirano T, Miki N, Chitnis AB, Huh TL (2000) Analysis of upstream elements in the HuC promoter leads to the establishment of transgenic zebrafish with fluorescent neurons. Dev Biol 227:279–293
14. Ahrens MB, Li JM, Orger MB, Robson DN, Schier AF, Engert F, Portugues R (2012) Brain-wide neuronal dynamics during motor adaptation in zebrafish. Nature 485:471–477

15. Fetcho JR, O'Malley DM (1995) Visualization of active neural circuitry in the spinal cord of intact zebrafish. J Neurophysiol 73:399–406
16. Gahtan E, Sankrithi N, Campos JB, O'Malley DM (2002) Evidence for a widespread brain stem escape network in larval zebrafish. J Neurophysiol 87:608–614
17. Sharma D, Kinsey WH (2008) Regionalized calcium signaling in zebrafish fertilization. Int J Dev Biol 52:561–570
18. Reinhard E, Yokoe H, Niebling KR, Allbritton NL, Kuhn MA, Meyer T (1995) Localized calcium signals in early zebrafish development. Dev Biol 170:50–61
19. Creton R, Speksnijder JE, Jaffe LF (1998) Patterns of free calcium in zebrafish embryos. J Cell Sci 111(Pt 12):1613–1622
20. Sharma D, Kinsey WH (2013) PYK2: a calcium-sensitive protein tyrosine kinase activated in response to fertilization of the zebrafish oocyte. Dev Biol 373:130–140
21. Bonneau B, Nougarede A, Prudent J, Popgeorgiev N, Peyrieras N, Rimokh R, Gillet G (2014) The Bcl-2 homolog Nrz inhibits binding of IP3 to its receptor to control calcium signaling during zebrafish epiboly. Sci Signal 7:ra14
22. Kreiling JA, Balantac ZL, Crawford AR, Ren Y, Toure J, Zchut S, Kochilas L, Creton R (2008) Suppression of the endoplasmic reticulum calcium pump during zebrafish gastrulation affects left-right asymmetry of the heart and brain. Mech Dev 125:396–410
23. Hollmann V, Lucks V, Kurtz R, Engelmann J (2015) Adaptation-induced modification of motion selectivity tuning in visual tectal neurons of adult zebrafish. J Neurophysiol 114:2893–2902
24. Brustein E, Marandi N, Kovalchuk Y, Drapeau P, Konnerth A (2003) "In vivo" monitoring of neuronal network activity in zebrafish by two-photon Ca^{2+} imaging. Pflugers Arch 446:766–773
25. Niell CM, Smith SJ (2005) Functional imaging reveals rapid development of visual response properties in the zebrafish tectum. Neuron 45:941–951
26. Bovo E, Dvornikov AV, Mazurek SR, de Tombe PP, Zima AV (2013) Mechanisms of Ca(2)+ handling in zebrafish ventricular myocytes. Pflugers Arch 465:1775–1784
27. Paavola J, Schliffke S, Rossetti S, Kuo IY, Yuan S, Sun Z, Harris PC, Torres VE, Ehrlich BE (2013) Polycystin-2 mutations lead to impaired calcium cycling in the heart and predispose to dilated cardiomyopathy. J Mol Cell Cardiol 58:199–208
28. Samson SC, Ferrer T, Jou CJ, Sachse FB, Shankaran SS, Shaw RM, Chi NC, Tristani-Firouzi M, Yost HJ (2013) 3-OST-7 regulates BMP-dependent cardiac contraction. PLoS Biol 11:e1001727
29. Robin G, Allard B (2015) Voltage-gated Ca^{2+} influx through L-type channels contributes to sarcoplasmic reticulum Ca^{2+} loading in skeletal muscle. J Physiol 593:4781–4797
30. Yaksi E, Judkewitz B, Friedrich RW (2007) Topological reorganization of odor representations in the olfactory bulb. PLoS Biol 5:e178
31. Shimomura O, Johnson FH, Saiga Y (1962) Extraction, purification and properties of aequorin, a bioluminescent protein from the luminous hydromedusan, Aequorea. J Cell Comp Physiol 59:223–239
32. Chiesa A, Rapizzi E, Tosello V, Pinton P, de Virgilio M, Fogarty KE, Rizzuto R (2001) Recombinant aequorin and green fluorescent protein as valuable tools in the study of cell signalling. Biochem J 355:1–12
33. Webb SE, Miller AL (2000) Calcium signalling during zebrafish embryonic development. BioEssays 22:113–123
34. Cheung CY, Webb SE, Meng A, Miller AL (2006) Transient expression of apoaequorin in zebrafish embryos: extending the ability to image calcium transients during later stages of development. Int J Dev Biol 50:561–569
35. Baubet V, Le Mouellic H, Campbell AK, Lucas-Meunier E, Fossier P, Brulet P (2000) Chimeric green fluorescent protein-aequorin as bioluminescent Ca^{2+} reporters at the single-cell level. Proc Natl Acad Sci U S A 97:7260–7265
36. Ashworth R, Brennan C (2005) Use of transgenic zebrafish reporter lines to study calcium signalling in development. Brief Funct Genomic Proteomic 4:186–193

37. Naumann EA, Kampff AR, Prober DA, Schier AF, Engert F (2010) Monitoring neural activity with bioluminescence during natural behavior. Nat Neurosci 13:513–520
38. Knafo S, Prendergast A, Thouvenin O, Figueiredo SN, Wyart C (2017) Bioluminescence monitoring of neuronal activity in freely moving Zebrafish larvae. Bio Protoc 7:e2550
39. Bakayan A, Domingo B, Miyawaki A, Llopis J (2015) Imaging Ca^{2+} activity in mammalian cells and zebrafish with a novel red-emitting aequorin variant. Pflugers Arch 467:2031–2042
40. Pologruto TA, Yasuda R, Svoboda K (2004) Monitoring neural activity and $[Ca^{2+}]$ with genetically encoded Ca^{2+} indicators. J Neurosci 24:9572–9579
41. Miyawaki A, Llopis J, Heim R, McCaffery JM, Adams JA, Ikura M, Tsien RY (1997) Fluorescent indicators for Ca^{2+} based on green fluorescent proteins and calmodulin. Nature 388:882–887
42. Tsuruwaka Y, Konishi T, Miyawaki A, Takagi M (2007) Real-time monitoring of dynamic intracellular Ca^{2+} movement during early embryogenesis through expression of yellow cameleon. Zebrafish 4:253–260
43. Mizuno H, Sassa T, Higashijima S, Okamoto H, Miyawaki A (2013) Transgenic zebrafish for ratiometric imaging of cytosolic and mitochondrial Ca^{2+} response in teleost embryo. Cell Calcium 54:236–245
44. Higashijima S, Masino MA, Mandel G, Fetcho JR (2003) Imaging neuronal activity during zebrafish behavior with a genetically encoded calcium indicator. J Neurophysiol 90:3986–3997
45. Tsuruwaka Y, Shimada E, Tsutsui K, Ogawa T (2017) Ca^{2+} dynamics in zebrafish morphogenesis. PeerJ 5:e2894
46. Nakai J, Ohkura M, Imoto K (2001) A high signal-to-noise Ca^{2+} probe composed of a single green fluorescent protein. Nat Biotechnol 19:137–141
47. Muto A, Ohkura M, Kotani T, Higashijima S, Nakai J, Kawakami K (2011) Genetic visualization with an improved GCaMP calcium indicator reveals spatiotemporal activation of the spinal motor neurons in zebrafish. Proc Natl Acad Sci U S A 108:5425–5430
48. Zhang RW, Li XQ, Kawakami K, Du JL (2016) Stereotyped initiation of retinal waves by bipolar cells via presynaptic NMDA autoreceptors. Nat Commun 7:12650
49. Lacoste AM, Schoppik D, Robson DN, Haesemeyer M, Portugues R, Li JM, Randlett O, Wee CL, Engert F, Schier AF (2015) A convergent and essential interneuron pathway for Mauthner-cell-mediated escapes. Curr Biol 25:1526–1534
50. Riley E, Kopotiyenko K, Zhdanova I (2015) Prenatal and acute cocaine exposure affects neural responses and habituation to visual stimuli. Front Neural Circuits 9:41
51. Dreosti E, Odermatt B, Dorostkar MM, Lagnado L (2009) A genetically encoded reporter of synaptic activity in vivo. Nat Methods 6:883–889
52. Dorostkar MM, Dreosti E, Odermatt B, Lagnado L (2010) Computational processing of optical measurements of neuronal and synaptic activity in networks. J Neurosci Methods 188:141–150
53. Esposti F, Johnston J, Rosa JM, Leung KM, Lagnado L (2013) Olfactory stimulation selectively modulates the OFF pathway in the retina of zebrafish. Neuron 79:97–110
54. Rosa JM, Ruehle S, Ding H, Lagnado L (2016) Crossover Inhibition Generates Sustained Visual Responses in the Inner Retina. Neuron 90:308–319
55. Walker AS, Burrone J, Meyer MP (2013) Functional imaging in the zebrafish retinotectal system using RGECO. Front Neural Circuits 7:34
56. Akerboom J, Carreras Calderon N, Tian L, Wabnig S, Prigge M, Tolo J, Gordus A, Orger MB, Severi KE, Macklin JJ, Patel R, Pulver SR, Wardill TJ, Fischer E, Schuler C, Chen TW, Sarkisyan KS, Marvin JS, Bargmann CI, Kim DS, Kugler S, Lagnado L, Hegemann P, Gottschalk A, Schreiter ER, Looger LL (2013) Genetically encoded calcium indicators for multi-color neural activity imaging and combination with optogenetics. Front Mol Neurosci 6:2
57. Dana H, Mohar B, Sun Y, Narayan S, Gordus A, Hasseman JP, Tsegaye G, Holt GT, Hu A, Walpita D, Patel R, Macklin JJ, Bargmann CI, Ahrens MB, Schreiter ER, Jayaraman V, Looger LL, Svoboda K, Kim DS (2016) Sensitive red protein calcium indicators for imaging neural activity. elife 5:pii: e12727

58. Shen Y, Dana H, Abdelfattah AS, Patel R, Shea J, Molina RS, Rawal B, Rancic V, Chang YF, Wu L, Chen Y, Qian Y, Wiens MD, Hambleton N, Ballanyi K, Hughes TE, Drobizhev M, Kim DS, Koyama M, Schreiter ER, Campbell RE (2018) A genetically encoded Ca^{2+} indicator based on circularly permutated sea anemone red fluorescent protein eqFP578. BMC Biol 16:9
59. Fosque BF, Sun Y, Dana H, Yang CT, Ohyama T, Tadross MR, Patel R, Zlatic M, Kim DS, Ahrens MB, Jayaraman V, Looger LL, Schreiter ER (2015) Neural circuits. Labeling of active neural circuits in vivo with designed calcium integrators. Science 347:755–760
60. Meng CL, Chang DC (1994) Study of calcium signaling in cell cleavage using confocal microscopy. Biol Bull 187:234–235
61. Kimmel CB, Ballard WW, Kimmel SR, Ullmann B, Schilling TF (1995) Stages of embryonic development of the zebrafish. Dev Dyn 203:253–310
62. Webb SE, Miller AL (2003) Imaging intercellular calcium waves during late epiboly in intact zebrafish embryos. Zygote 11:175–182
63. Webb SE, Fluck RA, Miller AL (2011) Calcium signaling during the early development of medaka and zebrafish. Biochimie 93:2112–2125
64. Gilland E, Miller AL, Karplus E, Baker R, Webb SE (1999) Imaging of multicellular large-scale rhythmic calcium waves during zebrafish gastrulation. Proc Natl Acad Sci U S A 96:157–161
65. Lee KW, Webb SE, Miller AL (1999) A wave of free cytosolic calcium traverses zebrafish eggs on activation. Dev Biol 214:168–180
66. Eno C, Gomez T, Slusarski DC, Pelegri F (2018) Slow calcium waves mediate furrow microtubule reorganization and germ plasm compaction in the early zebrafish embryo. Development 145:dev156604
67. Wu SY, Shin J, Sepich DS, Solnica-Krezel L (2012) Chemokine GPCR signaling inhibits beta-catenin during zebrafish axis formation. PLoS Biol 10:e1001403
68. Chen J, Xia L, Bruchas MR, Solnica-Krezel L (2017) Imaging early embryonic calcium activity with GCaMP6s transgenic zebrafish. Dev Biol 430:385–396
69. Chang DC, Meng C (1995) A localized elevation of cytosolic free calcium is associated with cytokinesis in the zebrafish embryo. J Cell Biol 131:1539–1545
70. Chang DC, Lu P (2000) Multiple types of calcium signals are associated with cell division in zebrafish embryo. Microsc Res Tech 49:111–122
71. Webb SE, Lee KW, Karplus E, Miller AL (1997) Localized calcium transients accompany furrow positioning, propagation, and deepening during the early cleavage period of zebrafish embryos. Dev Biol 192:78–92
72. Leung CF, Webb SE, Miller AL (1998) Calcium transients accompany ooplasmic segregation in zebrafish embryos. Develop Growth Differ 40:313–326
73. Webb SE, Li WM, Miller AL (2008) Calcium signalling during the cleavage period of zebrafish development. Philos Trans R Soc Lond Ser B Biol Sci 363:1363–1369
74. Yuen MY, Webb SE, Chan CM, Thisse B, Thisse C, Miller AL (2013) Characterization of Ca^{2+} signaling in the external yolk syncytial layer during the late blastula and early gastrula periods of zebrafish development. Biochim Biophys Acta 1833:1641–1656
75. Schneider I, Houston DW, Rebagliati MR, Slusarski DC (2008) Calcium fluxes in dorsal forerunner cells antagonize beta-catenin and alter left-right patterning. Development 135:75–84
76. Yuan S, Zhao L, Brueckner M, Sun Z (2015) Intraciliary calcium oscillations initiate vertebrate left-right asymmetry. Curr Biol 25:556–567
77. White RM, Sessa A, Burke C, Bowman T, LeBlanc J, Ceol C, Bourque C, Dovey M, Goessling W, Burns CE, Zon LI (2008) Transparent adult zebrafish as a tool for in vivo transplantation analysis. Cell Stem Cell 2:183–189
78. Antinucci P, Hindges R (2016) A crystal-clear zebrafish for in vivo imaging. Sci Rep 6:29490
79. O'Malley DM, Kao YH, Fetcho JR (1996) Imaging the functional organization of zebrafish hindbrain segments during escape behaviors. Neuron 17:1145–1155

80. Liu KS, Fetcho JR (1999) Laser ablations reveal functional relationships of segmental hindbrain neurons in zebrafish. Neuron 23:325–335
81. Granato M, van Eeden FJ, Schach U, Trowe T, Brand M, Furutani-Seiki M, Haffter P, Hammerschmidt M, Heisenberg CP, Jiang YJ, Kane DA, Kelsh RN, Mullins MC, Odenthal J, Nusslein-Volhard C (1996) Genes controlling and mediating locomotion behavior of the zebrafish embryo and larva. Development 123:399–413
82. Yamanaka I, Miki M, Asakawa K, Kawakami K, Oda Y, Hirata H (2013) Glycinergic transmission and postsynaptic activation of CaMKII are required for glycine receptor clustering in vivo. Genes Cells 18:211–224
83. Marsden KC, Granato M (2015) In vivo Ca^{2+} imaging reveals that decreased dendritic excitability drives startle habituation. Cell Rep 13:1733–1740
84. Marsden KC, Jain RA, Wolman MA, Echeverry FA, Nelson JC, Hayer KE, Miltenberg B, Pereda AE, Granato M (2018) A Cyfip2-dependent excitatory interneuron pathway establishes the innate startle threshold. Cell Rep 23:878–887
85. Takahashi M, Narushima M, Oda Y (2002) In vivo imaging of functional inhibitory networks on the mauthner cell of larval zebrafish. J Neurosci 22:3929–3938
86. O'Malley DM, Sankrithi NS, Borla MA, Parker S, Banden S, Gahtan E, Detrich HW 3rd (2004) Optical physiology and locomotor behaviors of wild-type and nacre zebrafish. Methods Cell Biol 76:261–284
87. Sankrithi NS, O'Malley DM (2010) Activation of a multisensory, multifunctional nucleus in the zebrafish midbrain during diverse locomotor behaviors. Neuroscience 166:970–993
88. Wang WC, McLean DL (2014) Selective responses to tonic descending commands by temporal summation in a spinal motor pool. Neuron 83:708–721
89. Thiele TR, Donovan JC, Baier H (2014) Descending control of swim posture by a midbrain nucleus in zebrafish. Neuron 83:679–691
90. Warp E, Agarwal G, Wyart C, Friedmann D, Oldfield CS, Conner A, Del Bene F, Arrenberg AB, Baier H, Isacoff EY (2012) Emergence of patterned activity in the developing zebrafish spinal cord. Curr Biol 22:93–102
91. Plazas PV, Nicol X, Spitzer NC (2013) Activity-dependent competition regulates motor neuron axon pathfinding via PlexinA3. Proc Natl Acad Sci U S A 110:1524–1529
92. Sternberg JR, Severi KE, Fidelin K, Gomez J, Ihara H, Alcheikh Y, Hubbard JM, Kawakami K, Suster M, Wyart C (2016) Optimization of a neurotoxin to investigate the contribution of excitatory interneurons to speed modulation in vivo. Curr Biol 26:2319–2328
93. Friedmann D, Hoagland A, Berlin S, Isacoff EY (2015) A spinal opsin controls early neural activity and drives a behavioral light response. Curr Biol 25:69–74
94. Kelu JJ, Webb SE, Galione A, Miller AL (2018) TPC2-mediated Ca^{2+} signaling is required for the establishment of synchronized activity in developing zebrafish primary motor neurons. Dev Biol 438:57–68
95. Menelaou E, VanDunk C, McLean DL (2014) Differences in the morphology of spinal V2a neurons reflect their recruitment order during swimming in larval zebrafish. J Comp Neurol 522:1232–1248
96. Bohm UL, Prendergast A, Djenoune L, Nunes Figueiredo S, Gomez J, Stokes C, Kaiser S, Suster M, Kawakami K, Charpentier M, Concordet JP, Rio JP, Del Bene F, Wyart C (2016) CSF-contacting neurons regulate locomotion by relaying mechanical stimuli to spinal circuits. Nat Commun 7:10866
97. Hubbard JM, Bohm UL, Prendergast A, Tseng PB, Newman M, Stokes C, Wyart C (2016) Intraspinal sensory neurons provide powerful inhibition to motor circuits ensuring postural control during locomotion. Curr Biol 26:2841–2853
98. Knafo S, Fidelin K, Prendergast A, Tseng PB, Parrin A, Dickey C, Bohm UL, Figueiredo SN, Thouvenin O, Pascal-Moussellard H, Wyart C (2017) Mechanosensory neurons control the timing of spinal microcircuit selection during locomotion. elife 6:pii: e25260
99. Odermatt B, Nikolaev A, Lagnado L (2012) Encoding of luminance and contrast by linear and nonlinear synapses in the retina. Neuron 73:758–773

100. Antinucci P, Nikolaou N, Meyer MP, Hindges R (2013) Teneurin-3 specifies morphological and functional connectivity of retinal ganglion cells in the vertebrate visual system. Cell Rep 5:582–592
101. Wei HP, Yao YY, Zhang RW, Zhao XF, Du JL (2012) Activity-induced long-term potentiation of excitatory synapses in developing zebrafish retina in vivo. Neuron 75:479–489
102. Zhao XF, Ellingsen S, Fjose A (2009) Labelling and targeted ablation of specific bipolar cell types in the zebrafish retina. BMC Neurosci 10:107
103. Smyth VA, Di Lorenzo D, Kennedy BN (2008) A novel, evolutionarily conserved enhancer of cone photoreceptor-specific expression. J Biol Chem 283:10881–10891
104. Ma EY, Lewis A, Barabas P, Stearns G, Suzuki S, Krizaj D, Brockerhoff SE (2013) Loss of Pde6 reduces cell body Ca^{2+} transients within photoreceptors. Cell Death Dis 4:e797
105. Fraser SE (1992) Patterning of retinotectal connections in the vertebrate visual system. Curr Opin Neurobiol 2:83–87
106. Nevin LM, Robles E, Baier H, Scott EK (2010) Focusing on optic tectum circuitry through the lens of genetics. BMC Biol 8:126
107. Nikolaou N, Lowe AS, Walker AS, Abbas F, Hunter PR, Thompson ID, Meyer MP (2012) Parametric functional maps of visual inputs to the tectum. Neuron 76:317–324
108. Hunter PR, Lowe AS, Thompson ID, Meyer MP (2013) Emergent properties of the optic tectum revealed by population analysis of direction and orientation selectivity. J Neurosci 33:13940–13945
109. Gabriel JP, Trivedi CA, Maurer CM, Ryu S, Bollmann JH (2012) Layer-specific targeting of direction-selective neurons in the zebrafish optic tectum. Neuron 76:1147–1160
110. Preuss SJ, Trivedi CA, vom Berg-Maurer CM, Ryu S, Bollmann JH (2014) Classification of object size in retinotectal microcircuits. Curr Biol 24:2376–2385
111. Naumann EA, Fitzgerald JE, Dunn TW, Rihel J, Sompolinsky H, Engert F (2016) From whole-brain data to functional circuit models: the zebrafish optomotor response. Cell 167:947–960.e920
112. Semmelhack JL, Donovan JC, Thiele TR, Kuehn E, Laurell E, Baier H (2014) A dedicated visual pathway for prey detection in larval zebrafish. elife 3:e04878
113. Shepherd GM (1993) Principles of specificity and redundancy underlying the organization of the olfactory system. Microsc Res Tech 24:106–112
114. Mombaerts P (2001) How smell develops. Nat Neurosci 4(Suppl):1192–1198
115. Shipley MT, Ennis M (1996) Functional organization of olfactory system. J Neurobiol 30:123–176
116. Blumhagen F, Zhu P, Shum J, Scharer YP, Yaksi E, Deisseroth K, Friedrich RW (2011) Neuronal filtering of multiplexed odour representations. Nature 479:493–498
117. Friedrich RW, Laurent G (2001) Dynamic optimization of odor representations by slow temporal patterning of mitral cell activity. Science 291:889–894
118. Yaksi E, von Saint Paul F, Niessing J, Bundschuh ST, Friedrich RW (2009) Transformation of odor representations in target areas of the olfactory bulb. Nat Neurosci 12:474–482
119. Friedrich RW, Korsching SI (1997) Combinatorial and chemotopic odorant coding in the zebrafish olfactory bulb visualized by optical imaging. Neuron 18:737–752
120. Bundschuh ST, Zhu P, Scharer YP, Friedrich RW (2012) Dopaminergic modulation of mitral cells and odor responses in the zebrafish olfactory bulb. J Neurosci 32:6830–6840
121. Scharer YP, Shum J, Moressis A, Friedrich RW (2012) Dopaminergic modulation of synaptic transmission and neuronal activity patterns in the zebrafish homolog of olfactory cortex. Front Neural Circuits 6:76
122. Jacobson GA, Rupprecht P, Friedrich RW (2018) Experience-dependent plasticity of odor representations in the telencephalon of zebrafish. Curr Biol 28:1–14.e13
123. Zhu P, Frank T, Friedrich RW (2013) Equalization of odor representations by a network of electrically coupled inhibitory interneurons. Nat Neurosci 16:1678–1686
124. Yabuki Y, Koide T, Miyasaka N, Wakisaka N, Masuda M, Ohkura M, Nakai J, Tsuge K, Tsuchiya S, Sugimoto Y, Yoshihara Y (2016) Olfactory receptor for prostaglandin F2alpha mediates male fish courtship behavior. Nat Neurosci 19:897–904

125. Wakisaka N, Miyasaka N, Koide T, Masuda M, Hiraki-Kajiyama T, Yoshihara Y (2017) An adenosine receptor for olfaction in fish. Curr Biol 27:1437–1447.e1434
126. Jetti SK, Vendrell-Llopis N, Yaksi E (2014) Spontaneous activity governs olfactory representations in spatially organized habenular microcircuits. Curr Biol 24:434–439
127. Mathuru AS, Kibat C, Cheong WF, Shui G, Wenk MR, Friedrich RW, Jesuthasan S (2012) Chondroitin fragments are odorants that trigger fear behavior in fish. Curr Biol 22:538–544
128. Hinz C, Namekawa I, Behrmann-Godel J, Oppelt C, Jaeschke A, Muller A, Friedrich RW, Gerlach G (2013) Olfactory imprinting is triggered by MHC peptide ligands. Sci Rep 3:2800
129. DeMaria S, Berke AP, Van Name E, Heravian A, Ferreira T, Ngai J (2013) Role of a ubiquitously expressed receptor in the vertebrate olfactory system. J Neurosci 33:15235–15247
130. Reiten I, Uslu FE, Fore S, Pelgrims R, Ringers C, Diaz Verdugo C, Hoffman M, Lal P, Kawakami K, Pekkan K, Yaksi E, Jurisch-Yaksi N (2017) Motile-cilia-mediated flow improves sensitivity and temporal resolution of olfactory computations. Curr Biol 27:166–174
131. Quirin S, Vladimirov N, Yang CT, Peterka DS, Yuste R, Ahrens MB (2016) Calcium imaging of neural circuits with extended depth-of-field light-sheet microscopy. Opt Lett 41:855–858
132. Ahrens MB, Orger MB, Robson DN, Li JM, Keller PJ (2013) Whole-brain functional imaging at cellular resolution using light-sheet microscopy. Nat Methods 10:413–420
133. Yang Z, Mei L, Xia F, Luo Q, Fu L, Gong H (2015) Dual-slit confocal light sheet microscopy for in vivo whole-brain imaging of zebrafish. Biomed Opt Express 6:1797–1811
134. Tomer R, Lovett-Barron M, Kauvar I, Andalman A, Burns VM, Sankaran S, Grosenick L, Broxton M, Yang S, Deisseroth K (2015) SPED light sheet microscopy: fast mapping of biological system structure and function. Cell 163:1796–1806
135. Dunn TW, Mu Y, Narayan S, Randlett O, Naumann EA, Yang CT, Schier AF, Freeman J, Engert F, Ahrens MB (2016) Brain-wide mapping of neural activity controlling zebrafish exploratory locomotion. elife 5:e12741
136. Vladimirov N, Mu Y, Kawashima T, Bennett DV, Yang CT, Looger LL, Keller PJ, Freeman J, Ahrens MB (2014) Light-sheet functional imaging in fictively behaving zebrafish. Nat Methods 11:883–884
137. Portugues R, Feierstein CE, Engert F, Orger MB (2014) Whole-brain activity maps reveal stereotyped, distributed networks for visuomotor behavior. Neuron 81:1328–1343
138. Winter MJ, Windell D, Metz J, Matthews P, Pinion J, Brown JT, Hetheridge MJ, Ball JS, Owen SF, Redfern WS, Moger J, Randall AD, Tyler CR (2017) 4-dimensional functional profiling in the convulsant-treated larval zebrafish brain. Sci Rep 7:6581
139. Turrini L, Fornetto C, Marchetto G, Mullenbroich MC, Tiso N, Vettori A, Resta F, Masi A, Mannaioni G, Pavone FS, Vanzi F (2017) Optical mapping of neuronal activity during seizures in zebrafish. Sci Rep 7:3025
140. Vanwalleghem G, Heap LA, Scott EK (2017) A profile of auditory-responsive neurons in the larval zebrafish brain. J Comp Neurol 525:3031–3043
141. Kim DH, Kim J, Marques JC, Grama A, Hildebrand DGC, Gu W, Li JM, Robson DN (2017) Pan-neuronal calcium imaging with cellular resolution in freely swimming zebrafish. Nat Methods 14:1107–1114
142. Cong L, Wang Z, Chai Y, Hang W, Shang C, Yang W, Bai L, Du J, Wang K, Wen Q (2017) Rapid whole brain imaging of neural activity in freely behaving larval zebrafish (Danio rerio). elife 6:pii: e28158
143. Muto A, Kawakami K (2016) Calcium imaging of neuronal activity in free-swimming larval zebrafish. Methods Mol Biol 1451:333–341
144. Muto A, Lal P, Ailani D, Abe G, Itoh M, Kawakami K (2017) Activation of the hypothalamic feeding centre upon visual prey detection. Nat Commun 8:15029
145. Stephenson DG, Lamb GD, Stephenson GM (1998) Events of the excitation-contraction-relaxation (E-C-R) cycle in fast- and slow-twitch mammalian muscle fibres relevant to muscle fatigue. Acta Physiol Scand 162:229–245
146. Wen H, Brehm P (2005) Paired motor neuron-muscle recordings in zebrafish test the receptor blockade model for shaping synaptic current. J Neurosci 25:8104–8111

147. Lin CY, Chen JS, Loo MR, Hsiao CC, Chang WY, Tsai HJ (2013) MicroRNA-3906 regulates fast muscle differentiation through modulating the target gene homer-1b in zebrafish embryos. PLoS One 8:e70187
148. Klatt Shaw D, Gunther D, Jurynec MJ, Chagovetz AA, Ritchie E, Grunwald DJ (2018) Intracellular calcium mobilization is required for sonic hedgehog signaling. Dev Cell 45:512–525.e515
149. Perni S, Marsden KC, Escobar M, Hollingworth S, Baylor SM, Franzini-Armstrong C (2015) Structural and functional properties of ryanodine receptor type 3 in zebrafish tail muscle. J Gen Physiol 145:173–184
150. Kelu JJ, Webb SE, Parrington J, Galione A, Miller AL (2017) Ca^{2+} release via two-pore channel type 2 (TPC2) is required for slow muscle cell myofibrillogenesis and myotomal patterning in intact zebrafish embryos. Dev Biol 425:109–129
151. Xiyuan Z, Fink RHA, Mosqueira M (2017) NO-sGC pathway modulates Ca^{2+} release and muscle contraction in zebrafish skeletal muscle. Front Physiol 8:607
152. Linsley JW, Hsu IU, Groom L, Yarotskyy V, Lavorato M, Horstick EJ, Linsley D, Wang W, Franzini-Armstrong C, Dirksen RT, Kuwada JY (2017) Congenital myopathy results from misregulation of a muscle Ca^{2+} channel by mutant Stac3. Proc Natl Acad Sci U S A 114:E228–E236
153. Horstick EJ, Linsley JW, Dowling JJ, Hauser MA, McDonald KK, Ashley-Koch A, Saint-Amant L, Satish A, Cui WW, Zhou W, Sprague SM, Stamm DS, Powell CM, Speer MC, Franzini-Armstrong C, Hirata H, Kuwada JY (2013) Stac3 is a component of the excitation-contraction coupling machinery and mutated in Native American myopathy. Nat Commun 4:1952
154. Smith LL, Gupta VA, Beggs AH (2014) Bridging integrator 1 (Bin1) deficiency in zebrafish results in centronuclear myopathy. Hum Mol Genet 23:3566–3578
155. Gibbs EM, Davidson AE, Telfer WR, Feldman EL, Dowling JJ (2014) The myopathy-causing mutation DNM2-S619L leads to defective tubulation in vitro and in developing zebrafish. Dis Model Mech 7:157–161
156. Jackson HE, Ono Y, Wang X, Elworthy S, Cunliffe VT, Ingham PW (2015) The role of Sox6 in zebrafish muscle fiber type specification. Skelet Muscle 5:2
157. Shahid M, Takamiya M, Stegmaier J, Middel V, Gradl M, Kluver N, Mikut R, Dickmeis T, Scholz S, Rastegar S, Yang L, Strahle U (2016) Zebrafish biosensor for toxicant induced muscle hyperactivity. Sci Rep 6:23768
158. Gore AV, Monzo K, Cha YR, Pan W, Weinstein BM (2012) Vascular development in the zebrafish. Cold Spring Harb Perspect Med 2:a006684
159. Bakkers J (2011) Zebrafish as a model to study cardiac development and human cardiac disease. Cardiovasc Res 91:279–288
160. Chi NC, Shaw RM, Jungblut B, Huisken J, Ferrer T, Arnaout R, Scott I, Beis D, Xiao T, Baier H, Jan LY, Tristani-Firouzi M, Stainier DY (2008) Genetic and physiologic dissection of the vertebrate cardiac conduction system. PLoS Biol 6:e109
161. Weber M, Scherf N, Meyer AM, Panakova D, Kohl P, Huisken J (2017) Cell-accurate optical mapping across the entire developing heart. elife 6:pii: e28307
162. Fukuda R, Gunawan F, Beisaw A, Jimenez-Amilburu V, Maischein HM, Kostin S, Kawakami K, Stainier DY (2017) Proteolysis regulates cardiomyocyte maturation and tissue integration. Nat Commun 8:14495
163. Liu J, Bressan M, Hassel D, Huisken J, Staudt D, Kikuchi K, Poss KD, Mikawa T, Stainier DY (2010) A dual role for ErbB2 signaling in cardiac trabeculation. Development 137:3867–3875
164. Shimizu H, Schredelseker J, Huang J, Lu K, Naghdi S, Lu F, Franklin S, Fiji HD, Wang K, Zhu H, Tian C, Lin B, Nakano H, Ehrlich A, Nakai J, Stieg AZ, Gimzewski JK, Nakano A, Goldhaber JI, Vondriska TM, Hajnoczky G, Kwon O, Chen JN (2015) Mitochondrial Ca^{2+} uptake by the voltage-dependent anion channel 2 regulates cardiac rhythmicity. elife 4:pii: e04801

165. Hong TT, Smyth JW, Chu KY, Vogan JM, Fong TS, Jensen BC, Fang K, Halushka MK, Russell SD, Colecraft H, Hoopes CW, Ocorr K, Chi NC, Shaw RM (2012) BIN1 is reduced and Cav1.2 trafficking is impaired in human failing cardiomyocytes. Heart Rhythm 9:812–820
166. Kirchmaier BC, Poon KL, Schwerte T, Huisken J, Winkler C, Jungblut B, Stainier DY, Brand T (2012) The Popeye domain containing 2 (popdc2) gene in zebrafish is required for heart and skeletal muscle development. Dev Biol 363:438–450
167. Ramspacher C, Steed E, Boselli F, Ferreira R, Faggianelli N, Roth S, Spiegelhalter C, Messaddeq N, Trinh L, Liebling M, Chacko N, Tessadori F, Bakkers J, Laporte J, Hnia K, Vermot J (2015) Developmental alterations in heart biomechanics and skeletal muscle function in desmin mutants suggest an early pathological root for desminopathies. Cell Rep 11:1564–1576
168. Heckel E, Boselli F, Roth S, Krudewig A, Belting HG, Charvin G, Vermot J (2015) Oscillatory flow modulates mechanosensitive klf2a expression through trpv4 and trpp2 during heart valve development. Curr Biol 25:1354–1361
169. Kaufman CK, White RM, Zon L (2009) Chemical genetic screening in the zebrafish embryo. Nat Protoc 4:1422–1432
170. Dunn TW, Gebhardt C, Naumann EA, Riegler C, Ahrens MB, Engert F, Del Bene F (2016) Neural circuits underlying visually evoked escapes in larval zebrafish. Neuron 89:613–628

Chapter 37
Stimulus-Secretion Coupling in Beta-Cells: From Basic to Bedside

Md. Shahidul Islam

Abstract Insulin secretion in humans is usually induced by mixed meals, which upon ingestion, increase the plasma concentration of glucose, fatty acids, amino acids, and incretins like glucagon-like peptide 1. Beta-cells can stay in the off-mode, ready-mode or on-mode; the mode-switching being determined by the open state probability of the ATP-sensitive potassium channels, and the activity of enzymes like glucokinase, and glutamate dehydrogenase. Mitochondrial metabolism is critical for insulin secretion. A sound understanding of the intermediary metabolism, electrophysiology, and cell signaling is essential for comprehension of the entire spectrum of the stimulus-secretion coupling. Depolarization brought about by inhibition of the ATP sensitive potassium channel, together with the inward depolarizing currents through the transient receptor potential (TRP) channels, leads to electrical activities, opening of the voltage-gated calcium channels, and exocytosis of insulin. Calcium- and cAMP-signaling elicited by depolarization, and activation of G-protein-coupled receptors, including the free fatty acid receptors, are intricately connected in the form of networks at different levels. Activation of the glucagon-like peptide 1 receptor augments insulin secretion by amplifying calcium signals by calcium induced calcium release (CICR). In the treatment of type 2 diabetes, use of the sulfonylureas that act on the ATP sensitive potassium channel, damages the beta cells, which eventually fail; these drugs do not improve the cardiovascular outcomes. In contrast, drugs acting through the glucagon-like peptide-1 receptor protect the beta-cells, and improve cardiovascular outcomes. The use of the glucagon-like peptide 1 receptor agonists is increasing and that of sulfonylurea is decreasing. A better understanding of the stimulus-secretion coupling may lead to the discovery of other molecular targets for development of drugs for the prevention and treatment of type 2 diabetes.

M. S. Islam (✉)
Department of Clinical Science and Education, Södersjukhuset, Karolinska Institutet, Stockholm, Sweden

Department of Emergency Care and Internal Medicine, Uppsala University Hospital, Uppsala, Sweden
e-mail: Shahidul.islam@ki.se

© Springer Nature Switzerland AG 2020

M. S. Islam (ed.), *Calcium Signaling*, Advances in Experimental Medicine and Biology 1131, https://doi.org/10.1007/978-3-030-12457-1_37

Keywords Insulin secretion · Beta-cells · Glucokinase · Glutamate dehydrogenase · Glucagon-like peptide-1 · Calcium induced calcium release · ATP-sensitive potassium channel · Stimulus-secretion coupling · Islets of Langerhans · Voltage-gated calcium channels and insulin secretion · Transient receptor potential channels and insulin secretion · Mitochondria and insulin secretion · Type 2 diabetes

Abbreviations

$[Ca^{2+}]_c$	Cytoplasmic free Ca^{2+} concentration
EPAC	Exchange protein directly activated by cAMP
K_{ATP}	ATP-sensitive potassium channel
NBD	Nucleotide binding domain
PKA	Protein kinase A
SCHAD	short chain 3-hydroxyacyl-CoA dehydrogenase
SUR1	Sulfonylurea receptor 1
VGCC	Voltage-gated Ca^{2+} channel

37.1 Introduction

The β-cells of the islets of Langerhans sense primarily the changes in the concentration of glucose in the plasma, and respond promptly and precisely by changing the rate of secretion of insulin to maintain the concentration of glucose within the physiologic range. They act as fuel sensors by virtue of possessing the glucose sensor glucokinase, and the ATP sensor, the ATP-sensitive potassium (K_{ATP}) channel. These cells are rendered "glucose-competent" by the incretin hormone glucagon-like peptide-1 (GLP-1) [1]. Strictly speaking, the physiological stimulus for insulin secretion is food, which is taken in the form of mixed meals, and which increases the concentration of not only glucose, but also of other nutrients, including the 20 amino acids, and many fatty acids, in the plasma. Foods increase not only the plasma concentrations of the nutrients, but also of several gut-derived peptides including GLP-1, and gastric inhibitory polypeptide (GIP). In their native environment in the normal human body, the β-cells do not have any resting state; they secrete insulin, albeit at low rates, even during the fasting state [2]. The concentration of glucose in the plasma must be maintained above a certain level to prevent hypoglycemia, which is life threatening. For this reason, insulin secretion is tightly coupled to the concentration of glucose. These cells switch between three overlapping modes of secretion: the "off-mode", the "ready-mode", and the "on-mode" [3]. At the molecular level, the open state probability of the ATP-sensitive potassium (K_{ATP}) channel partly determines the switching between the three modes.

In normal β-cells, when the concentration of glucose in the plasma plunges into the hypoglycemic range, the open state probability of the K_{ATP} channel increases, switching the cells rapidly to the off-mode, and shutting off the insulin secretion. Concentration of glucose in the hypoglycemic range minimizes metabolism of glucose because the enzyme glucokinase that phosphorylates glucose in the first step of the glycolytic pathway has high K_m for glucose. This reduces the concentration of ATP, and increases the concentration of MgADP in the cytoplasm, leading to an increase in the open state probability of K_{ATP} channel, hyperpolarization of membrane potential, closure of the voltage-gated Ca^{2+} channels (VGCC), reduction of the cytoplasmic free Ca^{2+} concentration ($[Ca^{2+}]_c$) to the resting level, and cessation of insulin secretion. In this scenario, the K_{ATP} channel acts as an emergency safety device for rapidly shutting off insulin secretion. If the K_{ATP} channel is absent, or if the channel do not open normally due to some mutations, then the channel does not open even when the glucose concentration reaches the hypoglycemic range. Inactivating mutations of K_{ATP} channel genes permanently switch the β-cells to the ready-mode or the on-mode. The *KCNJ11* gene encodes the pore forming $K_{ir}6.2$ subunit of the K_{ATP} channel. Activating mutations in the *KCNJ11* gene switches the β-cells to the off-mode [4]. When the β-cells are in the off-mode due to mutations in the K_{ATP} channel genes, glucose and GLP-1 cannot stimulate insulin secretion, and the result is permanent neonatal diabetes mellitus. In normal subjects, during the fasting condition, when the plasma glucose concentration is ~5.5 mmol/L, the β-cells are in the ready-mode.

37.2 Calcium Signaling in the β-Cells

Glucose enters the β-cells through glucose transporter 2 following the concentration gradient, and it is phosphorylated by glucokinase. Further metabolism through the glycolytic pathway, and the mitochondrial metabolism increases the concentration of ATP, and reduces the concentration of ADP, both of which remain partly in the free forms, and partly in the magnesium bound forms. The binding of these different forms of the nucleotides to the sulfonylurea receptor 1 (SUR1) and $K_{ir}6.2$ subunits of the K_{ATP} channel determines the open state probability of the channel.

Critical events in the stimulus-secretion coupling in the β-cells are membrane depolarization, increase of $[Ca^{2+}]_c$, and increase of cAMP. Membrane depolarization occurs due to the reduction in the open state probability of the K_{ATP} channel, and activation of inward depolarizing currents mediated by several ion channels, including some transient receptor potential (TRP) channels [3, 5]. When the membrane potential reaches the respective thresholds for the activation of the VGCCs, the channels open, and Ca^{2+} enters into the cytoplasm. Ca^{2+} entering through the VGCCs triggers further Ca^{2+} release from the endoplasmic reticulum (ER) Ca^{2+} stores, a process called Ca^{2+}-induced Ca^{2+}-release (CICR) [6]. Stimulations of the β-cells by nutrients and agonists that induce insulin secretion increase the $[Ca^{2+}]_c$, often in the form of oscillations, and an increase in the $[Ca^{2+}]_c$ in the β-cells in response to different stimuli is seen as a sign of β-cell activation [7–10]. Activation of the phospholipase-C-linked G-protein coupled receptors by agonists like the neurotransmitter acetylcholine increases $[Ca^{2+}]_c$ by releasing Ca^{2+} from the ER through activation of the inositol 1,4,5 trisphosphate receptors [11]. GLP-1 activates the GLP-1 receptor leading to the increases of cAMP, which has pleiotropic effects that enhance insulin secretion [12]. Cross talks between Ca^{2+}- and cAMP-signaling occurs at multiple levels, and it is nearly impossible to dissect between pure Ca^{2+} signaling, and pure cAMP signaling.

37.3 Hexokinases

Insulin secretion is "supply driven". Glucokinase (hexokinase 4), one of the four members of the hexokinase family of enzymes, is the primary glucose sensor. It phosphorylates glucose to glucose-6 phosphate. The affinity of glucokinase for glucose is low, compared to that of other hexokinases. The half saturation (the concentration of glucose at which the enzyme is half saturated, $S_{0.5}$) is about 8 mmol/L. The low affinity of glucokinase for glucose ensures that the concentration of glucose in the plasma is kept within the physiologic range.

Glucokinase is the only hexokinase expressed in the adult β-cells. In the neonatal period hexokinase 1 is also expressed in the β-cells, but its expression is down-regulated after birth. Some mutations in the non-coding region of HK1 can impair the process of normal suppression of HK1 expression [13]. Persistent expression of HK1 after birth leads to congenital hyperinsulinism in neonates and infants [14]. These subjects develop hypoglycemia during prolonged fasting. Some human insulinoma cells express hexokinase-1, which can explain excessive insulin secretion even during hypoglycemia [15]. Some of these patients can be treated by diazoxide for variable length of time, but those who have severe hypoglycemia, and those who do not respond to diazoxide, need near total pancreratectomy.

Activating mutations of glucokinase shifts the glucose-dependency curve to the left, reduces glucose $S_{0.5,}$ and increases the activity index of the enzyme. People who have activating mutations of glucokinase have hyperinsulinism, and variable degrees of hypoglycemia, ranging from severe hypoglycemia during neonatal period to milder hypoglycemia later in life. In vitro experiments with islets isolated from mice that express an activating mutation (A456V) of glucokinase show that the threshold for glucose stimulated insulin secretion is shifted to the left [16]. Hypoglycemia in some of these subjects can be treated by diazoxide that binds to the SUR1 subunit of the K_{ATP} channel, and increases the open state probability of the channel. Some subjects do not respond to diazoxide and may need the somatostatin analogs for reducing insulin secretion, and others may need near-total pancreatectomy.

Inactivating mutations of glucokinase increase the $S_{0.5}$ of the enzyme for glucose when the mutations are located in the glucose-binding site, and reduces the K_m for ATP when the mutations are located in the ATP-binding site. These mutations decrease the calculated activity index of glucokinase. Heterozygous inactivating mutations of glucokinase usually cause "maturity onset diabetes of the young type 2" (MODY2). MODY2 subjects have only a mild form of diabetes with mild fasting hyperglycemia (5.5–8 mmol/L), and HbA1C 40–60 mmol/mol. MODY2 subjects do not suffer from the long term complications of diabetes, and treatment is not recommended outside pregnancy [17].

Completely inactive glucokinase due to homozygous or compound heterozygous mutations, which occur rarely, causes permanent neonatal diabetes mellitus (PNDM), requiring lifelong insulin therapy for severe hyperglycemia [18].

37.4 K_{ATP} Channel

The K_{ATP} channel of the β-cells is composed of two proteins, the regulatory sulfonylurea receptor 1 (SUR1) encoded by the *ABCC8* gene, and the pore-forming inwardly rectifying (ir) potassium channel $K_{ir}6.2$ (encoded by the *KCNJ11* gene). SUR1 binds the sulfonylurea class of antidiabetic drugs that promote insulin secretion. $K_{ir}6.2$ is an inward-rectifier K^+ channel, (which means that it has a greater tendency to allow K^+ to enter into a cell rather than flow out of a cell. Each K_{ATP} channel consists of eight subunits, four $K_{ir}6.2$ in the center surrounded by four SUR1 in the periphery. *ABCC8* and *KCNJ11* are two adjacent genes located on the short arm (p) of the chromosome 11 at position 15.1 (cytogenetic location 11p15.1). Opening of the K_{ATP} channel polarizes, and closure of the channel depolarizes the membrane potential of β-cells.

ATP potently inhibits the K_{ATP} channel activity by binding to the $K_{ir}6.2$ subunit. This binding and inhibitory action of ATP is independent of Mg^{2+}. On the other hand Mg-nucleotides, including both MgATP and MgADP activate the K_{ATP} channel by binding to the nucleotide binding domain (NBD) of SUR1 subunit. Binding of MgADP to the SUR1 is more efficacious than binding of MgATP to SUR1 in activating the K_{ATP} channel [19]. Sulfonylurea drugs inhibit K_{ATP} channel activity mostly by direct binding to the SUR1, but also by preventing the MgATP binding to the SUR1 and MgATP-mediated activation of the channel. Like MgADP and MgATP, diazoxide also activates K_{ATP} channel by binding to SUR1, although the exact site of diazoxide action is not known.

37.4.1 *Inactivating Mutations of K_{ATP} Channel*

Some mutations of the two K_{ATP} channel genes inhibit the ability of the channels to open, and other mutations impair the biogenesis or trafficking of the channels to the plasma membrane. These "inactivating" mutations cause persistent depolarization of the membrane, and persistent insulin secretion, despite hypoglycemia. Such β-cells are unable to switch to the off-mode, which normal β-cells would do when there is hypoglycemia. Inactivating mutations of the K_{ATP} channel genes are the commonest causes of hypoglycemia due to congenital hyperinsulinism. Some mutations impair K_{ATP} channel activity, and the degree of the impairments determines the responsiveness of the mutant K_{ATP} channel to diazoxide [20].

Some of these mutations act in a recessive, and others act in a dominant fashion. Recessive mutations in the *ABCC8* and *KCNJ11* genes impair trafficking of the channels to the plasma membrane causing almost total loss of the channel in all of the β-cells. For this reason, the hyperinsulinism caused by the recessive mutations do not respond to the treatment by diazoxide. On the other hand the dominant mutations allow normal trafficking of the channels to the plasma membrane, but these mutations impair response to MgADP or diazoxide to variable degrees. Hyperinsulinism caused by some of these dominant mutations are diazoxide-responsive, and others are diazoxide-unresponsive [20]. Paradoxically, diazoxide can also stimulate insulin secretions in some of these cases, the mechanism of which remains unknown [14, 21].

There are two histological forms of K_{ATP} hyperinsulinism: the diffuse form, and the focal form. In the diffuse form, all the β-cells in the pancreas are affected. In the focal form, the β-cells of only parts of the pancreas are affected, but the β-cells located outside the affected area of the pancreas are normal. Diffuse form of K_{ATP} hyperinsulinism is due to recessive mutations in the K_{ATP} channel genes, *ABCC8 or KCNJ11.*

The focal form is caused by "two hits". The first is a mono-allelic recessive mutation in the *ABCC8* or *KCNJ11* gene that is paternally transmitted. The second "hit" is loss of heterozygosity due to a somatic loss of maternal 11p15.1 with paternal isodisomy at the same locus, inside the focal lesion in the pancreas [22, 23]. The 11p15.1 region contains not only the K_{ATP} channel genes, but also some imprinted genes including *H19*, *IGF2*, and *CDKN1C*, which are involved in the regulation of cell proliferation [24]. The *CDKN1C* gene encodes cyclin dependent kinase inhibitor 1C, which is a tumor suppressor. Absence of cyclin dependent kinase inhibitor 1C is one of the factors that probably contributes to the development of the β-cell hyperplasia in the focal form of the K_{ATP} hyperinsulinism [24]. In the focal form of K_{ATP}-hyperinsulinism, there is not only complete loss of K_{ATP}-channel function, but also formation of β-cell "tumors" [24]. The β-cells in the "unaffected" part of the pancreas, in the focal form of $_{KAP}$ hyperinsulinism have only mono-allelic recessive mutation of the K_{ATP} channel genes, but no loss of maternal 11p15.1. As expected, these β-cells function normally [25].

It is important to identify people who have the focal form of the disease, because this form can be completely cured by selective surgical resection of the pathologic area of the pancreas. The diffuse form, if they are not responsive to medical treatments, requires 95–98% pancreatectomy. Genetic analysis is useful in identifying the diffuse forms and the focal forms of K_{ATP} hyperinsulinism. If there are two recessive mutations in the K_{ATP} channel genes then the hyperinsulinism is more likely to be caused by the diffuse form. If there is only one heterozygous recessive mutation in either *ABCC8* or *KCNJ11* gene, then the hyperinsulinism is highly likely to be due to the focal form of the disease [26]. β-cells selectively incorporate 18-Fluorodihydroxy phenylalanine (18F-DOPA). 18F-DOPA PET scan is currently the most accurate and specific method for identifying the focal lesions, but the facility is available only in a few advanced centers [27].

Inactivating mutations of the K_{ATP} channels cause inappropriately increased insulin secretion even during the fasting state leading to hypoglycemia. At first sight it may appear that absence of the K_{ATP} channel, or defective opening of these channels leads to persistent depolarization of the membrane-potential of the β-cells, opening of the VGCCs, increase of $[Ca^{2+}]_c$, leading to exocytosis of insulin, but the situation is more complicated than that. These cells secrete insulin even when the concentration of glucose is <5 mmol/L, when glucose metabolism through the high K_m glucokinase is dramatically reduced, suggesting that these cells metabolize glucose through other low K_m hexokinases. In fact, it turns out that in these cells there is a 16-fold increase in the expression of the hexokinase 1 (*HK1*) gene, and marked reduction in the expression of the glucokinase (*GCK*) gene. Expression of other important glycolytic enzymes is also increased in these cells. As a result, there is increased glycolysis, as measured by ^{13}C tracer studies, in these cells [28].

While these β-cells secrete insulin under fasting state, their response to glucose challenge is reduced or abolished [25, 28]. The insulin content of these islets is lower than that of the normal islets. Basal insulin secretion from these islets is higher, but there is little or no increase in insulin secretion upon glucose stimulation [25, 28]. In normal β-cells, glucose stimulates insulin secretion by depolarizing the membrane potential leading to the opening of the VGCCs, and elevating the $[Ca^{2+}]_c$, but in the β-cells where the K_{ATP} channel is absent or inactivated, the membrane potential and the $[Ca^{2+}]_c$ are already elevated under basal conditions leading to high basal insulin secretion [25, 28]. Sulfonylureas that close K_{ATP} channel, increase insulin secretion from normal islets, but fail to do so in the islets from patients with inactivating mutations of K_{ATP} channel [25]. Normal glucose sensing in these cells is impaired because of reduced expression of glucokinase, and increased expression of hexokinase 1, as mentioned before [28]. In these islets, metabolism of glucose is abnormal in many other ways too. For instance, glucose is used for synthesis of the amino acids glycine, serine and glutamine [28]. The expression of 3-phosphoglycerate dehydrogenase (PHGDH), the enzyme important for serine/glycine biosynthesis, is increased 38 fold [28]. Expression of another key enzyme for serine/glycine biosynthesis, phosphoserine aminotransferase is also increased tenfold [28].

While the insulin response of these cells to glucose challenge is reduced, their response to amino acid is increased [29]. These patients develop hypoglycemia after ingesting protein-rich meals. The absence or the closure of the K_{ATP} channel switches the β-cells from off-mode to the ready-mode. When in the ready mode, the cells can be easily switched to the on-mode by many stimuli, including many amino acids. In in vitro experiments, stimulation of islets isolated from patients who have inactivating mutations of the K_{ATP} channel, by a mixture of amino acids, increases insulin secretion, whereas stimulation of the normal islets fails to do so [28]. This may be due to increased amino acid metabolism as a consequence of activation of the enzyme glutamate dehydrogenase (GDH) in the β-ells that have K_{ATP}-channel with inactivating mutations [28, 30].

Inactivation of the K_{ATP} during the development of the β-cells, leads to persistent increase of $[Ca^{2+}]_c$. This induces homeostatic, adaptive or compensatory changes in the components of the Ca^{2+} signaling toolkit, by mechanisms that are not well understood. This phenomenon has been named "Ca^{2+} homeostasome" [31]. Thus, the phenotypes that we see in the inactivating mutations of the K_{ATP} channel, are not primarily the results of the alterations of the function of the K_{ATP} channel, rather the net results of global homeostatic reorganization of the cellular signaling network. As a consequence of the chronic inactivation of the K_{ATP} channel, numerous genes are upregulated, and numerous other genes are downregulated, leading to alterations in glucose and amino acid metabolism, to name a few [28].

37.4.2 Activating Mutations of the KATP Channel Genes

Activating mutations of the *KCNJ11* and the *ABCC8* genes increase the open state probability of the K_{ATP} channel. The ability of ATP to inhibit the channel is reduced to a variable extent depending on different mutations. These mutations switch the β-cells to the off-mode. The membrane potential of these β-cells remain hyperpolarized, the VGCCs cannot open, $[Ca^{2+}]_c$ remains low, and insulin secretion is reduced. These mutations account for $\sim 50\%$ of neonatal diabetes mellitus, which are commonly treated with insulin in the beginning. Once the genetic diagnosis is established, $> 90\%$ of these infants, children and young adults can be treated effectively, and safely by the sulfonylurea class of anti-diabetic drugs, for many years, perhaps for rest of their lives [32]. The doses of sulfonylurea drugs needed for the treatment of these patients are usually high, and vary depending on the severity of the mutations. The sulfonylurea drugs reduce the open state probability of the K_{ATP} channels in these subjects. In this scenario, the sulfonylureas switch the β-cells from the off-mode to the ready-mode, when the insulin secretion can be stimulated further by glucose, and other nutrients, and by glucagon-like peptide 1 (GLP-1) [4].

37.5 Voltage-Gated Potassium Channel

Voltage-gated K^+ (K_v) channels are activated by depolarization; they mediate the repolarizing phase of the action potential, and limit insulin secretion [33]. Activation of GLP-1 receptor can reduce K_v currents, and may thereby extend action potential, leading to increased Ca^{2+} entry through the VGCCs. This may be one of the many mechanisms that contribute to the GLP-1-induced augmentation of insulin secretion.

The pore forming α-subunit of one of the K_v channels, the $K_v7.1$ is encoded by the *KCNQ1* gene, which is expressed both in the human heart, and in the human β-cells [34]. Some mutations in the *KCNQ1* gene cause one particular type of long QT syndrome (LQTS). These subjects have lower plasma glucose concentration, and they have increased insulin response to oral glucose load. They get episodes of hypoglycemia 3–5 h after meals [34]. As a sequelae of increased insulin secretion and hypoglycemia, these patients also tend to have hypokalemia, which further increases the risk of cardiac arrhythmia, and cardiac death. Variants in the *KCNQ1* gene are also associated with susceptibility to type 2 diabetes [35].

37.6 GLP-1

Oral glucose is more effective than intravenous glucose in stimulating insulin secretion, even when the plasma glucose concentration is maintained at the same level. This is called incretin effect. In healthy persons, 25–75% of insulin release after a glucose load is due to the incretin effect, which is mediated mostly by GLP-1, and gastric inhibitory polypeptide (GIP) [36]. In type 2 diabetes the incretin effect is almost abolished. Two active forms of endogenous GLP-1 are GLP1 (7-36)NH2, and GLP1 [7–37].

GLP-1 is secreted by the L-cells located in the small and large intestine, from duodenum to the colon. There are more L-cells particularly in the ileum. After ingestion of foods the concentration of GLP-1 in the plasma increases rapidly. If the β-cells are in the ready-mode, which is the case when the glucose concentration in the plasma is above the threshold for insulin secretion (~5 mmol/L), then GLP-1 switches the β-cells to the on-mode. When the glucose concentration in the plasma is <5 mmol/L, the β-cells are not in the ready-mode, and GLP-1 cannot switch the β-cells to the on-mode despite increasing the cAMP concentration in the cell.

GLP-1 binds to the glucagon-like-peptide-1 receptor (GLP-1 receptor), which is expressed at variable levels, in numerous tissues, including the human β-cells, δ-cells and α-cells. Activation of the GLP-1 receptor of human β-cells leads to activation of adenylyl cyclase (not clear which isoform), and increase of cAMP in the cells. cAMP activates protein kinase A (PKA) and EPAC (exchange protein directly activated by cAMP), both of which act on many target proteins, including ion channels, enzymes, and the proteins that mediate exocytosis. Activation of the GLP-1 receptor facilitates inhibition of K_{ATP} channel [37], activation of the transient receptor potential channel subfamily M member 4 (TRPM4) [38, 39], and transient receptor potential channel subfamily M member 2 [40, 41], facilitating membrane depolarization. The effects of GLP-1 on the changes in the $[Ca^{2+}]_c$ is particularly dramatic [42–46]. GLP-1 augments Ca^{2+} signals qualitatively and quantitatively. It increases Ca^{2+} current through the L-type voltage-gated Ca^{2+} channel by a cAMP dependent manner [47–50]. GLP-1, through PKA and EPAC facilitates CICR that amplifies Ca^{2+} signals [43, 51–53] and amplifies insulin secretion.

PKA-mediated phosphorylation of serine-103 of the essential Ca^{2+}-sensor synaptotagmin 7 increases exocytosis of insulin [54]. The precise mechanism is unclear because PKA phosphorylation of the serine-103 does not increase the Ca^{2+}-sensitivity of exocytosis [see figure S3 of Wu et al. [54]].

Excess glucagon contributes to the pathogenesis of type 2 diabetes [55]. While GLP-1 increases insulin secretion, it inhibits glucagon secretion. The expression of GLP-1 receptor in human α-cells is low, compared to that in the human β-cells and δ-cells [56]. GLP-1 inhibits glucagon secretion from the human α-cells through direct activation of the GLP-1 receptor, formation of cAMP, and activation of the protein kinase A, which leads to phosphorylation and inhibition of the P/Q type of VGCCs of the α-cells [57].

The half-life of GLP-1 in the plasma is only 1–2 min because of its rapid degradation by dipeptidyl peptidase 4 (DPP4), which is ubiquitously expressed. Inhibitors of DPP4 increase the concentration of GLP-1, and the inhibitors are used in the treatment of type 2 diabetes. GLP-1 analogs with long half-life have been developed by substitution of some of the amino acids, and by attaching fatty acids to the molecule. These changes promote self-aggregation, albumin-binding, slow absorption of the injected GLP-1 analog from the subcutaneous injection site, slow degradation, and slow renal elimination. The long half-life of semaglutide, a long-acting GLP-1 analog, means that it is enough to take one injection per week.

A large population-based study showed that GLP-1 secretion is reduced in prediabetes, and screen-detected diabetes, as well as in the obese and overweight individuals, suggesting that reduced secretion of GLP-1 leads to impaired regulation of glucose metabolism and appetite [58]. Other studies suggest that impaired insulin secretion in type 2 diabetes is not due to diminished secretion of GLP-1, but is due to diminished responsiveness of the β-cells to GLP-1 [42, 59, 60].

In addition to enhancing insulin secretion, GLP-1 inhibits, glucagon secretion, promotes satiety, delays gastric emptying, reduces weight, and promotes natriuresis. Antidiabetic drugs that act on the K_{ATP} channel do not reduce, rather they may increase the risk of cardiovascular events, and mortality [61]. On the other hand, some of the GLP-1 receptor agonists e.g. liraglutide, and semaglutide, reduce the risk of cardiovascular events and mortality [62, 63]. For this reason, the use of sulfonylureas in the treatment of type 2 diabetes is decreasing, and that of some GLP-1 analogs is increasing.

37.7 Somatostatin (Also Called Somatotropin Release-Inhibiting Factor, SRIF)

The δ-cells, which resemble small neurons with long slender processes and secretory granule-rich knob-like endings, secrete somatostatin. Somatostatin is also secreted from the pituitary gland, brain and gastrointestinal tract. Two forms of somatostatin, somatostatin 14 and somatostatin 28 are known; somatostatin-14 is the predominant form secreted from the δ-cells. Somatostatin inhibits secretion of many hormones, including insulin and specially glucagon. There are five somatostatin receptors (SST) 1–5 [64], all coupled to the G-proteins $G_{i/o}$. In rodent islets, somatostatin inhibits glucagon secretion by activating the SST2 receptor and inhibits insulin secretion by activating the SST5 receptor [65, 66]. In human islets SST2 is the dominant receptor in both the α-cells and the β-cells, but both cells express also other somatostatin receptors at low level [67, 68]. Selective inhibition of SST5 increases secretion of insulin, and also of GLP-1 [69]. Glucose stimulates secretion of insulin and inhibits secretion of glucagon in an oscillatory fashion, where the glucagon oscillations are anti-parallel to those of insulin and somatostatin [70]. Apparently, each pulse of somatostatin inhibits glucagon secretion by paracrine interactions.

Somatostatin inhibits insulin secretion by hyperpolarizing the membrane potential through activation of the G-protein coupled inwardly rectifying K (GIRK) channel, a process mediated by $G_{\beta\gamma}$. The GIRK channel of the islet cells consists of hetero-multimers of $K_{ir}3.1$, $K_{ir}3.2$, or $K_{ir}3.4$. Somatostatin inhibits the P/Q type VGCC of β-cells through G-protein mediated mechanisms. It activates $G_{i/o}$, which causes inhibition of adenylyl cyclase, and reduced formation of cAMP. In addition to these, somatostatin also inhibits exocytosis by mechanisms that are not fully clear [68].

Patients who have inoperable insulinoma, glucagonoma, or other carcinoid tumors, are often symptomatically treated, with variable degree of success, by one of the somastatin analogues e.g. octreotide, lanreotide and pasireotide that bind mostly to SST2 or SST5, [71].

37.8 Mitochondrial Uncoupling Protein 2 (UCP2)

UCP2 has only mild uncoupling activity, but it has other exchange/transport activities, which are more prominent. It exchanges four carbon (C4) intermediates e.g. malate, oxaloacetate, and aspartate, for cytoplasmic phosphate, and it exports C4 metabolites from mitochondria to the cytoplasm [72]. These processes are driven by electrical potential and pH gradient across the inner mitochondrial membrane. Glucose metabolism can be shifted towards aerobic glycolysis or mitochondrial oxidation, depending on the activity of UCP2. High activity of UCP2 reduces concentration of oxaloacetate in the mitochondria, and thereby reduces glucose oxidation through the citric acid cycle, lowers ATP:ADP ratio, production of reactive oxygen species (ROS), and increases glycolysis. Decreased activity of UCP2 increases mitochondrial glucose oxidation, and reduces glycolysis. Metformin, the most commonly used drug for the treatment of type 2 diabetes, induces UCP2 expression [73]. This reduces metabolically active citric acid cycle intermediates in the mitochondria, and cells utilize more glucose through aerobic glycolysis, which can partly explain how metformin increases glucose utilization in the peripheral tissues.

In the β-cells high activity of UCP2 reduces glucose oxidation in the mitochondria, lowers the ATP:ADP ratio, and ROS production. These effects reduce insulin secretion. On the other hand, inactivating mutations of the *UCP2* gene increase oxidation of glucose in the mitochondria, increase ATP:ADP ratio, increase insulin secretion, and cause congenital hyperinsulinism [72, 74]. Hypoglycemia in patients with inactivating mutations of UCP2 occurs usually after 24 h of fasting, and it can often be treated by diazoxide.

37.9 Glutamate Dehydrogenase

Glutamate dehydrogenase I (GDH-I) located in the mitochondrial matrix, catalyzes oxidative deamination of glutamate to α-ketoglutarate and ammonia, in a reversible manner. For insulin secretion, GDH is an important enzyme that is highly regulated. It is inhibited to near zero by GTP, and the inhibition can be relieved by ADP. [ADP] > 35 uM rapidly increases activity of GDH-1. The amino acid leucine allosterically activates GDH-I by promoting ADP binding. When glucose concentration falls below the threshold for insulin secretion, i.e. <4–5 mmol/L, the concentration of ADP increases, which increases GDH-I activity. Some mutations of the *GLUD1* gene that encodes GDH-I, activate the enzyme either by increasing the maximal rate or by reducing the sensitivity to inhibition by GTP. Subjects with such mutations suffer from episodes of hypoglycemia and hyperammonemia about 2–6 h after a protein-rich meal.

37.10 3-Hydroxyacyl CoA Dehydrogenase

3-Hydroxyacyl CoA dehydrogenase (also called short chain 3-hydroxyacyl-CoA dehydrogenase, SCHAD) is a mitochondrial enzyme that catalyzes the third step of β-oxidation of fatty acids, i.e. it catalyzes formation of 3-ketoacyl-CoA from 3-hydroxyacyl-CoA. Inactivating mutations in the *HADH* gene that encodes for SCHAD, cause deficiency of SCHAD. These subjects have increased insulin secretion and episodes of hypoglycemia in response to protein-rich meals. The increase in the insulin secretion is due to the increase in the activity of GDH. Normally SCHAD inhibits GDH; deficiency of SCHAD removes the inhibition, and GDH is activated by leucine present in the protein-rich meal, leading to increased insulin secretion [75]. The hypoglycemia caused by SCHAD can be prevented by diazoxide, which switches the β-cells to the off-mode by increasing the open state probability of the K_{ATP} channel.

37.11 Free Fatty Acid Receptors

The four free fatty acid receptors, FFA1, FFA2, FFA3 and FFA4 are encoded by *FFAR1*, *FFAR2*, *FFAR3*, and *FFAR4* genes respectively [76]. FFA1 and FFA4 are activated by long chain fatty acids; FFA2 and FFA3 are activated by short chain fatty acids.

FFA1 (GPR40, G-protein coupled receptor 40): FFA1 is activated by long chain fatty acids (myristic acid, palmitic acid, oleic acid, linoleic acid, alpha-linoleic acid, arachidonic acid, ecosapentanoic acid, and docosahexanoic acid). FFA1 is coupled to $G_{\alpha q}$, and activation of the receptor triggers Ca^{2+} signals through the

PLC-inositol 1,4,5 trisphosphate pathway [77, 78]. Activation of FFA1 enhances glucose-stimulated insulin secretion. FFA1 is also expressed in the intestinal L-cells, where its activation elicits Ca^{2+} signals leading to the secretion of GLP-1 [79]. FFA1 is a potential target for developing drugs that promote insulin secretion in a glucose-dependent manner [80].

FFA4 (GPR120, G-protein coupled receptor 120): FFA4 is activated by long chain fatty acids. It is coupled to $G_{q/11}$. Interestingly, activation of FFA4 increases secretion of GLP-1, which enhance insulin secretion in a glucose-dependent manner [81].

GPR119 (G-protein coupled receptor 119): GPR119 is not strictly a free fatty acid receptor, but it is activated by lipid compounds like N-oleoylethanolamide, and N-palmitoylethanolamide. This receptor is highly expressed in the β-cells and in the L-cells of small intestine that secrete GLP-1. Activation of GPR119 by orally active synthetic agonists increases GLP-1 level and enhances insulin secretion in a glucose-dependent manner [82].

37.12 Hepatocyte Nuclear Factor 4 Alpha

Loss of function mutations in the hepatocyte nuclear factor 4 alpha (*HNF4A*), (also called *MODY1*) gene reduces both glucose- and arginine-induced insulin secretion by mechanisms that are not fully clear [83]. It has been demonstrated that HNF4A directly induces expression of x-box protein 1 (XBP1). XBP1 protein is a transcription factor encoded by the *XBP1* gene. Loss of function mutations of *HNF4A* that cause “maturity onset diabetes of the young type 1” (MODY1), reduce expression of XBP1, which in turn reduces the area of the ER network, the concentration of Ca^{2+} in the ER, and the magnitude of the glucose-induced increase of $[Ca^{2+}]_c$ [84]. These patients present with diabetes during early childhood or adolescence, and are initially treated by insulin as in type 1 diabetes. Once the molecular diagnosis is established, many of these patients can be treated by sulfonylureas for variable period, but later on when sulfonylureas fail, they need insulin therapy.

37.13 Mitochondria

Mitochondrial ATP production is essential for stimulating normal insulin secretion. In β-cells, which lack lactate dehydrogenase, pyruvate, the end-product of glycolysis, is metabolized in the mitochondria, which is essential for robust insulin secretion in response to glucose [85].

Mitochondrial dysfunction caused by mutations in the mitochondrial DNA often cause impaired insulin secretion and diabetes. The A3243G mutation in the mitochondrially encoded tRNA leucine 1 (*MT-TL1*) gene causes maternally inherited diabetes and deafness (MIDD). The mutation impairs the ability of the β-cells to secrete insulin probably due to impaired generation of ATP by the mutant mitochondria [86].

"Transcription factor B1, mitochondrial" (*TFB1M*) gene encodes for a dimethyltransferase that dimethylates mitochondrial 12S rRNA. It is also necessary for mitochondrial gene expression. A common variant (rs950994) in the *TFB1M* gene is associated with impaired insulin secretion and increased risk for developing type 2 diabetes [87]. Studies using knockout mice and clonal β-cells, show that the risk SNP leads to reduced level of TFB1M protein, reduced mitochondrial translation, reduced oxidative phosphorylation, reduced production of ATP, and consequent impairment of insulin secretion [87].

37.14 Insulin Secretion in Type 2 Diabetes

Insulin secretion in response to intake of mixed meals is not biphasic [88, 89]. The so called biphasic insulin secretion is an experimental artifact generated by rapid and sustained stimulation of the β-cells with glucose [90]. In type 2 diabetes, insulin secretion, especially the first phase of the insulin secretion, in response to rapid and sustained stimulation by glucose, is impaired. This impairment is seen even in the early stages of the natural history of diabetes [91], and even in the normoglycemic first-degree relatives of people who have diabetes [92, 93]. There is strong evidence that the impairment of insulin secretion is essential for the development of type 2 diabetes. In fact, most of the genes that are strongly associated with type 2 diabetes, appear to be important for the development, and the function of β-cells [94, 95]. One of the cornerstone of the treatment of diabetes is to improve insulin secretion either by drugs that act on the K_{ATP} channel or the GLP-1 receptor.

Sulfonylureas and glinides reduce the open state probability of the K_{ATP} channel and modestly improve food induced increase in insulin secretion in people with type 2 diabetes. Apparently, in type 2 diabetes, a proportion of β-cells remain in the off-mode, and by binding to the SUR1 subunit of the K_{ATP} channel, these drugs switch these β-cells to the ready-mode. In normal subjects, these drugs shift the dose-response curve of glucose-induced insulin secretion to the left. In diabetes, the dose-response curves of glucose-induced insulin secretion, are flatter and shifted to the right, compared to those in the normal subjects. Sulfonylureas almost normalize, and shift the curve to left in people with prediabetes, but their effects in people with diabetes are only modest [96]. The insulin response to the sulfonylureas remains lower in people with diabetes compared to that in the normal subjects. Prolonged treatment with sulfonylureas makes the β-cells unresponsive to the treatment, usually permanently [97]. Apparently, persistent depolarization, and consequent chronic elevation of the $[Ca^{2+}]_c$ in the β-cells, lead to adaptive, or compensatory changes that ultimately make the β-cells unresponsive to glucose [31]. We see a similar "glucose-blindness" in the SUR1 knockout mice [98], and in patients who have inactivating mutations of the K_{ATP} channel genes [25, 28].

Drugs that act on the K_{ATP} channel have been used in the treatment of diabetes for decades, but their use is declining since they often induce hypoglycemia, increase weight, and most importantly, they do not improve the cardiovascular outcomes [61].

Impaired insulin secretion in type 2 diabetes can be treated preferably by drugs that act, directly or indirectly, on the GLP-1 receptor. These drugs increase secretion of insulin in a glucose-dependent manner, and at the same time reduce secretion of glucagon [99].

In contrast to the drugs that act on the K_{ATP} channel or the GLP-1 receptor, metformin, the first line drug for the treatment of diabetes, impairs glucose-induced $[Ca^{2+}]_c$ response in the β-cells, and thereby reduce insulin secretion [100, 101]. It is possible that, in the β-cells, metformin, by inducing UCP2 expression [73] reduces mitochondrial oxidation, and promotes glucose utilization through aerobic glycolysis. This may be seen as a way to achieve "β-cell rest".

In type 2 diabetes impairment of β-cells function probably appears first, which then leads to a decrease of β-cell mass as the disease progresses [102, 103]. It appears that, one population of the β-cells remains in the off-mode, and they can be switched to the ready-mode by the anti-diabetic sulfonylurea drugs. A second population of β-cells remains in the ready-mode, and they can be switched to the on-mode by drugs that act on the GLP-1 receptor. A third population of β-cells remains in the on-mode, and these can be granted "rest" by metformin. A fourth population of β-cells are in the process of dying because of many factors including amyloid deposition, and apoptosis. Drugs that act on the GLP-1 receptor appear to protect all the three population of β-cells by preventing apoptosis, and possibly even by supporting proliferation of the cells [12].

Acknowledgements Financial support was obtained from the Karolinska Institutet and the Uppsala County Council.

References

1. Holz GG, Kuhtreiber WM, Habener JF (1993) Pancreatic beta-cells are rendered glucose-competent by the insulinotropic hormone glucagon-like peptide-1(7-37). Nature 361(6410):362–365
2. Song SH, McIntyre SS, Shah H, Veldhuis JD, Hayes PC, Butler PC (2000) Direct measurement of pulsatile insulin secretion from the portal vein in human subjects. J Clin Endocrinol Metab 85(12):4491–4499
3. Islam MS (2011) TRP channels of islets. Adv Exp Med Biol 704:811–830
4. Gloyn AL, Pearson ER, Antcliff JF, Proks P, Bruining GJ, Slingerland AS et al (2004) Activating mutations in the gene encoding the ATP-sensitive potassium-channel subunit Kir6.2 and permanent neonatal diabetes. N Engl J Med 350(18):1838–1849
5. Islam MS (2014) Calcium signaling in the islets. In: Islam MS (ed) Islets of Langerhans, 2nd edn. Springer, Dordrecht, pp 605–632
6. Islam MS (2002) The ryanodine receptor calcium channel of beta-cells: molecular regulation and physiological significance. Diabetes 51(5):1299–1309

7. Martin F, Soria B (1996) Glucose-induced [Ca^{2+}]i oscillations in single human pancreatic islets. Cell Calcium 20(5):409–414
8. Quesada I, Todorova MG, Alonso-Magdalena P, Beltra M, Carneiro EM, Martin F et al (2006) Glucose induces opposite intracellular Ca^{2+} concentration oscillatory patterns in identified alpha- and beta-cells within intact human islets of Langerhans. Diabetes 55(9):2463–2469
9. Kindmark H, Kohler M, Arkhammar P, Efendic S, Larsson O, Linder S et al (1994) Oscillations in cytoplasmic free calcium concentration in human pancreatic islets from subjects with normal and impaired glucose tolerance. Diabetologia 37(11):1121–1131
10. Hellman B, Gylfe E, Bergsten P, Grapengiesser E, Lund PE, Berts A et al (1994) Glucose induces oscillatory Ca^{2+} signalling and insulin release in human pancreatic beta cells. Diabetologia 37(Suppl 2):S11–S20
11. Nordenskjöld F, Andersson B, Islam MS (2019) Expression of the inositol 1,4,5-trisphosphate receptor and the ryanodine receptor Ca^{2+}-release channels in the beta-cells and alpha- cells of the human islets of Langerhans. Adv Exp Med Biol 1131:271–281
12. Rowlands J, Heng J, Newsholme P, Carlessi R (2018) Pleiotropic effects of GLP-1 and analogs on cell signaling, metabolism, and function. Front Endocrinol (Lausanne) 9:672
13. Pinney SE, Ganapathy K, Bradfield J, Stokes D, Sasson A, Mackiewicz K et al (2013) Dominant form of congenital hyperinsulinism maps to HK1 region on 10q. Horm Res Paediatr 80(1):18–27
14. Henquin JC, Sempoux C, Marchandise J, Godecharles S, Guiot Y, Nenquin M et al (2013) Congenital hyperinsulinism caused by hexokinase I expression or glucokinase-activating mutation in a subset of beta-cells. Diabetes 62(5):1689–1696
15. Henquin JC, Nenquin M, Guiot Y, Rahier J, Sempoux C (2015) Human Insulinomas show distinct patterns of insulin secretion in vitro. Diabetes 64(10):3543–3553
16. Pino MF, Kim KA, Shelton KD, Lindner J, Odili S, Li C et al (2007) Glucokinase thermolability and hepatic regulatory protein binding are essential factors for predicting the blood glucose phenotype of missense mutations. J Biol Chem 282(18):13906–13916
17. Chakera AJ, Steele AM, Gloyn AL, Shepherd MH, Shields B, Ellard S et al (2015) Recognition and Management of Individuals with Hyperglycemia because of a heterozygous Glucokinase mutation. Diabetes Care 38(7):1383–1392
18. Bennett K, James C, Mutair A, Al-Shaikh H, Sinani A, Hussain K (2011) Four novel cases of permanent neonatal diabetes mellitus caused by homozygous mutations in the glucokinase gene. Pediatr Diabetes 12(3 Pt 1):192–196
19. Gribble FM, Tucker SJ, Haug T, Ashcroft FM (1998) MgATP activates the beta cell KATP channel by interaction with its SUR1 subunit. Proc Natl Acad Sci U S A 95(12):7185–7190
20. Macmullen CM, Zhou Q, Snider KE, Tewson PH, Becker SA, Aziz AR et al (2011) Diazoxide-unresponsive congenital hyperinsulinism in children with dominant mutations of the beta-cell sulfonylurea receptor SUR1. Diabetes 60(6):1797–1804
21. Ponmani C, Gannon H, Hussain K, Senniappan S (2013) Paradoxical hypoglycaemia associated with diazoxide therapy for hyperinsulinaemic hypoglycaemia. Horm Res Paediatr 80(2):129–133
22. de Lonlay P, Fournet JC, Rahier J, Gross-Morand MS, Poggi-Travert F, Foussier V et al (1997) Somatic deletion of the imprinted 11p15 region in sporadic persistent hyperinsulinemic hypoglycemia of infancy is specific of focal adenomatous hyperplasia and endorses partial pancreatectomy. J Clin Invest 100(4):802–807
23. Kapoor RR, Flanagan SE, Arya VB, Shield JP, Ellard S, Hussain K (2013) Clinical and molecular characterisation of 300 patients with congenital hyperinsulinism. Eur J Endocrinol 168(4):557–564
24. Sempoux C, Guiot Y, Dahan K, Moulin P, Stevens M, Lambot V et al (2003) The focal form of persistent hyperinsulinemic hypoglycemia of infancy: morphological and molecular studies show structural and functional differences with insulinoma. Diabetes 52(3):784–794
25. Henquin JC, Nenquin M, Sempoux C, Guiot Y, Bellanne-Chantelot C, Otonkoski T et al (2011) In vitro insulin secretion by pancreatic tissue from infants with diazoxide-resistant congenital hyperinsulinism deviates from model predictions. J Clin Invest 121(10):3932–3942

26. Snider KE, Becker S, Boyajian L, Shyng SL, MacMullen C, Hughes N et al (2013) Genotype and phenotype correlations in 417 children with congenital hyperinsulinism. J Clin Endocrinol Metab 98(2):E355–E363
27. Yorifuji T, Horikawa R, Hasegawa T, Adachi M, Soneda S, Minagawa M et al (2017) Clinical practice guidelines for congenital hyperinsulinism. Clin Pediatr Endocrinol 26(3):127–152
28. Li C, Ackermann AM, Boodhansingh KE, Bhatti TR, Liu C, Schug J et al (2017) Functional and metabolomic consequences of KATP channel inactivation in human islets. Diabetes 66(7):1901–1913
29. Thornton PS, MacMullen C, Ganguly A, Ruchelli E, Steinkrauss L, Crane A et al (2003) Clinical and molecular characterization of a dominant form of congenital hyperinsulinism caused by a mutation in the high-affinity sulfonylurea receptor. Diabetes 52(9):2403–2410
30. Wilson DF, Cember ATJ, Matschinsky FM (2018) Glutamate dehydrogenase: role in regulating metabolism and insulin release in pancreatic beta-cells. J Appl Physiol (1985) 125(2):419–428
31. Schwaller B (2012) The regulation of a cell's Ca^{2+} signaling toolkit: the Ca^{2+} homeostasome. Adv Exp Med Biol 740:1–25
32. Bowman P, Sulen A, Barbetti F, Beltrand J, Svalastoga P, Codner E et al (2018) Effectiveness and safety of long-term treatment with sulfonylureas in patients with neonatal diabetes due to KCNJ11 mutations: an international cohort study. Lancet Diabetes Endocrinol 6(8):637–646
33. MacDonald PE, Salapatek AM, Wheeler MB (2002) Glucagon-like peptide-1 receptor activation antagonizes voltage-dependent repolarizing K(+) currents in beta-cells: a possible glucose-dependent insulinotropic mechanism. Diabetes 51(Suppl 3):S443–S447
34. Torekov SS, Iepsen E, Christiansen M, Linneberg A, Pedersen O, Holst JJ et al (2014) KCNQ1 long QT syndrome patients have hyperinsulinemia and symptomatic hypoglycemia. Diabetes 63(4):1315–1325
35. Yasuda K, Miyake K, Horikawa Y, Hara K, Osawa H, Furuta H et al (2008) Variants in KCNQ1 are associated with susceptibility to type 2 diabetes mellitus. Nat Genet 40(9):1092–1097
36. Nauck MA, Meier JJ (2018) Incretin hormones: their role in health and disease. Diabetes Obes Metab 20(Suppl 1):5–21
37. Light PE, Manning Fox JE, Riedel MJ, Wheeler MB (2002) Glucagon-like peptide-1 inhibits pancreatic ATP-sensitive potassium channels via a protein kinase A- and ADP-dependent mechanism. Mol Endocrinol 16(9):2135–2144
38. Ma Z, Bjorklund A, Islam MS (2017) A TRPM4 inhibitor 9-Phenanthrol inhibits glucose- and glucagon-like peptide 1-induced insulin secretion from rat islets of Langerhans. J Diabetes Res 2017:5131785
39. Marabita F, Islam MS (2017) Expression of transient receptor potential channels in the purified human pancreatic beta-cells. Pancreas 46(1):97–101
40. Pang B, Kim S, Li D, Ma Z, Sun B, Zhang X et al (2017) Glucagon-like peptide-1 potentiates glucose-stimulated insulin secretion via the transient receptor potential melastatin 2 channel. Exp Ther Med 14(5):5219–5227
41. Bari MR, Akbar S, Eweida M, Kuhn FJ, Gustafsson AJ, Luckhoff A et al (2009) H2O2-induced Ca^{2+} influx and its inhibition by N-(p-amylcinnamoyl) anthranilic acid in the beta-cells: involvement of TRPM2 channels. J Cell Mol Med 13(9B):3260–3267
42. Hodson DJ, Mitchell RK, Bellomo EA, Sun G, Vinet L, Meda P et al (2013) Lipotoxicity disrupts incretin-regulated human beta cell connectivity. J Clin Invest 123(10):4182–4194
43. Islam MS (2010) Calcium signaling in the islets. Adv Exp Med Biol 654:235–259
44. Krishnan K, Ma Z, Bjorklund A, Islam MS (2015) Calcium signaling in a genetically engineered human pancreatic beta-cell line. Pancreas 44(5):773–777
45. Holz GG, Leech CA, Heller RS, Castonguay M, Habener JF (1999) cAMP-dependent mobilization of intracellular Ca^{2+} stores by activation of ryanodine receptors in pancreatic beta-cells. A Ca^{2+} signaling system stimulated by the insulinotropic hormone glucagon-like peptide-1-(7-37). J Biol Chem 274(20):14147–14156

46. Kang G, Joseph JW, Chepurny OG, Monaco M, Wheeler MB, Bos JL et al (2003) Epac-selective cAMP analog 8-pCPT-2′-O-Me-cAMP as a stimulus for Ca^{2+}-induced Ca^{2+} release and exocytosis in pancreatic beta-cells. J Biol Chem 278(10):8279–8285
47. Suga S, Kanno T, Nakano K, Takeo T, Dobashi Y, Wakui M (1997) GLP-I(7-36) amide augments Ba^{2+} current through L-type Ca^{2+} channel of rat pancreatic beta-cell in a cAMP-dependent manner. Diabetes 46(11):1755–1760
48. Jacobo SM, Guerra ML, Hockerman GH (2009) Cav1.2 and Cav1.3 are differentially coupled to glucagon-like peptide-1 potentiation of glucose-stimulated insulin secretion in the pancreatic beta-cell line INS-1. J Pharmacol Exp Ther 331(2):724–732
49. Gromada J, Bokvist K, Ding WG, Holst JJ, Nielsen JH, Rorsman P (1998) Glucagon-like peptide 1 (7-36) amide stimulates exocytosis in human pancreatic beta-cells by both proximal and distal regulatory steps in stimulus-secretion coupling. Diabetes 47(1):57–65
50. Yada T, Itoh K, Nakata M (1993) Glucagon-like peptide-1-(7-36)amide and a rise in cyclic adenosine 3′,5′-monophosphate increase cytosolic free Ca^{2+} in rat pancreatic beta-cells by enhancing Ca^{2+} channel activity. Endocrinology 133(4):1685–1692
51. Kang G, Chepurny OG, Rindler MJ, Collis L, Chepurny Z, Li WH et al (2005) A cAMP and Ca^{2+} coincidence detector in support of Ca^{2+}-induced Ca^{2+} release in mouse pancreatic beta cells. J Physiol 566(Pt 1):173–188
52. Dyachok O, Gylfe E (2004) Ca^{2+}-induced Ca^{2+} release via inositol 1,4,5-trisphosphate receptors is amplified by protein kinase A and triggers exocytosis in pancreatic beta-cells. J Biol Chem 279(44):45455–45461
53. Bruton JD, Lemmens R, Shi CL, Persson-Sjogren S, Westerblad H, Ahmed M et al (2003) Ryanodine receptors of pancreatic beta-cells mediate a distinct context-dependent signal for insulin secretion. FASEB J 17(2):301–303
54. Wu B, Wei S, Petersen N, Ali Y, Wang X, Bacaj T et al (2015) Synaptotagmin-7 phosphorylation mediates GLP-1-dependent potentiation of insulin secretion from beta-cells. Proc Natl Acad Sci U S A 112(32):9996–10001
55. Unger RH, Cherrington AD (2012) Glucagonocentric restructuring of diabetes: a pathophysiologic and therapeutic makeover. J Clin Invest 122(1):4–12
56. Zhang Y, Parajuli KR, Fava GE, Gupta R, Xu W, Nguyen LU et al (2018) GLP-1 receptor in pancreatic alpha cells regulates glucagon secretion in a glucose-dependent bidirectional manner. Diabetes 68(1):34–44
57. Ramracheya R, Chapman C, Chibalina M, Dou H, Miranda C, Gonzalez A et al (2018) GLP-1 suppresses glucagon secretion in human pancreatic alpha-cells by inhibition of P/Q-type Ca^{2+} channels. Physiol Rep 6(17):e13852
58. Faerch K, Torekov SS, Vistisen D, Johansen NB, Witte DR, Jonsson A et al (2015) GLP-1 response to Oral glucose is reduced in prediabetes, screen-detected type 2 diabetes, and obesity and influenced by sex: the ADDITION-PRO Study. Diabetes 64(7):2513–2525
59. Calanna S, Christensen M, Holst JJ, Laferrere B, Gluud LL, Vilsboll T et al (2013) Secretion of glucagon-like peptide-1 in patients with type 2 diabetes mellitus: systematic review and meta-analyses of clinical studies. Diabetologia 56(5):965–972
60. Ruetten H, Gebauer M, Raymond RH, Calle RA, Cobelli C, Ghosh A et al (2018) Mixed meal and intravenous L-arginine tests both stimulate incretin release across glucose tolerance in man: lack of correlation with beta cell function. Metab Syndr Relat Disord 16(8):406–415
61. Azoulay L, Suissa S (2017) Sulfonylureas and the risks of cardiovascular events and death: a methodological meta-regression analysis of the observational studies. Diabetes Care 40(5):706–714
62. Marso SP, Bain SC, Consoli A, Eliaschewitz FG, Jodar E, Leiter LA et al (2016) Semaglutide and cardiovascular outcomes in patients with type 2 diabetes. N Engl J Med 375(19):1834–1844
63. Marso SP, Daniels GH, Brown-Frandsen K, Kristensen P, Mann JF, Nauck MA et al (2016) Liraglutide and cardiovascular outcomes in type 2 diabetes. N Engl J Med 375(4):311–322
64. Gunther T, Tulipano G, Dournaud P, Bousquet C, Csaba Z, Kreienkamp HJ et al (2018) International Union of Basic and Clinical Pharmacology. CV. Somatostatin receptors: structure,

function, ligands, and new nomenclature. Pharmacol Rev 70(4):763–835
65. Strowski MZ, Parmar RM, Blake AD, Schaeffer JM (2000) Somatostatin inhibits insulin and glucagon secretion via two receptors subtypes: an in vitro study of pancreatic islets from somatostatin receptor 2 knockout mice. Endocrinology 141(1):111–117
66. Strowski MZ, Kohler M, Chen HY, Trumbauer ME, Li Z, Szalkowski D et al (2003) Somatostatin receptor subtype 5 regulates insulin secretion and glucose homeostasis. Mol Endocrinol 17(1):93–106
67. Braun M (2014) The somatostatin receptor in human pancreatic beta-cells. Vitam Horm 95:165–193
68. Kailey B, van de Bunt M, Cheley S, Johnson PR, MacDonald PE, Gloyn AL et al (2012) SSTR2 is the functionally dominant somatostatin receptor in human pancreatic beta- and alpha-cells. Am J Physiol Endocrinol Metab 303(9):E1107–E1116
69. Liu W, Shao PP, Liang GB, Bawiec J, He J, Aster SD et al (2018) Discovery and pharmacology of a novel somatostatin subtype 5 (SSTR5) antagonist: synergy with DPP-4 inhibition. ACS Med Chem Lett 9(11):1082–1087
70. Salehi A, Qader SS, Grapengiesser E, Hellman B (2007) Pulses of somatostatin release are slightly delayed compared with insulin and antisynchronous to glucagon. Regul Pept 144(1–3):43–49
71. Oberg K (2018) Management of functional neuroendocrine tumors of the pancreas. Gland Surg 7(1):20–27
72. Vozza A, Parisi G, De Leonardis F, Lasorsa FM, Castegna A, Amorese D et al (2014) UCP2 transports C4 metabolites out of mitochondria, regulating glucose and glutamine oxidation. Proc Natl Acad Sci U S A 111(3):960–965
73. Anedda A, Rial E, Gonzalez-Barroso MM (2008) Metformin induces oxidative stress in white adipocytes and raises uncoupling protein 2 levels. J Endocrinol 199(1):33–40
74. Ferrara CT, Boodhansingh KE, Paradies E, Fiermonte G, Steinkrauss LJ, Topor LS et al (2017) Novel hypoglycemia phenotype in congenital hyperinsulinism due to dominant mutations of uncoupling protein 2. J Clin Endocrinol Metab 102(3):942–949
75. Li C, Chen P, Palladino A, Narayan S, Russell LK, Sayed S et al (2010) Mechanism of hyperinsulinism in short-chain 3-hydroxyacyl-CoA dehydrogenase deficiency involves activation of glutamate dehydrogenase. J Biol Chem 285(41):31806–31818
76. Davenport AP, Alexander SP, Sharman JL, Pawson AJ, Benson HE, Monaghan AE et al (2013) International union of basic and clinical pharmacology. LXXXVIII. G protein-coupled receptor list: recommendations for new pairings with cognate ligands. Pharmacol Rev 65(3):967–986
77. Zhao Y, Wang L, Qiu J, Zha D, Sun Q, Chen C (2013) Linoleic acid stimulates $[Ca^{2+}]i$ increase in rat pancreatic beta-cells through both membrane receptor- and intracellular metabolite-mediated pathways. PLoS One 8(4):e60255
78. Fujiwara K, Maekawa F, Yada T (2005) Oleic acid interacts with GPR40 to induce Ca^{2+} signaling in rat islet beta-cells: mediation by PLC and L-type Ca^{2+} channel and link to insulin release. Am J Physiol Endocrinol Metab 289(4):E670–E677
79. Psichas A, Larraufie PF, Goldspink DA, Gribble FM, Reimann F (2017) Chylomicrons stimulate incretin secretion in mouse and human cells. Diabetologia 60(12):2475–2485
80. Guo B, Guo S, Huang J, Li J, Li J, Chen Q et al (2018) Design and optimization of 2,3-dihydrobenzo[b][1,4]dioxine propanoic acids as novel GPR40 agonists with improved pharmacokinetic and safety profiles. Bioorg Med Chem 26(22):5780–5791
81. Sundstrom L, Myhre S, Sundqvist M, Ahnmark A, McCoull W, Raubo P et al (2017) The acute glucose lowering effect of specific GPR120 activation in mice is mainly driven by glucagon-like peptide 1. PLoS One 12(12):e0189060
82. Matsumoto K, Yoshitomi T, Ishimoto Y, Tanaka N, Takahashi K, Watanabe A et al (2018) DS-8500a, an orally available G protein-coupled receptor 119 agonist, upregulates glucagon-like Peptide-1 and enhances glucose-dependent insulin secretion and improves glucose homeostasis in type 2 diabetic rats. J Pharmacol Exp Ther 367(3):509–517

83. Yamagata K, Senokuchi T, Lu M, Takemoto M, Fazlul Karim M, Go C et al (2011) Voltage-gated K^+ channel KCNQ1 regulates insulin secretion in MIN6 beta-cell line. Biochem Biophys Res Commun 407(3):620–625
84. Moore BD, Jin RU, Lo H, Jung M, Wang H, Battle MA et al (2016) Transcriptional regulation of X-box-binding protein one (XBP1) by hepatocyte nuclear factor 4alpha (HNF4Alpha) is vital to Beta-cell function. J Biol Chem 291(12):6146–6157
85. Malmgren S, Nicholls DG, Taneera J, Bacos K, Koeck T, Tamaddon A et al (2009) Tight coupling between glucose and mitochondrial metabolism in clonal beta-cells is required for robust insulin secretion. J Biol Chem 284(47):32395–32404
86. Fex M, Nicholas LM, Vishnu N, Medina A, Sharoyko VV, Nicholls DG et al (2018) The pathogenetic role of beta-cell mitochondria in type 2 diabetes. J Endocrinol 236(3):R145–RR59
87. Koeck T, Olsson AH, Nitert MD, Sharoyko VV, Ladenvall C, Kotova O et al (2011) A common variant in TFB1M is associated with reduced insulin secretion and increased future risk of type 2 diabetes. Cell Metab 13(1):80–91
88. Sorenson RL, Lindell DV, Elde RP (1980) Glucose stimulation of somatostatin and insulin release from the isolated, perfused rat pancreas. Diabetes 29(9):747–751
89. Grodsky GM (1972) A threshold distribution hypothesis for packet storage of insulin and its mathematical modeling. J Clin Invest 51(8):2047–2059
90. Henquin JC, Ishiyama N, Nenquin M, Ravier MA, Jonas JC (2002) Signals and pools underlying biphasic insulin secretion. Diabetes 51(Suppl 1):S60–S67
91. Davies MJ, Rayman G, Grenfell A, Gray IP, Day JL, Hales CN (1994) Loss of the first phase insulin response to intravenous glucose in subjects with persistent impaired glucose tolerance. Diabet Med 11(5):432–436
92. Nyholm B, Porksen N, Juhl CB, Gravholt CH, Butler PC, Weeke J et al (2000) Assessment of insulin secretion in relatives of patients with type 2 (non-insulin-dependent) diabetes mellitus: evidence of early beta-cell dysfunction. Metabolism 49(7):896–905
93. O'Rahilly S, Turner RC, Matthews DR (1988) Impaired pulsatile secretion of insulin in relatives of patients with non-insulin-dependent diabetes. N Engl J Med 318(19):1225–1230
94. Prasad RB, Groop L (2015) Genetics of type 2 diabetes-pitfalls and possibilities. Genes (Basel) 6(1):87–123
95. Wood AR, Jonsson A, Jackson AU, Wang N, van Leewen N, Palmer ND et al (2017) A genome-wide association study of IVGTT-based measures of first-phase insulin secretion refines the underlying physiology of type 2 diabetes variants. Diabetes 66(8):2296–2309
96. Cerasi E, Efendic S, Thornqvist C, Luft R (1979) Effect of two sulphonylureas on the dose kinetics of glucose-induced insulin release in normal and diabetic subjects. Acta Endocrinol 91(2):282–293
97. Delawter DE, Moss JM, Tyroler S, Canary JJ (1959) Secondary failure of response to tolbutamide treatment. J Am Med Assoc 171:1786–1792
98. Doliba NM, Qin W, Vatamaniuk MZ, Li C, Zelent D, Najafi H et al (2004) Restitution of defective glucose-stimulated insulin release of sulfonylurea type 1 receptor knockout mice by acetylcholine. Am J Physiol Endocrinol Metab 286(5):E834–E843
99. Gutniak M, Orskov C, Holst JJ, Ahren B, Efendic S (1992) Antidiabetogenic effect of glucagon-like peptide-1 (7-36)amide in normal subjects and patients with diabetes mellitus. N Engl J Med 326(20):1316–1322

100. Gelin L, Li J, Corbin KL, Jahan I, Nunemaker CS (2018) Metformin inhibits mouse islet insulin secretion and alters intracellular calcium in a concentration-dependent and duration-dependent manner near the circulating range. J Diabetes Res 2018:9163052
101. Leclerc I, Woltersdorf WW, da Silva XG, Rowe RL, Cross SE, Korbutt GS et al (2004) Metformin, but not leptin, regulates AMP-activated protein kinase in pancreatic islets: impact on glucose-stimulated insulin secretion. Am J Physiol Endocrinol Metab 286(6):E1023–E1031
102. Meier JJ, Bonadonna RC (2013) Role of reduced beta-cell mass versus impaired beta-cell function in the pathogenesis of type 2 diabetes. Diabetes Care 36(Suppl 2):S113–S119
103. Chen C, Cohrs CM, Stertmann J, Bozsak R, Speier S (2017) Human beta cell mass and function in diabetes: recent advances in knowledge and technologies to understand disease pathogenesis. Mol Metab 6(9):943–957

Chapter 38
Calcium Dynamics and Synaptic Plasticity

Pedro Mateos-Aparicio and Antonio Rodríguez-Moreno

Abstract Synaptic plasticity is a fundamental property of neurons referring to the activity-dependent changes in the strength and efficacy of synaptic transmission at preexisting synapses. Such changes can last from milliseconds to hours, days, or even longer and are involved in learning and memory as well as in development and response of the brain to injuries. Several types of synaptic plasticity have been described across neuronal types, brain regions, and species, but all of them share in one way or another capital importance of Ca^{2+}-mediated processes. In this chapter, we will focus on the Ca^{2+}-dependent events necessary for the induction and expression of multiple forms of synaptic plasticity.

Keywords Synaptic plasticity · NMDA · AMPA · calcium · LTP · LTD · Short-term plasticity · Second messengers · Transmitter release · Kinases

38.1 Introduction

Calcium (Ca^{2+}) is a divalent cation essential for all known forms of life. It is involved in the regulation of a myriad of cellular events such as metabolic control, mitochondrial function, apoptosis, intracellular signaling cascades, gene expression or cellular motility [1]. In neurons, Ca^{2+} acts as a second messenger controlling neuronal excitability, development of neuronal morphology, formation of synapses, synaptic release, gene expression, and synaptic plasticity [2]. While extracellular Ca^{2+} concentration ranges in the mM scale, intracellular free Ca^{2+} concentration is tightly maintained within the nM range (~100 nM). In order to achieve and maintain such steep concentration gradient, neurons have developed multiple strategies over the course of evolution to keep cytosolic Ca^{2+} ions compartmentalized or

P. Mateos-Aparicio (✉) · A. Rodríguez-Moreno (✉)
Laboratorio de Neurociencia Celular y Plasticidad, Departamento de Fisiología, Anatomía y Biología Celular, Universidad Pablo de Olavide, Sevilla, Spain
e-mail: pmatmor@upo.es; arodmor@upo.es

© Springer Nature Switzerland AG 2020

M. S. Islam (ed.), *Calcium Signaling*, Advances in Experimental Medicine and Biology 1131, https://doi.org/10.1007/978-3-030-12457-1_38

sequestered as well as extruded outside the cell. Most of the processes involved in Ca^{2+} function and regulation are carried by a vast array of membrane proteins (i.e. ion channels and exchanger pumps) that permit Ca^{2+} influx through the plasma membrane or release Ca^{2+} from intracellular organelles or stores. In addition, the specific functions of Ca^{2+} depend upon numerous Ca^{2+}-binding proteins (i.e. Ca^{2+} buffers and Ca^{2+} sensors) that regulate numerous neuronal intracellular cascades. When a signal such as membrane depolarization or an action potential occurs in neurons, it can trigger the opening of plasma membrane and organelle Ca^{2+} channels, resulting in a sudden increase in intracellular Ca^{2+} concentration which modifies the conformation of signaling proteins which in turn translate the Ca^{2+} signal into downstream cellular effects.

Synaptic plasticity is a fundamental property of neurons referring to the activity-dependent changes in the strength and efficacy of synaptic transmission at preexisting synapses [3]. Such changes can last from milliseconds to hours, days, or even longer and are involved in the correct development of the brain, learning and memory processes, and recovery of the brain after injuries [4, 5]. Multiple types of synaptic plasticity have been described across neuronal types, brain regions, and species, but all of them share in one way or another capital importance of Ca^{2+}-mediated processes. Here, we review the Ca^{2+}-dependent events necessary for the induction and expression of different forms of synaptic plasticity.

38.2 The Structure of Ca^{2+} Signaling During Synaptic Plasticity

In the plasma membrane, Ca^{2+} ions can flow into the neuron through voltage-gated Ca^{2+} channels (VGCCs) and the Ca^{2+}-permeable type of glutamate receptor N-methyl-D-aspartate receptors (NMDARs). In some synapses, an unconventional type of α-amino-3-hydroxy-5-methyl-4-isoxazolepropionic acid receptors (AMPARs), so-called Ca^{2+}-permeable AMPARs (CP-AMPARs) has been reported in some types of plasticity [6] as has been some types of Kainate receptors (KAR) [7–13]. Neuronal VGCCs are heteromultimeric proteins composed by a pore-forming $\alpha1$ subunit and auxiliary β, $\alpha2$, δ, and γ subunits [14]. Ten different $\alpha1$ subunits define three channel families that conduct Ca^{2+} currents with different physiological and pharmacological properties [14–16]. L-type Ca^{2+} currents, carried by Ca_V 1.1–1.4 channels, show high-voltage activation, large single channel conductance and slow voltage-dependent inactivation. In most cases, they are typically located in the postsynaptic membrane and provide the main Ca^{2+} signal that triggers postsynaptic forms of plasticity. N-, P/Q-, and R-type currents are carried by Ca_V 2.1, 2.2, and 2.3, respectively. In most cases, these channels are inserted in the presynaptic terminal, mediating fast synaptic transmission and providing the Ca^{2+} influx necessary for transmitter release. Finally, T-type currents carried by Ca_v 3.1–3.3 channels show the lowest voltage dependence (also called low voltage-gated Ca^{2+} channels). Given their low voltage-dependent activation, they can facilitate synaptic plasticity by depolarizing the postsynaptic membrane potential and therefore allowing the activation of other Ca^{2+} channels types and/or NMDA receptors.

Inside the neuron, Ca^{2+} can be released from intracellular stores including the endoplasmic reticulum (ER), mitochondria, lysosomes, endosomes, Golgi vesicles, and secretory granules [1]. The most prominent contribution of intracellular stores during plasticity is that from the ER. The lumen contains Ca^{2+}-binding proteins in a concentration of 3–4 orders of magnitude higher than the cytosol (luminal total $Ca^{2+} > 1$ mM; free Ca^{2+} 100–700 μM) [17]. Release of Ca^{2+} from the ER is mediated by the inositol 1,4,5-triphosphate receptors ($InsP_3Rs$) and ryanodine receptors (RyRs). Both types of receptors are highly Ca^{2+} sensitive so they can mediate Ca^{2+}-induced Ca^{2+} release upon Ca^{2+} influx from membrane VGCCs or NMDA receptors [2]. $InsP_3Rs$ are Ca^{2+}-selective cation channels that open upon binding the second messenger $InsP_3$ and Ca^{2+} [18, 19]. Therefore, following rises in cytoplasmic Ca^{2+} and production of the second messenger $InsP_3$, $InsP_3Rs$ open and allow Ca^{2+} movement to the cytoplasm. RyRs are sensitive to Ca^{2+} entering either from outside or neighboring receptors [1]. They are also responsible for ER Ca^{2+} release. There are three isoforms of RyRs and although all of them have been reported in neurons, the RyR2 and 3 isoforms are the most commonly found in the brain [20]. Ca^{2+} binding increases the RyRs sensitivity to other ligands such as caffeine or adenosine diphosphate ribose (rADP). When active, they interact closely with multiple Ca^{2+}-binding proteins, Ca^{2+}-dependent enzymes (protein kinases and phosphatases) or even Ca_v 1 channels [20].

Synaptic terminals are tightly packed with numerous proteins that can bind Ca^{2+} and activate multiple downstream signaling cascades, modify gene expression, or interact with the release machinery. Ca^{2+} entry produces a local and transient increase in cytosolic Ca^{2+} concentration with temporal and spatial properties that differ between neurons, depending on the type and distribution of Ca^{2+}-binding proteins that transduce the rise of Ca^{2+} levels into biochemical responses. If a Ca^{2+} source is tightly coupled to the release machinery, it is known as Ca^{2+} nano-domain, however loose coupling defines local Ca^{2+} micro-domains where the source of Ca^{2+} and release machinery are separated by μm [21]. Neurons express a vast number of Ca^{2+} sensor proteins that are essential for the multiple effects of Ca^{2+} [22]. Among the most studied are the family of synaptotagmins (Syt1-7), important Ca^{2+} sensors of fast kinetics that mediate fast synaptic release in presynaptic terminals such as the calyx of Held or inhibitory synapses [23–26]. In addition, a wide variety of Ca^{2+} sensors contains EF-hand motifs, a highly conserved Ca^{2+}-binding motif [22, 27, 28] as for example is the ubiquitous Ca^{2+} sensor protein calmodulin (CaM) [29–31]. CaM binds Ca^{2+}, resulting in a conformational change and subsequent regulation of multiple target proteins, such as Ca^{2+}/CaM-dependent protein kinases or phosphatases of key importance in long-lasting plasticity.

Finally, another important group of cytosolic Ca^{2+}-binding proteins are the so-called "Ca^{2+}-buffers" [32]. Ca^{2+}-buffers are a subset of intracellular EF-hand containing proteins that do not belong to the Ca^{2+} sensor family, including parvalbumins, calbindin-D9k, calbindin-D28k, and calretinin [32–34]. Ca^{2+}-buffers chelate Ca^{2+} signals whenever there is a rise in cytosolic Ca^{2+} concentration and shape the properties, kinetics, and signaling properties of Ca^{2+} currents.

As seen, there are a vast number of elements affecting Ca^{2+} regulation during basal conditions and synaptic plasticity (Table 38.1). The precise contribution of each element to synaptic transmission and plasticity is probably synapse-specific and although in some synapses some elements are well known, many others still

Table 38.1 Historical overview of important events related with calcium dynamics during synaptic plasticity

Year	Milestones in calcium dynamics and synaptic plasticity
1954–65	Early discoveries on Ca^{2+} signaling in myofibrils
1964–73	Discovery of long-term potentiation in rabbit hippocampus by Bliss and Lømo
1965	Eric Kandel discovered Short-term facilitation in Aplysia
1967	Hagiwara and Nakajima describe the pharmacology of calcium spikes
	Harald Reuter provides the first voltage-clamp recordings of Ca^{2+} currents in neurons
1968	Katz and Miledi formulate the residual calcium hypothesis
	Discovery of calmodulin
	Discovery of cAMP
	Donald Walsh and Ed Krebs discover PKA
1977	Lynch and colleagues uncover LTD
1978	Bert Sakmann and Erwin Neher develop the patch-clamp technique
1981	Llinás and Yarom discover low-threshold Ca^{2+} spikes (T-type)
1983	Collingridge and colleagues demonstrated the involvement of NMDARs in LTP
1985	Discovery of N-type Ca^{2+} channels
	Llinás and colleagues discover P-type Ca^{2+} channels in Purkinje neurons
	Neuronal L-type channels
1986	Regulation of CaMKII
	Ca^{2+}-permeability of NMDARs
1991	Ca^{2+}-permeable AMPARs
	Participation of PKC in LTD
1992–97	Erwin Neher studies on Ca^{2+} buffering
1993	First evidence of a role of N-type Ca^{2+} channels in neurotransmission
1995	Q-type Ca^{2+} current terminology
1995–97	First evidence of STDP in neocortex
1998	Pharmacological separation of R-type currents
	Intracellular stores required for LTD
2003	Neocortical NMDA-dependent presynaptic LTD
2005	NMDAR-independent LTP in mossy fibers
2007	CP-AMPARs can induce LTP

remain unknown. In the next section, we review representative examples of several forms of synaptic plasticity and how the elements listed before interact to give rise a complex chain of events leading to short or long-lasting increases or decreases of synaptic strength.

38.3 Calcium Dynamics in Short-Term Plasticity

Short-term plasticity (STP) is a form of potentiation or depression of synaptic transmission present in probably all types of synapses [3, 35]. STP lasts from milliseconds to several minutes and is thought to be important in fast adaptations to sensory inputs, transient changes in behavioral states, and short-lasting forms of memory [3]. An action potential arriving to the presynaptic terminal can evoke influx of Ca^{2+} ions which bind to specialized Ca^{2+} sensor proteins that mediate fusion of synaptic vesicles and exocytosis [36]. Most forms of STP rely on transient accumulation of Ca^{2+} triggered by a short burst of action potentials. Presynaptic Ca^{2+} accumulation then modifies the probability of neurotransmitter release by modulating the biochemical events that produce the exocytosis of synaptic vesicles. STP can be induced following paired-pulse protocols and repetitive or tetanic stimulation at high frequencies [35]. It is important to note that multiple forms of STP coexist in the majority of synapses and the net synaptic strength is the result of an interaction between these forms, although the relative contribution of each one is controlled by the initial release probability and presynaptic activity pattern [37].

Short-term facilitation (STF) describes the increase in transmitter release lasting from milliseconds up to several seconds or minutes. Several forms of STF have been described including paired-pulse facilitation in response to paired-pulse protocols, augmentation (lasting seconds) and post-tetanic potentiation (PTP, lasting minutes) following high-frequency repetitive stimulation protocols. The main presynaptic Ca^{2+} current involved in STF is the P/Q-type current, carried by Ca_v 2.1 channels [38–43]. Ca_v 2.2 and 2.3 channels, carrying N- and R-type Ca^{2+} currents respectively, also contribute to synaptic transmission but lack the unique high-affinity Ca^{2+}-induced facilitation properties of Ca_v 2.1 channels [44].

Over the years, a mechanistic explanation of STF has remained elusive. However, substantial advances have been done in the past 20 years and several non-exclusive mechanisms have been proposed to explain STF [37]. The residual Ca^{2+} hypothesis, initially suggested in 1968 [45], proposes that an action potential evokes a local Ca^{2+} signal from tens to hundreds of millimolar that triggers release, but then lower levels of Ca^{2+} persist in the presynaptic terminal (residual Ca^{2+}) [37]. A second action potential would evoke another Ca^{2+} signal that would build-up over the residual Ca^{2+} thus causing facilitation. This would be the case if residual Ca^{2+} is a significant fraction of the local Ca^{2+} signal [37]. However, estimations of local and residual Ca^{2+} suggested that residual Ca^{2+} is about 1% of the local Ca^{2+} signal [46], so the enhancement of transmission after the second action potential would be very small, insufficient to account for paired-pulse facilitation.

Synapses can overcome this limitation by expressing two types of Ca^{2+} sensors, slow high-affinity (*e.g.* Syt7) and fast low-affinity proteins (*e.g.* Syt1, 2, 5/9) [24]. Thus, presynaptic residual Ca^{2+} can activate the high-affinity sensor to produce facilitation [37]. Third, the existence of Ca^{2+} buffers such as calbindin can reduce the initial probability of release by chelating local Ca^{2+}, however if Ca^{2+} levels are high enough to saturate local Ca^{2+} buffers, further incoming Ca^{2+} following subsequent action potentials would reach the release site, contributing to facilitation [37]. Finally, a fourth mechanism is the use-dependent facilitation of VGCCs mediated by the activation of CaM after Ca^{2+} entry [47, 48].

Short-term depression (STD) refers to the chain of events causing short-lasting decreases in transmitter release probability. As for STF, several mechanisms have been proposed over the years to account for the properties of STD observed following paired-pulse or tetanic stimulation. The simplest depletion model of STD proposes that the first stimulus triggers the release of a large fraction of the readily releasable pool (RRP) [37]. If the second action potential arrives and released vesicles are not replaced, the RRP is depleted and the response is depressed. The model implies that the initial fraction of RRP released must be large. This model accounts for the basic properties of paired-pulse depression observed in many synapses, however, in some hippocampal synapses the extent of depression does not correlate with the initial release [49]. Other models including inactivation of release sites or inactivation of VGCCs have been suggested to explain differences between synapses [37].

38.4 Calcium Dynamics in Long-Term Plasticity

Long-term potentiation (LTP) and long-term depression (LTD), lasting from minutes to hours or days, are thought to be a main neural substrate for learning and memory processes. Nowadays, multiple forms of LTP and LTD have been described, many of them depending upon Ca^{2+} dynamics at the presynaptic or postsynaptic sites. In the recent years, significant advances in the understanding of the molecular mechanisms involved in long-lasting decrease of synaptic strength, called long-term depression (LTD), have also been made [50]. LTD is important in hippocampus-dependent learning and memory processes, fear conditioning in amygdala, recognition memory in perirhinal cortex, development of visual and somatosensory cortices, impairments in learning and memory induced by acute stress, Fragile X syndrome, psychiatric disorders, or drug-addiction and cortico-limbic-striatal circuits [50]. LTD is typically induced using low-frequency stimulation, pairing stimulation with depolarization, and using spike-timing dependent (STDP) protocols (typically post-pre protocols, although some forms of LTD can be induced following pre-post protocols, depending on the developmental stage). In this section we outline the role of Ca^{2+} dynamics in the induction and expression of different known forms of LTP and LTD at different synapses.

38.4.1 Long-Term Potentiation

38.4.1.1 Postsynaptic NMDAR-Dependent LTP (Fig. 38.1a)

The best studied example of LTP is the NMDA-dependent LTP at CA3-CA1 synapses (Schaffer collateral-CA1) of the hippocampus [51, 52]. Most of the knowledge about this type of plasticity has been obtained using the brain slice preparation. The induction of this form of LTP requires activation of NMDA receptors during strong postsynaptic depolarization, which leads to an increase in postsynaptic Ca^{2+} concentration subsequently activating downstream biochemical processes [51, 53, 54]. Postsynaptic NMDA-dependent LTP can be experimentally induced following high-frequency tetanic stimulation, using a so-called "pairing-protocol" in which the postsynaptic cell is persistently depolarized while low-frequency stimulation is applied [55], or by using STDP protocols [56]. Upon repetitive stimulation, activation of postsynaptic NMDA receptors leads to Ca^{2+} influx restricted to the dendritic spine, increasing postsynaptic Ca^{2+} concentration and subsequent activation of the calcium/calmodulin-dependent protein kinase II, also known as CaMKII [30, 55, 57–59]. The increase in postsynaptic Ca^{2+} concentration leads to the activation of other protein kinases. For example, Ca^{2+} ions can bind calmodulin and activate the membrane enzyme adenylyl cyclase, resulting in increased cytosolic concentration of cyclic adenosine monophosphate (cAMP) which ultimately can activate the protein kinase A (PKA). PKA has been shown to indirectly boost the activity of CAMKII [60–62]. Recently, it was shown that an isoform of protein kinase C (PKC), which is also activated by raises in spine Ca^{2+} through NMDARs, termed PKCα, is directly involved in the induction and maintenance of LTP [63]. The activity of the above-mentioned kinases results in augmented single-channel conductance of dendritic AMPARs [64, 65] as well as in the incorporation of a higher number of AMPARs to the postsynaptic membrane during the expression and maintenance of LTP [66–69].

38.4.1.2 Postsynaptic NMDAR-Independent LTP (Fig. 38.1b)

Although the classical form of NMDA-dependent LTP has been object of intense research effort since its discovery, other forms of LTP independent of postsynaptic activation of NMDARs have been demonstrated [15]. VGCCs, in particular L-type currents conducted by $Ca_v1.2$ and $Ca_v1.3$ channels, are involved in several forms of NMDA-independent LTP. Inhibition of L-type current reduced LTP induced chemically by blockade of voltage-gated K^+ (K_v) channels with tetraethylammonium (TEA) [70, 71]. Also, LTP induced in vivo is partly blocked by $Ca_v1.2$ inhibition [72]. Moreover, LTP induced after NMDAR blockade or low-level theta stimulation has been shown to depend on $Ca_v1.2$ channels [73, 74]. Therefore, there is accumulating evidence for some forms of NMDA-independent LTP which depend on Ca^{2+} influx through Ca_v1 channels.

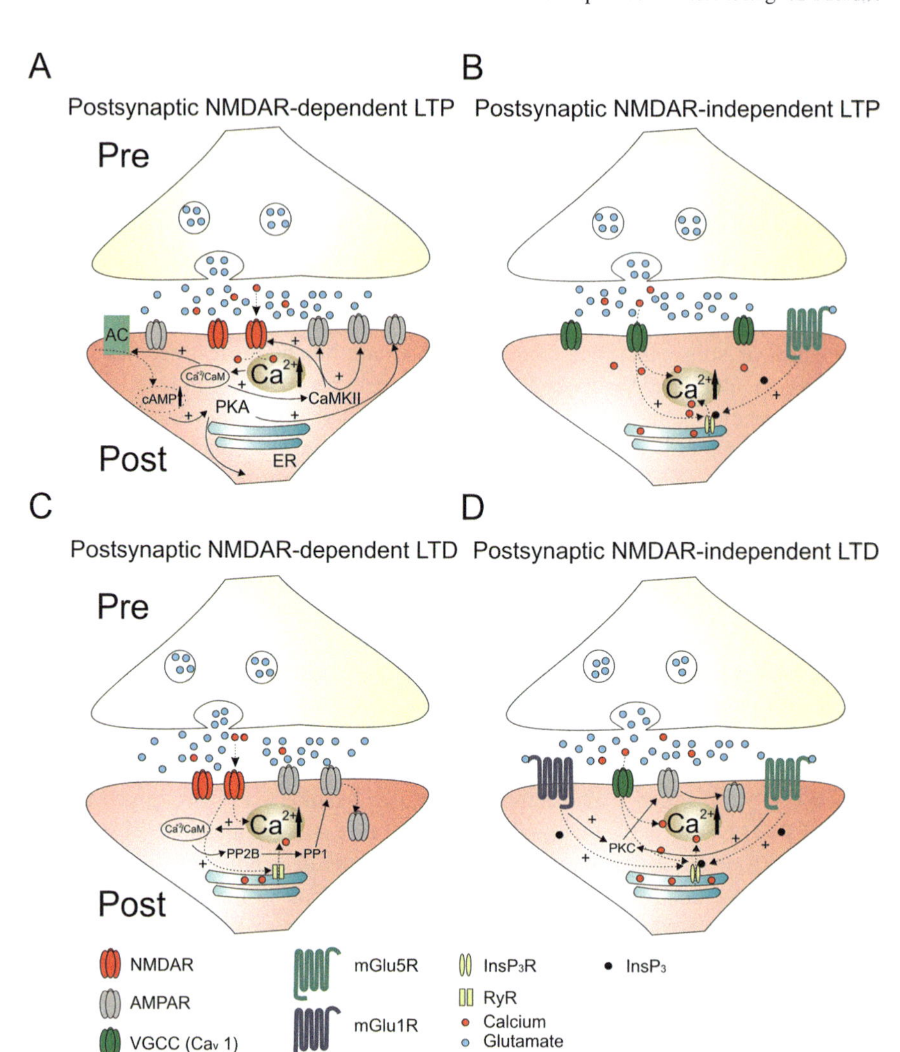

Fig. 38.1 Schematic examples of Ca^{2+}-related events during postsynaptic, NMDA-dependent and independent forms of LTP/LTD. (**a**) Ca^{2+} rise in the postsynaptic membrane through NMDA receptors activates the Ca^{2+}/CaM-activated AC and produces a subsequent increase in cAMP levels and PKA activation. Also, the Ca^{2+}/CaM complex can activate CaM-dependent kinases such as CaMKII. PKA and CaMKII activity lead to the insertion of additional AMPARs in the postsynaptic membrane, increasing synaptic strength. (**b**) Other forms of LTP include Ca^{2+} entry through postsynaptic VGCCs and activation of mGluRs, leading to the accumulation of second messenger and Ca^{2+}release from internal stores. (**c**) In prototypic cases of NMDA-dependent LTD, Ca^{2+} influx through postsynaptic NMDA receptors and ER-RyRs triggers the activation of a Ca^{2+}/CaM complex which in turn activates the phosphatase PP2B or calcineurin, leading to endocytosis of AMPAR. (**d**) In mGluR-dependent, NMDA-independent forms of LTD, activation of mGlu1R and mGlu5R and production of second messengers lead to activation of PKC-dependent pathways that will end in AMPAR endocytosis. In this form of LTD there is also Ca^{2+} influx from postsynaptic VGCCs. Solid arrow lines represent enzyme-dependent pathways whereas dashed arrow lines represent ion movement or production of second messengers

38.4.1.3 Presynaptic LTP (Fig. 38.2a)

Less studied than other forms of LTP, presynaptic LTP (preLTP) involves long-lasting modifications in the probability of neurotransmitter release [75–77]. This form of plasticity is ubiquitously expressed in the mammalian brain and may underlie behavioral responses occurring in vivo. In this section, we will review some of the best known examples of preLTP and their underlying Ca^{2+} dynamics.

In general, the mechanisms of induction of preLTP occur either in the presynaptic terminal or require the presence of a retrograde messenger produced by the postsynaptic neuron. The best characterized and prototypic form of preLTP can be found at the synapses established between the axons (mossy fibers, MF) of dentate gyrus granule cells and pyramidal CA3 neurons [78]. In this synapse, high-frequency stimulation or firing patterns occurring in vivo [79, 80] elicit an increase in Ca^{2+} concentration within the MF terminal [81–83] which results in a long-lasting increase of presynaptic release probability. Ca^{2+} influx mediated by Ca_v 2.3 channels (R-type currents) play a minor role in the overall Ca^{2+} influx during basal synaptic transmission (mostly mediated by N- and P/Q-type Ca^{2+} mediated by Ca_v 2.1 and 2.2 channels respectively) [84, 85], however Ca_v 2.3 channels contribute significantly to this type of preLTP. Presynaptic Ca^{2+} influx then activates a Ca^{2+}/CaM-dependent adenylyl cyclase, increasing presynaptic cAMP levels and subsequent activation of PKA, which ultimately phosphorylates other presynaptic substrates, resulting in a long-lasting increase of glutamate release probability [86–90].

Therefore, this type of preLTP is independent of NMDARs and cAMP/PKA-dependent. It has been found in other brain regions such as cerebellum [91], thalamus [92], subiculum [93], amygdala [94], and neocortex [95]. In MF-CA3 as well as MF-interneuron synapses [96–98], it has been proposed an additional mechanism for preLTP, consisting on the postsynaptic release of a retrograde messenger following postsynaptic Ca^{2+} increase mediated by Ca_v 1 channels and group I mGluR-dependent Ca^{2+} release from internal stores [99–101].

Although there is accumulating evidences pointing towards the presence of presynaptic NMDARs in some synapses, their roles in synaptic plasticity remain under debate [102]. NMDA-dependent preLTP has been described in some synapses of the amygdala and cerebellar cortex [75]. In this case, Ca^{2+} influx through NMDARs located in the presynaptic terminal is the main signal triggering downstream cascades that increase the presynaptic transmitter release probability during hours. The most common form of NMDA-dependent preLTP involves the increase of presynaptic levels of cAMP (presumably by activating the Ca^{2+}/CaM-dependent adenylyl cyclase) and subsequent PKA activation which phosphorylates target molecules of the presynaptic release machinery [103–106].

The expression mechanisms of preLTP also involve Ca^{2+} influx into presynaptic terminals. For example, at perforant path synapses contacting CA1 pyramidal neurons, a long-lasting increase of Ca_v 2.1 (N-type) channel activity has been demonstrated [107]. In addition, another study in the lateral amygdala showed a persistently increased Ca_v 1 (L-type) channel activity that mediates a long-lasting

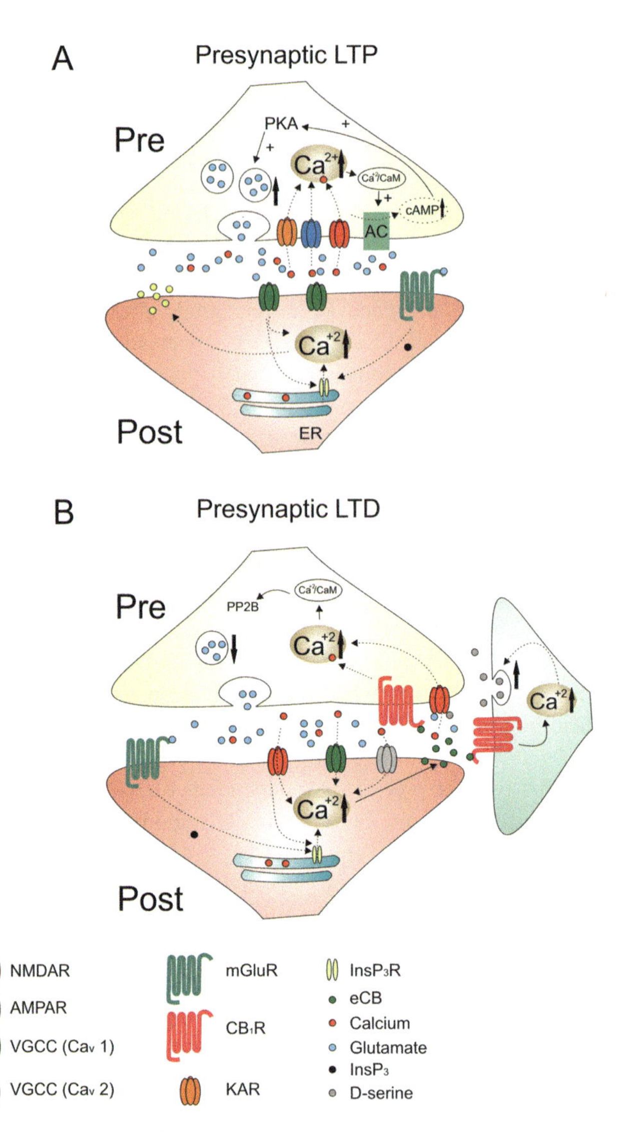

Fig. 38.2 Representation of Ca^{2+}-dependent events during presynaptic LTP/LTD. (**a**) In presynaptic forms of LTP, Ca^{2+} entry through NMDA (i.e. NMDA-dependent), VGCCs (Ca_v 2 channels), or KARs lead to presynaptic activation of Ca^{2+}/CaM complex, production of cAMP and activation of PKA, increasing the probability of transmitter release. In addition, coincident activation of postsynaptic mGluRs and Ca^{2+} influx through VGCCs and release from internal stores, and release of retrograde messengers (yellow) is required in some forms of preLTP. (**b**) During presynaptic LTD, presynaptic Ca^{2+} influx through NMDARs increase Ca^{2+} levels, activating Ca^{2+}/CaM-dependent phosphatases such as calcineurin, which in turn reduces the transmitter release probability. In addition, postsynaptic Ca^{2+} elevation through NMDARs, VGCCs, CP-AMPARs, or release from internal stores lead to eCB release, activating presynaptic eCB_1Rs (increasing presynaptic Ca^{2+} level) and astroglial eCB_1Rs. Glial CB_1R activation results in release of D-serine, boosting presynaptic NMDA activity. Solid arrow lines represent enzyme-dependent pathways whereas dashed arrow lines represent ion movement or production of second messengers

increase in glutamate release [108]. However, in the MF-CA3 of the hippocampus and the parallel fiber-Purkinje synapse in the cerebellum no changes in Ca^{2+} influx were observed following preLTP, suggesting that the expression mechanisms in these synapses occur downstream Ca^{2+} entry [75, 109–112]. Therefore, variability in the induction and expression of preLTP mechanisms indicate that further research is required to elucidate the molecular mechanisms more in detail. Given the increasing attention that presynaptic forms of LTP are receiving in the past few years, significant advances in this matter are expected to come in the upcoming future.

38.4.2 Long-Term Depression

38.4.2.1 Postsynaptic NMDAR-Dependent LTD (Fig. 38.1c)

As for LTP, NMDAR-dependency of LTD was first discovered at hippocampal CA3-CA1 synapses [113, 114]. One year later, in 1993, a similar form of LTD was also found in neocortical synapses [115]. In NMDAR-dependent LTD induced by low-frequency stimulation or STDP protocols, the main signal triggering downstream cascades that will end up in a sustained reduction in transmitter release is mainly postsynaptic Ca^{2+} influx through NMDA receptors [114]. The classical view is that LTD requires a modest increase in Ca^{2+} whereas LTP is elicited when Ca^{2+} levels increase beyond an induction threshold [51, 116]. However, further investigation has revealed that this is not always the case [117]. In this view, Ca^{2+} influx through NMDARs binds CaM, then activating a Ca^{2+} -dependent protein phosphatase cascade composed by Ca^{2+}/CaM-dependent phosphatase calcineurin which dephosphorylates inhibitor-1, leading to the activation of protein phosphatase 1 (PP1) [3, 50, 118–120], responsible for dephosphorilations either in AMPAR subunits or kinases that will end up in the endocytosis of postsynaptic AMPARs and LTD. The expression mechanisms of NMDAR-dependent LTD involve mainly endocytosis or removal of AMPARs from the postsynaptic membrane. This process involves different phosphatases that are directly or indirectly activated by Ca^{2+} influx. In addition, some studies suggest that protein kinases may be involved in the expression of this form of LTD [50]. Finally, the Ca^{2+} sensor hippocalcin has been suggested to mediate the coupling between Ca^{2+} signals and AMPAR endocytosis [121].

38.4.2.2 Postsynaptic NMDAR-Independent LTD (Fig. 38.1d)

As knowledge about LTD mechanisms in different synapses increased, new forms of LTD were uncovered across brain regions. NMDAR-independent LTD has been found in the hippocampus, cerebellum, ventral tegmental area, neostriatum, neocortex, or nucleus accumbens, for example.

The best studied form of NMDAR-independent LTD is that mediated by metabotropic glutamate receptors or mGluR-LTD. In the hippocampus [122], cerebellum [123], and perirhinal cortex [124], mGluR-LTD is dependent on Group I mGluRs (mGluR1 and 5) [125]. Since Group I mGluRs are located extrasynaptically, the induction of Group I mGluR-LTD is typically achieved by high-frequency trains (up to 300 Hz), burst or short trains at 66 Hz, paired-pulses of low frequency stimulation, or bath application of Group I mGluR agonists such as DHPG [126]. Afferent stimulation induces the opening of postsynaptic VGCCs and Group I mGluRs results in a postsynaptic elevation of second messengers such as $InsP_3$ pathway that will induce Ca^{2+} release from intracellular stores. Ca^{2+} influx activates the PKC pathway, eventually leading to removal of AMPARs in the postsynaptic membrane. In addition, DHPG-induced LTD can be observed in absence of Ca^{2+} and unaffected by PKC inhibition, suggesting that mGluR-LTD involves Ca^{2+}-dependent and -independent pathways [127, 128]. A role of Ca^{2+} entering into the cell trough calcium-permeable KARs has also been described.

Other forms of NMDAR-independent LTD involve postsynaptic release of endogenous endocannabinoids (eCBs) which act as retrograde messengers at the astrocyte or the presynaptic terminal (see below). In this form of LTD, neuronal activity triggers eCB release from the postsynaptic site, which binds to eCB_1Rs in the presynaptic site, suppressing neurotransmitter release at glutamatergic and GABAergic synapses [75]. Typically, presynaptic CB_1R activation is required during the induction of plasticity but not necessarily afterwards. In addition, postsynaptic Ca^{2+} rise through VGCCs is required for eCB release. At the presynaptic terminal, CB_1R activation inhibits presynaptic VGCCs, thereby reducing synaptic release.

38.4.2.3 Presynaptic LTD (Fig. 38.2b)

Conversely to preLTD, presynaptic LTD (preLTD) refers to long-lasting presynaptic changes that decrease synaptic strength [75]. As for preLTP, preLTD is widely expressed in many brain regions [75, 76]. There is evidence suggesting that preLTD may play important roles in certain behaviors, for example exploration of large spatial landmarks facilitates preLTD at MF-CA3 synapses [129]. In addition, fear extinction in the amygdala is linked to preLTD of prefrontal cortex-basolateral amygdala synapses [130]. At layer 4 to layer 2/3 synapses of developing barrel cortex, sensory evoked activity patterns induced preLTD potentially playing a role in synaptic pruning and refinement of cortical circuits during development [131].

NMDAR-Dependent PreLTD

Although currently subject of debate [102], accumulating evidence suggests that presynaptic NMDARs can mediate LTD in several synapses. A number of studies in visual and somatosensory cortices as well as in the hippocampus have provided the strongest evidence for a role of presynaptic NMDARs in LTD [102, 131–142]. Here, the induction of NMDA-dependent preLTD requires presynaptic Ca^{2+} influx through NMDA receptors and, within a narrow time window, coincident activation of postsynaptic Group I mGluRs and Ca_v 1 channels and the release of endocannabinoids, acting as retrograde messengers at the presynaptic terminal or astrocytes [137, 139, 143] producing a long-lasting reduction of transmitter release by activating CB_1Rs and presynaptic NMDARs. Precise details at molecular level beyond presynaptic NMDARs activation remains to be determined.

Recently, a similar form of preLTD was found in developing CA3-CA1 synapses [139]. In this case, the induction of preLTD involves presynaptic Ca^{2+} influx and activation of the Ca^{2+}-dependent phosphatase calcineurin.

NMDAR-Independent PreLTD

The MF-CA3 synapse not only expresses preLTP as discussed before, a form of preLTD independent of NMDARs has also been reported [144, 145]. In this case, although the precise mechanism is not clear, it potentially involves mGluRs and metabotropic kainate receptors. There is presynaptic Ca^{2+} influx through VGCCs and subsequent activation of CaM and Ca^{2+}/CaM-dependent kinases [11, 146–148]. Another form of preLTD has been described in synapses established between mossy fibers and *stratum lucidum* interneurons [149]. In this case, the induction of preLTD is mediated by postsynaptic Ca^{2+} increase through CP-AMPARs and presynaptic Ca_v 2.1 channels presumably through postsynaptic release of unidentified retrograde messenger. The reduction of presynaptic Ca^{2+} influx leads to long-lasting reduction of neurotransmitter release probability.

38.5 Concluding Remarks

Ca^{2+} dynamics and signaling in mammalian cells and especially in neurons is an extremely complex phenomenon producing variable effects in different neurons. The huge variability of mechanisms found in relation to synaptic plasticity is a major challenge for the future. Traditionally, it has been proposed that high Ca^{2+} influx is required for LTP whereas moderate Ca^{2+} rises induced LTD under some conditions. More realistic protocols based on firing patterns found in vivo or STDP protocols have revealed that the traditional assumption does not hold in some synapses [117]. Future research will be required to investigate in detail the chain of events involved in Ca^{2+} signaling at different synapses. For example, a detailed knowledge of receptor subunits composition, Ca^{2+} sensors and buffers involved, and Ca^{2+}-dependent enzymatic processes will be of great help in the quest for a general mechanistic framework of synaptic plasticity. In addition, novel roles of NMDA receptors must be investigated in detail to clarify the relevance of presynaptic and/or putative metabotropic NMDA receptors and relevant signaling molecules. How the complex machinery regulating Ca^{2+} dynamics controls the physiological aspects of neuronal communication in vivo and affects behavior in health and disease will be a major challenge for the future of the field.

Acknowlegments Work in the group is supported by the Ministerio de Economía y Competitividad (MINECO/FEDER) of Spain (Grant BFU2015-68655-P to A.R.-M.). P.M.-A. is supported by a postdoctoral "Juan de la Cierva-Formación" Fellowship from MINECO.

References

1. Clapham DE (2007) Calcium signaling. Cell 131(6):1047–1058
2. Berridge MJ (1998) Neuronal calcium signaling. Neuron 21(1):13–26
3. Citri A, Malenka RC (2008) Synaptic plasticity: multiple forms, functions, and mechanisms. Neuropsychopharmacology 33(1):18–41
4. Cramer SC, Sur M, Dobkin BH, O'Brien C, Sanger TD, Trojanowski JQ et al (2011) Harnessing neuroplasticity for clinical applications. Brain 134(6):1591–1609
5. Cooke SF, Bliss TVP (2006) Plasticity in the human central nervous system. Brain 129(7):1659–1673
6. Lalanne T, Oyrer J, Farrant M, Sjöström PJ (2018) Synapse type-dependent expression of calcium-permeable AMPA receptors. Front Synaptic Neurosci 10:34
7. Sihra TS, Flores G, Rodríguez-Moreno A (2014) Kainate receptors: multiple roles in neuronal plasticity. Neuroscientist 20(1):29–43
8. Negrete-Díaz JV, Duque-Feria P, Andrade-Talavera Y, Carrión M, Flores G, Rodríguez-Moreno A (2012) Kainate receptor-mediated depression of glutamatergic transmission involving protein kinase A in the lateral amygdala. J Neurochem 121(1):36–43
9. Rodríguez-Moreno A, Sihra TS (2011) Kainate receptors. Novel signaling insights. Adv Exp Med Biol 717:vii–xi, xiii
10. Rodríguez-Moreno A, Sihra TS (2007) Kainate receptors with a metabotropic modus operandi. Trends Neurosci 30(12):630–637
11. Negrete-Díaz JV, Sihra TS, Delgado-García JM, Rodríguez-Moreno A (2007) Kainate receptor-mediated presynaptic inhibition converges with presynaptic inhibition mediated by Group II mGluRs and long-term depression at the hippocampal mossy fiber-CA3 synapse. J Neural Transm (Vienna) 114(11):1425–1431
12. Negrete-Díaz JV, Sihra TS, Delgado-García JM, Rodríguez-Moreno A (2006) Kainate receptor-mediated inhibition of glutamate release involves protein kinase A in the mouse hippocampus. J Neurophysiol 96(4):1829–1837
13. Rodríguez-Moreno A, Lerma J (1998) Kainate receptor modulation of GABA release involves a metabotropic function. Neuron 20(6):1211–1218
14. Catterall WA (2011) Voltage-gated calcium channels. Cold Spring Harb Perspect Biol 3(8):a003947
15. Nanou E, Catterall WA (2018) Calcium Channels, Synaptic Plasticity, and Neuropsychiatric Disease. Neuron 98(3):466–481
16. Zamponi GW, Striessnig J, Koschak A, Dolphin AC (2015) The physiology, pathology, and pharmacology of voltage-gated calcium channels and their future therapeutic potential. Pharmacol Rev 67(4):821–870
17. Foskett JK, White C, Cheung K-H, Mak D-OD (2007) Inositol trisphosphate receptor Ca^{2+} release channels. Physiol Rev 87(2):593–658
18. Ferris CD, Huganir RL, Supattapone S, Snyder SH (1989) Purified inositol 1,4,5-trisphosphate receptor mediates calcium flux in reconstituted lipid vesicles. Nature 342(6245):87–89

19. Maeda N, Kawasaki T, Nakade S, Yokota N, Taguchi T, Kasai M et al (1991) Structural and functional characterization of inositol 1,4,5-trisphosphate receptor channel from mouse cerebellum. J Biol Chem 266(2):1109–1116
20. Lanner JT, Georgiou DK, Joshi AD, Hamilton SL (2010) Ryanodine receptors: structure, expression, molecular details, and function in calcium release. Cold Spring Harb Perspect Biol 2(11):a003996-a
21. Eggermann E, Bucurenciu I, Goswami SP, Jonas P (2011) Nanodomain coupling between Ca^{2+} channels and sensors of exocytosis at fast mammalian synapses. Nat Rev Neurosci 13(1):7–21
22. McCue HV, Haynes LP, Burgoyne RD (2010) The diversity of calcium sensor proteins in the regulation of neuronal function. Cold Spring Harb Perspect Biol 2(8):a004085
23. Chen C, Arai I, Satterfield R, Young SM Jr, Jonas P (2017) Synaptotagmin 2 Is the Fast Ca^{2+} Sensor at a Central Inhibitory Synapse. Cell Rep 18(3):723–736
24. Chen C, Jonas P (2017) Synaptotagmins: that's why so many. Neuron 94(4):694–696
25. Chen C, Satterfield R, Young SM Jr, Jonas P (2017) Triple function of synaptotagmin 7 ensures efficiency of high-frequency transmission at central GABAergic synapses. Cell Rep 21(8):2082–2089
26. Luo F, Südhof TC (2017) Synaptotagmin-7-mediated asynchronous release boosts high-fidelity synchronous transmission at a central synapse. Neuron 94(4):826–39 e3
27. Burgoyne RD (2007) Neuronal calcium sensor proteins: generating diversity in neuronal Ca^{2+} signalling. Nat Rev Neurosci 8(3):182–193
28. Burgoyne RD, Haynes LP (2012) Understanding the physiological roles of the neuronal calcium sensor proteins. Mol Brain 5(1):2
29. Malenka RC, Kauer JA, Perkel DJ, Mauk MD, Kelly PT, Nicoll RA et al (1989) An essential role for postsynaptic calmodulin and protein kinase activity in long-term potentiation. Nature 340(6234):554–557
30. Lledo PM, Hjelmstad GO, Mukherji S, Soderling TR, Malenka RC, Nicoll RA (1995) Calcium/calmodulin-dependent kinase II and long-term potentiation enhance synaptic transmission by the same mechanism. Proc Natl Acad Sci U S A 92(24):11175–11179
31. Lisman J, Malenka RC, Nicoll RA, Malinow R (1997) Learning mechanisms: the case for CaM-KII. Science 276(5321):2001–2002
32. Schwaller B (2010) Cytosolic Ca^{2+} buffers. Cold Spring Harb Perspect Biol 2(11):a004051
33. Blaustein MP (1988) Calcium transport and buffering in neurons. Trends Neurosci 11(10):438–443
34. Matthews EA, Dietrich D (2015) Buffer mobility and the regulation of neuronal calcium domains. Front Cell Neurosci 9:48
35. Zucker RS, Regehr WG (2002) Short-term synaptic plasticity. Annu Rev Physiol 64:355–405
36. Südhof TC (2013) Neurotransmitter release: the last millisecond in the life of a synaptic vesicle. Neuron 80(3):675–690
37. Regehr WG (2012) Short-term presynaptic plasticity. Cold Spring Harb Perspect Biol 4(7):a005702
38. Borst JG, Sakmann B (1998) Facilitation of presynaptic calcium currents in the rat brainstem. J Physiol 513(Pt 1):149–155
39. Cuttle MF, Tsujimoto T, Forsythe ID, Takahashi T (1998) Facilitation of the presynaptic calcium current at an auditory synapse in rat brainstem. J Physiol 512(Pt 3):723–729
40. Lee A, Scheuer T, Catterall WA (2000) Ca^{2+}/calmodulin-dependent facilitation and inactivation of P/Q-type Ca^{2+} channels. J Neurosci 20(18):6830–6838
41. Lee A, Westenbroek RE, Haeseleer F, Palczewski K, Scheuer T, Catterall WA (2002) Differential modulation of Ca(v)2.1 channels by calmodulin and Ca^{2+}-binding protein 1. Nat Neurosci 5(3):210–217
42. Lee A, Wong ST, Gallagher D, Li B, Storm DR, Scheuer T et al (1999) Ca^{2+}/calmodulin binds to and modulates P/Q-type calcium channels. Nature 399(6732):155–159
43. Lee A, Zhou H, Scheuer T, Catterall WA (2003) Molecular determinants of Ca^{2+}/calmodulin-dependent regulation of Ca(v)2.1 channels. Proc Natl Acad Sci U S A 100(26):16059–16064

44. Liang H, DeMaria CD, Erickson MG, Mori MX, Alseikhan BA, Yue DT (2003) Unified mechanisms of Ca^{2+} regulation across the Ca^{2+} channel family. Neuron 39(6):951–960
45. Katz B, Miledi R (1968) The role of calcium in neuromuscular facilitation. J Physiol 195(2):481–492
46. Schneggenburger R, Neher E (2005) Presynaptic calcium and control of vesicle fusion. Curr Opin Neurobiol 15(3):266–274
47. Catterall WA, Few AP (2008) Calcium channel regulation and presynaptic plasticity. Neuron 59(6):882–901
48. Mochida S, Few AP, Scheuer T, Catterall WA (2008) Regulation of presynaptic Ca(V)2.1 channels by Ca^{2+} sensor proteins mediates short-term synaptic plasticity. Neuron 57(2):210–216
49. Chen G, Harata NC, Tsien RW (2004) Paired-pulse depression of unitary quantal amplitude at single hippocampal synapses. Proc Natl Acad Sci U S A 101(4):1063–1068
50. Collingridge GL, Peineau S, Howland JG, Wang YT (2010) Long-term depression in the CNS. Nat Rev Neurosci 11:459–473
51. Malenka RC, Nicoll RA (1993) NMDA-receptor-dependent synaptic plasticity: multiple forms and mechanisms. Trends Neurosci 16(12):521–527
52. Nicoll RA (2017) A brief history of long-term potentiation. Neuron 93(2):281–290
53. Malenka RC (1991) Postsynaptic factors control the duration of synaptic enhancement in area CA1 of the hippocampus. Neuron 6(1):53–60
54. Malenka RC (1991) The role of postsynaptic calcium in the induction of long-term potentiation. Mol Neurobiol 5(2–4):289–295
55. Herring BE, Nicoll RA (2016) Long-term potentiation: from CaMKII to AMPA receptor trafficking. Annu Rev Physiol 78(1):351–365
56. Markram H, Lubke J, Frotscher M, Sakmann B (1997) Regulation of synaptic efficacy by coincidence of postsynaptic APs and EPSPs. Science 275(5297):213–215
57. Pettit DL, Perlman S, Malinow R (1994) Potentiated transmission and prevention of further LTP by increased CaMKII activity in postsynaptic hippocampal slice neurons. Science 266(5192):1881–1885
58. Silva AJ, Stevens CF, Tonegawa S, Wang Y (1992) Deficient hippocampal long-term potentiation in alpha-calcium-calmodulin kinase II mutant mice. Science 257(5067):201–206
59. Silva AJ, Wang Y, Paylor R, Wehner JM, Stevens CF, Tonegawa S (1992) Alpha calcium/calmodulin kinase II mutant mice: deficient long-term potentiation and impaired spatial learning. Cold Spring Harb Symp Quant Biol 57:527–539
60. Blitzer RD, Connor JH, Brown GP, Wong T, Shenolikar S, Iyengar R et al (1998) Gating of CaMKII by cAMP-regulated protein phosphatase activity during LTP. Science 280(5371):1940–1942
61. Lisman J (1989) A mechanism for the Hebb and the anti-Hebb processes underlying learning and memory. Proc Natl Acad Sci U S A 86(23):9574–9578
62. Makhinson M, Chotiner JK, Watson JB, O'Dell TJ (1999) Adenylyl cyclase activation modulates activity-dependent changes in synaptic strength and Ca^{2+}/calmodulin-dependent kinase II autophosphorylation. J Neurosci 19(7):2500–2510
63. Colgan LA, Hu M, Misler JA, Parra-Bueno P, Moran CM, Leitges M et al (2018) PKCα integrates spatiotemporally distinct Ca^{2+} and autocrine BDNF signaling to facilitate synaptic plasticity. Nat Neurosci 21:1027–1037
64. Benke TA, Luthi A, Isaac JT, Collingridge GL (1998) Modulation of AMPA receptor unitary conductance by synaptic activity. Nature 393(6687):793–797
65. Soderling TR, Derkach VA (2000) Postsynaptic protein phosphorylation and LTP. Trends Neurosci 23(2):75–80
66. Bredt DS, Nicoll RA (2003) AMPA receptor trafficking at excitatory synapses. Neuron 40(2):361–379
67. Collingridge GL, Isaac JT, Wang YT (2004) Receptor trafficking and synaptic plasticity. Nat Rev Neurosci 5(12):952–962

68. Sheng M, Kim MJ (2002) Postsynaptic signaling and plasticity mechanisms. Science 298(5594):776–780
69. Song I, Huganir RL (2002) Regulation of AMPA receptors during synaptic plasticity. Trends Neurosci 25(11):578–588
70. Huber KM, Mauk MD, Kelly PT (1995) LTP induced by activation of voltage-dependent Ca^{2+} channels requires protein kinase activity. Neuroreport 6(9):1281–1284
71. Huber KM, Mauk MD, Kelly PT (1995) Distinct LTP induction mechanisms: contribution of NMDA receptors and voltage-dependent calcium channels. J Neurophysiol 73(1):270–279
72. Freir DB, Herron CE (2003) Inhibition of L-type voltage dependent calcium channels causes impairment of long-term potentiation in the hippocampal CA1 region in vivo. Brain Res 967(1–2):27–36
73. Moosmang S, Haider N, Klugbauer N, Adelsberger H, Langwieser N, Muller J et al (2005) Role of hippocampal Cav1.2 Ca^{2+} channels in NMDA receptor-independent synaptic plasticity and spatial memory. J Neurosci 25(43):9883–9892
74. Staubli U, Lynch G (1987) Stable hippocampal long-term potentiation elicited by 'theta' pattern stimulation. Brain Res 435(1–2):227–234
75. Castillo PE (2012) Presynaptic LTP and LTD of excitatory and inhibitory synapses. Cold Spring Harb Perspect Biol 4(2):pii: a005728
76. Monday HR, Younts TJ, Castillo PE (2018) Long-term plasticity of neurotransmitter release: emerging mechanisms and contributions to brain function and disease. Annu Rev Neurosci 41(1):299–322
77. Yang Y, Calakos N (2013) Presynaptic long-term plasticity. Front Synaptic Neurosci 5:8
78. Nicoll RA, Schmitz D (2005) Synaptic plasticity at hippocampal mossy fibre synapses. Nat Rev Neurosci 6:863–876
79. Gundlfinger A, Breustedt J, Sullivan D, Schmitz D (2010) Natural spike trains trigger short- and long-lasting dynamics at hippocampal mossy fiber synapses in rodents. PLoS One 5(4):e9961
80. Mistry R, Dennis S, Frerking M, Mellor JR (2011) Dentate gyrus granule cell firing patterns can induce mossy fiber long-term potentiation in vitro. Hippocampus 21(11):1157–1168
81. Mellor J, Nicoll RA (2001) Hippocampal mossy fiber LTP is independent of postsynaptic calcium. Nat Neurosci 4(2):125–126
82. Zalutsky RA, Nicoll RA (1990) Comparison of two forms of long-term potentiation in single hippocampal neurons. Science 248(4963):1619–1624
83. Zalutsky RA, Nicoll RA (1991) Comparison of two forms of long-term potentiation in single hippocampus neurons. Correct Sci 251(4996):856
84. Breustedt J, Vogt KE, Miller RJ, Nicoll RA, Schmitz D (2003) Alpha1E-containing Ca^{2+} channels are involved in synaptic plasticity. Proc Natl Acad Sci U S A 100(21):12450–12455
85. Dietrich D, Kirschstein T, Kukley M, Pereverzev A, von der Brelie C, Schneider T et al (2003) Functional specialization of presynaptic Cav2.3 Ca^{2+} channels. Neuron 39(3):483–496
86. Huang YY, Li XC, Kandel ER (1994) cAMP contributes to mossy fiber LTP by initiating both a covalently mediated early phase and macromolecular synthesis-dependent late phase. Cell 79(1):69–79
87. Villacres EC, Wong ST, Chavkin C, Storm DR (1998) Type I adenylyl cyclase mutant mice have impaired mossy fiber long-term potentiation. J Neurosci 18(9):3186–3194
88. Wang H, Pineda VV, Chan GC, Wong ST, Muglia LJ, Storm DR (2003) Type 8 adenylyl cyclase is targeted to excitatory synapses and required for mossy fiber long-term potentiation. J Neurosci 23(30):9710–9718
89. Weisskopf MG, Castillo PE, Zalutsky RA, Nicoll RA (1994) Mediation of hippocampal mossy fiber long-term potentiation by cyclic AMP. Science 265(5180):1878–1882
90. Andrade-Talavera Y, Duque-Feria P, Negrete-Díaz JV, Sihra TS, Flores G, Rodríguez-Moreno A (2012) Presynaptic kainate receptor-mediated facilitation of glutamate release involves Ca^{2+} -calmodulin at mossy fiber-CA3 synapses. J Neurochem 122(5):891–899
91. Salin PA, Malenka RC, Nicoll RA (1996) Cyclic AMP mediates a presynaptic form of LTP at cerebellar parallel fiber synapses. Neuron 16(4):797–803

92. Castro-Alamancos MA, Calcagnotto ME (1999) Presynaptic long-term potentiation in corticothalamic synapses. J Neurosci 19(20):9090–9097
93. Behr J, Wozny C, Fidzinski P, Schmitz D (2009) Synaptic plasticity in the subiculum. Prog Neurobiol 89(4):334–342
94. López de Armentia M, Sah P (2007) Bidirectional synaptic plasticity at nociceptive afferents in the rat central amygdala. J Physiol 581(Pt 3):961–970
95. Chen HX, Jiang M, Akakin D, Roper SN (2009) Long-term potentiation of excitatory synapses on neocortical somatostatin-expressing interneurons. J Neurophysiol 102(6):3251–3259
96. Galván EJ, Calixto E, Barrionuevo G (2008) Bidirectional Hebbian plasticity at hippocampal mossy fiber synapses on CA3 interneurons. J Neurosci 28(52):14042–14055
97. Galván EJ, Cosgrove KE, Barrionuevo G (2011) Multiple forms of long-term synaptic plasticity at hippocampal mossy fiber synapses on interneurons. Neuropharmacology 60(5):740–747
98. Galván EJ, Cosgrove KE, Mauna JC, Card JP, Thiels E, Meriney SD et al (2010) Critical involvement of postsynaptic protein kinase activation in long-term potentiation at hippocampal mossy fiber synapses on CA3 interneurons. J Neurosci 30(8):2844–2855
99. Jaffe D, Johnston D (1990) Induction of long-term potentiation at hippocampal mossy-fiber synapses follows a Hebbian rule. J Neurophysiol 64(3):948–960
100. Kapur A, Yeckel MF, Gray R, Johnston D (1998) L-Type calcium channels are required for one form of hippocampal mossy fiber LTP. J Neurophysiol 79(4):2181–2190
101. Yeckel MF, Kapur A, Johnston D (1999) Multiple forms of LTP in hippocampal CA3 neurons use a common postsynaptic mechanism. Nat Neurosci 2(7):625–633
102. Bouvier G, Larsen RS, Rodríguez-Moreno A, Paulsen O, Sjöström PJ (2018) Towards resolving the presynaptic NMDA receptor debate. Curr Opin Neurobiol 51:1–7
103. Fourcaudot E, Gambino F, Humeau Y, Casassus G, Shaban H, Poulain B et al (2008) cAMP/PKA signaling and RIM1alpha mediate presynaptic LTP in the lateral amygdala. Proc Natl Acad Sci U S A 105(39):15130–15135
104. Lachamp PM, Liu Y, Liu SJ (2009) Glutamatergic modulation of cerebellar interneuron activity is mediated by an enhancement of GABA release and requires protein kinase A/RIM1alpha signaling. J Neurosci 29(2):381–392
105. Liu SJ, Lachamp P (2006) The activation of excitatory glutamate receptors evokes a long-lasting increase in the release of GABA from cerebellar stellate cells. J Neurosci 26(36):9332–9339
106. Samson RD, Pare D (2005) Activity-dependent synaptic plasticity in the central nucleus of the amygdala. J Neurosci 25(7):1847–1855
107. Ahmed MS, Siegelbaum SA (2009) Recruitment of N-Type Ca^{2+} channels during LTP enhances low release efficacy of hippocampal CA1 perforant path synapses. Neuron 63(3):372–385
108. Fourcaudot E, Gambino F, Casassus G, Poulain B, Humeau Y, Luthi A (2009) L-type voltage-dependent Ca^{2+} channels mediate expression of presynaptic LTP in amygdala. Nat Neurosci 12(9):1093–1095
109. Kamiya H, Umeda K, Ozawa S, Manabe T (2002) Presynaptic Ca^{2+} entry is unchanged during hippocampal mossy fiber long-term potentiation. J Neurosci 22(24):10524–10528
110. Regehr WG, Tank DW (1991) The maintenance of LTP at hippocampal mossy fiber synapses is independent of sustained presynaptic calcium. Neuron 7(3):451–459
111. Reid CA, Dixon DB, Takahashi M, Bliss TV, Fine A (2004) Optical quantal analysis indicates that long-term potentiation at single hippocampal mossy fiber synapses is expressed through increased release probability, recruitment of new release sites, and activation of silent synapses. J Neurosci 24(14):3618–3626
112. Chen C, Regehr WG (1997) The mechanism of cAMP-mediated enhancement at a cerebellar synapse. J Neurosci 17(22):8687–8694

113. Dudek SM, Bear MF (1992) Homosynaptic long-term depression in area CA1 of hippocampus and effects of N-methyl-D-aspartate receptor blockade. Proc Natl Acad Sci U S A 89(10):4363–4367
114. Mulkey RM, Malenka RC (1992) Mechanisms underlying induction of homosynaptic long-term depression in area CA1 of the hippocampus. Neuron 9(5):967–975
115. Kirkwood A, Dudek SM, Gold JT, Aizenman CD, Bear MF (1993) Common forms of synaptic plasticity in the hippocampus and neocortex in vitro. Science 260(5113):1518–1521
116. Cummings JA, Mulkey RM, Nicoll RA, Malenka RC (1996) Ca^{2+} signaling requirements for long-term depression in the hippocampus. Neuron 16(4):825–833
117. Evans RC, Blackwell KT (2015) Calcium: amplitude, duration, or location? Biol Bull 228(1):75–83
118. Isaac J (2001) Protein phosphatase 1 and LTD: synapses are the architects of depression. Neuron 32(6):963–966
119. Morishita W, Connor JH, Xia H, Quinlan EM, Shenolikar S, Malenka RC (2001) Regulation of synaptic strength by protein phosphatase 1. Neuron 32(6):1133–1148
120. Kirkwood A, Bear MF (1994) Homosynaptic long-term depression in the visual cortex. J Neurosci 14(5 Pt 2):3404–3412
121. Palmer CL, Lim W, Hastie PG, Toward M, Korolchuk VI, Burbidge SA et al (2005) Hippocalcin functions as a calcium sensor in hippocampal LTD. Neuron 47(4):487–494
122. Oliet SH, Malenka RC, Nicoll RA (1997) Two distinct forms of long-term depression coexist in CA1 hippocampal pyramidal cells. Neuron 18(6):969–982
123. Linden DJ, Connor JA (1991) Participation of postsynaptic PKC in cerebellar long-term depression in culture. Science 254(5038):1656–1659
124. Jo J, Heon S, Kim MJ, Son GH, Park Y, Henley JM et al (2008) Metabotropic glutamate receptor-mediated LTD involves two interacting Ca^{2+} sensors, NCS-1 and PICK1. Neuron 60(6):1095–1111
125. Lüscher C, Huber KM (2010) Group 1 mGluR-dependent synaptic long-term depression: mechanisms and implications for circuitry and disease. Neuron 65(4):445–459
126. Bellone C, Luscher C, Mameli M (2008) Mechanisms of synaptic depression triggered by metabotropic glutamate receptors. Cell Mol Life Sci 65(18):2913–2923
127. Fitzjohn SM, Palmer MJ, May JE, Neeson A, Morris SA, Collingridge GL (2001) A characterisation of long-term depression induced by metabotropic glutamate receptor activation in the rat hippocampus in vitro. J Physiol 537(Pt 2):421–430
128. Schnabel R, Kilpatrick IC, Collingridge GL (1999) An investigation into signal transduction mechanisms involved in DHPG-induced LTD in the CA1 region of the hippocampus. Neuropharmacology 38(10):1585–1596
129. Hagena H, Manahan-Vaughan D (2011) Learning-facilitated synaptic plasticity at CA3 mossy fiber and commissural-associational synapses reveals different roles in information processing. Cereb Cortex 21(11):2442–2449
130. Cho JH, Deisseroth K, Bolshakov VY (2013) Synaptic encoding of fear extinction in mPFC-amygdala circuits. Neuron 80(6):1491–1507
131. Rodríguez-Moreno A, Gonzalez-Rueda A, Banerjee A, Upton AL, Craig MT, Paulsen O (2013) Presynaptic self-depression at developing neocortical synapses. Neuron 77(1):35–42
132. Bender VA, Bender KJ, Brasier DJ, Feldman DE (2006) Two coincidence detectors for spike timing-dependent plasticity in somatosensory cortex. J Neurosci 26(16):4166–4177
133. Bender KJ, Allen CB, Bender VA, Feldman DE (2006) Synaptic basis for Whisker deprivation-induced synaptic depression in rat somatosensory cortex. J Neurosci 26(16):4155–4165
134. Larsen Rylan S, Smith Ikuko T, Miriyala J, Han Ji E, Corlew Rebekah J, Smith Spencer L et al (2014) Synapse-specific control of experience-dependent plasticity by presynaptic NMDA receptors. Neuron 83(4):879–893
135. Rodríguez-Moreno A, Paulsen O (2008) Spike timing-dependent long-term depression requires presynaptic NMDA receptors. Nat Neurosci 11(7):744–745

136. Nevian T, Sakmann B (2006) Spine Ca^{2+} signaling in spike-timing-dependent plasticity. J Neurosci 26(43):11001–11013
137. Sjöström PJ, Turrigiano GG, Nelson SB (2003) Neocortical LTD via coincident activation of presynaptic NMDA and cannabinoid receptors. Neuron 39(4):641–654
138. Rodríguez-Moreno A, Banerjee A, Paulsen O (2010) Presynaptic NMDA receptors and spike timing-dependent depression at cortical synapses. Front Synaptic Neurosci 2:18
139. Andrade-Talavera Y, Duque-Feria P, Paulsen O, Rodríguez-Moreno A (2016) Presynaptic spike timing-dependent long-term depression in the mouse hippocampus. Cereb Cortex 26(8):3637–3654
140. Banerjee A, Gonzalez-Rueda A, Sampaio-Baptista C, Paulsen O, Rodríguez-Moreno A (2014) Distinct mechanisms of spike timing-dependent LTD at vertical and horizontal inputs onto L2/3 pyramidal neurons in mouse barrel cortex. Phys Rep 2(3):e00271
141. Pérez-Rodríguez M, Arroyo-García LE, Prius-Mengual J, Andrade-Talavera Y, Armengol JA, Pérez-Villegas EM et al (2018) Adenosine receptor-mediated developmental loss of spike timing-dependent depression in the hippocampus. Cereb Cortex. https://doi.org/10.1093/cercor/bhy194
142. Rodríguez-Moreno A, Kohl MM, Reeve JE, Eaton TR, Collins HA, Anderson HL et al (2011) Presynaptic induction and expression of timing-dependent long-term depression demonstrated by compartment-specific photorelease of a use-dependent NMDA receptor antagonist. J Neurosci 31(23):8564–8569
143. Duguid I, Sjöström PJ (2006) Novel presynaptic mechanisms for coincidence detection in synaptic plasticity. Curr Opin Neurobiol 16(3):312–322
144. Yokoi M, Kobayashi K, Manabe T, Takahashi T, Sakaguchi I, Katsuura G et al (1996) Impairment of hippocampal mossy fiber LTD in mice lacking mGluR2. Science 273(5275):645–647
145. Kobayashi K, Manabe T, Takahashi T (1996) Presynaptic long-term depression at the hippocampal mossy fiber-CA3 synapse. Science 273(5275):648–650
146. Tzounopoulos T, Janz R, Südhof TC, Nicoll RA, Malenka RC (1998) A role for cAMP in long-term depression at hippocampal mossy fiber synapses. Neuron 21(4):837–845
147. Kobayashi K, Manabe T, Takahashi T (1999) Calcium-dependent mechanisms involved in presynaptic long-term depression at the hippocampal mossy fibre-CA3 synapse. Eur J Neurosci 11(5):1633–1638
148. Lyon L, Borel M, Carrion M, Kew JN, Corti C, Harrison PJ et al (2011) Hippocampal mossy fiber long-term depression in Grm2/3 double knockout mice. Synapse 65(9):945–954
149. Maccaferri G, Toth K, McBain CJ (1998) Target-specific expression of presynaptic mossy fiber plasticity. Science 279(5355):1368–1370

Chapter 39
Calcium Signaling During Brain Aging and Its Influence on the Hippocampal Synaptic Plasticity

Ashok Kumar

Abstract Calcium (Ca^{2+}) ions are highly versatile intracellular signaling molecules and are universal second messenger for regulating a variety of cellular and physiological functions including synaptic plasticity. Ca^{2+} homeostasis in the central nervous system endures subtle dysregulation with advancing age. Research has provided abundant evidence that brain aging is associated with altered neuronal Ca^{2+} regulation and synaptic plasticity mechanisms. Much of the work has focused on the hippocampus, a brain region critically involved in learning and memory, which is particularly susceptible to dysfunction during aging. The current chapter takes a specific perspective, assessing various Ca^{2+} sources and the influence of aging on Ca^{2+} sources and synaptic plasticity in the hippocampus. Integrating the knowledge of the complexity of age-related alterations in neuronal Ca^{2+} signaling and synaptic plasticity mechanisms will positively shape the development of highly effective therapeutics to treat brain disorders including cognitive impairment associated with aging and neurodegenerative disease.

Keywords Calcium homeostasis · Aging · Hippocampus · N-methyl-D-aspartate receptor · Voltage-dependent calcium channels · Intracellular calcium stores · Synaptic plasticity · LTP · LTD

39.1 Introduction

The Calcium (Ca^{2+}) ions are primary signaling molecules regulating a plethora of diverse and important cellular processes including apoptosis, energy production, gene regulation, cell proliferation, membrane excitability, and synaptic plasticity [1–5]. Due to the omnipresent environment of Ca^{2+} signaling, Ca^{2+} is one of the most highly regulated ions. The cytoplasmic concentration of Ca^{2+} is maintained

A. Kumar (✉)
Department of Neuroscience, McKnight Brain Institute, University of Florida, Gainesville, FL, USA
e-mail: kash@ufl.edu

© Springer Nature Switzerland AG 2020

M. S. Islam (ed.), *Calcium Signaling*, Advances in Experimental Medicine and Biology 1131, https://doi.org/10.1007/978-3-030-12457-1_39

considerably lower than the concentration in the extracellular space [6–8]. Therefore, any variation in Ca^{2+} signaling mechanisms, unless compensated by another mechanism, will result in a modification of cell function. Ca^{2+} signaling within a neuron is very complex, and an extremely precise and dynamic regulation of Ca^{2+} concentration is required for normal functioning. Ca^{2+} signaling depends primarily on a prompt and momentary increase in intracellular Ca^{2+} concentration through influx of Ca^{2+} from various sources including Ca^{2+} permeable plasma membrane receptors, ion channels, and internal Ca^{2+} sources. In most cells, multiple mechanisms exist whereby an augmentation in intracellular Ca^{2+} concentrations may ensue.

The major sources of intracellular Ca^{2+} include Ca^{2+} influx through various voltage-dependent Ca^{2+} channels (VDCCs), ligand-gated nonselective glutamate receptors, such as N-methyl-D-aspartate (NMDA) receptor, store-operated Ca^{2} channels (SOCs) as well as the release of Ca^{2+} from intracellular Ca^{2} stores (ICS) [9–13]. The transient receptor potential (TRP) channels, which are non-selective cation channels with high Ca^{2+} permeability, also contribute to influx of Ca^{2+} into the cell cytoplasm [14–19] (Fig. 39.1). The relative contribution of these sources will depend on the cell type: neuron, astrocyte, oligodendrocyte or microglia. In the case of neurons, Ca^{2+} sources will vary depending on their size, transmitter system, and location in excitatory or inhibitory neural circuits.

An elevation in intracellular concentration of free Ca^{2} plays an important role in the induction and maintenance of activity-dependent synaptic plasticity [20–22]. The level of Ca^{2} in response to synaptic activity determines the degree and direction of synaptic efficacy. Long-term potentiation (LTP) and long-term depression (LTD), the two major forms of activity-dependent synaptic plasticity, are the best cellular correlates of learning and memory, and are studied extensively across various brain regions [23–27]. LTP is a long lasting increase in synaptic strength in response to intense synaptic activity [20, 28, 29]. The induction of LTP requires activation of postsynaptic NMDA receptors resulting in a large, yet brief, influx of Ca^{2+} through the NMDA receptor channel. In turn, this large rise in intracellular Ca^{2+} activates Ca^{2+} sensitive kinases such Ca^{2+}/calmodulin-dependent kinase II (CaMKII). Kinase activity increases the strength of the synaptic response through phosphorylation of α-amino-3-hydroxy-5-methyl-4-isoxazolepropionic acid (AMPA) glutamate receptors, which leads to insertion of additional AMPA-glutamate receptors into the post-synaptic membrane [30–35]. In contrast to LTP, LTD is induced by low synaptic activity, which leads to a modest and prolonged rise in intracellular Ca^{2+} concentration [36–40]. The modest and sustained rise in intracellular Ca^{2+} concentration activates Ca^{2+}-sensitive phosphatases that decrease synaptic transmission through dephosphorylation of AMPA-glutamate receptors, resulting in their removal from the post-synaptic membrane [41] (Fig. 39.2). Due to the differential level of Ca^{2+} involved in the generation of the various forms of synaptic plasticity, any treatment that modifies Ca^{2+} influx into the cytoplasm can influence the direction and degree of synaptic plasticity. The dependence on differential intracellular Ca^{2+} levels in determining the form of synaptic plasticity underlies the observation that stimulation patterns for the induction of LTP and LTD tend toward high and low-

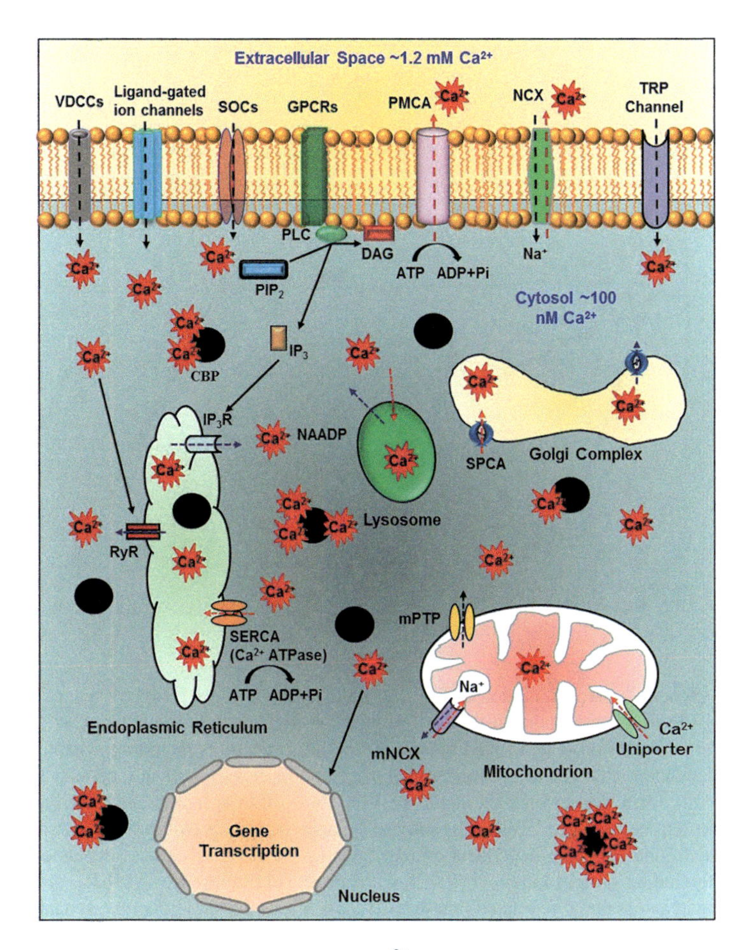

Fig. 39.1 The schematic illustration of various Ca^{2+} sources including ligand-gated ion channels (such as ionotropic glutamate NMDA receptor), voltage-dependent Ca^{2+} channels (*VDCC*), G protein-coupled receptors (*GPCRs*), store-operated Ca^{2} channels (SOCs), and transient receptor potential channels (TRP channels). Various channels permeable to Ca^{2+} exist and are expressed both on cell plasma membrane and on membrane of organelles. Alterations in cytosolic Ca^{2+} are dynamically regulated through interplay of various Ca^{2+}channels, exchange pumps, and transporters. Ca^{2+} (*red spark*) influxes into the cytosol (*black dashed arrows*) through various sources in a healthy cell. NMDA receptor and VDCCs are highly selective plasma proteins that mediate Ca^{2+} signals in response to membrane depolarization. The release of Ca^{2+} into the cytoplasm also occurs from the intracellular Ca^{2+} stores through ryanodine receptors (*RyR*), and inositol (1,4,5)-trisphosphate receptor (*IP3R*) involving phospholipase C (PLC), diacylglycerol (DAG), and inositol (1,4,5) trisphosphate (IP_3). Organelles, including endoplasmic reticulum, mitochondria, and lysosomes act as a Ca^{2+} buffering system, releasing, and sequestering Ca^{2+} in and out of the cytosol. Further, the model depicts Ca^{2+} buffering and extrusion pathways (*red dashed arrows*), involving plasma membrane Ca^{2+} ATPase (PMCA), sodium-calcium exchangers (NCX) located on plasma membrane and mitochondria (mNCX), sarcoplasmic reticulum Ca^{2+} ATPases (*SERCA*), various Ca^{2+} binding proteins (*CBP*), nicotinic acid adenine dinucleotide phosphate (NAADP), mitochondrial Ca^{2+} uniporter, mitochondrial permeability transition pore (*mPTP*), and secretory pathway Ca^{2+}-ATPases (*SPCA*)

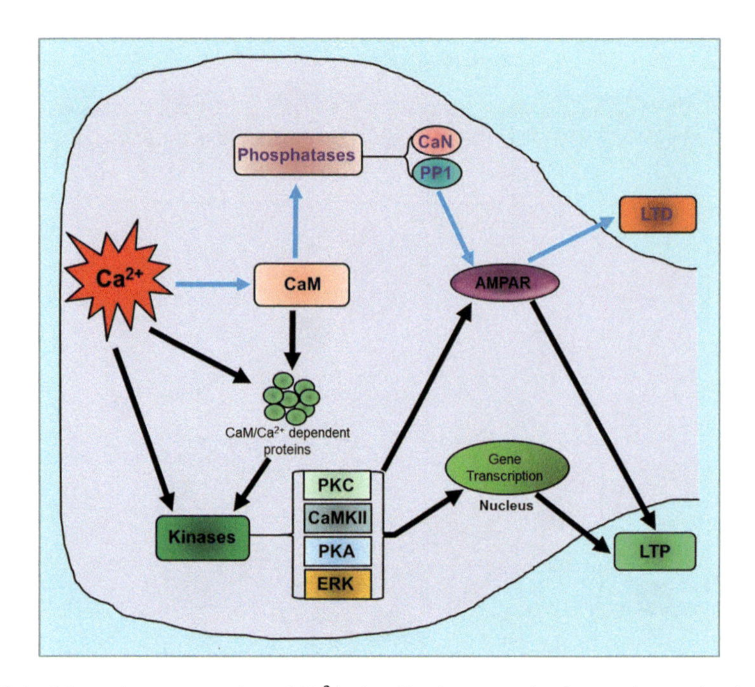

Fig. 39.2 Schematic representation of Ca^{2+} signaling in two major forms of synaptic plasticity, LTD and LTP. The homeostasis of intracellular Ca^{2+} is dynamically regulated. Ca^{2+} is required by a large number of proteins to mediate various cellular functions including synaptic plasticity. Activity-dependent changes in synaptic function results in elevation of intracellular Ca^{2+} levels through influx from various sources including VDCCs, NMDA receptor, and ICS. The large, but brief rise in intracellular Ca^{2+} following intense synaptic activity activates Ca^{2+} sensitive kinases and Ca^{2+} calmodulin (CaM)-dependent proteins. Activation of various kinase-dependent signaling cascades, including protein kinase C (PKC), CaM-dependent kinase II (CaMKII), protein kinase A (PKA), and extracellular regulated kinase (ERK), increases the strength of the synaptic response (LTP) through phosphorylation of AMPA type glutamate receptors (AMPAR), which leads to insertion of additional AMPA-glutamate receptors into the post-synaptic membrane (black arrows). LTD is induced by low synaptic activity, which leads to a modest and prolonged rise in intracellular Ca^{2+} concentration. The modest and sustained rise in intracellular Ca^{2+} activates Ca^{2+}-sensitive phosphatases including calcineurin (CaN) and protein phosphatase 1 (PP1) via CaM and induces a decrease in synaptic transmission (LTD) through dephosphorylation of AMPAR, resulting in their removal from the post-synaptic membrane (blue arrows)

frequency patterns, respectively. Theoretical models suggest that synaptic plasticity is a function of synaptic activity, such that low frequency stimulation induces LTD. As neural activity increases, there is a transition from net LTD to induction of LTP [40, 42]. The thresholds for induction of LTD and LTP, as defined by afferent activity, are thought to reflect activity-dependent changes in the level of intracellular Ca^{2+}, which in turn, activate Ca^{2+}-dependent enzymes. The current chapter centers on age-associated changes on various Ca^{2+} sources and how these alterations in Ca^{2+} signaling in the CA1 region of the hippocampus contributes to altered synaptic plasticity.

39.2 Ca^{2+} Hypothesis of Brain Aging and Influence of Aging on Various Ca^{2+} Sources

The Ca^{2+} 'dysregulation' hypothesis of brain aging and Alzheimer's disease formulated in the 1980s was based on discrete observations of alterations in processes that are regulated by Ca^{2+} [43–46]. Almost four decades of research, has accumulated substantial evidence for alterations in neuronal Ca^{2+} homeostasis in contributing to changes in various processes including cellular senescence and neurodegenerative diseases [47–60]. However, no single mechanism for Ca^{2+} dysregulation has been established. Rather, the causes and consequences of Ca^{2+} dysregulation vary across the central nervous system. As we increase our sophistication for identifying molecular and cellular processes, we are likely to find complex patterns of impaired/spared cellular function related to multiple Ca^{2+} regulating mechanisms. Normal aging is not associated with a loss of neurons [61, 62], so altered Ca^{2+} dependent physiology hypothesis including senescent synaptic plasticity is proposed. Age-associated changes in Ca^{2+} signaling mechanisms, including Ca^{2+} sources, contribute to alteration in synaptic function and presumably account for impaired cognitive function [21, 46, 55, 56, 60, 63–68].

39.3 Voltage-Dependent Ca^{2+} Channels

Voltage-dependent Ca^{2+} channels (VDCCs) are highly selective plasma membrane proteins, which open in response to membrane depolarization and allow Ca^{2+} influx into the cell from the extracellular space. These proteins, are heteromultimers composed of an α_1 subunit and three auxiliary subunits, $\alpha_2\delta$, β_{1-4}, and γ [69–72], provide one of the most effective sources of Ca^{2+} influx into the neuron [73]. The pore forming α_1 subunit (190 kDa) of VDCCs is the primary subunit essential for channel functioning. Each α_1 subunit has four homologous domains (I-IV), which are composed of six transmembrane helices. The fourth transmembrane helix of each domain contains the voltage-sensing motif. Two classes of VDCCs have been described; high-voltage-gated and low-voltage-gated channels, which are activated by strong and weak depolarization, respectively. Based on differential biophysical properties and sensitivity to pharmacological agents, high-voltage-gated channels are further classified into the L ($Ca_v1.1$-3), P/Q ($Ca_v2.1$), and N ($Ca_v2.2$) type channels. The low-voltage-gated channels include the T ($Ca_v3.1$) type channels. In addition, an intermediate-voltage-gated channel, R ($Ca_v2.3$) type is expressed throughout the central nervous system [74–80].

In aged animals, the whole-cell L-type Ca^{2+} currents in CA1 pyramidal neurons are enhanced [81, 82], and there is an increase in the density of functional L-type VDCCs [65]. The idea that L-channels are augmented in the hippocampus during aging is also supported by mRNA and protein expression studies indicating an increase in $Ca_v1.3$ [83–86]. Furthermore, posttranslational changes including the phosphorylation state of the $Ca_V1.2$ channel could contribute to age-associated increase in activity [87–89]. However, L-channel associated intracellular Ca^{2+} transients may demonstrate region specific variations within the hippocampus itself. For example, result indicates that CA3 interneurons in aged hippocampus exhibit no alterations in intracellular Ca^{2+} transients at resting state; however, larger Ca^{2+} transients are evident in the presence of external excitatory drive produced by kainate application [90].

It is not clearly elucidated why L-channels increase in the hippocampus during aging. The enhancement in L-channel function appears to be specific to hippocampal pyramidal cells. The expression of L-channels in the hippocampus is regulated by the sex steroid estrogen, such that an increased expression is associated with the decline of the hormone during aging [91]. Finally, it is possible that the increased L-channel function, enhanced the amplitude of afterhyperpolarization (AHP), and reduction in cell excitability represent compensatory mechanisms associated with Ca^{2+} dysregulation during senescence, which attempts to restrict depolarization and further influx of Ca^{2+} through NMDA receptors [21, 22, 63, 92]. Several cellular biomarkers of senescent physiology in the hippocampus are dependent on VDCC function, and L-type channel blockers can reverse age-related changes in the magnitude of the AHP and spike frequency adaptation [39, 44, 46, 65, 93, 94].

39.4 Ligand-Gated Ion Channels: N-Methyl-D-Aspartate (NMDA) Receptor

Various less-selective ligand-gated ion channels, including NMDA receptors, allow influx of Ca^{2+} into intracellular space. NMDA receptors are ionotropic non-selective cationic glutamate receptors, which play a critical role in the rapid regulation of synaptic plasticity. NMDA receptors are hetero-tetrameric protein complexes composed of two classes of subunits, the ubiquitously expressed and essential subunit (GluN1) and a modulatory subunit (GluN2A- GluN2D) [95–100]. The majority of NMDA receptors are assemblies of two GluN1 subunits, the ubiquitously expressed and obligatory subunit, and two GluN2A-D subunits, a modulatory subunit. In addition, GluN3 subunits (GluN3A and GluN3B), without involving GluN2 subunits, can assemble with GluN1 subunits to form functional receptors [101–105]. The activation of NMDA receptors requires binding of a ligand (glutamate), membrane depolarization (to remove the Mg^{2+} block of the channel), and binding of a co-agonist, glycine. Since NMDA receptor is a non-selective cation channel, its activation and opening leads to simultaneous influx of Na^{+} and Ca^{2+} ions [106]. However, between the two predominant ionotropic glutamate receptors (AMPA and NMDA) subtypes, the NMDA receptors are the most permeable to Ca^{2+} ions [107].

There is considerable evidence to indicate that aging is associated with a decline in NMDA receptor function within brain regions involved in higher brain function including learning and memory [108–115]. Perhaps the strongest evidence for a reduction in NMDA receptor function comes from electrophysiological studies, which indicate that the NMDA receptor mediated excitatory post synaptic potentials in the Schaeffer collateral pathway of the hippocampus are reduced by approximately 50% in aged animals [64, 110, 111, 116–120]. Several studies indicate a decrease in the level of NMDA receptor protein expression in the hippocampus during aging [114, 115, 121–129]; further, the decrease has primarily been localized to region CA1 [112, 130–132]. These studies report reduced binding of [^{3}H] glutamate (agonist site), [^{3}H] glycine (GluN1 site), [^{3}H] CPP (a competitive antagonist to the L-glutamate binding site), and [^{3}H] MK-801 (an open channel blocker) in the hippocampus of aged rats. However, others have reported no age-related change in antagonist binding [122, 125, 133, 134] or an increased MK-801 binding in animals with learning and retention deficits [135, 136]. It is interesting to note that MK-801 binds to the hydrophobic channel domain of NMDA receptor, exclusively labeling open channels. Thus, an apparent increase in NMDA receptor channel open time may act as a compensatory mechanism for the decrease in receptor number [137]. However, the majority of reports indicate that the net function of the NMDA receptors decreases at CA3-CA1 hippocampal synaptic contacts during senescence [116–120].

One of the potential mechanisms for the observed decrease in the NMDA receptor function is related to altered expression of specific NMDA receptor subunits [138]. Significant decreases have been observed in the expression of GluN1 protein [111, 139, 140] and GluN1 mRNA [141] levels in the aged hippocampus. In contrast, other studies report no age-related decrease in GluN1 protein expression in the whole hippocampus [115, 142]. Despite the lack of congruent changes in the expression levels in the hippocampus, other brain regions exhibit a decline in GluN1 mRNA expression during aging. Indeed, senescence-related decrease in the GluN1 mRNA expression has been observed in the medial basal hypothalamus-median eminence [113], in the medial and lateral prefrontal cortices [143], and in the insular, orbital, and somatosensory cortices [129]. Results indicate age-related changes in the modulatory GluN2 subunits. A decrease in the GluN2A protein expression has been observed in the hippocampus [114, 142]. Furthermore, GluN2A mRNA expression was reported to decline in the ventral hippocampus [141]. In contrast, results from other studies report no significant change in the GluN2A protein expression levels in the hippocampus [142, 144]. Age-related changes have also been reported for GluN2B subunit of the NMDA receptor; in particular the expression of GluN2B protein [115, 140] and GluN2B mRNA [141, 145] declines in the hippocampus.

From a physiological standpoint, the changes in the expression of specific GluN2 subunits could have dramatic influences on NMDA receptor function through the regulation of mean channel open time and conductance of the NMDA receptors. Studies on recombinant NMDA receptor expressed in *Xenopus* oocytes demonstrate that NMDA receptors containing the GluN2A subunit (GluN2A-NMDARs) have faster deactivation kinetics relative to GluN2B containing NMDARs (GluN2B-NMDARs) [95], such that smaller ion flux is observed for the GluN2A-NMDARs, relative to the GluN2B-NMDA receptors. Thus, a shift in the level of GluN2 subunit expression could modify the time course and magnitude of the Ca^{2+} signal leading to reduced Ca^{2+} influx associated with loss of GluN2B. A shift in GluN2A and GluN2B expression is thought to contribute to developmental changes in cognition and synaptic function [146].

Alternatively, it is possible that alterations in the NMDA receptor localization through the insertion of receptors into the membrane or recruitment of extra-synaptic receptors into the synapse will have important effects on NMDA receptor function during aging. It has been suggested that GluN2B containing receptors may be more prevalent at extra-synaptic sites [147], which could temporarily house the NMDA receptors, before being internalized into the cytoplasm [148, 149]. Results indicate that extra-synaptic NMDA receptors couple to different signaling cascades, initiate mechanisms that oppose synaptic potentiation, by shutting off the activity of cAMP response element binding protein and decreasing expression of brain-derived neurotropic factor [150, 151]. However, it remains to be clearly investigated whether altered localization of the NMDA receptors (specifically extra-synaptic localization) is the mechanism by which the NMDA receptor function declines during senescence.

Another likely candidate mechanism for regulating NMDA receptor function during aging is posttranslational modification of the receptor. In particular, the function of the NMDA receptor is influenced by its phosphorylation state. Activation of the tyrosine kinase [152, 153], protein kinase C [154, 155] and protein kinase A [156] increases NMDA receptor mediated currents. In contrast, protein phosphatases, including calcineurin (CaN) and protein phosphatase 1, decrease NMDA receptor currents [153, 156, 157]. Phosphorylation state of GluN2A and GluN2B subunits can rapidly regulate surface expression and localization of these receptors [158–161]. For example, phosphorylation of serine residues within the alternatively spliced cassettes of the C-terminal tail of GluN1 promotes receptor trafficking from the endoplasmic reticulum and insertion into the postsynaptic membrane [162, 163]. Finally, increased phosphatase activity has been linked to the internalization of NMDA receptors [164]. Thus, the kinases and phosphatases act as molecular switches which increase or decrease NMDA receptor function, respectively. Interestingly, aging is associated with a shift in the balance of kinase/phosphatase activity, favoring an increase in the phosphatase activity [22, 88, 165]. Thus, alterations in the phosphorylation state of the NMDA receptor could underlie the decrease in the NMDA receptor function during aging [166].

NMDA receptor function can be altered by the oxidation and reduction of sulfhydryl moieties on their structure. Previous research demonstrates that oxidizing agents like 5,5′-dithiobis(2-nitrobenzoic acid) [167], hydroxyl radicals generated by xanthine/xanthine oxidase [168] and oxidized glutathione [169] decrease NMDA receptor function in the neuronal cell cultures. The decrease in NMDA receptor function under oxidizing conditions is thought to result from the formation of disulfide bonds on the sulfhydryl group containing amino acid residues in NMDA receptors [170–172]. The aging brain is associated with an increase in the levels of oxidative stress and/or a decrease in redox buffering capacity [173–175], conditions that should promote a decrease in NMDA receptor function. Little or no effect of oxidizing agents was observed for older animals, suggesting that cells were already in an oxidized state. In contrast, reducing conditions enhanced NMDA receptor mediated synaptic responses in hippocampus of aged animals [64, 118–120, 176–179]. Results provide evidence for a link between the redox-mediated decrease in NMDA receptor function and the emergence of an age-related cognitive phenotype with impairment in the rapid acquisition and retention of novel spatial information [118, 119]. Further, results demonstrate that the age-related decrease in NMDA receptor-mediated synaptic responses at CA3-CA1 hippocampal synapses is related to redox state such that the reducing agent significantly enhanced the NMDA receptor component of the synaptic response to a greater extent in cognitively impaired animals relative to unimpaired animals [118, 180].

NMDA receptor function in neurons is regulated by local supporting cells, astrocytes and microglia, thus acting as an additional possible mechanism for the age-related changes to NMDA receptor function. Astrocytes are a major source of D-serine, an endogenous co-agonist for the NMDA receptor, which binds to the glycine site [181]. An age-related loss of D-serine is observed in the hippocampus of rats [182]. Furthermore, the age-related decline in the NMDA receptor function is rescued by D-cycloserine [128]. Microglia contribute to the brain's immune system and activated microglia can also release D-serine [183, 184]. In accordance with this idea, recent reports suggest that microglia can potentiate the NMDA receptor-mediated synaptic responses in cortical neurons [185, 186]. Finally, there is evidence for a feedback reduction in NMDA receptors due to excess synaptic glutamate activity during microglial activation [187, 188]. In light of the interactions of NMDA receptors and glial cells, it is important to consider the possibility that the decrease in NMDA receptor function might represent a compensatory neuro-protective mechanism associated with inappropriate receptor activity or increased Ca^{2+} due to other mechanisms. Thus, impaired NMDA receptor-dependent synaptic plasticity and memory decline may be epiphenomena due to processes for cell preservation [21, 22, 63, 92]. Indeed, upregulation of GluN2B subunits improves synaptic plasticity and memory in aged mice [117, 189] indicating that increased NMDA receptor function can ameliorate physiological aging.

39.5 Ca^{2+} Release from Intracellular Ca^{2+} Stores

Intracellular Ca^{2+} stores (ICS), in addition to Ca^{2+} influx from outside the cell, dynamically participate and play a significant role in regulating larger Ca^{2+} signals [190, 191]. Organelles, including the endoplasmic reticulum (ER), mitochondria, nuclear envelope, neurotransmitter vesicles, and lysosomes play a dual role by acting as a Ca^{2+} buffering-sequestering system for intracellular Ca^{2+} and releasing Ca^{2+} into the cytoplasm [58, 192–202]. Thus, there are at least two possible mechanisms by which ICS regulate Ca^{2+} homeostasis: (1) release of stored Ca^{2+} to enhance Ca^{2+} signals and (2) removing cytosolic Ca^{2+} following a large influx.

Two independent pathways, Ca^{2+}-induced Ca^{2+} release (CICR) and inositol (1,4,5)-trisphosphate (IP_3)-induced Ca^{2+} release (IICR) control Ca^{2+} release from intracellular stores. The CICR is a Ca^{2+} augmentation process initiated by Ca^{2+} influx through the plasma membrane that activates ryanodine receptors (RyRs), while IICR is IP_3-acivated release of Ca^{2+} from intracellular sources [193]. The release of Ca^{2+} from the ER is initiated by the activation of G protein-coupled receptors (GPCRs). GPCRs activate phospholipase C (PLC) to form diacylglycerol (DAG) and IP_3 that act on IP_3 receptors (IP_3Rs) to release Ca^{2+} from ICS. Several studies indicate age-related changes in GPCRs or PLC [203–205]. Despite a general decrease in the receptor, the literature suggests that a decrease in IP_3 induced Ca^{2+} release is either limited to cortical cells [206] or no age-related change is observed [206–210]. The disconnect between a reduction in IP_3R expression and the apparent absence of an effect of age on IP_3-induced Ca^{2+} release may be due to increased oxidation of the IP_3Rs which has been demonstrated to increase IP_3R function in brain cells [211, 212]. As such, reduced expression may act as compensation for an altered redox state, in order to maintain proper IP_3 signaling.

CICR is a Ca^{2+} amplification process that is initiated by influx of Ca^{2+} through membrane channels or from ICS through the activation of IP_3Rs. The intracellular Ca^{2+} binds RyRs to release additional Ca^{2+} into the cytosol from the ER. Accumulating evidence supports a role of altered CICR in contributing to altered physiology of normal aging. Rather, an age-related increase in oxidative stress and a shift in the intracellular redox state may enhance the responsiveness of RyRs to intracellular Ca^{2+} [68, 213–215]. Ca^{2+} release from ICS contributes towards enhancing NMDA-receptor mediated Ca^{2+} influx for induction of LTP [216–219]. Recent results reveal that activation of Ca^{2+} release from RyRs receptor facilitated LTD while attenuation of RyR-mediated Ca^{2+} release significantly prevented the induction of LTD in young animals [220]. Aging is associated with an increase RyRs protein content in hippocampal tissue and enhanced oxidation of cysteine residues located on RyRs [221]. Increased CICR appears to contribute to altered physiology in hippocampal neurons [38, 177, 222–224]. As noted above, hippocampal cells exhibit increase Ca^{2+} from L-type Ca^{2+} channels, which could provide a source of Ca^{2+} to fill ICS and activate CICR from ICS. Thus, the contribution of CICR to aging physiology in hippocampal cells may be due to a summation of various mechanisms.

39.6 Influence of Aging-Associated Altered Ca^{2+} Regulation on Synaptic Plasticity: LTP and LTD

Complex and synergistic molecular and cellular processes probably contribute to memory formation and storage in the brain. Several evidence suggest that multiple mechanisms including synaptic changes and intrinsic neuronal excitability are involved in information processing and storage [23–27, 225, 226]. Aging is a complex process; in general, aging is associated with a shift in synaptic plasticity favoring decreased synaptic transmission (i.e. LTD) and a reduced ability to enhance synaptic transmission through LTP. Indeed, the impairment in LTP may begin in middle age [227]. It has been suggested that the shift in synaptic plasticity, favoring LTD over LTP, contributes to the decrease in synaptic transmission observed in aged animals [21, 22, 63, 92]. In considering the mechanism for synaptic plasticity involvement in regulating synaptic transmission during aging, it is important to note that the shift in synaptic modifiability is not due to a change in the expression mechanisms. For example, there is no age-related difference in the maximal LTP magnitude observed under conditions in which strong burst of synaptic stimulation is delivered [228–231]. In addition, maximal LTP can be observed in aged animals when single pulses are combined with strong postsynaptic depolarization [232], or weak stimulation is combined with increased Ca^{2+} in the recording medium [233] suggesting that the NMDA receptor/Ca^{2+}-dependent signaling pathways for the induction of LTP are maintained.

Similarly, the increase in LTD in aged animals is not due to an enhancement in the expression mechanism or an increase in the maximum amplitude of LTD. Robust LTD can be observed in adults when Ca^{2+} levels are elevated in the recording media [228]. In addition, a substantial level of LTD can be observed by using more effective induction protocols involving paired-pulse low-frequency stimulation [234, 235]. In fact, administration of multiple episodes of the paired-pulse low-frequency stimulation produces similar levels of LTD in aged and adult animals [231]. Thus, it can be concluded that the age-related difference in synaptic modifiability is not due to a change in the expression mechanisms.

One possibility for a shift in the balance of LTP and LTD is an adjustment in the induction mechanisms, which are engaged to initiate synaptic modifications. The original reports of increased susceptibility to LTD in aged animals hypothesized that altered Ca^{2+} regulation resulted in a shift in the threshold for induction of synaptic plasticity over the life span [21, 60, 228]. Indeed, the ability to generate LTD using the weaker stimulation protocol decreases as animals mature from juvenile to adult [235, 236], and the propensity to induce LTD is once again augmented with advanced age [38, 228, 237, 238]. The dependence on Ca^{2+} regulation for LTD induction is readily demonstrated by altering the Ca^{2+}/Mg^{2+} ratio in the recording medium. While LTD is observed for aged animals and little or no LTD is observed for adult animals under Ca^{2+}/Mg^{2+} conditions that mimic cerebral spinal fluid, LTD can be readily observed in adults if the level of Ca^{2+} is increased relative to Mg^{2+}. Moreover, the elevation of Ca^{2+}/Mg^{2+} ratio may prevent LTD in the oldest animals due to a shift in intracellular Ca^{2+} levels beyond the range needed for LTD-induction or possibly due to increased susceptibility to toxic effects of higher Ca^{2+} in the oldest animals.

Interestingly, an age-related reduction in susceptibility to LTP is observed when stimulation parameters are set near the threshold for LTP-induction [223, 237, 239–241]. The induction mechanism for LTP involves the activation of NMDA receptors through the coincident binding of glutamate to the receptor and sufficient postsynaptic depolarization to remove the Mg^{2+} block from the NMDA receptor channel. In turn, weaken synaptic depolarization or disruption of NMDA receptors could contribute to impairments in induction of LTP. Evidence has been provided that NMDA receptor function may be compromised due to altered Ca^{2+} regulation leading to increased activity of the Ca^{2+}-dependent phosphatase, CaN. The activity of CaN depends on a modest rise in intracellular Ca^{2+} and aged memory impaired animals exhibit an increase in CaN activity [165]. In turn, CaN can act on NMDA receptors to reduce Ca^{2+} influx [157, 242].

The idea that induction of LTP is subdued as a result of a reduction in NMDA receptor activation is supported by research showing that induction deficits can be overcome by strong postsynaptic depolarization [232]. Indeed, there are several reasons to believe that an inability to achieve sufficient postsynaptic depolarization, a prerequisite for NMDA receptor activation, may be more problematic for LTP induction during aging. The reduced synaptic strength of aged animals may result in a reduced afferent cooperativity in depolarizing the postsynaptic neuron and an inability to reach the level of depolarization needed for NMDA receptor activation. The AHP, a Ca^{2+}-dependent, K^{+} mediated process, contributes to hyperpolarization and controls the cell excitability. Importantly, generation of the AHP depends on Ca^{2+} and all three AHP components, fast (fAHP), a medium (mAHP), and a slow (sAHP), rely on Ca^{2+} [243]. The amplitude and duration of AHP is significantly enhanced in aged animals [44, 56, 66, 93, 94, 177, 222, 226, 244–246]. The inability to depolarize the cell is compounded, during patterned stimulation, due to the larger AHP. In fact, results suggest that it is the large AHP which underlies much of the LTP impairment [21, 22, 63, 92] and may even mask the propensity for enhanced LTP-induction during aging [223]. Normally, there is a relationship between the frequency of afferent stimulation required for LTP induction and the level of depolarization [247]. However, as noted above, a large and long-lasting AHP is observed following an action potential in older animals [44, 56, 66, 93, 94, 177, 222, 244–246]. The large AHP disrupts the integration of depolarizing postsynaptic potentials and the duration of this disruption is a function of the extent and duration of the AHP [21, 60, 92]. The disruption would increase the level of stimulation needed for LTP (Fig. 39.3). Result shows that pharmacological manipulations that reduce the AHP shift the frequency response functions such that LTP can be observed for much lower stimulation frequencies that would normally not elicit LTP in young or aged animals. The reduction in the AHP amplitude permits increased activation of NMDA receptor, to shift the threshold for induction of synaptic plasticity [39, 229]. It should be noted that L-channel blockade does not completely ameliorate age-related differences. The AHP amplitude is reduced but not to the levels observed in young animals [248]. In aged rats, blockade of VDCCs [39] or inhibition of Ca^{2+} stores [223] reduces the AHP and facilitates induction of LTP. The process can be reversed such that an increase in the AHP following the addition of an L-channel agonist prevents LTP-induction by 5 Hz stimulation [223]. Thus, an interesting aspect of this work is the fact that while induction of LTP depends on a large rise in intracellular Ca^{2+}, LTP-induction is facilitated by

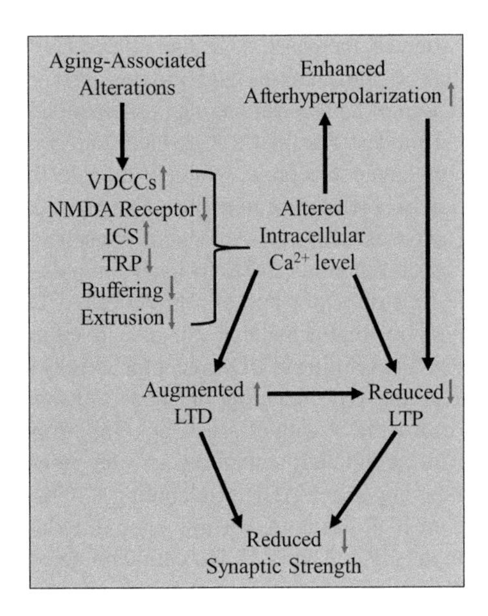

Fig. 39.3 Diagrammatic illustration of the relationship between altered Ca^{2+} regulation and synaptic plasticity during aging. An age-related alterations in Ca^{2+} regulation (increased release from voltage-dependent Ca^{2+} channels-VDCCs, intracellular stores-ICS, and transient receptor potential channels-TRP channels, and decreased NMDA receptor function) augments intracellular Ca^{2+} concentration during neural activity, which facilitates LTD-induction in response to low synaptic activity and increases the degree and duration of the afterhyperpolarization. In turn, a reduction in the level of depolarization due to a decline in cooperativity of decreased strength of synaptic contacts and an enhanced afterhyperpolarization acts to impair NMDA receptor activation and subsequent LTP-induction. The shift in the balance of LTP/LTD, preferring LTD, acts to decline synaptic strength with advanced age

blocking several Ca^{2+} sources that contribute to the AHP. The augmentation in the AHP amplitude diminishes the activation of NMDA channels, further contributing to impaired synaptic plasticity [60, 63]. The enhanced AHP with age is linked to increased involvement of VDCCs and internal Ca^{2+} stores [44, 46, 66, 93, 94, 177, 222, 223, 237, 248–254]. Thus, changes in Ca^{2+} from VDCCs or ICS can act through the AHP to impair NMDA receptor function.

A complimentary method for investigating the relationship between the AHP and LTP threshold is to reduce the AHP through manipulation of the potassium channels. For example, blockade of SK channels by apamin increases cell excitability and facilitates LTP-induction [39]. Moreover, deletion of the Kvβ1.1 subunit results in enhanced cell repolarization during repetitive firing by preventing A-type potassium channel inactivation. In turn, the normal spike broadening and increased Ca^{2+} influx through VDCCs is impaired by rapid repolarization. In aged, Kvβ1.1 knockout mice, the AHP is reduced, LTP is facilitated, and spatial memory is enhanced [255]. The results emphasize that the source of Ca^{2+} provides an overriding control of synaptic modifiability, shifting the threshold frequency of LTP-induction.

The shift in Ca^{2+} sources, increased AHP, and altered Ca^{2+} signaling, involving a shift in the activity of phosphatases and kinases also give rise to increased susceptibility to induction of LTD during aging. As described earlier, the induction of LTD depends on a modest rise in Ca^{2+}, which activates a signaling cascade to dephosphorylate glutamate receptors. Studies indicate that induction of LTD depends on suppression of NMDA receptors through a process in which Ca^{2+} influx through L-channels activates the Ca^{2+}-dependent phosphatase, CaN, resulting in reduced NMDA receptor function [247]. In aged animals, blockade of NMDA receptors can reduce, but does not necessarily prevent LTD [39]. In fact, LTD induction is facilitated by treatments that enhance potassium channel currents, hyperpolarizing the cell and limiting NMDA receptor activity [256]. Treatments that reduce the amplitude of the AHP including blockade of L-channels [39], depletion of Ca^{2+} stores [38], or treatment with estrogen [93, 238] impairs LTD induction in aged animals. The increase in CICR activates Ca^{2+}-dependent potassium channels in the membrane, inducing larger AHP. Attenuating CICR, by blocking RyR or depletion of Ca^{2+} from ICS, has a greater influence in reducing the amplitude of the AHP in aged animals [177, 222, 223], indicating an aging-specific mechanism. Thus, like LTP, the source of Ca^{2+} is important in determining synapse modifiability and LTD depends on Ca^{2+} from VDCCs and intracellular stores rather than NMDA receptors.

These results suggest a shift in Ca^{2+} homeostasis such that aging cells exhibit reduced Ca^{2+} influx from NMDA receptors and an increase contribution from VDCCs and ICS to intracellular Ca^{2+} during neural activity [21, 22, 63, 92]. In addition, it is likely that aged neurons exhibit changes in intracellular buffering and processes for extrusion of Ca^{2+} [257]. Collectively, this shift results in changes in Ca^{2+}-related physiology including an increase in the AHP amplitude and impaired LTP induction, at least under physiological Ca^{2+}/Mg^{2+} conditions. The change in Ca^{2} regulation, which shifts the cell away from Ca^{2+} influx through NMDA receptors, may be neuroprotective against Ca^{2+} mediated damage and thus act as compensation for increased vulnerability to neurotoxicity [258]. Physiological studies consistently indicate that NMDA receptor mediated excitatory postsynaptic potentials in the Schaffer collateral pathway of the hippocampus are significantly reduced in aged animals [110, 111, 117–120, 128, 176]. In turn, a decrease in NMDA receptor function is likely to influence induction of synaptic plasticity [259, 260] (i.e. metaplasticity [261]).

Alternatively, the shift in Ca^{2+} homeostasis could result from age-related increase in oxidative stress [262–264]. Reactive oxygen species could induce a rise in intracellular Ca^{2+} through release of Ca^{2+} from Ca^{2+} binding proteins (i.e. decreased buffering), and oxidation of Ca^{2+} regulatory proteins (calmodulin, SERCA, PMCA) would disrupt intracellular stores, and increase entry through Ca^{2+} channels [263, 265]. In the hippocampus, oxidative stress has effects which mimic aging; increasing Ca^{2+} influx through L-channels [266, 267], increasing the function of Ca^{2+}-dependent K^{+} channels [251, 268], and decreasing NMDA receptor function [269]. Furthermore, oxygen radicals can influence the activity of

Ca^{2+} dependent enzymes. The unstable super oxide ($O^{\cdot}$) or high levels of H_2O_2 (beyond the physiological range) inhibit CaN in tissue homogenates [270, 271]. However, in intact tissue, reactive oxygen species increase CaN activity, either through changes in the CaN inhibitory protein [272] or altered Ca^{2+} regulation involving increased Ca^{2+} from intracellular stores and VDCCs, leading to impaired induction of LTP [270]. Recent results provide evidence that enhanced oxidative stress also contributes to a reduction in Ca^{2+} influx [273].

Several recent studies have provided ample evidence that FK506-binding protein 12.6/1b (FKBP1b), a regulator of neuronal intracellular Ca^{2+}, negatively regulates Ca^{2+} release from RyRs, and attenuates influx of Ca^{2+} from L-type VDCCs. In addition, selective down regulation of FKBP1b in the hippocampus of young animals induced RyR-mediated dysregulation of Ca^{2+} homeostasis similar to that observed in aged animals [56, 274, 275]. Finally, viral-mediated upregulation of FKBP1b in the hippocampus reverses aging-associated dysregulation of Ca^{2+} signaling, reduces age-associated augmented AHP amplitude, restores genomic regulation, and ameliorates cognitive performance [55, 56]. Latest results demonstrated that age-associated enhanced oxidation of RyR contributes to decreased LTP induction and cognitive impairment [221].

39.7 Conclusion

The Ca^{2+} ion is a central signaling molecule in a wide range of cellular processes including synaptic plasticity, and the proper functioning of Ca^{2+} signaling machinery is essential for normal neuronal function. Alterations in neuronal Ca^{2+} homeostasis and Ca^{2+}-dependent physiological processes are the most consistent neurobiological manifestations of normal brain aging. Ca^{2+} dysregulation is not ubiquitous and mechanisms of altered regulation are restricted to specific cell populations. For example, an age-related increase in L-type Ca^{2+} channels is relatively specific to hippocampal pyramidal cells. Furthermore, an age-related decrease in NMDA receptor function, specifically in the hippocampus, suggests a possible compensatory mechanism to limit intracellular Ca^{2+} levels. However, such a mechanism may protect the cell at the expense of cell function. Cell specific susceptibility to Ca^{2+} dysregulation depends on environmental and genomic factors in addition to the availability of mechanisms for handling Ca^{2+}. The level of neural activity may render some regions more susceptible to oxidative stress, resulting in multiple changes to increase intracellular Ca^{2+} concentration including increased release of Ca^{2+} from ICS, impaired Ca^{2+} pumps, and weakened Ca^{2+} buffering. In turn, gene mutations may interact with age and cell specific alterations in Ca^{2+} regulation to produce the pattern of neuronal death, which characterizes neurodegenerative diseases. Due to the importance of Ca^{2+} as a crucial signaling molecule, selective regulation of Ca^{2+} in a particular set of neurons may be a daunting task for treating age-related diseases. Clearly, future studies are needed to delineate the contribution of several mechanisms in optimizing neuronal Ca^{2+} regulating machinery and the influence of abnormal Ca^{2+} signaling in age-associated altered synaptic plasticity processes and cognitive function.

Acknowledgements Supported by National Institute of Aging grants R37AG036800, RO1AG049711, RO1AG037984, and RO1AG052258 and the Evelyn F. McKnight Brain Research Foundation.

References

1. Carafoli E (2002) Calcium signaling: a tale for all seasons. Proc Natl Acad Sci U S A 99(3):1115–1122
2. Capiod T (2016) Extracellular calcium has multiple targets to control cell proliferation. Adv Exp Med Biol 898:133–156
3. Clapham DE (2007) Calcium signaling. Cell 131(6):1047–1058
4. Toth AB, Shum AK, Prakriya M (2016) Regulation of neurogenesis by calcium signaling. Cell Calcium 59(2–3):124–134
5. Berridge MJ (2012) Calcium signalling remodelling and disease. Biochem Soc Trans 40(2):297–309
6. Berridge MJ, Lipp P, Bootman MD (2000) The versatility and universality of calcium signalling. Nat Rev Mol Cell Biol 1(1):11–21
7. Orrenius S, Zhivotovsky B, Nicotera P (2003) Regulation of cell death: the calcium-apoptosis link. Nat Rev Mol Cell Biol 4(7):552–565
8. Rizzuto R (2001) Intracellular Ca^{2+} pools in neuronal signalling. Curr Opin Neurobiol 11(3):306–311
9. Berridge MJ (1998) Neuronal calcium signaling. Neuron 21(1):13–26
10. Geiger JR, Melcher T, Koh DS, Sakmann B, Seeburg PH, Jonas P et al (1995) Relative abundance of subunit mRNAs determines gating and Ca^{2+} permeability of AMPA receptors in principal neurons and interneurons in rat CNS. Neuron 15(1):193–204
11. Ghosh A, Ginty DD, Bading H, Greenberg ME (1994) Calcium regulation of gene expression in neuronal cells. J Neurobiol 25(3):294–303
12. Parekh AB, Putney JW Jr (2005) Store-operated calcium channels. Physiol Rev 85(2):757–810
13. Prakriya M, Lewis RS (2015) Store-operated calcium channels. Physiol Rev 95(4):1383–1436
14. Clapham DE (2003) TRP channels as cellular sensors. Nature 426(6966):517–524
15. Moran MM, Xu H, Clapham DE (2004) TRP ion channels in the nervous system. Curr Opin Neurobiol 14(3):362–369
16. Ramsey IS, Delling M, Clapham DE (2006) An introduction to TRP channels. Annu Rev Physiol 68:619–647
17. Bai JZ, Lipski J (2014) Involvement of TRPV4 channels in Abeta(40)-induced hippocampal cell death and astrocytic Ca^{2+} signalling. Neurotoxicology 41:64–72
18. Hartmann J, Henning HA, Konnerth A (2011) mGluR1/TRPC3-mediated synaptic transmission and calcium signaling in mammalian central neurons. Cold Spring Harb Perspect Biol 3(4):pii: a006726
19. Zhang H, Sun S, Wu L, Pchitskaya E, Zakharova O, Fon Tacer K et al (2016) Store-operated Calcium Channel complex in postsynaptic spines: a new therapeutic target for Alzheimer's disease treatment. J Neurosci 36(47):11837–11850
20. Cavazzini M, Bliss T, Emptage N (2005) Ca^{2+} and synaptic plasticity. Cell Calcium 38(3–4):355–367

21. Foster TC (1999) Involvement of hippocampal synaptic plasticity in age-related memory decline. Brain Res Rev 30(3):236–249
22. Foster TC (2007) Calcium homeostasis and modulation of synaptic plasticity in the aged brain. Aging Cell 6(3):319–325
23. Bear MF, Abraham WC (1996) Long-term depression in hippocampus. Annu Rev Neurosci 19:437–462
24. Malenka RC, Bear MF (2004) LTP and LTD: an embarrassment of riches. Neuron 44(1):5–21
25. Collingridge G (1987) Synaptic plasticity. The role of NMDA receptors in learning and memory. Nature 330(6149):604–605
26. Collingridge GL, Bliss TV (1995) Memories of NMDA receptors and LTP. Trends Neurosci 18(2):54–56
27. Collingridge GL, Peineau S, Howland JG, Wang YT (2010) Long-term depression in the CNS. Nat Rev Neurosci 11(7):459–473
28. Bliss TV, Collingridge GL (1993) A synaptic model of memory: long-term potentiation in the hippocampus. Nature 361(6407):31–39
29. Kumar A (2011) Long-term potentiation at CA3-CA1 hippocampal synapses with special emphasis on aging, disease, and stress. Front Aging Neurosci 3:7
30. Muller W, Connor JA (1991) Dendritic spines as individual neuronal compartments for synaptic Ca^{2+} responses. Nature 354(6348):73–76
31. Nicoll RA, Malenka RC (1999) Expression mechanisms underlying NMDA receptor-dependent long-term potentiation. Ann N Y Acad Sci 868:515–525
32. Conti R, Lisman J (2002) A large sustained Ca^{2+} elevation occurs in unstimulated spines during the LTP pairing protocol but does not change synaptic strength. Hippocampus 12(5):667–679
33. Matsuzaki M, Honkura N, Ellis-Davies GC, Kasai H (2004) Structural basis of long-term potentiation in single dendritic spines. Nature 429(6993):761–766
34. Lisman J, Yasuda R, Raghavachari S (2012) Mechanisms of CaMKII action in long-term potentiation. Nat Rev Neurosci 13(3):169–182
35. Chang JY, Parra-Bueno P, Laviv T, Szatmari EM, Lee SR, Yasuda R (2017) CaMKII autophosphorylation is necessary for optimal integration of Ca^{2+} signals during LTP induction, but not maintenance. Neuron 94(4):800–808 e4
36. Dudek SM, Bear MF (1992) Homosynaptic long-term depression in area CA1 of hippocampus and effects of N-methyl-D-aspartate receptor blockade. Proc Natl Acad Sci U S A 89(10):4363–4367
37. Dunwiddie T, Lynch G (1978) Long-term potentiation and depression of synaptic responses in the rat hippocampus: localization and frequency dependency. J Physiol 276:353–367
38. Kumar A, Foster TC (2005) Intracellular calcium stores contribute to increased susceptibility to LTD induction during aging. Brain Res 1031(1):125–128
39. Norris CM, Halpain S, Foster TC (1998) Reversal of age-related alterations in synaptic plasticity by blockade of L-type Ca^{2+} channels. J Neurosci 18(9):3171–3179
40. Artola A, Singer W (1993) Long-term depression of excitatory synaptic transmission and its relationship to long-term potentiation. Trends Neurosci 16(11):480–487
41. Lee HK, Kameyama K, Huganir RL, Bear MF (1998) NMDA induces long-term synaptic depression and dephosphorylation of the GluR1 subunit of AMPA receptors in hippocampus. Neuron 21(5):1151–1162
42. Bienenstock EL, Cooper LN, Munro PW (1982) Theory for the development of neuron selectivity: orientation specificity and binocular interaction in visual cortex. J Neurosci 2(1):32–48
43. Khachaturian ZS (1989) Calcium, membranes, aging, and Alzheimer's disease. Introduction and overview. Ann N Y Acad Sci 568:1–4
44. Landfield PW, Pitler TA (1984) Prolonged Ca^{2+}–dependent afterhyperpolarizations in hippocampal neurons of aged rats. Science 226(4678):1089–1092
45. Gibson GE, Peterson C (1987) Calcium and the aging nervous system. Neurobiol Aging 8(4):329–343

46. Disterhoft JF, Thompson LT, Moyer JR, Mogul DJ (1996) Calcium-dependent afterhyperpolarization and learning in young and aging hippocampus. Life Sci 59(5–6):413–420
47. Alzheimer's Association Calcium Hypothesis W (2017) Calcium hypothesis of Alzheimer's disease and brain aging: a framework for integrating new evidence into a comprehensive theory of pathogenesis. Alzheimers Dement 13(2):178–182 e17
48. Frazier HN, Maimaiti S, Anderson KL, Brewer LD, Gant JC, Porter NM et al (2017) Calcium's role as nuanced modulator of cellular physiology in the brain. Biochem Biophys Res Commun 483(4):981–987
49. Gibson GE, Thakkar A (2017) Interactions of mitochondria/metabolism and calcium regulation in Alzheimer's disease: a calcinist point of view. Neurochem Res 42(6):1636–1648
50. Pchitskaya E, Popugaeva E, Bezprozvanny I (2018) Calcium signaling and molecular mechanisms underlying neurodegenerative diseases. Cell Calcium 70:87–94
51. Sompol P, Norris CM (2018) Ca^{2+}, Astrocyte activation and calcineurin/NFAT signaling in age-related neurodegenerative diseases. Front Aging Neurosci 10:199
52. Vardjan N, Verkhratsky A, Zorec R (2017) Astrocytic pathological calcium homeostasis and impaired vesicle trafficking in neurodegeneration. Int J Mol Sci 18(2):358
53. Verkhratsky A, Rodriguez-Arellano JJ, Parpura V, Zorec R (2017) Astroglial calcium signalling in Alzheimer's disease. Biochem Biophys Res Commun 483(4):1005–1012
54. Zorec R, Parpura V, Verkhratsky A (2018) Astroglial vesicular network: evolutionary trends, physiology and pathophysiology. Acta Physiol (Oxford) 222(2)
55. Gant JC, Blalock EM, Chen KC, Kadish I, Thibault O, Porter NM et al (2018) FK506-binding protein 12.6/1b, a negative regulator of $[Ca^{2+}]$, rescues memory and restores genomic regulation in the Hippocampus of aging rats. J Neurosci 38(4):1030–1041
56. Gant JC, Chen KC, Kadish I, Blalock EM, Thibault O, Porter NM et al (2015) Reversal of aging-related neuronal Ca^{2+} dysregulation and cognitive impairment by delivery of a transgene encoding FK506-binding protein 12.6/1b to the Hippocampus. J Neurosci 35(30):10878–10887
57. Toescu EC, Verkhratsky A (2007) The importance of being subtle: small changes in calcium homeostasis control cognitive decline in normal aging. Aging Cell 6(3):267–273
58. Murchison D, Griffith WH (2007) Calcium buffering systems and calcium signaling in aged rat basal forebrain neurons. Aging Cell 6(3):297–305
59. Thibault O, Gant JC, Landfield PW (2007) Expansion of the calcium hypothesis of brain aging and Alzheimer's disease: minding the store. Aging Cell 6(3):307–317
60. Foster TC, Norris CM (1997) Age-associated changes in Ca^{2+} –dependent processes: relation to hippocampal synaptic plasticity. Hippocampus 7(6):602–612
61. Rapp PR, Gallagher M (1996) Preserved neuron number in the hippocampus of aged rats with spatial learning deficits. Proc Natl Acad Sci U S A 93(18):9926–9930
62. West MJ (1993) Regionally specific loss of neurons in the aging human hippocampus. Neurobiol Aging 14(4):287–293
63. Foster TC (2012) Dissecting the age-related decline on spatial learning and memory tasks in rodent models: N-methyl-D-aspartate receptors and voltage-dependent Ca(2)(+) channels in senescent synaptic plasticity. Prog Neurobiol 96(3):283–303
64. Kumar A, Foster TC (2018) Alteration in NMDA receptor mediated glutamatergic neurotransmission in the Hippocampus during senescence. Neurochem Res 44(1):38–48
65. Thibault O, Landfield PW (1996) Increase in single L-type calcium channels in hippocampal neurons during aging. Science 272(5264):1017–1020
66. Tombaugh GC, Rowe WB, Rose GM (2005) The slow afterhyperpolarization in hippocampal CA1 neurons covaries with spatial learning ability in aged Fisher 344 rats. J Neurosci 25(10):2609–2616
67. Murphy GG, Rahnama NP, Silva AJ (2006) Investigation of age-related cognitive decline using mice as a model system: behavioral correlates. Am J Geriatr Psychiatry 14(12):1004–1011
68. Abu-Omar N, Das J, Szeto V, Feng ZP (2018) Neuronal ryanodine receptors in development and aging. Mol Neurobiol 55(2):1183–1192

69. Catterall WA (2000) Structure and regulation of voltage-gated Ca^{2+} channels. Annu Rev Cell Dev Biol 16:521–555
70. Dolphin AC (2006) A short history of voltage-gated calcium channels. Br J Pharmacol 147(Suppl 1):S56–S62
71. Jones SW (1998) Overview of voltage-dependent calcium channels. J Bioenerg Biomembr 30(4):299–312
72. Kang MG, Chen CC, Felix R, Letts VA, Frankel WN, Mori Y et al (2001) Biochemical and biophysical evidence for gamma 2 subunit association with neuronal voltage-activated Ca^{2+} channels. J Biol Chem 276(35):32917–32924
73. Bertolino M, Llinas RR (1992) The central role of voltage-activated and receptor-operated calcium channels in neuronal cells. Annu Rev Pharmacol Toxicol 32:399–421
74. Veselovskii NS, Fedulova SA (1983) 2 types of calcium channels in the somatic membrane of spinal ganglion neurons in the rat. Dokl Akad Nauk SSSR 268(3):747–750
75. Nowycky MC, Fox AP, Tsien RW (1985) Three types of neuronal calcium channel with different calcium agonist sensitivity. Nature 316(6027):440–443
76. Bean BP (1989) Classes of calcium channels in vertebrate cells. Annu Rev Physiol 51:367–384
77. Carbone E, Lux HD (1984) A low voltage-activated, fully inactivating Ca channel in vertebrate sensory neurones. Nature 310(5977):501–502
78. Fedulova SA, Kostyuk PG, Veselovsky NS (1985) Two types of calcium channels in the somatic membrane of new-born rat dorsal root ganglion neurones. J Physiol 359:431–446
79. Soong TW, Stea A, Hodson CD, Dubel SJ, Vincent SR, Snutch TP (1993) Structure and functional expression of a member of the low voltage-activated calcium channel family. Science 260(5111):1133–1136
80. Nilius B, Hess P, Lansman JB, Tsien RW (1985) A novel type of cardiac calcium channel in ventricular cells. Nature 316(6027):443–446
81. Campbell LW, Hao SY, Thibault O, Blalock EM, Landfield PW (1996) Aging changes in voltage-gated calcium currents in hippocampal CA1 neurons. J Neurosci 16(19):6286–6295
82. Brewer LD, Dowling AL, Curran-Rauhut MA, Landfield PW, Porter NM, Blalock EM (2009) Estradiol reverses a calcium-related biomarker of brain aging in female rats. J Neurosci 29(19):6058–6067
83. Herman JP, Chen KC, Booze R, Landfield PW (1998) Up-regulation of alpha1D Ca^{2+} channel subunit mRNA expression in the hippocampus of aged F344 rats. Neurobiol Aging 19(6):581–587
84. Veng LM, Mesches MH, Browning MD (2003) Age-related working memory impairment is correlated with increases in the L-type calcium channel protein alpha1D (Cav1.3) in area CA1 of the hippocampus and both are ameliorated by chronic nimodipine treatment. Brain Res Mol Brain Res 110(2):193–202
85. Chen KC, Blalock EM, Thibault O, Kaminker P, Landfield PW (2000) Expression of alpha 1D subunit mRNA is correlated with L-type Ca^{2+} channel activity in single neurons of hippocampal "zipper" slices. Proc Natl Acad Sci U S A 97(8):4357–4362
86. Nunez-Santana FL, Oh MM, Antion MD, Lee A, Hell JW, Disterhoft JF (2014) Surface L-type Ca^{2+} channel expression levels are increased in aged hippocampus. Aging Cell 13(1):111–120
87. Davare MA, Hell JW (2003) Increased phosphorylation of the neuronal L-type Ca^{2+} channel Ca(v)1.2 during aging. Proc Natl Acad Sci U S A 100(26):16018–16023
88. Norris CM, Halpain S, Foster TC (1998) Alterations in the balance of protein kinase/phosphatase activities parallel reduced synaptic strength during aging. J Neurophysiol 80(3):1567–1570
89. Norris CM, Blalock EM, Chen KC, Porter NM, Landfield PW (2002) Calcineurin enhances L-type Ca^{2+} channel activity in hippocampal neurons: increased effect with age in culture. Neuroscience 110(2):213–225

90. Lu CB, Hamilton JB, Powell AD, Toescu EC, Vreugdenhil M (2009) Effect of ageing on CA3 interneuron sAHP and gamma oscillations is activity-dependent. Neurobiol Aging 32(5):956–965. [epub ahead of print]
91. Foster TC (2005) Interaction of rapid signal transduction cascades and gene expression in mediating estrogen effects on memory over the life span. Front Neuroendocrinol 26(2):51–64
92. Foster TC, Kumar A (2002) Calcium dysregulation in the aging brain. Neuroscientist 8(4):297–301
93. Kumar A, Foster TC (2002) 17beta-estradiol benzoate decreases the AHP amplitude in CA1 pyramidal neurons. J Neurophysiol 88(2):621–626
94. Moyer JR Jr, Thompson LT, Black JP, Disterhoft JF (1992) Nimodipine increases excitability of rabbit CA1 pyramidal neurons in an age- and concentration-dependent manner. J Neurophysiol 68(6):2100–2109
95. Cull-Candy S, Brickley S, Farrant M (2001) NMDA receptor subunits: diversity, development and disease. Curr Opin Neurobiol 11(3):327–335
96. Kutsuwada T, Kashiwabuchi N, Mori H, Sakimura K, Kushiya E, Araki K et al (1992) Molecular diversity of the NMDA receptor channel. Nature 358(6381):36–41
97. Meguro H, Mori H, Araki K, Kushiya E, Kutsuwada T, Yamazaki M et al (1992) Functional characterization of a heteromeric NMDA receptor channel expressed from cloned cDNAs. Nature 357(6373):70–74
98. Monyer H, Sprengel R, Schoepfer R, Herb A, Higuchi M, Lomeli H et al (1992) Heteromeric NMDA receptors: molecular and functional distinction of subtypes. Science 256(5060):1217–1221
99. Moriyoshi K, Masu M, Ishii T, Shigemoto R, Mizuno N, Nakanishi S (1991) Molecular cloning and characterization of the rat NMDA receptor. Nature 354(6348):31–37
100. Kumar A (2015) NMDA receptor function during senescence: implication on cognitive performance. Front Neurosci 9:473
101. Laube B, Kuhse J, Betz H (1998) Evidence for a tetrameric structure of recombinant NMDA receptors. J Neurosci 18(8):2954–2961
102. Al-Hallaq RA, Jarabek BR, Fu Z, Vicini S, Wolfe BB, Yasuda RP (2002) Association of NR3A with the N-methyl-D-aspartate receptor NR1 and NR2 subunits. Mol Pharmacol 62(5):1119–1127
103. Schuler T, Mesic I, Madry C, Bartholomaus I, Laube B (2008) Formation of NR1/NR2 and NR1/NR3 heterodimers constitutes the initial step in N-methyl-D-aspartate receptor assembly. J Biol Chem 283(1):37–46
104. Sucher NJ, Akbarian S, Chi CL, Leclerc CL, Awobuluyi M, Deitcher DL et al (1995) Developmental and regional expression pattern of a novel NMDA receptor-like subunit (NMDAR-L) in the rodent brain. J Neurosci 15(10):6509–6520
105. Low CM, Wee KS (2010) New insights into the not-so-new NR3 subunits of N-methyl-D-aspartate receptor: localization, structure, and function. Mol Pharmacol 78(1):1–11
106. Chen PE, Geballe MT, Stansfeld PJ, Johnston AR, Yuan H, Jacob AL et al (2005) Structural features of the glutamate binding site in recombinant NR1/NR2A N-methyl-D-aspartate receptors determined by site-directed mutagenesis and molecular modeling. Mol Pharmacol 67(5):1470–1484
107. Garaschuk O, Schneggenburger R, Schirra C, Tempia F, Konnerth A (1996) Fractional Ca^{2+} currents through somatic and dendritic glutamate receptor channels of rat hippocampal CA1 pyramidal neurones. J Physiol 491(Pt 3):757–772
108. Gonzales RA, Brown LM, Jones TW, Trent RD, Westbrook SL, Leslie SW (1991) N-methyl-D-aspartate mediated responses decrease with age in Fischer 344 rat brain. Neurobiol Aging 12(3):219–225
109. Pittaluga A, Fedele E, Risiglione C, Raiteri M (1993) Age-related decrease of the NMDA receptor-mediated noradrenaline release in rat hippocampus and partial restoration by D-cycloserine. Eur J Pharmacol 231(1):129–134

110. Barnes CA, Rao G, Shen J (1997) Age-related decrease in the N-methyl-D-aspartateR-mediated excitatory postsynaptic potential in hippocampal region CA1. Neurobiol Aging 18(4):445–452
111. Eckles-Smith K, Clayton D, Bickford P, Browning MD (2000) Caloric restriction prevents age-related deficits in LTP and in NMDA receptor expression. Brain Res Mol Brain Res 78(1–2):154–162
112. Magnusson KR (1998) The aging of the NMDA receptor complex. Front Biosci 3:e70–e80
113. Gore AC, Oung T, Woller MJ (2002) Age-related changes in hypothalamic gonadotropin-releasing hormone and N-methyl-D-aspartate receptor gene expression, and their regulation by oestrogen, in the female rat. J Neuroendocrinol 14(4):300–309
114. Liu P, Smith PF, Darlington CL (2008) Glutamate receptor subunits expression in memory-associated brain structures: regional variations and effects of aging. Synapse 62(11):834–841
115. Zhao X, Rosenke R, Kronemann D, Brim B, Das SR, Dunah AW et al (2009) The effects of aging on N-methyl-d-aspartate receptor subunits in the synaptic membrane and relationships to long-term spatial memory. Neuroscience 162(4):933–945
116. Bodhinathan K, Kumar A, Foster TC (2007) Oxidative stress decreases NMDA receptor function in the hippocampus of aged animals. Soc Neurosci Abstr:N18/256.8
117. Brim BL, Haskell R, Awedikian R, Ellinwood NM, Jin L, Kumar A et al (2013) Memory in aged mice is rescued by enhanced expression of the GluN2B subunit of the NMDA receptor. Behav Brain Res 238:211–226
118. Kumar A, Foster TC (2013) Linking redox regulation of NMDAR synaptic function to cognitive decline during aging. J Neurosci 33(40):15710–15715
119. Lee WH, Kumar A, Rani A, Foster TC (2014) Role of antioxidant enzymes in redox regulation of N-methyl-D-aspartate receptor function and memory in middle-aged rats. Neurobiol Aging 35(6):1459–1468
120. Kumar A, Rani A, Scheinert RB, Ormerod BK, Foster TC (2018) Nonsteroidal anti-inflammatory drug, indomethacin improves spatial memory and NMDA receptor function in aged animals. Neurobiol Aging 70:184–193
121. Bonhaus DW, Perry WB, McNamara JO (1990) Decreased density, but not number, of N-methyl-D-aspartate, glycine and phencyclidine binding sites in hippocampus of senescent rats. Brain Res 532(1–2):82–86
122. Kito S, Miyoshi R, Nomoto T (1990) Influence of age on NMDA receptor complex in rat brain studied by in vitro autoradiography. J Histochem Cytochem 38(12):1725–1731
123. Magnusson KR (1995) Differential effects of aging on binding sites of the activated NMDA receptor complex in mice. Mech Ageing Dev 84(3):227–243
124. Magnusson KR, Kresge D, Supon J (2006) Differential effects of aging on NMDA receptors in the intermediate versus the dorsal hippocampus. Neurobiol Aging 27(2):324–333
125. Miyoshi R, Kito S, Doudou N, Nomoto T (1991) Influence of age on N-methyl-D-aspartate antagonist binding sites in the rat brain studied by in vitro autoradiography. Synapse 8(3):212–217
126. Tamaru M, Yoneda Y, Ogita K, Shimizu J, Nagata Y (1991) Age-related decreases of the N-methyl-D-aspartate receptor complex in the rat cerebral cortex and hippocampus. Brain Res 542(1):83–90
127. Wenk GL, Walker LC, Price DL, Cork LC (1991) Loss of NMDA, but not GABA-A, binding in the brains of aged rats and monkeys. Neurobiol Aging 12(2):93–98
128. Billard JM, Rouaud E (2007) Deficit of NMDA receptor activation in CA1 hippocampal area of aged rats is rescued by D-cycloserine. Eur J Neurosci 25(8):2260–2268
129. Das SR, Magnusson KR (2008) Relationship between mRNA expression of splice forms of the zeta1 subunit of the N-methyl-D-aspartate receptor and spatial memory in aged mice. Brain Res 1207:142–154
130. Gazzaley AH, Weiland NG, McEwen BS, Morrison JH (1996) Differential regulation of NMDAR1 mRNA and protein by estradiol in the rat hippocampus. J Neurosci 16(21):6830–6838

131. Magnusson KR, Cotman CW (1993) Age-related changes in excitatory amino acid receptors in two mouse strains. Neurobiol Aging 14(3):197–206
132. Wenk GL, Barnes CA (2000) Regional changes in the hippocampal density of AMPA and NMDA receptors across the lifespan of the rat. Brain Res 885(1):1–5
133. Araki T, Kato H, Nagaki S, Shuto K, Fujiwara T, Itoyama Y (1997) Effects of vinconate on age-related alterations in [3H]MK-801, [3H]glycine, sodium-dependent D-[3H]aspartate, [3H]FK-506 and [3H]PN200-110 binding in rats. Mech Ageing Dev 95(1–2):13–29
134. Shimada A, Mukhin A, Ingram DK, London ED (1997) N-methyl-D-aspartate receptor binding in brains of rats at different ages. Neurobiol Aging 18(3):329–333
135. Ingram DK, Garofalo P, Spangler EL, Mantione CR, Odano I, London ED (1992) Reduced density of NMDA receptors and increased sensitivity to dizocilpine-induced learning impairment in aged rats. Brain Res 580(1–2):273–280
136. Topic B, Willuhn I, Palomero-Gallagher N, Zilles K, Huston JP, Hasenohrl RU (2007) Impaired maze performance in aged rats is accompanied by increased density of NMDA, 5-HT1A, and alpha-adrenoceptor binding in hippocampus. Hippocampus 17(1):68–77
137. Serra M, Ghiani CA, Foddi MC, Motzo C, Biggio G (1994) NMDA receptor function is enhanced in the hippocampus of aged rats. Neurochem Res 19(4):483–487
138. Magnusson KR (2000) Declines in mRNA expression of different subunits may account for differential effects of aging on agonist and antagonist binding to the NMDA receptor. J Neurosci 20(5):1666–1674
139. Liu F, Day M, Muniz LC, Bitran D, Arias R, Revilla-Sanchez R et al (2008) Activation of estrogen receptor-beta regulates hippocampal synaptic plasticity and improves memory. Nat Neurosci 11(3):334–343
140. Mesches MH, Gemma C, Veng LM, Allgeier C, Young DA, Browning MD et al (2004) Sulindac improves memory and increases NMDA receptor subunits in aged Fischer 344 rats. Neurobiol Aging 25(3):315–324
141. Adams MM, Morrison JH, Gore AC (2001) N-methyl-D-aspartate receptor mRNA levels change during reproductive senescence in the hippocampus of female rats. Exp Neurol 170(1):171–179
142. Sonntag WE, Bennett SA, Khan AS, Thornton PL, Xu X, Ingram RL et al (2000) Age and insulin-like growth factor-1 modulate N-methyl-D-aspartate receptor subtype expression in rats. Brain Res Bull 51(4):331–338
143. Magnusson KR, Bai L, Zhao X (2005) The effects of aging on different C-terminal splice forms of the zeta1(NR1) subunit of the N-methyl-d-aspartate receptor in mice. Brain Res Mol Brain Res 135(1–2):141–149
144. Martinez Villayandre B, Paniagua MA, Fernandez-Lopez A, Chinchetru MA, Calvo P (2004) Effect of vitamin E treatment on N-methyl-D-aspartate receptor at different ages in the rat brain. Brain Res 1028(2):148–155
145. Magnusson KR (2001) Influence of diet restriction on NMDA receptor subunits and learning during aging. Neurobiol Aging 22(4):613–627
146. Dumas TC (2005) Developmental regulation of cognitive abilities: modified composition of a molecular switch turns on associative learning. Prog Neurobiol 76(3):189–211
147. Massey PV, Johnson BE, Moult PR, Auberson YP, Brown MW, Molnar E et al (2004) Differential roles of NR2A and NR2B-containing NMDA receptors in cortical long-term potentiation and long-term depression. J Neurosci 24(36):7821–7828
148. Blanpied TA, Scott DB, Ehlers MD (2002) Dynamics and regulation of clathrin coats at specialized endocytic zones of dendrites and spines. Neuron 36(3):435–449
149. Lau CG, Zukin RS (2007) NMDA receptor trafficking in synaptic plasticity and neuropsychiatric disorders. Nat Rev Neurosci 8(6):413–426
150. Hardingham GE, Fukunaga Y, Bading H (2002) Extrasynaptic NMDARs oppose synaptic NMDARs by triggering CREB shut-off and cell death pathways. Nat Neurosci 5(5):405–414
151. Vanhoutte P, Bading H (2003) Opposing roles of synaptic and extrasynaptic NMDA receptors in neuronal calcium signalling and BDNF gene regulation. Curr Opin Neurobiol 13(3):366–371

152. Heidinger V, Manzerra P, Wang XQ, Strasser U, Yu SP, Choi DW et al (2002) Metabotropic glutamate receptor 1-induced upregulation of NMDA receptor current: mediation through the Pyk2/Src-family kinase pathway in cortical neurons. J Neurosci 22(13):5452–5461
153. Wang LY, Orser BA, Brautigan DL, MacDonald JF (1994) Regulation of NMDA receptors in cultured hippocampal neurons by protein phosphatases 1 and 2A. Nature 369(6477):230–232
154. Ben-Ari Y, Aniksztejn L, Bregestovski P (1992) Protein kinase C modulation of NMDA currents: an important link for LTP induction. Trends Neurosci 15(9):333–339
155. Chen L, Huang LY (1992) Protein kinase C reduces Mg2+ block of NMDA-receptor channels as a mechanism of modulation. Nature 356(6369):521–523
156. Raman IM, Tong G, Jahr CE (1996) Beta-adrenergic regulation of synaptic NMDA receptors by cAMP-dependent protein kinase. Neuron 16(2):415–421
157. Lieberman DN, Mody I (1994) Regulation of NMDA channel function by endogenous Ca^{2+}-dependent phosphatase. Nature 369(6477):235–239
158. Chung HJ, Huang YH, Lau LF, Huganir RL (2004) Regulation of the NMDA receptor complex and trafficking by activity-dependent phosphorylation of the NR2B subunit PDZ ligand. J Neurosci 24(45):10248–10259
159. Gardoni F, Schrama LH, Kamal A, Gispen WH, Cattabeni F, Di Luca M (2001) Hippocampal synaptic plasticity involves competition between Ca^{2+}/calmodulin-dependent protein kinase II and postsynaptic density 95 for binding to the NR2A subunit of the NMDA receptor. J Neurosci 21(5):1501–1509
160. Hallett PJ, Spoelgen R, Hyman BT, Standaert DG, Dunah AW (2006) Dopamine D1 activation potentiates striatal NMDA receptors by tyrosine phosphorylation-dependent subunit trafficking. J Neurosci 26(17):4690–4700
161. Lin Y, Jover-Mengual T, Wong J, Bennett MV, Zukin RS (2006) PSD-95 and PKC converge in regulating NMDA receptor trafficking and gating. Proc Natl Acad Sci U S A 103(52):19902–19907
162. Carroll RC, Zukin RS (2002) NMDA-receptor trafficking and targeting: implications for synaptic transmission and plasticity. Trends Neurosci 25(11):571–577
163. Scott DB, Blanpied TA, Swanson GT, Zhang C, Ehlers MD (2001) An NMDA receptor ER retention signal regulated by phosphorylation and alternative splicing. J Neurosci 21(9):3063–3072
164. Snyder EM, Nong Y, Almeida CG, Paul S, Moran T, Choi EY et al (2005) Regulation of NMDA receptor trafficking by amyloid-beta. Nat Neurosci 8(8):1051–1058
165. Foster TC, Sharrow KM, Masse JR, Norris CM, Kumar A (2001) Calcineurin links Ca^{2+} dysregulation with brain aging. J Neurosci 21(11):4066–4073
166. Coultrap SJ, Bickford PC, Browning MD (2008) Blueberry-enriched diet ameliorates age-related declines in NMDA receptor-dependent LTP. Age 30(4):263–272
167. Aizenman E, Lipton SA, Loring RH (1989) Selective modulation of NMDA responses by reduction and oxidation. Neuron 2(3):1257–1263
168. Aizenman E (1995) Modulation of N-methyl-D-aspartate receptors by hydroxyl radicals in rat cortical neurons in vitro. Neurosci Lett 189(1):57–59
169. Sucher NJ, Lipton SA (1991) Redox modulatory site of the NMDA receptor-channel complex: regulation by oxidized glutathione. J Neurosci Res 30(3):582–591
170. Aizenman E, Hartnett KA, Reynolds IJ (1990) Oxygen free radicals regulate NMDA receptor function via a redox modulatory site. Neuron 5(6):841–846
171. Choi Y, Chen HV, Lipton SA (2001) Three pairs of cysteine residues mediate both redox and zn2+ modulation of the nmda receptor. J Neurosci 21(2):392–400
172. Sullivan JM, Traynelis SF, Chen HS, Escobar W, Heinemann SF, Lipton SA (1994) Identification of two cysteine residues that are required for redox modulation of the NMDA subtype of glutamate receptor. Neuron 13(4):929–936
173. Foster TC (2006) Biological markers of age-related memory deficits: treatment of senescent physiology. CNS Drugs 20(2):153–166

174. Parihar MS, Kunz EA, Brewer GJ (2008) Age-related decreases in NAD(P)H and glutathione cause redox declines before ATP loss during glutamate treatment of hippocampal neurons. J Neurosci Res 86(10):2339–2352
175. Poon HF, Calabrese V, Calvani M, Butterfield DA (2006) Proteomics analyses of specific protein oxidation and protein expression in aged rat brain and its modulation by L-acetylcarnitine: insights into the mechanisms of action of this proposed therapeutic agent for CNS disorders associated with oxidative stress. Antioxid Redox Signal 8(3–4):381–394
176. Bodhinathan K, Kumar A, Foster TC (2010) Intracellular redox state alters NMDA receptor response during aging through Ca^{2+}/calmodulin-dependent protein kinase II. J Neurosci 30(5):1914–1924
177. Bodhinathan K, Kumar A, Foster TC (2010) Redox sensitive calcium stores underlie enhanced after hyperpolarization of aged neurons: role for ryanodine receptor mediated calcium signaling. J Neurophysiol 104(5):2586–2593
178. Haxaire C, Turpin FR, Potier B, Kervern M, Sinet PM, Barbanel G et al (2012) Reversal of age-related oxidative stress prevents hippocampal synaptic plasticity deficits by protecting d-serine-dependent NMDA receptor activation. Aging Cell 11(2):336–344
179. Robillard JM, Gordon GR, Choi HB, Christie BR, MacVicar BA (2011) Glutathione restores the mechanism of synaptic plasticity in aged mice to that of the adult. PLoS One 6(5):e20676
180. Kumar A, Yegla B, Foster TC (2018) Redox signaling in neurotransmission and cognition during aging. Antioxid Redox Signal 28(18):1724–1745
181. Schell MJ, Molliver ME, Snyder SH (1995) D-serine, an endogenous synaptic modulator: localization to astrocytes and glutamate-stimulated release. Proc Natl Acad Sci U S A 92(9):3948–3952
182. Williams SM, Diaz CM, Macnab LT, Sullivan RK, Pow DV (2006) Immunocytochemical analysis of D-serine distribution in the mammalian brain reveals novel anatomical compartmentalizations in glia and neurons. Glia 53(4):401–411
183. Wu S, Barger SW (2004) Induction of serine racemase by inflammatory stimuli is dependent on AP-1. Ann N Y Acad Sci 1035:133–146
184. Wu SZ, Bodles AM, Porter MM, Griffin WS, Basile AS, Barger SW (2004) Induction of serine racemase expression and D-serine release from microglia by amyloid beta-peptide. J Neuroinflammation 1(1):2
185. Hayashi Y, Ishibashi H, Hashimoto K, Nakanishi H (2006) Potentiation of the NMDA receptor-mediated responses through the activation of the glycine site by microglia secreting soluble factors. Glia 53(6):660–668
186. Moriguchi S, Mizoguchi Y, Tomimatsu Y, Hayashi Y, Kadowaki T, Kagamiishi Y et al (2003) Potentiation of NMDA receptor-mediated synaptic responses by microglia. Brain Res Mol Brain Res 119(2):160–169
187. Rosi S, Ramirez-Amaya V, Hauss-Wegrzyniak B, Wenk GL (2004) Chronic brain inflammation leads to a decline in hippocampal NMDA-R1 receptors. J Neuroinflammation 1(1):12
188. Rosi S, Vazdarjanova A, Ramirez-Amaya V, Worley PF, Barnes CA, Wenk GL (2006) Memantine protects against LPS-induced neuroinflammation, restores behaviorally-induced gene expression and spatial learning in the rat. Neuroscience 142(4):1303–1315
189. Cao X, Cui Z, Feng R, Tang YP, Qin Z, Mei B et al (2007) Maintenance of superior learning and memory function in NR2B transgenic mice during ageing. Eur J Neurosci 25(6):1815–1822
190. Ly CV, Verstreken P (2006) Mitochondria at the synapse. Neuroscientist 12(4):291–299
191. Mattson MP, LaFerla FM, Chan SL, Leissring MA, Shepel PN, Geiger JD (2000) Calcium signaling in the ER: its role in neuronal plasticity and neurodegenerative disorders. Trends Neurosci 23(5):222–229
192. Verkhratsky A, Toescu EC (1998) Calcium and neuronal ageing. Trends Neurosci 21(1):2–7
193. Verkhratsky AJ, Petersen OH (1998) Neuronal calcium stores. Cell Calcium 24(5–6):333–343
194. Petersen OH, Gerasimenko OV, Gerasimenko JV, Mogami H, Tepikin AV (1998) The calcium store in the nuclear envelope. Cell Calcium 23(2–3):87–90

195. Petersen OH, Michalak M, Verkhratsky A (2005) Calcium signalling: past, present and future. Cell Calcium 38(3–4):161–169
196. Toescu EC, Verkhratsky A (2004) Ca^{2+} and mitochondria as substrates for deficits in synaptic plasticity in normal brain ageing. J Cell Mol Med 8(2):181–190
197. Duchen MR (2000) Mitochondria and calcium: from cell signalling to cell death. J Physiol 529(Pt 1):57–68
198. Nicholls DG, Budd SL (2000) Mitochondria and neuronal survival. Physiol Rev 80(1):315–360
199. Solovyova N, Veselovsky N, Toescu EC, Verkhratsky A (2002) Ca^{2+} dynamics in the lumen of the endoplasmic reticulum in sensory neurons: direct visualization of Ca^{2+}-induced Ca^{2+} release triggered by physiological Ca^{2+} entry. EMBO J 21(4):622–630
200. McGuinness L, Bardo SJ, Emptage NJ (2007) The lysosome or lysosome-related organelle may serve as a Ca^{2+} store in the boutons of hippocampal pyramidal cells. Neuropharmacology 52(1):126–135
201. Toescu EC, Myronova N, Verkhratsky A (2000) Age-related structural and functional changes of brain mitochondria. Cell Calcium 28(5–6):329–338
202. Sanmartin CD, Paula-Lima AC, Garcia A, Barattini P, Hartel S, Nunez MT et al (2014) Ryanodine receptor-mediated Ca^{2+} release underlies iron-induced mitochondrial fission and stimulates mitochondrial Ca^{2+} uptake in primary hippocampal neurons. Front Mol Neurosci 7:13
203. Roth GS (1995) Changes in tissue responsiveness to hormones and neurotransmitters during aging. Exp Gerontol 30(3–4):361–368
204. Mizutani T, Nakashima S, Nozawa Y (1998) Changes in the expression of protein kinase C (PKC), phospholipases C (PLC) and D (PLD) isoforms in spleen, brain and kidney of the aged rat: RT-PCR and Western blot analysis. Mech Ageing Dev 105(1–2):151–172
205. Nicolle MM, Colombo PJ, Gallagher M, McKinney M (1999) Metabotropic glutamate receptor-mediated hippocampal phosphoinositide turnover is blunted in spatial learning-impaired aged rats. J Neurosci 19(21):9604–9610
206. Burnett DM, Daniell LC, Zahniser NR (1990) Decreased efficacy of inositol 1,4,5-trisphosphate to elicit calcium mobilization from cerebrocortical microsomes of aged rats. Mol Pharmacol 37(4):566–571
207. Stutzmann GE, Smith I, Caccamo A, Oddo S, Laferla FM, Parker I (2006) Enhanced ryanodine receptor recruitment contributes to Ca^{2+} disruptions in young, adult, and aged Alzheimer's disease mice. J Neurosci 26(19):5180–5189
208. Igwe OJ, Ning L (1993) Inositol 1,4,5-trisphosphate arm of the phosphatidylinositide signal transduction pathway in the rat cerebellum during aging. Neurosci Lett 164(1–2):167–170
209. Martini A, Battaini F, Govoni S, Volpe P (1994) Inositol 1,4,5-trisphosphate receptor and ryanodine receptor in the aging brain of Wistar rats. Neurobiol Aging 15(2):203–206
210. Simonyi A, Xia J, Igbavboa U, Wood WG, Sun GY (1998) Age differences in the expression of metabotropic glutamate receptor 1 and inositol 1,4,5-trisphosphate receptor in mouse cerebellum. Neurosci Lett 244(1):29–32
211. Long LH, Liu J, Liu RL, Wang F, Hu ZL, Xie N et al (2009) Differential effects of methionine and cysteine oxidation on $[Ca^{2+}]$ i in cultured hippocampal neurons. Cell Mol Neurobiol 29(1):7–15
212. Peuchen S, Duchen MR, Clark JB (1996) Energy metabolism of adult astrocytes in vitro. Neuroscience 71(3):855–870
213. Bull R, Finkelstein JP, Humeres A, Behrens MI, Hidalgo C (2007) Effects of ATP, Mg2+, and redox agents on the Ca^{2+} dependence of RyR channels from rat brain cortex. Am J Physiol Cell Physiol 293(1):C162–C171
214. Gokulrangan G, Zaidi A, Michaelis ML, Schoneich C (2007) Proteomic analysis of protein nitration in rat cerebellum: effect of biological aging. J Neurochem 100(6):1494–1504
215. Hidalgo C, Bull R, Behrens MI, Donoso P (2004) Redox regulation of RyR-mediated Ca^{2+} release in muscle and neurons. Biol Res 37(4):539–552

216. Alford S, Frenguelli BG, Schofield JG, Collingridge GL (1993) Characterization of Ca^{2+} signals induced in hippocampal CA1 neurones by the synaptic activation of NMDA receptors. J Physiol 469:693–716
217. Matias C, Dionisio JC, Quinta-Ferreira ME (2002) Thapsigargin blocks STP and LTP related calcium enhancements in hippocampal CA1 area. Neuroreport 13(18):2577–2580
218. Yamazaki Y, Fujii S, Goto JI, Fujiwara H, Mikoshiba K (2015) Activation of inositol 1,4,5-trisphosphate receptors during preconditioning low-frequency stimulation suppresses subsequent induction of long-term potentiation in hippocampal CA1 neurons. Neuroscience 311:195–206
219. Sugita M, Yamazaki Y, Goto JI, Fujiwara H, Aihara T, Mikoshiba K et al (2016) Role of postsynaptic inositol 1, 4, 5-trisphosphate receptors in depotentiation in guinea pig hippocampal CA1 neurons. Brain Res 1642:154–162
220. Arias-Cavieres A, Adasme T, Sanchez G, Munoz P, Hidalgo C (2018) Raynodine receptor-mediated calcium release has a key role in hippocampal LTD induction. Front Cell Neurosci 12:403
221. Arias-Cavieres A, Adasme T, Sanchez G, Munoz P, Hidalgo C (2017) Aging impairs hippocampal- dependent recognition memory and LTP and prevents the associated RyR up-regulation. Front Aging Neurosci 9:111
222. Gant JC, Sama MM, Landfield PW, Thibault O (2006) Early and simultaneous emergence of multiple hippocampal biomarkers of aging is mediated by Ca^{2+}–induced Ca^{2+} release. J Neurosci 26(13):3482–3490
223. Kumar A, Foster TC (2004) Enhanced long-term potentiation during aging is masked by processes involving intracellular calcium stores. J Neurophysiol 91(6):2437–2444
224. Paula-Lima AC, Adasme T, Hidalgo C (2014) Contribution of Ca^{2+} release channels to hippocampal synaptic plasticity and spatial memory: potential redox modulation. Antioxid Redox Signal 21(6):892–914
225. Disterhoft JF, Oh MM (2006) Learning, aging and intrinsic neuronal plasticity. Trends Neurosci 29(10):587–599
226. Disterhoft JF, Oh MM (2007) Alterations in intrinsic neuronal excitability during normal aging. Aging Cell 6(3):327–336
227. Rex CS, Kramar EA, Colgin LL, Lin B, Gall CM, Lynch G (2005) Long-term potentiation is impaired in middle-aged rats: regional specificity and reversal by adenosine receptor antagonists. J Neurosci 25(25):5956–5966
228. Norris CM, Korol DL, Foster TC (1996) Increased susceptibility to induction of long-term depression and long- term potentiation reversal during aging. J Neurosci 16(17):5382–5392
229. Shankar S, Teyler TJ, Robbins N (1998) Aging differentially alters forms of long-term potentiation in rat hippocampal area CA1. J Neurophysiol 79(1):334–341
230. Diana G, Domenici MR, Loizzo A (1994) Scotti de Carolis A, Sagratella S. Age and strain differences in rat place learning and hippocampal dentate gyrus frequency-potentiation. Neurosci Lett 171(1–2):113–116
231. Kumar A, Thinschmidt JS, Foster TC, King MA (2007) Aging effects on the limits and stability of Long-term synaptic potentiation and depression in rat hippocampal area CA1. J Neurophysiol 98(2):594–601
232. Barnes CA, Rao G, McNaughton BL (1996) Functional integrity of NMDA-dependent LTP induction mechanisms across the lifespan of F-344 rats. Learn Mem 3(2–3):124–137
233. Watabe AM, O'Dell TJ (2003) Age-related changes in theta frequency stimulation-induced long-term potentiation. Neurobiol Aging 24(2):267–272
234. Zamani MR, Desmond NL, Levy WB (2000) Estradiol modulates long-term synaptic depression in female rat hippocampus. J Neurophysiol 84(4):1800–1808
235. Kemp N, McQueen J, Faulkes S, Bashir ZI (2000) Different forms of LTD in the CA1 region of the hippocampus: role of age and stimulus protocol. Eur J Neurosci 12(1):360–366
236. Dudek SM, Bear MF (1993) Bidirectional long-term modification of synaptic effectiveness in the adult and immature hippocampus. J Neurosci 13(7):2910–2918

237. Hsu KS, Huang CC, Liang YC, Wu HM, Chen YL, Lo SW et al (2002) Alterations in the balance of protein kinase and phosphatase activities and age-related impairments of synaptic transmission and long-term potentiation. Hippocampus 12(6):787–802
238. Vouimba RM, Foy MR, Foy JG, Thompson RF (2000) 17beta-estradiol suppresses expression of long-term depression in aged rats. Brain Res Bull 53(6):783–787
239. Moore CI, Browning MD, Rose GM (1993) Hippocampal plasticity induced by primed burst, but not long-term potentiation, stimulation is impaired in area CA1 of aged Fischer 344 rats. Hippocampus 3(1):57–66
240. Deupree DL, Bradley J, Turner DA (1993) Age-related alterations in potentiation in the CA1 region in F344 rats. Neurobiol Aging 14(3):249–258
241. Rosenzweig ES, Rao G, McNaughton BL, Barnes CA (1997) Role of temporal summation in age-related long-term potentiation- induction deficits. Hippocampus 7(5):549–558
242. Tong G, Jahr CE (1994) Regulation of glycine-insensitive desensitization of the NMDA receptor in outside-out patches. J Neurophysiol 72(2):754–761
243. Sah P, Faber ES (2002) Channels underlying neuronal calcium-activated potassium currents. Prog Neurobiol 66(5):345–353
244. Kumar A, Foster T (2007) Environmental enrichment decreases the afterhyperpolarization in senescent rats. Brain Res 1130(1):103–107
245. Disterhoft JF, Oh MM (2006) Pharmacological and molecular enhancement of learning in aging and Alzheimer's disease. J Physiol Paris 99(2–3):180–192
246. Kumar A, Rani A, Tchigranova O, Lee WH, Foster TC (2012) Influence of late-life exposure to environmental enrichment or exercise on hippocampal function and CA1 senescent physiology. Neurobiol Aging 33(4):828 e1–e17
247. Froemke RC, Poo MM, Dan Y (2005) Spike-timing-dependent synaptic plasticity depends on dendritic location. Nature 434(7030):221–225
248. Power JM, Wu WW, Sametsky E, Oh MM, Disterhoft JF (2002) Age-related enhancement of the slow outward calcium-activated potassium current in hippocampal CA1 pyramidal neurons in vitro. J Neurosci 22(16):7234–7243
249. Kerr DS, Campbell LW, Hao SY, Landfield PW (1989) Corticosteroid modulation of hippocampal potentials: increased effect with aging. Science 245(4925):1505–1509
250. Pitler TA, Landfield PW (1990) Aging-related prolongation of calcium spike duration in rat hippocampal slice neurons. Brain Res 508(1):1–6
251. Gong LW, Gao TM, Huang H, Zhou KX, Tong Z (2002) ATP modulation of large conductance Ca^{2+}-activated K(+) channels via a functionally associated protein kinase A in CA1 pyramidal neurons from rat hippocampus. Brain Res 951(1):130–134
252. Disterhoft JF, Moyer JR Jr, Thompson LT, Kowalska M (1993) Functional aspects of calcium-channel modulation. Clin Neuropharmacol 16(Suppl 1):S12–S24
253. Power JM, Oh MM, Disterhoft JF (2001) Metrifonate decreases sI(AHP) in CA1 pyramidal neurons in vitro. J Neurophysiol 85(1):319–322
254. Moyer JR Jr, Power JM, Thompson LT, Disterhoft JF (2000) Increased excitability of aged rabbit CA1 neurons after trace eyeblink conditioning. J Neurosci 20(14):5476–5482
255. Murphy GG, Fedorov NB, Giese KP, Ohno M, Friedman E, Chen R et al (2004) Increased neuronal excitability, synaptic plasticity, and learning in aged Kvbeta1.1 knockout mice. Curr Biol 14(21):1907–1915
256. Azad SC, Eder M, Simon W, Hapfelmeier G, Dodt HU, Zieglgansberger W et al (2004) The potassium channel modulator flupirtine shifts the frequency-response function of hippocampal synapses to favour LTD in mice. Neurosci Lett 370(2–3):186–190
257. Kumar A, Bodhinathan K, Foster TC (2009) Susceptibility to calcium dysregulation during brain aging. Front Aging Neurosci 1:2
258. Phillips RG, Meier TJ, Giuli LC, McLaughlin JR, Ho DY, Sapolsky RM (1999) Calbindin D28K gene transfer via herpes simplex virus amplicon vector decreases hippocampal damage in vivo following neurotoxic insults. J Neurochem 73(3):1200–1205
259. Dore K, Stein IS, Brock JA, Castillo PE, Zito K, Sjostrom PJ (2017) Unconventional NMDA receptor signaling. J Neurosci 37(45):10800–10807

260. Zorumski CF, Izumi Y (2012) NMDA receptors and metaplasticity: mechanisms and possible roles in neuropsychiatric disorders. Neurosci Biobehav Rev 36(3):989–1000
261. Abraham WC, Williams JM (2008) LTP maintenance and its protein synthesis-dependence. Neurobiol Learn Mem 89(3):260–268
262. Annunziato L, Amoroso S, Pannaccione A, Cataldi M, Pignataro G, D'Alessio A et al (2003) Apoptosis induced in neuronal cells by oxidative stress: role played by caspases and intracellular calcium ions. Toxicol Lett 139(2–3):125–133
263. Squier TC (2001) Oxidative stress and protein aggregation during biological aging. Exp Gerontol 36(9):1539–1550
264. Serrano F, Klann E (2004) Reactive oxygen species and synaptic plasticity in the aging hippocampus. Ageing Res Rev 3(4):431–443
265. Suzuki K, Nakamura M, Hatanaka Y, Kayanoki Y, Tatsumi H, Taniguchi N (1997) Induction of apoptotic cell death in human endothelial cells treated with snake venom: implication of intracellular reactive oxygen species and protective effects of glutathione and superoxide dismutases. J Biochem (Tokyo) 122(6):1260–1264
266. Lu C, Chan SL, Fu W, Mattson MP (2002) The lipid peroxidation product 4-hydroxynonenal facilitates opening of voltage-dependent Ca^{2+} channels in neurons by increasing protein tyrosine phosphorylation. J Biol Chem 277(27):24368–24375
267. Akaishi T, Nakazawa K, Sato K, Saito H, Ohno Y, Ito Y (2004) Modulation of voltage-gated Ca^{2+} current by 4-hydroxynonenal in dentate granule cells. Biol Pharm Bull 27(2):174–179
268. Gong L, Gao TM, Huang H, Tong Z (2000) Redox modulation of large conductance calcium-activated potassium channels in CA1 pyramidal neurons from adult rat hippocampus. Neurosci Lett 286(3):191–194
269. Lu C, Chan SL, Haughey N, Lee WT, Mattson MP (2001) Selective and biphasic effect of the membrane lipid peroxidation product 4-hydroxy-2,3-nonenal on N-methyl-D-aspartate channels. J Neurochem 78(3):577–589
270. Kamsler A, Segal M (2004) Hydrogen peroxide as a diffusible signal molecule in synaptic plasticity. Mol Neurobiol 29(2):167–178
271. Ullrich V, Namgaladze D, Frein D (2003) Superoxide as inhibitor of calcineurin and mediator of redox regulation. Toxicol Lett 139(2–3):107–110
272. Lin CH, Yeh SH, Leu TH, Chang WC, Wang ST, Gean PW (2003) Identification of calcineurin as a key signal in the extinction of fear memory. J Neurosci 23(5):1574–1579
273. Gorlach A, Bertram K, Hudecova S, Krizanova O (2015) Calcium and ROS: a mutual interplay. Redox Biol 6:260–271
274. Gant JC, Blalock EM, Chen KC, Kadish I, Porter NM, Norris CM et al (2014) FK506-binding protein 1b/12.6: a key to aging-related hippocampal Ca^{2+} dysregulation? Eur J Pharmacol 739:74–82
275. Gant JC, Chen KC, Norris CM, Kadish I, Thibault O, Blalock EM et al (2011) Disrupting function of FK506-binding protein 1b/12.6 induces the Ca(2)+−dysregulation aging phenotype in hippocampal neurons. J Neurosci 31(5):1693–1703

Chapter 40
Calcium Signaling in Endothelial Colony Forming Cells in Health and Disease

Francesco Moccia

Abstract Endothelial colony forming cells (ECFCs) represent the only known truly endothelial precursors. ECFCs are released in peripheral circulation to restore the vascular networks dismantled by an ischemic insult or to sustain the early phases of the angiogenic switch in solid tumors. A growing number of studies demonstrated that intracellular Ca^{2+} signaling plays a crucial role in driving ECFC proliferation, migration, homing and neovessel formation. For instance, vascular endothelial growth factor (VEGF) triggers intracellular Ca^{2+} oscillations and stimulates angiogenesis in healthy ECFCs, whereas stromal derived factor-1α promotes ECFC migration through a biphasic Ca^{2+} signal. The Ca^{2+} toolkit endowed to circulating ECFCs is extremely plastic and shows striking differences depending on the physiological background of the donor. For instance, inositol-1,4,5-trisphosphate-induced Ca^{2+} release from the endoplasmic reticulum is downregulated in tumor-derived ECFCs, while agonists-induced store-operated Ca^{2+} entry is up-regulated in renal cellular carcinoma and is unaltered in breast cancer and reduced in infantile hemangioma. This remodeling of the Ca^{2+} toolkit prevents VEGF-induced pro-angiogenic Ca^{2+} oscillations in tumor-derived ECFCs. An emerging theme of research is the dysregulation of the Ca^{2+} toolkit in primary myelofibrosis-derived ECFCs, as this myeloproliferative disorder may depend on a driver mutation in the calreticulin gene. In this chapter, I provide a comprehensive, but succinct, description on the architecture and role of the intracellular Ca^{2+} signaling toolkit in ECFCs derived from umbilical cord blood and from peripheral blood of healthy donors, cancer patients and subjects affected by primary myelofibrosis.

Keywords Endothelial colony forming cells · Ca^{2+} signaling · Inositol-1 · 4 · 5-trisphosphate receptors · Two-pore channel 1 · STIM1 · Orai1 · TRPC1 · TRPC3 · VEGF · SDF-1α

F. Moccia (✉)
Laboratory of General Physiology, Department of Biology and Biotechnology "L. Spallanzani", University of Pavia, Pavia, Italy
e-mail: francesco.moccia@unipv.it

© Springer Nature Switzerland AG 2020

M. S. Islam (ed.), *Calcium Signaling*, Advances in Experimental Medicine and Biology 1131, https://doi.org/10.1007/978-3-030-12457-1_40

40.1 Introduction

Endothelial progenitor cells (EPCs) are released in the bloodstream to replace senescent or injured endothelial cells and to restore the vascular networks disrupted by an ischemic insult [1, 2]. Moreover, EPCs sustain tumor angiogenesis and metastasis by providing the building blocks for nascent vasculature and by stimulating local angiogenesis in a paracrine manner [3]. Circulating EPCs can be subdivided into two main cellular populations based on their hematopoietic or endothelial lineage [1]. Myeloid angiogenic cells (MACs), also known as circulating angiogenic cells (CACs) or "early" EPCs, represent a population of hematopoietic EPCs that are released from bone marrow and promote vascular repair in a paracrine manner, even though they are not able to physically engraft within neovessels and show limited proliferative potential. Endothelial colony forming cells (ECFCs), also known as blood outgrowth endothelial cells (BOECs) or "late" EPCs, constitute the only known truly endothelial precursors with an enormous potential for vascular reconstruction [1, 4]. ECFCs express the surface antigens CD31, CD105, CD144, CD146, von Willebrand factor (vWF), VEGFR-2 (KDR) and display the ability to ingest the ability to ingest acetylated low density lipoprotein (AcLDL). Furthermore, ECFCs lack the hematopoietic or monocytes/macrophage surface antigens CD14, CD45, or CD115 [5]. ECFCs mainly reside within vascular stem cell niches, display a robust clonogenic potential, assemble into capillary-like structures in vitro and give raise to durable and functional blood vessels in multiple murine models of cardiovascular disorders [1, 4]. Moreover, ECFCs drive the angiogenic switch during the early phases of vascular growth in many types of solid cancers, including melanoma and breast cancer [4, 6]. Therefore, ECFCs are regarded as the most suitable cellular candidates for regenerative stem cell therapy of cardiovascular disorders [7] and the most promising target for anti-angiogenic therapy [3, 8].

Intracellular Ca^{2+} signaling has long been known to stimulate angiogenesis under both physiological and pathological conditions [9]. Growth factors and cytokines, such as vascular endothelial growth factor (VEGF), basic fibroblast growth factor (bFGF), epidermal growth factor (EGF), angiopoietins, and stromal cell-derived factor-1α (SDF-1α), stimulate endothelial proliferation, migration, and tube formation through an increase in intracellular Ca^{2+} concentration ($[Ca^{2+}]_i$) [10–17]. Recent studies demonstrated that Ca^{2+} signaling plays a key role in ECFC proliferation, tube formation and neovessel formation [18, 19]. This review discusses how insulin-like growth factor 2 (IGF2), SDF-1α and VEGF impinge on distinct Ca^{2+} waveforms to regulate the angiogenic behavior of ECFCs derived from human peripheral blood and umbilical cord blood (PB-ECFCs and UCB-ECFCs, respectively). Furthermore, this review discusses how the remodeling of the Ca^{2+} toolkit in PB-ECFCs may contribute to the development of resistance to anti-angiogenic therapies in renal cellular carcinoma (RCC), breast cancer (BC), and infantile hemangioma (IH) patients. Finally, this review discusses how the Ca^{2+} signaling machinery undergoes an extensive rewiring also in ECFCs deriving from subjects affected by primary myelofibrosis (PMF-ECFCs), a myeloproliferative disorder that may include a driver mutation in the calreticulin gene [20].

40.2 The Intracellular Ca^{2+} Toolkit in Peripheral Blood-Derived Endothelial Colony Forming Cells

The concerted interaction of multiple Ca^{2+}-transporting systems contributes to maintain the resting $[Ca^{2+}]_i$ in ECFCs and to clear cytosolic Ca^{2+} upon extracellular stimulation [18, 19]. A thorough transcriptomic analysis revealed that ECFCs express some of the most widespread endothelial transporter and pump isoforms [21]. These include Sarco-Endoplasmic Reticulum Ca^{2+}-ATPase 2B (SERCA2B), which sequesters cytosolic Ca^{2+} into the endoplasmic reticulum (ER) lumen, the most abundant intracellular Ca^{2+} reservoir; Plasma Membrane Ca^{2+}-ATPase 1B (PMCA1B) and PMCA4B, which extrude Ca^{2+} towards the extracellular space. Surprisingly, ECFCs lack the endothelial isoforms of the Na^+/Ca^{2+} exchanger (NCX), e.g. NCX1.3 and NCX1.7, which sustain PMCA-mediated Ca^{2+} extrusion across the plasma membrane in vascular endothelial cells [22]. In agreement with this observation, reverting the Na^+ gradient through Na^+ substitution, which induces NCX to switch from the forward (3 Na^+ in: 1 Ca^{2+} out) to the reverse (3 Na^+ out: 1 Ca^{2+} in) Ca^{2+} entry mode [23], did not cause any detectable increase in $[Ca^{2+}]_i$ in ECFCs [18].

ECFCs impinge on two main Ca^{2+} sources to generate intracellular Ca^{2+} signals in response to extracellular stimulation [18, 19], namely the ER and the extracellular milieu (Fig. 40.1). Growth factors and cytokines bind, respectively, to their specific Tyrosine Kinase Receptors (TRKs), such as VEGFR2 (for VEGF), and G-Protein Coupled Receptors (GPCRs), such as IGF2 receptor (IGFR2) and CXCR4 (for SDF-1α), thereby causing an increase in $[Ca^{2+}]_i$ through the activation of multiple isoforms of phospholipase C (PLC) [19, 24–26]. Accordingly, GPCRs and TKRs recruit, respectively, PLCβ and PLCγ, which cleave phosphatidylinositol 4,5-bisphosphate (PIP_2), a minor phospholipid component of the plasma membrane, into the intracellular second messengers, inositol-1,4,5-trisphosphate ($InsP_3$) and diacylglycerol (DAG) [19]. ECFCs are endowed with PLCβ2, which may be recruited by IGF2 and SDF-1α through the $G_{\beta\gamma}$ dimer [24], while it is still unclear which PLCγ isoforms (PLCγ1–4) they express. In addition, ECFCs express all the three $InsP_3$ receptor isoforms ($InsP_3Rs$), namely $InsP_3R1$, $InsP_3R2$ and $InsP_3R3$ (Table 40.1) [26]. $InsP_3$, therefore, binds to ER-embedded $InsP_3Rs$ and causes transient or rhythmical Ca^{2+} release into the cytosol, leading to a massive reduction in ER Ca^{2+} concentration [24, 26, 27]. Conversely, ECFCs lack ryanodine receptors (RyRs) (Table 40.1) and do not produce Ca^{2+} signals in response to caffeine [27, 28]. In addition, endogenous Ca^{2+} may be accumulated within the acidic vesicles of the endolysosomal (EL) compartment [29]. ECFCs express high protein levels of the two-pore channel 1 (TPC1) (Table 40.1), which is gated by the newly discovered

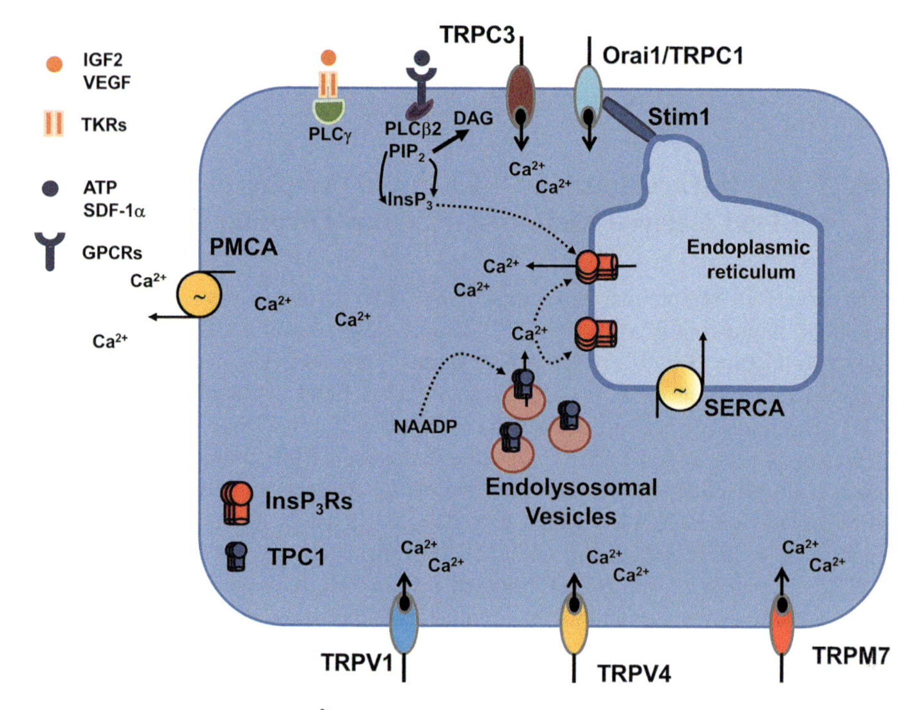

Fig. 40.1 The intracellular Ca^{2+} toolkit in human endothelial colony forming cells. Extracellular autacoids bind to specific G-protein Coupled Receptors (GPCRs) or Tyrosin Kinase Receptors (TKRs), thereby activating PLCβ2 and PLCγ, respectively, which cleave PIP_2 into $InsP_3$ and DAG. $InsP_3$ triggers ER-dependent Ca^{2+} release through $InsP_3Rs$, while DAG gates TRPC3 in UCB-ECFCs. The newly discovered second messenger NAADP stimulates TPC1 to cause Ca^{2+} release from the acidic vesicles of the endolysosomal (EL) system. NAADP-induced Ca^{2+} release could, in turn, recruit $InsP_3Rs$ through the Ca^{2+}-induced Ca^{2+} release process. The $InsP_3$-dependent drop in ER Ca^{2+} levels induces SOCE, which is mediated by the interaction between STIM1, Orai1 and TRPC1. Moreover, extracellular Ca^{2+} entry may occur through TRPV1, TRPV4 and TRPM7. TKRs and GPCRs are representative of IGFR2 and VEGFRs and of P_{2Y} receptors and CXCR4, respectively

second messenger, nicotinic acid adenine dinucleotide phosphate (NAADP), and mediates massive EL Ca^{2+} release into the cytoplasm [28, 30]. Mitochondria may also modulate intracellular Ca^{2+} signals by buffering $InsP_3$-dependent Ca^{2+} release and extracellular Ca^{2+} influx in ECFCs [21, 31]. For instance, $InsP_3$-driven mitochondrial Ca^{2+} signals sustain ECFC bioenergetics [32], as reported in vascular endothelium [33].

Extracellular Ca^{2+} entry in ECFCs is mediated by store-operated Ca^{2+} entry (SOCE) and members of the Transient Receptor Potential (TRP) superfamily of non-selective cation channels (Table 40.1) [19, 34]. SOCE is physiologically activated by the $InsP_3$-induced depletion of the ER Ca^{2+} store and is accomplished

Table 40.1 Composition of the Ca^{2+} toolkit in different types of endothelial colony forming cells

Parameter	PB-ECFCs	RCC-ECFCs	BC-ECFCs	IH-ECFCs	PMF-ECFCs	UCB-ECFCs
$[Ca^{2+}]_{ER}$	/	No [21]	No [76]	No [43]	No [99]	ND
SOCE	No [27]	No [35]	=	=	No [99]	No [27]
VEGF sensitivity	No [26]	No [35]	No [76]	No [43]	No [101]	No [48]
$InsP_3R1–3$	No [26], mRNA	No [35], mRNA	No [76]	No [43]	No [99]	No [48], mRNA
RyR1–3	No [27], mRNA	ND	ND	ND	ND	ND
TPC1	No [28], mRNA	ND	ND	ND	ND	ND
STIM1	No [27], mRNA & protein	No [35], mRNA & protein	No [76], mRNA & protein	No [43], mRNA & protein	No [99], mRNA & protein	No [27], mRNA & protein
STIM2	No [27], mRNA & protein	No [35], mRNA & protein	No [76], mRNA & protein	No [43], mRNA & protein	No [99], mRNA & protein	No [27], mRNA & protein
Orai1	No [27], mRNA & protein	No [35], mRNA & protein	No [76], mRNA & protein	No [43], mRNA & protein	No [99], mRNA & protein	No [27], mRNA & protein
Orai2	No [27], mRNA	No [35], mRNA	No [76], mRNA	No [43], mRNA	No [99], mRNA	No [27], mRNA
Orai3	No [27], mRNA & protein	No [35], mRNA	No [76], mRNA	No [43], mRNA	No [99], mRNA	No [27], mRNA
TRPC1	No [27], mRNA & protein	No [35], mRNA & protein	No [76], mRNA & protein	No [43], mRNA & protein	No [99], mRNA & protein	No [27], mRNA & protein
TRPC3	No [27, 48], mRNA & protein	No [35], mRNA	No [76], mRNA	No [43], mRNA	No, [99], mRNA	No [48], mRNA & protein
TRPC4	No [27], mRNA & protein	No [35], mRNA & protein	No [76], mRNA & protein	No [43], mRNA & protein	No [99], mRNA & protein	No [27], mRNA & protein
TRPC5–7	No [27], mRNA	No [35], mRNA	No [76], mRNA	No [43], mRNA	No [99], mRNA	No [27], mRNA
TRPV1	No [27], Protein	ND	ND	ND	ND	ND
TRPV4	No [52], mRNA & protein	ND	ND	ND	ND	ND
TRPM7	No [54], mRNA & protein	ND	ND	ND	ND	ND

Abbreviations: *$InsP_3R$* inositol-1,4,5-trisphosphate receptors, *ND* not defined, *RyR* ryanodine receptors, *SOCE* store-operated Ca^{2+} entry, *TPC* two-pore channels, *TRPC* Canonical Transient Receptor Potential, *TRPM* Melastatin Transient Receptor Potential, *TRPV* Vanilloid Transient Receptor Potential

by the interaction between Stromal Interaction Molecule 1 (STIM1), a sensor of ER Ca^{2+} levels, and Orai1 and Transient Receptor Potential (TRP) Canonical 1 (TRPC1), which form the Ca^{2+}-permeable channel(s) on the plasma membrane [19, 27, 35, 36]. $InsP_3$-evoked ER-dependent Ca^{2+} release induced STIM1 aggregation into ER-plasma membrane clusters, known as *puncta*, in which STIM1 interacts with and activates Orai1 and TRPC1 [19, 36]. It is, however, still unknown whether Orai1 and TRPC1 assemble into a unique supermolecular complex, as observed in human megakaryocytes [37] and platelets [38], or mediate two distinct STIM1-dependent channels, as described in human submandibular gland cells [39], mouse salivary acinar cells [40] and mouse pancreatic acinar cells [41]. TRPC1 has been shown to contribute to SOCE also in mouse bone marrow-derived ECFCs [42]. SOCE in ECFCs is blocked by 10 μM La^{3+} or Gd^{3+}, YM-58483/BTP-2, which lacks side effects in ECFCs [34], Pyr6 and carboxyamidotriazole (CAI) [27, 35, 43].

The TRP superfamily consists of 28 (27 human) members which fall into six subfamilies based on amino acid sequence similarity and function [44]: TRPC for canonical (TRPC1–7), TRPV for vanilloid (TRPV1–6), TRPM for melastatin (TRPM1–8), TRPA for ankyrin (TRPA1), TRPML for mucolipin (TRPML1–3), and TRPP for polycystin (TRPP; TRPP2, TRPP3 and TRPP5) [44]. Little information is available regarding TRP channel expression in human ECFCs. Apart from TRPC1, PB-ECFCs express TRPC4, while they lack TRPC3, TRPC5 and TRPC6, which play a crucial role in angiogenesis and blood pressure regulation in vascular endothelial cells [45–47]. Nevertheless, TRPC3 is abundantly expressed in UCB-ECFCs (Table 40.1) [48]. On the other hand, PB-ECFCs possess multiple TRP channel isoforms, i.e. TRPV1, TRPV4 and TRPM7 (Table 40.1) [19]. TRPV1 is a polymodal non-selective cation channel that is gated by noxious heat (>43 °C), protons (pH < 5.9), endovanilloids, and many natural products, including dietary compounds, such as capsaicin, eugenol, and gingerol [49]. A recent study showed that TRPV1 mediates cell proliferation and tube formation in ECFCs by promoting anandamide uptake in a Ca^{2+}-independent manner [50]. This effect was further enhanced by capsaicin [50]. Similar to TRPV1, TRPV4 is a polymodal Ca^{2+}-permeable channel that is activated by a disparate array of stimuli, including heat (>25 °C), hypotonicity, acidic pH, arachidonic acid (AA) and phospholipase A2-mediated AA metabolites [51]. AA stimulates TRPV4 to stimulate ECFC proliferation by recruiting the endothelial nitric oxide (NO) synthase (eNOS) and triggering robust NO release [28, 52]. Finally, TRPM7 is a constitutively active non-selective cation channel which is highly permeable to extracellular divalent cations, i.e. Ca^{2+} and Mg^{2+}, and is physiologically regulated by multiple mechanisms, including: a reduction in intracellular Mg^{2+} and Mg-ATP, depletion of PIP_2 in the plasma membrane upon PLC activation, laminar shear stress and membrane stretch [53]. Recent work showed that TRPM7 has no impact on ECFC growth, although it stimulates proliferation in human microvascular endothelial cells [54].

40.3 IGF2 and SDF-1α Stimulate ECFC Homing Through an Increase in $[Ca^{2+}]_i$

Ischemic tissues release a multitude of cytokines, including IGF2 and SDF-1α, which recruit circulating ECFCs or stimulate ECFC mobilization from the vascular niches and promote their physical incorporation within nascent neovessels [24, 55]. IGF2 stimulated IGFR2 to cause an increase in $[Ca^{2+}]_i$ through the recruitment of PLCβ2 [24]. Intracellular Ca^{2+} mobilization was, in turn, required to drive UCB-ECFC chemotaxis and adhesion to a fibronectin matrix in vitro. Furthermore, IGF2-evoked Ca^{2+} signals induced UCB-ECFC recruitment to injured vessels and incorporation into newly formed vessels in vivo, thereby promoting vascular repair in a murine model of hindlimb ischemia [24]. Subsequent work demonstrated that also SDF-1α stimulates PB-ECFC migration in vitro and neovessel formation in vivo through intracellular Ca^{2+} signaling [25]. SDF-1α-evoked increase in $[Ca^{2+}]_i$ was mediated by CXCR4 and achieved the largest magnitude at 10 ng/mL, whereas it desensitized at higher concentrations (i.e. 50 ng/mL). The Ca^{2+} response to SDF-1α consisted in an initial Ca^{2+} peak, which was mediated by $InsP_3$-dependent ER Ca^{2+} release, followed by a prolonged plateau phase, which was due to SOCE activation [25]. SDF-1α induced Ca^{2+} signaling, in turn, engaged the phosphoinositide 3-kinase (PI3K)/Akt and extracellular signal–regulated kinase (ERK) signaling pathways [25], which are both necessary for ECFC migration and neovessel formation [25, 56]. Therefore, intracellular Ca^{2+} signaling plays a crucial role in promoting ECFC homing to hypoxic tissues by mediating the chemotactic effect of multiple cytokines.

40.4 VEGF-Induced Intracellular Ca^{2+} Oscillations Stimulate ECFC Proliferation and Tube Formation

VEGF is massively released from hypoxic cells to stimulate sprouting angiogenesis and promote proliferation in EPCs (i.e. MACs and ECFCs) recruited to injured vessels [2, 57]. VEGF promotes the angiogenic activity of ECFCs by binding to VEGFR-2 [57, 58], which activates multiple downstream signaling pathways [59], including an increase in $[Ca^{2+}]_i$ [10, 14, 60]. A recent investigation demonstrated that VEGF-induced intracellular Ca^{2+} oscillations drive cell proliferation and tube formation in PB-ECFCs (Table 40.1) [26, 36]. VEGF-induced Ca^{2+} spikes were asynchronous between individual PB-ECFCs from the same coverslip and displayed a dose-response activity. The magnitude of the first Ca^{2+} spike increases while latency of the oscillatory response decreases when VEGF concentration was increased from 1 ng/ml to 50 ng/ml, with maximal frequency of Ca^{2+} oscillations attained at 10 ng/mL [26]. Pharmacological manipulation revealed that VEGF-induced intracellular Ca^{2+} oscillations were triggered by rhythmical $InsP_3$-dependent Ca^{2+} release from the ER and maintained over time by SOCE in a SERCA-dependent manner [26]. The transcription factor, Nuclear factor kappa enhancer binding protein (NF-κB), represents the Ca^{2+}-dependent decoder which translates intracellular Ca^{2+} oscillations into a mitogenic output in vascular endothelial cells [61, 62]. The NF-κB family of transcription factors comprises five members: p50, p52, p65 (also known as REL-A), c-REL and REL-B, which assemble with each other into homo- or heterodimeric complexes [63]. Under resting conditions, NF-κB dimers are retained within the cytosol through the physical interaction with the inhibitory IkB proteins, which masks their nuclear localization sequence (NLS). Intracellular Ca^{2+} oscillations recruit the Ca^{2+}/Calmodulin (CaM)-dependent protein kinase IV (CaMKIV), which phosphorylates and activates two IκB kinases, i.e. IKKα and IKKβ. In turn, these kinases phosphorylate IκB at S32 and S36, respectively [19, 61, 62]. In particular, IKKβ-induced phosphorylation leads to IκB ubiquitination and subsequent proteosomal degradation, thereby exposing the previously masked NLS of NF-κB, which is free to translocate into the nucleus and initiate the transcription of target genes [63, 64]. VEGF-induced intracellular Ca^{2+} oscillations promoted IκB phosphorylation, whereby thimoquinone, a rather selective NF-κB inhibitor, prevented VEGF-dependent PB-ECFC proliferation and tube formation [26].

Earlier work demonstrated that UCB-ECFCs are more sensitive to VEGF compared to their circulating counterparts [65], although there is no significant difference in VEGFR-2 expression between the two cell types [27, 48]. This finding suggests that the pro-angiogenic signaling pathway recruited downstream of VEGFR-2 is more robustly activated in UCB-ECFCs [2, 66]. In agreement with this hypothesis, the oscillatory Ca^{2+} activity induced by VEGF is significantly enhanced in UCB-ECFCs compared to PB-ECFCs (Table 40.1) [2, 48]. The oscillatory response to VEGF was triggered by TRPC3, which is gated by DAG and is selectively expressed in UCB-, but not PB-ECFCs (Fig. 40.1) [48]. TRPC3-mediated extracellular Ca^{2+} entry, in turn, triggered the dynamic interplay between rhythmical $InsP_3$-dependent ER Ca^{2+} release and SOCE, which maintains the oscillations over time [48]. This observation showed that the Ca^{2+} toolkit endowed to ECFCs is extremely plastic and varies depending on the source (Table 40.1), i.e. PB vs. UCB. Furthermore, it led to the hypothesis that introducing TRPC3 in autologous aging or dysfunctional ECFCs could rejuvenate or partially rescue their reparative phenotype [66].

40.5 The Intracellular Ca^{2+} Toolkit Is Remodeled in Tumor-Derived ECFCs

The Ca^{2+} toolkit is dramatically altered not only in cancer cells [67, 68], but also in tumor stromal cells [69–74], thereby contributing to establish many cancer hallmarks, such as aberrant angiogenesis, invasion and metastasis. Tumor-derived ECFCs were isolated from peripheral blood of cancer patients, as described in the pioneering study by Yoder and coworkers [65]. The frequency of ECFCs is significantly increased in peripheral blood of RCC and BC patients [75], which is consistent with their role in tumor vascularization [4, 6]. Furthermore, RCC- and BC-derived ECFCs (termed, respectively, RCC-ECFCs and BC-ECFCs) show significant ultrastructural differences when compared to healthy ECFCs. For instance, the mitochondria were more numerous and displayed an elongated and branched shape, which reflects fusion or incomplete separation, in tumor-derived ECFCs [21, 76]. Moreover, cisternae of rough ER (rER) were more widely spaced and occupied a wider area in RCC- and BC-ECFCs compared to healthy ECFCs. Similarly, smooth ER (sER) vesicles occupied a large part of the cytoplasm in tumor-derived, but not healthy, ECFCs [21, 76]. Ca^{2+} imaging recordings revealed that these major morphological differences were accompanied by a dramatic remodeling of the Ca^{2+} toolkit. An earlier report showed that $InsP_3$-dependent ER Ca^{2+} release was significantly reduced in RCC-ECFCs [35], due to the reduction in the ER Ca^{2+} concentration ($[Ca^{2+}]_{ER}$) (Table 40.1) [21] and the downregulation of $InsP_3R$ transcripts (Table 40.1) [35]. Conversely, SOCE amplitude was enhanced due to the up-regulation of STIM1, Orai1 and TRPC1 expression (Table 40.1) [35]. The reduction in $[Ca^{2+}]_{ER}$ could be due to the slowing down of SERCA pump activity [21], which is associated with the up-regulation of TMTC1, a tetratricopeptide repeats (TPR)-containing protein that physically interacts with and inhibits SERCA [77]. On the other hand, there was no difference in the expression levels of SERCA2B, PMCA1B, PMCA4B and of the two major ER Ca^{2+} buffering proteins, calreticulin and calnexin, between normal and RCC-derived ECFCs [21]. Remodeling of the Ca^{2+} toolkit has the potential to modulate ECFC resistance to a number of anticancer therapies [18, 69]. For instance, the reduction in $[Ca^{2+}]_{ER}$ and $InsP_3R$ expression resulted in lower sensitivity to pro-apoptotic stimulation in RCC-ECFCs [21], as routinely observed in cancer cells [68, 78], due to the impairment of $InsP_3$-driven ER-to-mitochondria Ca^{2+} transfer [21]. Conversely, there was no difference in the expression levels of the anti-apoptotic protein, Bcl-2, and of the pro-apoptotic protein, Bak [21]. Furthermore, RCC-ECFCs became insensitive to VEGF (Table 40.1), which failed to trigger $InsP_3$-dependent intracellular Ca^{2+} oscillations [18, 35]. This finding could explain why anti-VEGF therapy does not improve the overall survival of RCC patients and why a remarkable fraction of these subjects manifest resistance at the start of treatment [3, 18, 79]. Nevertheless, the pharmacological blockade of SOCE with 10 μM La^{3+} or Gd^{3+}, YM-58483/BTP-2 or CAI, inhibited RCC-ECFC proliferation and tube formation [35]. This finding suggested that SOCE could provide a reliable target for anti-angiogenic therapies in RCC [18, 34, 79].

A subsequent investigation showed that the intracellular Ca^{2+} toolkit is also remodeled in BC-ECFCs [76]. These cells displayed a significant reduction in the $InsP_3$-regulated ER Ca^{2+} pool (Table 40.1), although $InsP_3Rs$ were normally expressed. Although $[Ca^{2+}]_{ER}$ has not been directly measured in BC-ECFCs [76], TMTC1 was overexpressed in these cells and may be responsible for the down-regulation of $InsP_3$-sensitive Ca^{2+} store [75]. Conversely, the amplitude of SOCE was not enhanced as only STIM1, but not Orai1 and TRPC1, was up-regulated (Table 40.1) [76]. Similar to RCC-ECFCs, the reduction in the $InsP_3$-releasable ER

Ca^{2+} pool prevented VEGF from triggering robust intracellular Ca^{2+} oscillations in BC-ECFCs (Table 40.1), although there was no difference in VEGFR-2 expression compared to healthy cells. As a consequence, VEGF failed to promote proliferation and tube formation in BC-ECFCs [76]. As for RCC, this feature could explain why anti-VEGF drugs do not significantly improve the overall survival of BCC patients [80]. It should, however, be pointed out that the pharmacological blockade of SOCE did suppress BC-ECFC proliferation and tube formation.

Finally, the Ca^{2+} toolkit is also subtly remodeled in ECFCs derived from children suffering from infantile hemangioma (IH-ECFCs) [43]. IH is a benign vascular tumor that arises a few weeks after birth and can undergo spontaneous regression over the next 7–10 years [81]. Unfortunately, 24–38% of IH patients can develop complications that induce severe comorbidities and require medical interventions in 24–38% of the patients [82]. The aberrant vascular growth that results in IH is exacerbated by a population of resident endothelial precursors [81, 83], that are likely to be ECFCs. A recent study revealed that the $InsP_3$-releasable ER Ca^{2+} pool is dramatically reduced in IH-ECFCs, although $InsP_3Rs$ are normally expressed (Table 40.1). Interestingly, the amplitude of SOCE was dampened, although there was no change in the expression levels of STIM1, Orai1 and TRPC1 (Table 40.1) [43]. These findings led to the hypothesis that SOCE was constitutively activated in IH-ECFCs, so that STIM1, Orai1 and TRPC1 could not be fully recruited by extracellular stimulation [43]. The Mn^{2+} quenching technique is a widely employed strategy to evaluate resting Ca^{2+} permeability in vascular endothelial cells [84, 85]. This approach revealed that IH-ECFCs, but not healthy ECFCs, exhibit a constitutive SOCE which refills the ER and maintains the Ca^{2+} response to $InsP_3$ in a SERCA-dependent manner [43]. Similar to agonists-induced SOCE [35, 76], this constitutive Ca^{2+} entry was blocked by YM-58483/BTP-2 and 10 μM La^{3+} [43]. Therefore, it has been suggested that the chronic reduction in $[Ca^{2+}]_{ER}$ leads to the basal activation of store-operated channels to prevent excessive ER Ca^{2+} depletion. Constitutive SOCE, in turn, drives IH-ECFC proliferation by recruiting eNOS and enhancing basal NO release [43]. Of note, VEGF was not able to stimulate intracellular Ca^{2+} oscillations in IH-ECFCs (Table 40.1) [28].

Overall, these findings confirm that the Ca^{2+} toolkit in ECFCs is differently remodeled depending on the tumor type (Table 40.1). Nevertheless, tumor-derived ECFCs share three common features: (1) they display a significant reduction in the $InsP_3$-releasable ER Ca^{2+} pool, which could be due to a reduction in $[Ca^{2+}]_{ER}$ (and to $InsP_3Rs$ down-regulation in RCC-ECFCs); (2) they are insensitive to VEGF, which does not trigger intracellular Ca^{2+} oscillations and do not promote in vitro angiogenesis; and (3) they require SOCE to proliferate and assemble into capillary-like tubular structures. The molecular determinants that remodel the Ca^{2+} toolkit in tumor-derived ECFCs are not clear. It has been shown that BC-derived TRPC5-containing exosomes determine the up-regulation of TRPC5 expression in human dermal microvascular endothelial cells [86]. Moreover, the angiogenic behavior of several EPC subsets, including ECFCs, may be reprogrammed by circulating microRNAs (miRNAs) resident in exosomes and microvesicles [58, 87–89].

40.6 The Intracellular Ca^{2+} Toolkit Is Remodeled in Primary Myelofibrosis-Derived ECFCs

Primary myelofibrosis (PMF) is a Philadelphia chromosome-negative (Ph-neg) chronic myeloproliferative neoplasm which is characterized by bone marrow fibrosis, myeloid metaplasia, and splenomegaly. Splenomegaly, in turn, is sustained by enhanced mobilization of bone marrow-derived hematopoietic stem cells and circulating ECFCs, which support neovessel formation in the spleen [90, 91]. PMF is caused by somatic mutation in the JAK2 gene, resulting in a valine to phenylalanine substitution at position 617 (JAK2-V617F) in $\approx$60% of patients, in the calreticulin gene in 25% of patients, or in the gene encoding for the receptor for thrombopoietin (myeloproliferative leukemia virus oncogene, MPL) in 5% of patients [92]. Conversely, $\approx$10% of patients do not harbor any recognized mutation in their hematopoietic cells [20]. Calreticulin is a major ER Ca^{2+}-binding protein that exerts a multilevel control of intracellular Ca^{2+} dynamics. For instance, calreticulin determines the amount of ER-releasable Ca^{2+} [93], controls SOCE [94], and finely modulates SERCA2B activity [95]. The CALR mutation in PMF consists of either a 52-bp deletion (type 1) or a 5 bp TGTC insertion at the COOH-terminal, which contains the Ca^{2+}-binding site and the ER-retention domain [96]. Since the high capacity Ca^{2+}-binding domain of calreticulin is also able to impede SOCE activation [94], a mutation in this domain causes a decrease in the ER Ca^{2+} storage ability and an increase in SOCE amplitude in megakaryocytes from PMF patients, which are more evident for type 1 CALR mutation [97]. Long before the discovery of the driver role of calreticulin in PMF, a dramatic remodeling of the Ca^{2+} toolkit was described in ECFCs isolated from subjects carrying the JAK2 mutation (PMF-ECFCs). Although PMF-ECFCs do not belong to the neoplastic clone [98], they displayed a remarkable increase in the ER-releasable Ca^{2+} pool (Table 40.1) [99]. Nevertheless, $InsP_3$-dependent Ca^{2+} mobilization was attenuated (Table 40.1), although all the $InsP_3R$ isoforms were up-regulated [99]. SOCE was up-regulated and displayed additional molecular, pharmacological and functional differences when compared to healthy ECFCs (Table 40.1). Expression of STIM1, Orai1, Orai3, TRPC1 and TRPC4 proteins were also sup-regulated in PMF-ECFCs (Table 40.1). Intriguingly, two distinct store-operated channels were activated upon ER Ca^{2+} depletion in PMF-ECFCs. One channel was activated by $InsP_3$-dependent Ca^{2+} mobilization and was inhibited by YM-58483/BTP-2 and 10 μM La^{3+} and Gd^{3+}. The other channel was stimulated by pharmacological depletion of the ER Ca^{2+} pool with cyclopiazonic acid (CPA), a selective inhibitor of SERCA activity, but was insensitive to Gd^{3+}. The pharmacological profile of the $InsP_3$-sensitive SOCE was compatible with that described in RCC-ECFCs [35], BC-ECFCs [76] and IH-ECFCs [43] and strongly suggests that this pathway is mediated by STIM1, Orai1 and TRPC1. Molecular identity of the CPA-induced, Gd^{3+}-resistant SOCE is far less clear. Nevertheless, it has been shown that Orai1 may interact with TRPC1 and TRPC4 to form a STIM1-regulated channel that is insensitive to Gd^{3+} in a heterologous expression system, such as HEK-293 cells

[100]. Of note, YM-58483/BTP-2 only weakly affected PMF-ECFC proliferation, while Gd^{3+} was ineffective [99]. A subsequent report demonstrated that VEGF is able to trigger intracellular Ca^{2+} oscillations in PMF-ECFCs (Table 40.1), but this oscillatory activity is down-regulated compared to normal ECFCs and does not promote proliferation and tube formation [101]. Furthermore, the dynamic interplay between $InsP_3$-dependent ER Ca^{2+} release and SOCE, which shapes the periodic Ca^{2+} transients, was initiated by TRPC1-mediated Ca^{2+} entry. TRPC1 was, in turn, gated by DAG in a store-independent manner [52]. Notably, 1-oleoyl-2-acetyl-sn-glycerol, a membrane permeable DAG analogue, did not cause any detectable increase in $[Ca^{2+}]_i$ in PB-ECFCs [45]. The data described above can be attributed to the changes observed in the intracellular Ca^{2+} toolkits of differently sourced ECFCs (as shown in Table 40.1).

40.7 Conclusion

ECFCs represent a sharp double-edged sword. They are released in the blood to restore local blood perfusion by regenerating the vascular network injured by an ischemic event. Hence, these cells are regarded as one of the most suitable candidates for regenerative treatment of multiple cardiovascular disorders. On the other hand, ECFCs sustain the early phases of the angiogenic switch in a growing number of solid tumors and provide a promising target for developing alternative anti-angiogenic therapies. Intracellular Ca^{2+} signaling plays a key role in promoting proliferation, tube formation, homing and neovessel formation in both healthy and tumor-derived ECFCs. Therefore, it has been suggested that the Ca^{2+} toolkit could be genetically manipulated to rejuvenate or restore the reparative phenotype of autologous ECFCs [2, 66], or to interfere with tumor vascularization and prevent cancer growth and metastasis [18, 79]. For instance, autologous senescent ECFCs could be engineered to overexpress TRPC3, thereby increasing their sensitivity to VEGF, as reported in umbilical cord blood-derived ECFCs [2, 66, 102]. The molecular components of SOCE can be pharmacologically targeted to either prevent or minimize neovascularization in RCC, BC and IH [18, 79]. A fascinating feature of the Ca^{2+} toolkit expressed by ECFCs resides in its sensitivity to the pathophysiological conditions of the donors. Peripheral blood-derived ECFCs differ from UCB-ECFCs and from tumor-derived ECFCs. For instance, they lack $InsP_3R1$, but are endowed with TRPC3 which boosts their sensitivity to VEGF [48]. Furthermore, RCC-, BC-, IH- and PMF-derived ECFCs display distinct Ca^{2+} signaling machineries, although they all show a reduction in $[Ca^{2+}]_{ER}$, which prevents them from responding to VEGF. Understanding how solid cancers and myeloproliferative disorders, such as PMF, differently alter the Ca^{2+} toolkit in ECFCs will be a challenging field of research. Furthermore, it is predictable that the Ca^{2+} toolkit is also remodeled in ECFCs deriving from individuals affected by cardiovascular disorders, as recently shown in mouse MACs [103].

References

1. Medina RJ, Barber CL, Sabatier F, Dignat-George F, Melero-Martin JM, Khosrotehrani K et al (2017) Endothelial progenitors: a consensus statement on nomenclature. Stem Cells Transl Med 6(5):1316–1320
2. Moccia F, Ruffinatti FA, Zuccolo E (2015) Intracellular Ca(2)(+) signals to reconstruct a broken heart: still a theoretical approach? Curr Drug Targets 16(8):793–815
3. Moccia F, Zuccolo E, Poletto V, Cinelli M, Bonetti E, Guerra G et al (2015) Endothelial progenitor cells support tumour growth and metastatisation: implications for the resistance to anti-angiogenic therapy. Tumour Biol 36(9):6603–6614
4. Banno K, Yoder MC (2018) Tissue regeneration using endothelial colony-forming cells: promising cells for vascular repair. Pediatr Res 83(1–2):283–290
5. Yoder MC (2012) Human endothelial progenitor cells. Cold Spring Harb Perspect Med 2(7):a006692
6. Naito H, Wakabayashi T, Kidoya H, Muramatsu F, Takara K, Eino D et al (2016) Endothelial side population cells contribute to tumor angiogenesis and antiangiogenic drug resistance. Cancer Res 76(11):3200–3210
7. Tasev D, Koolwijk P, van Hinsbergh VW (2016) Therapeutic potential of human-derived endothelial colony-forming cells in animal models. Tissue Eng Part B Rev 22(5):371–382
8. Laurenzana A, Margheri F, Chilla A, Biagioni A, Margheri G, Calorini L et al (2016) Endothelial progenitor cells as shuttle of anticancer agents. Hum Gene Ther 27:784–791
9. Moccia F, Tanzi F, Munaron L (2014) Endothelial remodelling and intracellular calcium machinery. Curr Mol Med 14(4):457–480
10. Noren DP, Chou WH, Lee SH, Qutub AA, Warmflash A, Wagner DS et al (2016) Endothelial cells decode VEGF-mediated Ca^{2+} signaling patterns to produce distinct functional responses. Sci Signal 9(416):ra20
11. Pafumi I, Favia A, Gambara G, Papacci F, Ziparo E, Palombi F et al (2015) Regulation of angiogenic functions by angiopoietins through calcium-dependent signaling pathways. Biomed Res Int 2015:965271
12. Moccia F, Berra-Romani R, Tritto S, Signorelli S, Taglietti V, Tanzi F (2003) Epidermal growth factor induces intracellular Ca^{2+} oscillations in microvascular endothelial cells. J Cell Physiol 194:139–150
13. Munaron L, Fiorio Pla A (2000) Calcium influx induced by activation of tyrosine kinase receptors in cultured bovine aortic endothelial cells. J Cell Physiol 185(3):454–463
14. Yokota Y, Nakajima H, Wakayama Y, Muto A, Kawakami K, Fukuhara S et al (2015) Endothelial Ca^{2+} oscillations reflect VEGFR signaling-regulated angiogenic capacity in vivo. Elife 4:e08817
15. Sameermahmood Z, Balasubramanyam M, Saravanan T, Rema M (2008) Curcumin modulates SDF-1alpha/CXCR4-induced migration of human retinal endothelial cells (HRECs). Invest Ophthalmol Vis Sci 49(8):3305–3311
16. Fiorio Pla A, Gkika D (2013) Emerging role of TRP channels in cell migration: from tumor vascularization to metastasis. Front Physiol 4:311
17. Antoniotti S, Lovisolo D, Fiorio Pla A, Munaron L (2002) Expression and functional role of bTRPC1 channels in native endothelial cells. FEBS Lett 510(3):189–195
18. Moccia F, Poletto V (2015) May the remodeling of the Ca(2)(+) toolkit in endothelial progenitor cells derived from cancer patients suggest alternative targets for anti-angiogenic treatment? Biochim Biophys Acta 1853(9):1958–1973
19. Moccia F, Guerra G (2016) Ca^{2+} Signalling in endothelial progenitor cells: friend or foe? J Cell Physiol 231(2):314–327
20. Rumi E, Pietra D, Pascutto C, Guglielmelli P, Martínez-Trillos A, Casetti I et al (2014) Clinical effect of driver mutations of JAK2, CALR, or MPL in primary myelofibrosis. Blood 124(7):1062–1069

21. Poletto V, Dragoni S, Lim D, Biggiogera M, Aronica A, Cinelli M et al (2016) Endoplasmic reticulum Ca^{2+} handling and apoptotic resistance in tumor-derived endothelial colony forming cells. J Cell Biochem 117(10):2260–2271
22. Moccia F, Berra-Romani R, Tanzi F (2012) Update on vascular endothelial Ca^{2+} signalling: a tale of ion channels, pumps and transporters. World J Biol Chem 3(7):127–158
23. Berra-Romani R, Raqeeb A, Guzman-Silva A, Torres-Jacome J, Tanzi F, Moccia F (2010) Na^{+}-Ca^{2+} exchanger contributes to Ca^{2+} extrusion in ATP-stimulated endothelium of intact rat aorta. Biochem Biophys Res Commun 395(1):126–130
24. Maeng YS, Choi HJ, Kwon JY, Park YW, Choi KS, Min JK et al (2009) Endothelial progenitor cell homing: prominent role of the IGF2-IGF2R-PLCbeta2 axis. Blood 113(1):233–243
25. Zuccolo E, Di Buduo C, Lodola F, Orecchioni S, Scarpellino G, Kheder DA et al (2018) Stromal cell-derived factor-1alpha promotes endothelial colony-forming cell migration through the Ca^{2+}-dependent activation of the extracellular signal-regulated kinase 1/2 and phosphoinositide 3-kinase/AKT pathways. Stem Cells Dev 27(1):23–34
26. Dragoni S, Laforenza U, Bonetti E, Lodola F, Bottino C, Berra-Romani R et al (2011) Vascular endothelial growth factor stimulates endothelial colony forming cells proliferation and tubulogenesis by inducing oscillations in intracellular Ca^{2+} concentration. Stem Cells 29(11):1898–1907
27. Sanchez-Hernandez Y, Laforenza U, Bonetti E, Fontana J, Dragoni S, Russo M et al (2010) Store-operated Ca^{2+} entry is expressed in human endothelial progenitor cells. Stem Cells Dev 19(12):1967–1981
28. Zuccolo E, Dragoni S, Poletto V, Catarsi P, Guido D, Rappa A et al (2016) Arachidonic acid-evoked Ca^{2+} signals promote nitric oxide release and proliferation in human endothelial colony forming cells. Vasc Pharmacol 87:159–171
29. Morgan AJ, Platt FM, Lloyd-Evans E, Galione A (2011) Molecular mechanisms of endolysosomal Ca^{2+} signalling in health and disease. Biochem J 439(3):349–374
30. Di Nezza F, Zuccolo E, Poletto V, Rosti V, De Luca A, Moccia F et al (2017) Liposomes as a putative tool to investigate NAADP signaling in vasculogenesis. J Cell Biochem 118: 3722–3729
31. Wang YW, Zhang JH, Yu Y, Yu J, Huang L (2016) Inhibition of store-operated calcium entry protects endothelial progenitor cells from H2O2-induced apoptosis. Biomol Ther (Seoul) 24(4):371–379
32. Choi JW, Son SM, Mook-Jung I, Moon YJ, Lee JY, Wang KC et al (2017) Mitochondrial abnormalities related to the dysfunction of circulating endothelial colony-forming cells in moyamoya disease. J Neurosurg 129:1–9
33. Kluge MA, Fetterman JL, Vita JA (2013) Mitochondria and endothelial function. Circ Res 112(8):1171–1188
34. Moccia F, Dragoni S, Lodola F, Bonetti E, Bottino C, Guerra G et al (2012) Store-dependent Ca^{2+} entry in endothelial progenitor cells as a perspective tool to enhance cell-based therapy and adverse tumour vascularization. Curr Med Chem 19(34):5802–5818
35. Lodola F, Laforenza U, Bonetti E, Lim D, Dragoni S, Bottino C et al (2012) Store-operated Ca^{2+} entry is remodelled and controls in vitro angiogenesis in endothelial progenitor cells isolated from tumoral patients. PLoS One 7(9):e42541
36. Li J, Cubbon RM, Wilson LA, Amer MS, McKeown L, Hou B et al (2011) Orai1 and CRAC channel dependence of VEGF-activated Ca^{2+} entry and endothelial tube formation. Circ Res 108(10):1190–1198
37. Di Buduo CA, Balduini A, Moccia F (2016) Pathophysiological significance of store-operated calcium entry in megakaryocyte function: opening new paths for understanding the role of calcium in thrombopoiesis. Int J Mol Sci 17(12):2055
38. Jardin I, Lopez JJ, Salido GM, Rosado JA (2008) Orai1 mediates the interaction between STIM1 and hTRPC1 and regulates the mode of activation of hTRPC1-forming Ca^{2+} channels. J Biol Chem 283(37):25296–25304

39. Cheng KT, Liu X, Ong HL, Swaim W, Ambudkar IS (2011) Local Ca(2)+ entry via Orai1 regulates plasma membrane recruitment of TRPC1 and controls cytosolic Ca(2)+ signals required for specific cell functions. PLoS Biol 9(3):e1001025
40. Pani B, Liu X, Bollimuntha S, Cheng KT, Niesman IR, Zheng C et al (2013) Impairment of TRPC1-STIM1 channel assembly and AQP5 translocation compromise agonist-stimulated fluid secretion in mice lacking caveolin1. J Cell Sci 126(Pt 2):667–675
41. Hong JH, Li Q, Kim MS, Shin DM, Feske S, Birnbaumer L et al (2011) Polarized but differential localization and recruitment of STIM1, Orai1 and TRPC channels in secretory cells. Traffic 12(2):232–245
42. Du LL, Shen Z, Li Z, Ye X, Wu M, Hong L et al (2018) TRPC1 deficiency impairs the endothelial progenitor cell function via inhibition of calmodulin/eNOS pathway. J Cardiovasc Transl Res 11:339–345
43. Zuccolo E, Bottino C, Diofano F, Poletto V, Codazzi AC, Mannarino S et al (2016) Constitutive store-operated Ca^{2+} entry leads to enhanced nitric oxide production and proliferation in infantile hemangioma-derived endothelial Colony-forming cells. Stem Cells Dev 25(4):301–319
44. Gees M, Colsoul B, Nilius B (2010) The role of transient receptor potential cation channels in Ca^{2+} signaling. Cold Spring Harb Perspect Biol 2(10):a003962
45. Guerra G, Lucariello A, Perna A, Botta L, De Luca A, Moccia F (2018) The role of endothelial Ca^{2+} signaling in neurovascular coupling: a view from the lumen. Int J Mol Sci 19(4). https://doi.org/10.3390/ijms19040938
46. Hamdollah Zadeh MA, Glass CA, Magnussen A, Hancox JC, Bates DO (2008) VEGF-mediated elevated intracellular calcium and angiogenesis in human microvascular endothelial cells in vitro are inhibited by dominant negative TRPC6. Microcirculation 15(7):605–614
47. Antigny F, Girardin N, Frieden M (2012) Transient receptor potential canonical channels are required for in vitro endothelial tube formation. J Biol Chem 287(8):5917–5927
48. Dragoni S, Laforenza U, Bonetti E, Lodola F, Bottino C, Guerra G et al (2013) Canonical transient receptor potential 3 channel triggers vascular endothelial growth factor-induced intracellular Ca^{2+} oscillations in endothelial progenitor cells isolated from umbilical cord blood. Stem Cells Dev 22(19):2561–2580
49. Vriens J, Appendino G, Nilius B (2009) Pharmacology of vanilloid transient receptor potential cation channels. Mol Pharmacol 75(6):1262–1279
50. Hofmann NA, Barth S, Waldeck-Weiermair M, Klec C, Strunk D, Malli R et al (2014) TRPV1 mediates cellular uptake of anandamide and thus promotes endothelial cell proliferation and network-formation. Biol Open 3(12):1164–1172
51. White JP, Cibelli M, Urban L, Nilius B, McGeown JG, Nagy I (2016) TRPV4: molecular conductor of a diverse orchestra. Physiol Rev 96(3):911–973
52. Dragoni S, Guerra G, Fiorio Pla A, Bertoni G, Rappa A, Poletto V et al (2015) A functional transient receptor potential vanilloid 4 (TRPV4) channel is expressed in human endothelial progenitor cells. J Cell Physiol 230(1):95–104
53. Fleig A, Chubanov V (2014) Trpm7. Handb Exp Pharmacol 222:521–546
54. Baldoli E, Maier JA (2012) Silencing TRPM7 mimics the effects of magnesium deficiency in human microvascular endothelial cells. Angiogenesis 15(1):47–57
55. Tu TC, Nagano M, Yamashita T, Hamada H, Ohneda K, Kimura K et al (2016) A chemokine receptor, CXCR4, which is regulated by hypoxia-inducible factor 2alpha, is crucial for functional endothelial progenitor cells migration to ischemic tissue and wound repair. Stem Cells Dev 25(3):266–276
56. Oh BJ, Kim DK, Kim BJ, Yoon KS, Park SG, Park KS et al (2010) Differences in donor CXCR4 expression levels are correlated with functional capacity and therapeutic outcome of angiogenic treatment with endothelial colony forming cells. Biochem Biophys Res Commun 398(4):627–633
57. Joo HJ, Song S, Seo HR, Shin JH, Choi SC, Park JH et al (2015) Human endothelial colony forming cells from adult peripheral blood have enhanced sprouting angiogenic potential through up-regulating VEGFR2 signaling. Int J Cardiol 197:33–43

58. Su SH, Wu CH, Chiu YL, Chang SJ, Lo HH, Liao KH et al (2017) Dysregulation of vascular endothelial growth factor receptor-2 by multiple miRNAs in endothelial colony-forming cells of coronary artery disease. J Vasc Res 54(1):22–32
59. Koch S, Claesson-Welsh L (2012) Signal transduction by vascular endothelial growth factor receptors. Cold Spring Harb Perspect Med 2(7):a006502
60. Potenza DM, Guerra G, Avanzato D, Poletto V, Pareek S, Guido D et al (2014) Hydrogen sulphide triggers VEGF-induced intracellular Ca^{2+} signals in human endothelial cells but not in their immature progenitors. Cell Calcium 56:225–234
61. Zhu L, Song S, Pi Y, Yu Y, She W, Ye H et al (2011) Cumulated Ca2(+) spike duration underlies Ca2(+) oscillation frequency-regulated NFkappaB transcriptional activity. J Cell Sci 124(Pt 15):2591–2601
62. Zhu LP, Luo YG, Chen TX, Chen FR, Wang T, Hu Q (2008) Ca^{2+} oscillation frequency regulates agonist-stimulated gene expression in vascular endothelial cells. J Cell Sci 121(15):2511–2518
63. Chen ZJ (2005) Ubiquitin signalling in the NF-kappaB pathway. Nat Cell Biol 7(8):758–765
64. Oeckinghaus A, Ghosh S (2009) The NF-kappaB family of transcription factors and its regulation. Cold Spring Harb Perspect Biol 1(4):a000034
65. Ingram DA, Mead LE, Tanaka H, Meade V, Fenoglio A, Mortell K et al (2004) Identification of a novel hierarchy of endothelial progenitor cells using human peripheral and umbilical cord blood. Blood 104(9):2752–2760
66. Moccia F, Lucariello A, Guerra G (2018) TRPC3-mediated Ca^{2+} signals as a promising strategy to boost therapeutic angiogenesis in failing hearts: the role of autologous endothelial colony forming cells. J Cell Physiol 233(5):3901–3917
67. Monteith GR, Prevarskaya N, Roberts-Thomson SJ (2017) The calcium-cancer signalling nexus. Nat Rev Cancer 17(6):367–380
68. Prevarskaya N, Ouadid-Ahidouch H, Skryma R, Shuba Y (2014) Remodelling of Ca^{2+} transport in cancer: how it contributes to cancer hallmarks? Philos Trans R Soc Lond Ser B Biol Sci 369(1638):20130097
69. Moccia F (2018) Endothelial Ca^{2+} signaling and the resistance to anticancer treatments: partners in crime. Int J Mol Sci 19(1):E217
70. Pupo E, Pla AF, Avanzato D, Moccia F, Cruz JE, Tanzi F et al (2011) Hydrogen sulfide promotes calcium signals and migration in tumor-derived endothelial cells. Free Radic Biol Med 51(9):1765–1773
71. Fiorio Pla A, Ong HL, Cheng KT, Brossa A, Bussolati B, Lockwich T et al (2012) TRPV4 mediates tumor-derived endothelial cell migration via arachidonic acid-activated actin remodeling. Oncogene 31(2):200–212
72. Avanzato D, Genova T, Fiorio Pla A, Bernardini M, Bianco S, Bussolati B et al (2016) Activation of P2X7 and P2Y11 purinergic receptors inhibits migration and normalizes tumor-derived endothelial cells via cAMP signaling. Sci Rep 6:32602
73. Fiorio Pla A, Grange C, Antoniotti S, Tomatis C, Merlino A, Bussolati B et al (2008) Arachidonic acid-induced Ca^{2+} entry is involved in early steps of tumor angiogenesis. Mol Cancer Res 6(4):535–545
74. Genova T, Grolez GP, Camillo C, Bernardini M, Bokhobza A, Richard E et al (2017) TRPM8 inhibits endothelial cell migration via a non-channel function by trapping the small GTPase Rap1. J Cell Biol 216(7):2107–2130
75. Moccia F, Fotia V, Tancredi R, Della Porta MG, Rosti V, Bonetti E et al (2017) Breast and renal cancer-derived endothelial colony forming cells share a common gene signature. Eur J Cancer 77:155–164
76. Lodola F, Laforenza U, Cattaneo F, Ruffinatti FA, Poletto V, Massa M et al (2017) VEGF-induced intracellular Ca^{2+} oscillations are down-regulated and do not stimulate angiogenesis in breast cancer-derived endothelial colony forming cells. Oncotarget 8:95223–95246
77. Sunryd JC, Cheon B, Graham JB, Giorda KM, Fissore RA, Hebert DN (2014) TMTC1 and TMTC2 are novel endoplasmic reticulum tetratricopeptide repeat-containing adapter proteins involved in calcium homeostasis. J Biol Chem 289(23):16085–16099

78. Vanoverberghe K, Vanden Abeele F, Mariot P, Lepage G, Roudbaraki M, Bonnal JL et al (2004) Ca^{2+} homeostasis and apoptotic resistance of neuroendocrine-differentiated prostate cancer cells. Cell Death Differ 11(3):321–330
79. Moccia F, Dragoni S, Poletto V, Rosti V, Tanzi F, Ganini C et al (2014) Orai1 and transient receptor potential channels as novel molecular targets to impair tumor neovascularisation in renal cell carcinoma and other malignancies. Anti Cancer Agents Med Chem 14(2):296–312
80. Carmeliet P, Jain RK (2011) Molecular mechanisms and clinical applications of angiogenesis. Nature 473(7347):298–307
81. Greenberger S, Bischoff J (2013) Pathogenesis of infantile haemangioma. Br J Dermatol 169(1):12–19
82. Grzesik P, Wu JK (2017) Current perspectives on the optimal management of infantile hemangioma. Pediatr Health Med Ther 8:107–116
83. Bischoff J (2009) Progenitor cells in infantile hemangioma. J Craniofac Surg 20(Suppl 1):695–697
84. Zuccolo E, Lim D, Kheder DA, Perna A, Catarsi P, Botta L et al (2017) Acetylcholine induces intracellular Ca^{2+} oscillations and nitric oxide release in mouse brain endothelial cells. Cell Calcium 66:33–47
85. Moccia F, Berra-Romani R, Baruffi S, Spaggiari S, Adams DJ, Taglietti V et al (2002) Basal nonselective cation permeability in rat cardiac microvascular endothelial cells. Microvasc Res 64(2):187–197
86. Ma X, Chen Z, Hua D, He D, Wang L, Zhang P et al (2014) Essential role for TrpC5-containing extracellular vesicles in breast cancer with chemotherapeutic resistance. Proc Natl Acad Sci U S A 111(17):6389–6394
87. Plummer PN, Freeman R, Taft RJ, Vider J, Sax M, Umer BA et al (2013) MicroRNAs regulate tumor angiogenesis modulated by endothelial progenitor cells. Cancer Res 73(1):341–352
88. Katoh M (2013) Therapeutics targeting angiogenesis: genetics and epigenetics, extracellular miRNAs and signaling networks (review). Int J Mol Med 32(4):763–767
89. Chang TY, Tsai WC, Huang TS, Su SH, Chang CY, Ma HY et al (2017) Dysregulation of endothelial colony-forming cell function by a negative feedback loop of circulating miR-146a and -146b in cardiovascular disease patients. PLoS One 12(7):e0181562
90. Rosti V, Bonetti E, Bergamaschi G, Campanelli R, Guglielmelli P, Maestri M et al (2010) High frequency of endothelial colony forming cells marks a non-active myeloproliferative neoplasm with high risk of splanchnic vein thrombosis. PLoS One 5(12):e15277
91. Barosi G, Rosti V, Massa M, Viarengo GL, Pecci A, Necchi V et al (2004) Spleen neoangiogenesis in patients with myelofibrosis with myeloid metaplasia. Br J Haematol 124(5):618–625
92. Szuber N, Tefferi A (2018) Driver mutations in primary myelofibrosis and their implications. Curr Opin Hematol 25(2):129–135
93. Mesaeli N, Nakamura K, Zvaritch E, Dickie P, Dziak E, Krause KH et al (1999) Calreticulin is essential for cardiac development. J Cell Biol 144(5):857–868
94. Fasolato C, Pizzo P, Pozzan T (1998) Delayed activation of the store-operated calcium current induced by calreticulin overexpression in RBL-1 cells. Mol Biol Cell 9(6):1513–1522
95. John LM, Lechleiter JD, Camacho P (1998) Differential modulation of SERCA2 isoforms by calreticulin. J Cell Biol 142(4):963–973
96. Klampfl T, Gisslinger H, Harutyunyan AS, Nivarthi H, Rumi E, Milosevic JD et al (2013) Somatic mutations of calreticulin in myeloproliferative neoplasms. N Engl J Med 369(25):2379–2390
97. Pietra D, Rumi E, Ferretti VV, Di Buduo CA, Milanesi C, Cavalloni C et al (2016) Differential clinical effects of different mutation subtypes in CALR-mutant myeloproliferative neoplasms. Leukemia 30(2):431–438
98. Piaggio G, Rosti V, Corselli M, Bertolotti F, Bergamaschi G, Pozzi S et al (2009) Endothelial colony-forming cells from patients with chronic myeloproliferative disorders lack the disease-specific molecular clonality marker. Blood 114(14):3127–3130

99. Dragoni S, Laforenza U, Bonetti E, Reforgiato M, Poletto V, Lodola F et al (2014) Enhanced expression of Stim, orai, and TRPC transcripts and proteins in endothelial progenitor cells isolated from patients with primary myelofibrosis. PLoS One 9(3):e91099
100. Liao Y, Plummer NW, George MD, Abramowitz J, Zhu MX, Birnbaumer L (2009) A role for Orai in TRPC-mediated Ca^{2+} entry suggests that a TRPC:Orai complex may mediate store and receptor operated Ca^{2+} entry. Proc Natl Acad Sci U S A 106(9):3202–3206
101. Dragoni S, Reforgiato M, Zuccolo E, Poletto V, Lodola F, Ruffinatti FA et al (2015) Dysregulation of VEGF-induced proangiogenic Ca^{2+} oscillations in primary myelofibrosis-derived endothelial colony-forming cells. Exp Hematol 43(12):1019–30 e3
102. Moccia F, Dragoni S, Cinelli M, Montagnani S, Amato B, Rosti V et al (2013) How to utilize Ca^{2+} signals to rejuvenate the repairative phenotype of senescent endothelial progenitor cells in elderly patients affected by cardiovascular diseases: a useful therapeutic support of surgical approach? BMC Surg 13(Suppl 2):S46
103. Wang LY, Zhang JH, Yu J, Yang J, Deng MY, Kang HL et al (2015) Reduction of store-operated Ca^{2+} entry correlates with endothelial progenitor cell dysfunction in atherosclerotic mice. Stem Cells Dev 24(13):1582–1590

Chapter 41
Sensing Extracellular Calcium – An Insight into the Structure and Function of the Calcium-Sensing Receptor (CaSR)

Sergei Chavez-Abiega, Iris Mos, Patricia P. Centeno, Taha Elajnaf, Wolfgang Schlattl, Donald T. Ward, Joachim Goedhart, and Enikö Kallay

---

Author contributed equally with all other contributors. Sergei Chavez-Abiega, Iris Mos and Patricia P. Centeno

---

S. Chavez-Abiega
Systems Bioinformatics, Amsterdam Institute for Molecules, Medicines, and Systems, VU University, Amsterdam, The Netherlands

Section of Molecular Cytology, van Leeuwenhoek Centre for Advanced Microscopy, Swammerdam Institute for Life Sciences, University of Amsterdam, Amsterdam, The Netherlands
e-mail: s.chavezabiega@vu.nl

I. Mos
Department of Drug Design and Pharmacology, Faculty of Health and Medical Sciences, University of Copenhagen, Copenhagen, Denmark
e-mail: iris.mos@sund.ku.dk

P. P. Centeno · D. T. Ward
Faculty of Biology Medicine and Health, The University of Manchester, Manchester, UK
e-mail: patricia.pacioscenteno@postgrad.manchester.ac.uk; d.ward@manchester.ac.uk

T. Elajnaf · E. Kallay (✉)
Department of Pathophysiology and Allergy Research, Center of Pathophysiology, Infectiology & Immunology, Medical University of Vienna, Vienna, Austria
e-mail: taha.elajnaf@meduniwien.ac.at; enikoe.kallay@meduniwien.ac.at

W. Schlattl
Computer Science Department, University of Torino; S.A.F.AN. BIOINFORMATICS, Torino, Italy
e-mail: wolfgang.schlattl@unito.it

J. Goedhart
Section of Molecular Cytology, van Leeuwenhoek Centre for Advanced Microscopy, Swammerdam Institute for Life Sciences, University of Amsterdam, Amsterdam, The Netherlands
e-mail: j.goedhart@uva.nl

© Springer Nature Switzerland AG 2020

M. S. Islam (ed.), *Calcium Signaling*, Advances in Experimental Medicine and Biology 1131, https://doi.org/10.1007/978-3-030-12457-1_41

Abstract The calcium-sensing receptor (CaSR) is a G protein-coupled receptor that plays a key role in calcium homeostasis, by sensing free calcium levels in blood and regulating parathyroid hormone secretion in response. The CaSR is highly expressed in parathyroid gland and kidney where its role is well characterised, but also in other tissues where its function remains to be determined. The CaSR can be activated by a variety of endogenous ligands, as well as by synthetic modulators such as Cinacalcet, used in the clinic to treat secondary hyperparathyroidism in patients with chronic kidney disease. The CaSR couples to multiple G proteins, in a tissue-specific manner, activating several signalling pathways and thus regulating diverse intracellular events. The multifaceted nature of this receptor makes it a valuable therapeutic target for calciotropic and non-calciotropic diseases. It is therefore essential to understand the complexity behind the pharmacology, trafficking, and signalling characteristics of this receptor. This review provides an overview of the latest knowledge about the CaSR and discusses future hot topics in this field.

Keywords Extracellular calcium · Parathyroid hormone · G protein-coupled receptor · G proteins · Biased signalling · Calcimimetics · Calcilytics · Allosteric modulators · Orthosteric ligands · Cellular trafficking

Abbreviations

1,25D3	1α,25-dihydroxyvitamin D3
AC	Adenylate cyclase
ADH	Autosomal dominant hypocalcaemia
AP2	Adaptor protein-2
cAMP	Cyclic adenosine monophosphate
Ca^{2+}	Calcium
CaSR	Calcium-sensing receptor
CR	Cysteine rich domain
$[Ca^{2+}]_o$	Extracellular calcium concentration
$Ca^{2+}{}_i$	Intracellular calcium
$[Ca^{2+}]_i$	Intracellular calcium concentration
CKD	Chronic kidney disease
DAG	Diacylglycerol
ECD	Extracellular domain
ER	Endoplasmic reticulum
ERK	Extracellular signal-regulated kinase
FHH	Familial hypocalciuric hypercalcaemia
GABA	Gamma-aminobutyric acid
GAPs	GTPase-activating proteins
GEFs	Guanine nucleotide exchange factors
GDIs	Guanine nucleotide dissociation inhibitors
GPCR	G protein-coupled receptor

GRKs	G protein-coupled receptor kinases
GSK3	Glycogen synthase kinase-3
HEK	Human embryonic kidney
HEK-CaSR	HEK293 cells stably expressing the CaSR
ICD	Intracellular domain
IGF-1	Insulin-like growth factor 1
IP3	Inositol 1,4,5-trisphosphate
JNK	C-Jun amino-terminal kinases
mGlu	Metabotropic glutamate receptor
NAM	Negative allosteric modulator
LB	Lobe-shaped domain
MAPKs	Mitogen-activated protein kinases
NAM	Negative allosteric modulator
NSHPT	Neonatal severe hyperparathyroidism
NKCC2	Na-K-Cl cotransporter 2
PA	Phosphatidic acid
PAM	Positive allosteric modulator
PDEs	Phosphodiesterases
Pi	Inorganic phosphate
PI3Ks	Phosphoinositide 3-kinases
PIP2	Phosphatidylinositol 4,5-bisphosphate
PKA	Protein Kinase A
PKB	Protein Kinase B
PKC	Protein Kinase C
PLA2	Phospholipase A2
PLC	Phospholipase C
PLD	Phospholipase D
PreProPTH	Prepro-parathyroid hormone
PT	Parathyroid
PTH	Parathyroid hormone
PTHrP	Parathyroid hormone-related protein
PTx	Pertussis-toxin
RGS	Regulator of G protein signalling
RAMPs	Receptor activity-modifying proteins
TMD	Transmembrane domain
TAS1R	Taste 1 receptors
VFD	Venus flytrap domain

41.1 Introduction

Calcium (Ca^{2+}) is a macro element representing 1.5–2% of an adult's total body weight and is mostly found in bones and teeth. Only 1% of the body's Ca^{2+} is located in cells and tissues, where it regulates numerous critical cellular responses.

The extracellular calcium concentration $[Ca^{2+}]_o$ is much higher (20,000-fold) than in the cytosol and changes in this balance trigger various signalling pathways. This gradient allows Ca^{2+} to act as a second messenger in intracellular signalling [1]. Many tissues are equipped with a cell-surface sensor for Ca^{2+} that extends the signalling properties of Ca^{2+} to being an extracellular first messenger also. This receptor is known as the extracellular calcium-sensing receptor (CaSR) and is a G protein-coupled receptor (GPCR).

CaSR is a member of the class C GPCRs which also includes the metabotropic glutamate (mGlu) receptors, the gamma-aminobutyric acid (GABA) receptors, the taste 1 receptors (TAS1R) and 8 orphan receptors [2]. The class C orphan receptor GPRC6 shares the highest sequence similarity with the CaSR (Fig. 41.1).

The CaSR plays an essential role in calcium homeostasis and its existence was confirmed by cloning in 1993 [3]. CaSR senses changes in $[Ca^{2+}]_o$ and also interacts with other multivalent cations (Mg^{2+}, Gd^{3+}), organic cations (neomycin), polyamines (spermine, polyarginine and polylysine) and possibly even beta amyloid [4]. It is highly expressed in the parathyroid glands, pancreas, duodenum and kidney and less in the digestive system, stomach and respiratory system [5]. This review will give an overview of the (patho)physiological roles, structure, ligands, trafficking, signalling pathways and tissue specific functions of the CaSR.

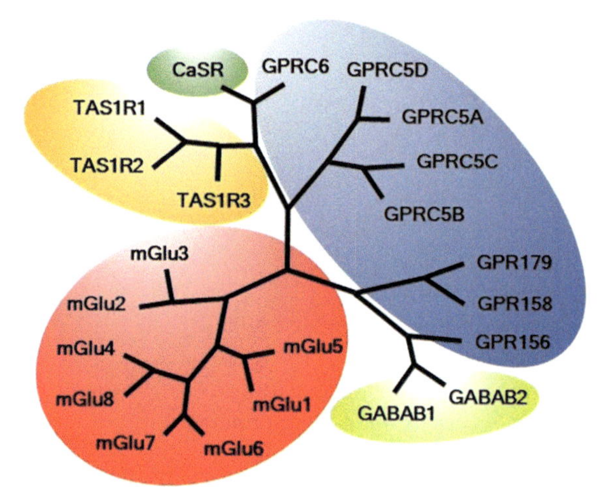

Fig. 41.1 Phylogenetic tree of the class C GPCRs generated with the neighbour-joining method of the full receptor sequences; The different colors represent the receptor families based on the endogenous ligand affiliation: CaSR family: calcium-sensing receptor (green), GPR family: class C orphan receptors with unknown endogenous ligands (blue), GABA family: gamma-aminobutyric acid receptors (yellow), mGlu family: metabotropic glutamate receptors (red), TAS1R family: taste 1 receptors (orange)

41.1.1 Physiological Role of the CaSR in Calcium Regulation

The blood $[Ca^{2+}]_o$ is influenced by parathyroid hormone (PTH), 1α,25-dihydroxyvitamin D_3 (1,25D_3) and calcitonin [4, 6]. PTH is expressed by and secreted from chief cells of the parathyroid gland and acts upon the kidneys, bones, and intestines. In the kidney, CaSR stimulates Ca^{2+} and Mg^{2+} reabsorption in the distal tubules whereas in the proximal tubules it promotes the excretion of hydrogen phosphate and dihydrogen phosphate. PTH induces the 25-hydroxyvitamin D_3 1α-hydroxylase to produce 1,25D3 which is essential for intestinal Ca^{2+} absorption [7]. Hypercalcaemia is prevented by CaSR, which inhibits PTH secretion and suppresses the transcription of PreProPTH and cell proliferation (Fig. 41.2) [8]. PTH stimulates bone remodelling. Calcitonin protects against hypercalcaemia and inhibits osteoclast activity and consequently the release of Ca^{2+} from bones. The transcription of calcitonin in thyroidal C-cells is inhibited by increasing 1,25D3 concentrations [6]. However, the impact of calcitonin in maintaining systemic

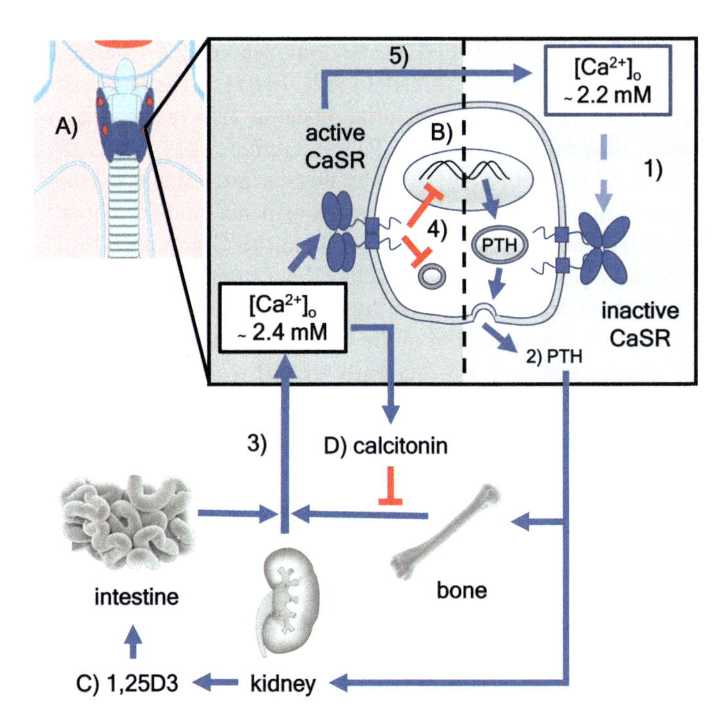

Fig. 41.2 Overview of the calcium homeostasis. (**a**) Location of the parathyroid glands (red dots); (**b**) Chief cell of the parathyroid gland, 1) CaSR is inactive at low $[Ca^{2+}]_o$ and PTH is secreted; 2) PTH stimulates Ca^{2+} release from bone, the reabsorption from the kidney and the 1,25D_3 synthesis which induces Ca^{2+} uptake from the intestine (**c**); 3) The resulting increase in blood Ca^{2+} activates CaSR and at high concentrations calcitonin secretion; (**d**) Calcitonin inhibits the osteoclast activity and transiently the Ca^{2+} release from bone; 4) The active CaSR inhibits the PTH expression and secretion and consequently lowers the blood Ca^{2+} level; 5) Until the process starts again at 1)

blood Ca^{2+} is still contradictory because its absence or excess does not result in any significant metabolic abnormalities.

The physiological range of serum $[Ca^{2+}]_o$ is tightly regulated between 2.2 and 2.4 mM by the CaSR, facilitated by the high cooperativity of Ca^{2+} on the receptor. About half of the $[Ca^{2+}]_o$ is free, and the rest of it is bound mainly to albumin. PTH secretion is induced when the free $[Ca^{2+}]_o$ drops below 1.2 mM (to ~2.2 mM total Ca^{2+}) and it is effectively suppressed by CaSR activation when free $[Ca^{2+}]_o$ rises above 1.2 mM (towards 2.5 mM) [9]. High free $[Ca^{2+}]_o$ activates renal CaSR leading to inhibition of Ca^{2+} reabsorption resulting in elevated renal Ca^{2+} excretion [10, 11].

41.1.2 Pathophysiological Role of the CaSR

After the successful cloning of the bovine parathyroid CaSR [3], a number of diseases were identified which are caused by CaSR mutations.

Heterozygous loss-of-function mutations in CaSR are associated with familial hypocalciuric hypercalcaemia (FHH1) and homozygous CaSR mutations to neonatal severe hyperparathyroidism (NSHPT) [12]. FHH1 is characterised by disabled Ca^{2+} reabsorption causing hypocalciuria, moderate hypercalcaemia, hypermagnesaemia and a disabled inhibition of PTH secretion which leads to an elevated steady-state $[Ca^{2+}]_o$ level [4, 13]. Usually this disease remains asymptomatic over one's lifetime, but a few patients show signs of pancreatitis or chondrocalcinosis. FHH2 and FHH3 are the result of mutations in the G protein GNA11 and APS1 gene, respectively [13], these proteins acting downstream of CaSR signalling.

NSHPT is characterised by severe hypercalcaemia and very high PTH levels. The defective feedback regulation of the CaSR leads to bone demineralisation and to pathological fractures [4]. It is currently treated with bisphosphonates, dialysis, calcimimetics or by total parathyroidectomy [13, 14].

In contrast, autosomal dominant hypocalcaemia type 1 (ADH1) and type 2 (ADH2) are caused by gain-of-function mutations of the CaSR and the GNA11 gene, respectively. ADH1 results in a reduced steady-state of blood $[Ca^{2+}]_o$ and causes low PTH levels, hypercalciuria, hypomagnesaemia, and hyperphosphataemia. Symptoms of type 1 are paraesthesia, tetany, epilepsy, severe hypocalcaemia and basal ganglia calcification which are the same for ADH2 but without hypercalciuria and hypomagnesaemia [12]. Another gain-of-function disease is connected to a renal salt-wasting form called Bartter Syndrome type-5. It is the result of unrestrained CaSR activity which leads to dysfunctional Na-K-Cl cotransporter (NKCC2)-dependent NaCl reabsorption [4].

Autoimmune diseases of the CaSR have also been described in rare cases due to the presence of anti-CaSR antibodies. These antibodies can have CaSR-stimulating or CaSR-blocking effects causing a form of acquired autoimmune hypoparathyroidism or autoimmune hypocalciuric hypercalcaemia, respectively [13].

Mutations of the CaSR are also observed in a variety of non-calciotropic diseases, for example the R990G variant is associated with an elevated risk for hypercalciuria and nephrolithiasis [15]. Other diseases are connected to changed expression levels of the receptor. In colorectal and parathyroid cancer CaSR expression is decreased or lost, attenuating its tumour preventive effect. In breast and prostate tumours, CaSR is overexpressed which correlates with an increasing risk for metastases to the bone [16]. There is also evidence that changes in CaSR activity or expression are associated with alterations in cardiac function, insulin secretion, postprandial blood glucose regulation, lipolysis and inhibition of myocardial cell proliferation. In the digestive tract, CaSR shows anti-inflammatory, anti-secretory, pro-absorbent, and obstructive properties while in the respiratory tract CaSR activation is associated with inflammation and nonspecific hyperresponsiveness in asthma [13, 17].

41.2 Structure of the CaSR

The CaSR functions as a disulphide-tethered homodimer composed of three main domains: an extracellular domain (ECD), a heptahelical transmembrane domain (TMD) and an intracellular C-terminal domain (ICD) [18] (Fig. 41.3).

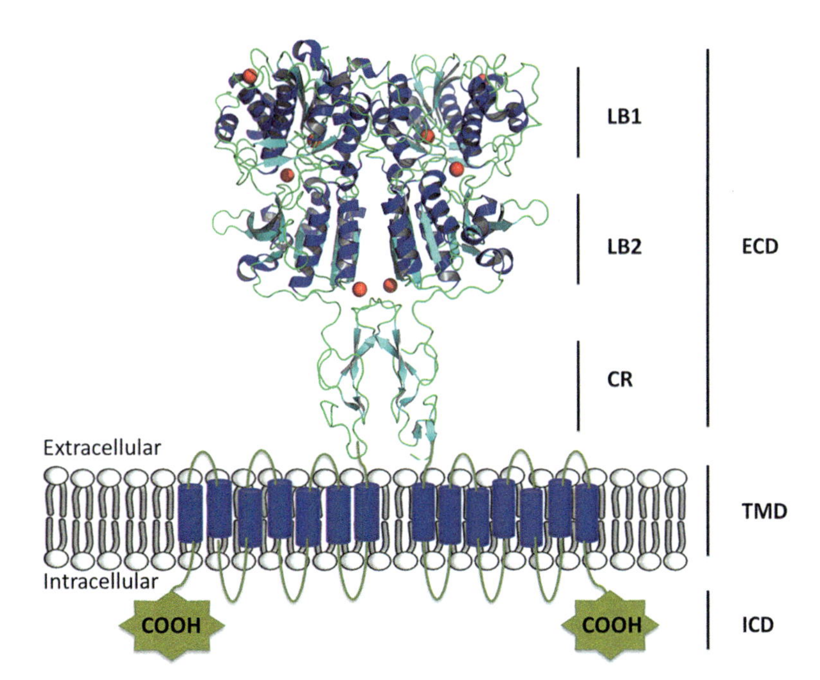

Fig. 41.3 Structure of the CaSR including the crystal structure of the ECD from the human CaSR, formed by LB1, LB2 and CR (PDB: 5k5s), and a schematic representation of the TMD formed by the seven transmembrane helixes followed by the ICD. Calcium ions are represented as red spheres in the ECD

The human CaSR ECD contains 612 amino acids and consists of two lobe-shaped domains (LB1 and LB2) that form the N-terminal Venus Fly Trap (VFT) domain, and the cysteine-rich (CR) domain [19]. The VFT is the ligand-binding region, reminiscent of gram-negative bacterial periplasmic binding proteins [20]. Both LB domains are formed by typical β-sheets and α-helices, where the central parallel β-sheets are sandwiched by α-helices [21]. The CR region, located between the ECD and the TMD contains nine-conserved cysteines. It transmits and amplifies signals from the VFT domain to the intracellular loops of the TMD [22]. The CR region is present in all class C GPCRs except in GABA B receptors and is required for receptor activation [21–23].

CaSR is expressed on the cell surface as a homodimer formed by direct interactions involving the ECD and the TMD. The ECDs of both monomers interact in a side-by-side fashion by a covalent disulphide bridge involving residues Cys-129 and Cys-131, whereas the TMDs establish hydrophobic interactions between them [24, 25]. However, there is also evidence suggesting heterodimerisation with other class C GPCRs. These heterodimers are considered new types of receptors that lead to changes in CaSR expression, signalling and sensitivity. For instance, CaSR may form dimers with mGlu1a or mGlu5, in hippocampal and cerebellar neurons, and with GABA B receptors [26, 27].

The CaSR ECD also includes 20–40 kDa of either high mannose or complex carbohydrates. These glycosylations are believed to be important for cell-surface localisation of the CaSR, intracellular trafficking, protein folding and secretion [25].

Recently, two different groups have simultaneously resolved the crystal structures of the human CaSR ECD in resting and active conformations [28, 29]. Zhang and partners crystallised the ECD in the active conformation and identified two Ca^{2+} binding sites plus an additional orthosteric binding site for L-Trp. The Ca^{2+} binding sites can also be occupied by other divalent metals such as Mg^{2+} whereas, the additional orthosteric binding site was occupied by a L-Trp derivate, L-1,2,3,4,-tetrahydronorharman-3-carboxylic acid and it was located in the hinge region between the two subdomains. The L-Trp binding site was described as crucial for receptor activation and stabilisation of the active conformation [29]. Meanwhile, Geng and colleagues crystallised the receptor in its active and inactive conformations. The active structure was obtained in the presence of 10 mM Ca^{2+} and 10 mM L-Trp, when the receptor is in its closed conformation (active state, closed-closed) (Fig. 41.4 left). They also identified the same orthosteric binding site described by Zhang and partners, located in the ligand-binding cleft of each protomer and also occupied by L-Trp. In this model, the authors defined four different Ca^{2+}-binding sites in the active structure, including one Ca^{2+}-binding site in each protomer that is common in the active and inactive structures suggesting an integral part of the receptor. On the other hand, the inactive CaSR ECD structure was obtained in the presence and absence of 2 mM Ca^{2+}, when the receptor is in open conformation (inactive state, open-open) and the interdomain cleft is empty (Fig. 41.4 right). In this model, they also defined three anion-binding sites. The authors proposed a CaSR activation model where L-Trp facilitates the CaSR-ECD closure by contacting LB1 and LB2 domains of the VFT module to bring the CR domains

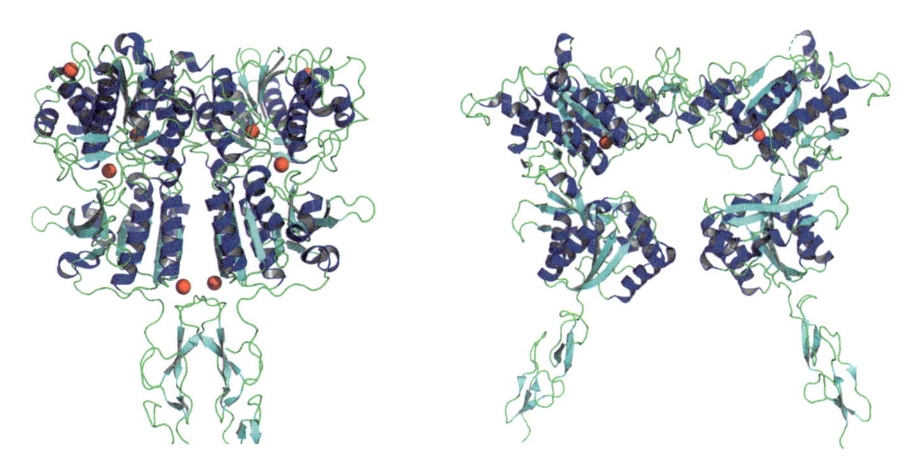

Fig. 41.4 Crystal structure of the human CaSR ECD in its active (left) and inactive (right) conformations. In the active conformation the VFT is closed and LB1 and LB2 interact to bring the CR domains closer together. In the inactive conformation the VFT is open and the interactions between LB1 and LB2 are minimal, therefore the CR domains do not interact. Calcium ions are shown as red spheres. PDB accession numbers: 5k5s and 5k5t [28]

closer together. These interactions form a large homodimer interface that is unique for the active state, reduce the distance between the C-terminal tails and might cause a rearrangement of the TMD [28] (Fig. 41.4).

Results from both groups suggest that the CaSR follows a universal activation mechanism similar for all class C GPCRs, despite the low sequence similarity (20–30%) [30]. This mechanism can be summarised in three steps. First, agonist binding causes the closure of the VFT. Second, membrane-proximal domains associate forming a homodimer interface between LB2 and CR domains. Third, agonist binding is accompanied by an approach between the C-terminal ends of ECDs of both protomers suggesting rearrangement of the TMD [31].

The ICD allows accurate receptor-specific control of diverse downstream signalling pathways. It represents the most diverse region of the class C GPCRs and determines selectivity of CaSR by coupling to different G proteins through the intracellular loops [32]. The ICD is exposed to the cytoplasm and begins with Lys-863 [33]. The amino acid sequence of the ICD is well conserved among species, although amino acids in the C-terminal tail are quite diverse [32]. Until now, two residues (Phe-706 and Leu-703) in intracellular loop two and eight residues (including Leu-797 and Phe-801) in intracellular loop three have been shown to be important for activating phospholipase C (PLC), the major pathway of the CaSR intracellular signalling [34]. Furthermore, there are several well-defined phosphorylation sites, especially Thr-888, in the ICD that are important for protein kinase C (PKC)-dependent inhibition of CaSR [22, 35, 36]. Also, this inhibitory effect may be counteracted by a protein phosphatase (most likely PP2A) that dephosphorylates Thr-888, restoring CaSR responsiveness [37].

41.3 CaSR Modulation

GPCRs can recognise diverse extracellular stimuli and are one of the most successful pharmaceutical target classes for different disorders. The ligands for GPCRs are typically polypeptides, amino acids and/or other small biological molecules that bind in well-defined pockets [38]. The CaSR, as a multifaceted receptor, is able to bind a broad range of molecules in addition to Ca^{2+}, its primary ligand. CaSR modulators can be divided into two groups: type I or orthosteric modulators, which bind to the active site, and type II or allosteric modulators, which bind elsewhere in the receptor.

41.3.1 Orthosteric Modulators of the CaSR

Orthosteric modulators are type I CaSR agonists and include all ligands that are thought to compete with Ca^{2+} for the same binding sites on the receptor. In addition, they are sufficient to activate the CaSR on their own, in the absence of Ca^{2+}.

Although Ca^{2+} is crucial for CaSR function, many other organic cations can activate CaSR in vitro for instance the divalent cations Mg^{2+} and Sr^{2+} and trivalent cations such as Gd^{3+}, as well as heavy metals such as Pb^{2+} and Co^{2+} which are more potent than Ca^{2+} [39]. In fact, the order of agonist potency for inositol metabolism in bovine parathyroid cells depends on two factors; the charge of the ion and the ionic radii. Thus, among ions with the same charge, those with greater radius have a greater potency and among ions of a similar size those with greater charge have a greater potency [40]. The order of potency for the main orthosteric modulators is as follows: $Gd^{3+} > La^{3+} > Ca^{2+} = Ba^{2+} > Sr^{2+} > Mg^{2+}$ [39]. Many organic polycations, such as the poly-amino acids poly-L-lysine or poly-arginine and aminoglycoside antibiotics such as neomycin, are also orthosteric modulators of the CaSR [41–43]. Polyamines produced in the gut and in the synaptic cleft in vivo, are also CaSR agonists. Spermine is the most potent polyamine followed by spermidine and putrescine. In this case, potency is linked to the number of amine groups in the ligand [44].

41.3.2 Allosteric Modulators of the CaSR

In addition to orthosteric agonists, the CaSR can also be activated by allosteric modulators, sometimes referred to as type II CaSR agonists, and these do not compete for the same binding sites as Ca^{2+}, instead they allosterically modify the endogenous affinity of the receptor for $Ca^{2+}{}_o$ [45]. The allosteric modulators affect the conformational equilibrium of the receptor and they can be divided into two groups: activators or positive allosteric modulators (PAM) if they shift the equilibrium towards the active state, and inhibitors or negative allosteric modulators (NAM), if they stabilise the inactive state. Both types of modulators include compounds that can be found in the body under physiological conditions like L-aromatic amino acids, glutathione, ionic strength and alkalinisation [46], but also synthetic compounds like calcimimetic drugs [47].

L-aromatic amino acids were the first endogenous PAMs identified and these include L-Phe, L-Tyr, L-His and L-Trp, with the short aliphatic amino acids L-Thr and L-Ala also effective [48]. L-amino acids increase CaSR sensitivity in the presence of other agonists, such as Ca^{2+} or Gd^{3+}. This demonstrates that CaSR is able to sense a broad range of nutrients having special relevance in the gastrointestinal tract where the CaSR has been identified as an L-amino acid sensor for macronutrient-dependent hormone secretion [49, 50]. In addition, increased aromatic L-amino acid concentration suppresses PTH secretion stereoselectively by activating endogenous CaSR [51]. Therefore, L-amino acids may play an important role as physiological regulators of PTH secretion and calcium metabolism via CaSR modulation.

Interestingly, pH and ionic strength play a double role in modulating CaSR sensitivity. In the case of pH, CaSR sensitivity can be enhanced when pH is elevated (>7.5), but also reduced when pH is low (<7.3) [52]. Decreasing blood pH by only 0.2–0.4 units significantly increases PTH secretion, suggesting a functionally less active CaSR [53]. In contrast, moderate alkalinisation equivalent to that seen in metabolic alkalosis significantly inhibits PTH secretion independently of a change in $[Ca^{2+}]_o$, suggestive of a more sensitive CaSR [54]. This has been confirmed in vitro, whereby small pathophysiologic pH changes (0.2 units) significantly inhibit CaSR-induced intracellular calcium $Ca^{2+}{}_i$ mobilisation (and also extracellular signal-regulated kinase (ERK1/2) phosphorylation and actin polymerisation [55]) in CaSR-HEK cells and in bovine parathyroid cells [34]. Similarly, increasing the ionic strength of the surrounding buffer can also reduce CaSR sensitivity, whereas reducing the buffer's ionic strength enhances CaSR sensitivity [46]. This suggests that protons and Na^+ can both act as NAMs of the CaSR.

41.3.3 Synthetic Modulators of the CaSR

Over the last 20 years, scientists have been looking for drugs to alleviate pathological abnormalities in plasma PTH and Ca^{2+} levels. As the secretion of PTH is mainly regulated by CaSR, compounds that affect this receptor are good candidates to treat PTH disorders. Thus, new synthetic allosteric modulators with higher potency and specificity have been developed.

Nemeth and colleagues at NPS Pharmaceuticals Inc. successfully identified two small organic molecules that caused a leftward shift in the concentration-response curve of the CaSR for $[Ca^{2+}]_o$. They named them calcimimetics. These compounds are able to potentiate the effects of $[Ca^{2+}]_o$ probably by stabilising the active conformation of the receptor by binding to the TMD [56–58]. Calcimimetics are considered type II CaSR agonists and most of them are phenylalkylamines and derivatives of Ca^{2+} channel blockers [57, 59]. Some Ca^{2+} channel blockers can also activate the CaSR, worsening the effects in pulmonary arterial hypertension [60].

Cinacalcet, a calcimimetic molecule more easily absorbed than the initially identified analogue NPS R-568, was the first PAM acting on a GPCR to receive FDA approval and enter the clinic. It represents a targeted therapy for the treatment of disorders linked to hyperparathyroidism, including chronic kidney disease (CKD), life-threatening NSHPT, and parathyroid carcinoma [61–63]. In patients with end-stage CKD, treatment with Cinacalcet lowers PTH levels after 2–4 h [64]. However, calcimimetics can evoke significant side effects including adverse gastrointestinal effects, due to the fact that the CaSR is expressed in many other tissues, where it activates different signalling pathways [64, 65]. Apart from nausea, the main side effect of Cinacalcet is hypocalcaemia [66]. Recently, a new peptide calcimimetic called Etelcalcetide (Parsabiv) has just received FDA approval for the treatment of secondary hyperparathyroidism in adult haemodialysis patients with CKD [66, 67]. Other calcimimetics in use either as research tools or as potential clinical agents include Calindol (AC265347) and Velcalcetide (AMG416) [68].

In contrast, synthetic CaSR NAMs called calcilytics have opposite effects to calcimimetics. Calcilytics include the substituted phenyl-O-alkylamine NPS 2143 and NPS 89636 [56]. Their binding site is located within the CaSR TMD and is partly overlapping with the calcimimetic binding site. Two other structural types of compounds, amino alcohols (e.g. Ronacaleret) and quinazolinones (e.g. ATF936), were identified by high-throughput screening and shown to reduce CaSR affinity for $[Ca^{2+}]_o$ [56, 69]. As calcilytics can increase endogenous PTH secretion by inhibiting CaSR they were initially developed to treat osteoporosis by delivering endogenous, anabolic pulses of PTH but they had insufficient efficacy [65]. Currently, calcilytics are being studied in different drug repurposing projects, including asthma and other lung-related diseases [70].

41.4 CaSR Trafficking

Receptor trafficking plays a critical role in GPCR activity through tight regulation of GPCR expression levels at the cellular surface. This regulation can be divided into two opposing routes: (1) trafficking of newly synthesised GPCRs to the cellular surface (i.e. exocytic trafficking) and (2) removal of GPCRs from the cell surface to intracellular compartments (i.e. endocytic trafficking) [71]. This section will focus on the processes and interacting partners involved in CaSR trafficking.

41.4.1 From Protein Synthesis to the Cellular Surface

To initiate downstream signalling, a GPCR is required to be present at the cellular surface where the agonist binding site is accessible to ligand stimulation and its intracellular part can interact with G proteins or other binding partners [71–73]. The outward motion of newly synthesised GPCRs to the cellular surface is driven by exocytic receptor trafficking. In this section, the term exocytic trafficking will be used in the broadest sense to refer to protein synthesis, protein maturation and the transport of newly synthesised GPCRs from the endoplasmic reticulum (ER) and Golgi system to the cellular surface [74].

To date, the processes and binding partners involved in exocytic CaSR trafficking are poorly understood. In humans, the gene that encodes for CaSR is located on chromosome 3q13.3-21 [75]. The CaSR protein is transcribed from six out of the eight mapped exons in this gene and transcription can be initiated from two different promoter sites (i.e. promoter P1 or P2) [76, 77]. In an investigation into the regulation of CaSR transcription, Canaff and Hendy have identified functional vitamin D and NF-κB response elements within both promoters of the CaSR gene [78, 79]. In agreement with these findings, vitamin D and several proinflammatory cytokines have been reported to upregulate rodent and human CaSR expression [79–82].

Correct protein folding and protein maturation through post-translational modifications are essential for cell-surface targeting of GPCRs. Protein folding into the GPCR's functional three-dimensional conformation is assisted by chaperones. To date, a large number of chaperones or GPCR-interacting proteins with chaperone function have been identified, but none of these proteins have been associated with CaSR folding [72, 83]. CaSR maturation involves extensive N-linked glycosylation in the ECD. A total of 11 potential N-linked glycosylation sites have been identified in the CaSR protein. Glycosylation of at least three sites have been found crucial for cell-surface expression. Moreover, western blot analyses of cell lysates containing CaSR demonstrate immunoreactive bands at approximately 140–160 kDa corresponding to immature monomeric CaSR and fully mature monomeric CaSR respectively [25, 84–86].

As mentioned earlier, the CaSR predominantly exists on the cell surface as a homodimer, but with evidence suggesting potential heterodimerisation with other class C GPCR members including mGlu and GABA B receptors [26, 87]. The CaSR homodimerisation process takes place in the ER and is directed by the formation of disulphide linkages and non-covalent interactions at the dimer interface, as confirmed by the recently resolved crystal structures of the CaSR ECD [28, 29].

Protein synthesis and maturation are strictly regulated by the cell to ensure that only correctly folded and fully matured GPCRs are targeted for trafficking towards the cellular surface. This quality control system is proposed to be regulated by GPCR-interacting proteins such as the previously mentioned chaperones as well as by recognition of conserved retention or export motifs [71, 83]. Bouschet et al. have investigated the involvement of receptor activity-modifying proteins (RAMPs) in exocytic CaSR trafficking. According to Bouschet et al., CaSR interaction with RAMP subtype 1 or 3 facilitates delivery to the cellular surface [88]. This view is supported by Desai and co-workers who demonstrated direct interactions between CaSR and both RAMP subtypes at the cellular surface using FRET-based stoichiometry [89]. The CaSR-interacting protein dorfin mediates ER-associated degradation of the receptor, while filamin A, another interacting protein, protects CaSR from degradation [90–93]. Furthermore, an extended phosphorylation-regulated arginine-rich region was identified in the carboxyl terminus of CaSR which has been shown to be involved in intracellular retention through interaction with 14-3-3 proteins [94–97].

In general, the number of receptors expressed at the cellular surface influences the magnitude of downstream signalling responses. Multiple studies have demonstrated that differences in cell surface expression levels influence CaSR-mediated signalling. Cell surface expression levels of CaSR can be influenced by multiple factors [84, 98, 99]. First, CaSR expression is affected by numerous naturally occurring mutations and polymorphisms. Interestingly, cell surface expression levels of most CaSR mutants could be effectively rectified towards wild-type expression levels upon treatment with calcimimetics or calcilytics [90, 94, 100–102]. Second, phosphorylation at residue Ser-899, a protein kinase A (PKA) phosphorylation site located next to the extended arginine-rich region, has been reported to increase CaSR surface localisation by disruption of 14-3-3 protein binding [94, 95]. Third, a novel trafficking mechanism, referred to as agonist-driven insertional signalling, has been proposed to regulate cell surface expression in response to CaSR activation. According to this mechanism, agonist binding promotes an increase in the forward trafficking of newly synthesised CaSR to the cellular surface from a consistently present intracellular CaSR pool [95].

41.4.2 From the Cellular Surface to Protein Degradation

Endocytic receptor trafficking, also commonly referred to as receptor endocytosis or receptor internalisation, regulates the duration and magnitude of GPCR-activated G protein signalling responses by effective removal of GPCRs from the cellular surface. Besides its crucial role in the termination of GPCR activity, multiple studies have linked receptor endocytosis to the initiation of non-canonical G protein-independent signalling pathways [71, 103]. Endocytic trafficking of CaSR was described to play a role in parathyroid hormone-related protein (PTHrP) secretion and ERK1/2 activation [104, 105].

The molecular mechanism underlying endocytic CaSR trafficking is still poorly investigated. The reported experimental data is rather controversial, and there is no general agreement about the endocytic trafficking route of CaSR. One of the main findings related to CaSR endocytosis is the ability to initiate endocytic trafficking independently of ligand activation [95, 104, 106]. Pi and colleagues stated that phosphorylation preferentially by GRK4 promotes ß-arrestin binding

[107]. However, Lorenz et al. argue that phosphorylation by PKC rather than GRKs mediates ß-arrestin recruitment [108]. This disagreement could potentially be linked to receptor origin as Pi et al. measured desensitisation of rat CaSR while the studies of Lorenz et al. were conducted with human CaSR.

The internalised CaSR can be either recycled or degraded [95, 104, 106, 109]. Similarly as for the exocytic trafficking pathway, the endocytic trafficking pathway is strongly regulated by GPCR-interacting proteins and conserved motifs [103, 110, 111]. In 2012, a presumed internalisation motif linked to lysosomal degradation has been discovered at the CaSR carboxyl terminus [106]. Interestingly, this motif shows an overlap with the filamin A binding site indicating that filamin A might be involved in both exocytic and endocytic trafficking of the CaSR [91]. This hypothesis is supported by the finding that filamin A contributes to the localisation of CaSR to caveolae, a specialised cell membrane region known to be involved in clathrin-independent endocytosis [92, 112, 113]. Furthermore, the CaSR-interacting protein AMSH-1 (associated molecule with the SH3 domain of STAM) has been reported to promote ubiquitin-mediated degradation of internalised CaSR [104, 114].

The ability to rectify the expression of disease-related mutants by calcimimetics and calcilytics highlights the therapeutic potential of modulating endocytic CaSR trafficking in the treatment of CaSR-related diseases. However, further research is needed to fully understand the molecular mechanism underlying CaSR trafficking and its potential as therapeutic target.

41.5 Overview of Signalling Pathways Activated by the CaSR

This section will focus on the signalling pathways mediated by the CaSR, with special attention to the diversity of responses elicited upon activation of the receptor in different tissues. Figure 41.5 shows a simplified overview of what is known to date about CaSR signalling.

41.5.1 G Proteins Activated by the CaSR

G proteins, also called guanine nucleotide-binding proteins, can bind the guanine nucleotides GDP and GTP. GTP-bound, active G proteins have GTPase activity, hydrolysing GTP into GDP and inorganic phosphate (Pi), returning the G protein to its inactive GDP-bound form. The equilibrium of GDP- and GTP-bound forms of the G proteins is a result of the activities of three groups of molecules. Guanine nucleotide exchange factors (GEFs) activate the G proteins by exchanging GDP for GTP. GTPase-activating proteins (GAPs) accelerate the GTPase activity of the G protein and thus terminate its activity. Guanine nucleotide dissociation inhibitors (GDIs) bind GDP-bound G proteins and inhibit activation by the GEFs. G proteins

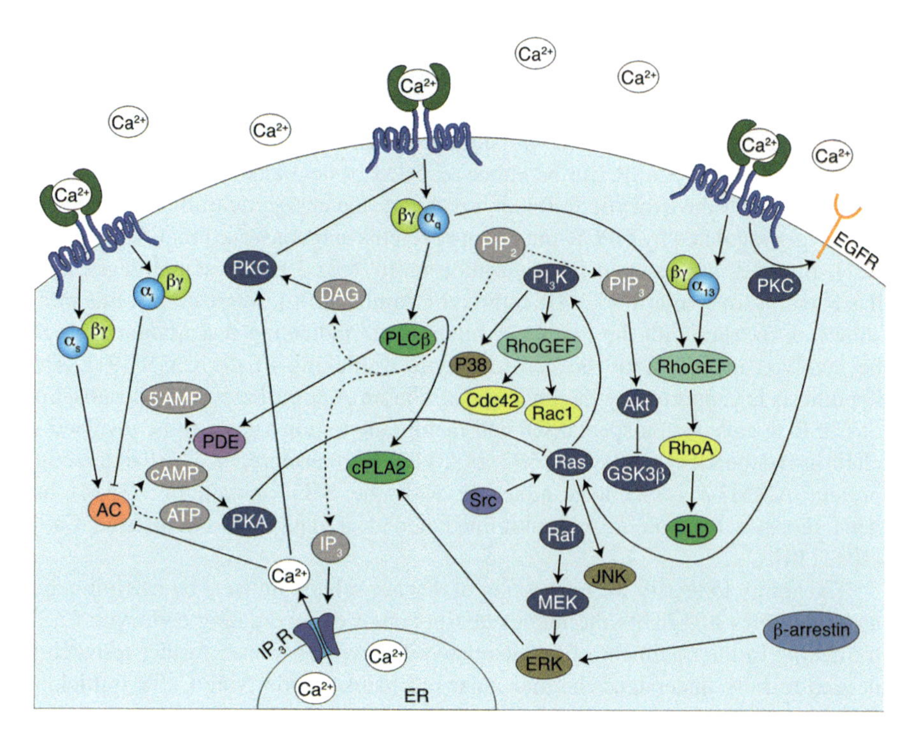

Fig. 41.5 General representation of the signalling pathways activated by CaSR. Arrows, bar-headed lines, and dashed arrows represent activation, inhibition, and a chemical reaction respectively. In green and blue, a CaSR homodimer; in light blue and yellow, the α and βγ subunits of the G proteins; in dark blue, kinases; in light green, phospholipases; in light brown, MAP kinases; in light yellow, Rho GTPases; in grey, second messengers and derivatives

can be either heterotrimeric or monomeric. In this section we will refer to them as "G proteins" and "small GTPases" respectively.

G proteins are heterotrimers composed of α, β, and γ subunits. To date, a total of 21 α, 6 β, and 12 different γ subunits have been identified in humans. G proteins are activated by GPCRs, such as the CaSR, which act as GEFs. The exchange of GDP for GTP, bound to the α subunit, results in the dissociation of the α subunit from the βγ dimer. Subsequently, both the GTP-bound α subunit and the βγ heterodimer activate signalling pathways until the GTP is hydrolysed into GDP and the heterotrimer is reassembled. GAPs accelerate the hydrolysis and are also known as regulators of G protein signalling (RGS). These events are illustrated in Fig. 41.6.

Among the well-accepted effects of the βγ dimer are the regulation of K^+ and voltage-dependent Ca^{2+} channels, adenylyl cyclases (ACs), phospholipases C (PLCs), and phosphoinositide 3-kinases (PI_3Ks) [115]. In addition, βγ heterodimers have been suggested to affect transcription, trafficking and signalling at different subcellular locations [116].

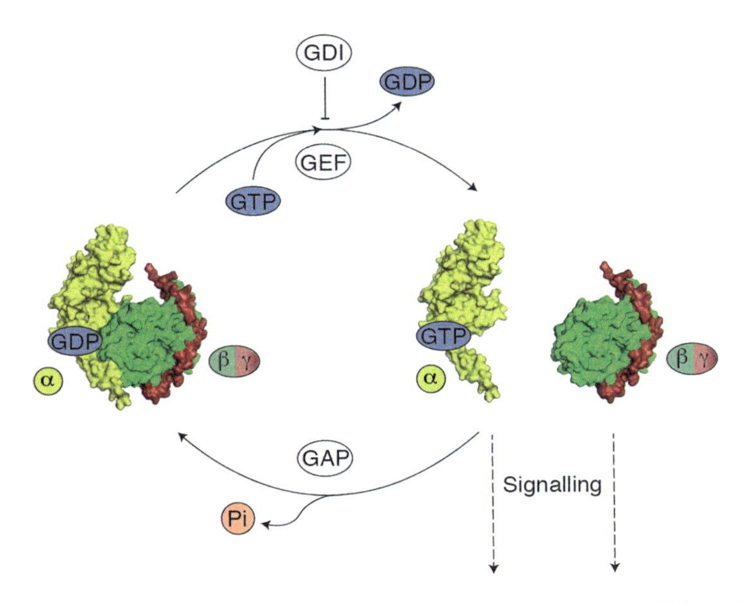

Fig. 41.6 Simplified representation of the cycles of activation and deactivation of G proteins

As for many GPCRs, the CaSR couples to more than one family of heterotrimeric G proteins, especially to $G_{q/11}$ and $G_{i/o}$. However several studies also suggest that the CaSR may couple to members of the $G_{12/13}$ family, and also to G_s in cancer-derived cell lines [18].

41.5.1.1 $G_{q/11}$

G_q and G_{11} share 90% sequence homology, are ubiquitously expressed, and have similar functions. For historic reasons, most studies focus on G_q and to a lesser extent on G_{11} [117]. The α subunits of $G_{q/11}$ activate PLCβ, which cleaves membrane-located phosphatidylinositol 4,5-bisphosphate (PIP_2) into the second messengers 1,2-diacylglycerol (DAG) and inositol 1,4,5-trisphosphate (IP_3). IP_3 diffuses into the cytosol and binds the IP_3 receptors that reside in the ER, causing the release of $Ca^{2+}{}_i$. IP_3 is then metabolised to IP_2 and IP_1. Increased $[Ca^{2+}]_i$ together with DAG, localised in the plasma membrane, results in the recruitment and activation of multiple isoforms of PKC. PKC phosphorylates numerous other proteins [118], including CaSR at Thr-888 to regulate $Ca^{2+}{}_i$ oscillations [119]. CaSR-induced IP_3 generation and $Ca^{2+}{}_i$ mobilisation was first shown in *Xenopus laevis* oocytes expressing the bovine parathyroid CaSR in the original cloning paper [3] but is more commonly investigated in HEK293 cells stably expressing the CaSR (HEK-CaSR) [18] [120] thus confirming the central role of the $G_{q/11}$-PLCβ pathway in CaSR signalling.

In addition to the classic $G_{q/11}$-PLCβ pathway, recent studies have shown that $G_{q/11}$ can also activate other signalling pathways via RhoGEFs such as RhoA [121–123], although the relevance of this pathway for the CaSR has not yet been determined.

41.5.1.2 $G_{i/o}$

The members of the $G_{i/o}$ family are characterised by their sensitivity to pertussis toxin (PTx), which inhibits their interaction with the GPCR, the only exception being G_z [124]. G_{i1}, G_{i2}, G_{i3} and G_o subtypes share a high sequence homology and probably have overlapping functions, although G_o is localised predominantly in the central nervous system [125]. Due to the relatively high abundance of this family of G proteins compared to the others, and since the majority of signalling events activated by $\beta\gamma$ are sensitive to PTx [126], the signalling by $\beta\gamma$ dimers is often attributed to activation of $G_{i/o}$ [127].

Activation of G_i inhibits several types of ACs. ACs increase cytosolic levels of cAMP and therefore G_i activation lowers cAMP levels. Studies in HEK-CaSR cells show that increased $[Ca^{2+}]_o$ decreases forskolin-induced increase in cAMP, suggesting the activation of G_i [120]. In bovine parathyroid and HEK-CaSR cells, CaSR stimulates ERK1/2 phosphorylation via $G_{q/11}$ and G_i pathways [128]. A later study suggested that G_{i2} is the $G_{i/o}$ subtype responsible for ERK1/2 activation [105].

41.5.1.3 $G_{12/13}$

The activation of $G_{12/13}$ proteins recruit to the membrane and activate RhoGEFs that specifically activate RhoA, such as p115-RhoGEF, which also acts as a GAP through its RGS domain terminating the activity of $G_{12/13}$ [129]. Using Madin-Darby canine kidney cells stably overexpressing CaSR, Miller and collaborators found that $[Ca^{2+}]_o$ activated phospholipase D via RhoA, and that this was mediated by $G_{12/13}$ and independent of $G_{q/11}$ and G_i [130]. Another study, suggested a pathway specifically activated by L-Phe via $G_{12/13}$ in mouse embryonic fibroblasts that resulted in $Ca^{2+}{}_i$ oscillations [131].

41.5.1.4 G_s

A few studies have reported G_s coupling to the CaSR. Wysolmerski and collaborators first showed that CaSR couples to G_i in healthy mammary epithelial cells, but then switches to G_s in both MCF-7 human breast cancer cells, and in Comma-D immortalised murine mammary cells. Surprisingly, no IP_1 accumulation was observed, whereas the levels of active mitogen-activated protein kinases (MAPKs) were increased upon stimulation with high $[Ca^{2+}]_o$. They also found that cAMP regulated the secretion of PTHrP via PKA [132], which was corroborated in a recent study [133]. In mouse pituitary gland tumour derived AtT-20 cells, the same group showed that CaSR activation stimulated PTHrP via the same mechanism, G_s-cAMP-PKA, independently of PLC or PKC [134]. In a previous study using the same cell line, increases in IP_1 concentrations were sensitive to PTx, showing simultaneous coupling both to G_i and G_s [135]. G protein switching has been observed also for the β2-adrenergic receptor, where PKA phosphorylates the receptor, increasing its affinity for G_i versus G_s. As a result, it switches signalling from cAMP/PKA to MAPK activation [136].

41.5.2 Rho GTPases

Rho GTPases belong to the Ras family of GTPases, which are the most known small monomeric GTPases. Among the Rho GTPases, the best characterised are RhoA, Rac1 and Cdc42. Rho GTPases play a central role in cell migration, cell polarity, and cell cycle progression, by regulating cell adhesion and actin cytoskeleton dynamics [137]. The activation of RhoA has been traditionally associated exclusively with $G_{12/13}$ signalling, however there is increasing evidence of activation by $G_{q/11}$ via RhoGEFs and independent of PLCβ [138]. It has been suggested that the CaSR activates PI_4-kinase via Rho [139]. CaSR activation produced actin stress fibre assembly in HEK-CaSR, in a Rho kinase-dependent mechanism. This phenomenon was PTx-insensitive and the PLCβ inhibitor U73122 showed no effect [140]. Since U73122 can activate ion channels at the concentrations used to inhibit PLCβ [141] the recently available potent and specific $G_{q/11}$ inhibitors FR900359 and YM-254890 may prove better reagents for the investigation of $G_{q/11}$ signalling [142].

A study in human keratinocytes showed that CaSR-dependent activation of RhoA plays a role in cell-cell adhesion [143], whereas experiments in human podocytes showed that CaSR activated RhoA via $Ca^{2+}{}_i$ mobilisation, in a mechanism dependent of the ion channel TRPC6 [144].

The activation of Rac and Cdc42 by G proteins is less clearly defined. In highly motile cells βγ-mediated activation of PI_3K and the GEF PRex resulted in Rac1 activity. Whether these signalling modules play a role in CaSR signal transduction remains to be demonstrated [145]. In primary human monocyte-derived macrophages CaSR activated Rac and/or Cdc42, but no RhoA, to regulate membrane ruffling via a mechanism dependent on PI_3K [146]. A study in a human T cell line found that CaSR can promote cell migration by activating Cdc42, also via a PI_3K-dependent mechanism [147]. A study in HEK-CaSR cells showed that membrane ruffling is $G_{q/11}$-dependent and $G_{12/13}$-independent, suggesting activation of Rho GTPases by $G_{q/11}$ [148].

41.5.3 *β-Arrestins*

In addition to their key role in terminating G protein signalling pathways activated by GPCRs, β-arrestins can also activate signalling events [149]. Specifically in HEK-CaSR cells, β-arrestin 1 is involved in CaSR-induced plasma membrane ruffling [150] while β-arrestins 1 and 2 are involved in CaSR-induced ERK1/2 activation [151].

41.5.4 *CaSR-Induced Protein Kinase Activation*

The CaSR activates a number of protein kinase families including glycogen synthase kinase-3 (GSK3), Akt, and the MAPKs, and these will be detailed in turn.

41.5.4.1 Akt and GSK-3β

Akt, or protein kinase B, is a protein kinase that regulates multiple functions such as growth, proliferation and transcription. The first step for Akt activation is binding to PIP_3 in the membrane. PIP_3-bound Akt is sequentially phosphorylated first at Thr-308 and then at Ser-473 for full activation [152]. GSK3 is involved in the phosphorylation of over a hundred substrates, and it interacts with multiple types of receptors. It exists in two isoforms, α and β, and it can be phosphorylated by PKA, PKC, and Akt, among others. Phosphorylation of GSK3-β at Ser-9 results in inhibition of the binding to certain substrates that require binding to a domain in the protein prior to phosphorylation [153].

Studies in fetal rat calvarial cells, murine osteoblast 2T3 cells, and human osteoblasts, show that CaSR activation results in phosphorylation of Akt at Thr-308 and Ser-473, and of GSK3-β at Ser-9 [154, 155]. Further, in proximal tubular opossum kidney cells, the CaSR ligands neomycin and gentamicin elicit phosphorylation of Akt and GSK3-β in a PI_3K-dependent fashion [42].

41.5.4.2 MAPKs

Several studies have recently explored the role played by CaSR in the phosphorylation of protein kinases, both in healthy tissue and in disease models. The MAPKs include ERK, c-Jun amino-terminal kinases (JNK), and P38. These proteins are activated by phosphorylation and thus we will refer to the active phosphorylated forms as p-ERK1/2, p-JNK and p-P38.

Activation of ERK1/2 can be Ras- or PKC-dependent. Ras-dependent activation involves PI_3K, Src family kinases, and receptor tyrosine kinases such as the epidermal growth factor receptor. A study in HEK-CaSR cells showed that ERK1/2 activation by CaSR was Ras-dependent, relied largely on PI_3K activity, and was independent of tyrosine kinase activity [156]. In contrast, another study in HEK-CaSR cells and in bovine parathyroid cells, showed that the cytoplasmic tyrosine kinase inhibitor, herbimycin, inhibited ERK1/2 phosphorylation [128]. A similar result was observed for ERK1 in Rat-1 fibroblasts [157]. In proximal tubular opossum-kidney cells, activation of CaSR by neomycin induced P38 activation via a PI_3K-mediated mechanism [41].

Across different tissues, increased CaSR expression and activation correlates positively with an increase in p-ERK1/2 levels [41, 154, 158–162], except for a study on hearts of a rat epilepsy model where p-ERK1/2 levels decreased [163]. A similar positive correlation was observed for p-JNK [161–164], whereas one study showed no effect [160]. As for p-P38, a similar number of studies show a positive correlation [41, 161, 163, 164] or no effect [158, 160, 162].

Overall, a prolonged exposure to CaSR agonists increases mRNA or protein expression levels of CaSR, and this phenomenon often correlates positively with an increase in active ERK1/2, JNK, and P38. Differences in phosphorylation are observed reflecting the different signalling profiles of CaSR in different tissues.

41.6 Ligand-Biased Signalling Through the CaSR

Ligand-biased signalling is a relatively new concept based on the idea that a receptor can exist in multiple active conformations, each stabilised by a specific ligand, with characteristic binding kinetics, and therefore with a particular signalling profile [165]. For GPCRs, this would translate into different coupling behaviours towards G proteins and β-arrestins [166]. For allosteric modulators, the concept extends to how these molecules affect positively or negatively each of the pathways activated by the orthosteric ligands. Exploiting this phenomenon offers great potential for the discovery and development of new drugs with increased efficacy and safety, which is of course of interest for the pharmaceutical industry [167].

Several studies have used the CaSR as a model to study ligand-biased signalling, given that it can activate multiple signalling pathways and it can be modulated by a wide range of different ligands [168]. This phenomenon has been studied using pharmacological assays, applied in a high-throughput manner, and using multiple agonists. The receptor readouts usually follow changes in $Ca^{2+}{}_i$ mobilisation, IP_1 accumulation, cAMP levels, phosphorylation of ERK1/2, and plasma membrane ruffling. The first two provide information on the $G_{q/11}$-PLCβ pathway; cAMP on G_i and G_s activity; p-ERK1/2 on G_q, G_i, and β-arrestins; and PM ruffling on Rho GTPases and β-arrestins.

These studies often rely on obtaining concentration-response curves for different ligands, and comparing calculated values such as EC_{50}, dissociation constant, maximum response, or cooperativity αβ for allosteric modulators. Additionally, receptor expression levels are often also determined, and in fact regulation of cell surface expression has been proposed as a mechanism of bias by allosteric

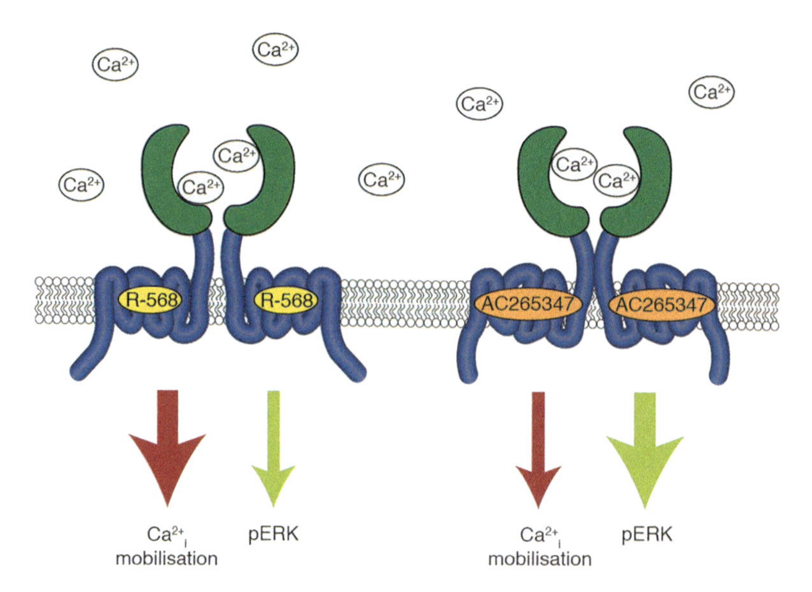

Fig. 41.7 Representation of bias by the positive allosteric modulators R-568 and AC265347 in $Ca^{2+}{}_{i}$ mobilisation and phosphorylation of ERK1/2 upon activation of the CaSR by $Ca^{2+}{}_{o}$

modulators [100, 169]. Multiple studies in HEK-CaSR cells have addressed ligand-bias by orthosteric ligands, as well as positive and negative allosteric modulators [151, 170–173], including one that explored the effect of naturally occurring CaSR mutations on ligand-bias [102].

These systematic in vitro studies provide valuable information to understand the differences in the effects of the ligands in vivo. For example, Leach and collaborators found that the calcimimetic AC265347 tunes the effect of $[Ca^{2+}]_{o}$ to favour phosphorylation of ERK1/2 and accumulation of IP_1, as compared to $Ca^{2+}{}_{i}$ mobilisation. Interestingly, AC265347 did not increase trafficking of loss-of-expression CaSR mutants, an effect observed by other calcimimetics, suggesting that it may act via a new mechanism [170]. Figure 41.7 shows an example of the signalling bias caused by two of the allosteric modulators used in this study.

41.7 Tissue-Specific Signalling of the CaSR

The pleiotropy of the CaSR arises as a result of its ability to couple to various G proteins and thus to mediate distinct signalling pathways. Consequently, the CaSR may fine-tune several physiological processes in a tissue-specific manner. The ability of GPCRs to mediate tissue-specific signalling is dictated by the cellular environment, as evidenced by recombinant systems where the same GPCR can have different pharmacological profiles in different cellular backgrounds. This phenomenon is termed tissue-specific signalling or system bias. It arises when ligands favour the interaction of a receptor ensemble to auxiliary proteins, or when receptors form heterodimers with distinct pharmacological signatures [174]. The capacity of CaSR ligands to promote coupling to multiple G proteins and to differing extents was previously discussed in the context of biased signalling. Here, tissue-specific signalling is discussed in light of the evidence showing interaction of the CaSR with other proteins and the formation of heterodimers in different cellular environments.

The CaSR interacts with various proteins that influence its signalling signature. Several CaSR interacting proteins have been identified and these include inwardly-rectifying potassium channels [175] and the previously described filamin A [92, 93, 176] as well as the RAMPs [88]. In addition, the CaSR may form heterodimers with other class C GPCRs including mGlu1a, mGlu5 and GABA B receptors, as shown in endogenous and recombinant systems [87]. Such heterodimerisation could thus provide the CaSR another mechanism for tissue-specific signalling.

41.8 Future Topics and Concluding Remarks

The recently published crystallographic data of the CaSR ECD structure has shed some light on CaSR ligand recognition, receptor activation, allosteric modulation, as well as on the structural basis of dimerisation. The medicinal importance of CaSR modulation is clear and therefore obtaining the ECD structure will facilitate structure-based drug discovery and might open up further therapeutic approaches. These data also raise the question of whether L-aromatic amino acids and relevant anions should be added to the experimental buffers when studying CaSR function to preserve the receptor's native conformation. The full structure of the CaSR has yet to be determined and thus obtaining the crystal structures of the CaSR's TMD and ICD is a high priority in the field as this will help understanding effector interactions. In addition, the current structural data provides only a snapshot of the receptor in a fixed conformation, whereas in physiology the receptor is a dynamic molecule wobbling between multiple conformations. Thus, we need a new and more dynamic approach able to reveal a protein's structure in its transition states, ideally allowing us to see conformational changes upon agonist/antagonist binding at the receptor.

The CaSR field would also benefit from new reagents such as a CaSR-selective radioligand, as calcium itself is a too low-affinity ligand to be of use in binding studies. Such a radioligand would allow researchers to investigate whether CaSR biased signalling might be driven by ligand binding kinetics. Moreover, the radioligand could be used to study CaSR expression in native tissues and cells as CaSR expression analysis is currently hampered by nonspecific staining of commercially available CaSR antibodies. Next, the newly emerging FRET-based biosensors can be used to observe G protein activation of the CaSR directly, providing dynamic information on the first step in the signalling cascade [177]. Indeed by taking advantage of the plethora of fluorescent biosensors available, given particular spectroscopic properties, it is possible to measure activation of multiple proteins simultaneously and in real-time [178]. This would be of particular value when studying biased signalling, as current protocols are susceptible to time- and assay-specific artefacts.

Finally, a recent publication suggested that internalised CaSR could have a role in sustained signalling [148]. This phenomenon has been proposed before for some class A GPCRs [179], and implies a new signalling mechanism by CaSR. The relevance of the observations for internalised CaSR signalling needs to be addressed in follow-up studies.

We can conclude therefore that CaSR activation results in a wide range of downstream signals and functions at different timescales and across a variety of tissues. To make sense of this complexity will require better understanding of a range of factors including differential ligand affinity and bias, receptor heterodimerisation, as well as downstream effector selection. The benefit of such information could be the rational development of novel drugs with improved efficacy and safety.

Acknowledgements The authors thank Hans Bräuner-Osborne for critical review of the manuscript. The authors of this chapter have received funding from the European Union's Horizon 2020 research and innovation programme under grant agreement No 675228 (CaSR Biomedicine).

References

1. Crichton RR (2008) Biological inorganic chemistry. Biological inorganic chemistry
2. Alexander SPH, Christopoulos A, Davenport AP, Kelly E, Marrion NV, Peters JA et al (2017) The concise guide to pharmacology 2017/18: G protein-coupled receptors. Br J Pharmacol 174:S17–S129
3. Brown EM, Gamba G, RIccardi D, Lombardi M, Butters R, Kifor O et al (1993) Cloning and characterization of an extracellular Ca^{2+}-sensing receptor from bovine parathyroid. Nature 366:461–464
4. Conigrave AD (2016) The calcium-sensing receptor and the parathyroid: past, present, future. Front Physiol 7:563
5. Uhlen M, Fagerberg L, Hallstrom BM, Lindskog C, Oksvold P, Mardinoglu A et al (2015) Tissue-based map of the human proteome. Science 347(6220):1260419–1260419
6. Felsenfeld AJ, Levine BS (2015) Calcitonin, the forgotten hormone: does it deserve to be forgotten? Clin Kidney J 8(2):180–187
7. Mutschler E, Geisslinger G, Kroemer HK, Menzel S, Ruth P Mutschler Arzneimittelwirkungen – Pharmakologie, Klinische Pharmakologie, Toxikologie. Lehrbuch der Pharmakologie, der klinischen Pharmakologie und Toxikologie; mit einführenden Kapiteln in die Anatomie, Physiologie und Pathophysiologie; mit 257 Tabellen und 1417 Strukturformeln. 2013. XXIII, 1197 Seiten
8. Brown EM, MacLeod RJ (2001) Extracellular calcium sensing and extracellular calcium signaling. Physiol Rev 81(1):239–297
9. Conigrave AD, Quinn SJ, Brown EM (2000) Cooperative multi-modal sensing and therapeutic implications of the extracellular Ca^{2+} sensing receptor. Trends Pharmacol Sci 21:401–407

10. Kantham L, Quinn SJ, Egbuna OI, Baxi K, Butters R, Pang JL et al (2009) The calcium-sensing receptor (CaSR) defends against hypercalcemia independently of its regulation of parathyroid hormone secretion. Am J Physiol Endocrinol Metab 297(4):E915–E923
11. Loupy A, Ramakrishnan SK, Wootla B, Chambrey R, De La Faille R, Bourgeois S et al (2012) PTH-independent regulation of blood calcium concentration by the calcium-sensing receptor. J Clin Invest 122(9):3355–3367
12. Ward BK, Magno AL, Davis EA, Hanyaloglu AC, Stuckey BGA, Burrows M et al (2004) Functional deletion of the calcium-sensing receptor in a case of neonatal severe hyperparathyroidism. J Clin Endocrinol Metab 89:3721–3730
13. Vahe C, Benomar K, Espiard S, Coppin L, Jannin A, Odou MF et al (2017) Diseases associated with calcium-sensing receptor. Orphanet J Rare Dis 12:19
14. Pallan S, Rahman MO, Khan AA (2012) Diagnosis and management of primary hyperparathyroidism. BMJ 344(7849):181–192
15. Assimos DG (2015) The G allele of CaSR R990G polymorphism increases susceptibility to urolithiasis and hypercalciuria: evidences from a comprehensive meta-analysis. J Urol 194:1014
16. Tennakoon S, Aggarwal A, Kallay E (2016) The calcium-sensing receptor and the hallmarks of cancer. Biochim Biophys Acta 1863(6 Pt B):1398–1407
17. Yarova PL, Stewart AL, Sathish V, Britt RD, Thompson MA, Lowe APP et al (2015) Calcium-sensing receptor antagonists abrogate airway hyperresponsiveness and inflammation in allergic asthma. Sci Transl Med 7(284):284ra58
18. Conigrave AD, Ward DT (2013) Calcium-sensing receptor (CaSR): pharmacological properties and signaling pathways. Best Pract Res Clin Endocrinol Metab 27(3):315–331
19. Muto T, Tsuchiya D, Morikawa K, Jingami H (2007) Structures of the extracellular regions of the group II/III metabotropic glutamate receptors. Proc Natl Acad Sci 104(10):3759–3764
20. Silve C, Petrel C, Leroy C, Bruel H, Mallet E, Rognan D et al (2005) Delineating a Ca^{2+} binding pocket within the venus flytrap module of the human calcium-sensing receptor. J Biol Chem 280(45):37917–37923
21. Hendy GN, Canaff L, Cole DEC (2013) The CASR gene: alternative splicing and transcriptional control, and calcium-sensing receptor (CaSR) protein: structure and ligand binding sites. Best Pract Res Clin Endocrinol Metab 27:285–301
22. Bai M (2004) Structure-function relationship of the extracellular calcium-sensing receptor. Cell Calcium 35:197–207
23. Hauache OM, Hu J, Ray K, Spiegel AM (2000 Aug) Functional interactions between the extracellular domain and the seven-transmembrane domain in Ca^{2+} receptor activation. Endocrine 13(1):63–70
24. Ward DT, Brown EM, Harris HW (1998) Disulfide bonds in the extracellular calcium-polyvalent cation-sensing receptor correlate with dimer formation and its response to divalent cations in vitro. J Biol Chem 273(23):14476–14483
25. Ray K, Clapp P, Goldsmith PK, Spiegel AM (1998) Identification of the sites of N-linked glycosylation on the human calcium receptor and assessment of their role in cell surface expression and signal transduction. J Biol Chem 273(51):34558–34567
26. Gama L, Wilt SG, Breitwieser GE (2001) Heterodimerization of calcium sensing receptors with metabotropic glutamate receptors in neurons. J Biol Chem 276(42):39053–39059
27. Chang W, Tu C, Cheng Z, Rodriguez L, Chen T-H, Gassmann M et al (2007) Complex formation with the Type B gamma-aminobutyric acid receptor affects the expression and signal transduction of the extracellular calcium-sensing receptor. Studies with HEK-293 cells and neurons. J Biol Chem 282(34):25030–25040
28. Geng Y, Mosyak L, Kurinov I, Zuo H, Sturchler E, Cheng TC et al (2016) Structural mechanism of ligand activation in human calcium-sensing receptor. elife 5(July):1–25
29. Zhang C, Zhang T, Zou J, Miller CL, Gorkhali R, Yang J et al (2016) Structural basis for regulation of human calcium-sensing receptor by magnesium ions and an unexpected tryptophan derivative co-agonist. Sci Adv 2(5):e1600241

30. Kunishima N, Shimada Y, Tsuji Y, Sato T, Yamamoto M, Kumasaka T et al (2000) Structural basis of glutamate recognition by a dimeric metabotropic glutamate receptor. Nature 407(6807):971–977
31. Matsushita S, Nakata H, Kubo Y, Tateyama M (2010) Ligand-induced rearrangements of the GABAB receptor revealed by fluorescence resonance energy transfer. J Biol Chem 285(14):10291–10299
32. Tfelt-Hansen J, Brown EM (2005) The calcium-sensing receptor in normal physiology and pathophysiology: a review. Crit Rev Clin Lab Sci 42:35–70
33. Garrett JE, Capuano IV, Hammerland LG, Hung BC, Brown EM, Hebert SC et al (1995) Molecular cloning and functional expression of human parathyroid calcium receptor cDNAs. J Biol Chem 270(21):12919–12925
34. Chang W, Chen TH, Pratt S, Shoback D (2000) Amino acids in the second and third intracellular loops of the parathyroid Ca^{2+}-sensing receptor mediate efficient coupling to phospholipase C. J Biol Chem 275(26):19955–19963
35. Brown EM, Lian JB (2008) New insights in bone biology: unmasking skeletal effects of the extracellular calcium-sensing receptor. Sci Signal 1:pe40
36. Bai M, Trivedi S, Brown EM (1998) Dimerization of the extracellular calcium-sensing receptor (CaR) on the cell surface of CaR-transfected HEK293 cells. J Biol Chem 273(36):23605–23610
37. McCormick WD, Atkinson-Dell R, Campion KL, Mun HC, Conigrave AD, Ward DT (2010) Increased receptor stimulation elicits differential calcium-sensing receptorT888 dephosphorylation. J Biol Chem 285(19):14170–14177
38. Granier S, Kobilka B (2012) A new era of GPCR structural and chemical biology. Nat Chem Biol 8:670–673
39. Handlogten ME, Shiraishi N, Awata H, Huang CF, Miller RT (2000) Extracellular Ca^{2+}-sensing receptor is a promiscuous divalent cation sensor that responds to lead. Am J Physiol Physiol 279(6):F1083–F1091
40. Quinn SJ, Ye CP, Diaz R, Kifor O, Bai M, Vassilev P et al (1997) The Ca^{2+}-sensing receptor: a target for polyamines. Am J Phys 273:C1315–C1323
41. Ward DT, McLarnon SJ, Riccardi D (2002) Aminoglycosides increase intracellular calcium levels and ERK activity in proximal tubular OK cells expressing the extracellular calcium-sensing receptor. J Am Soc Nephrol 13(6):1481–1489
42. Ward DT, Maldonado-Pérez D, Hollins L, Riccardi D (2005) Aminoglycosides induce acute cell signaling and chronic cell death in renal cells that express the calcium-sensing receptor. J Am Soc Nephrol 16(5):1236–1244
43. Brauner-Osborne H, Wellendorph P, Jensen A (2007) Structure, pharmacology and therapeutic prospects of family C G-protein coupled receptors. Curr Drug Targets 8(1):169–184
44. Chang W, Shoback D (2004) Extracellular Ca^{2+}-sensing receptors – an overview. Cell Calcium 35(3):183–196
45. Nemeth EF (2004) Calcimimetic and calcilytic drugs: just for parathyroid cells? Cell Calcium 35(3):283–289
46. Bandyopadhyay S, Tfelt-Hansen J, Chattopadhyay N (2010) Diverse roles of extracellular calcium-sensing receptor in the central nervous system. J Neurosci Res 88:2073–2082
47. Ward DT, Riccardi D (2012) New concepts in calcium-sensing receptor pharmacology and signalling. Br J Pharmacol 165:35–48
48. Conigrave AD, Quinn SJ, Brown EM (2000 Apr) L-amino acid sensing by the extracellular Ca^{2+}-sensing receptor. Proc Natl Acad Sci U S A 97(9):4814–4819
49. Liou AP, Sei Y, Zhao X, Feng J, Lu X, Thomas C et al (2011) The extracellular calcium-sensing receptor is required for cholecystokinin secretion in response to L-phenylalanine in acutely isolated intestinal I cells. Am J Physiol Gastrointest Liver Physiol 300(4):G538–G546
50. Feng J, Petersen CD, Coy DH, Jiang J-K, Thomas CJ, Pollak MR et al (2010 Oct) Calcium-sensing receptor is a physiologic multimodal chemosensor regulating gastric G-cell growth and gastrin secretion. Proc Natl Acad Sci U S A 107(41):17791–17796

51. Conigrave AD, Mun H-C, Delbridge L, Quinn SJ, Wilkinson M, Brown EM (2004 Sep) L-amino acids regulate parathyroid hormone secretion. J Biol Chem 279(37):38151–38159
52. Quinn SJ, Bai M, Brown EM (2004) pH sensing by the calcium-sensing receptor. J Biol Chem 279(36):37241–37249
53. López I, Aguilera-Tejero E, Estepa JC, Rodriguez M, Felsenfeld AJ (2004) Role of acidosis-induced increases in calcium on PTH secretion in acute metabolic and respiratory acidosis in the dog. Am J Physiol Endocrinol Metab 286(5):E780–E785
54. Lopez I, Rodriguez M, Felsenfeld AJ, Estepa JC, Aguilera-Tejero E (2003) Direct suppressive effect of acute metabolic and respiratory alkalosis on parathyroid hormone secretion in the dog. J Bone Miner Res 18(8):1478–1485
55. Campion KL, McCormick WD, Warwicker J, Khayat MEB, Atkinson-Dell R, Steward MC et al (2015) Pathophysiologic changes in extracellular pH modulate parathyroid calcium-sensing receptor activity and secretion via a histidine-independent mechanism. J Am Soc Nephrol 26(9):2163–2171
56. Nemeth EF (2002) The search for calcium receptor antagonists (calcilytics). J Mol Endocrinol 29(1):15–21
57. Nemeth EF, Steffey ME, Hammerland LG, Hung BC, Van Wagenen BC, DelMar EG et al (1998) Calcimimetics with potent and selective activity on the parathyroid calcium receptor. Proc Natl Acad Sci U S A 95(March):4040–4045
58. Miedlich S, Gama L, Breitwieser GE (2002) Calcium sensing receptor activation by a calcimimetic suggests a link between cooperativity and intracellular calcium oscillations. J Biol Chem 277(51):49691–49699
59. Saidak Z, Brazier M, Kamel S, Mentaverri R (2009) Agonists and allosteric modulators of the calcium-sensing receptor and their therapeutic applications. Mol Pharmacol 76(6):1131–1144
60. Yamamura A (2016) Molecular mechanism of Dihydropyridine Ca^{2+} channel blockers in pulmonary hypertension. Yakugaku Zasshi 136(10):1373–1377
61. Block GA, Zaun D, Smits G, Persky M, Brillhart S, Nieman K et al (2010) Cinacalcet hydrochloride treatment significantly improves all-cause and cardiovascular survival in a large cohort of hemodialysis patients. Kidney Int 78(6):578–589
62. Gannon AW, Monk HM, Levine MA (2014) Cinacalcet monotherapy in neonatal severe hyperparathyroidism: a case study and review. J Clin Endocrinol Metab 99(1):7–11
63. Silverberg SJ, Rubin MR, Faiman C, Peacock M, Shoback DM, Smallridge RC et al (2007) Cinacalcet hydrochloride reduces the serum calcium concentration in inoperable parathyroid carcinoma. J Clin Endocrinol Metab 92(10):3803–3808
64. Nemeth EF, Goodman WG (2016) Calcimimetic and calcilytic drugs: feats, flops, and futures. Calcif Tissue Int 98:341–358
65. Steddon SJ, Cunningham J (2005) Calcimimetics and calcilytics – fooling the calcium receptor. Lancet 365(9478):2237–2239
66. Block GA, Bushinsky DA, Cheng S, Cunningham J, Dehmel B, Drueke TB et al (2017) Effect of etelcalcetide vs cinacalcet on serum parathyroid hormone in patients receiving hemodialysis with secondary hyperparathyroidism. JAMA 317(2):156–164
67. Eidman KE, Wetmore JB (2018) Managing hyperparathyroidism in hemodialysis: role of etelcalcetide. Int J Nephrol Renovasc Dis 11:69–80
68. Ma J-N, Owens M, Gustafsson M, Jensen J, Tabatabaei A, Schmelzer K et al (2011) Characterization of highly efficacious allosteric agonists of the human calcium-sensing receptor. J Pharmacol Exp Ther 337(1):275–284
69. Nemeth EF, Delmar EG, Heaton WL, Miller MA, Lambert LD, Conklin RL et al (2001) Calcilytic compounds: potent and selective Ca^{2+} receptor antagonists that stimulate secretion of parathyroid hormone. J Pharmacol Exp Ther 299(1):323–331
70. Lembrechts R, Brouns I, Schnorbusch K, Pintelon I, Kemp PJ, Timmermans J-P et al (2013) Functional expression of the multimodal extracellular calcium-sensing receptor in pulmonary neuroendocrine cells. J Cell Sci 126(Pt 19):4490–4501

71. Drake MT, Shenoy SK, Lefkowitz RJ (2006) Trafficking of G protein-coupled receptors. Circ Res 99(6):570–582
72. Ritter SL, Hall RA (2009) Fine-tuning of GPCR activity by receptor-interacting proteins. Nat Rev Mol Cell Biol 10(12):819–830
73. Pierce KL, Premont RT, Lefkowitz RJ (2002) Seven-transmembrane receptors. Nat Rev Mol Cell Biol 3(9):639–650
74. Tokarev AA, Alfonso A, Segev N (2009) Overview of intracellular compartments and trafficking pathways. In: Trafficking inside cells: pathways, mechanisms and regulation. Springer, New York, pp 3–14
75. Janicic N, S E, Pausova Z, Seldin MF, Rivi M, Szpirer J et al (1995) Mapping of the calcium-sensing receptor gene (CASR) to human chromosome 3q13. 3-21 by fluorescence in situ hybridization, vol 801. Springer, New York, pp 798–801
76. Chikatsu N, Fukumoto S, Takeuchi Y, Suzawa M, Obara T, Matsumoto T et al (2000) Cloning and characterization of two promoters for the human calcium-sensing receptor (CaSR) and changes of CaSR expression in parathyroid adenomas. J Biol Chem 275(11):7553–7557
77. Yun FHJ, Wong BYL, Chase M, Shuen AY, Canaff L, Thongthai K et al (2007) Genetic variation at the calcium-sensing receptor (CASR) locus: implications for clinical molecular diagnostics. Clin Biochem 40(8):551–561
78. Canaff L, Hendy GN (2002) Human calcium-sensing receptor gene. Vitamin D response elements in promoters P1 and P2 confer transcriptional responsiveness to 1,25-dihydroxyvitamin D. J Biol Chem 277(33):30337–30350
79. Canaff L, Hendy GN (2005) Calcium-sensing receptor gene transcription is up-regulated by the proinflammatory cytokine, interleukin-1β: role of the NF-κB pathway and κB elements. J Biol Chem 280(14):14177–14188
80. Brown AJ, Zhong M, Finch J, Ritter C, McCracken R, Morrissey J et al (1996) Rat calcium-sensing receptor is regulated by vitamin D but not by calcium. Am J Phys 270(3 Pt 2):F454–F460
81. Yao JJ (2005) Regulation of renal calcium receptor gene expression by 1,25-dihydroxyvitamin D3 in genetic hypercalciuric stone-forming rats. J Am Soc Nephrol 16(5):1300–1308
82. Fetahu IS, Hummel DM, Manhardt T, Aggarwal A, Baumgartner-Parzer S, Kállay E (2014) Regulation of the calcium-sensing receptor expression by 1,25-dihydroxyvitamin D3, interleukin-6, and tumor necrosis factor alpha in colon cancer cells. J Steroid Biochem Mol Biol 144:228–231
83. Dong C, Filipeanu CM, Duvernay MT, Wu G (2007) Regulation of G protein-coupled receptor export trafficking. Biochim Biophys Acta Biomembr 1768(4):853–870
84. Ray K, Fan GF, Goldsmith PK, Spiegel AM (1997) The carboxyl terminus of the human calcium receptor. Requirements for cell-surface expression and signal transduction. J Biol Chem 272(50):31355–31361
85. Fan G, Goldsmith PK, Collins R, Dunn CK, Krapcho KJ, Rogers KV et al (1997) N-linked glycosylation of the human Ca^{2+} receptor is essential for its expression at the cell surface. Endocrinology 138(5):1916–1922
86. Aida K, Koishi S, Tawata M, Onaya T (1995) Molecular cloning of a putative Ca^{2+}-sensing receptor cDNA from human kidney. Biochem Biophys Res Commun 214(2):524–529
87. Chang W, Tu C, Cheng Z, Rodriguez L, Chen TH, Gassmann M et al (2007) Complex formation with the type B γ-aminobutyric acid receptor affects the expression and signal transduction of the extracellular calcium-sensing receptor: studies with HEK-293 cells and neurons. J Biol Chem 282(34):25030–25040
88. Bouschet T, Stéphane M, Henley JM (2005) Receptor-activity-modifying proteins are required for forward trafficking of the calcium-sensing receptor to the plasma membrane. J Cell Sci 118(20):4709–4720
89. Desai AJ, Roberts DJ, Richards GO, Skerry TM (2014) Role of receptor activity modifying protein 1 in function of the calcium sensing receptor in the human TT thyroid carcinoma cell line. PLoS One 9(1):e85237

90. Huang Y, Breitwieser GE (2007) Rescue of calcium-sensing receptor mutants by allosteric modulators reveals a conformational checkpoint in receptor biogenesis. J Biol Chem 282(13):9517–9525
91. Zhang M, Breitwieser GE (2005) High affinity interaction with filamin A protects against calcium-sensing receptor degradation. J Biol Chem 280(12):11140–11146
92. Hjälm G, MacLeod RJ, Kifor O, Chattopadhyay N, Brown EM (2001) Filamin-A binds to the carboxyl-terminal tail of the calcium-sensing receptor, an interaction that participates in CaR-mediated activation of mitogen-activated protein kinase. J Biol Chem 276(37):34880–34887
93. Awata H, Huang C, Handlogten ME, Miller RT (2001) Interaction of the calcium-sensing receptor and filamin, a potential scaffolding protein. J Biol Chem 276(37):34871–34879
94. Stepanchick A, McKenna J, McGovern O, Huang Y, Breitwieser GE (2010) Calcium sensing receptor mutations implicated in pancreatitis and idiopathic epilepsy syndrome disrupt an arginine-rich retention motif. Cell Physiol Biochem 26(3):363–374
95. Grant MP, Stepanchick A, Cavanaugh A, Breitwieser GE (2011) Agonist-driven maturation and plasma membrane insertion of calcium-sensing receptors dynamically control signal amplitude. Sci Signal 4(200):1–9
96. Grant MP, Cavanaugh A, Breitwieser GE (2015) 14-3-3 proteins buffer intracellular calcium sensing receptors to constrain signaling. PLoS One 10(8):1–20
97. Arulpragasam A, Magno AL, Ingley E, Brown SJ, Conigrave AD, Ratajczak T et al (2012) The adaptor protein 14-3-3 binds to the calcium-sensing receptor and attenuates receptor-mediated Rho kinase signalling. Biochem J 441(3):995–1007
98. Brennan SC, Mun H-C, Leach K, Kuchel PW, Christopoulos A, Conigrave AD (2015) Receptor expression modulates calcium-sensing receptor mediated intracellular Ca^{2+} mobilization. Endocrinology 156(4):1330–1342
99. Chang W, Pratt S, Chen TH, Bourguignon L, Shoback D (2001) Amino acids in the cytoplasmic C terminus of the parathyroid Ca^{2+}–sensing receptor mediate efficient cell-surface expression and phospholipase C activation. J Biol Chem 276(47):44129–44136
100. Leach K, Wen A, Cook AE, Sexton PM, Conigrave AD, Christopoulos A (2013) Impact of clinically relevant mutations on the pharmacoregulation and signaling bias of the calcium-sensing receptor by positive and negative allosteric modulators. Endocrinology 154(3):1105–1116
101. White E, McKenna J, Cavanaugh A, Breitwieser GE (2009) Pharmacochaperone-mediated rescue of calcium-sensing receptor loss-of-function mutants. Mol Endocrinol 23(7):1115–1123
102. Leach K, Wen A, Davey AE, Sexton PM, Conigrave AD, Christopoulos A (2012) Identification of molecular phenotypes and biased signaling induced by naturally occurring mutations of the human calcium-sensing receptor. Endocrinology 153(9):4304–4316
103. Moore CAC, Milano SK, Benovic JL (2007) Regulation of receptor trafficking by GRKs and arrestins. Annu Rev Physiol 69(1):451–482
104. Reyes-Ibarra AP, García-Regalado A, Ramírez-Rangel I, Esparza-Silva AL, Valadez-Sánchez M, Vázquez-Prado J et al (2007) Calcium-sensing receptor endocytosis links extracellular calcium signaling to parathyroid hormone-related peptide secretion via a Rab11a-dependent and AMSH-sensitive mechanism. Mol Endocrinol 21(6):1394–1407
105. Holstein DM, Berg KA, Leeb-Lundberg LMF, Olson MS, Saunders C (2004) Calcium-sensing receptor-mediated ERK1/2 activation requires Gα i2 coupling and dynamin-independent receptor internalization. J Biol Chem 279(11):10060–10069
106. Zhuang X, Northup JK, Ray K (2012) Large putative PEST-like sequence motif at the carboxyl tail of human calcium receptor directs lysosomal degradation and regulates cell surface receptor level. J Biol Chem 287(6):4165–4176
107. Pi M, Oakley RH, Gesty-Palmer D, Cruickshank RD, Spurney RF, Luttrell LM et al (2005) β-arrestin- and G protein receptor kinase-mediated calcium-sensing receptor desensitization. Mol Endocrinol 19(4):1078–1087

108. Lorenz S, Frenzel R, Paschke R, Breitwieser GE, Miedlich SU (2007) Functional desensitization of the extracellular calcium-sensing receptor is regulated via distinct mechanisms: role of G protein-coupled receptor kinases, protein kinase C and β-arrestins. Endocrinology 148(5):2398–2404
109. Huang Y, Niwa JI, Sobue G, Breitwieser GE (2006) Calcium-sensing receptor ubiquitination and degradation mediated by the E3 ubiquitin ligase dorfin. J Biol Chem 281(17):11610–11617
110. Marchese A, Paing MM, Temple BRS, Trejo J (2008) G protein–coupled receptor sorting to endosomes and lysosomes. Annu Rev Pharmacol Toxicol 48(1):601–629
111. Magalhaes AC, Dunn H, Ferguson SSG (2012) Regulation of GPCR activity, trafficking and localization by GPCR-interacting proteins. Br J Pharmacol 165(6):1717–1736
112. Kifor O, Diaz R, Butters R, Kifor I, Brown EM (1998) The calcium-sensing receptor is localized in caveolin-rich plasma membrane domains of bovine parathyroid cells. J Biol Chem 273(34):21708–21713
113. Chini B, Parenti M (2004) G-protein coupled receptors in lipid rafts and caveolae: how, when and why do they go there? J Mol Endocrinol 32(2):325–338
114. Herrera-Vigenor F, Hernández-García R, Valadez-Sánchez M, Vázquez-Prado J, Reyes-Cruz G (2006) AMSH regulates calcium-sensing receptor signaling through direct interactions. Biochem Biophys Res Commun 347(4):924–930
115. Khan SM, Sleno R, Gora S, Zylbergold P, Laverdure J-P, Labbe J-C et al (2013) The expanding roles of Gβγ subunits in G protein-coupled receptor signaling and drug action. Pharmacol Rev 65(2):545–577
116. Khan SM, Sung JY, Hébert TE (2016) Gβγ subunits—different spaces, different faces. Pharmacol Res 111:434–441
117. Hubbard KB, Hepler JR (2006) Cell signalling diversity of the Gqα family of heterotrimeric G proteins. Cell Signal 18(2):135–150
118. Newton AC (2018) Protein kinase C: perfectly balanced. Crit Rev Biochem Mol Biol 53(2):208–230
119. Davies SL, Ozawa A, McCormick WD, Dvorak MM, Ward DT (2007) Protein kinase C-mediated phosphorylation of the calcium-sensing receptor is stimulated by receptor activation and attenuated by calyculin-sensitive phosphatase activity. J Biol Chem 282(20):15048–15056
120. Chang W, Pratt S, Chen TH, Nemeth E, Huang Z, Shoback D (1998) Coupling of calcium receptors to inositol phosphate and cyclic AMP generation in mammalian cells and Xenopus laevis oocytes and immunodetection of receptor protein by region-specific antipeptide antisera. J Bone Miner Res 13(4):570–580
121. Rojas RJ, Yohe ME, Gershburg S, Kawano T, Kozasa T, Sondek J (2007) Gαq directly activates p63RhoGEF and trio via a conserved extension of the Dbl homology-associated pleckstrin homology domain. J Biol Chem 282(40):29201–29210
122. Lutz S, Shankaranarayanan A, Coco C, Ridilla M, Nance MR, Vettel C et al (2007) Structure of Gaq-p63RhoGEF-RhoA complex reveals a pathway for the activation of RhoA by GPCRs. Science 318:1923–1927
123. Van Unen J, Reinhard NR, Yin T, Wu YI, Postma M, Gadella TWJ et al (2015) Plasma membrane restricted RhoGEF activity is sufficient for RhoA-mediated actin polymerization. Sci Rep 5(October):1–16
124. Gagnon AW, Manning DR, Catani L, Gewirtz A, Poncz M, Brass LF (1991) Identification of Gz as a pertussis toxin-insensitive G protein in human platelets and megakaryocytes. Blood 78:1247–1254
125. Jiang M, Bajpayee NS (2009) Molecular mechanisms of go signaling. Neurosignals 17(1):23–41
126. Smrcka AV (2008) G protein βγ subunits: central mediators of G protein-coupled receptor signaling. Cell Mol Life Sci 65:2191–2214
127. Wettschureck N, Offermanns S (2005) Mammalian G proteins and their cell type specific functions. Physiol Rev 85:1159–1204

128. Kifor O, MacLeod RJ, Diaz R, Bai M, Yamaguchi T, Yao T et al (2001) Regulation of MAP kinase by calcium-sensing receptor in bovine parathyroid and CaR-transfected HEK293 cells. Am J Physiol Renal Physiol 280(2):F291–F302
129. Kozasa T, Hajicek N, Chow CR, Suzuki N (2011) Signalling mechanisms of RhoGTPase regulation by the heterotrimeric G proteins G12 and G13. J Biochem 150(4):357–369
130. Huang CF, Hujer KM, Wu ZZ, Miller RT (2004) The Ca^{2+}-sensing receptor couples to G alpha(12/13) to activate phospholipase D in Madin-Darby canine kidney cells. Am J Physiol Physiol 286(1):C22–C30
131. Rey O, Young SH, Yuan J, Slice L, Rozengurt E (2005) Amino acid-stimulated Ca^{2+} oscillations produced by the Ca^{2+}-sensing receptor are mediated by a phospholipase C/inositol 1,4,5-trisphosphate-independent pathway that requires G12, Rho, filamin-A, and the actin cytoskeleton. J Biol Chem 280(24):22875–22882
132. Mamillapalli R, VanHouten J, Zawalich W, Wysolmerski J (2008) Switching of G-protein usage by the calcium-sensing receptor reverses its effect on parathyroid hormone-related protein secretion in normal versus malignant breast cells. J Biol Chem 283(36):24435–24447
133. Kim W, Takyar FM, Swan K, Jeong J, Vanhouten J, Sullivan C et al (2016) Calcium-sensing receptor promotes breast cancer by stimulating intracrine actions of parathyroid hormone-related protein. Cancer Res 76(18):5348–5360
134. Mamillapalli R, Wysolmerski J (2010) The calcium-sensing receptor couples to Galpha(s) and regulates PTHrP and ACTH secretion in pituitary cells. J Endocrinol 204(3):287–297
135. Emanuel RL, Adler GK, Kifor O, Quinn SJ, Fuller F, Krapcho K et al (1996) CaSR expression and regulation by extracellular calcium in the AtT-20 pituitary cell line. Mol Endocrinol 10(5):555–565
136. Daaka Y, Luttrell LM, Lefkowitz RJ (1997) Switching of the coupling of the beta2-adrenergic receptor to different G proteins by protein kinase A. Nature 390(November):88–91
137. Hodge RG, Ridley AJ (2016) Regulating Rho GTPases and their regulators. Nat Rev Mol Cell Biol 17(8):496–510
138. Vogt S, Grosse R, Schultz G, Offermanns S (2003) Receptor-dependent RhoA activation in G12/G13-deficient cells. Genetic evidence for an involvement of Gq/G11. J Biol Chem 278(31):28743–28749
139. Huang C, Handlogten ME, Tyler Miller R (2002) Parallel activation of phosphatidylinositol 4-kinase and phospholipase C by the extracellular calcium-sensing receptor. J Biol Chem 277(23):20293–20300
140. Davies SL, Gibbons CE, Vizard T, Ward DT, Sarah L, Gibbons CE et al (2006) CaSR induces Rho kinase-mediated actin stress fiber assembly and altered cell morphology, but not in response to aromatic amino acids. Am J Physiol Cell Physiol 290:1543–1551
141. Leitner MG, Michel N, Behrendt M, Dierich M, Dembla S, Wilke BU et al (2016) Direct modulation of TRPM4 and TRPM3 channels by the phospholipase C inhibitor U73122. Br J Pharmacol 173:2555–2569
142. Kukkonen JP (2016) G-protein inhibition profile of the reported Gq/11 inhibitor UBO-QIC. Biochem Biophys Res Commun 469(1):101–107
143. Tu CL, Chang W, Bikle DD (2011) The calcium-sensing receptor-dependent regulation of cell-cell adhesion and keratinocyte differentiation requires rho and filamin a. J Invest Dermatol 131(5):1119–1128
144. Zhang L, Ji T, Wang Q, Meng K, Zhang R, Yang H et al (2017) Calcium-sensing receptor stimulation in cultured glomerular podocytes induces TRPC6-dependent calcium entry and RhoA activation. Cell Physiol Biochem 43:1777–1789
145. Vazquez-Prado J, Bracho-Valdes I, Cervantes-Villagrana RD, Reyes-Cruz G (2016) G pathways in cell polarity and migration linked to oncogenic GPCR signaling: potential relevance in tumor microenvironment. Mol Pharmacol 90(5):573–586
146. Canton J, Schlam D, Breuer C, Gütschow M, Glogauer M, Grinstein S (2016) Calcium-sensing receptors signal constitutive macropinocytosis and facilitate the uptake of NOD2 ligands in macrophages. Nat Commun 7(11284):1–12

147. Chang F, Kim JM, Choi Y, Park K (2018) MTA promotes chemotaxis and chemokinesis of immune cells through distinct calcium-sensing receptor signaling pathways. Biomaterials 150:14–24
148. Gorvin CM, Rogers A, Hastoy B, Tarasov AI, Frost M, Sposini S et al (2018) AP2σ mutations impair calcium-sensing receptor trafficking and signaling, and show an endosomal pathway to spatially direct G-protein selectivity. Cell Rep 22(4):1054–1066
149. Lefkowitz R (2007) Introduction to special section on β-Arrestins. Annu Rev Physiol 69(1)
150. Bouschet T, Martin S, Kanamarlapudi V, Mundell S, Henley JM (2007) The calcium-sensing receptor changes cell shape via a beta-arrestin-1 ARNO ARF6 ELMO protein network. J Cell Sci 120(Pt):2489–2497
151. Thomsen ARB, Hvidtfeldt M, Bräuner-Osborne H (2012) Biased agonism of the calcium-sensing receptor. Cell Calcium 51(2):107–116
152. Hemmings BA, Restuccia DF, Wrana JL, Alto NM, Orth K, Kopan R et al (2012) PI3K-PKB/Akt pathway. Cold Spring Harb Perspect Biol 4:a011189
153. Beurel E, Grieco SF, Jope RS (2015) Glycogen synthase kinase-3 (GSK3): regulation, actions, and diseases. Pharmacol Ther 148:114–131
154. Dvorak MM, Siddiqua A, Ward DT, Carter DH, Dallas SL, Nemeth EF et al (2004) Physiological changes in extracellular calcium concentration directly control osteoblast function in the absence of calciotropic hormones. Proc Natl Acad Sci U S A 101(14):5140–5145
155. Rybchyn MS, Slater M, Conigrave AD, Mason RS (2011) An Akt-dependent increase in canonical Wnt signaling and a decrease in sclerostin protein levels are involved in strontium ranelate-induced osteogenic effects in human osteoblasts. J Biol Chem 286(27):23771–23779
156. Hobson SA, Wright J, Lee F, McNeil SE, Bilderback T, Rodland KD (2003) Activation of the MAP kinase cascade by exogenous calcium-sensing receptor. Mol Cell Endocrinol 200:189–198
157. McNeil SE, Hobson SA, Nipper V, Rodland KD (1998) Functional calcium-sensing receptors in rat fibroblasts are required for activation of SRC kinase and mitogen-activated protein kinase in response to extracellular calcium. J Biol Chem 273(2):1114–1120
158. Wang P, Wang L, Wang S, Li S, Li Y, Zhang L (2015) Effects of calcium-sensing receptors on apoptosis in rat hippocampus during hypoxia/reoxygenation through the ERK1/2 pathway. Int J Clin Exp Pathol 8(9):10808–10815
159. Mizumachi H, Yoshida S, Tomokiyo A, Hasegawa D, Hamano S, Yuda A et al (2017) Calcium-sensing receptor-ERK signaling promotes odontoblastic differentiation of human dental pulp cells. Bone 101:191–201
160. Li T, Sun M, Yin X, Wu C, Wu Q, Feng S et al (2013) Expression of the calcium sensing receptor in human peripheral blood T lymphocyte and its contribution to cytokine secretion through MAPKs or NF-κB pathways. Mol Immunol 53(4):414–420
161. Kong W-Y, Tong L-Q, Zhang H-J, Cao Y-G, Wang G-C, Zhu J-Z et al (2016) The calcium-sensing receptor participates in testicular damage in streptozotocin-induced diabetic rats. Asian J Androl 18:803–808
162. Qi H, Cao Y, Huang W, Liu Y, Wang Y, Li L et al (2013) Crucial role of calcium-sensing receptor activation in cardiac injury of diabetic rats. PLoS One 8(5):e65147
163. Li L, Chen F, Cao YG, Qi HP, Huang W, Wang Y et al (2015) Role of calcium-sensing receptor in cardiac injury of hereditary epileptic rats. Pharmacology 95:10–21
164. Zhen Y, Ding C, Sun J, Wang Y, Li S, Dong L (2016) Activation of the calcium-sensing receptor promotes apoptosis by modulating the JNK/p38 MAPK pathway in focal cerebral ischemia-reperfusion in mice. Am J Transl Res 8(2):911–921
165. Herenbrink CK, Sykes DA, Donthamsetti P, Canals M, Coudrat T, Shonberg J et al (2016) The role of kinetic context in apparent biased agonism at GPCRs. Nat Commun 7:1–14
166. Kenakin T (2011) Functional selectivity and biased receptor signaling. J Pharmacol Exp Ther 336(2):296–302
167. Kenakin T, Christopoulos A (2013) Signalling bias in new drug discovery: detection, quantification and therapeutic impact. Nat Rev Drug Discov 12(3):205–216

168. Leach K, Conigrave AD, Sexton PM, Christopoulos A (2015) Towards tissue-specific pharmacology: insights from the calcium-sensing receptor as a paradigm for GPCR (patho)physiological bias. Trends Pharmacol Sci 36(4):215–225
169. Cavanaugh A, Huang Y, Breitwieser GE (2012) Behind the curtain: cellular mechanisms for allosteric modulation of calcium-sensing receptors. Br J Pharmacol 165(6):1670–1677
170. Cook AE, Mistry SN, Gregory KJ, Furness SGB, Sexton PM, Scammells PJ et al (2015) Biased allosteric modulation at the CaS receptor engendered by structurally diverse calcimimetics. Br J Pharmacol 172(1):185–200
171. Davey AE, Leach K, Valant C, Conigrave AD, Sexton PM, Christopoulos A (2012) Positive and negative allosteric modulators promote biased signaling at the calcium-sensing receptor. Endocrinology 153(3):1232–1241
172. Leach K, Gregory KJ, Kufareva I, Khajehali E, Cook AE, Abagyan R et al (2016) Towards a structural understanding of allosteric drugs at the human calcium-sensing receptor. Cell Res 26(5):574–592
173. Thomsen ARB, Worm J, Jacobsen SE, Stahlhut M, Latta M, Brauner-Osborne H (2012) Strontium is a biased agonist of the calcium-sensing receptor in rat medullary thyroid carcinoma 6-23 cells. J Pharmacol Exp Ther 343(3):638–649
174. Kenakin T (2002) Drug efficacy at G protein–coupled receptors. Annu Rev Pharmacol Toxicol 42(1):349–379
175. Huang C, Sindic A, Hill CE, Hujer KM, Chan KW, Sassen M et al (2007) Interaction of the Ca^{2+}-sensing receptor with the inwardly rectifying potassium channels Kir4.1 and Kir4.2 results in inhibition of channel function. Am J Physiol Renal Physiol 44106:1073–1081
176. Pi M, Spurney RF, Tu Q, Hinson T, Quarles LD (2018) Calcium-sensing receptor activation of Rho involves filamin and Rho-guanine nucleotide exchange factor. Endocrinology 143(January):3830–3838
177. Adjobo-Hermans MJW, Goedhart J, van Weeren L, Nijmeijer S, Manders EMM, Offermanns S et al (2011) Real-time visualization of heterotrimeric G protein Gq activation in living cells. BMC Biol 9(1):32
178. Depry C, Mehta S, Zhang J (2013) Multiplexed visualization of dynamic signaling networks using genetically encoded fluorescent protein-based biosensors. Pflugers Arch Eur J Physiol 465(3):373–381
179. Jean-Alphonse F, Bowersox S, Chen S, Beard G, Puthenveedu MA, Hanyaloglu AC (2014) Spatially restricted G protein-coupled receptor activity via divergent endocytic compartments. J Biol Chem 289(7):3960–3977

Chapter 42
Extracellular Ca^{2+} in Bone Marrow

Ryota Hashimoto

Abstract Our blood serum Ca^{2+} levels are maintained within a narrow range (Ca^{2+} homeostasis) through a complex feedback system. However, local bone marrow Ca^{2+} levels can reach high concentrations, at least transiently, due to bone resorption, which is one of the notable features of the bone marrow stroma. Bone homeostasis is maintained by both the balance between osteoblastic bone formation and osteoclastic bone resorption and the balance of mesenchymal stem cell differentiation into osteoblasts and adipocytes. It has been reported that under culture conditions of infrequent adipocyte differentiation (no treatment with insulin or dexamethasone), high extracellular Ca^{2+} enhances osteoblast but not adipocyte accumulation in bone marrow stromal cells. In contrast, under culture conditions of predominant adipocyte differentiation (treatment with insulin and dexamethasone), high extracellular Ca^{2+} enhances adipocyte but not osteoblast accumulation in bone marrow stromal cells. Thus, the increased extracellular Ca^{2+} caused by bone resorption might enhance osteoblast development to reform missing bone under conditions of infrequent adipocyte differentiation (such as the normal physiological state) and might accelerate adipocyte accumulation instead of osteoblastic bone formation under conditions of predominant adipocyte differentiation (such as aging, obesity, use of glucocorticoids, and postmenopause). Moreover, increased adipocyte accumulation in bone marrow suppresses lymphohematopoiesis and contributes to a dysfunction of osteogenesis.

Keywords Extracellular Ca^{2+} · Hypercalcemia · Hypocalcemia · Bone marrow · Mesenchymal stem cells · Osteoblasts · Adipocytes

---

R. Hashimoto (✉)
Department of Physiology, Juntendo University Faculty of Medicine, Bunkyo-ku, Tokyo, Japan
e-mail: hryota@juntendo.ac.jp

© Springer Nature Switzerland AG 2020

M. S. Islam (ed.), *Calcium Signaling*, Advances in Experimental Medicine and Biology 1131, https://doi.org/10.1007/978-3-030-12457-1_42

42.1 Introduction

The concentration of calcium ions (Ca^{2+}) is well-controlled both intracellularly and extracellularly to maintain homeostasis in various vertebrates. Although the extracellular concentration of Ca^{2+} is on the order of 1 mM, the intracellular Ca^{2+} concentration ($[Ca^{2+}]_i$) is maintained on the order of 100 nM. Cells can increase their $[Ca^{2+}]_i$ to approximately 1 μM after stimulation. Increased intracellular Ca^{2+} works as a second messenger to control various cell functions (including fertilization, proliferation, differentiation, muscle contraction, and secretion) [1, 2]. The reader can reference other chapters in this book for more information regarding intracellular Ca^{2+}. Changes in extracellular Ca^{2+} concentration also seem to be important in cell functions. The molecular identification of the extracellular calcium-sensing receptor (CaSR), which is a monitor for extracellular Ca^{2+}, has opened up the possibility that Ca^{2+} might also function as a first messenger [3–5]. In this chapter, I focus on extracellular Ca^{2+}, particularly in bone marrow.

42.2 Significance of Maintaining Extracellular Ca^{2+} Levels within a Narrow Range

There is approximately 1.2 kg of total body Ca in adult humans, and approximately 99% of total body Ca exists in bones and teeth, where it is important for supporting the three-dimensional structure of the body. Approximately 1% of total body Ca is present in cells, and 0.1% is present in the extracellular fluid [6–8]. These numbers appear contradictory because the volume of intracellular fluid is approximately twice as great as that of the extracellular fluid,and the intracellular concentration of Ca^{2+} is approximately 10,000 times lower than the extracellular Ca^{2+} concentration. Although the cytoplasmic Ca^{2+} concentration is low, a substantial amount of Ca is intracellularly stored in the endoplasmic reticulum (ER; the ER in striated muscle is structurally and functionally different from that in other cells and is named the sarcoplasmic reticulum (SR)) and mitochondria [9], which results in more Ca content in cells than in the extracellular fluid. Despite the low amount of extracellular Ca^{2+}, the concentration of extracellular Ca^{2+} in blood serum must be strictly regulated within a narrow range. The normal range of blood serum Ca concentrations is 2.2–2.6 mmol/L (8.5–10.5 mg/dL). Blood serum Ca levels greater than 2.6 mmol/L are defined as hypercalcemia, whereas levels less than 2.2 mmol/L are defined as hypocalcemia [10, 11]. Approximately 40% of plasma Ca is protein-bound, and 10% of Ca is in a complex with anions, such as phosphate and citrate; thus, approximately half of plasma Ca is in its free form (ionized form; Ca^{2+}), which is physiologically important [1, 2].

The symptoms of hypercalcemia include drowsiness, constipation, nausea, cardiac arrhythmias, etc., while the typical symptom of hypocalcemia is a tetany, which is a continuous tonic muscle spasm. As Ca^{2+} blocks sodium channels and inhibits the depolarization of neurons, hypercalcemia raises the threshold for depolarization, while hypocalcemia lowers the threshold for depolarization [12–14]. Thus, high levels of Ca^{2+} decrease the conductance of the neuronal membrane to Na^{+}, which leads to decreased excitability, resulting in hypotonus of muscles. This relation can explain the drowsiness, constipation and nausea in hypercalcemia and tetany in hypocalcemia. This mechanism can also explain bradycardia and atrioventricular block in hypercalcemia. On the other hand, QT shortening in hypercalcemia is explained by a rapid influx of Ca^{2+} through L-type Ca^{2+} channels in cardiomyocytes. The rapid influx of Ca^{2+} quickly reaches the threshold to close the L-type Ca^{2+} channels, thus reducing the duration of phase 2 of the myocardial action potential, which results in QT shortening in hypercalcemia [15, 16]. It has also been suggested that reducing the duration of both phases 2 and 3, during which time Ca^{2+}-activated K^{+} channels increase outward K^{+} currents and accelerate the process of repolarization, contributes to shortening the QT interval in hypercalcemia [17].

42.3 Mechanisms to Maintain Extracellular Ca^{2+} Levels Within a Narrow Range

To prevent these unfavorable symptoms, our blood serum Ca^{2+} levels are maintained within a narrow range (Ca^{2+} homeostasis) through a complex feedback system that involves hormones. First, the CaSR is a G protein-coupled receptor and a monitor of extracellular $[Ca^{2+}]$ [3–5]. In response to increased serum Ca^{2+} levels, the C cells (parafollicular cells) of the thyroid gland [18, 19] increase the secretion of calcitonin (CT). In response to decreased serum Ca^{2+} levels, the chief cells of the parathyroid gland increase parathyroid hormone (PTH) secretion [20].

The main target organs of these hormones are bone, the small intestine, and the kidney (Fig. 42.1). Bone is a vast reservoir of Ca, and the resorption of bone mineral releases Ca^{2+} into the blood [21]. The small intestine is the site where dietary Ca^{2+} is absorbed [22], and the kidney regulates Ca^{2+} excretion [23]. The input of Ca^{2+} into the blood can originate from bone mineral resorption and intestinal Ca^{2+} absorption. Moreover, the output of Ca^{2+} from the blood can occur through bone formation and renal Ca^{2+} excretion. The Ca^{2+} balance is the net sum of these Ca^{2+} inputs and outputs in the body.

PTH increases Ca^{2+} levels in the blood through bone and the kidneys (Figs. 42.2 and 42.4):

1. PTH promotes the absorption of Ca^{2+} from the bone in 2 ways. The rapid phase is an increase in serum Ca^{2+} within minutes. Bone surfaces are covered with a cellular layer containing both osteocytes and osteoblasts that is referred to as the osteocytic membrane. Between the osteocytic membrane and the solid bone, there is bone material, which is not fully crystallized and is referred to as the bone fluid. When PTH binds to receptors on osteoblasts and osteocytes,

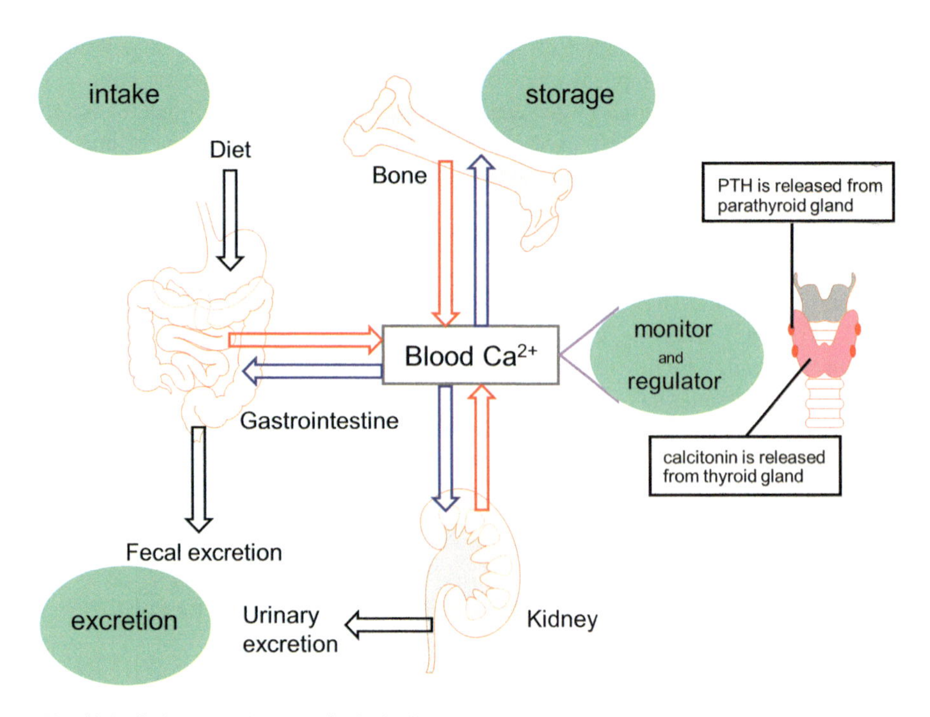

Fig. 42.1 Ca inputs and outputs in the body
Arrows show Ca flows; arrows indicating the increasing and decreasing of blood Ca^{2+} are red and blue, respectively. Bone is a vast reservoir of Ca, and resorption of bone mineral releases Ca^{2+} into the blood. The small intestine is the site where dietary Ca^{2+} is absorbed, and the kidney regulates Ca^{2+} excretion. Input of Ca^{2+} into the blood can originate from bone mineral resorption, intestinal Ca^{2+} absorption, and reabsorption from the kidney. Output of Ca^{2+} from the blood can occur through bone formation, secretion in the colon, and filtration in the kidney. The Ca^{2+} balance is the net sum of these Ca^{2+} inputs and outputs in the body. In response to increased blood Ca^{2+} levels, the C cells of the thyroid gland increase the secretion of calcitonin. In response to decreased blood Ca^{2+} levels, the chief cells of the parathyroid gland increase parathyroid hormone (PTH) secretion

the osteocytic membrane pumps Ca^{2+} from the bone fluid into the extracellular fluid. The slow phase of bone resorption occurs over several days. When osteoblasts and osteocytes are stimulated by PTH, they upregulate the expression of RANK ligand (RANKL) on the plasma membrane. RANKL binds to the receptor activator of the nuclear factor-kappa B (RANK) of osteoclast precursors, which activates signaling pathways that promote osteoclast differentiation and activation. Osteoclasts are bone-resorbing cells and produce a release of Ca^{2+} [24–27].

2. PTH increases the reabsorption of Ca^{2+} in the kidney (i.e., PTH suppresses renal Ca^{2+} excretion) [23].

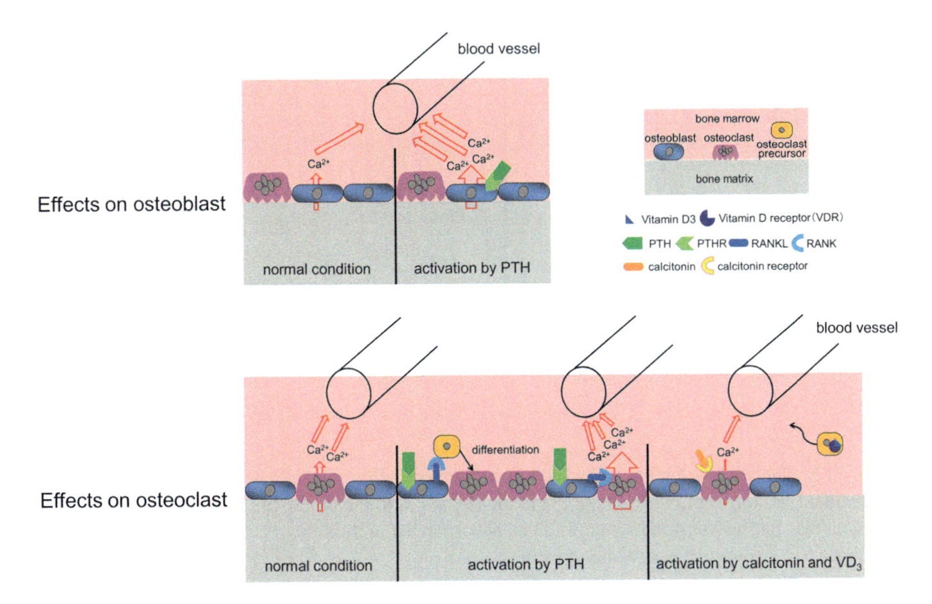

Fig. 42.2 Bone is a vast reservoir of Ca
Bone is a vast reservoir of Ca, and absorption of bone mineral releases Ca^{2+} into the blood. PTH promotes the absorption of Ca^{2+} from the bone in 2 ways. The rapid phase is an increase in serum Ca^{2+} within minutes, which appears to occur through the activation of osteoblasts. The slow phase of bone resorption occurs over several days. When osteoblasts are stimulated by PTH, they upregulate the expression of RANK ligand (RANKL). RANKL binds to receptor activator of nuclear factor-kappa B (RANK) of osteoclast precursors, which activates signaling pathways that promote osteoclast differentiation and activation. Osteoclasts are bone-resorbing cells and produce a release of Ca^{2+}. Calcitonin inhibits the activity of osteoclasts. Active vitamin D enhances the mobilization of osteoclast precursors from the bone to the blood. Thus, calcitonin and vitamin D suppress the absorption of Ca^{2+} from the bone

PTH also converts 25-hydroxyvitamin D to its most active metabolite, 1,25-dihydroxyvitamin D-3 [1,25-$(OH)_2D_3$], in the kidney. Active vitamin D, in turn, acts to suppress PTH production in the parathyroid gland, which is a mild inhibitory system of PTH release [28, 29]. The inhibition of PTH release primarily occurs by a direct effect of the Ca^{2+} concentration at the parathyroid gland as previously described.

Active vitamin D works as a hormone and increases Ca^{2+} levels in the blood through the small intestine and kidney (Figs. 42.2, 42.3 and 42.4):

1. Active vitamin D increases intestinal Ca^{2+} absorption from ingested food [30, 31].
2. Active vitamin D increases the reabsorption of Ca^{2+} in the kidney (i.e., active vitamin D suppresses renal Ca^{2+} excretion) [23].

Active vitamin D has been widely used in the treatment of rickets, osteomalacia, and osteoporosis to prevent bone loss. Active vitamin D enhances the mobilization of osteoclast precursors from the bone to the blood through the suppression of

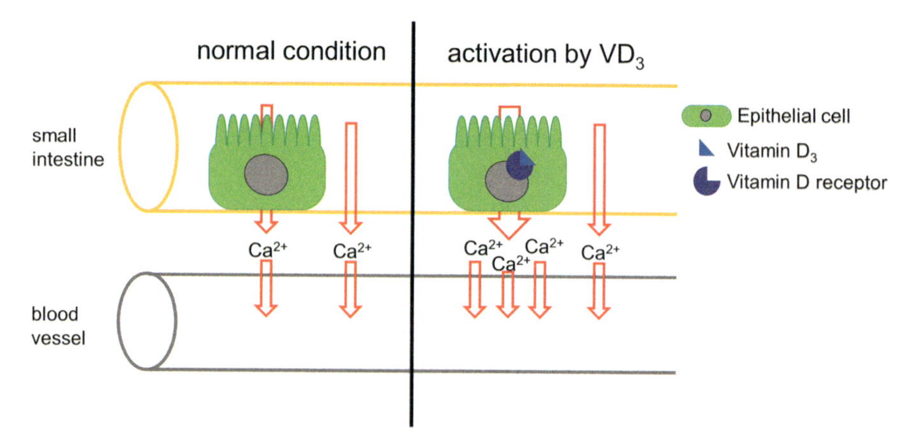

Fig. 42.3 The small intestine is the site where dietary Ca is absorbed
Intestinal Ca^{2+} absorption results in the input of Ca^{2+} into the blood through the transcellular and paracellular pathways. Active vitamin D causes changes in the function of epithelial cells, which enhance Ca^{2+} transport across the intestine

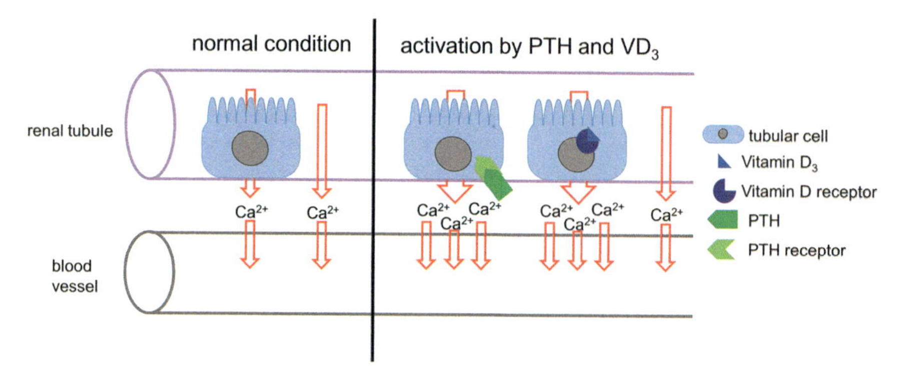

Fig. 42.4 The kidney regulates Ca excretion
The ionized and complexed form of blood Ca (excluding the protein bound form) is freely filtered through the renal glomerulus. Approximately 99% of filtered Ca is reabsorbed in the normal condition into the blood along the renal tubules through the transcellular and paracellular pathways. PTH and active vitamin D change the function of tubular cells, which increase the reabsorption of Ca^{2+} in the kidney (i.e., PTH and active vitamin D suppress renal Ca excretion)

sphingosine-1-phosphate receptor-2 (S1PR2) expression, thereby contributing to limiting osteoclastic bone resorption [32]. This report resolves the discrepancy of in vitro studies that have shown active vitamin D increases the expression of RANKL in osteoblasts and osteocytes, thereby acting as osteoclastogenic, bone-resorbing factors, which is inconsistent with the action of vitamin D on bone in vivo [33].

Calcitonin reduces Ca^{2+} levels in the blood through bone (Fig. 42.2):

1. Calcitonin inhibits the activity of osteoclasts. The osteoclast inhibition reduces the amount of Ca^{2+} released into the blood from bone [34].

The importance of human calcitonin for Ca^{2+} homeostasis in humans has not been established. However, calcitonin of fish (particularly salmon) actively affects human Ca^{2+} homeostasis [34]. In fact, salmon calcitonin has been used therapeutically for the treatment of hypercalcemia and osteoporosis. However, the use of salmon calcitonin is limited because of its involvement in prostate cancer and bone metastasis [35]. Thus, in various ways, these hormones and organs participate in supplying Ca^{2+} to the blood and removing it from the blood when necessary (Figs. 42.1, 42.2, 42.3 and 42.4).

42.4 Extracellular Ca^{2+} in Bone Marrow

42.4.1 In the Developing Skeletal Tissue

Our body length rapidly increases during fetal life and early childhood; however, linear growth progressively slows and eventually ceases during adolescence. Thus, skeletal development begins in the early stages of embryogenesis and continues postnatally until the peak bone mass is achieved in early adulthood. Longitudinal bone growth occurs at the growth plate of the long bone by endochondral ossification, a two-step process in which cartilage is first formed and then remodeled into bone. In the growth plate, chondrocytes proliferate, mature, hypertrophy, reach terminal differentiation and then deposit Ca^{2+}/phosphate-containing mineral in the surrounding matrix. Within this mineralized matrix, chondrocytes release growth factors to induce vascular invasion and guide the differentiation of incoming osteoclast progenitors and osteoblasts, which have respective bone resorbing and bone forming activities that replace cartilage with bone [24, 36, 37]. It has also been reported that hypertrophic chondrocytes can survive and become osteoblasts and osteocytes during endochondral bone formation [38]. Growth plates exist throughout adolescence to support longitudinal bone growth by repeating the previously described cell differentiation programs until the time of growth plate closure in early adulthood.

Accelerated or delayed chondrocyte differentiation leads to disorganized growth plates and retarded bone growth. Various factors, such as PTH-related protein (PTHrP) and insulin-like growth factor 1 (IGF1), and their receptors control the pace of chondrocyte differentiation [39, 40]. Here, I introduce Ca^{2+} and the CaSR as key modulators of chondrocyte differentiation. As previously described, blood serum Ca^{2+} levels are maintained within a narrow range. However, local bone marrow Ca^{2+} levels can reach high concentrations due to bone resorption [41], which is one of the notable features of the bone marrow stroma, indicating that bone marrow stroma can have elevated extracellular Ca^{2+} in vivo, at least transiently. It has been reported that high Ca^{2+} reduces the expression of early differentiation markers of chondrocytes, increases the expression of late differentiation markers, and increases mineral accumulation [42, 43]. In the growth plates, the CaSR is detected in chondrocytes, and its expression increases as cells hypertrophy [44]. In addition, mice with chondrocyte-specific ablation of the CaSR gene exhibit a shorter, undermineralized skeleton due to delayed differentiation of hypertrophic chondrocytes [45]. These reports suggest that Ca^{2+} promotes chondrocyte differentiation.

42.4.2 In Adulthood – Under a Normal Physiological State

The mature skeleton in adulthood (following growth plate closure) is subsequently maintained by continuous bone remodeling; approximately 10% of the total adult bone mass turns over each year, resulting in a complete regeneration of the adult skeleton every 10 years [46, 47]. Bone remodeling is the replacement of old bone tissue by new bone tissue. It involves the processes of bone resorption by osteoclasts and bone deposition by osteoblasts. Bone homeostasis is maintained by the balance between osteoclastic bone resorption and osteoblastic bone formation [48].

Bone resorption by osteoclasts, which are derived from hematopoietic stem cells, gives rise to the release of Ca^{2+}. Increased Ca^{2+} enhances the proliferation [49, 50], differentiation [51, 52], and chemotaxis [49, 53] of osteoblasts to reform missing bone under a normal physiological state. In bone, the CaSR is detected in osteoblasts and osteocytes [44]. In mice that lack CaSR in immature osteoblasts, bone defects, which result from decreased osteoblast numbers, abnormal mineralizing activities, and increased osteoclast numbers and activities, are observed [45, 54]. These results indicate that the CaSR is directly essential for the proliferation, survival, and maturation of immature osteoblasts and that both the numbers and activities of osteoclasts are regulated through the CaSR of osteoblasts.

The increased Ca^{2+} also directly affects osteoclasts to prevent their further expansion and aberrant bone resorption. High Ca^{2+} inhibits the differentiation [55, 56] and bone-resorbing functions in osteoclasts [57–59] and increases the apoptosis of osteoclasts [56]. CaSR expression has been detected in osteoclasts, and the CaSR is involved in these functions [55, 56, 59]. These balanced bone-resorbing and bone-forming activities through Ca^{2+} sustain a steady bone turnover rate to continuously remodel the skeleton without bone loss [24, 36, 37].

42.4.3 In Adulthood – Under Aging, Obesity, Use of Glucocorticoids, and Postmenopause

Bone marrow stromal cells include mesenchymal stem cells, which are capable of differentiating into osteoblasts, chondrocytes, and adipocytes [60]. Bone homeostasis is maintained by both the balance between osteoblastic bone formation and osteoclastic bone resorption [48] and the balance of mesenchymal stem cell differentiation into osteoblasts and adipocytes. As described in (42.4.2), high Ca^{2+} enhances the chemotaxis, proliferation, and differentiation of osteoblasts. However, despite the increased Ca^{2+} that results from bone resorption, bone mass declines and adipocyte levels in the marrow increase during aging [61], obesity [61], the use of glucocorticoids [62, 63], and postmenopause [64, 65]. Adipocyte accumulation in bone marrow suppresses lymphohematopoiesis [66, 67] and contributes to a dysfunction of osteogenesis [68–71]. Aging impairs the osteogenic lineage, and high-fat diet feeding activates the expansion of the adipogenic lineage in bone marrow [61]. Glucocorticoids are used in the treatment of a wide range of diseases, and glucocorticoid-induced osteoporosis is the most common secondary cause of osteoporosis. Glucocorticoids commit the differentiation of stem cells to adipocytes in preference to osteoblasts, which results in decreased numbers of osteoblasts. In addition, glucocorticoids enhance the differentiation and activities of osteoclasts [62, 63]. Postmenopausal osteoporosis is associated with estrogen deficiency because the loss of estrogens leads to a reduction in bone formation, an increase in bone resorption, and higher levels of marrow adipogenesis [72–75]. Estrogen promotes osteoblast differentiation and inhibits adipocyte differentiation in bone marrow stromal cells [76]. Osteoporotic fractures are common in patients with chronic kidney disease [77]. However, there are no significant differences in the gene expression of stemness markers or the morphology of adipocytes and osteoblasts differentiated from mesenchymal stem cells between chronic kidney disease and control groups [78].

We have reported that under culture conditions of infrequent adipocyte differentiation (no treatment with insulin or dexamethasone), high extracellular Ca^{2+} enhances osteoblast but not adipocyte accumulation in bone marrow stromal cells [79]. It is important to note that under culture conditions of predominant adipocyte differentiation (treatment with insulin and dexamethasone), high extracellular Ca^{2+} enhances adipocyte but not osteoblast accumulation in bone marrow stromal cells (Fig. 42.5, [79]). In addition, we have reported that another CaSR agonist, Sr^{2+}, enhances adipocyte accumulation under culture conditions of predominant adipocyte differentiation [80]. We have also reported that high extracellular Ca^{2+} enhances the proliferation of bone marrow stromal cells by increasing $[Ca^{2+}]_i$ and enhances their differentiation into adipocytes by decreasing intracellular cAMP (Fig. 42.5, [80, 81]). I propose that the increased extracellular Ca^{2+} caused by bone resorption might enhance osteoblast development to form bone under conditions of infrequent adipocyte differentiation (such as the normal physiological state) and might accelerate adipocyte accumulation instead of osteoblastic bone formation under conditions of predominant adipocyte differentiation (such as aging, obesity, the use of glucocorticoids, and postmenopause). Moreover, increased adipocyte accumulation in bone marrow suppresses lymphohematopoiesis and contributes to a dysfunction of osteogenesis.

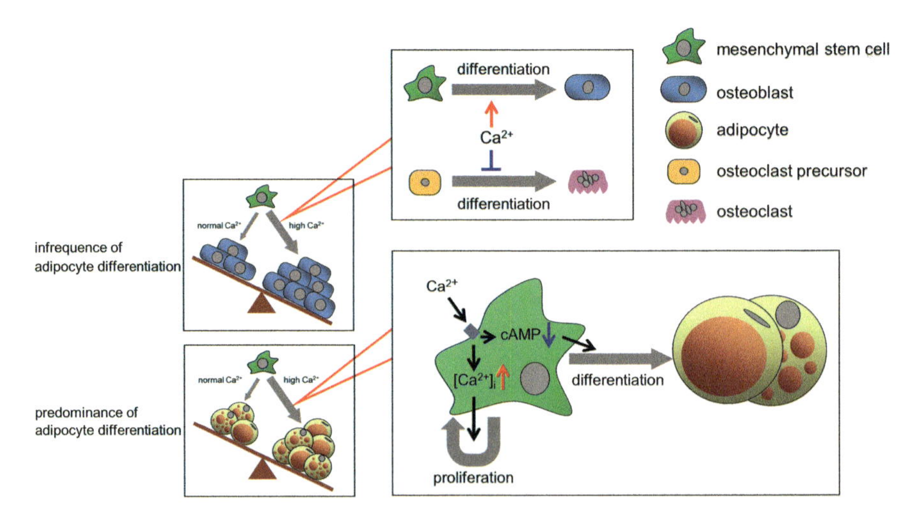

Fig. 42.5 Effects of extracellular Ca^{2+} on bone marrow cells
Bone marrow contains mesenchymal stem cells, which are capable of differentiating into osteoblasts, chondrocytes, and adipocytes. Under conditions of infrequent adipocyte differentiation, high Ca^{2+} enhances the proliferation and osteochondrogenesis of mesenchymal stem cells and suppresses the differentiation of osteoclast precursors. Under conditions of predominant adipocyte differentiation, high Ca^{2+} enhances the proliferation and adipogenesis of mesenchymal stem cells, which might accelerate adipocyte accumulation instead of osteoblastic bone formation. High Ca^{2+} enhances the proliferation of bone marrow stromal cells by increasing $[Ca^{2+}]_i$ and enhances their differentiation into adipocytes by decreasing intracellular cAMP

References

1. Berridge MJ, Bootman MD, Lipp P (1998) Calcium – a life and death signal. Nature 395:645–648
2. Mikoshiba K (2015) Role of IP3 receptor signaling in cell functions and diseases. Adv Biol Reg 57:217–227
3. Brown EM, Gamba G, Riccardi D, Lombardi M, Butters R, Kifor O, Sun A, Hediger MA, Lytton J, Hebert SC (1993) Cloning and characterization of an extracellular Ca^{2+}-sensing receptor from bovine parathyroid. Nature 366:575–580
4. Bouschet T, Henley JM (2005) Calcium as an extracellular signalling molecule: perspectives on the calcium sensing receptor in the brain. C R Biol 328:691–700
5. Hofer AM, Brown EM (2003) Extracellular calcium sensing and signalling. Nat Rev Mol Cell Biol 4:530–538
6. Weaver CM, Gordon CM, Janz KF, Kalkwarf HJ, Lappe JM, Lewis R, O'Karma M, Wallace TC, Zemel BS (2016) The National Osteoporosis Foundation's position statement on peak bone mass development and lifestyle factors: a systematic review and implementation recommendations. Osteoporos Int 27:1281–1386
7. Bushinsky, D. A. (2010) Contribution of intestine, bone, kidney, and dialysis to extracellular fluid calcium content, Clin J Am Soc Nephrol*: CJASN.* 5 Suppl 1, S12-S22
8. Peacock M (2010) Calcium metabolism in health and disease. Clin J Am Soc Nephrol 5(Suppl 1):S23–S30
9. Pizzo P, Pozzan T (2007) Mitochondria-endoplasmic reticulum choreography: structure and signaling dynamics. Trends Cell Biol 17:511–517

10. Stokes VJ, Nielsen MF, Hannan FM, Thakker RV (2017) Hypercalcemic disorders in children. J Bone Miner Res 32:2157–2170
11. Mirrakhimov AE (2015) Hypercalcemia of malignancy: an update on pathogenesis and management. N Am J Med Sci 7:483–493
12. Frankenhaeuser B, Hodgkin AL (1957) The action of calcium on the electrical properties of squid axons. J Physiol 137:218–244
13. Yamamoto D, Yeh JZ, Narahashi T (1984) Voltage-dependent calcium block of normal and tetramethrin-modified single sodium channels. Biophys J 45:337–344
14. Armstrong CM, Cota G (1999) Calcium block of Na^{+} channels and its effect on closing rate. Proc Natl Acad Sci U S A 96:4154–4157
15. Ahmed R, Hashiba K (1988) Reliability of QT intervals as indicators of clinical hypercalcemia. Clin Cardiol 11:395–400
16. Grandi E, Pasqualini FS, Pes C, Corsi C, Zaza A, Severi S (2009) Theoretical investigation of action potential duration dependence on extracellular Ca^{2+} in human cardiomyocytes. J Mol Cell Cardiol 46:332–342
17. Kazama I (2017) High-calcium exposure to frog heart: a simple model representing hypercalcemia-induced ECG abnormalities. J Vet Med Sci 79:71–75
18. Freichel M, Zink-Lorenz A, Holloschi A, Hafner M, Flockerzi V, Raue F (1996) Expression of a calcium-sensing receptor in a human medullary thyroid carcinoma cell line and its contribution to calcitonin secretion. Endocrinology 137:3842–3848
19. McGehee DS, Aldersberg M, Liu KP, Hsuing S, Heath MJ, Tamir H (1997) Mechanism of extracellular Ca^{2+} receptor-stimulated hormone release from sheep thyroid parafollicular cells. J Physiol 502. (Pt 1:31–44
20. Nemeth EF, Steffey ME, Hammerland LG, Hung BC, Van Wagenen BC, DelMar EG, Balandrin MF (1998) Calcimimetics with potent and selective activity on the parathyroid calcium receptor. Proc Natl Acad Sci U S A 95:4040–4045
21. Blair HC, Robinson LJ, Huang CL, Sun L, Friedman PA, Schlesinger PH, Zaidi M (2011) Calcium and bone disease. BioFactors (Oxford, England) 37:159–167
22. Bronner F (2003) Mechanisms of intestinal calcium absorption. J Cell Biochem 88:387–393
23. Jeon US (2008) Kidney and calcium homeostasis. Electrolyte Blood Press 6:68–76
24. Santa Maria C, Cheng Z, Li A, Wang J, Shoback D, Tu CL, Chang W (2016) Interplay between CaSR and PTH1R signaling in skeletal development and osteoanabolism. Semin Cell Dev Biol 49:11–23
25. Levine BS, Rodriguez M, Felsenfeld AJ (2014) Serum calcium and bone: effect of PTH, phosphate, vitamin D and uremia. Nefrologia Publ oficial de la Sociedad Espanola Nefrologia 34:658–669
26. Nakashima T, Hayashi M, Fukunaga T, Kurata K, Oh-Hora M, Feng JQ, Bonewald LF, Kodama T, Wutz A, Wagner EF, Penninger JM, Takayanagi H (2011) Evidence for osteocyte regulation of bone homeostasis through RANKL expression. Nat Med 17:1231–1234
27. Xiong J, Onal M, Jilka RL, Weinstein RS, Manolagas SC, O'Brien CA (2011) Matrix-embedded cells control osteoclast formation. Nat Med 17:1235–1241
28. Henry HL (2011) Regulation of vitamin D metabolism. Best Pract Res Clin Endocrinol Metab 25:531–541
29. Brenza HL, DeLuca HF (2000) Regulation of 25-hydroxyvitamin D3 1alpha-hydroxylase gene expression by parathyroid hormone and 1,25-dihydroxyvitamin D3. Arch Biochem Biophys 381:143–152
30. Veldurthy V, Wei R, Oz L, Dhawan P, Jeon YH, Christakos S (2016) Vitamin D, calcium homeostasis and aging. Bone research 4:16041
31. Bouillon R, Suda T (2014) Vitamin D: calcium and bone homeostasis during evolution. BoneKEy reports 3:480
32. Kikuta J, Kawamura S, Okiji F, Shirazaki M, Sakai S, Saito H, Ishii M (2013) Sphingosine-1-phosphate-mediated osteoclast precursor monocyte migration is a critical point of control in antibone-resorptive action of active vitamin D. Proc Natl Acad Sci U S A 110:7009–7013

33. Suda T, Takahashi F, Takahashi N (2012) Bone effects of vitamin D – discrepancies between in vivo and in vitro studies. Arch Biochem Biophys 523:22–29
34. Naot D, Cornish J (2008) The role of peptides and receptors of the calcitonin family in the regulation of bone metabolism. Bone 43:813–818
35. Warrington JI, Richards GO, Wang N (2017) The role of the calcitonin peptide family in prostate cancer and bone metastasis. Curr Mol Biol Rep 3:197–203
36. Goltzman D, Hendy GN (2015) The calcium-sensing receptor in bone–mechanistic and therapeutic insights. Nat Rev Endocrinol 11:298–307
37. Riccardi D, Brennan SC, Chang W (2013) The extracellular calcium-sensing receptor, CaSR, in fetal development. Best Pract Res Clin Endocrinol Metab 27:443–453
38. Yang L, Tsang KY, Tang HC, Chan D, Cheah KS (2014) Hypertrophic chondrocytes can become osteoblasts and osteocytes in endochondral bone formation. Proc Natl Acad Sci U S A 111:12097–12102
39. Kronenberg HM (2003) Developmental regulation of the growth plate. Nature 423:332–336
40. Wang Y, Cheng Z, Elalieh HZ, Nakamura E, Nguyen MT, Mackem S, Clemens TL, Bikle DD, Chang W (2011) IGF-1R signaling in chondrocytes modulates growth plate development by interacting with the PTHrP/Ihh pathway. J Bone Miner Res 26:1437–1446
41. Silver IA, Murrills RJ, Etherington DJ (1988) Microelectrode studies on the acid microenvironment beneath adherent macrophages and osteoclasts. Exp Cell Res 175:266–276
42. Bonen DK, Schmid TM (1991) Elevated extracellular calcium concentrations induce type X collagen synthesis in chondrocyte cultures. J Cell Biol 115:1171–1178
43. Rodriguez L, Tu C, Cheng Z, Chen TH, Bikle D, Shoback D, Chang W (2005) Expression and functional assessment of an alternatively spliced extracellular Ca^{2+}–sensing receptor in growth plate chondrocytes. Endocrinology 146:5294–5303
44. Chang W, Tu C, Chen TH, Komuves L, Oda Y, Pratt SA, Miller S, Shoback D (1999) Expression and signal transduction of calcium-sensing receptors in cartilage and bone. Endocrinology 140:5883–5893
45. Chang W, Tu C, Chen TH, Bikle D, Shoback D (2008) The extracellular calcium-sensing receptor (CaSR) is a critical modulator of skeletal development. Sci Signal 1:ra1
46. Manolagas SC (2000) Birth and death of bone cells: basic regulatory mechanisms and implications for the pathogenesis and treatment of osteoporosis. Endocr Rev 21:115–137
47. van Schaick E, Zheng J, Perez Ruixo JJ, Gieschke R, Jacqmin P (2015) A semi-mechanistic model of bone mineral density and bone turnover based on a circular model of bone remodeling. J Pharmacokinet Pharmacodyn 42:315–332
48. Sharan K, Siddiqui JA, Swarnkar G, Chattopadhyay N (2008) Role of calcium-sensing receptor in bone biology. Indian J Med Res 127:274–286
49. Yamaguchi, T., Chattopadhyay, N., Kifor, O., Butters, R. R., Jr., Sugimoto, T. & Brown, E. M. (1998) Mouse osteoblastic cell line (MC3T3-E1) expresses extracellular calcium ($Ca^{2+}o$)-sensing receptor and its agonists stimulate chemotaxis and proliferation of MC3T3-E1 cells, J Bone Miner Res 13, 1530-1538
50. Quarles, L. D., Hartle, J. E., 2nd, Siddhanti, S. R., Guo, R. & Hinson, T. K. (1997) A distinct cation-sensing mechanism in MC3T3-E1 osteoblasts functionally related to the calcium receptor, J Bone Miner Res 12, 393-402
51. Dvorak MM, Siddiqua A, Ward DT, Carter DH, Dallas SL, Nemeth EF, Riccardi D (2004) Physiological changes in extracellular calcium concentration directly control osteoblast function in the absence of calciotropic hormones. Proc Natl Acad Sci U S A 101:5140–5145
52. Yamauchi M, Yamaguchi T, Kaji H, Sugimoto T, Chihara K (2005) Involvement of calcium-sensing receptor in osteoblastic differentiation of mouse MC3T3-E1 cells. Am J Physiol Endocrinol Metab 288:E608–E616
53. Godwin SL, Soltoff SP (1997) Extracellular calcium and platelet-derived growth factor promote receptor-mediated chemotaxis in osteoblasts through different signaling pathways. J Biol Chem 272:11307–11312

54. Dvorak-Ewell MM, Chen TH, Liang N, Garvey C, Liu B, Tu C, Chang W, Bikle DD, Shoback DM (2011) Osteoblast extracellular Ca^{2+} –sensing receptor regulates bone development, mineralization, and turnover. J Bone Miner Res 26:2935–2947
55. Kanatani M, Sugimoto T, Kanzawa M, Yano S, Chihara K (1999) High extracellular calcium inhibits osteoclast-like cell formation by directly acting on the calcium-sensing receptor existing in osteoclast precursor cells. Biochem Biophys Res Commun 261:144–148
56. Mentaverri R, Yano S, Chattopadhyay N, Petit L, Kifor O, Kamel S, Terwilliger EF, Brazier M, Brown EM (2006) The calcium sensing receptor is directly involved in both osteoclast differentiation and apoptosis. FASEB J 20:2562–2564
57. Datta HK, MacIntyre I, Zaidi M (1989) The effect of extracellular calcium elevation on morphology and function of isolated rat osteoclasts. Biosci Rep 9:747–751
58. Zaidi M, Kerby J, Huang CL, Alam T, Rathod H, Chambers TJ, Moonga BS (1991) Divalent cations mimic the inhibitory effect of extracellular ionised calcium on bone resorption by isolated rat osteoclasts: further evidence for a "calcium receptor". J Cell Physiol 149:422–427
59. Kameda T, Mano H, Yamada Y, Takai H, Amizuka N, Kobori M, Izumi N, Kawashima H, Ozawa H, Ikeda K, Kameda A, Hakeda Y, Kumegawa M (1998) Calcium-sensing receptor in mature osteoclasts, which are bone resorbing cells. Biochem Biophys Res Commun 245:419–422
60. Friedenstein AJ, Gorskaja JF, Kulagina NN (1976) Fibroblast precursors in normal and irradiated mouse hematopoietic organs. Exp Hematol 4:267–274
61. Ambrosi TH, Scialdone A, Graja A, Gohlke S, Jank AM, Bocian C, Woelk L, Fan H, Logan DW, Schurmann A, Saraiva LR, Schulz TJ (2017) Adipocyte accumulation in the bone marrow during obesity and aging impairs stem cell-based hematopoietic and bone regeneration. Cell Stem Cell 20:771–784.e6
62. Compston J (2018) Glucocorticoid-induced osteoporosis: an update. Endocrine 61:7–16
63. Hachemi Y, Rapp AE, Picke AK, Weidinger G, Ignatius A, Tuckermann J (2018) Molecular mechanisms of glucocorticoids on skeleton and bone regeneration after fracture. J Mol Endocrinol 61:R75–r90
64. Beekman KM, Veldhuis-Vlug AG, den Heijer M, Maas M, Oleksik AM, Tanck MW, Ott SM, van't Hof RJ, Lips P, Bisschop PH, Bravenboer N (2017) The effect of raloxifene on bone marrow adipose tissue and bone turnover in postmenopausal women with osteoporosis. Bone
65. Syed FA, Oursler MJ, Hefferanm TE, Peterson JM, Riggs BL, Khosla S (2008) Effects of estrogen therapy on bone marrow adipocytes in postmenopausal osteoporotic women. Osteoporos Int 19:1323–1330
66. Naveiras O, Nardi V, Wenzel PL, Hauschka PV, Fahey F, Daley GQ (2009) Bone-marrow adipocytes as negative regulators of the haematopoietic microenvironment. Nature 460:259–263
67. Payne MW, Uhthoff HK, Trudel G (2007) Anemia of immobility: caused by adipocyte accumulation in bone marrow. Med Hypotheses 69:778–786
68. Elbaz A, Wu X, Rivas D, Gimble JM, Duque G (2010) Inhibition of fatty acid biosynthesis prevents adipocyte lipotoxicity on human osteoblasts in vitro. J Cell Mol Med 14:982–991
69. Maurin AC, Chavassieux PM, Frappart L, Delmas PD, Serre CM, Meunier PJ (2000) Influence of mature adipocytes on osteoblast proliferation in human primary cocultures. Bone 26:485–489
70. Maurin AC, Chavassieux PM, Vericel E, Meunier PJ (2002) Role of polyunsaturated fatty acids in the inhibitory effect of human adipocytes on osteoblastic proliferation. Bone 31:260–266
71. Lecka-Czernik B, Moerman EJ, Grant DF, Lehmann JM, Manolagas SC, Jilka RL (2002) Divergent effects of selective peroxisome proliferator-activated receptor-gamma 2 ligands on adipocyte versus osteoblast differentiation. Endocrinology 143:2376–2384
72. Eastell R, O'Neill TW, Hofbauer LC, Langdahl B, Reid IR, Gold DT, Cummings SR (2016) Postmenopausal osteoporosis. Nat Rev Dis Primers 2:16069
73. Levin VA, Jiang X, Kagan R (2018) Estrogen therapy for osteoporosis in the modern era. Osteoporos Int 29:1049–1055

74. Nuttall ME, Gimble JM (2000) Is there a therapeutic opportunity to either prevent or treat osteopenic disorders by inhibiting marrow adipogenesis? Bone 27:177–184
75. Zallone A (2006) Direct and indirect estrogen actions on osteoblasts and osteoclasts. Ann N Y Acad Sci 1068:173–179
76. Okazaki R, Inoue D, Shibata M, Saika M, Kido S, Ooka H, Tomiyama H, Sakamoto Y, Matsumoto T (2002) Estrogen promotes early osteoblast differentiation and inhibits adipocyte differentiation in mouse bone marrow stromal cell lines that express estrogen receptor (ER) alpha or beta. Endocrinology 143:2349–2356
77. Lips P, Goldsmith D, de Jongh R (2017) Vitamin D and osteoporosis in chronic kidney disease. J Nephrol 30:671–675
78. Yamada A, Yokoo T, Yokote S, Yamanaka S, Izuhara L, Katsuoka Y, Shimada Y, Shukuya A, Okano HJ, Ohashi T, Ida H (2014) Comparison of multipotency and molecular profile of MSCs between CKD and healthy rats. Hum Cell 27:59–67
79. Hashimoto R, Katoh Y, Nakamura K, Itoh S, Iesaki T, Daida H, Nakazato Y, Okada T (2012) Enhanced accumulation of adipocytes in bone marrow stromal cells in the presence of increased extracellular and intracellular [Ca(2)(+)]. Biochem Biophys Res Commun 423:672–678
80. Hashimoto R, Katoh Y, Miyamoto Y, Itoh S, Daida H, Nakazato Y, Okada T (2015) Increased extracellular and intracellular Ca(2)(+) lead to adipocyte accumulation in bone marrow stromal cells by different mechanisms. Biochem Biophys Res Commun 457:647–652
81. Hashimoto R, Katoh Y, Miyamoto Y, Nakamura K, Itoh S, Daida H, Nakazato Y, Okada T (2017) High extracellular Ca^{2+} enhances the adipocyte accumulation of bone marrow stromal cells through a decrease in cAMP. Cell Calcium 67:74–80

Chapter 43
Calcium in Cell-Extracellular Matrix Interactions

Sandeep Gopal, Hinke A. B. Multhaupt, and John R. Couchman

Abstract In multicellular organisms, the cells are surrounded by persistent, dynamic extracellular matrix (ECM), the largest calcium reservoir in animals. ECM regulates several aspects of cell behavior including cell migration and adhesion, survival, gene expression and differentiation, thus playing a significant role in health and disease. Calcium is reported to be important in the assembly of ECM, where it binds to many ECM proteins. While serving as a calcium reservoir, ECM macromolecules can directly interact with cell surface receptors resulting in calcium transport across the membrane. This chapter mainly focusses on the role of cell-ECM interactions in cellular calcium regulation and how calcium itself mediates these interactions.

Keywords Extracellular matrix · Cell adhesion · Calcium · Proteoglycans · Integrins · Syndecans · TRP channels · Focal adhesions · Actin cytoskeleton · Mechanosensing

43.1 Introduction

Extracellular matrix (ECM) is a complex network of proteins and polysaccharides that provides mechanical stability for tissues while maintaining its dynamic nature. Major structural proteins in the ECM include collagens, elastin and their associated

---

S. Gopal (✉)
Development and Stem Cells Program, Monash Biomedicine Discovery Institute, Department of Anatomy and Developmental Biology, Monash University, Clayton, VIC, Australia
e-mail: sandeep.gopal@monash.edu

H. A. B. Multhaupt · J. R. Couchman
Biotech Research & Innovation Center, University of Copenhagen, Copenhagen, Denmark
e-mail: hinke.multhaupt@bric.ku.dk; john.couchman@bric.ku.dk

© Springer Nature Switzerland AG 2020

M. S. Islam (ed.), *Calcium Signaling*, Advances in Experimental Medicine and Biology 1131, https://doi.org/10.1007/978-3-030-12457-1_43

glycoproteins, laminins, fibronectin, proteoglycans and matricellular proteins (e.g. tenascins, osteopontin, thrombospondins). The proportion of structural proteins in the ECM varies between tissue types, thus providing a varying degree of stiffness and elasticity to ECM as required [1]. ECM composition is known to change both spatially and temporally as well as in response to inflammation and disease. ECM, initially purported to be an inert support structure, in fact regulates many functions in animals including cell adhesion, durotaxis, gene expression and cell differentiation.

ECM is the largest source of free calcium ions in a multicellular organism, with a 10,000 fold higher concentration than cytosol [2]. The free calcium ion concentration in ECM is approximately 1.2 mM. From the ECM, calcium enters cells via voltage gated or ligand gated calcium channels present in the plasma membrane. As with many interactions, ECM-calcium cross-talk is a two-way conversation that plays a significant role in cell signaling pathways. While ECM regulates the calcium entry into the cell by various means, the calcium maintains a strong influence over several cell-ECM interactions. The entry of calcium from ECM into the cells is a tightly regulated process that involves several calcium channels and regulators, and sometimes cell structures such as focal adhesions and cytoskeleton [3–5].

Calcium can affect most of the aspects of a cell's life from fertilization to apoptosis [6, 7]. At cellular level, calcium ions are a key second messenger in many signaling pathways. It can bind to hundreds of molecules, in some cases to trigger various cellular signaling pathways and in others to regulate intracellular calcium level itself. Binding affinities of calcium to cellular proteins range from nM to mM range with many protein conformational properties dependent on the interaction. This chapter focusses on the role of calcium in regulating cell-ECM interactions and the role of ECM in controlling cellular calcium.

43.2 Role of Calcium in ECM and Cells

In a multicellular mammalian system, cells are always under persistent, dynamic mechanical tension [8–10]. Cells exert tension not only on extracellular matrix but also on neighboring cells. The mechanical tension exerted on the matrix is mediated through junctions such as focal adhesions that are connected to the actin cytoskeleton. Intra- and extracellular calcium levels are powerful regulators of cytoskeletal dynamics and cell-matrix adhesion [11]. Such effects of calcium on cytoskeletal assembly and contractility are due to the results of calcium interaction with several cytoskeletal proteins [12–14] . Roles for calcium in cytoskeletal maintenance are well known, though a comprehensive picture of the mechanisms remains unclear. This is mainly due to the presence of a multitude of protein networks in the cytoskeleton and junctions, which are subject to both rapid and long-term dynamics.

43.2.1 Matrix proteins

Mechanotransduction is an important component of cell survival, where mechanical forces influence signaling cascades and cell behavior. As ECM directly interacts with cells, there will be a significant tension applied to the cell membrane. While mechanotransduction pathways involve cell surface receptors, ion channels, actin cytoskeleton and tyrosine protein kinases, the role of ECM proteins appears to be essential. Previous reports have shown that the several of the ECM components regulate cytosolic calcium levels. At the same time, they have the ability to bind to calcium in the ECM. Depleting or enriching ECM components can significantly alter the mechanical properties of the matrix, which in turn affects the mechanically sensitive channels. For instance, TRP channels, also known as stretch activated channels, are regulated by extracellular tension derived from cell-ECM interactions. This section lists some ECM proteins and how they are connected with calcium homeostasis in the tissues.

43.2.1.1 Fibronectin

Fibronectin is a well characterized and abundant ECM proteins of vertebrates [15]. It can interact with transmembrane proteoglycans and integrins, two major cell surface receptor families that can initiate an array of downstream signaling. Early reports using *Entamoeba histolytica* trophozoites, an active feeding stage of a sporozoan parasite, showed that direct contact with fibronectin induces the cytosolic calcium spikes [16]. Similar results were observed in multiple cell types where fibronectin influenced the cytosolic calcium level [17]. The effect of fibronectin on calcium was later shown to be the result of engagement of its integrin-binding Arg-Gly-Asp motif with subsequent activation of stretch-activated calcium channels (Fig. 43.1) [18]. On the other hand, the significant increase in the fibronectin deposition in the ECM during wound healing is achieved through the TRPC3-induced transcription of fibronectin [19]. Recently identified role of syndecan-4 proteoglycan in cytosolic calcium regulation also appeared to be initiated by fibronectin. It has been demonstrated that the fibroblasts plated on the integrin-binding region of fibronectin (FN110-KDa, Fig. 43.1a) have an elevated calcium level, which can be reduced to the levels observed on whole fibronectin by the addition of solution containing the heparin-binding region, Hep II (fibronectin type III repeats 12–15, Fig. 43.1) [20]. This addition of Hep II domain also rectified the defects in focal adhesions and cytoskeleton observed in the fibroblasts plated on FN110-KDa [21]. Other syndecans also regulate cytosolic calcium, though their connection with fibronectin is unclear. This role of proteoglycans in calcium regulation will be discussed below.

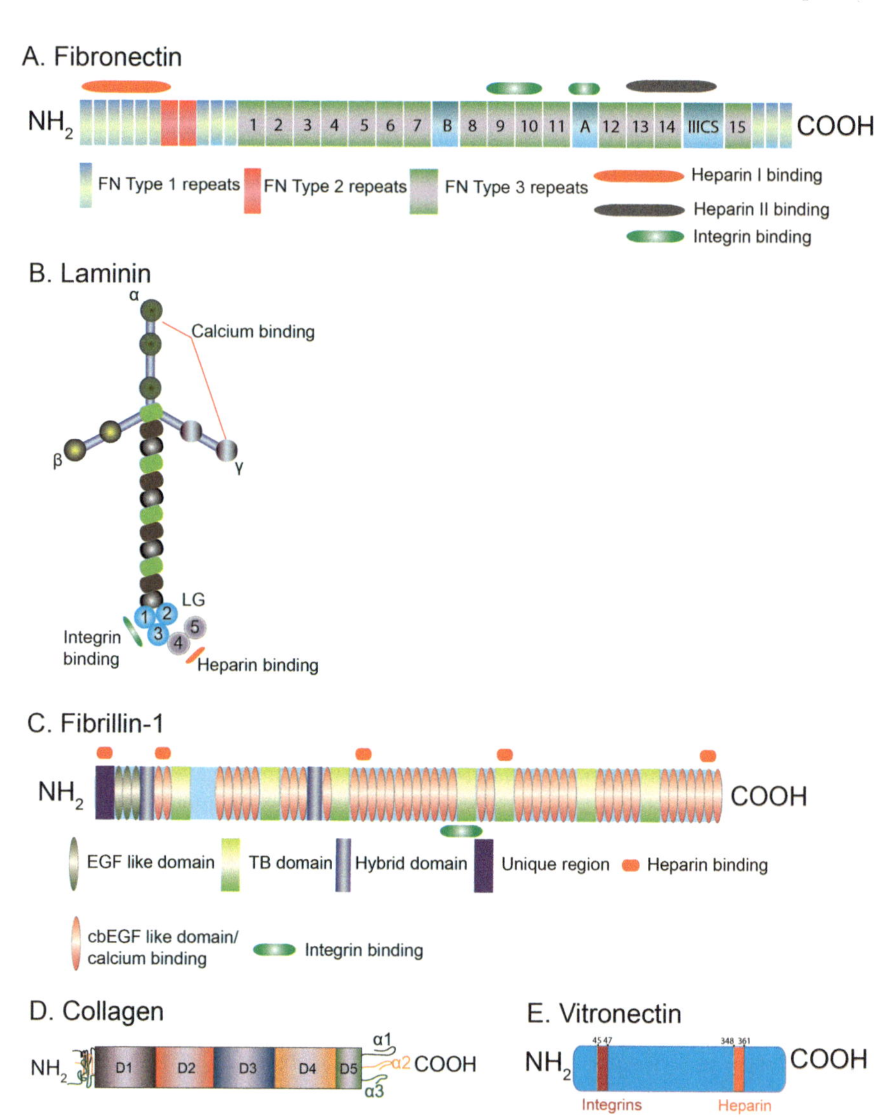

Fig. 43.1 Schematics of major calcium responsive/ regulating ECM molecules. (**a**) Fibronectin structure showing the integrin and heparin (heparan sulfate analogue) interacting domains. (**b**) Laminins are formed of one α, one β and one γ chains. The integrin and heparin interacting regions are shown in the figure. Calcium binding domains of laminins are located on the α and γ arms. (**c**) Fibrillin-1 structure showing the integrin and heparin interacting domains and high calcium affinity cbEGF like domains. (**d**) Collagen I monomer is comprised of three alpha chains forming a supercoiled triple helix. Final structure of collagen monomer assumes a rod-like structure of 300 nm length. Like other ECM molecules, collagen interacts with integrins and heparan sulfate chains. *E* Vitronectin interacts with integrins and heparan sulfate chains

43.2.1.2 Laminins

Laminins are the second most abundant structural component of basement membranes, with an ability to interact with integrins, dystroglycan receptors and negatively charged moieties such as heparan sulfate chains and sulfatides. A total 15 laminin polypeptides are expressed in vertebrates and several of them appear to have calcium binding sites or calcium-induced self-assembly mechanisms [22–24]. Laminins are composed of three distinct subunits (α, β, and γ) (Fig. 43.1b) that self-assemble in a calcium-dependent manner [24]. The self-assembly of laminin is further enhanced during their aggregation onto lipid bilayers where both aggregation and self-assembly depends on calcium binding [25]. There is a limited knowledge regarding the role of laminins in regulating cytosolic calcium. Addition of soluble laminin-211 (α2β1γ1), but not laminin-111 (α1β1γ1) to osteoclast precursor cells resulted in the release of calcium from intracellular calcium stores [26]. The more recent reports suggest that the laminin β2 chain interacts directly with voltage-gated calcium channels in neuromuscular junctions [27–29] where they control calcium sensitivity during neurotransmission [30]. This interaction between laminin β2 and voltage gated calcium channels may involve cytoskeletal proteins such as spectrins, plectin 1, dystrophin, myosin-1and α3 integrin cell surface receptor, though the details remain unknown [29, 31]. It is unclear how widespread the interaction between laminin β2 and voltage gated calcium channel property may be. However, a mutation in laminin α1 subunit in patients with cerebellar ataxia resulted in defective cellular calcium homeostasis [32]. Taken together, it is clear that laminins require calcium for structural integrity, but in turn may influence cytosolic calcium levels.

43.2.1.3 Fibrillin

Fibrillins are large ECM proteins that polymerize to form microfibrils, which can contribute to the architecture of extracellular matrix, including elastic fiber assemblies. Mutations in fibrillins are known to be associated with Marfan syndrome [33], a genetic disease leading to structural defects in connective tissue. At least four isoforms (fibrillin 1–4) of fibrillins are identified so far, with fibrillin-1 being the most studied. The role of calcium in microfibril organization was identified when incubation with EGTA or EDTA resulted in the disintegration of microfibril structure [34, 35]. Fibrillins possess multiple tandem repeats of epidermal growth factor-like motifs that can bind calcium with high affinity, an essential requirement in microfibril formation [36, 37]. Reports suggest that fibrillins contain calcium-binding epidermal growth factor (cbEGF) like domains that are interspersed with TGF-β binding protein like domains [38]. The interdomain interactions between cbEGF and TGF-β binding protein like domain appear to be strongly dependent on calcium [39]. It was shown that the calcium binding to cbEGF like domain plays a major role in maintaining fibrillin-1 structure by restricting the mobility of the interdomain regions [40]. In addition, calcium also appears to be essential for the homotypic and heterotypic interactions between fibrillin isoforms [41, 42].

Finally, calcium plays important role in fibrillin interactions with several matrix proteins including versican, aggrecan, fibulins and HSPGs [43–46]. The presence of calcium ions in the ECM embued fibrillin with resistance to proteolytic enzymes such as trypsin, matrix metalloproteinases and plasmin [47, 48]. While fibrillins can initiate cell signaling through cell surface receptors such as integrins and syndecan HSPGs [43, 49, 50], currently there is no direct evidence to suggest they can regulate cytosolic calcium, but seems likely given other data on these receptor systems.

43.2.1.4 Collagen

Collagens are the most abundant proteins of the ECM. There are at least 28 types of chronologically named (I–XXVIII, Fig. 43.1d) collagens identified in vertebrates. While collagens are important in calcification of cartilage during aging, this section of the chapter focuses only on the role of collagen in calcium mobility in cells. It has been debated for a long time whether the collagen can induce a calcium rise in cells. Most of the information about the role of collagen in calcium regulation comes from platelets. Some reports suggest that collagen raises cytosolic calcium levels in human platelets, possibly by activating calcium channels via phospholipase C and inositol 1,4,5 trisphosphate [51–53]. This observation was later supported when the platelets showed a steep increase in calcium levels when they were in contact with type 1 Collagen [54]. As the most abundant protein in ECM, the collagen contributes to the stiffness and tensile strength of the matrix. As mentioned previously, the ECM-mediated tension can alter calcium levels in cells. Studies using human mesenchymal stem cells plated on the collagen showed that calcium oscillations in the cells vary depending on the rigidity of the collagen matrix [55]. The average calcium spikes in the cells were increased with the increase in rigidity of the collagen matrix, indicating the calcium oscillations are sensitive to mechanical cues. While the study addressed the role of collagen rigidity, it may well be true for several other proteins since nearly all structural ECM proteins can contribute to matrix stiffness.

43.2.2 Focal Adhesions

Aside from hemidesmosomes, the major points of contacts with cells and ECM are focal adhesions (Fig. 43.2), a group of mechanosensitive, macromolecular assemblies that anchor cells to ECM. While multiple studies collectively identified a total of 2000 proteins in the focal adhesions depending on the cell types and the analysis methods, the core adhesome formed of approximately 60 proteins [56]. Focal adhesions are formed at a specific range when the plasma membrane is at a distance of 15 nm from the ECM. Focal adhesions are present at the termini of

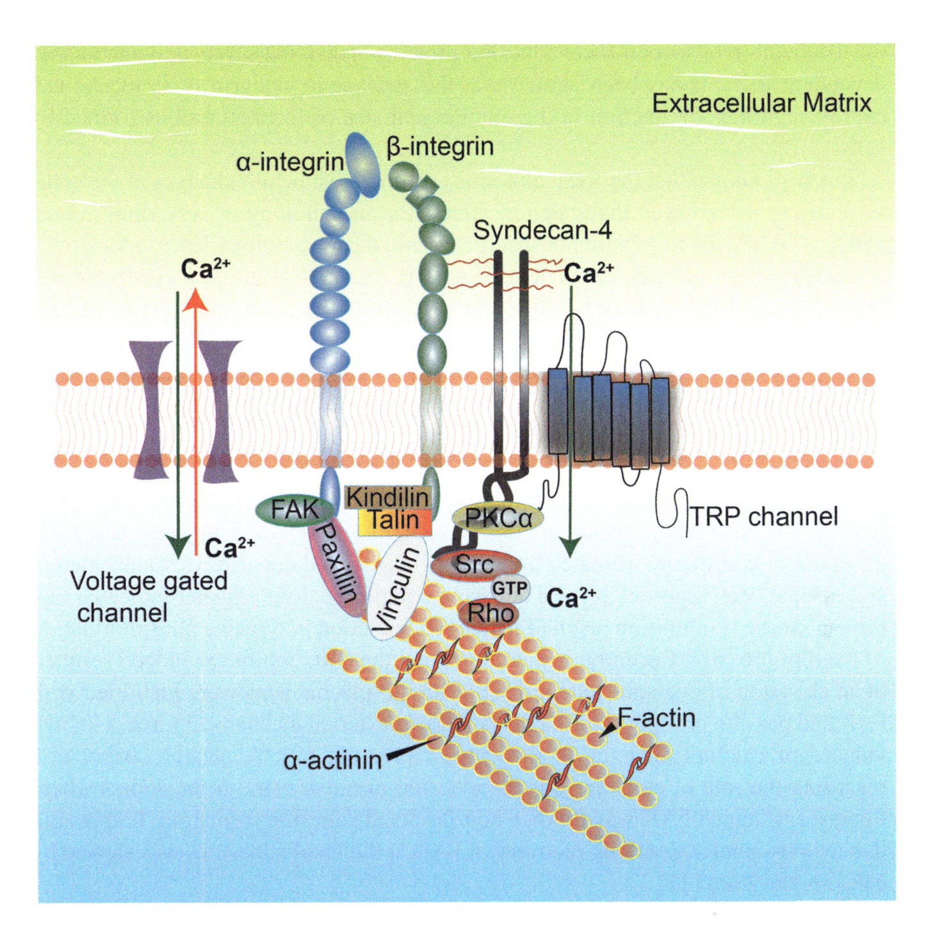

Fig. 43.2 Cell surface receptors in the focal adhesion formation, cytoskeletal organization and calcium regulation. The cell surface receptors integrins and syndecan-4 bind with ECM proteins to activate a large group of downstream signaling molecules to control focal adhesion and cytoskeletal dynamics. Integrins bind directly or indirectly to FAK, talin, kindlin and paxillin through its cytoplasmic domain to trigger focal adhesion formation, whereas syndecan-4 interacts with PKCα and trigger further downstream signaling. The focal adhesions are attached to the actin cytoskeleton formed by F-actin bundles connected by α-actinin, where the vinculin acts as a mechano-transducer between the cytoskeleton and focal adhesions. The synergistic action of integrins and syndecan-4 is essential for the formation of focal adhesions and organized cytoskeleton. Mechano-sensitive (TRP channels) and/or voltage gated channels are regulated by both integrins and syndecans, where they form complexes and appear to co-localize with other focal adhesion molecules

actin bundles (stress fibers), where, in a protein-dense plaque, they are connected to receptors, notably integrins (Fig. 43.2). Key protein components of the plaque include talin, vinculin, paxillin, kindlins and kinases (Fig. 43.2) [56–58]. Focal adhesion dynamics are maintained by localized cooperative signaling between integrins and syndecan-4 proteoglycans [59, 60]. While integrins and syndecans may be mechanoreceptors and have limited effects on transcription, they control the localization of several focal adhesion proteins and signal survival in anchorage-dependent cells. It has been shown that the absence of integrin or syndecan can cause a significant reduction in the number and size of focal adhesions formed by cells.

It is well known that the focal adhesions respond to calcium changes in the cells. An early report showed that calcium affects the morphology of spreading human platelets. While the report did not specifically use the terminology ‘focal adhesions’, it mentioned the calcium dependent formation of small discrete, optically denser regions suggesting the presence of focal specializations at the ventral cell surface [61]. Later reports showed cells attach to fibronectin in calcium dependent manner. Even though the cell surface receptors involved was unknown at the time, one mechanism of attachment to fibronectin was shown to be mediated by glycosaminoglycan sugar chains, specifically heparan sulfate chains [62].

The role of calcium in adherent cells was further emphasized when enhanced calcium influx was shown to induce phosphorylation of focal adhesion-associated tyrosine kinase, FAK in adherent cells [63, 64]. Later reports suggested that FAK phosphorylation was coordinated by protein kinase C and integrins [65] and calcium spikes have been reported in focal adhesion complexes. Reports later indicated that protein kinase C inhibition resulted in marked reduction in tyrosine phosphorylation of paxillin [66, 67]. Recently, it was found that the PKC inhibition indeed resulted in an elevation of cytosolic calcium [20]. Similar mechanisms were identified with FAK and paxillin phosphorylation in response to external mechanical force, a known inducer of calcium changes in the cytosol [18, 68, 69]. A clearer observation regarding the role of calcium in focal adhesion turnover came from studies where fluorescent-tagged FAK was followed during focal adhesion formation. It appeared that the post photo-bleaching recovery of FAK at the focal adhesions was slowed by calcium elevation [11].

Until recently, no calcium channels were identified in any focal adhesion specific proteomic analysis (see review by [70]. However, it could be due to the low expression of channels in the focal adhesions making them difficult to detect. One of the latest mass spectrometry analyses of focal adhesions revealed that the presence of SLC9A1, SLC16A3, KCNH2, PKD-1 and TRPM7 in the focal adhesions. Recent studies by immunofluorescence showed that focal adhesions contain stretch-activated channels TRPC7 (Fig. 43.3) and TRPM4 whose expression changes appears to affect focal adhesion size [20, 71]. Currently, there is no evidence indicating that these channels form heterodimers in the focal adhesions. The observation that these channels are present in focal adhesions is very recent and their function in the focal adhesions remains unclear. TRPM4 itself is not a calcium channel, though it can be activated by calcium. Given the presence of TRPC7, it is possible the TRPM4 is activated by the calcium entry through TRPC7. The transient silencing of TRPM4 appears to lead to larger focal adhesions, leading to the speculation that TRPM4 is involved in focal adhesion turnover [71].

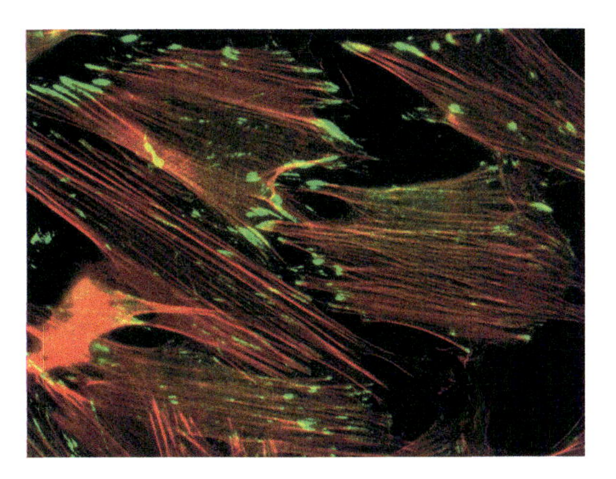

Fig. 43.3 The cells showing staining for TRPC7 channel (green) and filamentous actin (red). The TRPC7 channels appears to localize at the end of actin filaments, where the actin cytoskeleton attaches to the focal adhesions

43.2.3 The Actin Cytoskeleton

Cytoskeleton provides rigidity to the cells and facilitates cell shape maintenance. In addition, the actin cytoskeleton plays important roles in cell migration and tension sensing [72–74]. Actin forms filamentous bundles that are bridged by a second protein called α-actinin [75, 76]. In non-muscle cells, these structures are mostly visible in activated (myo)fibroblasts. Over 20 years ago the ability of the cells to contract in response to external mechanical forces was shown to be dependent on calcium influx from the ECM [77]. One of the initial observations showed that the addition of calcium to *Dictyostelium*, an eukaryotic, phagotrophic bacterivore led to restricted depolymerization of actin filaments where ends of the actin filaments were resistant to depolymerization [78]. Later, α-actinin, a 34 kDa actin-bundling protein and gelation factor (ABP-120) were identified as the components that bind to calcium in order to promote this process [79–81]. Studies using mutant α-actinin, 34 kDa actin-bundling protein, and knock-out models showed that the actin bundling was reduced in the cells and leading to a fragile, loose cytoskeleton [82]. Research from Edwards and Booth suggested a number of cytoskeletal proteins of molecular mass 72, 69, 38, 36 and 32 kDa were regulated by calcium concentration. Depletion of calcium using EGTA suggested the release of these proteins from their complexes led to a more open meshwork of disorganized cytoskeleton. This research also provided the first evidence for a possible effect of local calcium concentration on cytoskeletal proteins [83]. Previously, it was shown that increase in calcium concentration induced degeneration of monkey and human peripheral nerves due to cytoskeletal collapse [84]. Peripheral nerves already have comparatively high calcium concentrations and further increase in this calcium initiated the cytoskeletal collapse. However, lowering the calcium concentration did not produce significant changes during the examined time period [84]. These findings were further consolidated when the role of calcium in actin cytoskeletal organization was identified in osteoblasts [85]. Parathyroid hormone is necessary for maintaining optimum blood calcium concentration level by controlling calcium release from bone. The treatment of osteoblasts with parathyroid hormone resulted in changes in osteoblast cytoskeletal organization due to an elevation in cytosolic calcium. This increase in the intracellular calcium concentration resulted in an impaired cytoskeleton leading to the retraction of the cells and a reduction of triton X-100 insoluble cytoskeletal components, suggesting a decrease in polymerized actin and tubulin in the cytoskeleton [85]. Furthermore, later research showed that the force-induced actin reorganization depended on cytosolic calcium levels and tyrosine phosphorylation of several cytoskeletal proteins including paxillin [86]. Another output of calcium-cytoskeleton cross-talk is the rapid actin assembly and disassembly caused by calcium oscillations in response to local flux through the membrane [87]. These observations essentially led to the idea that the integrity of cytoskeleton depends on maintaining optimum intracellular calcium concentrations.

Eukaryotic cells respond to a calcium-mediated effect on cytoskeleton mainly through calmodulin (**CAL**cium-**MODUL**ated prote**IN,** CaM), a calcium binding messenger protein [88]. Calmodulin is a small protein (17 kDa) with four helix-loop-helix structural domains called EF hands. Similar to another cytoskeleton protein α-actinin, calmodulin can sequester calcium ions in the EF hands which are essential for their structure [89, 90]. Calmodulin also possesses a hinged region giving structural flexibility to the backbone that helps to wrap around the target molecule. The calcium interaction property of calmodulin is interesting since it can target several cellular proteins that are not able to interact directly with calcium, thus acting as a signal transducer. The role of calcium/calmodulin in cell signaling is best exemplified by the major effector protein, calcium/calmodulin-dependent protein kinase II (CaMKII), a family of serine/threonine-specific protein kinases. CaMKII is often identified as a 'structural kinase' that can interact with cytoskeleton, a process that can be influenced by laminin interactions at the cell surface [91]. For instance, the variable region of oligomeric CaMKIIβ has an F-actin binding domain that directs CaMKIIβ to stress fibers. This promotes the localization of CaMKIIβ to the cytoskeleton [92, 93]. Calcium/calmodulin interactions triggers CaMKII phosphorylation by direct binding. As the calcium/calmodulin and F-actin binding site in CaMKII are located in proximity, it is possible that calcium/calmodulin binding may block CaMKII-F-actin interaction leading to cytoskeletal changes [93]. This observation was supported by using CaMKII mutant that was unable to interact with calcium/calmodulin, where the CaMKII interaction with filamentous actin was enhanced. Taken together, current knowledge suggests that the increase in calcium in cells may lead to calcium/calmodulin binding to CaMKII that governs the removal of CaMKII from actin cytoskeleton. Since binding of CaMKII to F-actin is required for stability of F-actin bundles [94], the increase in calcium may eventually results in a disassembled cytoskeleton. These observations were confirmed by biochemical assays suggesting CaMKII indeed regulates actin assembly, structure and remodeling [95]. In addition, in vitro studies suggested that actin could be phosphorylated exclusively by CaMKIIβ but not by other isoforms. This could be enhanced by the depletion of calcium/calmodulin complexes [96]. A more recent observation aligns with the findings where elevation of calcium through TRPC7 channel resulted in disorganized cytoskeleton [20]. The GTPase activating protein,

IQGAP was identified as another modulator between calmodulin and cytoskeleton. Knocking down of IQGAP1 led to the reduction of calmodulin in the actin rich cortex of mast cells. Cells exhibited changes in local calcium concentration followed by reorganization of cytoskeleton [97].

Importantly, cytosolic calcium regulates not only the cytoskeletal organization but also contractility. In this respect, myosin is the key protein. Myosins are family of at least 20 motor proteins involved in actin-based eukaryotic cell motility [98–100]. The head of the myosin protein contains the actin- and nucleotide-binding sites and the catalytic domain consisting of an α-helical strand, which is stabilized by the binding of calmodulin or calmodulin-like light chain subunits [101]. The C-terminal region, which is distinct for each myosin class, mediates the association of myosins with each other and anchoring them for movement relative to actin filaments [102]. Intracellular calcium levels heavily influence the activity of specific myosins. Most of the members of myosin family are inactive in the absence of calcium. For example, myosin V and VII have been shown by electron microscopy and analytical ultracentrifugation to fold into a more compact, inactive structure in the absence of calcium [103]. On the other hand increased cytosolic calcium leads to reductions in the progressive movement of motor proteins, which is speculated to result from the dissociation of calmodulin light chain from heavy chain affecting the stability of myosin [104]. Moreover, calcium levels have a major influence on interactions mediated by myosin head and tail [105]. Another calcium binding molecule that controls cytoskeletal contractility is troponin, which consists of a calcium binding subunit (troponin C), inhibitory subunit (troponin L) and tropomyosin-binding subunit (troponin T). Troponin mainly appears on the actin bundles of striated muscles, where they activate muscle contraction. Like several other cytoskeletal proteins, troponin C contains multiple EF hand domains. The two EF hands at the C-terminus troponin C harbor two high affinity calcium binding sites, whereas the single EF hands on the N-terminus have a low affinity calcium binding site [106, 107]. Depending on calcium levels, troponin regulates the interaction between myosin and actin filaments in striated muscles, thus affecting the contractility of the cytoskeleton [108–111].

The ability of calcium to alter cytoskeletal organization and contractility may have a broad impact on the life of an organism, emanating from the control of cell morphology. For example, the organization of dendritic spines depends on the local calcium concentration in neurons [112, 113]. Dendritic spines are highly motile structures whose morphology is cytoskeleton-dependent. Rapid reorganization of spines is crucial for the synapses residing on it, which eventually affects the neural transmission [114]. Observations from starfish oocytes suggest that calcium signals are increased with the increase in the expression of cofilin, an actin-depolymerizing factor that modulates cytoskeletal dynamics, cell motility and cytokinesis. Increased cofilin expression promotes hormone-induced oocyte maturation and fertilization [115]. More important aspects of cytosolic calcium levels came from the studies of cardiac physiology. The impaired diastolic function of the heart is believed be the result of dysfunctional calcium regulation, which can be improved through modulation of myofilament calcium sensitivity [116]. At this point it is worthy to note that none of the studies indicated a change in protein level expression of any of the cytoskeletal proteins in response to change in cytosolic calcium. This means the changes in cytoskeletal structures are simply due to altered localization and function of various cytoskeletal proteins. For example, studies on roles of the cytoskeleton in exocytosis suggested that increased cytosolic calcium concentration directly affecting F-actin organization might cause the relocation of actin from the bundles, rather than affecting protein synthesis [117].

43.2.4 Cell Surface Receptors

43.2.4.1 Proteoglycans

Proteoglycans are composed of a core protein with glycosaminoglycan sugar chains covalently attached to the core protein. Proteoglycans are an important component of the ECM, but others at the cell surface function as receptors. Proteoglycans interact with ECM molecules, mainly through their glycosaminoglycan chains. Over 30 proteoglycans are encoded in mammals where many maintain tissue specific expression [118]. Proteoglycans of the syndecan family control multiple cellular functions, including cell adhesion, cytoskeletal organization, cell migration and protein transport [118–120]. Syndecans are the only proteoglycans identified to regulate cytosolic calcium (Fig. 43.4). They are transmembrane receptors that can interact with the extracellular ligands such as fibronectin, laminin and collagen [22]. Recently, syndecans were identified as regulators of mechanosensitive transient receptor potential (TRP) channels. Syndecans-1 and -4 were shown to regulate transient receptor potential (TRP) channels. Syndecan-1, a prominent syndecan of epithelial cells appeared to control TRPC4 channels [20]. Similarly, ubiquitously expressed syndecan-4 controls TRPC7 channels in mesenchymal cells [20]. While these two syndecans can associate with TRPC channels, it is highly likely that other syndecans regulate other members from TRP family, not least since similar properties could be shown in the single syndecan of *Caenorhabditis elegans* [20]. Syndecan-4 mediated regulation of TRP channel seems to be maintained through PKC activation and direct complex formation between syndecan, the channels and the cytoskeletal protein α-actinin [20]. This agrees with a previous observation made in goldfish retinal bipolar cells where calcium was altered with the activation of PKC by PMA. Both researches, performed over a decade apart in two different systems, found that the PKC induced calcium modification results in F-actin filament reorganization.

Syndecan-4, perhaps through ligand-induced clustering, can activate PKCα through direct interaction. It has been reported that TRP channels are negatively regulated by a PKC mediated phosphorylation of a regulatory serine residue situated near the C- terminus of the channel [121]. It is speculated that the active PKCα near the cytoplasmic domain of syndecan-4 can phosphorylate this regulatory serine residue of the channel, thus leading to the closure of the channel (Fig. 43.4).

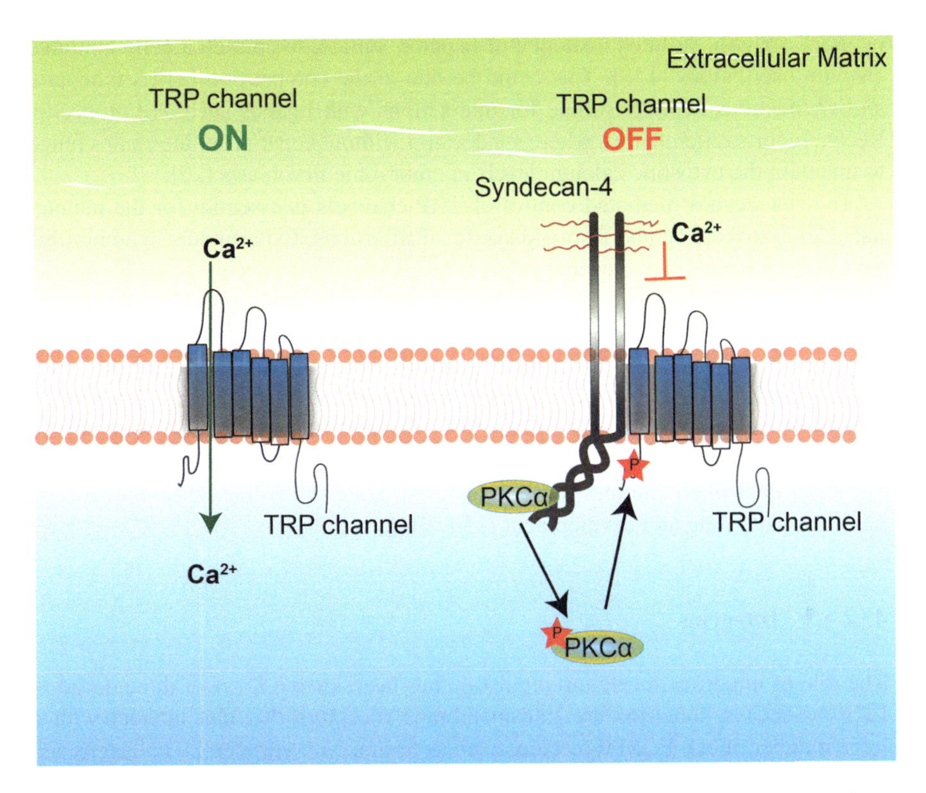

Fig. 43.4 Syndecan mediated regulation of TRP channels: In the presence of syndecan-4, the active PKCα phosphorylates TRP channels. This leads to the closing of the channel

The complex formed between TRP channels and syndecan-4 should enable the channel to be in the close proximity of the PKCα. Interestingly, the complex formation between syndecan-4 and the TRP channels appears to be independent of heparan sulfate chains on the ectodomain of syndecan-4, whereas the channel regulation results from heparan sulfate chain interactions with extracellular matrix [20]. Syndecan-4 interacts with the fibronectin, specifically the 31kD HepII domain of fibronectin through heparan sulfate chains. This classic interaction is known to trigger PKCα activation in the presence of the membrane phospholipid, PtdIns4,5P_2 [122–125]. Therefore, it can be speculated that the active PKCα phosphorylates the TRP channel and other cytoskeletal regulatory proteins [121, 126, 127]. Overall, these findings support the idea that the syndecan mediated calcium control is initiated by the heparan sulfate chain engagement with ECM, that transduces PKCα activation and channel regulation. Currently, fibronectin is the paradigmatic ECM component that was shown to have the ability to initiate syndecan-4 mediated control of TRP channels. This property could, however, be common to many ECM molecules and awaits investigation. Previous observations support the idea that heparan sulfate chains are essential for this process. One report showed that adding heparin, a highly sulfated analogue of heparan sulfate, to epithelial cells inhibited calcium oscillations [128]. This could be due to the competition between heparin and endogenous heparan sulfate for interactions with ligands such as fibronectin. Recent reports reiterate this, where syndecan-4 without heparan sulfate chains failed to maintain the cytosolic calcium levels in embryonic fibroblasts [20].

The syndecan-4 mediated control of TRP channels is essential for the maintenance of cytoskeleton and formation of focal adhesions. Experiments in embryonic fibroblasts revealed that elevated calcium levels seen in syndecan-4 null fibroblasts result in disorganized cytoskeleton and smaller focal adhesions [20]. Currently, several syndecan-dependent phenotypes may be explained as secondary effects of their role in cytosolic calcium regulation. However, there is no significant information available regarding the organismal effect of syndecan-TRP channel cross talk in mammals. However, one report suggests that syndecan-4 and TRPC6 may regulate glomerular permeability [129]. Recent reviews also highlighted that syndecan controlled calcium may be of importance in cardiac diseases, cancers, neuronal patterning and development [130–134].

43.2.4.2 Integrins

The role of integrins in calcium regulation has been known for over three decades. Like syndecans, integrins are transmembrane receptors that can interact with a large number of cell ECM molecules such as fibronectin, vitronectin, collagens and laminins, and participate in signal transduction (Figs. 43.1 and 43.2). Integrins are composed of transmembrane α- and a β- subunits, where they can assume active or inactive conformations. There are 24 integrin heterodimers in vertebrates formed by different combinations of 18 α- and 8 β- subunits [135, 136]. Structural analysis revealed that calcium ions compete with manganese ions for integrin binding [137–139], in regulating the switch between active and inactive integrin conformations. This was further confirmed when removal of calcium appeared to increase ligand binding affinity of the integrin [140]. The divalent ion interaction with integrin also known to stabilize the integrin interaction with its ligands [139], though this is attributed to manganese rather than calcium ions.

A number of integrins can facilitate calcium entry into cells, thus affecting cellular processes including cell-ECM adhesion and cytoskeletal regulation. On the other hand, integrin -mediated cell adhesion was also shown to be controlled by cytosolic calcium levels. A subset of integrins interact with ECM proteins through an Arg-Gly-Asp motif on the ECM protein (e.g. fibronectin and vitronectin interactions with αV or α5β1 integrins) [141–144]. It has been shown that supplying cells with Arg-Gly-Asp peptides elicited a transient change in intracellular calcium levels [145–147], likely due to the integrin engagement with the peptide ligand.

Integrin-induced calcium changes are mainly associated with voltage-gated calcium channels. For instance, integrins differentially control L-type calcium channels in vascular smooth cells, where αVβ3 integrin may negatively regulate the channel while α5β1 integrin may enhance calcium entry [148]. Subsequent research showed that the α5β1 integrin mediated control of L-type channels may be maintained via a tyrosine phosphorylation cascade, signaling between focal adhesion kinase, vinculin, paxillin and c-Src [12, 149]. This signaling appeared to be initiated by the interaction of integrins with extracellular proteins such as fibronectin and vitronectin, probably through the Arg-Gly-Asp motif. Currently the reason behind differential regulation of channels in response to Arg-Gly-Asp interaction is unclear. One possibility is that other calcium channels or internal calcium stores might be compensating the effects of integrin-ECM induced calcium changes. Alternatively, each integrin may associate with distinct channels, an area in need of further research. A more recent report showed that α4β1 integrin can similarly influence L-type calcium channels as a result of their interaction with the Leu-Asp-Val sequence of an alternately spliced fibronectin variant [150]. Finally, the interaction of α7 integrin with laminin in skeletal muscle resulted in not only the activation of L-type channels, but also calcium release from intracellular calcium stores [151]. As mentioned before, integrins and laminins may form complexes with voltage gated channel Ca(v)α in other systems such as synapses. However, the direct role of integrin in calcium release from intracellular calcium stores remains unclear. Finally, integrins are also involved in controlling mechanosensitive channels either directly or indirectly. Applying mechanical forces directly to integrins evoked an increase in cytosolic calcium levels, which seems to require a reciprocal force from the cytoskeleton [152, 153]. However, the application of the force may lead to more global effects on the plasma membrane. Therefore, the increase in calcium might be a combined result of the force applied on the plasma membrane and cytoskeleton instead of, or in addition to, an integrin-specific one.

While the role of integrin in calcium regulation is well established, the directionality of the regulation appeared to depend on cell and mode of integrin involvement. For instance, αVβ3 integrin appears to negatively regulate calcium entry in vascular smooth cells [148] whereas it increases calcium levels in rat osteoclasts [154]. These experiments were, however, different in terms of integrin presentation to cells. The experiments with vascular smooth cells were based on the integrin present on the cell surface whereas the αVβ3 integrin was supplied in solution to rat osteoclasts. This may have resulted in a competition for ligand binding between cell surface and exogenous integrins. Overall, the consensus is that integrin-controlled calcium regulation is initiated by the clustering of integrins on the cell surface following their interaction with ECM proteins [155].

43.3 Conclusions

The ability of ECM proteins to regulate cytosolic calcium levels in addition to signaling through kinase networks is an indicator of the complexity of cell-ECM interactions. Over last two decades the role of ECM in controlling cell behavior has been increasingly elucidated and some of the pathways are controlled through calcium. On the other hand, the ability of calcium to control cell-ECM interactions, mainly though focal adhesions is also well known. This two-way interaction indicates feedback mechanisms, though the precise control mechanisms remain elusive. Since cell surface receptors are the gatekeepers of these interactions, it is highly likely the decisive factor is the localization of these receptors on the cell surface and the formation of localized and functional signaling complexes. The recent addition of syndecans as calcium regulators along with integrins reveals a complex, yet a delicately controlled mechanism, subject to the balance of cell surface receptor expression and function.

43.4 Future Research

As expected, the role of calcium in cell-ECM cross talk is a complex research area. While ECM and tissue fluids remain the largest calcium storage sites, intracellular calcium stores play a significant role in calcium regulation that impacts cell adhesion and morphology. The most challenging aspect of calcium studies is the identification of relevant calcium sources that are involved in specific cell signaling. Secondly, there is a need to identify exactly how the localized calcium is controlled. Microdomains within the cell with regard to ions, mRNAs and junctions are apparent. For example, calcium dynamics in focal adhesions may have a decisive effect on the focal adhesion turnover, yet there is no significant information available on its regulation. In addition, further information needs to be elucidated on the mechanosensitive calcium channels where cell-ECM interactions play a major role. It is not yet fully clear which TRP channels are regulated by which syndecans and how much redundancy may be in the system. It is also possible that other TRP subfamilies could be similarly regulated, but this is currently unknown. Applying computational models may be in the realm of possibilities to elucidate mechanotransduction between cell and ECM. By incorporating calcium regulation in the computational models may help to understand the complexity of calcium in cell-ECM interactions.

Acknowledgments Authors thank Ms. Ioli Mitsou for images for Fig. 43.3. SG is supported by Senior Postdoctoral Fellowship from Monash University.

References

1. Janson IA, Putnam AJ (2015) Extracellular matrix elasticity and topography: material-based cues that affect cell function via conserved mechanisms. J Biomed Mater Res A 103(3):1246–1258
2. Carafoli E, Krebs J (2016) Why calcium? How calcium became the best communicator. J Biol Chem 291(40):20849–20857
3. Avila-Medina J, Mayoral-Gonzalez I, Dominguez-Rodriguez A, Gallardo-Castillo I, Ribas J, Ordonez A et al (2018) The complex role of store operated calcium entry pathways and

related proteins in the function of cardiac, skeletal and vascular smooth muscle cells. Front Physiol 9:257
4. Zhang AH, Sharma G, Undheim EAB, Jia X, Mobli M (2018) A complicated complex: ion channels, voltage sensing, cell membranes and peptide inhibitors. Neurosci Lett
5. Kobayashi T, Sokabe M (2010) Sensing substrate rigidity by mechanosensitive ion channels with stress fibers and focal adhesions. Curr Opin Cell Biol 22(5):669–676
6. Whitaker M (2006) Calcium at fertilization and in early development. Physiol Rev 86(1):25–88
7. Mattson MP, Chan SL (2003) Calcium orchestrates apoptosis. Nat Cell Biol 5(12):1041–1043
8. Orr AW, Helmke BP, Blackman BR, Schwartz MA (2006) Mechanisms of mechanotransduction. Dev Cell 10(1):11–20
9. Vogel V, Sheetz M (2006) Local force and geometry sensing regulate cell functions. Nat Rev Mol Cell Biol 7(4):265–275
10. Ringer P, Colo G, Fassler R, Grashoff C (2017) Sensing the mechano-chemical properties of the extracellular matrix. Matrix Biol 64:6–16
11. Giannone G, Ronde P, Gaire M, Beaudouin J, Haiech J, Ellenberg J et al (2004) Calcium rises locally trigger focal adhesion disassembly and enhance residency of focal adhesion kinase at focal adhesions. J Biol Chem 279(27):28715–28723
12. Wu X, Davis GE, Meininger GA, Wilson E, Davis MJ (2001) Regulation of the L-type calcium channel by alpha 5beta 1 integrin requires signaling between focal adhesion proteins. J Biol Chem 276(32):30285–30292
13. Ciobanasu C, Faivre B, Le Clainche C (2014) Actomyosin-dependent formation of the mechanosensitive talin-vinculin complex reinforces actin anchoring. Nat Commun 5:3095
14. Drmota Prebil S, Slapsak U, Pavsic M, Ilc G, Puz V, de Almeida RE et al (2016) Structure and calcium-binding studies of calmodulin-like domain of human non-muscle alpha-actinin-1. Sci Rep 6:27383
15. Naba A, Clauser KR, Hoersch S, Liu H, Carr SA, Hynes RO (2012) The matrisome: in silico definition and in vivo characterization by proteomics of normal and tumor extracellular matrices. Mol Cell Proteomics 11(4):M111.014647
16. Carbajal ME, Manning-Cela R, Pina A, Franco E, Meza I (1996) Fibronectin-induced intracellular calcium rise in Entamoeba histolytica trophozoites: effect on adhesion and the actin cytoskeleton. Exp Parasitol 82(1):11–20
17. Nebe B, Rychly J, Knopp A, Bohn W (1995) Mechanical induction of beta 1-integrin-mediated calcium signaling in a hepatocyte cell line. Exp Cell Res 218(2):479–484
18. Lee HS, Millward-Sadler SJ, Wright MO, Nuki G, Salter DM (2000) Integrin and mechanosensitive ion channel-dependent tyrosine phosphorylation of focal adhesion proteins and beta-catenin in human articular chondrocytes after mechanical stimulation. J Bone Miner Res 15(8):1501–1509
19. Ishise H, Larson B, Hirata Y, Fujiwara T, Nishimoto S, Kubo T et al (2015) Hypertrophic scar contracture is mediated by the TRPC3 mechanical force transducer via NFkB activation. Sci Rep 5:11620
20. Gopal S, Sogaard P, Multhaupt HA, Pataki C, Okina E, Xian X et al (2015) Transmembrane proteoglycans control stretch-activated channels to set cytosolic calcium levels. J Cell Biol 210(7):1199–1211
21. Gopal S, Bober A, Whiteford JR, Multhaupt HA, Yoneda A, Couchman JR (2010) Heparan sulfate chain valency controls syndecan-4 function in cell adhesion. J Biol Chem 285(19):14247–14258
22. Xian X, Gopal S, Couchman JR (2010) Syndecans as receptors and organizers of the extracellular matrix. Cell Tissue Res 339(1):31–46
23. Yurchenco PD, Cheng YS (1993) Self-assembly and calcium-binding sites in laminin. A three-arm interaction model. J Biol Chem 268(23):17286–17299
24. Cheng YS, Champliaud MF, Burgeson RE, Marinkovich MP, Yurchenco PD (1997) Self-assembly of laminin isoforms. J Biol Chem 272(50):31525–31532

25. Kalb E, Engel J (1991) Binding and calcium-induced aggregation of laminin onto lipid bilayers. J Biol Chem 266(28):19047–19052
26. Colucci S, Giannelli G, Grano M, Faccio R, Quaranta V, Zallone AZ (1996) Human osteoclast-like cells selectively recognize laminin isoforms, an event that induces migration and activates Ca^{2+} mediated signals. J Cell Sci 109(Pt 6):1527–1535
27. Nishimune H (2012) Molecular mechanism of active zone organization at vertebrate neuromuscular junctions. Mol Neurobiol 45(1):1–16
28. Nishimune H, Sanes JR, Carlson SS (2004) A synaptic laminin-calcium channel interaction organizes active zones in motor nerve terminals. Nature 432(7017):580–587
29. Sunderland WJ, Son YJ, Miner JH, Sanes JR, Carlson SS (2000) The presynaptic calcium channel is part of a transmembrane complex linking a synaptic laminin (alpha4beta2gamma1) with non-erythroid spectrin. J Neurosci 20(3):1009–1019
30. Chand KK, Lee KM, Schenning MP, Lavidis NA, Noakes PG (2015) Loss of beta2-laminin alters calcium sensitivity and voltage-gated calcium channel maturation of neurotransmission at the neuromuscular junction. J Physiol 593(1):245–265
31. Carlson SS, Valdez G, Sanes JR (2010) Presynaptic calcium channels and alpha3-integrins are complexed with synaptic cleft laminins, cytoskeletal elements and active zone components. J Neurochem 115(3):654–666
32. Cali T, Lopreiato R, Shimony J, Vineyard M, Frizzarin M, Zanni G et al (2015) A novel mutation in isoform 3 of the plasma membrane Ca^{2+} pump impairs cellular Ca^{2+} homeostasis in a patient with cerebellar ataxia and laminin subunit 1alpha mutations. J Biol Chem 290(26):16132–16141
33. Dietz HC, Pyeritz RE (1995) Mutations in the human gene for fibrillin-1 (FBN1) in the Marfan syndrome and related disorders. Hum Mol Genet 4 Spec No:1799–1809
34. Handford PA (2000) Fibrillin-1, a calcium binding protein of extracellular matrix. Biochim Biophys Acta 1498(2–3):84–90
35. Kielty CM, Shuttleworth CA (1993) The role of calcium in the organization of fibrillin microfibrils. FEBS Lett 336(2):323–326
36. Sakai LY, Keene DR, Engvall E (1986) Fibrillin, a new 350-kD glycoprotein, is a component of extracellular microfibrils. J Cell Biol 103(6 Pt 1):2499–2509
37. Reinhardt DP, Mechling DE, Boswell BA, Keene DR, Sakai LY, Bachinger HP (1997) Calcium determines the shape of fibrillin. J Biol Chem 272(11):7368–7373
38. Corson GM, Chalberg SC, Dietz HC, Charbonneau NL, Sakai LY (1993) Fibrillin binds calcium and is coded by cDNAs that reveal a multidomain structure and alternatively spliced exons at the 5′ end. Genomics 17(2):476–484
39. Jensen SA, Corbett AR, Knott V, Redfield C, Handford PA (2005) Ca^{2+}−dependent interface formation in fibrillin-1. J Biol Chem 280(14):14076–14084
40. Smallridge RS, Whiteman P, Werner JM, Campbell ID, Handford PA, Downing AK (2003) Solution structure and dynamics of a calcium binding epidermal growth factor-like domain pair from the neonatal region of human fibrillin-1. J Biol Chem 278(14):12199–12206
41. Lin G, Tiedemann K, Vollbrandt T, Peters H, Batge B, Brinckmann J et al (2002) Homo- and heterotypic fibrillin-1 and -2 interactions constitute the basis for the assembly of microfibrils. J Biol Chem 277(52):50795–50804
42. Marson A, Rock MJ, Cain SA, Freeman LJ, Morgan A, Mellody K et al (2005) Homotypic fibrillin-1 interactions in microfibril assembly. J Biol Chem 280(6):5013–5021
43. Cain SA, Baldock C, Gallagher J, Morgan A, Bax DV, Weiss AS et al (2005) Fibrillin-1 interactions with heparin. Implications for microfibril and elastic fiber assembly. J Biol Chem 280(34):30526–30537
44. Tiedemann K, Batge B, Muller PK, Reinhardt DP (2001) Interactions of fibrillin-1 with heparin/heparan sulfate, implications for microfibrillar assembly. J Biol Chem 276(38):36035–36042
45. Isogai Z, Aspberg A, Keene DR, Ono RN, Reinhardt DP, Sakai LY (2002) Versican interacts with fibrillin-1 and links extracellular microfibrils to other connective tissue networks. J Biol Chem 277(6):4565–4572

46. El-Hallous E, Sasaki T, Hubmacher D, Getie M, Tiedemann K, Brinckmann J et al (2007) Fibrillin-1 interactions with fibulins depend on the first hybrid domain and provide an adaptor function to tropoelastin. J Biol Chem 282(12):8935–8946
47. Reinhardt DP, Ono RN, Sakai LY (1997) Calcium stabilizes fibrillin-1 against proteolytic degradation. J Biol Chem 272(2):1231–1236
48. Ashworth JL, Murphy G, Rock MJ, Sherratt MJ, Shapiro SD, Shuttleworth CA et al (1999) Fibrillin degradation by matrix metalloproteinases: implications for connective tissue remodelling. Biochem J 340(Pt 1):171–181
49. Mariko B, Ghandour Z, Raveaud S, Quentin M, Usson Y, Verdetti J et al (2010) Microfibrils and fibrillin-1 induce integrin-mediated signaling, proliferation and migration in human endothelial cells. Am J Physiol Cell Physiol 299(5):C977–C987
50. Massam-Wu T, Chiu M, Choudhury R, Chaudhry SS, Baldwin AK, McGovern A et al (2010) Assembly of fibrillin microfibrils governs extracellular deposition of latent TGF beta. J Cell Sci 123(Pt 17):3006–3018
51. Poole AW, Watson SP (1995) Regulation of cytosolic calcium by collagen in single human platelets. Br J Pharmacol 115(1):101–106
52. Shiraishi M, Ikeda M, Ogawa H, Tu CH, Ito K (1998) Impaired cytosolic calcium mobilization and aggregation in response to collagen in platelets from Japanese black cattle with Chediak-Higashi syndrome. Am J Vet Res 59(6):744–749
53. Roberts DE, McNicol A, Bose R (2004) Mechanism of collagen activation in human platelets. J Biol Chem 279(19):19421–19430
54. Nesbitt WS, Giuliano S, Kulkarni S, Dopheide SM, Harper IS, Jackson SP (2003) Intercellular calcium communication regulates platelet aggregation and thrombus growth. J Cell Biol 160(7):1151–1161
55. Shih YR, Tseng KF, Lai HY, Lin CH, Lee OK (2011) Matrix stiffness regulation of integrin-mediated mechanotransduction during osteogenic differentiation of human mesenchymal stem cells. J Bone Miner Res 26(4):730–738
56. Horton ER, Byron A, Askari JA, DHJ N, Millon-Fremillon A, Robertson J et al (2015) Definition of a consensus integrin adhesome and its dynamics during adhesion complex assembly and disassembly. Nat Cell Biol 17(12):1577–1587
57. Byron A, Humphries JD, Craig SE, Knight D, Humphries MJ (2012) Proteomic analysis of alpha4beta1 integrin adhesion complexes reveals alpha-subunit-dependent protein recruitment. Proteomics 12(13):2107–2114
58. Carisey A, Tsang R, Greiner AM, Nijenhuis N, Heath N, Nazgiewicz A et al (2013) Vinculin regulates the recruitment and release of core focal adhesion proteins in a force-dependent manner. Curr Biol 23(4):271–281
59. Morgan MR, Humphries MJ, Bass MD (2007) Synergistic control of cell adhesion by integrins and syndecans. Nat Rev Mol Cell Biol 8(12):957–969
60. Roper JA, Williamson RC, Bass MD (2012) Syndecan and integrin interactomes: large complexes in small spaces. Curr Opin Struct Biol 22(5):583–590
61. Zobel CR, Woods A (1983) Effect of calcium on the morphology of human platelets spread on glass substrates. Eur J Cell Biol 30(1):83–92
62. Laterra J, Norton EK, Izzard CS, Culp LA (1983) Contact formation by fibroblasts adhering to heparan sulfate-binding substrata (fibronectin or platelet factor 4). Exp Cell Res 146(1):15–27
63. Hamawy MM, Mergenhagen SE, Siraganian RP (1993) Tyrosine phosphorylation of pp125FAK by the aggregation of high affinity immunoglobulin E receptors requires cell adherence. J Biol Chem 268(10):6851–6854
64. Schaller MD, Borgman CA, Cobb BS, Vines RR, Reynolds AB, Parsons JT (1992) pp125FAK a structurally distinctive protein-tyrosine kinase associated with focal adhesions. Proc Natl Acad Sci USA 89(11):5192–5196
65. Shattil SJ, Haimovich B, Cunningham M, Lipfert L, Parsons JT, Ginsberg MH et al (1994) Tyrosine phosphorylation of pp125FAK in platelets requires coordinated signaling through integrin and agonist receptors. J Biol Chem 269(20):14738–14745

66. Zachary I, Sinnett-Smith J, Turner CE, Rozengurt E (1993) Bombesin, vasopressin, and endothelin rapidly stimulate tyrosine phosphorylation of the focal adhesion-associated protein paxillin in Swiss 3T3 cells. J Biol Chem 268(29):22060–22065
67. Turner CE (2000) Paxillin and focal adhesion signalling. Nat Cell Biol 2(12):E231–E236
68. Gopal S, Multhaupt HA, Pocock R, Couchman JR (2016) Cell-extracellular matrix and cell-cell adhesion are linked by syndecan-4. Matrix Biol
69. Sachs F (2010) Stretch-activated ion channels: what are they? Physiology (Bethesda) 25(1):50–56
70. Mitsou I, Multhaupt HAB, Couchman JR (2017) Proteoglycans, ion channels and cell-matrix adhesion. Biochem J 474(12):1965–1979
71. Caceres M, Ortiz L, Recabarren T, Romero A, Colombo A, Leiva-Salcedo E et al (2015) TRPM4 Is a novel component of the adhesome required for focal adhesion disassembly, migration and contractility. PloS One 10(6):e0130540
72. Tang DD, Gerlach BD (2017) The roles and regulation of the actin cytoskeleton, intermediate filaments and microtubules in smooth muscle cell migration. Respir Res 18(1):54
73. Oishi T, Kimura K, Hanada M, Samejima M (1982) Assistance extended to patients with poor prognosis and their families – a lesson in role playing. Kango Gijutsu 28(16):2183–2188
74. Hayakawa K, Tatsumi H, Sokabe M (2011) Actin filaments function as a tension sensor by tension-dependent binding of cofilin to the filament. J Cell Biol 195(5):721–727
75. Hampton CM, Taylor DW, Taylor KA (2007) Novel structures for alpha-actinin:F-actin interactions and their implications for actin-membrane attachment and tension sensing in the cytoskeleton. J Mol Biol 368(1):92–104
76. Meyer RK, Aebi U (1990) Bundling of actin filaments by alpha-actinin depends on its molecular length. J Cell Biol 110(6):2013–2024
77. Sharp WW, Simpson DG, Borg TK, Samarel AM, Terracio L (1997) Mechanical forces regulate focal adhesion and costamere assembly in cardiac myocytes. Am J Phys 273(2 Pt 2):H546–H556
78. Brown SS, Yamamoto K, Spudich JA (1982) A 40,000-dalton protein from Dictyostelium discoideum affects assembly properties of actin in a Ca^{2+}–dependent manner. J Cell Biol 93(1):205–210
79. Fechheimer M, Taylor DL (1984) Isolation and characterization of a 30,000-dalton calcium-sensitive actin cross-linking protein from Dictyostelium discoideum. J Biol Chem 259(7):4514–4520
80. Rivero F, Furukawa R, Fechheimer M, Noegel AA (1999) Three actin cross-linking proteins, the 34 kDa actin-bundling protein, alpha-actinin and gelation factor (ABP-120), have both unique and redundant roles in the growth and development of Dictyostelium. J Cell Sci 112(Pt 16):2737–2751
81. Witke W, Hofmann A, Koppel B, Schleicher M, Noegel AA (1993) The Ca^{2+}-binding domains in non-muscle type alpha-actinin: biochemical and genetic analysis. J Cell Biol 121(3):599–606
82. Furukawa R, Maselli A, Thomson SA, Lim RW, Stokes JV, Fechheimer M (2003) Calcium regulation of actin crosslinking is important for function of the actin cytoskeleton in Dictyostelium. J Cell Sci 116(Pt 1):187–196
83. Edwards HC, Booth AG (1987) Calcium-sensitive, lipid-binding cytoskeletal proteins of the human placental microvillar region. J Cell Biol 105(1):303–311
84. Badalamente MA, Hurst LC, Stracher A (1986) Calcium-induced degeneration of the cytoskeleton in monkey and human peripheral nerves. J Hand Surg (Br) 11(3):337–340
85. Lomri A, Marie PJ (1990) Distinct effects of calcium- and cyclic AMP-enhancing factors on cytoskeletal synthesis and assembly in mouse osteoblastic cells. Biochim Biophys Acta 1052(1):179–186
86. Glogauer M, Arora P, Yao G, Sokholov I, Ferrier J, McCulloch CA (1997) Calcium ions and tyrosine phosphorylation interact coordinately with actin to regulate cytoprotective responses to stretching. J Cell Sci 110(Pt 1):11–21

87. Veksler A, Gov NS (2009) Calcium-actin waves and oscillations of cellular membranes. Biophys J 97(6):1558–1568
88. Stevens FC (1983) Calmodulin: an introduction. Can J Biochem Cell Biol 61(8):906–910
89. Chin D, Means AR (2000) Calmodulin: a prototypical calcium sensor. Trends Cell Biol 10(8):322–328
90. Chou JJ, Li S, Klee CB, Bax A (2001) Solution structure of Ca^{2+}-calmodulin reveals flexible hand-like properties of its domains. Nat Struct Biol 8(11):990–997
91. Easley CA, Faison MO, Kirsch TL, Lee JA, Seward ME, Tombes RM (2006) Laminin activates CaMK-II to stabilize nascent embryonic axons. Brain Res 1092(1):59–68
92. Shen K, Teruel MN, Subramanian K, Meyer T (1998) CaMKIIbeta functions as an F-actin targeting module that localizes CaMKIIalpha/beta heterooligomers to dendritic spines. Neuron 21(3):593–606
93. Lin YC, Redmond L (2008) CaMKIIbeta binding to stable F-actin in vivo regulates F-actin filament stability. Proc Natl Acad Sci USA 105(41):15791–15796
94. Okamoto K, Narayanan R, Lee SH, Murata K, Hayashi Y (2007) The role of CaMKII as an F-actin-bundling protein crucial for maintenance of dendritic spine structure. Proc Natl Acad Sci USA 104(15):6418–6423
95. Sanabria H, Swulius MT, Kolodziej SJ, Liu J, Waxham MN (2009) {beta}CaMKII regulates actin assembly and structure. J Biol Chem 284(15):9770–9780
96. O'Leary H, Lasda E, Bayer KU (2006) CaMKIIbeta association with the actin cytoskeleton is regulated by alternative splicing. Mol Biol Cell 17(11):4656–4665
97. Psatha MI, Razi M, Koffer A, Moss SE, Sacks DB, Bolsover SR (2007) Targeting of calcium:calmodulin signals to the cytoskeleton by IQGAP1. Cell Calcium 41(6):593–605
98. Sellers JR (2000) Myosins: a diverse superfamily. Biochim Biophys Acta 1496(1):3–22
99. Hartman MA, Spudich JA (2012) The myosin superfamily at a glance. J Cell Sci 125(Pt 7):1627–1632
100. Syamaladevi DP, Spudich JA, Sowdhamini R (2012) Structural and functional insights on the Myosin superfamily. Bioinf Biol Insights 6:11–21
101. Preller M, Manstein DJ (2013) Myosin structure, allostery, and mechano-chemistry. Structure 21(11):1911–1922
102. Foth BJ, Goedecke MC, Soldati D (2006) New insights into myosin evolution and classification. Proc Natl Acad Sci USA 103(10):3681–3686
103. Wang F, Thirumurugan K, Stafford WF, Hammer JA 3rd, Knight PJ, Sellers JR (2004) Regulated conformation of myosin V. J Biol Chem 279(4):2333–2336
104. Lu H, Krementsova EB, Trybus KM (2006) Regulation of myosin V processivity by calcium at the single molecule level. J Biol Chem 281(42):31987–31994
105. Jung HS, Komatsu S, Ikebe M, Craig R (2008) Head-head and head-tail interaction: a general mechanism for switching off myosin II activity in cells. Mol Biol Cell 19(8):3234–3242
106. Li MX, Saude EJ, Wang X, Pearlstone JR, Smillie LB, Sykes BD (2002) Kinetic studies of calcium and cardiac troponin I peptide binding to human cardiac troponin C using NMR spectroscopy. Eur Biophys J 31(4):245–256
107. Herzberg O, Moult J, James MN (1986) Calcium binding to skeletal muscle troponin C and the regulation of muscle contraction. Ciba Found Symp 122:120–144
108. Murakami K, Yumoto F, Ohki SY, Yasunaga T, Tanokura M, Wakabayashi T (2005) Structural basis for Ca^{2+}–regulated muscle relaxation at interaction sites of troponin with actin and tropomyosin. J Mol Biol 352(1):178–201
109. Lehman W, Hatch V, Korman V, Rosol M, Thomas L, Maytum R et al (2000) Tropomyosin and actin isoforms modulate the localization of tropomyosin strands on actin filaments. J Mol Biol 302(3):593–606
110. McKillop DF, Geeves MA (1993) Regulation of the interaction between actin and myosin subfragment 1: evidence for three states of the thin filament. Biophys J 65(2):693–701
111. Sequeira V, Nijenkamp LL, Regan JA, van der Velden J (2014) The physiological role of cardiac cytoskeleton and its alterations in heart failure. Biochim Biophys Acta 1838(2):700–722

112. Engert F, Bonhoeffer T (1999) Dendritic spine changes associated with hippocampal long-term synaptic plasticity. Nature 399(6731):66–70
113. Toni N, Buchs PA, Nikonenko I, Bron CR, Muller D (1999) LTP promotes formation of multiple spine synapses between a single axon terminal and a dendrite. Nature 402(6760):421–425
114. Oertner TG, Matus A (2005) Calcium regulation of actin dynamics in dendritic spines. Cell Calcium 37(5):477–482
115. Nusco GA, Chun JT, Ercolano E, Lim D, Gragnaniello G, Kyozuka K et al (2006) Modulation of calcium signalling by the actin-binding protein cofilin. Biochem Biophys Res Commun 348(1):109–114
116. Lovelock JD, Monasky MM, Jeong EM, Lardin HA, Liu H, Patel BG et al (2012) Ranolazine improves cardiac diastolic dysfunction through modulation of myofilament calcium sensitivity. Circ Res 110(6):841–850
117. Yoneda M, Nishizaki T, Tasaka K, Kurachi H, Miyake A, Murata Y (2000) Changes in actin network during calcium-induced exocytosis in permeabilized GH3 cells: calcium directly regulates F-actin disassembly. J Endocrinol 166(3):677–687
118. Couchman JR, Pataki CA (2012) An introduction to proteoglycans and their localization. J Histochem Cytochem 60(12):885–897
119. Afratis NA, Nikitovic D, Multhaupt HA, Theocharis AD, Couchman JR, Karamanos NK (2017) Syndecans - key regulators of cell signaling and biological functions. FEBS J 284(1):27–41
120. Couchman JR (2010) Transmembrane signaling proteoglycans. Annu Rev Cell Dev Biol 26:89–114
121. Trebak M, Hempel N, Wedel BJ, Smyth JT, Bird GS, Putney JW Jr (2005) Negative regulation of TRPC3 channels by protein kinase C-mediated phosphorylation of serine 712. Mol Pharmacol 67(2):558–563
122. Oh ES, Woods A, Couchman JR (1997) Syndecan-4 proteoglycan regulates the distribution and activity of protein kinase C. J Biol Chem 272(13):8133–8136
123. Keum E, Kim Y, Kim J, Kwon S, Lim Y, Han I et al (2004) Syndecan-4 regulates localization, activity and stability of protein kinase C-alpha. Biochem J 378(Pt 3):1007–1014
124. Horowitz A, Murakami M, Gao Y, Simons M (1999) Phosphatidylinositol-4,5-bisphosphate mediates the interaction of syndecan-4 with protein kinase C. Biochemistry 38(48):15871–15877
125. Oh ES, Woods A, Lim ST, Theibert AW, Couchman JR (1998) Syndecan-4 proteoglycan cytoplasmic domain and phosphatidylinositol 4,5-bisphosphate coordinately regulate protein kinase C activity. J Biol Chem 273(17):10624–10629
126. Tu LC, Chou CK, Chen HC, Yeh SF (2001) Protein kinase C-mediated tyrosine phosphorylation of paxillin and focal adhesion kinase requires cytoskeletal integrity and is uncoupled to mitogen-activated protein kinase activation in human hepatoma cells. J Biomed Sci 8(2):184–190
127. Fogh BS, Multhaupt HA, Couchman JR (2014) Protein kinase C, focal adhesions and the regulation of cell migration. J Histochem Cytochem 62(3):172–184
128. Trinkaus-Randall V, Kewalramani R, Payne J, Cornell-Bell A (2000) Calcium signaling induced by adhesion mediates protein tyrosine phosphorylation and is independent of pHi. J Cell Physiol 184(3):385–399
129. Liu Y, Echtermeyer F, Thilo F, Theilmeier G, Schmidt A, Schulein R et al (2012) The proteoglycan syndecan 4 regulates transient receptor potential canonical 6 channels via RhoA/Rho-associated protein kinase signaling. Arterioscler Thromb Vasc Biol 32(2):378–385
130. Lunde IG, Herum KM, Carlson CC, Christensen G (2016) Syndecans in heart fibrosis. Cell Tissue Res 365(3):539–552
131. Couchman JR, Multhaupt H, Sanderson RD (2016) Recent insights into cell surface heparan sulphate proteoglycans and cancer. F1000Res 5

132. Saied-Santiago K, Bulow HE (2018) Diverse roles for glycosaminoglycans in neural patterning. Dev Dyn 247(1):54–74
133. Christensen G, Herum KM, Lunde IG (2018) Sweet, yet underappreciated: Proteoglycans and extracellular matrix remodeling in heart disease. Matrix Biol
134. Gopal S, Couchman J, Pocock R (2016) Redefining the role of syndecans in C. elegans biology. Worm 5(1):e1142042
135. Humphries JD, Byron A, Humphries MJ (2006) Integrin ligands at a glance. J Cell Sci 119(Pt 19):3901–3903
136. Hynes RO (2002) Integrins: bidirectional, allosteric signaling machines. Cell 110(6):673–687
137. Xiong JP, Stehle T, Goodman SL, Arnaout MA (2003) Integrins, cations and ligands: making the connection. J Thromb Haemost 1(7):1642–1654
138. Luo BH, Carman CV, Springer TA (2007) Structural basis of integrin regulation and signaling. Annu Rev Immunol 25:619–647
139. Craig D, Gao M, Schulten K, Vogel V (2004) Structural insights into how the MIDAS ion stabilizes integrin binding to an RGD peptide under force. Structure 12(11):2049–2058
140. Zhang K, Chen J (2012) The regulation of integrin function by divalent cations. Cell Adhes Migr 6(1):20–29
141. Stefansson S, Su EJ, Ishigami S, Cale JM, Gao Y, Gorlatova N et al (2007) The contributions of integrin affinity and integrin-cytoskeletal engagement in endothelial and smooth muscle cell adhesion to vitronectin. J Biol Chem 282(21):15679–15689
142. Cherny RC, Honan MA, Thiagarajan P (1993) Site-directed mutagenesis of the arginine-glycine-aspartic acid in vitronectin abolishes cell adhesion. J Biol Chem 268(13):9725–9729
143. Charo IF, Nannizzi L, Smith JW, Cheresh DA (1990) The vitronectin receptor alpha v beta 3 binds fibronectin and acts in concert with alpha 5 beta 1 in promoting cellular attachment and spreading on fibronectin. J Cell Biol 111(6 Pt 1):2795–2800
144. Chen J, Maeda T, Sekiguchi K, Sheppard D (1996) Distinct structural requirements for interaction of the integrins alpha 5 beta 1, alpha v beta 5, and alpha v beta 6 with the central cell binding domain in fibronectin. Cell Adhes Commun 4(4–5):237–250
145. Sjaastad MD, Angres B, Lewis RS, Nelson WJ (1994) Feedback regulation of cell-substratum adhesion by integrin-mediated intracellular Ca^{2+} signaling. Proc Natl Acad Sci USA 91(17):8214–8218
146. Sarin V, Gaffin RD, Meininger GA, Muthuchamy M (2005) Arginine-glycine-aspartic acid (RGD)-containing peptides inhibit the force production of mouse papillary muscle bundles via alpha 5 beta 1 integrin. J Physiol 564(Pt 2):603–617
147. Lin CY, Hilgenberg LG, Smith MA, Lynch G, Gall CM (2008) Integrin regulation of cytoplasmic calcium in excitatory neurons depends upon glutamate receptors and release from intracellular stores. Mol Cell Neurosci 37(4):770–780
148. Wu X, Mogford JE, Platts SH, Davis GE, Meininger GA, Davis MJ (1998) Modulation of calcium current in arteriolar smooth muscle by alphav beta3 and alpha5 beta1 integrin ligands. J Cell Biol 143(1):241–252
149. Gui P, Wu X, Ling S, Stotz SC, Winkfein RJ, Wilson E et al (2006) Integrin receptor activation triggers converging regulation of Cav1.2 calcium channels by c-Src and protein kinase A pathways. J Biol Chem 281(20):14015–14025
150. Waitkus-Edwards KR, Martinez-Lemus LA, Wu X, Trzeciakowski JP, Davis MJ, Davis GE et al (2002) alpha(4)beta(1) Integrin activation of L-type calcium channels in vascular smooth muscle causes arteriole vasoconstriction. Circ Res 90(4):473–480
151. Kwon MS, Park CS, Choi K, Ahnn J, Kim JI, Eom SH et al (2000) Calreticulin couples calcium release and calcium influx in integrin-mediated calcium signaling. Mol Biol Cell 11(4):1433–1443
152. Jiao R, Cui D, Wang SC, Li D, Wang YF (2017) Interactions of the mechanosensitive channels with extracellular matrix, integrins, and cytoskeletal network in osmosensation. Front Mol Neurosci 10:96

153. Matthews BD, Thodeti CK, Tytell JD, Mammoto A, Overby DR, Ingber DE (2010) Ultra-rapid activation of TRPV4 ion channels by mechanical forces applied to cell surface beta1 integrins. Integr Biol (Camb) 2(9):435–442
154. Zimolo Z, Wesolowski G, Tanaka H, Hyman JL, Hoyer JR, Rodan GA (1994) Soluble alpha v beta 3-integrin ligands raise $[Ca^{2+}]i$ in rat osteoclasts and mouse-derived osteoclast-like cells. Am J Phys 266(2 Pt 1):C376–C381
155. Bhattacharya S, Ying X, Fu C, Patel R, Kuebler W, Greenberg S et al (2000) alpha(v)beta(3) integrin induces tyrosine phosphorylation-dependent Ca^{2+} influx in pulmonary endothelial cells. Circ Res 86(4):456–462

Index

© Springer Nature Switzerland AG 2020

M. S. Islam (ed.), *Calcium Signaling*, Advances in Experimental Medicine and Biology 1131, https://doi.org/10.1007/978-3-030-12457-1